Compendium of Pumped Storage Plants in the United States

Prepared by the Task Committee on
Pumped Storage of the Hydropower Committee of the
Energy Division of the
American Society of Civil Engineers

Published by the
American Society of Civil Engineers
345 East 47th Street
New York, NY 10017-2398

ABSTRACT

The *Compendium of Pumped Storage Plants in the United States* was prepared by the Pumped Storage Task Committee of the Hydropower Committee of the ASCE Energy Division. It is intended to be a reference document to describe the characteristics of existing pumped storage plants in the United States. Through the use of tables and lists, the *Compendium* summarizes the major civil engineering features of 36 pumped storage plants in order to make the process of analyzing or developing new sites more efficient; to make the preliminary design of new pumped storage projects with similar characteristics more efficient; and to provide a useful educational tool for those new to the field of pumped storage.

Library of Congress Cataloging-in-Publication Data

Compendium of pumped storage plants in the United States/prepared by the
  Task Committee on Pumped Storage of the Hydropower Committee of the
  Energy Division of the American Society of Civil Engineers.
    p.cm.
   Includes index.
   ISBN 0-87262-991-0
    1.Pumped storage power plants—United States—Directories. 2.Pumped storage power plants—United States—Statistics. 3.Pumped storage power plants—United State—Design and construction. I.American Society of Civil Engineers. Task Committee on Pumped Storage.
TK1423.C66 1993   93-33564
621.31'2134—dc20   CIP

The material presented in this publication has been prepared in accordance with generally recognized engineering principles and practices, and is for general information only. This information should not be used without first securing competent advice with respect to its suitability for any general or specific application.

The contents of this publication are not intended to be and should not be construed to be a standard of the American Society of Civil Engineers (ASCE) and are not intended for use as a reference in purchase specifications, contracts, regulations, statutes, or any other legal document.

No reference made in this publication to any specific method, product, process or service constitutes or implies an endorsement, recommendation, or warranty thereof by ASCE.

ASCE makes no representation or warranty of any kind, whether express or implied, concerning the accuracy, completeness, suitability or utility of any information, apparatus, product, or process discussed in this publication, and assumes no liability therefor.

Anyone utilizing this information assumes all liability arising from such use, including but not limited to infringement of any patent or patents.

Photocopies. Authorization to photocopy material for internal or personal use under circumstances not falling within the fair use provisions of the Copyright Act is granted by ASCE to libraries and other users registered with the Copyright Clearance Center (CCC) Transactional Reporting Service, provided that the base fee of $2.00 per article plus $.25 per page copied is paid directly to CCC, 27 Congress Street, Salem, MA 01970. The identification for ASCE Books is 0-87262-991-0/93 $2.00 + $.25. Requests for special permission or bulk copying should be addressed to Permissions & Copyright Dept., ASCE.

Copyright © 1993 by the American Society of Civil Engineers,
All Rights Reserved.
Library of Congress Catalog Card No: 93-33564
ISBN 0-87262-991-0
Manufactured in the United States of America.

FOREWORD

The *Compendium* of Pumped Storage Plants in the United States was prepared by the Pumped Storage Task Committee of the Hydropower Committee of the ASCE Energy Division. Work began in 1980, and, initially, a subcommittee planned to collect information on pumped storage plants worldwide. However, after two attempts at collecting information in 1981 and 1984, the scope of the work was reduced to include only those plants located in the United States. This made the effort achievable and left the job of investigating plants outside of the United States to a future task committee, if appropriate.

The need for the work was driven by the lack of a single document that summarizes the major civil engineering features of existing pumped storage plants and the desire for a document that could serve as an early reference for conceptual studies and design of future pumped storage plants. The objective of the *Compendium* is to meet that need.

The *Compendium* was computerized in 1985 and 1986. However, activity was suspended in 1987 to concentrate on completion of the "Civil Engineering Guidelines for Planning and Designing Hydroelectric Developments." These *Guidelines* were published in 1989. The *Compendium* was envisioned as an extension of these guidelines. Activity on the *Compendium* resumed in 1990. Collection of information for the *Compendium* was completed in February 1992.

Recognition is due the Hydropower Committee members, its Control Group members who supported the development of the *Compendium*, the Chairman and Vice-Chairman of the Task Committee who led the effort, and to the Task

Committee members, especially those who remained active through the completion of the *Compendium* and dedicated many hours of personal time. Recognition is also due the organizations who supported the participation of the Task Committee members, allowing them to attend meetings and providing to them the office assistance necessary for completion of their activities. Recognition is also due the individuals who did research to develop and provide the information in the format requested by the Task Committee, and last, but not least, to the many plant managers and their staff who responded with the information requested.

ASCE ENERGY DIVISION
EXECUTIVE COMMITTEE CONTACT MEMBERS

Donald Matchett, P.E.　　　　　　　　Stone & Webster Engineering Corp.
　　　　　　　　　　　　　　　　　　Denver, Colorado (1990)

Arvids Zagars, P.E.　　　　　　　　　Harza Engineering Company
　　　　　　　　　　　　　　　　　　Chicago, Illinois

ASCE HYDROPOWER DEVELOPMENT COMMITTEE
CONTROL GROUP MEMBERS (1988-1992)

Edgar T. Moore, Jr., P.E.　　　　　　　Chairman, ASCE Hydropower
　　　　　　　　　　　　　　　　　　Development Committee
　　　　　　　　　　　　　　　　　　Harza Engineering Company
　　　　　　　　　　　　　　　　　　Chicago, Illinois

Thomas G. Gebhart, Jr., Ph.D., P.E.　　Gebhart Engineers, Inc.
　　　　　　　　　　　　　　　　　　Austin, Texas

Dr. Antonio Ferreira, P.E.　　　　　　 6 Jeane Drive
　　　　　　　　　　　　　　　　　　Holyoke, Massachusetts 01040

Thomas H. Logan, P.E.　　　　　　　　Thomas L. Logan Associates
　　　　　　　　　　　　　　　　　　1310 Wadsworth Boulevard
　　　　　　　　　　　　　　　　　　Suite 100
　　　　　　　　　　　　　　　　　　Lakewood, Colorado 80215

ASCE PUMPED STORAGE TASK COMMITTEE
FOR PUMPED STORAGE COMPENDIUM

Throughout this process, many individuals and organizations have supported the development of the *Compendium*. Task Committee members were active in both data collection and review. They did research to determine who had information on the pumped storage plant or plants they were assigned and then contacted the organizations and individuals. Many Task Committee members used either their own or their company's resources to attend review meetings to develop the content and the format of the *Compendium* and to collect the information on each plant. For some members, this work required considerable time and diligence. Many searched through the literature themselves to find the data needed for the *Compendium*. Others worked with several different individuals, departments, or companies to collect the needed information. Without the dedicated work of these committee members, the *Compendium* could not have been completed. Members of the Task Committee who attended meetings and lead the collection of information for one or more plants and their organizations are:

Task Committee Member	Organization
Ralph R. Roza, Jr., P.E.	Chairman
	Corps of Engineers
A. Hassan Makarechian	Vice Chairman
	Stone & Webster Corp.
David Burgoine	Acres International Corp.
Friedrich Fahlbusch	Morrison Knudson Engineers
Dr. Antonio Ferreira	Northeast Utilities

Task Committee Member	Organization
Gary Grinnell	HDR Engineering, Inc.
George Kanakaris	Ebasco Services, Inc.
Dr. Gabor Karadi	University of Wisconsin
Richard Mascolo	Retired, Stone & Webster Corp.
Hugh McKay	Duke Power Company
Divyendu Narayan	New York Power Authority
Lee Nash	Tennessee Valley Authority
David Sveum	Bureau of Reclamation
Bill Thompson	Bureau of Reclamation
Arvids Zagars	Harza Engineering Company

INDIVIDUALS COLLECTING INFORMATION FOR COMMITTEE MEMBERS

In addition, many people worked to provide information. Because many plants have been in existence for some time, this was not always a straightforward process. In many cases, boxes of design documents had to be retrieved from storage, and much research had to be completed to identify sources of information. This often involved contacting several people or departments within one or more organizations. The individuals known to have provided this type of information to the *Compendium* include:

Name	Plants
F.E. Adkins	Hiwassee
David Bakhoum	Edward Hyatt, Thermalito
John Brooks	Grand Coulee, Mt. Elbert
David Burgoine	Salina, Carters
Robert A. Burks	Balsam Meadow
Rodney J. Clark	Castaic
Jerry A. Davis	Helms
J.M. Epps	Raccoon Mountain
Friedrich E. Fahlbusch	Helms, Taum Saulk
Dennis Fenske	Clarence Cannon
David Ferguson	Edward Hyatt, Thermalito
Antonio Ferreira	Kinzua, Rocky River, Yards Creek, Bear Swamp, Muddy Run Northfield Mountain
Paul J. Flanagan	Giannelli
Richard Gerkowski	Ludington
Fred Harty	Northfield Mountain
Kenneth L. Hern	Smith Mountain
Gina L. Holland	Cabin Creek
George Kanakaris	Ludington, Smith Mountain, Yards Creek
Dr. Gabor Karadi	Wallace, Castaic
D.M. Keith	Hiwassee
Thomas H. Logan	Bear Swamp, Northfield Mountain
Bill Lynch	Richard B. Russell
A. Hassan Makarechian	Cabin Creek, Horse Mesa, Mormon Flat, Northfield Mountain
Dick Mascolo	Balsam Meadow, Smith Mountain, Flat Iron
R.L. Maurel	Balsam Meadow
Scott McDonald	Smith Mountain
Hugh G. McKay, III	Bad Creek, Jocassee
W.E. Moore	Fairfield
Divyendu Narayan	Blenheim Gilboa, Lewiston
Lee A. Nash	Hiwassee, Raccoon Mountain
Luther Newton	De Gray
Dan Pellouchoud	Horse Mesa, Mormon Flat
B.G. Ragsdale	Hiwassee, Raccoon Mountain
Ralph R. Roza, Jr.	De Gray, Clarence Cannon, others

Name	Plants
Vedula Sarma	Salina
Charles H. Snow	Carters
David Starks	Edward Hyatt, Thermalito
Sam Stockman	Fairfield
David Sveum	Grand Coulee, others
Bill Thompson	Grand Coulee, Mount Elbert, Flat Iron
J.E. Throckmorton	Balsam Meadow
A.M. Waddell	Hiwassee, Raccoon Mountain
C.D. Wagner	Hiwassee, Raccoon Mountain
Arvids Zagars	Bath County, Rocky Mountain, Kinzua

SPECIAL THANKS

As Chairman of the Task Committee, I would like to extend special thanks to my wife, Jo, and my children, Chad, Devin, Andy, and Miriam, for their sustained patience and support. Thanks also to Linda Howell who typed much of this document and kept me organized. Thanks to Connie Carman for her editorial suggestions. Thanks to Hassan Makarechian, Vice-Chairman; David Burgoine; Fred Fahlbusch; Tony Ferriera; George Kanakaris; Dr. Gabor Karadi; Divyendu Narayan; Bill Thompson; and Arvids Zagars for being the team that really made this project work. Special thanks also to Dick Mascolo who collected information for the Smith Mountain project, even though he was retired and used his own time and financial resources. Dick provided additional information to the subcommittee even after having had major heart surgery. Thanks also to ASCE and to all of the Task Committee members for believing that this document would, indeed, be completed and published.

Respectfully Submitted,

Ralph R. Roza, P.E.
Chairman, ASCE Task Committee

SPECIAL RECOGNITION

Special credit is given to the Chairman of the Task Committee, Ralph Roza, for organizing all the necessary related activities and for spending uncounted hours of his personal time in publishing the *Compendium*. As a result of his efforts, the *Compendium* became a reality and the highly technical information is presented in a manner that is beneficial to the profession.

Respectfully Submitted,

Arvids Zagars, P.E.
ASCE Energy Division
Contract Member

MEASUREMENT CONVERSIONS

Conversion formulas used in the development of the *Compendium* are as follows:

To Convert:	Multiply by:
Meters to feet	3.28084
Meters^3 per second to feet^3 per second	35.31466
Meters^3 to yards^3	1.30795
Meters^2 x 10^6 to acres	247.1
Meters^3 x 10^6 to acre-feet	810.7085
Meters^2 x 10^6 to miles^2	0.3861
Meters^2 to feet^2	10.764
Millimeters to inches	0.03937
Grams x 10^6 to ton (short 2,000 lb.)	1.10231
Newton x Meters^2 to pound-force x feet^2	2.41927
Kilometers to miles	0.62136

ABBREVIATIONS AND SYMBOLS

The following table presents a summary of the abbreviations used in this document.

Abbreviation	Full Text
A	Ampères
eff	efficiency
ft	feet
G	Giga or times one billion
g	grams
GWh	Gigawatt-hours
h	hours
hp	horsepower
in	inches
k	kilo (or times 1,000)
kV	kilo-Volts
kVA	kilo-Volts-Amperes
L/H	Length to Head ratio
lbf	pound-force
m	meters
M	Mega or times one million
Max	Maximum
mi	miles
Min	Minimum
min	minutes
mm	millimeters
msl	mean sea level
MVA	Mega Volt Ampères
MW	Megawatts
MWh	Megawatt-hours
PMF	Probable Maximum Flood
rpm	revolutions per minute
V	Volts
vel	velocity
s	seconds
W	Watt
yd	yard

TABLE OF CONTENTS

List of Sections

Section	Page
Introduction	1
Purpose and Scope of the Compendium	1
Organization and Content of the Document	3
Section 1–Tables of Information by Plant Feature	9
Section 2–Information by Plant	85

List of Figures

Figure No.	Item	Page
Figure 1	Map of Pumped Storage Plants in the United States	4
Figure 2	Growth in the Installed Capacity of Pumped Storage Units	5
Figure 3	Efficiency of Pumped Storage Plants vs Time	6

TABLE OF CONTENTS

List of Sections

Section 1—Tables of Information by Plant Feature

Tables in English Units:

    Table 1–Summary of Plant Data 11
    Table 2–Summary of Data on Upper Dams, Reservoirs,
        and Spillways..19
    Table 3–Summary of Data on Upper Intakes 24
    Table 4–Summary of Data on Conduits 28
    Table 5–Summary of Data on Pump-Turbines 31
    Table 6–Summary of Data on Lower Dams, Reservoirs,
        and Spillways..34
    Table 7–Summary of Data on Lower Intakes...............................40
    Table 8–Data on Environmental Features 43

Tables in Metric Units:

    Table 9–Summary of Plant Data 54
    Table 10–Summary of Data on Upper Dams, Reservoirs,
        and Spillways..62
    Table 11–Summary of Data on Upper Intakes 67
    Table 12–Summary of Data on Conduits 71
    Table 13–Summary of Data on Pump-Turbines 74
    Table 14–Summary of Data on Lower Dams, Reservoirs,
        and Spillways..77
    Table 15–Summary of Data on Lower Intakes............................82

TABLE OF CONTENTS

List of Sections

Section 2—Information by Plant

Plant Name

Bad Creek	87
Balsam Meadow	106
Bath County	126
Bear Swamp	155
Blenheim-Gilboa	175
Cabin Creek	191
Carters	210
Castaic	237
Clarence Cannon	258
De Gray	282
Edward Hyatt	306
Fairfield	335
Flatiron	370
Gianelli San Luis	386
Grand Coulee	400
Helms	415
Hiwassee-Unit 2	439
Horse Mesa #4	452

Plant Name

Jocassee	461
Kinzua (Seneca)	473
Lewiston	501
Ludington	517
Mormon Flat #2	531
Mount Elbert	540
Muddy Run	559
Northfield Mountain	575
Raccoon Mountain	602
Rocky Mountain	619
Rocky River	640
Russell Dam	648
Salina	655
Smith Mountain	675
Taum Sauk	689
Thermalito	700
Wallace Dam	709
Yards Creek	727

Index 743

INTRODUCTION

PURPOSE AND SCOPE OF THE COMPENDIUM

The pumped storage compendium is intended to be a reference document to describe the characteristics of existing pumped storage plants in the United States. The *Compendium* focuses on the civil engineering aspects of the plants and has the following main objectives:

- Make the process of analyzing or developing new sites more efficient. Information on the designs and layouts presented in the compendium can be viewed as a collection of previous solutions to the problem of how engineers and others used available site resources to optimize power benefits and provide other benefits to society. Because the layout and design of each pumped storage plant is unique and depends to a large extent on the physical characteristics of the site, the *Compendium* provides engineers with many examples of conceptual layouts for a wide range of pumped storage sites.

- Make the preliminary design of new pumped storage projects with similar characteristics more efficient. In addition to presenting conceptual drawings and descriptions of existing plants, the *Compendium* presents many tables and listings of the civil engineering characteristics of each plant. This information is intended to facilitate the analysis of development trends, the comparison of the characteristics of various plants, and the examination of the close relationship between the components of a particular plant. In addition, it provides a handy source of information regarding the dimensions of components of existing plants for comparison during project conceptual studies.

- Provide a useful educational tool for those new to the field of pumped storage. The *Compendium* should be used in conjunction with Volume 5, "Pumped Storage and Tidal Power," of the recently completed "Civil Engineering Guidelines for Planning and Designing Hydroelectric Developments." This volume presents information on the history of pumped storage, development trends, and the planning and design of pumped storage plants.

The *Compendium* is believed to be the most comprehensive source of civil engineering-related information on pumped storage plants in the United States. The engineers who collected the information made every attempt to obtain accurate information. In addition, the information has been reviewed by a committee of engineers who were active in the data collection process and familiar with the *Compendium* data needs and its intended uses. However, no guarantees can be made about the accuracy of the data. The data for each plant were collected either from plant owners or from the company or companies that designed the plant, or both. In a few cases, only limited information was available.

Figure 1 shows the location of each of the pumped storage plants in the United States. Figures 2 and 3 illustrate only two of many relationships that can be developed from the compendium data. Figure 2 shows the growth of pumped storage capacity from 1929, the year of the first pumped storage plant in the United States, Rocky River, through 1995, the year the Rocky Mountain project, which is currently under construction, is scheduled to be online. Other pumped storage projects are in the planning stages or in the early phases of construction, but these were not included in the compendium. Figure 3 shows the steady increase in the efficiency of pumped storage plants. This reflects the increased attention to the design of the hydraulic and mechanical components of newer plants, a result of steady increases in energy costs.

ORGANIZATION AND CONTENT OF THE DOCUMENT

The *Compendium* information is presented in two sections. **Section 1** consists of tables that include information organized around the major features of the pumped storage plants. Each table presents this information for all of the plants in the *Compendium*. As such, this section facilitates comparison of the dimensions and various characteristics of the plants and includes the following tables:

Title	Table Number	
	English	Metric
Summary of Plant Data	1	9
Summary of Data on Upper Dams, Reservoirs, and Spillways	2	10
Summary of Data on Upper Intakes	3	11
Summary of Data on Conduits	4	12
Summary of Data on Pump-Turbines	5	13
Summary of Data on Lower Dams, Reservoirs, and Spillways	6	14
Summary of Data on Lower Intakes	7	15
Data on Environmental Features	8	8

FIGURE 1
Map of Pumped Storage Plants in the United States

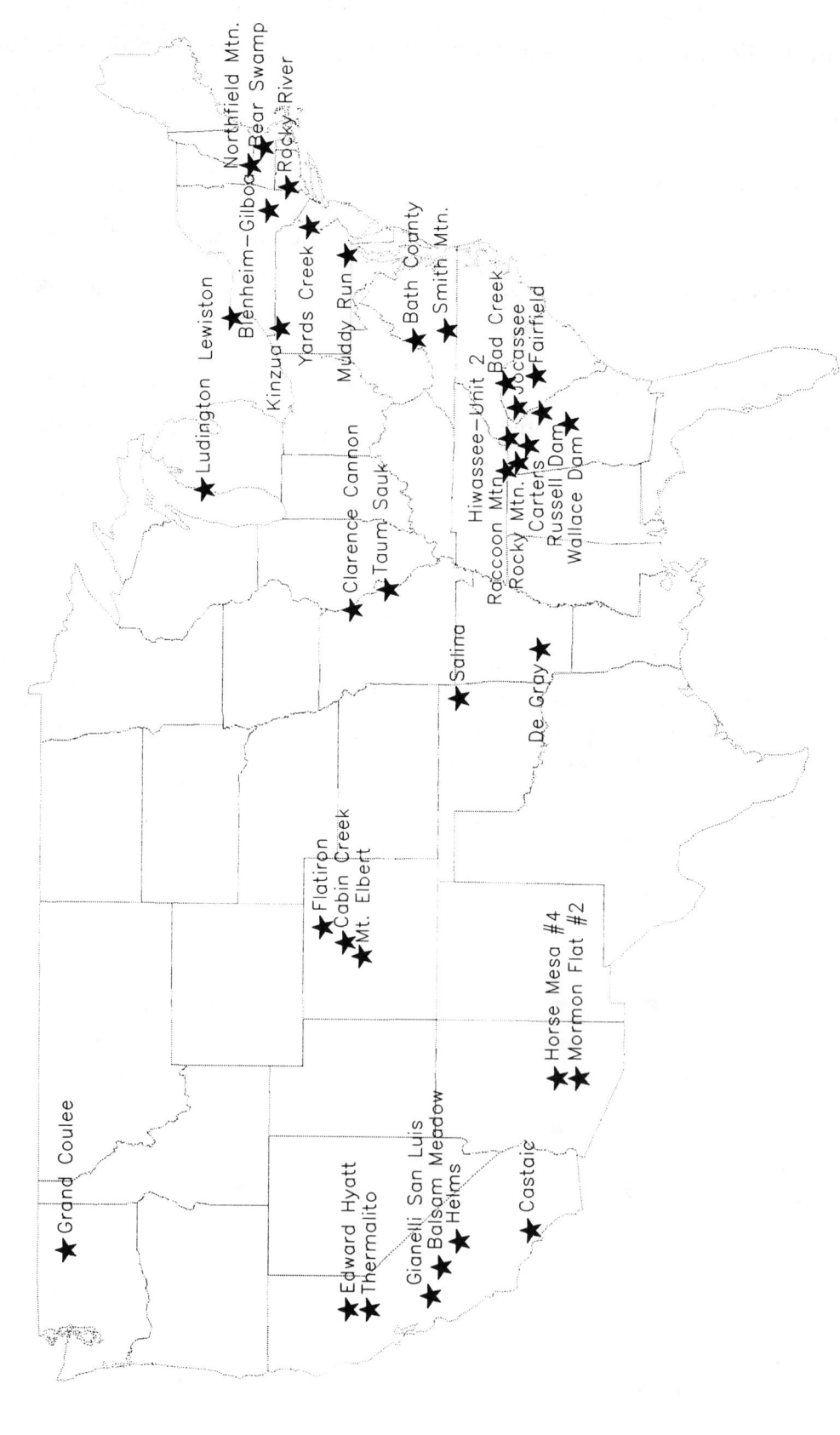

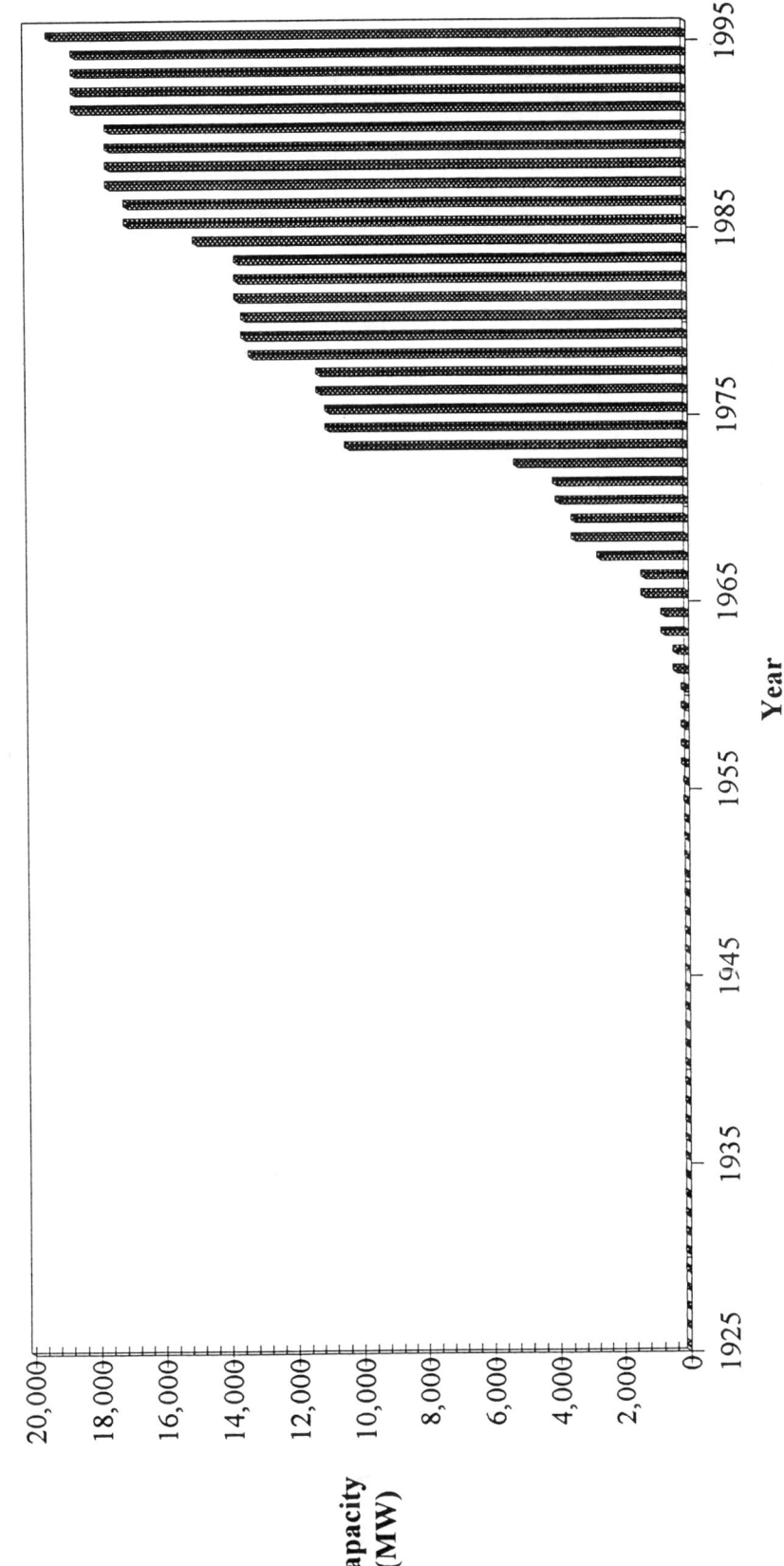

Figure 2
Growth in the Installed Capacity of Pumped Storage Units

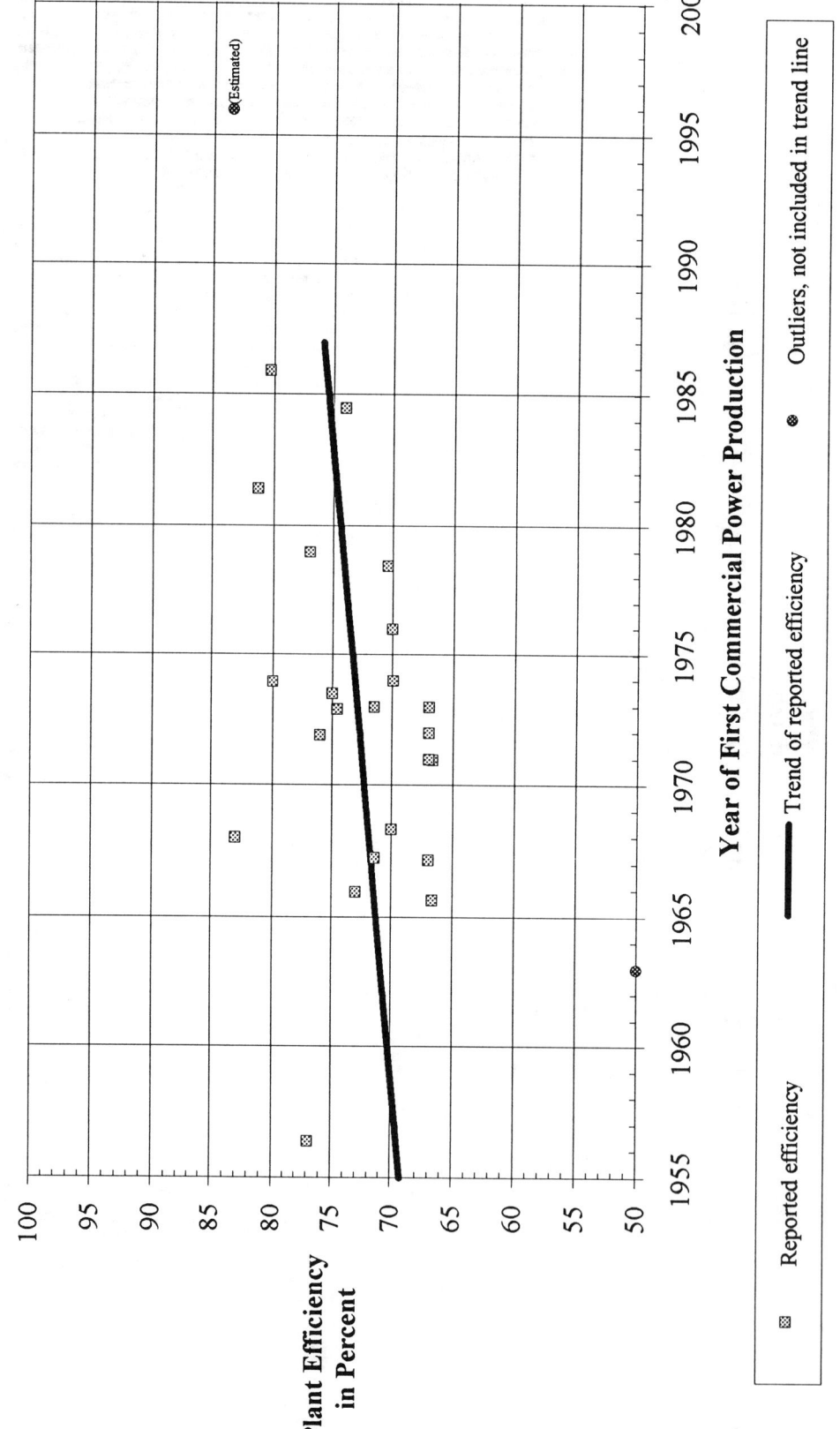

This section includes summary data only. More detailed and comprehensive information is found in section 2.

Section 2, which is organized by plant, presents all the information that has been collected for each plant and can be used to find information about a particular pumped storage plant. Section 2 includes:

- Four pages of technical information describing plant features. These pages contain over 250 items of information on each plant. Pages 1, 2, and 3 are presented twice, first in English units and then in Metric units. Page 4, which is primarily text descriptions, is presented once. Page 4 is presented in Metric units with English units in parenthesis where appropriate.

Page 1 includes information on the plant location, size, number of units, owner, designers, and so forth, plus information describing the upper and lower dams and reservoirs. **Page 2** includes information on the upper and lower intakes and the water passages. **Page 3** includes information on the powerhouse and related features such as pump-turbines and generators. **Page 4** includes information on annual energy generation, plant availability, maintenance outage times, and so forth, as well as miscellaneous notes, information describing any significant or unique problems encountered at the plant, and information regarding the recreation, fish and wildlife, and social aspects of the plant.

- One descriptive article (if available). The *Compendium* includes one or two short articles on each plant. Articles used are those that best describe the civil engineering and physical characteristics and layout of the plant.

- Drawings (if available). A limited number of drawings are provided. Where possible, simplified line drawings with limited detail have been used to provide an overview of the layout of the plant and its various components.

- One or two photographs (if available). Photographs have been included to provide an oblique aerial view showing an overall perspective of the plant or of major plant features.

- A bibliography of pertinent articles for the plant (if available).

SECTION 1
TABLES 1-15

ASCE Pumped Storage Compendium
Table 1 - Summary of Plant Data

Plant Name	Plant Location: Nearest Town, State	Owner/Address	Plant Capacity MW	First Commercial Power	Number of Reversible Units	Number of Main Conduits	Surge Tanks[1]	Static Head: Max, Min (ft) Max/Min Ratio	Storage Capacity Generation Hours[2]	Plant Eff. [3] %	Length of Waterways ft	L/H Ratio[4]
BAD CREEK PUMPED STORAGE PROJECT	Salem, SC	Duke Power Company P.O. Box 33189 Charlotte, North Carolina 28201-1006	1,000	1991	4	1	none	1230, 1040 Ratio: 1.18	24		9,519	8.3
BALSAM MEADOW	Shaver Lake, CA	Southern California Edison Company P.O. Box 800 2244 Walnut Grove Av Rosemead, California 91770	200	1987	1	1	downstream	1352, 1316 Ratio: 1.03	8		12,488	10.0
BATH COUNTY	Warm Springs, VA	Virginia Power Company and Allegheny Generating Company P.O. Box 26666 Richmond, Virginia 23261	2,100	1985	6	3	upstream	1260, 1080 Ratio: 1.17	11	80.30	9,117	7.7
BEAR SWAMP	Rowe, MA	New England Power Company 25 Research Drive Westborough, Massachusetts 01581	600	1974	2	2	none	770, 680 Ratio: 1.13	6	69.90	1,834	2.4
BLENHEIM-GILBOA PUMPED STORAGE PROJECT	Blenheim and Gilboa, NY	New York Power Authority 123 Main Street White Plains, New York 10601	1,000	1973	4	4	none	1143, 1097 Ratio: 1.04	12	75.00	3,895	3.5

[1] Entry indicates location of surge tank(s) relative to powerhouse; upstream, downstream, or both; or that none exists.
[2] Energy stored in upper reservoir divided by the plant's rated capacity.
[3] Plant efficiency is the energy generated divided by the energy used in pumping.
[4] Total length of waterways (upper tunnel, shaft, lower tunnel, penstocks, and tailrace tunnel) divided by the average static head.

Table 1, Page 1

ASCE Pumped Storage Compendium
Table 1 - Summary of Plant Data

Plant Name	Plant Location: Nearest Town, State	Owner/Address	Plant Capacity MW	First Commercial Power	Number of Reversible Units	Number of Main Conduits	Surge Tanks[1]	Static Head: Max, Min (ft) Max/Min Ratio	Storage Capacity Generation Hours[2]	Plant Eff.[3] %	Length of Waterways ft	L/H Ratio[4]
CABIN CREEK	Georgetown, CO	Public Service Company of Colorado 5900 E. 39th Avenue Denver, Colorado 80207	300	1967	2	2	none	1226, 1095 Ratio: 1.12	5	67.00	4,300	3.7
CARTERS	Chatsworth, GA	Mobile District, Corps of Engineers P.O. Box 2288 Mobile, Alabama 36628-0001	258	1976	2	2	none	406, 315 Ratio: 1.29	7	70.00	838	2.4
CASTAIC POWER PROJECT	Castaic, CA	Los Angeles Department of Water and Power 111 North Hope Street Los Angeles, California 90012-2694	1,250	1973	6	1	upstream	1098, 800 Ratio: 1.37	10	67.00	41,275	38.6
CLARENCE CANNON	Center, MO	St. Louis District, Corps of Engineers Box 20 B Monroe City, Missouri 63456	31	1984	1	1	none	117, 59 Ratio: 1.98	9		70	0.9
DE GRAY DAM AND RESERVOIR	Arkadelphia, AK	Vicksburgh District Corps of Engineers P.O. Box 60 Vicksburg, Mississippi 39180-0060	28	1971	1	1	none	206, 146 Ratio: 1.41	7	76.00	1,570	9.2

[1] Entry indicates location of surge tank(s) relative to powerhouse; upstream, downstream, or both; or that none exists.
[2] Energy stored in upper reservoir divided by the plant's rated capacity.
[3] Plant efficiency is the energy generated divided by the energy used in pumping.
[4] Total length of waterways (upper tunnel, shaft, lower tunnel, penstocks, and tailrace tunnel) divided by the average static head.

Table 1, Page 2

ASCE Pumped Storage Compendium
Table 1 - Summary of Plant Data

Plant Name	Plant Location: Nearest Town, State	Owner/Address	Plant Capacity MW	First Commercial Power	Number of Reversible Units	Number of Main Conduits	Surge Tanks[1]	Static Head: Max, Min (ft) Max/Min Ratio	Storage Capacity Generation Hours[2]	Plant Eff. [3] %	Length of Waterways ft	L/H Ratio[4]
EDWARD HYATT POWER PLANT	Oroville, CA	California Department of Water Resesources 1416 Ninth Street Sacramento, California 95814	293	1967	3	2	none	675, 500 Ratio: 1.35			581	0.9
FAIRFIELD	Jenkinsville, SC	South Carolina Electric & Gas Company 1426 Main Street Columbia, South Carolina 29201	512	1978	8	4	none	169, 155 Ratio: 1.09	8	70.40	1,095	7.3
FLATIRON POWERPLANT	Loveland, CO	Bureau of Reclamation Attn: D-3000 P.O. Box 25007 Denver, Colorado 80225	9	1954	1	1	upstream	298, 153 Ratio: 1.95			7,392	25.5
GIANELLI SAN LUIS PUMP-GENERATE PLANT	Los Banos, CA	Bureau of Reclamation Attn: D-3000 P.O. Box 25007 Denver, Colorado 80225	400	1968	8	4	none	324, 117 Ratio: 2.77	1274	83.00	2,146	10.9
GRAND COULEE	Grand Coulee, WA	Bureau of Reclamation Attn: D-3000 P.O. Box 25007 Denver, Colorado 80225	314	1973	6	6	none	363, 267 Ratio: 1.36	35		716	2.7

[1] Entry indicates location of surge tank(s) relative to powerhouse; upstream, downstream, or both; or that none exists.
[2] Energy stored in upper reservoir divided by the plant's rated capacity.
[3] Plant efficiency is the energy generated divided by the energy used in pumping.
[4] Total length of waterways (upper tunnel, shaft, lower tunnel, penstocks, and tailrace tunnel) divided by the average static head.

Table 1, Page 3

ASCE Pumped Storage Compendium
Table 1 - Summary of Plant Data

Plant Name	Plant Location: Nearest Town, State	Owner/Address	Plant Capacity MW	First Commercial Power	Number of Reversible Units	Number of Main Conduits	Surge Tanks[1]	Static Head: Max, Min (ft) Max/Min Ratio	Storage Capacity Generation Hours[2]	Plant Eff. [3] %	Length of Waterways ft	L/H Ratio[4]
HELMS	Shaver Lake, CA	Pacific Gas & Electric Company 77 Beale Street San Francisco, California 94536	1,206	1984	3	1	upstream and downstream	1745, 1470 Ratio: 1.19	153	74.00	19,803	12.3
HIWASSEE-UNIT 2	Murphy, NC	Tennessee Valley Authority 818 Power Building Chattanooga, Tennessee 37401	68	1956	1	1	none	255, 173 Ratio: 1.47	4228	76.90	190	1.0
HORSE MESA #4	Tortilla Flat, AZ	Bureau of Reclamation Attn: D-3000 P.O. Box 25007 Denver, Colorado 80225	97	1972	1	1	none	258, 236 Ratio: 1.10	245	67.00	187	0.8
JOCASSEE PUMPED STORAGE STATION	Salem, SC	Duke Power Company P.O. Box 1006 Charlotte, North Carolina 28201-1006	610	1973	4	4	none	335, 310 Ratio: 1.08	94	80.00	1,636	5.0
KINZUA	Warren, PA	Pennsylvania Electric Company 1001 Broad Street Johnstown, Pennsylvania 15907	350	1970	2	2	none	813, 644 Ratio: 1.26	11	66.70	2,802	4.3

[1] Entry indicates location of surge tank(s) relative to powerhouse; upstream, downstream, or both; or that none exists.
[2] Energy stored in upper reservoir divided by the plant's rated capacity.
[3] Plant efficiency is the energy generated divided by the energy used in pumping.
[4] Total length of waterways (upper tunnel, shaft, lower tunnel, penstocks, and tailrace tunnel) divided by the average static head.

Table 1, Page 4

ASCE Pumped Storage Compendium
Table 1 - Summary of Plant Data

Plant Name	Plant Location: Nearest Town, State	Owner/Address	Plant Capacity MW	First Commercial Power	Number of Reversible Units	Number of Main Conduits	Surge Tanks [1]	Static Head: Max, Min (ft) Max/Min Ratio	Storage Capacity Generation Hours [2]	Plant Eff. [3] %	Length of Waterways ft	L/H Ratio [4]
LEWISTON	Lewiston, NY	New York Power Authority, 123 Main Street, White Plains, New York 10601	240	1961	12	12	none	113, 66 Ratio: 1.71	20		152	2.0
LUDINGTON	Ludington, MI	Consumers Power (51%) / Detroit Edison (49%), 212 W. Michigan Avenue, Ludington, Michigan 49201	1,979	1973	6	6	none	364, 295 Ratio: 1.23	9	71.50	1,299	3.6
MORMON FLAT #2	Tortilla Flat, AZ	Bureau of Reclamation, Attn: D-3000, P.O. Box 25007, Denver, Colorado 80225	47	1971	1	1	none	139, 126 Ratio: 1.10	59	67.00	175	1.4
MT. ELBERT	Leadville, CO	Bureau of Reclamation, Attn: D-3000, P.O. Box 25007, Denver, Colorado 80225	200	1981	2	2	upstream	475, 400 Ratio: 1.19	12	81.30	3,000	6.7
MUDDY RUN	Drumore, PA	Philadelphia Electric Company, 2301 Market Street, Philadelphia, Pennsylvania 19101	800	1967	8	8	none	415, 361 Ratio: 1.15	14	71.40	1,290	3.7

[1] Entry indicates location of surge tank(s) relative to powerhouse; upstream, downstream, or both; or that none exists.
[2] Energy stored in upper reservoir divided by the plant's rated capacity.
[3] Plant efficiency is the energy generated divided by the energy used in pumping.
[4] Total length of waterways (upper tunnel, shaft, lower tunnel, penstocks, and tailrace tunnel) divided by the average static head.

Table 1, Page 5

ASCE Pumped Storage Compendium
Table 1 - Summary of Plant Data

Plant Name	Plant Location: Nearest Town, State	Owner/Address	Plant Capacity MW	First Commercial Power	Number of Reversible Units	Number of Main Conduits	Surge Tanks[1]	Static Head: Max, Min (ft) Max/Min Ratio	Storage Capacity Generation Hours[2]	Plant Eff.[3] %	Length of Waterways ft	L/H Ratio[4]
NORTHFIELD MOUNTAIN	Northfield & Erving, MA	Connecticut Light and Power Co./Western Mass. Electric Co. c/o Northeast Utilities P.O. Box 270 Hartford, Connecticut 06101-0270	1,080	1972	4	4	upstream and downstream	828, 735 Ratio: 1.13	10	74.60	6,320	8.5
RACCOON MOUNTAIN	Chattanooga, TN	Tennessee Valley Authority Mr 6N 36A Chattanooga, Tennessee 37401	1,530	1978	4	4	downstream	1042, 900 Ratio: 1.16	21	76.90	3,600	4.0
ROCKY MOUNTAIN	Armuchee, GA	Oglethorpe Power Corporation 2100 East Exchange Place P.O. Box 1349 Tucker, Georgia 30085	760	1995	3	3	none	690, 613 Ratio: 1.13	8	83.50	2,984	4.6
ROCKY RIVER	New Milford, CT	Connecticut Light and Power Company P.O. Box 270 Hartford, Connecticut 06101	32	1929	1	1	upstream	230, 210 Ratio: 1.10	837		1,677	7.5
RUSSEL DAM	Elberton, GA	Savannah District, Corps of Engineers 100 W. Oglethorpe Avenue Savannah, Georgia 31401	360	1987	4	4	none					

[1] Entry indicates location of surge tank(s) relative to powerhouse; upstream, downstream, or both; or that none exists.
[2] Energy stored in upper reservoir divided by the plant's rated capacity.
[3] Plant efficiency is the energy generated divided by the energy used in pumping.
[4] Total length of waterways (upper tunnel, shaft, lower tunnel, penstocks, and tailrace tunnel) divided by the average static head.

Table 1, Page 6

ASCE Pumped Storage Compendium
Table 1 - Summary of Plant Data

Plant Name	Plant Location: Nearest Town, State	Owner/Address	Plant Capacity MW	First Commercial Power	Number of Reversible Units	Number of Main Conduits	Surge Tanks[1]	Static Head: Max, Min (ft) Max/Min Ratio	Storage Capacity Generation Hours[2]	Plant Eff. [3] %	Length of Waterways ft	L/H Ratio[4]
SALINA	Salina, OK	Grand River Dam Authority P.O. Box 409 Vanita, Oklahoma 74301-0409	260	1968	6	6	none	245, 225 Ratio: 1.09	10	70.00	640	2.8
SMITH MOUNTAIN	Sandy Level, VA	Appalacian Power Company P.O. Box 2021 Roanoke, Virginia 24022-2121	240	1965	3	3	none	195, 174 Ratio: 1.12	14	73.00	170	0.9
TAUM SAUK	St. Louis, MO	Union Electric Company 1901 Chateau P.O. Box 149 St. Louis, Missouri 63166	350	1963	2	2	none	875, 764 Ratio: 1.15	8	50.00	7,250	9.2
THERMALITO POWER PLANT	Oroville, CA	California Department of Water Resources 1416 Ninth Street Sacramento, California 95965	115	1968	4	4	none	101, 85 Ratio: 1.19			118	1.3
WALLACE DAM	Eatonton, GE	Georgia Power Company P.O. Box 4545 Atlanta, Georgia 30302	209	1979	4	6	none	96, 89 Ratio: 1.08	1020		172	1.8

[1] Entry indicates location of surge tank(s) relative to powerhouse; upstream, downstream, or both; or that none exists.
[2] Energy stored in upper reservoir divided by the plant's rated capacity.
[3] Plant efficiency is the energy generated divided by the energy used in pumping.
[4] Total length of waterways (upper tunnel, shaft, lower tunnel, penstocks, and tailrace tunnel) divided by the average static head.

Table 1, Page 7

ASCE Pumped Storage Compendium
Table 1 - Summary of Plant Data

Plant Name	Plant Location: Nearest Town, State	Owner/Address	Plant Capacity MW	First Commercial Power	Number of Reversible Units	Number of Main Conduits	Surge Tanks[1]	Static Head: Max, Min (ft) Max/Min Ratio	Storage Capacity Generation Hours[2]	Plant Eff. [3] %	Length of Waterways ft	L/H Ratio[4]
YARDS CREEK STATION	Blairstown, NJ	Jersey Central P&L Co. & Public Service Electric & Gas Co. Madison Avenue At Punch Bowl Road Morristown, New Jersey 07960	360	1965	3	1	none	760, 688 Ratio: 1.11	8	66.67	3,500	4.8

[1] Entry indicates location of surge tank(s) relative to powerhouse; upstream, downstream, or both; or that none exists.
[2] Energy stored in upper reservoir divided by the plant's rated capacity.
[3] Plant efficiency is the energy generated divided by the energy used in pumping.
[4] Total length of waterways (upper tunnel, shaft, lower tunnel, penstocks, and tailrace tunnel) divided by the average static head.

Table 1, Page 8

ASCE Pumped Storage Compendium
Table 2 - Summary of Data on Upper Dams, Reservoirs, and Spillways

Plant Name	Dams Type	Dams Height ft	Dams Length ft	Dams Volume yd³ x10³	Reservoirs Type	Drainage Area mi²	Surface Area acres	Max/Min Elevations ft msl	Usable Storage acre-ft	Power Pool Fluctuation ft	Spillways Type	Peak Discharge ft³/s	Spillways Gates Type	Gates Number	Gates Height ft	Gates Width ft
BAD CREEK PUMPED STORAGE PROJECT	Zoned earthfill	361	2,602	466	Constructed	1.4	368	Max: 2,309.7 Min: 2,149.6	31,374	160.1	None					
BALSAM MEADOW	Rockfill embankment	120	1,460	354	Constructed	0.1	60	Max: 6,668.0 Min: 6,638.0	1,680	40.0	Open channel	2,000				
BATH COUNTY	Zoned fill	460	2,400	17,003	On stream	2.4	265	Max: 3,320.0 Min: 3,215.0	22,500	105.0	Fuse plug	18,000				
BEAR SWAMP	Rockfill central core	155		124	Constructed	1,250.0	1,120	Max: 1,600.0 Min: 1,550.0	5,260	50.0	Overflow weir	6,700				
BLENHEIM-GILBOA PUMPED STORAGE PROJECT	Earth & rockfill	162	11,900	5,875	Constructed	0.6	390	Max: 2,003.0 Min: 1,955.0	15,000	48.0	Emergency spill channel	10,000				
CABIN CREEK	Concrete faced, rockfill	210	1,460	1,194	On stream, constructed	1.0	23	Max: 11,196.0 Min: 11,105.0	1,400	91.0	Overflow weir, ungated					
CARTERS	Rockfill	445	2,053	15,000	Constructed	376.0	3,220	Max: 1,072.0 Min: 1,022.0	134,900	52.0	Concrete gravity	194,200	Tainter	5	36.50	42.00
CASTAIC POWER PROJECT	Earth and rockfill	4,000	1,090	6,860	Constructed, Pyramid Reservoir	293.0	1,357	Max: 2,578.0 Min: 2,340.0	20,000	18.0	Broad crested weir			1		
CLARENCE CANNON	Compacted earth, Clarence Cannon Dam	106	1,690	2,170	Constructed, Mark Twain Lake	2,318.0	38,400	Max: 638.0 Min: 567.0	355,000	71.0	Concrete ogee, gated	276,500	Tainter	4	39.00	50.00

Table 2, Page 1

ASCE Pumped Storage Compendium
Table 2 - Summary of Data on Upper Dams, Reservoirs, and Spillways

Plant Name	Dams Type	Dams Height ft	Dams Length ft	Dams Volume yd³×10³	Reservoirs Type	Drainage Area mi²	Surface Area acres	Max/Min Elevations ft msl	Power Pool Usable Storage acre-ft	Power Pool Fluctuation ft	Spillways Type	Peak Discharge ft³/s	Spillways Type	Gates Number	Gates Height ft	Gates Width ft
DE GRAY DAM AND RESERVOIR	Rolled earthfill, De Gray Dam	243	3,400	6,831	Constructed, De Grey Lake	453.0	13,400	Max: 402.0 Min: 367.0	393,200	41.0	Rock saddle, ungated	67,000				
EDWARD HYATT POWER PLANT	Zoned earthfill, Oroville Dam	770	6,920	80,000	Constructed, Oroville Reservoir	3,611.0	15,800	Max: 900.0 Min: 730.0	2,685,000	170.0	Concrete channel	440,000	Radial	8	33.50	17.58
FAIRFIELD	Random-filled	180	11,001	9,996	Constructed, Monticello Reservoir		6,795	Max: 425.0 Min: 420.5	29,023	4.5	None					
FLATIRON POWERPLANT	Zoned earthfill	214	1,959	4,264	Constructed, Carter Lake	5.7	1,142	Max: 5,760.0 Min: 5,625.0		135.0	None					
GIANELLI SAN LUIS PUMP-GENERATE PLANT	Zoned earthfill (B.F. Sisk-San Luis)	314	18,600	80,285	Constructed	84.7	13,000	Max: 544.0 Min: 326.0	1,909,219	218.0	Morning Glory 2.89 m (9.5 ft) diameter.	1,030				
GRAND COULEE	Two zoned earthfill, North Dam	145	1,170	1,659	Constructed, Feeder Canal from Banks Lake		27,000	Max: 1,571.0 Min: 1,557.0	715,000	6.0	None					
HELMS	Rockfill concrete faced, Courtright Dam	315	863	1,560	Existing, constructed, Courtright Lake	39.3	1,646	Max: 8,182.4 Min: 8,018.4	119,174	164.0	Overflow ogee	12,007				

ASCE Pumped Storage Compendium
Table 2 - Summary of Data on Upper Dams, Reservoirs, and Spillways

Plant Name	Dams				Reservoirs				Power Pool			Spillways				
	Type	Height ft	Length ft	Volume yd³x10³	Type	Drainage Area mi²	Surface Area acres	Max/Min Elevations ft msl	Usable Storage acre-ft	Fluctuation ft	Type	Peak Discharge ft³/s	Type	Gates Number	Height ft	Width ft
HIWASSEE-UNIT 2	Concrete, Hiwassee Dam	307	1,375	7,200	On stream	968.3	5,856	Max: 1,528.2 Min: 1,450.5	421,568	73.8	Concrete	129,993	Radial	7	23.00	31.99
HORSE MESA #4	Existing concrete thin arch dam	305	660	162	Existing Apache Lake constructed		2,669	Max: 1,913.6 Min: 1,896.7	116,985	18.0	One overflow and one tunnel spillway		9 radial & 1 fixed wheel	9	22.97	25.92
JOCASSEE PUMPED STORAGE STATION	Zoned fill, Jocassee Dam	385	1,750	13,000	Constructed, Lake Jocassee	148.0	7,565	Max: 1,110.0 Min: 1,080.0	215,700	30.0	Concrete chute		Taintor	2	33.00	38.00
KINZUA	Rolled fill - sandstone	115	7,800	260,805	Constructed	0.2	110	Max: 2,073.3 Min: 2,003.3	5,756	70.0	Fuse plug for over pumping	7,910				
LEWISTON	Earth and rockfill	55	34,500	9,250	Constructed, Lewiston Reservoir	3.0	1,900	Max: 655.0 Min: 620.0	57,000	35.0	None					
LUDINGTON	Asphalt faced, zoned earth embankment	170	31,824	37,669	Constructed	1.3	840	Max: 942.3 Min: 875.3	52,210	66.9	None					
MORMON FLAT #2	Existing concrete thin arch dam	224	380	60	Existing Canyon Lake constructed		939	Max: 1,660.9 Min: 1,645.0	25,699	15.0	Chute spillway		Fixed wheel	2	49.87	49.87

Table 2, Page 3

ASCE Pumped Storage Compendium
Table 2 - Summary of Data on Upper Dams, Reservoirs, and Spillways

Plant Name	Dams Type	Height ft	Length ft	Volume yd³×10³	Reservoirs Type	Drainage Area mi²	Surface Area acres	Max/Min Elevations ft msl	Power Pool Usable Storage acre-ft	Fluctuation ft	Spillways Type	Peak Discharge ft³/s	Type	Gates Number	Height ft	Width ft
MT. ELBERT	Zoned earthfill embankment, Mt. Elbert forebay	92	2,650	846	Constructed	1.0	282	Max: 9,645.7 Min: 9,590.0	7,395	30.0	Morning glory	9,069	None - stored inflow			
MUDDY RUN	Earth and rockfill	250	4,400	5,179	Constructed	9.2	960	Max: 520.0 Min: 470.0	33,158	50.0	Concrete, ungated	13,000				
NORTHFIELD MOUNTAIN	Central core rockfill	145	5,600	2,700	Constructed	0.5	275	Max: 1,004.5 Min: 920.0	14,350	84.5	Overflow	11,400	None			
RACCOON MOUNTAIN	Earth and rockfill	230	8,500	9,351	Constructed	1.4	528	Max: 1,672.0 Min: 1,530.0	36,340	142.0	None					
ROCKY MOUNTAIN	Zoned earthfill	80	12,800	10,738	Constructed		210	Max: 1,392.0 Min: 1,341.0	10,200	51.0	Fuse plug					
ROCKY RIVER	Earth with wooden core wall	100	952	368	Constructed	35.0	5,420	Max: 430.0 Min: 400.0	53,480	30.0	None					
RUSSEL DAM	Concrete	195	1,834	4,499	Constructed	283.0	26,653		126,864	5.0	Ogee concrete gravity	801,500	Tainter	10	45.00	50.00
SALINA	Rockfill and earth	200	2,300	3,700	Constructed		785	Max: 865.0 Min: 845.0		20.0	None		None			
SMITH MOUNTAIN	Concrete arch	235	820	191	Constructed, Smith Mountain Reservoir	1,023.9	20,608	Max: 795.0 Min: 787.0	157,764	8.0	Free overflow	200,000				

Table 2, Page 4

ASCE Pumped Storage Compendium
Table 2 - Summary of Data on Upper Dams, Reservoirs, and Spillways

Plant Name	Dams				Reservoirs				Power Pool			Spillways				
	Type	Height ft	Length ft	Volume yd³×10³	Type	Drainage Area mi²	Surface Area acres	Max/Min Elevations ft msl	Usable Storage acre-ft	Fluctuation ft	Type	Peak Discharge ft³/s	Type	Gates Number	Height ft	Width ft
TAUM SAUK	Rockfill, concrete face, asphalt floor	92	6,660	3,750	Constructed	0.1	55	Max: 1,597.0 Min: 1,505.0	4,350	88.0						
THERMALITO POWER PLANT	Zoned earthfill	71	16,000	1,580	Constructed	0.1	630	Max: 226.0 Min: 222.0	9,936	4.0						
WALLACE DAM	Concrete, earth, non-overflow	117	2,395	990	Constructed, Oconee Lake	1,830.0	19,050	Max: 435.0 Min: 430.0	756,694	5.0	Gated ogee	268,000	Radial lift	5	44.00	42.00
YARDS CREEK STATION	Rockfill with earth core	70	9,600	26,700	Constructed	0.2	154	Max: 1,555.0 Min: 1,506.0	4,650	49.0	None					

ASCE Pumped Storage Compendium
Table 3 - Summary of Data on Upper Intakes

Plant Name	Number of Intakes	Type of Intake	Area of Each Inlet ft²	Bar Racks Shape	Bar Racks Spacing in	Bar Racks Depth in	Bar Racks Thickness in	Bar Racks Diameter in	Service Gates Type	Service Gates Number	Service Gates Height ft	Service Gates Width ft	Emergency Gates Type	Emergency Gates Number	Emergency Gates Height ft	Emergency Gates Width ft
BAD CREEK PUMPED STORAGE PROJECT	1	Concrete submerged bellmouth	1,636						Spherical valves used as service gates	4			Spherical valves, 2.5 m (8.2 ft) diameter	4		
BALSAM MEADOW	1	Flaired concrete structure	1,150		5.00	3.00	0.38		Slide gate	2	10.0	17.5		None		
BATH COUNTY	3	Horizontal	4,408	Rectangular	9.50	5.00	0.63		Fixed wheel	6	28.7	12.8	Bulkhead	6	28.7	12.8
BEAR SWAMP	1	Glory hole	1,965							None				None		
BLENHEIM-GILBOA PUMPED STORAGE PROJECT	1	Ungated submerged circular weir (morning glory)	3,390		6.00	2.00	0.75			None				None		
CABIN CREEK	1	Vertical shaft with cover	5,736	Rectangular	4.00	4.50	0.44			None				None		
CARTERS	2	Concrete	7,064						Tractor	4	20.5	14.0		None		
CASTAIC POWER PROJECT	1	Reinforced concrete gravity structure, multiple compartmented	352							None			Flat-leaf coaster type	1	32.0	25.0
CLARENCE CANNON	1	Flaired, integral with powerhouse	1,728	Rectangular	8.00	3.00	0.69		Vertical, slide type - 3 bulkheads	1	37.3	17.5		None		
DE GRAY DAM AND RESERVOIR	1	Vertical, octagonal tower, with selective temperature control	1,890	Rectangular	6.00	3.00	0.63		Cylinder, 8.83 m (29 ft) diameter	1			Trashrack, bulkhead	4	23.5	22.5
EDWARD HYATT POWER PLANT	2	Inclined screen	380	Elliptical	4.50	3.50	0.75			None			Hydraulic, inclined slide gate	2	32.8	22.5

Table 3, Page 1

ASCE Pumped Storage Compendium
Table 3 - Summary of Data on Upper Intakes

Plant Name	Number of Intakes	Type of Intake	Area of Each Inlet ft²	Shape	Bar Racks Spacing in	Bar Racks Depth in	Bar Racks Thickness in	Bar Racks Diameter in	Service Gates Type	Service Gates Number	Service Gates Height ft	Service Gates Width ft	Emergency Gates Type	Emergency Gates Number	Emergency Gates Height ft	Emergency Gates Width ft
FAIRFIELD	4	Gated intake structure	1,501	Rectangular	7.00	3.00	1.00		Roller	1	30.5	18.3	Roller	4	30.5	18.3
FLATIRON POWERPLANT	1	Reinforced concrete	162		2.00	0.50	0.50			None			Butterfly	1	8.0	
GIANELLI SAN LUIS PUMP-GENERATE PLANT	4	Roller gate	2,151	Rectangular	6.00	2.00	0.63			None			Roller	4	22.0	17.5
GRAND COULEE	2	Canal	288		6.25	3.00	0.75			None			76.2 cm (30 in) spool valve	1		
HELMS	1	Submerged, horizontal	4,596	Circular	8.50			1.50	Slide	2	26.9	11.8	Fixed wheel	1	32.2	21.3
HIWASSEE-UNIT 2	1	Semicircular tower	3,050	Rectangular	6.00	6.00	1.00			None			Roller train lift gate	2	26.0	19.0
HORSE MESA #4	1	Penstock intake	715	Rectangular	7.25	3.50	0.75		N/A	None			Fixed wheel	1	20.3	17.1
JOCASSEE PUMPED STORAGE STATION	2	Circular, 8 sided	3,840	Rectangular	6.50	4.00	0.75		Cylindrical	2	27.5	39.6	Stop logs (steel)	24	9.5	19.0
KINZUA	1	Bell mouth with vortex	792							None				None		
LEWISTON	12	Concrete gravity with rectangular openings	880	Rectangular	6.00	3.00	1.00		Head gates	12	25.0	15.2	Stop log	None	25.0	16.2
LUDINGTON	1	Horizontal, reinforced concrete	817						Vertical lift, fixed wheel	6	28.9	28.9		None		
MORMON FLAT #2	1	Penstock intake	805	Rectangular	5.25	3.50	0.75			None			Fixed wheel	1	16.4	20.0

Table 3, Page 2

ASCE Pumped Storage Compendium
Table 3 - Summary of Data on Upper Intakes

Plant Name	Number of Intakes	Type of Intake	Area of Each Inlet ft²	Shape	Bar Racks Spacing in	Depth in	Thickness in	Diameter in	Service Gates Type	Number	Height ft	Width ft	Emergency Gates Type	Number	Height ft	Width ft
MT. ELBERT	2	Concrete structure	1,200	Elliptical	6.00	2.00	0.62			None			Wheel mounted	2	12.5	15.0
MUDDY RUN	4	Concrete structure, tower	2,224		6.00	3.50	0.63		Circular gate, 7.16 m (23.5 ft) diameter	1				None		
NORTHFIELD MOUNTAIN	1	Horizontal, side intake	4,055		6.00	0.75	0.75			None				None		
RACCOON MOUNTAIN	1	Horseshoe-shaped vertical concrete tower	27,648							None			Spherical valves 3.048 meters diameter	1	10.0	
ROCKY MOUNTAIN	1	Covered morning glory	3,848							None				None		
ROCKY RIVER	1	Circular concrete tower intake	1,170		3.00				Broom gate	1	18.0	18.5		None		
RUSSEL DAM	4															
SALINA	6		154	Rectangular	3.75	5.00	0.50			None				None		
SMITH MOUNTAIN	3	Bell mouth		Rectangular	4.00	3.50	0.37			None			Wheel	3	24.4	22.6
TAUM SAUK	1	Cylindrical vertical reinforced concrete	581							None				None		
THERMALITO POWER PLANT	4			Rectangular	6.00	2.25	0.50			None				None		
WALLACE DAM	6	Concrete	1,863	V-round H-elliptical	14.50	22.00	1.00	4.00 slipe		2	32.4	17.1		None		

Table 3, Page 3

ASCE Pumped Storage Compendium
Table 3 - Summary of Data on Upper Intakes

Plant Name	Number of Intakes	Type of Intake	Area of Each Inlet ft²	Shape	Bar Racks				Service Gates				Emergency Gates		
					Spacing in	Depth in	Thickness in	Diameter in	Type	Number	Height ft	Width ft	Type	Number Height ft	Width ft
YARDS CREEK STATION	1	Conventional bell mouth concrete	2,900	Rectangular	3.00	3.00	0.38		Steel-faced stop log panels	6	10.0	12.0		None	

Table 3, Page 4

ASCE Pumped Storage Compendium
Table 4 - Summary of Data on Conduits

Plant Name	Upper Tunnel				Shaft				Lower Tunnel				Penstocks				Tailrace Tunnel			
	Number	Length ft	Diameter ft	Vel. ft/s	Number	Length ft	Diameter ft	Vel. ft/s	Number	Length ft	Diameter ft	Vel. ft/s	Number	Length ft	Diameter ft	Vel. ft/s	Number	Length ft	Diameter ft	Vel. ft/s
BAD CREEK PUMPED STORAGE PROJECT	1	3,701	29.53	21.0	1	856	29.53	23.0	1	3,701	29.53	23.0	4	386	13.78	26.9	2	875	24.61	16.7
BALSAM MEADOW	1	2,854	18.00	9.1	1	939	13.50	16.2	1	875	18.00	9.1	1	320	12.00	20.5	1	7,500	18.00	9.1
BATH COUNTY	3	3,618	28.50	16.9	3	990	28.50	16.9	3	3,459	28.50	16.9	6	1,050	18.00	21.2				
BEAR SWAMP					1	749	25.00	5.4	1	410	25.00	5.4	2	175	17.50	5.5	2	500	22.01	3.5
BLENHEIM-GILBOA PUMPED STORAGE PROJECT					1	1,035	28.00	24.0	1	900	28.00	24.0	4	1,960	12.00	33.0				
CABIN CREEK					1	1,019	15.00	25.0	1	3,093	12.00	38.9	2	188	8.50	38.8				
CARTERS													2	838	18.00	21.2				
CASTAIC POWER PROJECT	1	37,775	30.00	26.0									6	3,500	12.50					
CLARENCE CANNON													1	70	26.00	11.3				
DE GRAY DAM AND RESERVOIR					1	118	29.00	15.4	1	1,258	29.00	15.4	1	194	13.17	20.5				
EDWARD HYATT POWER PLANT	2	385	22.00										6	196	12.00	30.0	2		35.00	10.0
FAIRFIELD	4	375	24.00										4	720	25.98	22.3				
FLATIRON POWERPLANT	1	6,912	10.00										1	480	8.00					

Table 4, Page 1

ASCE Pumped Storage Compendium
Table 4 - Summary of Data on Conduits

Plant Name	Upper Tunnel				Shaft				Lower Tunnel				Penstocks				Tailrace Tunnel			
	Number	Length ft	Diameter ft	Vel. ft/s	Number	Length ft	Diameter ft	Vel. ft/s	Number	Length ft	Diameter ft	Vel. ft/s	Number	Length ft	Diameter ft	Vel. ft/s	Number	Length ft	Diameter ft	Vel. ft/s
GIANELLI SAN LUIS PUMP-GENERATE PLANT													4	2,146	17.49	17.7				
GRAND COULEE	6	220	18.00	9.4	6	226	18.00	9.4					6	181	12.00	21.2	6	89	14.00	15.5
HELMS	1	13,215	27.00	17.3	1	1,745	27.00	17.3	1	4,068	27.00	17.3	3	505	15.42	38.1	3	270	15.49	17.4
HIWASSEE-UNIT 2													1	190	18.00	19.3				
HORSE MESA #4													1	187	15.42	30.2				
JOCASSEE PUMPED STORAGE STATION					2	276	33.50	18.0	2	1,200	33.50	18.0	4	160	20.50					
KINZUA					1	482	22.33	24.0	1	1,990	22.33	24.0	2	330	14.00	24.0				
LEWISTON													12	152	24.00	14.5				
LUDINGTON													6	1,299	28.54	27.9				
MORMON FLAT #2													1	175	18.01	21.3				
MT.ELBERT													2	3,000	15.00	20.3				
MUDDY RUN					4	350	24.50	18.5	4	440	24.50	18.5	8	500	14.00	28.5				
NORTHFIELD MOUNTAIN					1	850	31.00	26.5					4	340	14.79	33.5	1	5,130	33.00	22.6
RACCOON MOUNTAIN					1	900	35.00	21.0	1	975	35.00	21.0	4	150	17.80	20.3	1	1,575	35.00	21.0
ROCKY MOUNTAIN					1	567	35.01	19.2	1	1,935	35.01	19.2	3	482	19.00	21.8				
ROCKY RIVER													1	1,677	15.00		1		8.00	

Table 4, Page 2

ASCE Pumped Storage Compendium
Table 4 - Summary of Data on Conduits

Plant Name	Upper Tunnel			Shaft			Lower Tunnel			Penstocks			Tailrace Tunnel		
	Number	Length ft	Diameter ft	Vel. ft/s	Number	Length ft	Diameter ft	Vel. ft/s	Number	Length ft	Diameter ft	Vel. ft/s	Number	Length ft	Diameter ft	Vel. ft/s

Plant Name	Upper Tunnel Number	Length ft	Diameter ft	Vel. ft/s	Shaft Number	Length ft	Diameter ft	Vel. ft/s	Lower Tunnel Number	Length ft	Diameter ft	Vel. ft/s	Penstocks Number	Length ft	Diameter ft	Vel. ft/s	Tailrace Tunnel Number	Length ft	Diameter ft	Vel. ft/s
RUSSEL DAM													4		26.00					
SALINA													6	640	14.00					
SMITH MOUNTAIN													3	170	26.00					
TAUM SAUK					1	431	27.17	15.0	1	6,571	25.50	17.2	2	248	13.50	31.0				
THERMALITO POWER PLANT													4	118	21.00	11.3				
WALLACE DAM													6	95	25.50	15.7	6	77		21.2
YARDS CREEK STATION	1	1,500	20.00	26.0									3	2,000	10.00	29.0				

Table 4, Page 3

ASCE Pumped Storage Compendium
Table 5 - Summary of Data on Pump-Turbines

Plant Name	Plant Capacity MW	Number of Units	Diameter ft	Rotational Speed Synchronous/Overspeed rpm	Unit Submergence ft	Valves Number/type	Diameter ft	Generating at Best Gate Rated Head ft	Capacity MW	Discharge ft³/s	Efficiency %	Pumping at Best Gate Rated Head ft	Capacity MW	Discharge ft³/s	Efficiency %
BAD CREEK PUMPED STORAGE PROJECT	1,000	4	17.59	300.0/375.0	157.5	4 Spherical	8.4	1,013	270.0	3,143	91.60	1,246	318.0	1,783	92.60
BALSAM MEADOW	200	1	13.78	400.0/560.0	98.0	1 Spherical	7.0	1,266	221.0	2,103	91.60	1,324	173.0	1,429	93.60
BATH COUNTY	2,100	6	20.94	257.0/415.0	65.0	6 Spherical	9.5	1,080	380.0	4,700	90.70	1,100	420.0	4,200	92.70
BEAR SWAMP	600	2	19.16	225.0/NA	70.0	2 Spherical	11.0	720	298.0		92.50	685		4,430	91.50
BLENHEIM-GILBOA PUMPED STORAGE PROJECT	1,000	4	19.90	257.0/NA	50.0	4 Spherical	9.2	1,110	270.0	3,120	92.00	1,173	264.0	2,300	84.00
CABIN CREEK	300	2	13.75	360.0/485.0	35.0	2 Spherical	5.5	1,190	160.0			1,230		840	
CARTERS	258	2	20.66	150.0/NA	16.5	None		345	129.0	5,200		347	110.0	4,435	
CASTAIC POWER PROJECT	1,250	6	19.25	257.0/NA	50.0	6 Spherical	8.7	1,078	255.3	3,500		1,065	227.0	2,250	
CLARENCE CANNON	31	1	22.20	75.0/NA	6.0	None		75	26.5	4,750	90.00	75	33.8	4,500	
DE GRAY DAM AND RESERVOIR	28	1	16.20	128.5/182.0		1 Butterfly	14.0	171	28.9	2,150	93.00	175	30.5	1,850	88.00
EDWARD HYATT POWER PLANT	293	3	19.08	189.5/284.0	30.0	6 Ball valve	9.5	615	98.0		91.60	592	77.0	1,870	
FAIRFIELD	512	8	16.17	150.0/241.0	32.0	None		162	61.9	6,003	90.40	162		4,803	92.30

All plants have reversible Francis turbines, except Rocky River, which was built in 1929.
NA = Information is not available.

Table 5, Page 1

ASCE Pumped Storage Compendium
Table 5 - Summary of Data on Pump-Turbines

Plant Name	Plant Capacity MW	Number of Units	Diameter ft	Rotational Speed Synchronous/Overspeed rpm	Unit Submergence ft	Valves Number/type	Valves Diameter ft	Generating at Best Gate - Rated Head ft	Generating at Best Gate - Capacity MW	Generating at Best Gate - Discharge ft³/s	Generating at Best Gate - Efficiency %	Pumping at Best Gate - Rated Head ft	Pumping at Best Gate - Capacity MW	Pumping at Best Gate - Discharge ft³/s	Pumping at Best Gate - Efficiency %
FLATIRON POWERPLANT	9	1	8.42	300.0/555.0	9.0	1 Butterfly	6.3	290	8.5	470		240	9.7	370	
GIANELLI SAN LUIS PUMP-GENERATE PLANT	400	8	17.72	150.0/206.0	15.0	8 Butterfly	13.0	304	45.1	2,031	86.10	341	48.3	1,487	89.70
GRAND COULEE	314	6	14.32	200.0/NA	17.0	None		280	44.7	2,200	91.50	292	50.6	1,813	91.60
HELMS	1,206	3	16.83	360.0/532.0	200.1	3 Spherical	8.0	1,624	358.0	2,818	91.80	1,499	350.0	2,511	91.90
HIWASSEE-UNIT 2	68	1	15.17	105.9/161.0	1.0	None		190	60.0	4,180		205	76.0	3,900	
HORSE MESA #4	97	1	18.04	150.0/187.5	20.4	None		246	80.0			260		3,740	
JOCASSEE PUMPED STORAGE STATION	610	4	24.00	120.0/138.0	16.0	None		294	170.0		93.00	294	170.0	6,200	92.00
KINZUA	350	2	18.67	225.0/NA	72.0	2 Spherical	9.5	646	162.0			700	195.0	3,200	
LEWISTON	240	12	17.17	112.5/202.0	4.0	None		111	28.0	3,500	86.00	118	26.0	3,000	86.00
LUDINGTON	1,979	6	27.49	112.5/163.0	26.5	None		353	308.0	11,509	92.20	286	323.0	11,124	92.70
MORMON FLAT #2	47	1	17.06	138.5/173.0	13.3	None		135	41.8			143	45.5	4,070	
MT.ELBERT	200	2	18.83	180.0/270.0	38.0	None		400	103.0	3,325	92.00	430	132.4	3,400	
MUDDY RUN	800	8	17.91	180.0/NA	20.0	None		353	103.0			427	115.0	2,610	
NORTHFIELD MOUNTAIN	1,080	4	17.21	257.0/390.0	106.0	4 Spherical	9.5	745	259.5			740	267.7	3,600	

All plants have reversible Francis turbines, except Rocky River, which was built in 1929.
NA = Information is not available.

Table 5, Page 2

ASCE Pumped Storage Compendium
Table 5 - Summary of Data on Pump-Turbines

Plant Name	Plant Capacity MW	Number of Units	Diameter ft	Rotational Speed Synchronous/Overspeed rpm	Unit Submergence ft	Valves Number/type	Diameter ft	Generating at Best Gate				Pumping at Best Gate			
								Rated Head ft	Capacity MW	Discharge ft³/s	Efficiency %	Rated Head ft	Capacity MW	Discharge ft³/s	Efficiency %
RACCOON MOUNTAIN	1,530	4	16.58	300.0/455.0	107.0	--- 4 --- Spherical	10.0	1,020	400.0	5,020		1,000	360.0	3,850	
ROCKY MOUNTAIN	760	3	18.70	225.0/337.5	71.0	--- 3 --- Spherical	10.7	613	253.3	5,700		613	263.2	5,170	
ROCKY RIVER	32	1	4.50	327.0/634.0	5.8	--- 2 --- Butterfly	4.5	240	32.0	1,870	70.00	240	6.5	250	78.00
RUSSEL DAM	360	4		120.0/NA		-- None --		157	93.0			161	88.0		
SALINA	260	6	15.00	171.4/197.0	5.0	--- 6 --- Butterfly	13.4	225	42.0	2,800	83.00	245	48.0	1,900	86.00
SMITH MOUNTAIN	240	3	22.31	105.9/NA	4.0	-- None --		180	70.0		90.00	197	75.0	4,220	93.00
TAUM SAUK	350	2	21.25	200.0/220.0	32.0	--- 2 --- Spherical	9.0	790	220.0			764	179.0	2,650	
THERMALITO POWER PLANT	115	4	17.17	112.5/169.0		-- None --		85	26.0						
WALLACE DAM	209	4	21.92	85.7/105.0	7.0	-- None --		96	52.5	7,200	91.80	96	63.4	7,800	93.00
YARDS CREEK STATION	360	3	18.42	240.0/364.0	25.0	--- 3 --- Spherical	7.0	656	112.0			732		2,145	

All plants have reversible Francis turbines, except Rocky River, which was built in 1929.
NA = Information is not available.

Table 5, Page 3

ASCE Pumped Storage Compendium
Table 6 - Summary of Data on Lower Dams, Reservoirs, and Spillways

Plant Name	Dam Type	Height ft	Length ft	Volume yd³ x10³	Reservoir Type	Drainage Area mi²	Surface Area acres	Max/Min Elevations ft msl	Usable Storage acre-ft	Power Pool Fluctuation ft	Spillway Type	Peak Discharge ft³/s	Spillway Gates Type	No.	Height ft	Width ft
BAD CREEK PUMPED STORAGE PROJECT	Zoned fill	385	1,750	13,000	Existing, Jocassee Reservoir	148	7,565	Max: 1110.00 Min: 1080.00	215,700	30.0			Tainter	2	33.00	38.00
BALSAM MEADOW	Existing concrete gravity	185	1,760		Existing, Shaver Lake			Max: 5372.00 Min: 5330.00	60,000	42.0						
BATH COUNTY	Zoned fill	135	2,300	4,002	On stream	73	555	Max: 2118.00 Min: 2058.00	22,500	60.0	Chute	60,000	Radial	2	30.00	32.00
BEAR SWAMP	Earth and rockfill	130	900	567	On stream	1,250	1,440	Max: 870.00 Min: 830.00	5,900	40.0	Gated	73,900	Tainter	2	40.00	36.00
BLENHEIM-GILBOA PUMPED STORAGE PROJECT	Earth and rockfill	165	18,000	1,159	Constructed	40	430	Max: 900.00 Min: 860.00	15,000	40.0	Ogee, concrete gravity	150,000	Radial	3	46.50	38.00
CABIN CREEK	Compacted earth and rockfill, impervious core	95	1,150	1,043	Constructed	14	52	Max: 10012.0 Min: 9970.00	1,850	42.0	Ogee crest, 44.2 m (145 ft) wide + tunnel spillway.	13,600				
CARTERS	Concrete	35	208	766	Constructed	530	870	Max: 707.00 Min: 665.50	17,210	15.0		197,800	Tainter	4	36.50	42.00
CASTAIC POWER PROJECT	Earth and rockfill	170	1,960	5,903	Constructed, Castaic Reservoir	76	493	Max: 1540.00 Min: 1480.00	20,000	60.0	Broad crested weir					

ASCE Pumped Storage Compendium
Table 6 - Summary of Data on Lower Dams, Reservoirs, and Spillways

Plant Name	Dam Type	Height ft	Length ft	Volume yd³ ×10³	Reservoir Type	Drainage Area mi²	Surface Area acres	Max/Min Elevations ft msl	Power Pool Usable Storage acre-ft	Fluctuation ft	Type	Peak Discharge ft³/s	Spillway Gates Type	No.	Height ft	Width ft
CLARENCE CANNON	Rolled earth	31	1,550		Constructed, reregulating dam	29	1,200	Max: 528.00 Min: 521.00	4,960	7.0	Concrete, gravity type		Tainter	2	31.00	30.00
DE GRAY DAM AND RESERVOIR	Rolled earthfill	40	1,071	56	Constructed, reregulating dam	27	430	Max: 221.00 Min: 217.00	1,600	4.0	Ogee, ungated 94.5 m (310 ft) crest length	45,000				
EDWARD HYATT POWER PLANT	Concrete gravity	143	1,300	154	Constructed		323	Max: 225.00 Min: 221.00	13,000	4.0	Gated ogee crest		Radial	14	23.00	40.00
FAIRFIELD	Concrete, Crest Gates	46	2,001	62	Existing, Parr Reservoir	5	4,398	Max: 266.01 Min: 256.00	29,024	10.0	Overflow	228,000	Bascule	10	8.99	200.00
FLATIRON POWERPLANT	Zoned earthfill	86	1,726	382	Constructed, Flatiron Afterbay	7	47	Max: 5472.80 Min: 5462.00	440	10.8	Concrete, ungated	23,625				
GIANELLI SAN LUIS PUMP-GENERATE PLANT	Earth and rockfill (Forebay Dam - O'Neill)	87	14,301	2,877	Constructed - no inflow drainage area.		2,250	Max: 225.00 Min: 217.00			Morning Glory - 3.6 m (11.8 ft) diameter	3,300				
GRAND COULEE	Concrete gravity, Grand Coulee Dam	550	5,223	11,977	On stream, Franklin D. Roosevelt Lake	74,100	82,304	Max: 1290.00 Min: 1208.00	5,185,400	82.0	Overflow chute	1,000,000	Drum	11	30.00	135.00

Table 6, Page 2

ASCE Pumped Storage Compendium
Table 6 - Summary of Data on Lower Dams, Reservoirs, and Spillways

Plant Name	Dam Type	Height ft	Length ft	Volume yd³×10³	Reservoir Type	Drainage Area mi²	Surface Area acres	Max/Min Elevations ft msl	Power Pool Usable Storage acre-ft	Fluctuation ft	Type	Spillway Peak Discharge ft³/s	Type	Gates No.	Height ft	Width ft
HELMS	Rockfill concrete faced	260	3,330	3,700	Existing, constructed, Lake Wishon	177	1,166	Max: 6548.56 Min: 6440.29	89,989	109.9	Overflow, ogee	30,017	Radial	6	11.52	40.03
HIWASSEE-UNIT 2	Concrete	150	1,309	4,055	On stream	50	2,142	Max: 1276.57 Min: 1272.31	8,837	8.0	Concrete	156,585	Radial	10	23.00	31.99
HORSE MESA #4	Existing concrete thin arch dam.	224	380	125	Existing Canyon Lake, constructed		939	Max: 1660.93 Min: 1655.41	25,699	5.5	Chute spillway		Fixed wheel	2	49.87	49.87
JOCASSEE PUMPED STORAGE STATION	Homogenous earthfill, Keowee Dam Station	170	3,500	2,043	Constructed, Lake Keowee	439	18,372	Max: 800.00 Min: 775.00	391,700	25.0	Concrete chute		Taintor	4	35.00	38.00
KINZUA	Concrete and earthfill				On stream, Allegheney Reservoir	21,080	20,992	Max: 1365.00 Min: 1260.00	4,650	105.0						
LEWISTON	Concrete gravity	389	1,840	1,104	Constructed			Max: 565.00 Min: 548.00		28.0						
LUDINGTON	None				Lake Michigan		14,336,248	Max: 579.72 Min: 579.72			None					
MORMON FLAT #2	Existing concrete thin arch dam	207	1,531	120	Existing Saguaro Lake, constructed		1,255	Max: 1529.53 Min: 1522.31	8,521	7.0	2 spillways		Radial	9	22.97	27.00

ASCE Pumped Storage Compendium
Table 6 - Summary of Data on Lower Dams, Reservoirs, and Spillways

Plant Name	Dam Type	Dam Height ft	Dam Length ft	Dam Volume yd³×10³	Reservoir Type	Drainage Area mi²	Surface Area acres	Max/Min Elevations ft msl	Usable Storage acre-ft	Power Pool Fluctuation ft	Power Pool Type	Peak Discharge ft³/s	Spillway Type	Gates No.	Gates Height ft	Gates Width ft
MT. ELBERT	Zoned earthfill embankment	55	3,140	624	On stream, Twin Lakes	105	2,817	Max: 9200.00 Min: 9168.70	67,930	4.0	Morning glory 12.11 m (39.7 ft)	12,841	None			
MUDDY RUN	Concrete gravity	102	4,649	667	On stream, Conowingo Pond	27,000	2,656	Max: 109.00 Min: 105.50	80,990	4.5		1,170,000	Stoney type crest	50	22.50	38.00
NORTHFIELD MOUNTAIN	Existing concrete Turner Falls Dam	55			On stream, on the Connecticut River, Turner Falls Pond	7,163	2,000	Max: 185.00 Min: 176.00	12,745	9.0	Gated	210,000	3 Radial, 4 Bascule	7	13.25	120.00
RACCOON MOUNTAIN	Concrete with earthfill embankments, Nickajack Dam	81	3,766		On stream, constructed, Nickajack Reservoir	21,873	10,371	Max: 634.00 Min: 632.00	20,673	2.0		500,000	Radial	10	40.00	40.00
ROCKY MOUNTAIN	Concrete gravity, zoned earthfill	72	650	571	Constructed	15	600	Max: 710.50 Min: 681.00	15,690	20.0	Overflow	24,900	Radial	2	23.55	22.50

Table 6, Page 4

ASCE Pumped Storage Compendium
Table 6 - Summary of Data on Lower Dams, Reservoirs, and Spillways

Plant Name	Dam Type	Dam Height ft	Dam Length ft	Dam Volume yd³×10³	Reservoir Type	Drainage Area mi²	Surface Area acres	Max/Min Elevations ft msl	Power Pool Usable Storage acre-ft	Power Pool Fluctuation ft	Power Pool Type	Spillway Peak Discharge ft³/s	Spillway Type	Gates No.	Gates Height ft	Gates Width ft
ROCKY RIVER	No lower reservoir, on Housatonic River				On stream			Max: 200.00 Min: 200.00								
RUSSEL DAM	Zoned fill, concrete and earth	170	5,680	9,550	Constructed	6,144	71,100		1,045,000	18.0		1,015,000	Tainter	23	35.00	40.00
SALINA	Rockfill earth concrete constructed				Constructed, Lake Hudson			Max: 636.00 Min: 619.00		17.0						
SMITH MOUNTAIN	Concrete gravity	94	981	109	Constructed, Leesville Reservoir	1,505	3,262	Max: 613.00 Min: 600.00	37,779	13.0	Gated	242,000	Radial	4	35.10	49.87
TAUM SAUK	Ogee weir, concrete, gravity	60	390	22	Constructed	91	200	Max: 750.00 Min: 735.00	6,350	15.0	Ogee weir	120,000				
THERMALITO POWER PLANT	Zoned earthfill	39	42,000	5,020	Constructed	5	4,500	Max: 136.50 Min: 123.00	2,888	13.5						
WALLACE DAM	Concrete and earth, non-overflow.	100	2,766	947	Existing, Lake Sinclair	2,910	15,400	Max: 340.00 Min: 338.00	724,120	2.0	Gated ogee	479,000	Radial lift	24	18.00	30.00

Table 6, Page 5

ASCE Pumped Storage Compendium
Table 6 - Summary of Data on Lower Dams, Reservoirs, and Spillways

Plant Name	Dam				Reservoir				Power Pool			Spillway			
	Type	Height ft	Length ft	Volume yd³×10³	Type	Drainage Area mi²	Surface Area acres	Max/Min Elevations ft msl	Usable Storage acre-ft	Fluctuation ft	Type	Peak Discharge ft³/s	Gates Type	No.	Height ft	Width ft
YARDS CREEK STATION	Earthfill, till core	55	1,300	8,700	On stream, constructed	4	282	Max: 818.50 Min: 795.00	4,650	23.5	Concrete ogee	11,000				

Table 6, Page 6

ASCE Pumped Storage Compendium
Table 7 - Summary of Data on Lower Intakes

Plant Name	Number of Intakes	Type of Intake	Area of Each Inlet ft²	Bar Racks Shape	Bar Racks Spacing in	Bar Racks Depth in	Bar Racks Thickness in	Bar Racks Diameter in	Service Gates Type	Service Gates Number	Service Gates Height ft	Service Gates Width ft	Emergency Gates Type	Emergency Gates Number	Emergency Gates Height ft	Emergency Gates Width ft
BAD CREEK PUMPED STORAGE PROJECT	1	Vertical, 1 structure with 4 openings	1,647	Rectangular	0.59	0.39	0.10		Draft tube, bonnet side	4	19.3	18.2		None		
BALSAM MEADOW	1	Flaired concrete structure	1,150		5.00	3.00	0.38		Slide gate	2	10.0	17.5		None		
BATH COUNTY	6	Draft tube	1,420	Rectangular	9.75	4.00	0.75		Slide gate	12	19.5	14.2		None		
BEAR SWAMP	2		600		6.00	2.50	0.94		Slide	2	16.0	20.0		None		
BLENHEIM-GILBOA PUMPED STORAGE PROJECT	4	Draft tube	1,350		6.00	1.25	0.75		Bulkhead	4	12.5	13.5		None		
CABIN CREEK	2	Draft tube type								None				None		
CARTERS																
CASTAIC POWER PROJECT	2	West Branch of California Aqueduct and Castiac Creek.								None				None		
CLARENCE CANNON	1	Draft tube	1,620	Rectangular	8.00	3.00	0.69		Vertical lift, slide type - 3 bulkheads	1	69.0	19.5		None		
DE GRAY DAM AND RESERVOIR	1	Draft tube	240						Vertical lift, sliding	2	16.9	17.7		None		
EDWARD HYATT POWER PLANT																
FAIRFIELD	1	16 Draft tube openings	567	Rectangular	7.00	3.00	1.00		None				Slide	16	15.1	25.0

Table 7, Page 1

ASCE Pumped Storage Compendium
Table 7 - Summary of Data on Lower Intakes

Plant Name	Number of Intakes	Type of Intake	Area of Each Inlet ft²	Bar Racks Shape	Bar Racks Spacing in	Bar Racks Depth in	Bar Racks Thickness in	Bar Racks Diameter in	Service Gates Type	Service Gates Number	Service Gates Height ft	Service Gates Width ft	Emergency Gates Type	Emergency Gates Number	Emergency Gates Height ft	Emergency Gates Width ft
FLATIRON POWERPLANT									Butterfly	1	6.3			None		
GIANELLI SAN LUIS PUMP-GENERATE PLANT	8	Suction tube	628	Rectangular	6.50	2.00	0.63			None			Stop log	8	15.7	12.8
GRAND COULEE	2	Concrete	1,380		6.25	2.50	0.63			None			Reverse flow wheel	2	20.1	13.0
HELMS	1	Submerged, horizontal	3,671	Circular	8.50			1.50	Slide	2	15.4	33.8		None		
HIWASSEE-UNIT 2	1	Concrete draft tube	1,641		6.00	3.00	0.63		Slide	3	13.0	15.3		None		
HORSE MESA #4	1	Draft tube	723	Rectangular	7.25	3.50	0.75		N/A	None			N/A	None		
JOCASSEE PUMPED STORAGE STATION	4	Rectangular	4,122	Circular	6.50			1.50	Rectangular	6	22.0	14.0		None		
KINZUA	1	Tower designed for temperature control	615		1.50	0.38			Fixed wheel - vertical	2	23.5	16.5	Closure bulkhead	1	25.3	18.5
LEWISTON	12	Concrete gravity with recangular openings	667	Rectangular	6.00	3.00	1.00			None			Stop log		25.0	20.7
LUDINGTON	6	Draft tube	1,324	Circular	14.00	2.00			Slide	12	21.0	32.2		None		
MORMON FLAT #2	1	Draft tube	807	Rectangular	5.25	3.50	0.75			None				None		
MT. ELBERT	4	Draft tube	1,572	Elliptical	6.50	2.00	0.62			None				None		
MUDDY RUN	8	Draft tube								None				None		
NORTHFIELD MOUNTAIN	1	Horizontal	3,300		6.00	0.75	0.75			None				None		

Table 7, Page 2

ASCE Pumped Storage Compendium
Table 7 - Summary of Data on Lower Intakes

Plant Name	Number of Intakes	Type of Intake	Area of Each Inlet ft²	Shape	Bar Racks Spacing in	Bar Racks Depth in	Bar Racks Thickness in	Bar Racks Diameter in	Type	Service Gates Number	Service Gates Height ft	Service Gates Width ft	Type	Emergency Gates Number	Emergency Gates Height ft	Emergency Gates Width ft
RACCOON MOUNTAIN	1	Reinforced concrete	3,080						Slide	4	5.0	5.6		None		
ROCKY MOUNTAIN	3	Draft tube	724	Rectangular	9.00	4.00	0.75		Vertical lift	6	16.8	17.8		None		
ROCKY RIVER	2	Treadwell	119		4.00				Steel, slide	2	8.0	5.0		None		
RUSSEL DAM																
SALINA																
SMITH MOUNTAIN	4	Bell mouth								None			Wheel	4		
TAUM SAUK																
THERMALITO POWER PLANT		Afterbay pool								None				None		
WALLACE DAM	4	Concrete	1,662	V-round, H-elliptical	14.50	22.00	1.00	4.00	Slipe	3	16.1	23.5		None		
YARDS CREEK STATION	3	Divided draft tube	1,000	Rectangular	6.00	3.00	0.50		Steel panels	6	12.0	12.0		None		

ASCE PUMPED STORAGE COMPENDIUM
Table 8 - Data on Environmental Features

Plant Name	Fish and Wildlife	Recreation	Social
BAD CREEK PUMPED STORAGE PROJECT	Wildlife mitigation includes food plots and reforestation. The project has created a wildlife refuge.	The project provides a 69 Km (43-mi) hiking trail and several visitor access points. Fishing is available in the lower pool, which is Lake Jocassee.	The project has attracted some tourism. Currently there is only minimal development around the project area.
BALSAM MEADOW		Picnic and day use area (walk in only) at the Balsam Meadow Forebay Dam. Snow play area is located in the vicinity of the dam.	
BATH COUNTY	Fish stocked in recreation pools attract anglers. A number of material waste pile areas and excavation slopes provide open grazing areas for deer.	Downstream from the lower reservoir, two recreation ponds, campgrounds, and hiking trails have been provided. Nonpowered boating, a fishing pier, and a swimming beach are also present. Picnic areas, playground areas, camping areas, and one hunting area are available.	The project provides improved flood control for downstream communities. It also provides, improvement in minimum flows in Little Back Creek and Back Creek.
BEAR SWAMP			

Table 8, Page 1

ASCE PUMPED STORAGE COMPENDIUM
Table 8 - Data on Environmental Features

Plant Name	Fish and Wildlife	Recreation	Social
BLENHEIM-GILBOA PUMPED STORAGE PROJECT	The Power Authority built the 2.63 Mm² (650 acre) Mine Kill State Park in connection with the power project, acquired land for a wildlife management program in the project area, nd provided a 64 Km² (16 acre) fish pond about 25.9 Km (10 mi) northwest of the project near Summit.	An admission-free visitor's center, housed in a restored 19th century barn, is part of an educational-historical-conservation complex created by the Power Authority and is operated as a museum by the Schoharie County Historical Society.	The historic Lansing Manor, Visitor's Bureau exhibits, and a weather station are located near the plant. The conservation, recreation, and cultural facilities earned the Power Authority the Department of Interior's highest outdoor-recreation award of 1976.
CABIN CREEK	Goats and sheep can be seen on the slopes near the lower reservoir. An abundance of marmots are on the site. Elk can be seen above the upper reservoir.	There is no recreation on the Cabin Creek site. Fishing from the bank is permitted on Clear Lake immediately downstream from the lower reservoir. No boating is permitted on Clear Lake.	None.

Table 8, Page 2

ASCE PUMPED STORAGE COMPENDIUM
Table 8 - Data on Environmental Features

Plant Name	Fish and Wildlife	Recreation	Social
CARTERS	The project does not have any fish and wildlife issues.	Significant recreational opportunities were developed as part of the project. These included facilities for boating, hiking, camping, fishing, swimming, picnicking, hunting, bird watching, and a visitor center. Visitation exceeds 1 million people per year.	Positive socioeconomic impact has occurred. Tourism has increased, new residential developments have been started, and new time share houseboats are present at the marina. Land use has changed from only rural with timber production, to now including housing. Historic DeSoto camp area was inundated. The reservoir will be used for water supply by the city of Chatsworth.
CASTAIC POWER PROJECT		None	
CLARENCE CANNON	Improved water quality below the re-regulating dam, increased fishery and increased wildlife habitat have resulted.	The re-regulating dam and canyon dam offer fishing, boating, swimming, camping, waterskiing, hunting, wildlife habitat, and so forth.	Benefits include community water supply, flood control, navigation, and hydropower.

Table 8, Page 3

ASCE PUMPED STORAGE COMPENDIUM
Table 8 - Data on Environmental Features

Plant Name	Fish and Wildlife	Recreation	Social
DE GRAY DAM AND RESERVOIR	There are over 121 Mm² (30,000 acres) of land and water for fisheries, forestry, and wildlife management. Natural resources are indigenous to this region of Arkansas. This is a 546 Km² (1,350 acre) waterfowl refuge near the re-regulation pool. Bald eagles are winter residents.	There are 21 recreation sites, including 6 Class A camp areas, 19 boat-launching ramps, and 10 swimming beaches. Visitation in 1990 was 9,252,000 visitor hours.	A total of 1.726 Mm³ (1,400 acre-feet) in the re-regulation pool is maintained for continuous discharge downstream for water supply, fish and wildlife purposes and pollution abatement.
EDWARD HYATT POWER PLANT	King salmon, trout, bass, catfish, and panfish, in addition to deer, ducks, geese and turkeys are present.	In addition to boating, fishing, camping, waterskiing, swimming, and hunting, there are public boat ramps, picnic areas, campgrounds, and beaches available.	
FAIRFIELD	Waterfowl impoundments on the Broad river and the Enoree river.	Facilities include a 1.21 Mm² (300-acre) recreational lake, five boat landings, two Fairfield County parks and a scenic overlook.	The local fire department has a facility for filling tankers from the upper reservoir.
FLATIRON POWERPLANT		Recreation activities include day use and overnight camping.	The twin penstocks are considered landmarks by many of the residents of Loveland, Colorado.

Table 8, Page 4

ASCE PUMPED STORAGE COMPENDIUM
Table 8 - Data on Environmental Features

Plant Name	Fish and Wildlife	Recreation	Social
GIANELLI SAN LUIS PUMP-GENERATE PLANT	Fish and wildlife at the project include deer, squirrels, wild boar (feral pigs), eagles, rabbits, fish (shad, sucker, carp, balackfish, hitch, hardhead, catfish, perch, mosquito fish, crappie, bass, war mouth, sunfish, bluegill, salmon, flounder, sturgeon, spittail, and striped bass), pheasant, quail, pigeon, pelican, egret, heron, kit, red fox, and coyote.	Recreation activities include powerboating and sailboating, wind surfing, swimming, picnicing, camping, biking, fishing, waterskiing, hiking, horseback riding, nature study, bird watching and wild flower viewing.	Domestic water from the plant is supplied to the Departments of Forestry and Parks and Recreation. Improvements include 563 Km (350 miles) of roads, 40.5 Km² (10 acres) of parking lots, irrigation.
GRAND COULEE	Fish enhancement includes two net pens. Additionally, plans are being negotiated to enhance fish and wildlife programs.	All types of recreation occurs including swimming, boating, hiking, camping, and fishing. The area has about 2 million visitors per year including 100,000 campers, 50,000 boat launches, 485,000 visitor-center visitors. Archeologists look for artifacts during low water.	Relatively few people visited this area prior to dam construction. Grand Coulee and the associated irrigation project created two large reservoirs and four permanent towns whose primary industry is project maintenance and recreation.
HELMS	Plant fish to maintain pre-project catch rate. Wildlife enhancement in selected areas.	Boat launching & docking, camping, fishing access & picnic areas.	Rebuilt and paved USFS and county roads. Yearly snow removal on 33 miles of USFS and county roads.
HIWASSEE-UNIT 2			

Table 8, Page 5

ASCE PUMPED STORAGE COMPENDIUM
Table 8 - Data on Environmental Features

Plant Name	Fish and Wildlife	Recreation	Social
HORSE MESA #4		Recreation opportunities include fishing, waterskiing, and camping at the campgrounds.	
JOCASSEE PUMPED STORAGE STATION	Fish present at the lake include small- and large-mouth bass, rainbow and brown trout, catfish, crappie, shalle cracker, and pike. Wildlife include turkey, deer, bear, wildcat, fox, eagles, falcons, boar, mountain lion, quail, rabbit, mink, and beaver.	Four paved lake access areas with nine ramps and primitive campgrounds. Activities include swimming, waterskiing, and power sail-boating. Numerous fishing tournaments have been held.	The project includes 500 acres with a bathhouse, county store, beach, and camping and rental cabins which are leased to State of South Carolina, which performs operation and maintainance. The area is a South Carolina State Park and is part of the mitigation for project which is known as Keowee-Toxaway.
KINZUA		Recreation includes boating, fishing, and camping at Kinzua Dam.	

ASCE PUMPED STORAGE COMPENDIUM
Table 8 - Data on Environmental Features

Plant Name	Fish and Wildlife	Recreation	Social
LEWISTON	The Power Authority has built a fishing facility at the upper reservoir with an approach and parking for 20 cars.	Hyde Park was enlarged by 583 Km² (144 acres) deeded to the City of Niagara Falls from the Power Authority, and an existing 9-hole golf course was expanded to 18 holes. Another 18-hole course was rearranged. A modern clubhouse was built and an underpass conects the two courses. An artificial hill for coasting near the clubhouse was constructed. Also, at Lewiston Historical Park, 174 Km² (43 acres) were provided to the Niagra Frontier State Park Commission. Recreational facilities were developed there while historic sites were preserved and reconstructed.	The Power Authority has provided financial assistance to the town of Lewiston to improve and upgrade its water supply.
LUDINGTON	Yellow perch and forage species such as sculpins and darters are abundant near the plant offshore structures. Fishing near the plant provides for trout and salmon catches.	Recreational facilities include the Lake Michigan Vista Point Scenic Overlook; a picnic area with pavillion, restrooms, and playground; and a trailer park with 48 campsites, showers, restrooms and playground. Periodic "open houses" and special tours are held.	The plant is a landmark for boating navigation.

Table 8, Page 7

ASCE PUMPED STORAGE COMPENDIUM
Table 8 - Data on Environmental Features

Plant Name	Fish and Wildlife	Recreation	Social
MORMON FLAT #2		Recreation includes fishing, waterskiing, and the use of campgrounds.	
MT. ELBERT	Twin Lakes supports a healthy trout (especially Lake Trout) population. The forebay reservoir also supports trout.	The Mt. Elbert Powerplant was built in an existing reservoir. This lower reservoir was originally two natural lakes, that were enlarged twice by constructing two separate dams. The latest reservoir enlargement increased tourism and recreational activities including boating, hiking, camping, fishing, and swimming. Over 20,000 people tour the powerplant's visitor center each year. Recreational development includes two boat ramps, two picnic areas, a campground and parking areas.	Prior to the latest reservoir enlargement, most of the buildings of the old towns of Interlaken and Twin Lakes were moved and restored. This project is part of an irrigation project.
MUDDY RUN		Recreation incudes a park with fishing and camping facilities.	
NORTHFIELD MOUNTAIN		Hiking/skiing trails, skating, riverboat tours, and a visitor center are available.	

Table 8, Page 8

ASCE PUMPED STORAGE COMPENDIUM
Table 8 - Data on Environmental Features

Plant Name	Fish and Wildlife	Recreation	Social
RACCOON MOUNTAIN	The top lake is closed to the public. No hunting, fishing, and access to lower pool is allowed.	A boat ramp, handicapped ramps at the lower pool, and a picnic area are available. Visitor tours are also conducted.	None.
ROCKY MOUNTAIN	Preservation area is at the project site.	Recreationists enjoy the picnic areas, beach, boat ramps, fishing, swimming, playground, and trails and a camp in the upper reservoir area.	Highway improvement was included as a part of the project.
ROCKY RIVER	The lake has many species of fish and the open areas have deer, wild turkeys, and many other forms of wildlife.	The lake is ringed with homes and is heavily used for boating and fishing. Picknicking, fishing and nature trails are provided at Dikes Point.	
RUSSEL DAM	Bar screens being designed for fish protection.	Recreation facilities include a fishing pier, boat ramps, and a visitor center.	
SALINA		Boat ramps are readily available and the lake provide good fishing for bass, crappie, and other fish. No powered motor boats are allowed.	

Table 8, Page 9

ASCE PUMPED STORAGE COMPENDIUM
Table 8 - Data on Environmental Features

Plant Name	Fish and Wildlife	Recreation	Social
SMITH MOUNTAIN	The lake is stocked with gamefish, such as striped largemouth and smallmouth bass and muskel. The 20.23 Mm² (5,000 acre) wildlife refuge is managed by the Department of Game and Fisheries.	Recreational facilities include numerous picnic areas, over 25 marinas, a visitor center, wildlife trails, and a 20.23 Mm² (5,000-acre) wildlife management area.	
TAUM SAUK			
THERMALITO POWER PLANT	Trout, bass, catfish, panfish, and ducks are present in the area.	Facilities include a public boat ramp, picnic area, and a swimming beach. Recreationists enjoy boating, fishing, waterskiing, swimming, and hunting.	
WALLACE DAM	Active Osprey are nesting in the reservoir and Bald Eagles nest at the Dam. A wildlife refuge was created from the project. There are no fishery issues.	Recreational usage exceeds 600,000 visitors per year. Includes boating, hiking, camping, fishing, and swimming. project has caused extensive recreational development.	Project has increased tourism and rec development. Taxes collected have increased. Farmland converted to residential and recreatonal land near lake.

Table 8, Page 10

ASCE PUMPED STORAGE COMPENDIUM
Table 8 - Data on Environmental Features

Plant Name	Fish and Wildlife	Recreation	Social
YARDS CREEK STATION	There is no fishing in the reservoir and hunting is limited. The lower reservoir area is a wildlife refuge.	Recreation is passive, occurs during the daytime, and is year around. Facilities include a visitor center, nature overlooks, interpretive panels, hiking trails, a picnic area, and restrooms. No reservoir boating, swimming, or fishing is allowed. Hunting is allowed in specified areas per State game laws.	Group guided tours of the station are conducted.

ASCE Pumped Storage Compendium
Table 9 - Summary of Plant Data

Plant Name	Plant Location: Nearest Town, State	Owner/Address	Plant Capacity MW	First Commercial Power	Number of Reversible Units	Number of Main Conduits	Surge Tanks [1]	Static Head: Max, Min (m) Max/Min Ratio	Storage Capacity Generation Hours [2]	Plant Eff. [3] %	Length of Waterways m	L/H Ratio [4]
BAD CREEK PUMPED STORAGE PROJECT	Salem, SC	Duke Power Company P.O. Box 33189 Charlotte, North Carolina 28201-1006	1,000	1991	4	1	none	375, 317 Ratio: 1.2	24		2,901	8.3
BALSAM MEADOW	Shaver Lake, CA	Southern California Edison Company P.O. Box 800 2244 Walnut Grove Av Rosemead, California 91770	200	1987	1	1	downstream	412, 401 Ratio: 1.0	8		3,806	10.0
BATH COUNTY	Warm Springs, VA	Virginia Power Company and Allegheny Generating Company P.O. Box 26666 Richmond, Virginia 23261	2,100	1985	6	3	upstream	384, 329 Ratio: 1.2	11	80.30	2,779	7.7
BEAR SWAMP	Rowe, MA	New England Power Company 25 Research Drive Westborough, Massachusetts 01581	600	1974	2	2	none	235, 207 Ratio: 1.1	6	69.90	559	2.4
BLENHEIM-GILBOA PUMPED STORAGE PROJECT	Blenheim and Gilboa, NY	New York Power Authority 123 Main Street White Plains, New York 10601	1,000	1973	4	4	none	348, 334 Ratio: 1.0	12	75.00	1,187	3.5

[1] Entry indicates location of surge tank(s) relative to powerhouse; upstream, downstream, or both; or that none exists.
[2] Energy stored in upper reservoir divided by the plant's rated capacity.
[3] Plant efficiency is the energy generated divided by the energy used in pumping.
[4] Total length of waterways (upper tunnel, shaft, lower tunnel, penstocks, and tailrace tunnel) divided by the average static head.

Table 9, Page 1

ASCE Pumped Storage Compendium
Table 9 - Summary of Plant Data

Plant Name	Plant Location: Nearest Town, State	Owner/Address	Plant Capacity MW	First Commercial Power	Number of Reversible Units	Number of Main Conduits	Surge Tanks[1]	Static Head: Max, Min (m) Max/Min Ratio	Storage Capacity Generation Hours[2]	Plant Eff.[3] %	Length of Waterways m	L/H Ratio[4]
CABIN CREEK	Georgetown, CO	Public Service Company of Colorado, 5900 E. 39th Avenue, Denver, Colorado 80207	300	1967	2	2	none	374, 334 Ratio: 1.1	5	67.00	1,311	3.7
CARTERS	Chatsworth, GA	Mobile District, Corps of Engineers, P.O. Box 2288, Mobile, Alabama 36628-0001	258	1976	2	2	none	124, 96.0 Ratio: 1.3	7	70.00	255	2.4
CASTAIC POWER PROJECT	Castaic, CA	Los Angeles Department of Water and Power, 111 North Hope Street, Los Angeles, California 90012-2694	1,250	1973	6	1	upstream	335, 244 Ratio: 1.4	10	67.00	12,581	38.6
CLARENCE CANNON	Center, MO	St. Louis District, Corps of Engineers, Box 20 B, Monroe City, Missouri 63456	31	1984	1	1	none	35.7, 18.0 Ratio: 2.0	9		21	0.9
DE GRAY DAM AND RESERVOIR	Arkadelphia, AK	Vicksburgh District Corps of Engineers, P.O. Box 60, Vicksburg, Mississippi 39180-0060	28	1971	1	1	none	62.8, 44.5 Ratio: 1.4	7	76.00	479	9.2

[1] Entry indicates location of surge tank(s) relative to powerhouse; upstream, downstream, or both; or that none exists.
[2] Energy stored in upper reservoir divided by the plant's rated capacity.
[3] Plant efficiency is the energy generated divided by the energy used in pumping.
[4] Total length of waterways (upper tunnel, shaft, lower tunnel, penstocks, and tailrace tunnel) divided by the average static head.

Table 9, Page 2

ASCE Pumped Storage Compendium
Table 9 - Summary of Plant Data

Plant Name	Plant Location: Nearest Town, State	Owner/Address	Plant Capacity MW	First Commercial Power	Number of Reversible Units	Number of Main Conduits	Surge Tanks[1]	Static Head: Max, Min (m) Max/Min Ratio	Storage Capacity Generation Hours[2]	Plant Eff. [3] %	Length of Waterways m	L/H Ratio[4]
EDWARD HYATT POWER PLANT	Oroville, CA	California Department of Water Resoources 1416 Ninth Street Sacramento, California 95814	293	1967	3	2	none	206, 152 Ratio: 1.4			177	0.9
FAIRFIELD	Jenkinsville, SC	South Carolina Electric & Gas Company 1426 Main Street Columbia, South Carolina 29201	512	1978	8	4	none	51.5, 47.1 Ratio: 1.1	8	70.40	334	7.3
FLATIRON POWERPLANT	Loveland, CO	Bureau of Reclamation Attn: D-3000 P.O. Box 25007 Denver, Colorado 80225	9	1954	1	1	upstream	90.8, 46.6 Ratio: 1.9			2,253	25.5
GIANELLI SAN LUIS PUMP-GENERATE PLANT	Los Banos, CA	Bureau of Reclamation Attn: D-3000 P.O. Box 25007 Denver, Colorado 80225	400	1968	8	4	none	98.8, 35.7 Ratio: 2.8	1274	83.00	654	10.9
GRAND COULEE	Grand Coulee, WA	Bureau of Reclamation Attn: D-3000 P.O. Box 25007 Denver, Colorado 80225	314	1973	6	6	none	111, 81.4 Ratio: 1.4	35		218	2.7

[1] Entry indicates location of surge tank(s) relative to powerhouse; upstream, downstream, or both; or that none exists.
[2] Energy stored in upper reservoir divided by the plant's rated capacity.
[3] Plant efficiency is the energy generated divided by the energy used in pumping.
[4] Total length of waterways (upper tunnel, shaft, lower tunnel, penstocks, and tailrace tunnel) divided by the average static head.

Table 9, Page 3

ASCE Pumped Storage Compendium
Table 9 - Summary of Plant Data

Plant Name	Plant Location: Nearest Town, State	Owner/Address	Plant Capacity MW	First Commercial Power	Number of Reversible Units	Number of Main Conduits	Surge Tanks [1]	Static Head: Max, Min (m) Max/Min Ratio	Storage Capacity Generation Hours [2]	Plant Eff. [3] %	Length of Waterways m	L/H Ratio [4]
HELMS	Shaver Lake, CA	Pacific Gas & Electric Company 77 Beale Street San Francisco, California 94536	1,206	1984	3	1	upstream and downstream	532, 448 Ratio: 1.2	153	74.00	6,036	12.3
HIWASSEE-UNIT 2	Murphy, NC	Tennessee Valley Authority 818 Power Building Chattanooga, Tennessee 37401	68	1956	1	1	none	77.7, 52.7 Ratio: 1.5	4228	76.90	58	1.0
HORSE MESA #4	Tortilla Flat, AZ	Bureau of Reclamation Attn: D-3000 P.O. Box 25007 Denver, Colorado 80225	97	1972	1	1	none	78.7, 71.8 Ratio: 1.1	245	67.00	57	0.8
JOCASSEE PUMPED STORAGE STATION	Salem, SC	Duke Power Company P.O. Box 1006 Charlotte, North Carolina 28201-1006	610	1973	4	4	none	102, 94.5 Ratio: 1.1	94	80.00	499	5.0
KINZUA	Warren, PA	Pennsylvania Electric Company 1001 Broad Street Johnstown, Pennsylvania 15907	350	1970	2	2	none	248, 196 Ratio: 1.3	11	66.70	854	4.3

[1] Entry indicates location of surge tank(s) relative to powerhouse; upstream, downstream, or both; or that none exists.
[2] Energy stored in upper reservoir divided by the plant's rated capacity.
[3] Plant efficiency is the energy generated divided by the energy used in pumping.
[4] Total length of waterways (upper tunnel, shaft, lower tunnel, penstocks, and tailrace tunnel) divided by the average static head.

Table 9, Page 4

ASCE Pumped Storage Compendium
Table 9 - Summary of Plant Data

Plant Name	Plant Location: Nearest Town, State	Owner/Address	Plant Capacity MW	First Commercial Power	Number of Reversible Units	Number of Main Conduits	Surge Tanks[1]	Static Head: Max, Min (m) Max/Min Ratio	Storage Capacity Generation Hours[2]	Plant Eff. [3] %	Length of Waterways m	L/H Ratio[4]
LEWISTON	Lewiston, NY	New York Power Authority 123 Main Street White Plains, New York 10601	240	1961	12	12	none	34.4, 20.1 Ratio: 1.7	20		46	2.0
LUDINGTON	Ludington, MI	Consumers Power (51%) / Detroit Edison (49%) 212 W. Michigan Avenue Ludington, Michigan 49201	1,979	1973	6	6	none	111, 90.0 Ratio: 1.2	9	71.50	396	3.6
MORMON FLAT #2	Tortilla Flat, AZ	Bureau of Reclamation Attn: D-3000 P.O. Box 25007 Denver, Colorado 80225	47	1971	1	1	none	42.3, 38.4 Ratio: 1.1	59	67.00	53	1.4
MT.ELBERT	Leadville, CO	Bureau of Reclamation Attn: D-3000 P.O. Box 25007 Denver, Colorado 80225	200	1981	2	2	upstream	145, 122 Ratio: 1.2	12	81.30	914	6.7
MUDDY RUN	Drumore, PA	Philadelphia Electric Company 2301 Market Street Philadelphia, Pennsylvania 19101	800	1967	8	8	none	126, 110 Ratio: 1.1	14	71.40	393	3.7

[1] Entry indicates location of surge tank(s) relative to powerhouse; upstream, downstream, or both; or that none exists.
[2] Energy stored in upper reservoir divided by the plant's rated capacity.
[3] Plant efficiency is the energy generated divided by the energy used in pumping.
[4] Total length of waterways (upper tunnel, shaft, lower tunnel, penstocks, and tailrace tunnel) divided by the average static head.

Table 9, Page 5

ASCE Pumped Storage Compendium
Table 9 - Summary of Plant Data

Plant Name	Plant Location: Nearest Town, State	Owner/Address	Plant Capacity MW	First Commercial Power	Number of Reversible Units	Number of Main Conduits	Surge Tanks[1]	Static Head: Max, Min (m) Max/Min Ratio	Storage Capacity Generation Hours[2]	Plant Eff. [3] %	Length of Waterways m	L/H Ratio[4]
NORTHFIELD MOUNTAIN	Northfield & Erving, MA	Connecticut Light and Power Co./Western Mass. Electric Co. c/o Northeast Utilities P.O. Box 270 Hartford, Connecticut 06101-0270	1,080	1972	4	4	upstream and downstream	252, 224 Ratio: 1.1	10	74.60	1,926	8.5
RACCOON MOUNTAIN	Chattanooga, TN	Tennessee Valley Authority Mr 6N 36A Chattanooga, Tennessee 37401	1,530	1978	4	4	downstream	318, 274 Ratio: 1.2	21	76.90	1,097	4.0
ROCKY MOUNTAIN	Armuchee, GA	Oglethorpe Power Corporation 2100 East Exchange Place P.O. Box 1349 Tucker, Georgia 30085	760	1995	3	3	none	210, 187 Ratio: 1.1	8	83.50	910	4.6
ROCKY RIVER	New Milford, CT	Connecticut Light and Power Company P.O. Box 270 Hartford, Connecticut 06101	32	1929	1	1	upstream	70.1, 64.0 Ratio: 1.1	837		511	7.5
RUSSEL DAM	Elberton, GA	Savannah District, Corps of Engineers 100 W. Oglethorpe Avenue Savannah, Georgia 31401	360	1987	4	4	none					

[1] Entry indicates location of surge tank(s) relative to powerhouse; upstream, downstream, or both; or that none exists.
[2] Energy stored in upper reservoir divided by the plant's rated capacity.
[3] Plant efficiency is the energy generated divided by the energy used in pumping.
[4] Total length of waterways (upper tunnel, shaft, lower tunnel, penstocks, and tailrace tunnel) divided by the average static head.

Table 9, Page 6

ASCE Pumped Storage Compendium
Table 9 - Summary of Plant Data

Plant Name	Plant Location: Nearest Town, State	Owner/Address	Plant Capacity MW	First Commercial Power	Number of Reversible Units	Number of Main Conduits	Surge Tanks [1]	Static Head: Max, Min (m) Max/Min Ratio	Storage Capacity Generation Hours [2]	Plant Eff. [3] %	Length of Waterways m	L/H Ratio [4]
SALINA	Salina, OK	Grand River Dam Authority P.O. Box 409 Vanita, Oklahoma 74301-0409	260	1968	6	6	none	74.7, 68.6 Ratio: 1.1	10	70.00	195	2.8
SMITH MOUNTAIN	Sandy Level, VA	Appalacian Power Company P.O. Box 2021 Roanoke, Virginia 24022-2121	240	1965	3	3	none	59.5, 53.0 Ratio: 1.1	14	73.00	52	0.9
TAUM SAUK	St. Louis, MO	Union Electric Company 1901 Chateau P.O. Box 149 St. Louis, Missouri 63166	350	1963	2	2	none	267, 233 Ratio: 1.1	8	50.00	2,210	9.2
THERMALITO POWER PLANT	Oroville, CA	California Department of Water Resources 1416 Ninth Street Sacramento, California 95965	115	1968	4	4	none	30.8, 25.9 Ratio: 1.2			36	1.3
WALLACE DAM	Eatonton, GE	Georgia Power Company P.O. Box 4545 Atlanta, Georgia 30302	209	1979	4	6	none	29.3, 27.1 Ratio: 1.1	1020		52	1.8

[1] Entry indicates location of surge tank(s) relative to powerhouse; upstream, downstream, or both; or that none exists.
[2] Energy stored in upper reservoir divided by the plant's rated capacity.
[3] Plant efficiency is the energy generated divided by the energy used in pumping.
[4] Total length of waterways (upper tunnel, shaft, lower tunnel, penstocks, and tailrace tunnel) divided by the average static head.

Table 9, Page 7

ASCE Pumped Storage Compendium
Table 9 - Summary of Plant Data

Plant Name	Plant Location: Nearest Town, State	Owner/Address	Plant Capacity MW	First Commercial Power	Number of Reversible Units	Number of Main Conduits	Surge Tanks[1]	Static Head: Max, Min (m) Max/Min Ratio	Storage Capacity Generation Hours[2]	Plant Eff. [3] %	Length of Waterways m	L/H Ratio[4]
YARDS CREEK STATION	Blairstown, NJ	Jersey Central P&L Co. & Public Service Electric & Gas Co. Madison Avenue At Punch Bowl Road Morristown, New Jersey 07960	360	1965	3	1	none	232, 210 Ratio: 1.1	8	66.67	1,067	4.8

[1] Entry indicates location of surge tank(s) relative to powerhouse; upstream, downstream, or both; or that none exists.
[2] Energy stored in upper reservoir divided by the plant's rated capacity.
[3] Plant efficiency is the energy generated divided by the energy used in pumping.
[4] Total length of waterways (upper tunnel, shaft, lower tunnel, penstocks, and tailrace tunnel) divided by the average static head.

Table 9, Page 8

ASCE Pumped Storage Compendium
Table 10 - Summary of Data on Upper Dams, Reservoirs, and Spillways

Plant Name	Dams Type	Height m	Length m	Volume Mm³ [1]	Reservoirs Type	Drainage Area Mm² [2]	Surface Area Mm² [2]	Max/Min Elevations m msl	Power Pool Usable Storage Mm³ [1]	Fluctuation m	Spillways Type	Peak Discharge m³/s	Spillways Type	Gates Number	Height m	Width m
BAD CREEK PUMPED STORAGE PROJECT	Zoned earthfill	110.0	793	0.4	Constructed	3.50	1.49	Max: 704.0 Min: 655.2	38.7	48.8	None					
BALSAM MEADOW	Rockfill embankment	36.6	445	0.3	Constructed	0.26	0.24	Max: 2032.4 Min: 2023.3	2.1	12.2	Open channel	57				
BATH COUNTY	Zoned fill	140.2	732	13.0	On stream	6.22	1.07	Max: 1011.9 Min: 979.9	27.8	32.0	Fuse plug	510				
BEAR SWAMP	Rockfill central core	47.2		0.1	Constructed	3,237.50	4.53	Max: 487.7 Min: 472.4	6.5	15.2	Overflow weir	190				
BLENHEIM-GILBOA PUMPED STORAGE PROJECT	Earth & rockfill	49.4	3,627	4.5	Constructed	1.45	1.58	Max: 610.5 Min: 595.9	18.5	14.6	Emergency spill channel	283				
CABIN CREEK	Concrete faced, rockfill	64.0	445	0.9	On stream, constructed	2.59	0.09	Max: 3412.5 Min: 3384.8	1.7	27.7	Overflow weir, ungated					
CARTERS	Rockfill	135.6	626	11.5	Constructed	973.84	13.03	Max: 326.7 Min: 311.5	166.4	15.9	Concrete gravity	5,499	Tainter	5	11.1	12.8
CASTAIC POWER PROJECT	Earth and rockfill	1,219.2	332	5.2	Constructed, Pyramid Reservoir	758.87	5.49	Max: 785.8 Min: 713.2	24.7	5.5	Broad crested weir			1		
CLARENCE CANNON	Compacted earth, Clarence Cannon Dam	32.3	515	1.7	Constructed, Mark Twain Lake	6,003.63	155.40	Max: 194.5 Min: 172.8	437.9	21.6	Concrete ogee, gated	7,830	Tainter	4	11.9	15.2

[1] Mm³ is million cubic meters.
[2] Mm² is million square meters.

Table 10, Page 1

ASCE Pumped Storage Compendium
Table 10 - Summary of Data on Upper Dams, Reservoirs, and Spillways

Plant Name	Dams Type	Height m	Length m	Volume Mm³ [1]	Reservoirs Type	Drainage Area Mm² [2]	Surface Area Mm² [2]	Max/Min Elevations m msl	Usable Storage Mm³ [1]	Power Pool Fluctuation m	Spillways Type	Peak Discharge m³/s	Gates Type	Gates Number	Height m	Width m
DE GRAY DAM AND RESERVOIR	Rolled earthfill, De Gray Dam	74.1	1,036	5.2	Constructed, De Gray Lake	1,173.27	54.23	Max: 122.5 Min: 111.9	485.0	12.5	Rock saddle, ungated	1,897				
EDWARD HYATT POWER PLANT	Zoned earthfill, Oroville Dam	234.7	2,109	61.2	Constructed, Oroville Reservoir	9,352.50	63.94	Max: 274.3 Min: 222.5	3,311.9	51.8	Concrete channel	12,459	Radial	8	10.2	5.4
FAIRFIELD	Random-filled	54.9	3,353	7.6	Constructed, Monticello Reservoir	0.04	27.50	Max: 129.5 Min: 128.2	35.8	1.4	None					
FLATIRON POWERPLANT	Zoned earthfill	65.2	597	3.3	Constructed, Carter Lake	14.76	4.62	Max: 1755.6 Min: 1714.5		41.1	None					
GIANELLI SAN LUIS PUMP-GENERATE PLANT	Zoned earthfill (B.F. Sisk-San Luis)	95.7	5,669	61.4	Constructed	219.37	52.61	Max: 165.8 Min: 99.4	2,355.0	66.4	Morning Glory 2.89 m (9.5 ft) diameter.	29				
GRAND COULEE	Two zoned earthfill, North Dam	44.2	357	1.3	Constructed, Feeder Canal from Banks Lake		109.27	Max: 478.8 Min: 474.6	881.9	1.8	None					
HELMS	Rockfill concrete faced, Courtright Dam	96.0	263	1.2	Existing, constructed, Courtright Lake	101.80	6.66	Max: 2494.0 Min: 2444.0	147.0	50.0	Overflow ogee	340				
HIWASSEE-UNIT 2	Concrete, Hiwassee Dam	93.6	419	5.5	On stream	2,508.00	23.70	Max: 465.8 Min: 442.1	520.0	22.5	Concrete	3,681	Radial	7	7.0	9.8

[1] Mm³ is million cubic meters.
[2] Mm² is million square meters.

Table 10, Page 2

ASCE Pumped Storage Compendium
Table 10 - Summary of Data on Upper Dams, Reservoirs, and Spillways

Plant Name	Dams Type	Height m	Length m	Volume Mm³ [1]	Reservoirs Type	Drainage Area Mm² [2]	Surface Area Mm² [2]	Max/Min Elevations m msl	Power Pool Usable Storage Mm³ [1]	Fluctuation m	Spillways Type	Peak Discharge m³/s	Spillways Type	Gates Number	Height m	Width m
HORSE MESA #4	Existing concrete thin arch dam	93.0	201	0.1	Existing Apache Lake constructed		10.80	Max: 583.3 Min: 578.1	144.3	5.5	One overflow and one tunnel spillway		9 radial & 1 fixed wheel	9	7.0	7.9
JOCASSEE PUMPED STORAGE STATION	Zoned fill, Jocassee Dam	117.3	533	9.9	Constructed, Lake Jocassee	383.32	30.62	Max: 338.3 Min: 329.2	266.1	9.1	Concrete chute		Taintor	2	10.1	11.6
KINZUA	Rolled fill - sandstone	35.1	2,377	199.4	Constructed	0.45	0.45	Max: 631.9 Min: 610.6	7.1	21.3	Fuse plug for over pumping	224				
LEWISTON	Earth and rockfill	16.8	10,516	7.1	Constructed, Lewiston Reservoir	7.69	7.69	Max: 199.6 Min: 189.0	70.3	10.7	None					
LUDINGTON	Asphalt faced, zoned earth embankment	51.8	9,700	28.8	Constructed	3.37	3.40	Max: 287.2 Min: 266.8	64.4	20.4	None					
MORMON FLAT #2	Existing concrete thin arch dam	68.3	116		Existing Canyon Lake constructed		3.80	Max: 506.3 Min: 501.4	31.7	4.6	Chute spillway		Fixed wheel	2	15.2	15.2
MT. ELBERT	Zoned earthfill embankment, Mt. Elbert forebay	28.0	808	0.6	Constructed	2.59	1.14	Max: 2940.0 Min: 2923.0	9.1	9.1	Morning glory	257	None - stored inflow			
MUDDY RUN	Earth and rockfill	76.2	1,341	4.0	Constructed	23.83	3.89	Max: 158.5 Min: 143.3	40.9	15.2	Concrete, ungated	368				

[1] Mm³ is million cubic meters.
[2] Mm² is million square meters.

Table 10, Page 3

ASCE Pumped Storage Compendium
Table 10 - Summary of Data on Upper Dams, Reservoirs, and Spillways

Plant Name	Dams Type	Height m	Length m	Volume Mm³ [1]	Reservoirs Type	Drainage Area Mm² [2]	Surface Area Mm² [2]	Max/Min Elevations m msl	Power Pool Usable Storage Mm³ [1]	Power Pool Fluctuation m	Spillways Type	Peak Discharge m³/s	Gates Type	Gates Number	Height m	Width m
NORTHFIELD MOUNTAIN	Central core rockfill	44.2	1,707	2.1	Constructed	1.30	1.11	Max: 306.2 Min: 280.4	17.7	25.8	Overflow	323	None			
RACCOON MOUNTAIN	Earth and rockfill	70.1	2,591	7.1	Constructed	3.63	2.14	Max: 509.6 Min: 466.3	44.8	43.3	None					
ROCKY MOUNTAIN	Zoned earthfill	24.4	3,901	8.2	Constructed		0.85	Max: 424.3 Min: 408.7	12.6	15.5	Fuse plug					
ROCKY RIVER	Earth with wooden core wall	30.5	290	0.3	Constructed	90.65	21.93	Max: 131.1 Min: 121.9	66.0	9.1	None					
RUSSEL DAM	Concrete	59.4	559	3.4	Constructed	732.97	107.86		156.5	1.5	Ogee concrete gravity	22,696	Tainter	10	13.7	15.2
SALINA	Rockfill and earth	61.0	701	2.8	Constructed		3.18	Max: 263.7 Min: 257.6		6.1	None					
SMITH MOUNTAIN	Concrete arch	71.6	250	0.1	Constructed, Smith Mountain Reservoir	2,652.00	83.40	Max: 242.3 Min: 239.9	194.6	2.4	Free overflow	5,663				
TAUM SAUK	Rockfill, concrete face, asphalt floor	28.0	2,030	2.9	Constructed	0.22	0.22	Max: 486.8 Min: 458.7	5.4	26.8						
THERMALITO POWER PLANT	Zoned earthfill	21.6	4,877	1.2	Constructed	0.26	2.55	Max: 68.9 Min: 67.7	12.3	1.2						

[1] Mm³ is million cubic meters.
[2] Mm² is million square meters.

Table 10, Page 4

ASCE Pumped Storage Compendium
Table 10 - Summary of Data on Upper Dams, Reservoirs, and Spillways

Plant Name	Dams				Reservoirs				Power Pool		Spillways				
	Type	Height m	Length m	Volume Mm³ [1]	Type	Drainage Area Mm² [2]	Surface Area Mm² [2]	Max/Min Elevations m msl	Usable Storage Mm³ [1]	Fluctuation m	Type	Peak Discharge m³/s	Gates Type	Number	Height m / Width m
WALLACE DAM	Concrete, earth, non-overflow	35.7	730	0.8	Constructed, Oconee Lake	4,739.71	77.09	Max: 132.6 Min: 131.1	933.4	1.5	Gated ogee	7,589	Radial lift	5	13.4 / 12.8
YARDS CREEK STATION	Rockfill with earth core	21.3	2,926	20.4	Constructed	0.62	0.62	Max: 474.0 Min: 459.0	5.7	14.9	None				

[1] Mm³ is million cubic meters.
[2] Mm² is million square meters.

Table 10, Page 5

ASCE Pumped Storage Compendium
Table 11 - Summary of Data on Upper Intakes

Plant Name	Number of Intakes	Type of Intake	Area of Each Inlet m²	Bar Racks Shape	Bar Racks Spacing mm	Bar Racks Depth mm	Bar Racks Thickness mm	Bar Racks Diameter mm	Service Gates Type	Service Gates Number	Service Gates Height m	Service Gates Width m	Emergency Gates Type	Emergency Gates Number	Emergency Gates Height m	Emergency Gates Width m
BAD CREEK PUMPED STORAGE PROJECT	1	Concrete submerged bellmouth	152.0						Spherical valves used as service gates	4			Spherical valves, 2.5 m (8.2 ft) diameter	4		
BALSAM MEADOW	1	Flaired concrete structure	106.8		127	76	10		Slide gate	2	3.0	5.3		None		
BATH COUNTY	3	Horizontal	409.5	Rectangular	241	127	16		Fixed wheel	6	8.7	3.9	Bulkhead	6	8.7	3.9
BEAR SWAMP	1	Glory hole	182.6							None				None		
BLENHEIM-GILBOA PUMPED STORAGE PROJECT	1	Ungated submerged circular weir (morning glory)	314.9		152	51	19			None				None		
CABIN CREEK	1	Vertical shaft with cover	532.9	Rectangular	102	114	11			None				None		
CARTERS	2	Concrete	656.2						Tractor	4	6.2	4.3		None		
CASTAIC POWER PROJECT	1	Reinforced concrete gravity structure, multiple compartmented	32.7							None			Flat-leaf coaster type	1	9.8	7.6
CLARENCE CANNON	1	Flaired, integral with powerhouse	160.5	Rectangular	203	76	17		Vertical, slide type - 3 bulkheads	1	11.4	5.3		None		
DE GRAY DAM AND RESERVOIR	1	Vertical, octagonal tower, with selective temperature control	175.6	Rectangular	152	76	16		Cylinder, 8.83 m (29 ft) diameter	1			Trashrack, bulkhead	4	7.2	6.9

ASCE Pumped Storage Compendium
Table 11 - Summary of Data on Upper Intakes

Plant Name	Number of Intakes	Type of Intake	Area of Each Inlet m²	Bar Racks Shape	Bar Racks Spacing mm	Bar Racks Depth mm	Bar Racks Thickness mm	Bar Racks Diameter mm	Service Gates Type	Service Gates Number	Service Gates Height m	Service Gates Width m	Emergency Gates Type	Emergency Gates Number	Emergency Gates Height m	Emergency Gates Width m
EDWARD HYATT POWER PLANT	2	Inclined screen	35.3	Elliptical	114	89	19			None			Hydraulic, inclined slide gate	2	10.0	6.9
FAIRFIELD	4	Gated intake structure	139.4	Rectangular	178	76	25		Roller	1	9.3	5.6	Roller	4	9.3	5.6
FLATIRON POWERPLANT	1	Reinforced concrete	15.1		51	13	13			None			Butterfly	1	2.4	
GIANELLI SAN LUIS PUMP-GENERATE PLANT	4	Roller gate	199.8	Rectangular	152	51	16			None			Roller	4	6.7	5.3
GRAND COULEE	2	Canal	26.8		159	76	19			None			76.2 cm (30 in) spool valve	1		
HELMS	1	Submerged, horizontal	427.0	Circular	216			38	Slide	2	8.2	3.6	Fixed wheel	1	9.8	6.5
HIWASSEE-UNIT 2	1	Semicircular tower	283.4	Rectangular	152	152	25			None			Roller train lift gate	2	7.9	5.8
HORSE MESA #4	1	Penstock intake	66.4		184	89	19		N/A	None			Fixed wheel	1	6.2	5.2
JOCASSEE PUMPED STORAGE STATION	2	Circular, 8 sided	356.7	Rectangular	165	102	19		Cylindrical	2	8.4	12.1	Stop logs (steel)	24	2.9	5.8
KINZUA	1	Bell mouth with vortex	73.6							None				None		
LEWISTON	12	Concrete gravity with rectangular openings	81.8	Rectangular	152	76	25		Head gates	12	7.6	4.6	Stop log	None	7.6	4.9

Table 11, Page 2

ASCE Pumped Storage Compendium
Table 11 - Summary of Data on Upper Intakes

Plant Name	Number of Intakes	Type of Intake	Area of Each Inlet m²	Bar Racks Shape	Bar Racks Spacing mm	Bar Racks Depth mm	Bar Racks Thickness mm	Bar Racks Diameter mm	Service Gates Type	Service Gates Number	Service Gates Height m	Service Gates Width m	Emergency Gates Type	Emergency Gates Number	Emergency Gates Height m	Emergency Gates Width m
LUDINGTON	1	Horizontal, reinforced concrete	75.9						Vertical lift, fixed wheel	6	8.8	8.8		None		
MORMON FLAT #2	1	Penstock intake	74.8	Rectangular	133	89	19			None			Fixed wheel	1	5.0	6.1
MT. ELBERT	2	Concrete structure	111.5	Elliptical	152	51	16			None			Wheel mounted	2	3.8	4.6
MUDDY RUN	4	Concrete structure, tower	206.6		152	89	16		Circular gate, 7.16 m (23.5 ft) diameter	1				None		
NORTHFIELD MOUNTAIN	1	Horizontal, side intake	376.7		152	19	19			None				None		
RACCOON MOUNTAIN	1	Horseshoe-shaped vertical concrete tower	2568.6							None			Spherical valves 3.048 meters diameter	1	3.0	
ROCKY MOUNTAIN	1	Covered morning glory	357.5							None				None		
ROCKY RIVER	1	Circular concrete tower intake	108.7		76				Broom gate	1	5.5	5.6		None		
RUSSEL DAM	4															
SALINA	6		14.3	Rectangular	95	127	13			None				None		
SMITH MOUNTAIN	3	Bell mouth		Rectangular	102	89	9			None			Wheel	3	7.4	6.9
TAUM SAUK	1	Cylindrical vertical reinforced concrete	54.0							None				None		

Table 11, Page 3

ASCE Pumped Storage Compendium
Table 11 - Summary of Data on Upper Intakes

Plant Name	Number of Intakes	Type of Intake	Area of Each Inlet m²	Bar Racks Shape	Bar Racks Spacing mm	Bar Racks Depth mm	Bar Racks Thickness mm	Bar Racks Diameter mm	Service Gates Type	Service Gates Number	Service Gates Height m	Service Gates Width m	Emergency Gates Type	Emergency Gates Number	Emergency Gates Height m	Emergency Gates Width m
THERMALITO POWER PLANT	4			Rectangular	152	57	13			None				None		
WALLACE DAM	6	Concrete	173.1	V-round H-elliptical	368	559	25	102	Slipe	2	9.9	5.2		None		
YARDS CREEK STATION	1	Conventional bell mouth concrete	269.4	Rectangular	76	76	10		Steel-faced stop log panels	6	3.0	3.7		None		

Table 11, Page 4

ASCE Pumped Storage Compendium
Table 12 - Summary of Data on Conduits

Plant Name	Upper Tunnel Number	Length m	Diameter m	Vel. m/s	Shaft Number	Length m	Diameter m	Vel. m/s	Lower Tunnel Number	Length m	Diameter m	Vel. m/s	Penstocks Number	Length m	Diameter m	Vel. m/s	Tailrace Tunnel Number	Length m	Diameter m	Vel. m/s
BAD CREEK PUMPED STORAGE PROJECT	1	1,128.0	9.0	6.4	1	261.0	9.0	7.0	1	1,128.0	9.0	7.0	4	117.7	4.2	8.2	2	266.7	7.5	5.1
BALSAM MEADOW	1	869.9	5.5	2.8	1	286.2	4.1	4.9	1	266.7	5.5	2.8	1	97.5	3.7	6.2	1	2,285.9	5.5	2.8
BATH COUNTY	3	1,102.8	8.7	5.2	3	301.8	8.7	5.2	3	1,054.3	8.7	5.2	6	320.0	5.5	6.5				
BEAR SWAMP					1	228.3	7.6	1.6	1	125.0	7.6	1.6	2	53.3	5.3	1.7	2	152.4	6.7	1.1
BLENHEIM-GILBOA PUMPED STORAGE PROJECT					1	315.5	8.5	7.3	1	274.3	8.5	7.3	4	597.4	3.7	10.1				
CABIN CREEK					1	310.6	4.6	7.6	1	942.7	3.7	11.9	2	57.3	2.6	11.8				
CARTERS													2	255.4	5.5	6.5				
CASTAIC POWER PROJECT	1	11,513.8	9.1	7.9									6	1,066.8	3.8					
CLARENCE CANNON													1	21.3	7.9	3.4				
DE GRAY DAM AND RESERVOIR					1	36.0	8.8	4.7	1	383.4	8.8	4.7	1	59.1	4.0	6.2				
EDWARD HYATT POWER PLANT	2	117.3	6.7										6	59.7	3.7	9.1	2		10.7	3.0
FAIRFIELD	4	114.5	7.3										4	219.5	7.9	6.8				
FLATIRON POWERPLANT	1	2,106.8	3.0										1	146.3	2.4					

Table 12, Page 1

ASCE Pumped Storage Compendium
Table 12 - Summary of Data on Conduits

Plant Name	Upper Tunnel				Shaft				Lower Tunnel				Penstocks				Tailrace Tunnel			
	Number	Length m	Diameter m	Vel. m/s	Number	Length m	Diameter m	Vel. m/s	Number	Length m	Diameter m	Vel. m/s	Number	Length m	Diameter m	Vel. m/s	Number	Length m	Diameter m	Vel. m/s
GIANELLI SAN LUIS PUMP-GENERATE PLANT													4	654.0	5.3	5.4				
GRAND COULEE	6	67.1	5.5	2.9	6	68.9	5.5	2.9					6	55.2	3.7	6.5	6	27.0	4.3	4.7
HELMS	1	4,028.0	8.2	5.3	1	532.0	8.2	5.3	1	1,240.0	8.2	5.3	3	154.0	4.7	11.6	3	82.3	4.7	5.3
HIWASSEE-UNIT 2													1	57.8	5.5	5.9				
HORSE MESA #4													1	57.1	4.7	9.2				
JOCASSEE PUMPED STORAGE STATION					2	84.1	10.2	5.5	2	365.8	10.2	5.5	4	48.8	6.2					
KINZUA					1	146.9	6.8	7.3	1	606.6	6.8	7.3	2	100.6	4.3	7.3				
LEWISTON													12	46.3	7.3	4.4				
LUDINGTON													6	396.0	8.7	8.5				
MORMON FLAT #2													1	53.3	5.5	6.5				
MT. ELBERT													2	914.4	4.6	6.2				
MUDDY RUN					4	106.7	7.5	5.6	4	134.1	7.5	5.6	8	152.4	4.3	8.7				
NORTHFIELD MOUNTAIN					1	259.1	9.4	8.1					4	103.6	4.5	10.2	1	1,563.6	10.1	6.9
RACCOON MOUNTAIN					1	274.3	10.7	6.4	1	297.2	10.7	6.4	4	45.7	5.4	6.2	1	480.1	10.7	6.4
ROCKY MOUNTAIN					1	172.8	10.7	5.9	1	589.8	10.7	5.9	3	147.0	5.8	6.6				
ROCKY RIVER													1	511.2	4.6		1		2.4	

Table 12, Page 2

ASCE Pumped Storage Compendium
Table 12 - Summary of Data on Conduits

Plant Name	Upper Tunnel Number	Upper Tunnel Length m	Upper Tunnel Diameter m	Upper Tunnel Vel. m/s	Shaft Number	Shaft Length m	Shaft Diameter m	Shaft Vel. m/s	Lower Tunnel Number	Lower Tunnel Length m	Lower Tunnel Diameter m	Lower Tunnel Vel. m/s	Penstocks Number	Penstocks Length m	Penstocks Diameter m	Penstocks Vel. m/s	Tailrace Tunnel Number	Tailrace Tunnel Length m	Tailrace Tunnel Diameter m	Tailrace Tunnel Vel. m/s
RUSSEL DAM													4		7.9					
SALINA													6	195.1	4.3					
SMITH MOUNTAIN													3	51.8	7.9					
TAUM SAUK					1	131.4	8.3	4.6					2	75.6	4.1	9.4				
THERMALITO POWER PLANT													4	36.0	6.4	3.4				
WALLACE DAM													6	29.0	7.8	4.8	6		23.5	6.5
YARDS CREEK STATION	1	457.2	6.1	7.9					1	2,002.8	7.8	5.2	3	609.6	3.0	8.8				

ASCE Pumped Storage Compendium
Table 13 - Summary of Data on Pump-Turbines

Plant Name	Plant Capacity MW	Number of Units	Diameter m	Rotational Speed Synchronous/Overspeed rpm	Unit Submergence m	Valves Number/type	Diameter m	Generating at Best Gate				Pumping at Best Gate		
								Rated Head m	Rated Capacity MW	Discharge m³/s	Efficiency %	Rated Head m	Capacity MW	Discharge m³/s	Efficiency %
BAD CREEK PUMPED STORAGE PROJECT	1,000	4	5.4	300.0/375.0	48.0	4 Spherical	2.57	308.8	270.0	89.0	91.60	379.9	318.0	50.5	92.60
BALSAM MEADOW	200	1	4.2	400.0/560.0	29.9	1 Spherical	2.13	385.9	221.0	59.6	91.60	403.6	173.0	40.5	93.60
BATH COUNTY	2,100	6	6.4	257.0/415.0	19.8	6 Spherical	2.90	329.2	380.0	133.1	90.70	335.3	420.0	118.9	92.70
BEAR SWAMP	600	2	5.8	225.0/NA	21.3	2 Spherical	3.35	219.5	298.0		92.50	208.8		125.4	91.50
BLENHEIM-GILBOA PUMPED STORAGE PROJECT	1,000	4	6.1	257.0/NA	15.2	4 Spherical	2.82	338.3	270.0	88.3	92.00	357.5	264.0	65.1	84.00
CABIN CREEK	300	2	4.2	360.0/485.0	10.7	2 Spherical	1.68	362.7	160.0			374.9		23.8	
CARTERS	258	2	6.3	150.0/NA	5.0	None		105.2	129.1	147.3		105.8	110.0	125.6	
CASTAIC POWER PROJECT	1,250	6	5.9	257.0/NA	15.2	6 Spherical	2.64	328.6	255.3	99.1		324.6	227.0	63.7	
CLARENCE CANNON	31	1	6.8	75.0/NA	1.8	None		22.9	26.5	134.5	90.00	22.9	33.8	127.4	
DE GRAY DAM AND RESERVOIR	28	1	4.9	128.5/182.0		1 Butterfly	4.27	52.1	28.9	60.9	93.00	53.3	30.5	52.4	88.00
EDWARD HYATT POWER PLANT	293	3	5.8	189.5/284.0	9.1	6 Ball valve	2.90	187.5	98.0		91.60	180.4	77.0	53.0	
FAIRFIELD	512	8	4.9	150.0/241.0	9.8	None		49.4	61.9	170.0	90.40	49.4		136.0	92.30

All plants have reversible Francis turbines, except Rocky River, which was built in 1929.
NA = Information is not available.

Table 13, Page 1

ASCE Pumped Storage Compendium
Table 13 - Summary of Data on Pump-Turbines

Plant Name	Plant Capacity MW	Number of Units	Diameter m	Rotational Speed Synchronous/Overspeed rpm	Unit Submergence m	Valves Number/type	Valves Diameter m	Generating at Best Gate Rated Head m	Generating at Best Gate Capacity MW	Generating at Best Gate Discharge m³/s	Generating at Best Gate Efficiency %	Pumping at Best Gate Rated Head m	Pumping at Best Gate Capacity MW	Pumping at Best Gate Discharge m³/s	Pumping at Best Gate Efficiency %
FLATIRON POWERPLANT	9	1	2.6	300.0/555.0	2.7	1 Butterfly	1.93	88.4	8.5	13.3		73.2	9.7	10.5	
GIANELLI SAN LUIS PUMP-GENERATE PLANT	400	8	5.4	150.0/206.0	4.6	8 Butterfly	3.96	92.7	45.1	57.5	86.10	103.9	48.3	42.1	89.70
GRAND COULEE	314	6	4.4	200.0/NA	5.2	None		85.3	44.7	62.3	91.50	89.0	50.6	51.3	91.60
HELMS	1,206	3	5.1	360.0/532.0	61.0	3 Spherical	2.44	495.0	358.0	79.8	91.80	457.0	350.0	71.1	91.90
HIWASSEE-UNIT 2	68	1	4.6	105.9/161.0	0.3	None		57.9	60.0	118.4		62.5	76.0	110.4	
HORSE MESA #4	97	1	5.5	150.0/187.5	6.2	None		75.1	80.0			79.4		105.9	
JOCASSEE PUMPED STORAGE STATION	610	4	7.3	120.0/138.0	4.9	None		89.6	170.0		93.00	89.6	170.0	175.6	92.00
KINZUA	350	2	5.7	225.0/NA	21.9	2 Spherical	2.90	196.9	162.0			213.4	195.0	90.6	
LEWISTON	240	12	5.2	112.5/202.0	1.2	None		33.8	28.0	99.1	86.00	36.0	26.0	85.0	86.00
LUDINGTON	1,979	6	8.4	112.5/163.0	8.1	None		107.6	308.0	325.9	92.20	87.2	323.0	315.0	92.70
MORMON FLAT #2	47	1	5.2	138.5/173.0	4.1	None		41.1	41.8			43.6	45.5	115.3	
MT.ELBERT	200	2	5.7	180.0/270.0	11.6	None		121.9	103.0	94.2	92.00	131.1	132.4	96.3	
MUDDY RUN	800	8	5.5	180.0/NA	6.1	None		107.6	103.0			130.2	115.0	73.9	
NORTHFIELD MOUNTAIN	1,080	4	5.2	257.0/390.0	32.3	4 Spherical	2.90	227.1	259.5			225.6	267.7	101.9	

All plants have reversible Francis turbines, except Rocky River, which was built in 1929.
NA = Information is not available.

Table 13, Page 2

ASCE Pumped Storage Compendium
Table 13 - Summary of Data on Pump-Turbines

Plant Name	Plant Capacity MW	Number of Units	Diameter m	Rotational Speed Synchronous/Overspeed rpm	Unit Submergence m	Valves Number/type	Rated Head m	Generating at Best Gate Capacity MW	Discharge m³/s	Efficiency %	Rated Head m	Pumping at Best Gate Capacity MW	Discharge m³/s	Efficiency %
RACCOON MOUNTAIN	1,530	4	5.1	300.0/455.0	32.6	4 Spherical	310.8	400.0	142.2		304.8	360.0	109.0	
ROCKY MOUNTAIN	760	3	5.7	225.0/337.5	21.6	3 Spherical	186.8	253.3	161.4		186.8	263.2	146.4	
ROCKY RIVER	32	1	1.4	327.0/634.0	1.8	2 Butterfly	73.2	32.0	53.0	70.00	73.2	6.5	7.1	78.00
RUSSEL DAM	360	4	4.6	120.0/NA		None	47.9	93.0			49.1	88.0		
SALINA	260	6	4.6	171.4/197.0	1.5	6 Butterfly	68.6	42.0	79.3	83.00	74.7	48.0	53.8	86.00
SMITH MOUNTAIN	240	3	6.8	105.9/NA	1.2	None	54.9	70.0		90.00	60.0	75.0	119.5	93.00
TAUM SAUK	350	2	6.5	200.0/220.0	9.8	2 Spherical	240.8	220.0			232.9	179.0	75.0	
THERMALITO POWER PLANT	115	4	5.2	112.5/169.0		None	25.9	26.0						
WALLACE DAM	209	4	6.7	85.7/105.0	2.1	None	29.3	52.5	203.9	91.80	29.3	63.4	220.9	93.00
YARDS CREEK STATION	360	3	5.6	240.0/364.0	7.6	3 Spherical	199.9	112.0			223.1		60.7	

All plants have reversible Francis turbines, except Rocky River, which was built in 1929.
NA = Information is not available.

ASCE Pumped Storage Compendium
Table 14 - Summary of Data on Lower Dams, Reservoirs, and Spillways

Plant Name	Dam Type	Height m	Length m	Volume Mm³ [1]	Reservoir Type	Drainage Area Mm² [2]	Surface Area Mm² [2]	Max/Min Elevations m msl	Power Pool Usable Storage Mm³ [1]	Power Pool Fluctuation m	Spillway Type	Peak Discharge m³/s	Spillway Gates Type	Number	Height m	Width m
BAD CREEK PUMPED STORAGE PROJECT	Zoned fill	117.3	533	9.94	Existing, Jocassee Reservoir	383.3	30.62	Max: 338.3 Min: 329.2	266.1	9.1			Tainter	2	10.1	11.6
BALSAM MEADOW	Existing concrete gravity	56.4	536		Existing, Shaver Lake			Max: 1637.4 Min: 1624.6	74.0	12.8						
BATH COUNTY	Zoned fill	41.1	701	3.06	On stream	190.1	2.25	Max: 645.6 Min: 627.3	27.8	18.3	Chute	1,699	Radial	2	9.1	9.8
BEAR SWAMP	Earth and rockfill	39.6	274	0.43	On stream	3,237.5	5.83	Max: 265.2 Min: 253.0	7.3	12.2	Gated	2,093	Tainter	2	12.2	11.0
BLENHEIM-GILBOA PUMPED STORAGE PROJECT	Earth and rockfill	50.3	5,486	0.89	Constructed	103.6	1.74	Max: 274.3 Min: 262.1	18.5	12.2	Ogee, concrete gravity	4,248	Radial	3	14.2	11.6
CABIN CREEK	Compacted earth and rockfill, impervious core	29.0	351	0.80	Constructed	36.3	0.21	Max: 3051.7 Min: 3038.9	2.3	12.8	Ogee crest, 44.2 m (145 ft) wide + tunnel spillway.	385				
CARTERS	Concrete	10.7	63	0.59	Constructed	1,372.7	3.52	Max: 215.5 Min: 202.8	21.2	4.6		5,601	Tainter	4	11.1	12.8
CASTAIC POWER PROJECT	Earth and rockfill	51.8	597	4.51	Constructed, Castaic Reservoir	196.8	2.00	Max: 469.4 Min: 451.1	24.7	18.3	Broad crested weir					

[1] Mm³ is million cubic meters.
[2] Mm² is million square meters.

Table 14, Page 1

ASCE Pumped Storage Compendium
Table 14 - Summary of Data on Lower Dams, Reservoirs, and Spillways

Plant Name	Dam Type	Dam Height m	Dam Length m	Dam Volume Mm³ [1]	Reservoir Type	Drainage Area Mm² [2]	Surface Area Mm² [2]	Max/Min Elevations m msl	Power Pool Usable Storage Mm³ [1]	Power Pool Fluctuation m	Spillway Type	Peak Discharge m³/s	Spillway Gates Type	Gates Number	Gates Height m	Gates Width m
CLARENCE CANNON	Rolled earth	9.4	472		Constructed, reregulating dam	75.1	4.86	Max: 160.9 Min: 158.8	6.1	2.1	Concrete, gravity type		Tainter	2	9.4	9.1
DE GRAY DAM AND RESERVOIR	Rolled earthfill	12.2	326	0.04	Constructed, reregulating dam	69.9	1.74	Max: 67.4 Min: 66.1	2.0	1.2	Ogee, ungated 94.5 m (310 ft) crest length	1,274				
EDWARD HYATT POWER PLANT	Concrete gravity	43.6	396	0.12	Constructed	0.3	1.31	Max: 68.6 Min: 67.4	16.0	1.2	Gated ogee crest		Radial	14	7.0	12.2
FAIRFIELD	Concrete, Crest Gates	14.0	610	0.05	Existing, Parr Reservoir	12.2	17.80	Max: 81.1 Min: 78.0	35.8	3.1	Overflow	6,456	Bascule	10	2.7	61.0
FLATIRON POWERPLANT	Zoned earthfill	26.2	526	0.29	Constructed, Flatiron Afterbay	19.2	0.19	Max: 1668.1 Min: 1664.8	0.5	3.3	Concrete, ungated	669				
GIANELLI SAN LUIS PUMP-GENERATE PLANT	Earth and rockfill (Forebay Dam - O'Neill)	26.5	4,359	2.20	Constructed - no inflow drainage area.		9.11	Max: 68.6 Min: 66.1			Morning Glory - 3.6 m (11.8 ft) diameter	93				
GRAND COULEE	Concrete gravity, Grand Coulee Dam	167.6	1,592	9.16	On stream, Franklin D. Roosevelt Lake	191,919.2	333.08	Max: 393.2 Min: 368.2	6,396.1	25.0	Overflow chute	28,317	Drum	11	9.1	41.1

[1] Mm³ is million cubic meters.
[2] Mm² is million square meters.

Table 14, Page 2

ASCE Pumped Storage Compendium
Table 14 - Summary of Data on Lower Dams, Reservoirs, and Spillways

Plant Name	Dam Type	Dam Height m	Dam Length m	Dam Volume Mm³ [1]	Reservoir Type	Drainage Area Mm² [2]	Surface Area Mm² [2]	Max/Min Elevations m msl	Power Pool Usable Storage Mm³ [1]	Power Pool Fluctuation m	Spillway Type	Peak Discharge m³/s	Spillway Gates Type	Gates Number	Gates Height m	Gates Width m
HELMS	Rockfill concrete faced	79.2	1,015	2.83	Existing, constructed, Lake Wishon	459.0	4.72	Max: 1996.0 Min: 1963.0	111.0	33.5	Overflow, ogee	850	Radial	6	3.5	12.2
HIWASSEE-UNIT 2	Concrete	45.7	399	3.10	On stream	129.0	8.67	Max: 389.1 Min: 387.8	10.9	2.4	Concrete	4,434	Radial	10	7.0	9.8
HORSE MESA #4	Existing concrete thin arch dam.	68.3	116	0.10	Existing Canyon Lake, constructed		3.80	Max: 506.3 Min: 504.6	31.7	1.7	Chute spillway		Fixed wheel	2	15.2	15.2
JOCASSEE PUMPED STORAGE STATION	Homogenous earthfill, Keowee Dam Station	51.8	1,067	1.56	Constructed, Lake Keowee	1,137.0	74.35	Max: 243.8 Min: 236.2	483.2	7.6	Concrete chute		Taintor	4	10.7	11.6
KINZUA	Concrete and earthfill				On stream, Allegheney Reservoir	54,597.3	84.95	Max: 416.1 Min: 384.0	5.7	32.0						
LEWISTON	Concrete gravity	118.6	561	0.84	Constructed			Max: 172.2 Min: 167.0		8.5						
LUDINGTON	None				Lake Michigan		58,018.00	Max: 176.7 Min: 176.7			None					
MORMON FLAT #2	Existing concrete thin arch dam	63.1	467	0.09	Existing Saguaro Lake, constructed		5.08	Max: 466.2 Min: 464.0	10.5	2.1	2 spillways		Radial	9	7.0	8.2

[1] Mm³ is million cubic meters.
[2] Mm² is million square meters.

Table 14, Page 3

ASCE Pumped Storage Compendium
Table 14 - Summary of Data on Lower Dams, Reservoirs, and Spillways

Plant Name	Dam Type	Dam Height m	Dam Length m	Dam Volume Mm³ [1]	Reservoir Type	Drainage Area Mm² [2]	Surface Area Mm² [2]	Max/Min Elevations m msl	Power Pool Usable Storage Mm³ [1]	Power Pool Fluctuation m	Spillway Type	Peak Discharge m³/s	Spillway Gates Type	Number	Height m	Width m
MT. ELBERT	Zoned earthfill embankment	16.8	957	0.48	On stream, Twin Lakes	272.0	11.40	Max: 2804.2 Min: 2794.6	83.8	1.2	Morning glory 12.11 m (39.7 ft)	364	None			
MUDDY RUN	Concrete gravity	31.1	1,417	0.51	On stream, Conwingo Pond	69,930.1	10.75	Max: 33.2 Min: 32.2	99.9	1.4		33,131	Stoney type crest	50	6.9	11.6
NORTHFIELD MOUNTAIN	Existing concrete Turner Falls Dam	16.8			On stream, on the Connecticut River, Turner Falls Pond	18,552.2	8.09	Max: 56.4 Min: 53.6	15.7	2.7	Gated	5,947	3 Radial, 4 Bascule	7	4.0	36.6
RACCOON MOUNTAIN	Concrete with earthfill embankments, Nickajack Dam	24.7	1,148		On stream, constructed, Nickajack Reservoir	56,650.0	41.97	Max: 193.2 Min: 192.6	25.5	0.6		14,158	Radial	10	12.2	12.2
ROCKY MOUNTAIN	Concrete gravity, zoned earthfill	22.0	198	0.44	Constructed	38.9	2.43	Max: 216.6 Min: 207.6	19.4	6.1	Overflow	705	Radial	2	7.2	6.9
ROCKY RIVER	No lower reservoir, on Housatonic River				On stream			Max: 61.0 Min: 61.0								

[1] Mm³ is million cubic meters.
[2] Mm² is million square meters.

Table 14, Page 4

ASCE Pumped Storage Compendium
Table 14 - Summary of Data on Lower Dams, Reservoirs, and Spillways

Plant Name	Dam Height m	Dam Length m	Dam Volume Mm³ [1]	Dam Type	Reservoir Drainage Area Mm² [2]	Reservoir Surface Area Mm² [2]	Reservoir Max/Min Elevations m msl	Power Pool Usable Storage Mm³ [1]	Power Pool Fluctuation m	Spillway Type	Peak Discharge m³/s	Spillway Gates Type	Gates Number	Gates Height m	Gates Width m
RUSSEL DAM	51.8	1,731	7.30	Zoned fill, concrete and earth	15,913.0	287.74		1,289.0	5.5		28,742	Tainter	23	10.7	12.2
SALINA				Rockfill earth concrete constructed			Max: 193.9 Min: 188.7		5.2						
SMITH MOUNTAIN	28.6	299	0.08	Constructed, Leesville Reservoir	3,898.0	13.20	Max: 186.8 Min: 182.9	46.6	4.0	Gated	6,853	Radial	4	10.7	15.2
TAUM SAUK	18.3	119	0.02	Constructed	235.7	0.81	Max: 228.6 Min: 224.0	7.8	4.6	Ogee weir	3,398				
THERMALITO POWER PLANT	11.9	12,802	3.84	Constructed	13.0	18.21	Max: 41.6 Min: 37.5	3.6	4.1						
WALLACE DAM	30.5	843	0.72	Existing, Lake Sinclair	7,536.9	62.32	Max: 103.6 Min: 103.0	893.2	0.6	Gated ogee	13,564	Radial lift	24	5.5	9.1
YARDS CREEK STATION	16.8	396	6.65	On stream, constructed	11.1	1.14	Max: 249.5 Min: 242.3	5.7	7.2	Concrete ogee	311				

[1] Mm³ is million cubic meters.
[2] Mm² is million square meters.

Table 14, Page 5

ASCE Pumped Storage Compendium
Table 15 - Summary of Data on Lower Intakes

Plant Name	Number of Intakes	Type of Intake	Area of Each Inlet m²	Bar Racks Shape	Bar Racks Spacing mm	Bar Racks Depth mm	Bar Racks Thickness mm	Bar Racks Diameter mm	Service Gates Type	Service Gates Number	Service Gates Height m	Service Gates Width m	Emergency Gates Type	Emergency Gates Number	Emergency Gates Height m	Emergency Gates Width m
BAD CREEK PUMPED STORAGE PROJECT	1	Vertical, 1 structure with 4 openings	153.0	Rectangular	15		2		Draft tube, bonnet side	4	5.9	5.6		None		
BALSAM MEADOW	1	Flaired concrete structure	106.8		127	76	10		Slide gate	2	3.0	5.3		None		
BATH COUNTY	6	Draft tube	131.9	Rectangular	248	102	19		Slide gate	12	5.9	4.3		None		
BEAR SWAMP	2		55.7		152	64	24		Slide	2	4.9	6.1		None		
BLENHEIM-GILBOA PUMPED STORAGE PROJECT	4	Draft tube	125.4		152	32	19		Bulkhead	4	3.8	4.1		None		
CABIN CREEK	2	Draft tube type								None				None		
CARTERS																
CASTAIC POWER PROJECT	2	West Branch of California Aqueduct and Castiac Creek.								None				None		
CLARENCE CANNON	1	Draft tube	150.5	Rectangular	203	76	17		Vertical lift, slide type - 3 bulkheads	1	21.0	5.9		None		
DE GRAY DAM AND RESERVOIR	1	Draft tube	22.3						Vertical lift, sliding	2	5.2	5.4		None		
EDWARD HYATT POWER PLANT																
FAIRFIELD	1	16 Draft tube openings	52.7	Rectangular	178	76	25			None			Slide	16	4.6	7.6

Table 15, Page 1

ASCE Pumped Storage Compendium
Table 15 - Summary of Data on Lower Intakes

Plant Name	Number of Intakes	Type of Intake	Area of Each Inlet m²	Bar Racks Shape	Spacing mm	Depth mm	Thickness mm	Diameter mm	Service Gates Type	Number	Height m	Width m	Emergency Gates Type	Number	Height m	Width m
FLATIRON POWERPLANT									Butterfly	1	1.9			None		
GIANELLI SAN LUIS PUMP-GENERATE PLANT	8	Suction tube	58.3	Rectangular	165	51	16			None			Stop log	8	4.8	3.9
GRAND COULEE	2	Concrete	128.2		159	64	16			None			Reverse flow wheel	2	6.1	4.0
HELMS	1	Submerged, horizontal	341.0	Circular	216			38	Slide	2	4.7	10.3		None		
HIWASSEE-UNIT 2	1	Concrete draft tube	152.5		152	76	16		Slide	3	4.0	4.7		None		
HORSE MESA #4	1	Draft tube	67.2	Rectangular	184	89	19		N/A	None			N/A	None		
JOCASSEE PUMPED STORAGE STATION	4	Rectangular	382.9	Circular	165			38	Rectangular	6	6.7	4.3		None		
KINZUA	1	Tower designed for temperature control	57.1		38	10			Fixed wheel - vertical	2	7.2	5.0	Closure bulkhead	1	7.7	5.6
LEWISTON	12	Concrete gravity with rectangular openings	62.0	Rectangular	152	76	25			None			Stop log		7.6	6.3
LUDINGTON	6	Draft tube	123.0	Circular	356	51			Slide	12	6.4	9.8		None		
MORMON FLAT #2	1	Draft tube	75.0	Rectangular	133	89	19			None				None		
MT.ELBERT	4	Draft tube	146.0	Elliptical	165	51	16			None				None		
MUDDY RUN	8	Draft tube								None				None		
NORTHFIELD MOUNTAIN	1	Horizontal	306.6		152	19	19			None				None		

Table 15, Page 2

ASCE Pumped Storage Compendium
Table 15 - Summary of Data on Lower Intakes

Plant Name	Number of Intakes	Type of Intake	Area of Each Inlet m²	Bar Racks Shape	Bar Racks Spacing mm	Bar Racks Depth mm	Bar Racks Thickness mm	Bar Racks Diameter mm	Service Gates Type	Service Gates Number	Service Gates Height m	Service Gates Width m	Emergency Gates Type	Emergency Gates Number	Emergency Gates Height m	Emergency Gates Width m
RACCOON MOUNTAIN	1	Reinforced concrete	286.1						Slide	4	1.5	1.7		None		
ROCKY MOUNTAIN	3	Draft tube	67.3	Rectangular	229	102	19		Vertical lift	6	5.1	5.4		None		
ROCKY RIVER	2	Treadwell	11.1		102				Steel, slide	2	2.4	1.5		None		
RUSSEL DAM																
SALINA																
SMITH MOUNTAIN	4	Bell mouth								None			Wheel	4		
TAUM SAUK																
THERMALITO POWER PLANT		Afterbay pool								None				None		
WALLACE DAM	4	Concrete	154.4	V-round, H-elliptical	368	559	25	102	Slipe	3	4.9	7.2		None		
YARDS CREEK STATION	3	Divided draft tube	92.9	Rectangular	152	76	13		Steel panels	6	3.7	3.7		None		

Table 15, Page 3

SECTION 2
INFORMATION BY PLANT

Plant Name: **BAD CREEK PUMPED STORAGE PROJECT**

Plant location:
  Salem, SC
  Oconee County

Owner: Duke Power Company
  P.O. Box 33189
  Charlotte, North Carolina 28201-1006

Rated capacity   1,000 MW
Average static head   1,148 ft
Plant efficiency   %
Stored energy   24,000 MWh
Number of units   4

Designers:
  Duke Power Company, Design and Engineering Department

Construction time: 7 years, 2 months
Construction cost:
Price level:
First commercial power: May 1991
FERC project number: 2740

Plant Manager/Superintendent:
  Mike Cloninger
  Bad Creek Pumped Storage Project
  HC 76 Box 170
  Salem, South Carolina 29676
  (803) 944-2990

River or water source: Lake Jocassee (lower reservoir) + inflow from Bad Creek and tributaries

	UPPER RESERVOIR	LOWER RESERVOIR
DAM		
Type	Zoned earthfill	Zoned fill
Height (ft)	361	385
Crest length (ft)	2,602	1,750
Volume (yd³)	466,000	13,000,000
RESERVOIR		
Type	Constructed	Existing, Jocassee Reservoir
Surface area (acres)	368	7,565
Usable power storage (acre-ft)	31,374	215,700
Power pool fluctuation (ft)	160.1	30.0
Operating levels		
Maximum (ft)	2,309.7	1,110.0
Minimum (ft)	2,149.6	1,080.0
Drainage area (miles²)	1.4	148.0
Seepage (ft³/s)	4.944	
SPILLWAY		
Design flood		
Return period (years)		
Flow (ft³/s)		
Capacity (ft³/s)		46,000
Type	None	
Gates		
Number	None	2
Type		Tainter
Width (ft)		38.00
Height (ft)		33.00
OUTLET WORKS		
Discharge capacity (ft³/s)		
Number of water passages	None	None
Dimensions of water passages		
Height (ft)		
Width (ft)		
Diameter (ft)		
Type of gates		
Number of gates	None	None

Plant Name: **BAD CREEK PUMPED STORAGE PROJECT**

Plant location
  Salem, SC
  Oconee County

Owner: Duke Power Company
  P.O. Box 33189
  Charlotte, North Carolina 28201-1006

Rated capacity   1,000 MW
Average static head   350.0 m
Plant efficiency   %
Stored energy   24,000 MWh
Number of units   4

Designers:
  Duke Power Company, Design and Engineering Department

Construction time: 7 years, 2 months
Construction cost:
Price level:
First commercial power: May 1991
FERC project number: 2740

Plant Manager/Superintendent:
  Mike Cloninger
  Bad Creek Pumped Storage Project
  HC 76 Box 170
  Salem, South Carolina 29676
  (803) 944-2990

River or water source: Lake Jocassee (lower reservoir) + inflow from Bad Creek and tributaries

	UPPER RESERVOIR	LOWER RESERVOIR
DAM		
Type	Zoned earthfill	Zoned fill
Height (m)	110.0	117.3
Crest length (m)	793.0	533.4
Volume (m³)	356,000	9,939,218
RESERVOIR		
Type	Constructed	Existing, Jocassee Reservoir
Surface area (Mm²)	1.49	30.62
Usable power storage (Mm³)	38.700	266.064
Power pool fluctuation (m)	48.80	9.14
Operating levels		
Maximum (m)	704.00	338.33
Minimum (m)	655.20	329.18
Drainage area (Mm²)	3.500	383.320
Seepage (m³/s)	0.1400	
SPILLWAY		
Design flood		
Return period (years)		
Flow (m³/s)		
Capacity (m³/s)		1,303
Type	None	
Gates		
Number	None	2
Type		Tainter
Width (m)		11.582
Height (m)		10.058
OUTLET WORKS		
Discharge capacity (m³/s)		
Number of water passages	None	None
Dimensions of water passages		
Height (m)		
Width (m)		
Diameter (m)		
Type of gates		
Number of gates	None	None

Bad Creek Pumped Storage Project - Page 1 (Metric)

BAD CREEK PUMPED STORAGE PROJECT

INTAKES	UPPER INTAKE	LOWER INTAKE
Number	1	1
Type	Concrete submerged bellmouth	Vertical, 1 structure with 4 openings
Design discharge (ft³/s)	15,998	15,998
Gross inlet area (ft²) (at trash racks)	1,636	1,647
Bar racks		
spacing (in)		0.59
shape		Rectangular
depth/thickness (in)		0.39 / 0.10
diameter (in)		
Emergency gates		
number	4	None
height/width (ft)		
type	Spherical valves, 2.5 m (8.2 ft) diameter	
Service gates		
number	4	4
height/width (ft)		19.29 / 18.21
type	Spherical valves used as service gates	Draft tube, bonnet side
Bulkhead/stop logs (Y or N)	N	Y
number of units serviced		4
Hoists		
number	None	1
capacity (tons)		30
type		Outdoor gantry crane

WATER PASSAGES	Upper Tunnel	Shaft	Lower Tunnel	Surge Tanks Upper	Surge Tanks Lower	Penstocks	Tailrace Tunnel
Number	1	1	1			4	2
Diameter (ft)	29.5	29.5	29.5			13.8	24.6
Length (ft)	3,701	856	3,701			386	875
Maximum velocity (ft/s)	21.0	23.0	23.0			26.9	16.7
Concrete lining thickness (in)	19.68	11.76	11.76			23.64	11.76
Total length of concrete sections (ft)		856	3,701			173	875
Steel liner Thickness							
Minimum (in)						1.61	
Maximum (in)						2.13	
Material grade						A516 Grade 70	
Total length of steel-lined sections (ft)						213	

Notes:
Penstocks: Diameter varies from 2.57 m (8.43 ft) to 4.20 m (13.78 ft).

BAD CREEK PUMPED STORAGE PROJECT

INTAKES	UPPER INTAKE	LOWER INTAKE
Number	1	1
Type	Concrete submerged bellmouth	Vertical, 1 structure with 4 openings
Design discharge (m³/s)	453.0	453.0
Gross inlet area (m²) (at trash racks)	152.0	153.0
Bar Racks:		
spacing (mm)		15
shape		Rectangular
depth/thickness (mm)		10 / 2
diameter (mm)		
Emergency gates		
number	4	None
height/width (m)		
type	Spherical valves, 2.5 m (8.2 ft) diameter	
Service gates		
number	4	4
height/width (m)		5.880 / 5.550
type	Spherical valves used as service gates	Draft tube, bonnet side
Bulkhead/stop logs (Y or N)	N	Y
number of units serviced		4
Hoists		
number	None	1
capacity (Mg)		27
type		Outdoor gantry crane

WATER PASSAGES	Upper Tunnel	Shaft	Lower Tunnel	Surge Tanks Upper	Surge Tanks Lower	Penstocks	Tailrace Tunnel
Number	1	1	1			4	2
Diameter (m)	9.00	9.00	9.00			4.20	7.50
Length (m)	1,128.0	261.0	1,128.0			117.7	266.7
Maximum velocity (m/s)	6.40	7.00	7.00			8.20	5.10
Concrete lining thickness (m)	0.500	0.300	0.300			0.600	0.300
Total length of concrete sections (m)		261.0	1,128.0			52.8	266.7
Steel liner Thickness							
Minimum (mm)						41	
Maximum (mm)						54	
Material grade						A516 Grade 70	
Total length of steel-lined sections (m)						64.9	

Notes:
Penstocks: Diameter varies from 2.57 m (8.43 ft) to 4.20 m (13.78 ft).

BAD CREEK PUMPED STORAGE PROJECT

POWERHOUSE and RELATED FEATURES

Powerhouse Structure
Type: Underground
Length: 433 ft   Width: 74 ft   Height: 164 ft

Guard Valves
Number: 4   Diameter: 8.4 ft
Type: Spherical

Transformers
Number: 3 single phase
Ratings:
Voltages: (kV) 525 / 19 / 19   WYE-DELTA-DELTA

Generator
Rating generating (MVA): 313.0   Rating pumping (MVA): 293.0
Insulation type: Class F
Starting method: Synchronous (3 units w/1 across line, reduced voltage)
Starting equipment: Pony motors

Runners
Material: Stainless steel
Minimum unit submergence: 157.5 ft
WR²:
Manufacturer: Voith Hydro
Model test by: Voith Hydro

	Reversible Runners	Reversible Motor/Generator
Number	4	4
Diameter (ft)	17.60	Rotor 20.67 Stator 20.73
rpm synchronous	300.0	300.0
rpm overspeed	375.0	375.0
Type	Francis	Umbrella, vertical shaft

Information on Runners

Condition:	Gross Head (ft) Generating	Gross Head (ft) Pumping	Capacity (MW) Generating	Capacity (MW) Pumping	Discharge (ft³/s) Generating	Discharge (ft³/s) Pumping	Turbine/Pump Eff.(%) Generating	Turbine/Pump Eff.(%) Pumping
Maximum head & maximum power	1,230	1,230	390	295	4,308	2,560	89.0	92.0
Minimum head & maximum power	1,040	1,040	295	334	3,885	3,426	89.4	92.6

Note: Data in the above table are based on model tests.

Condition:	Net Head (ft) Generating	Net Head (ft) Pumping	Capacity (MW) Generating	Capacity (MW) Pumping	Discharge (ft³/s) Generating	Discharge (ft³/s) Pumping	Turbine/Pump Eff.(%) Generating	Turbine/Pump Eff.(%) Pumping
Rated head @ best gate	1,013	1,246	270	318	3,143	1,783	91.6	92.6

Note: Data in the above table are based on model tests.

Bad Creek Pumped Storage Project - Page 3 (English)

BAD CREEK PUMPED STORAGE PROJECT

POWERHOUSE and RELATED FEATURES

Powerhouse Structure
Type: Underground
Length: 132.0 m   Width: 22.5 m   Height: 50.0 m

Guard Valves
Number: 4   Diameter: 2.57 m
Type: Spherical

Transformers
Number: 3 single phase
Ratings:
Voltages: (kV) 525 / 19 / 19   WYE-DELTA-DELTA

Generator
Rating generating (MVA): 313.0   Rating pumping (MVA): 293.0
Insulation type: Class F
Starting method: Synchronous (3 units w/1 across line, reduced voltage)
Starting equipment: Pony motors

Runners
Material: Stainless steel
Minimum unit submergence: 48.00 m
WR^2:
Manufacturer: Voith Hydro
Model test by: Voith Hydro

	Reversible Runners	Reversible Motor/Generator		
Number	4			4
Diameter m	5.36	Rotor 6.300	Stator	6.320
rpm synchronous	300.0			300.0
rpm overspeed	375.0			375.0
Type	Francis	Umbrella, vertical shaft		

Information on Runners

Condition:	Gross Head (m)		Capacity (MW)		Discharge (m³/s)		Turbine/Pump Eff.(%)	
	Generating	Pumping	Generating	Pumping	Generating	Pumping	Generating	Pumping
Maximum head & maximum power	375.0	375.0	390	295	122.0	72.5	89.0	92.0
Minimum head & maximum power	317.0	317.0	295	334	110.0	97.0	89.4	92.6

Note: Data in the above table are based on model tests.

Condition:	Net Head (m)		Capacity (MW)		Discharge (m³/s)		Turbine/Pump Eff.(%)	
	Generating	Pumping	Generating	Pumping	Generating	Pumping	Generating	Pumping
Rated head @ best gate	308.8	379.9	270	318	89.0	50.5	91.6	92.6

Note: Data in the above table are based on model tests.

Bad Creek Pumped Storage Project - Page 3 (Metric)

BAD CREEK PUMPED STORAGE PROJECT

Plant Data:
   Average GWh generating per year:
   Average GWh pumping per year:
   Starting time from standstill (s):   120
   Changeover time pumping to generating (min):   7
   Planned/scheduled time
      between major overhauls (years):
   Outage time required per unit
      during major overhauls (weeks):
   Representative plant availability (%):
   Representative planned outages (weeks per year):

Miscelleneous Notes:
The upper reservoir is formed from three dams, having the following characteristics:

Type	Main Dam Zoned fill	West Dam Zoned fill	East Dike Zoned fill
Height (m)	110	52	27.4
Crest length (m)	793	277	274
Volume (y^3)	356,000	28,348	12,600

Type	Main Dam Zoned fill	West Dam Zoned fill	East Dike Zoned fill
Height (ft)	360	170	90
Crest length (ft)	2,600	909	890
Volume (y^3)	466,000	37,100	16,480

The penstock and tailrace tunnel lengths of 118 m (387 ft) and 266 m (873 ft), respectively, are averages.

The reinforced concrete sections of the tailrace tunnels, shafts, and upper tunnel vary from 0.3 m (1 ft) to 0.5 m (1.6 ft). The thickness of the reinforced concrete in the penstocks varies from 0.5 m (1.6 ft) to 0.6 m (2 ft).

Cavitation Experience:
Because the plant only recently came online, litte operating history is available.

Significant or Unique Problems:
No significant problems exist. Plant operation has increased turbidity in the lower reservoir.

List of Licenses Required:
FERC.

ENVIRONMENTAL FEATURES

Recreation:
The project provides a 69 Km (43-mi) hiking trail and several visitor

Bad Creek Pumped Storage Project - Page 4

BAD CREEK PUMPED STORAGE PROJECT

access points. Fishing is available in the lower pool, which is Lake Jocassee.

Fish and Wildlife:
Wildlife mitigation includes food plots and reforestation. The project has created a wildlife refuge.

Social:
The project has attracted some tourism. Currently there is only minimal development around the project area.

Page 4 (Continued)

DUKE POWER COMPANY
BAD CREEK PUMPED STORAGE PROJECT
PROJECT DESCRIPTION

PROJECT LOCATION

Bad Creek Pumped Storage Project is located in Oconee County, approximately 8 miles north northwest of Salem and 35 miles west northwest of Greenville, South Carolina. The existing Jocassee Reservoir will function as the lower reservoir for the Bad Creek project. The upper reservoir will be located on the Bad Creek and West Bad Creek project. The upper reservoir will be located on the Bad Creek and West Bad Creek tributaries of Howard Creek, approximately 1 mile west of the Whitewater River Arm of Lake Jocassee and within several thousand feet of the North Carolina State line.

The project will require approximately 1650 acres of land. All required lands are presently owned by Crescent Land and Timber Corporation, a wholly-owned subsidiary of Duke Power Company, but will be transferred to Duke Power Company prior to starting construction.

PLANT OPERATION

When completed the Bad Creek plant will be capable of producing 1000 Megawatts (MW) of electric power. This power will be used to help meet Duke Power Company's peak load demands. During periods of peak loads the plant will operate in the generating mode where water flows from the upper reservoir, through the turbine-generators, and out to the lower reservoir producing electricity. During lower demand periods (nights and weekends) the plant will operate in the pumping mode where the turbine-generators are reversed to become pumps which will pump water from the lower reservoir up to the upper reservoir.

The Bad Creek plan is designed for 60 hours of operation at full load during the peak load week. A typical week's operation would begin on Monday with the upper reservoir at full pond. Each day power would be generated and a portion of the water pumped back into the reservoir during off peak hours. Over the weekend, the upper reservoir would be filled back to full pond elevation.

The Bad Creek project is expected to operate on the average for the equivalent of 2500 hours per year at its rated generating capacity. However, operation may vary considerably from year to year, depending on the ability of the project to produce energy at a lower cost than the generation from other plants serving the system. This will be determined on a day-to-day basis.

SITE LAYOUT AND FACILITIES

ABOVE GROUND

UPPER RESERVOIR AND DAMS

The upper reservoir will be created by two large dams (Main and West Dams) and a saddledike (East Dike). The reservoir will have a surface area of 318 acres and a storage capacity of 33,303 acre-feet, of which 30,228 acre-feet will be usable storage between minimum elevation of 2150 feet and full pond elevation of 2310 feet. Maximum drawdown will be 160 feet with 3095 acre-feet of dead storage below elevation 2150 feet. Due to the extreme water level fluctuations neither fishing nor public access to the upper reservoir will be allowed.

1. <u>Main Dam</u>. The main dam, across Bad Creek, will consist of an impervious central core surrounded by a rockfill shell. Utilizing approximately 11,400,000 cubic yards of material, this dam will have a crest width of 30 feet, maximum base width of 1550 feet, maximum height of 355 feet, and will be 2600 feet long. The crest elevation of 2315 feet allows for 5.0 feet freeboard.

2. <u>West Dam</u>. The west dam is similar in structure to the main dam. The west dam is approximately 900-feet-long and 170-feet-high and will be constructed across West Bad Creek. Crest width will be 30 feet, maximum base width 350 feet, and crest elevation of 2315 feet. Approximately 1,363,000 cubic yards of material will be needed for this dam.

3. <u>East Dike</u>. The east dike is approximately 1000-feet-long and 95-feet-high across a natural depression on the eastern rim of the reservoir. Using approximately 479,000 cubic yards of material, this dike will have a maximum base width of 450 feet and a crest width of 30 feet. A low section of the dike with crest elevation of 2313 feet, 2 feet lower than the crests of the other dams, allows that portion of the dike to act as an emergency spillway.

INTAKE STRUCTURE

The intake structure for the power tunnel will be located in the southeast portion of the upper reservoir. The structure will consist of a submerged reinforced concrete drop inlet with one 41.5-foot diameter opening. Anticipated maximum flows through the structure are 14,479 cubic feet per second (c.f.s.) generating and 12,500 c.f.s. pumping.

OUTLET WORKS

A reinforced concrete outlet structure will be located in the discharge channel on the west shore of the Whitewater River Arm of Lake Jocassee. The structure will be equipped with 4 20-foot-wide by 30-foot-high bulkhead gates and 4 trashracks of the same size. A movable hoist will be provided to lift the gates. The bottom elevation of the openings will be 1032 feet. Rock excavated from the tunnels will be used to construct a submerged weir with a crest elevation of 1070 feet located approximately 1800 feet downstream from the discharge structure in Lake Jocassee. The weir will be used to prevent mixing of warmer water from the pumped storage discharge with the cooler water in the lower layer of the main lake in order to prevent adverse effects to the cold water fish habitat of Lake Jocassee.

LOWER RESERVOIR

Lake Jocassee, the upper pool of the Jocassee Pumped Storage Development, will serve as the lower pool of the Bad Creek project. At full pool elevation 1110 feet, Lake Jocassee has a water surface area of 7505 acres and a storage capacity of 1,160,300 acre-feet. At full pool, the Lake Jocassee water surface will be approximately 40 feet above the top of the Bad Creek project discharge structure openings, and a minimum pool elevation 1080 feet, the reservoir will be approximately 18 feet above the top of the opening.

ACCESS ROADS

Access is provided by means of a 4.8 mile paved road. Road alignment is based on a maximum 10 percent grade and a minimum 100 foot radius of curvature. The pavement design will meet South Carolina Highway specifications for secondary roads and will be designed to safely accommodate the movement of the heavy equipment necessary for powerhouse construction.

EQUIPMENT BUILDING

The equipment building is located above ground approximately 143 meters (469.2 feet) above the underground powerhouse and houses the control complex and diesel generators as well as other major electrical and Heating Ventilation and Air Conditioning (HVAC) equipment. The building is of steel construction and measures 18 meters (59.1 feet) wide by 42 meters (137.8 feet) long. About half of the building is 12.5 meters (41.0 feet) high (2 floors). The remaining portion is 6.0 meters (19.7 feet) high (1 floor).

TRANSFORMER YARD AND SWITCHYARD

The transformer yard and switchyard are located adjacent to the equipment building and contain the equipment necessary to transmit the electric power from the generators to the Duke Power system grid.

BELOW GROUND

METRIC SYSTEM

Duke has designed and will construct the underground portion of the Bad Creek project (plus the above ground Equipment Building) using the metric system of measurement. In this description both Metric and English dimensions are given to facilitate comparison and understanding.

WATER CONVEYANCE TUNNELS AND SHAFTS

In the generating mode, water is taken from Bad Creek reservoir via the submerged, drop inlet intake structure into a 9.0 meter (29.5 feet) diameter shaft. This shaft drops vertically 260 meters (853.0 feet) and then elbows into the power tunnel which is sloped toward the powerhouse at 7 percent. Near the powerhouse the power tunnel curves into a manifold tunnel from which four 4.2 meter (13.8 feet) diameter penstock tunnels emerge. The flow passes from these tunnels through the turbines and out of the powerhouse cavity by way of four 5.0 meter (16.4 feet) diameter draft tube tunnels. The draft tube tunnels bifurcate into two 7.5 meter (24.6 feet) diameter trailrace tunnels which discharge into Lake Jocassee through the discharge structure.

The water conveyance tunnels and shafts, plus the Vertical Access Shaft will be lined with cast-in-place concrete. The four penstock tunnels will be steel lined with cast-in-place concrete. The four penstock tunnels will be steel lined for 65 meters (213.3 feet) upstream from the powerhouse. The other tunnels and shafts will be unlined.

MAIN ACCESS TUNNEL AND VERTICAL ACCESS SHAFT

Main access to the powerhouse is supplied by a 9 meter (29.5 feet) wide by 8 meter (26.2 feet) high access tunnel. The tunnel is approximately 365-meters-long (1198 feet) and enters the powerhouse at elevation 309.5 meters. The bottom of the tunnel will accommodate a 2-land paved road to be used both during and after construction.

Access is also gained to the powerhouse by way of a stairway and elevator in the vertical access shaft. The vertical shaft is recessed in the downstream face of the powerhouse chamber. The shaft services all four floors in the powerhouse above. The shaft also houses the electric bus lines from the generators and major HVAC ducts.

CONSTRUCTION SHAFTS

Shafts are added to facilitate the construction effort. Both Duke and the tunneling contractor may add relatively small diameter shafts for ventilation, concrete transport, etc.

POWERHOUSE

The powerhouse chamber is located approximately 165 meters (541.3 feet) underground and 350 meters (1142.3 feet) upstream from the main access tunnel entrance. The chamber is 22.5 meters (73.8 feet) wide by 43 meters (141.1 feet) high by 131 meters (429.8 feet) long. It will contain a service bay and four turbine-generators. The powerhouse is constructed of reinforced concrete up to and including the operating floor at elevation 309.5. There are intermediate that house miscellaneous mechanical and electrical equipment. The four turbine-generators rated at 250 MW each are supported on mass concrete foundations that transfer the operating loads to the surrounding rock.

Major equipment is serviced by one 475 ton overhead bridge crane. The crane rails are supported by reinforced concrete beams measuring 1 meter (3.3 feet) wide by 2 meters (6.6 feet) deep anchored to the powerhouse chamber walls using rock anchors.

MISCELLANEOUS PROJECT ITEMS

FOOTHILLS HIKING TRAIL

A 35-mile (approximate) hiking trail will be built above Lake Jocassee to link with a State Park trail to the east, and U.S. Forest Service trail to the west. This trail system will then reach from Table Rock State Park westward to Oconee State Park.

TRANSMISSION FACILITIES

The transmission facilities consist of a single circuit 100 KV line to serve as a source of construction power, and a 525 KV line for transmitting generated power.

The 525 KV line will extend from Bad Creek to Oconee, while the 100 KV line will extend only to Jocassee. The two lines will share a common corridor of 7.36 miles from near Jocassee to Bad Creek.

The transmission line towers will be placed near the top of ridges and hills in the rugged topography between Jocassee and Bad Creek. This will allow the conductors to span the ravines. The natural vegetation in the ravines will be left intact wherever it is below the minimum safety space to the conductors.

In addition, wherever the rare plant Oconee bell occurs along the route, the Oconee bells and the surrounding vegetation will not be disturbed.

Access to the tower sites will be by logging roads which already exist at most locations. These roads are near the top of ridges, and thus their improvement and use will not disturb the streams and vegetation in adjacent ravine bottoms.

ENVIRONMENTAL STUDIES AND PROGRAMS

Extensive environmental studies, monitoring programs and protection activities have been or are being conducted with the Bad Creek project. Others will be performed as the project progresses. These include the following:

1. Howard Creek Water Quality Monitoring
2. Lake Jocassee Water Quality Monitoring
3. Mercury Monitoring in Howard Creek
4. Monitoring of Benthos in Howard Creek
5. Monitoring of Sediments in Howard Creek
6. Monitoring of Trout in Howard Creek
7. Stocking of Trout in Howard Creek
8. Studies of Entrainment and Mortality of Fish
9. Endangered Species Survey, and Plans for Endangered Species Protection
10. Providing Flow Augmentation to Howard Creek
11. Flow Augmentation Analysis
12. Implementation of Erosion Control Program
13. Wildlife Mitigation Program
14. Restricted Use of the Whitewater River Tract

CONSTRUCTION SCHEDULE

Construction work on the Bad Creek project will begin in August 1981. The first phase will consist of site preparation activities such as access roads, erosion control measures, and other initial above ground site work. Major work on the dams and underground excavation will not being until 1984 or 1985.

The project completion date will be dependent on economic and power need considerations and will be assessed on a year-by-year basis.

This project description was provided by Hugh G. McKay.

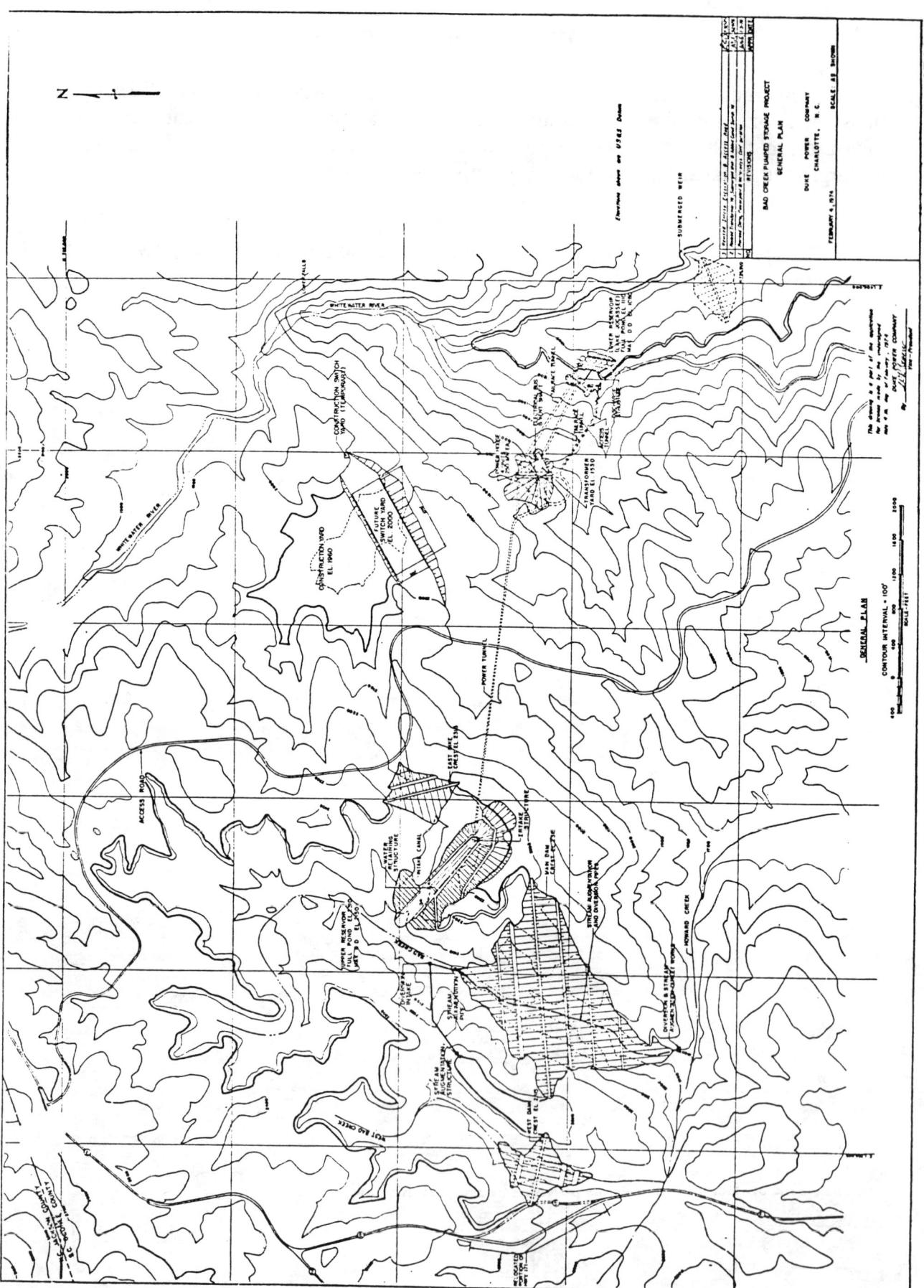

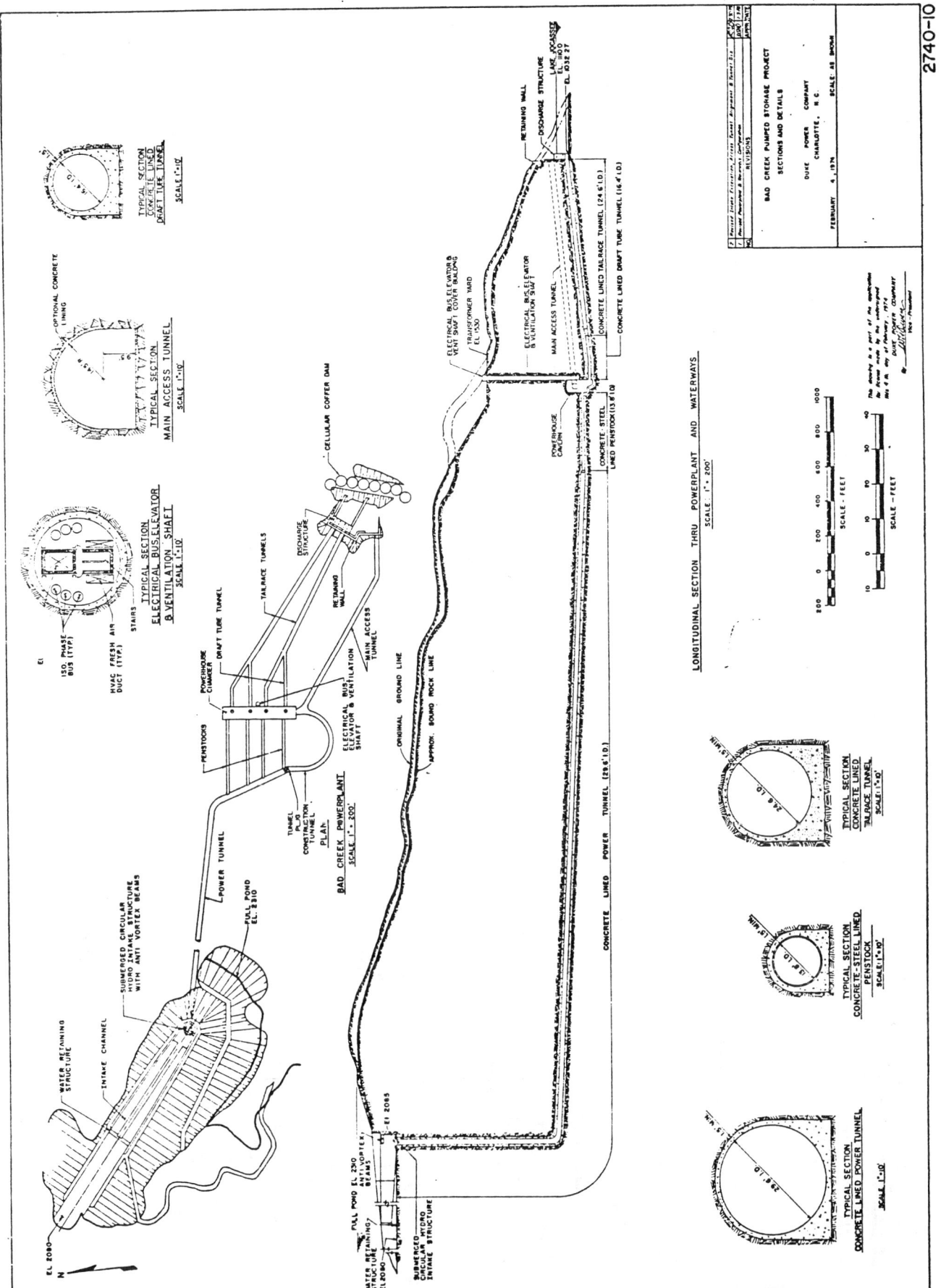

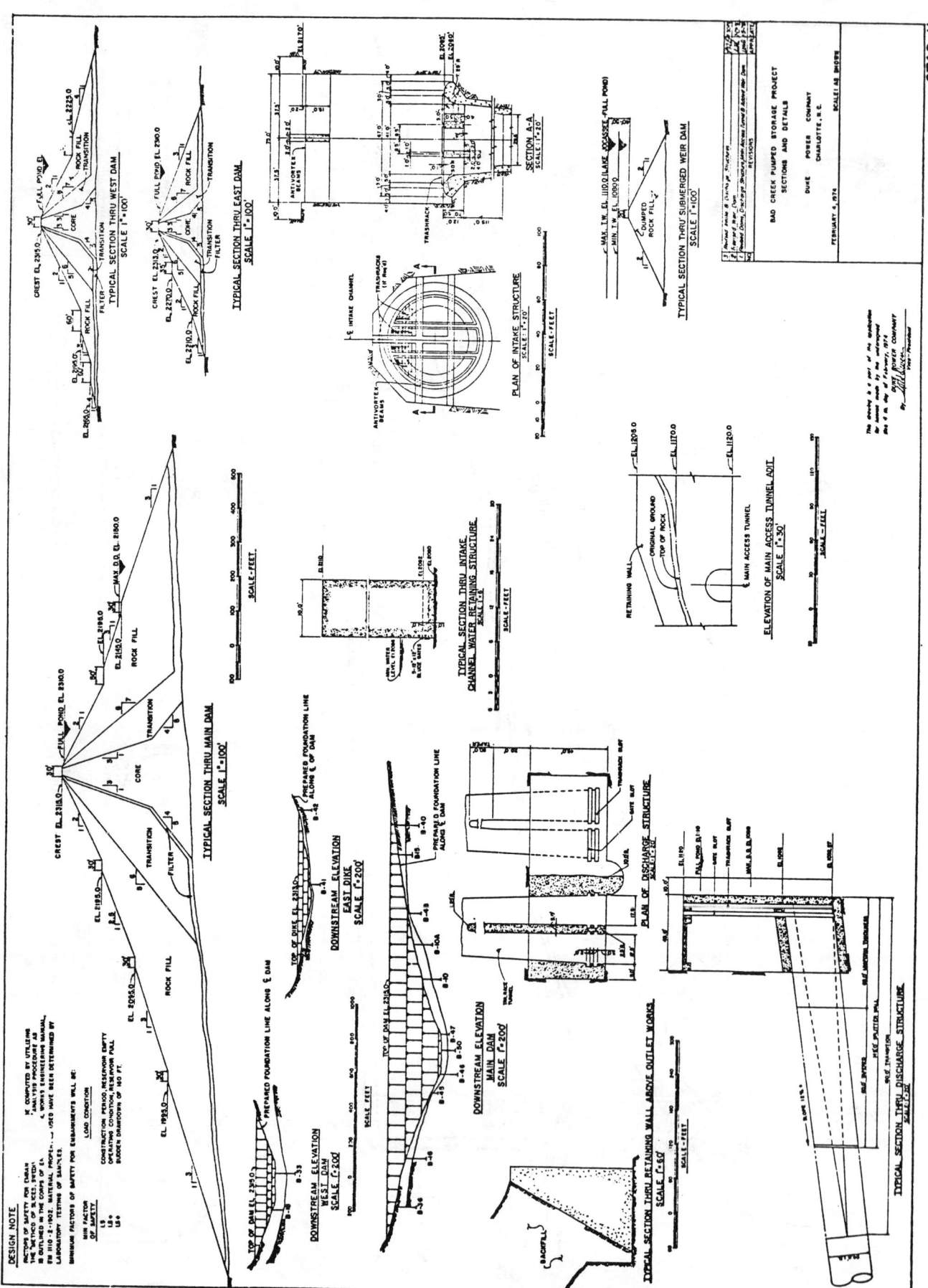

PUBLICATIONS RELATED TO THE BAD CREEK PUMPED STORAGE PROJECT

1. Hester, J.G. and Edmonds, R.F., 1974, Bad Creek Pumped Storage Project, Electric Power and the Civil Engineers, ASCE.

2. Schaeffer, M.F., and Steffems, R.E., 1979, Geology of the Bad Creek Pilot Tunnel, Northwestern South Carolina, Geologic Notes, South Carolina Geol. Survey, V. 23, p. 117-128.

3. Schaeffer, M.F., Steffens, R.E., and Hatcher, R.D., Jr., 1979, Insitu and Its Relationship to Joint Formation in the Toxaway Gneiss, Northwestern South Carolina, Southwestern Geology, v. 20, no. 3, p. 129-143.

4. Steffens, R.E., 1984, The Bad Creek Exploratory Programs, Proceedings, Tunneling in Soil and Rock, Geotech 1984, ASCE, p. 156-181.

5. Holley, T.H., 1978, Tunnel Excavation at the Bad Creek Project, Proceedings, RETC, v. 2, p. 839-849.

6. Steffens, R.E., 1989, The Underground Powerhouse and Tunnels for the Bad Creek Pumped Storage Project - A Case History, Proceedings, RETC, p. 291-310.

Plant Name: **BALSAM MEADOW**

Plant location: 　Shaver Lake, CA	Owner:　Southern California Edison Company 　　　　　P.O. Box 800 　　　　　2244 Walnut Grove Av 　　　　　Rosemead, California 91770
Rated capacity　　　　　　　200 MW Average static head　　　　1,250 ft Plant efficiency　　　　　　　　% Stored energy　　　　　　1,600 MWh Number of units　　　　　　　　1	Designers: 　　　　　Southern California Edison Company
Construction time: 3 years, 11 months Construction cost: Price level: 1986 First commercial power: December 1987 FERC project number: 67	Plant Manager/Superintendent: 　　　　　Southern California Edison Company 　　　　　P.O. Box 800 　　　　　2244 Walnut Grove Av 　　　　　Rosemead, California 91770 　　　　　(　)　-
River or water source: Huntington Lake and Pitman Creek	

DAM	UPPER RESERVOIR	LOWER RESERVOIR
Type	Rockfill embankment	Existing concrete gravity
Height (ft)	120	185
Crest length (ft)	1,460	1,760
Volume (yd³)	354,000	
RESERVOIR		
Type	Constructed	Existing, Shaver Lake
Surface area (acres)	60	
Usable power storage (acre-ft)	1,680	60,000
Power pool fluctuation (ft)	40.0	42.0
Operating levels		
Maximum (ft)	6,668.0	5,372.0
Minimum (ft)	6,638.0	5,330.0
Drainage area (miles²)	0.1	
Seepage (ft³/s)		
SPILLWAY		
Design flood		
Return period (years)		
Flow (ft³/s)	2,000	
Capacity (ft³/s)		
Type	Open channel	
Gates		
Number	None	None
Type		
Width (ft)		
Height (ft)		
OUTLET WORKS		
Discharge capacity (ft³/s)		
Number of water passages	None	None
Dimensions of water passages		
Height (ft)		
Width (ft)		
Diameter (ft)		
Type of gates		
Number of gates	None	None

Plant Name: **BALSAM MEADOW**

Plant location
  Shaver Lake, CA
	Owner: Southern California Edison Company
	       P.O. Box 800
	       2244 Walnut Grove Av
	       Rosemead, California 91770

Rated capacity 200 MW Average static head 381.0 m Plant efficiency % Stored energy 1,600 MWh Number of units 1	Designers: Southern California Edison Company

Construction time: 3 years, 11 months
Construction cost:
Price level: 1986
First commercial power: December 1987
FERC project number: 67
	Plant Manager/Superintendent:
	       Southern California Edison Company
	       P.O. Box 800
	       2244 Walnut Grove Av
	       Rosemead, California 91770
	       ( ) -

River or water source: Huntington Lake and Pitman Creek	

DAM	UPPER RESERVOIR	LOWER RESERVOIR
Type	Rockfill embankment	Existing concrete gravity
Height (m)	36.6	56.4
Crest length (m)	445.0	536.5
Volume (m^3)	270,653	
RESERVOIR		
Type	Constructed	Existing, Shaver Lake
Surface area (Mm^2)	0.24	
Usable power storage (Mm^3)	2.072	74.009
Power pool fluctuation (m)	12.19	12.80
Operating levels		
Maximum (m)	2,032.41	1,637.39
Minimum (m)	2,023.26	1,624.58
Drainage area (Mm^2)	0.260	
Seepage (m^3/s)		
SPILLWAY		
Design flood		
Return period (years)		
Flow (m^3/s)	57	
Capacity (m^3/s)		
Type	Open channel	
Gates		
Number	None	None
Type		
Width (m)		
Height (m)		
OUTLET WORKS		
Discharge capacity (m^3/s)		
Number of water passages	None	None
Dimensions of water passages		
Height (m)		
Width (m)		
Diameter (m)		
Type of gates		
Number of gates	None	None

Balsam Meadow - Page 1 (Metric)

BALSAM MEADOW

INTAKES	UPPER INTAKE	LOWER INTAKE
Number	1	1
Type	Flaired concrete structure	Flaired concrete structure
Design discharge (ft³/s)	2,100	2,100
Gross inlet area (ft²)	1,150	1,150
(at trash racks)		
Bar racks		
spacing (in)	5.00	5.00
shape		
depth/thickness (in)	3.00 / 0.38	3.00 / 0.38
diameter (in)		
Emergency gates		
number	None	None
height/width (ft)		
type		
Service gates		
number	2	2
height/width (ft)	10.00 / 17.50	10.00 / 17.50
type	Slide gate	Slide gate
Bulkhead/stop logs(Y or N)		
number of units serviced		
Hoists		
number	2	2
capacity (tons)		
type	Electric	Electric

WATER PASSAGES	Upper Tunnel	Shaft	Lower Tunnel	Surge Tanks Upper	Surge Tanks Lower	Penstocks	Tailrace Tunnel
Number	1	1	1		1	1	1
Diameter (ft)	18.0	13.5	18.0		30.0	12.0	18.0
Length (ft)	2,854	939	875		274	320	7,500
Maximum velocity (ft/s)	9.1	16.2	9.1			20.5	9.1
Concrete lining thickness (in)					72.00	24.00	15.60
Total length of concrete sections (ft)					156		425
Steel liner Thickness							
Minimum (in)		0.94	1.38			1.38	
Maximum (in)		1.38	1.38			2.25	
Material grade		A537	A537			A537	
Total length of steel-lined sections (ft)		440	875			320	

Notes:
Upper Tunnel: Paved invert only.
Lower Surge Tank: Liner is shotcrete.

Balsam Meadow - Page 2 (English)

BALSAM MEADOW

INTAKES	UPPER INTAKE	LOWER INTAKE
Number	1	1
Type	Flaired concrete structure	Flaired concrete structure
Design discharge (m³/s)	59.5	59.5
Gross inlet area (m²) (at trash racks)	106.8	106.8
Bar Racks:		
spacing (mm)	127	127
shape		
depth/thickness (mm)	76 / 10	76 / 10
diameter (mm)		
Emergency gates		
number	None	None
height/width (m)		
type		
Service gates		
number	2	2
height/width (m)	3.048 / 5.334	3.048 / 5.334
type	Slide gate	Slide gate
Bulkhead/stop logs (Y or N)		
number of units serviced		
Hoists		
number	2	2
capacity (Mg)		
type	Electric	Electric

WATER PASSAGES	Upper Tunnel	Shaft	Lower Tunnel	Surge Tanks Upper	Surge Tanks Lower	Penstocks	Tailrace Tunnel
Number	1	1	1		1	1	1
Diameter (m)	5.49	4.12	5.49		9.14	3.66	5.49
Length (m)	869.9	286.2	266.7		83.5	97.5	2,285.9
Maximum velocity (m/s)	2.77	4.94	2.77			6.25	2.77
Concrete lining thickness (m)					1.829	0.610	0.396
Total length of concrete sections (m)					47.5		129.5
Steel liner Thickness							
Minimum (mm)		24	35			35	
Maximum (mm)		35	35			57	
Material grade		A537	A537			A537	
Total length of steel-lined sections (m)		134.1	266.7			97.5	

Notes:
Upper Tunnel: Paved invert only.
Lower Surge Tank: Liner is shotcrete.

Balsam Meadow - Page 2 (Metric)

BALSAM MEADOW

POWERHOUSE and RELATED FEATURES

Powerhouse Structure
Type: Underground
Length: 188 ft   Width: 80 ft   Height: 77 ft

Guard Valves
Number: 1   Diameter: 7.0 ft
Type: Spherical

Transformers
Number: 1
Ratings: 235
Voltages: (kV) 230 / 13.8

Generator
Rating generating (MVA): 200.0   Rating pumping (MVA): 180.0
Insulation type: Class B
Starting method: Reduced voltage-current limiting reactor
Starting equipment:

Runners
Material: Stainless steel casting 13cr-4ni
Minimum unit submergence: 98.0 ft
WR^2:
Manufacturer: Mitsubishi
Model test by: Mitsubishi

	Reversible Runners	Reversible Motor/Generator
Number	1	1
Diameter (ft)	13.80	Rotor 16.62 Stator 16.85
rpm synchronous	400.0	400.0
rpm overspeed	560.0	560.0
Type	Francis	Vertical, air cooler

Information on Runners

Condition:	Gross Head (ft)		Capacity (MW)		Discharge (ft³/s)		Turbine/Pump Eff.(%)	
	Generating	Pumping	Generating	Pumping	Generating	Pumping	Generating	Pumping
Maximum head & maximum power	1,352	1,324	235	152	2,306	1,650	91.4	91.1
Minimum head & maximum power	1,316	1,258	205	179	2,196	1,788	91.2	92.6

Note: Data in the above table are based on warranty info.

Condition:	Net Head (ft)		Capacity (MW)		Discharge (ft³/s)		Turbine/Pump Eff.(%)	
	Generating	Pumping	Generating	Pumping	Generating	Pumping	Generating	Pumping
Rated head @ best gate	1,266	1,324	221	173	2,103	1,429	91.6	93.6

Note: Data in the above table are based on imperical data.

BALSAM MEADOW

POWERHOUSE and RELATED FEATURES

Powerhouse Structure
Type: Underground
Length: 57.3 m   Width: 24.4 m   Height: 23.5 m

Guard Valves
Number: 1   Diameter: 2.13 m
Type: Spherical

Transformers
Number: 1
Ratings: 235
Voltages: (kV) 230 / 13.8

Generator
Rating generating (MVA): 200.0   Rating pumping (MVA): 180.0
Insulation type: Class B
Starting method: Reduced voltage-current limiting reactor
Starting equipment:

Runners
Material: Stainless steel casting 13cr-4ni
Minimum unit submergence: 29.87 m
WR^2:
Manufacturer: Mitsubishi
Model test by: Mitsubishi

	Reversible Runners	Reversible Motor/Generator		
Number	1			1
Diameter m	4.20	Rotor 5.065	Stator	5.136
rpm synchronous	400.0			400.0
rpm overspeed	560.0			560.0
Type	Francis	Vertical, air cooler		

Information on Runners

Condition:	Gross Head (m)		Capacity (MW)		Discharge (m³/s)		Turbine/Pump Eff.(%)	
	Generating	Pumping	Generating	Pumping	Generating	Pumping	Generating	Pumping
Maximum head & maximum power	412.1	403.6	235	152	65.3	46.7	91.4	91.1
Minimum head & maximum power	401.1	383.4	205	179	62.2	50.6	91.2	92.6

Note: Data in the above table are based on warranty info.

Condition:	Net Head (m)		Capacity (MW)		Discharge (m³/s)		Turbine/Pump Eff.(%)	
	Generating	Pumping	Generating	Pumping	Generating	Pumping	Generating	Pumping
Rated head @ best gate	385.9	403.6	221	173	59.6	40.5	91.6	93.6

Note: Data in the above table are based on imperical data.

BALSAM MEADOW

Plant Data:
```
  Average GWh generating per year:
  Average GWh pumping per year:
  Starting time from standstill (s):                        360
  Changeover time pumping to generating (min):               15
  Planned/scheduled time
     between major overhauls (years):                         5
  Outage time required per unit
     during major overhauls (weeks):                          4
  Representative plant availability (%):                   90.0
  Representative planned outages (weeks per year):            2
```

Miscelleneous Notes:
  The upper tunnel is an unlined horseshoe shape.

Cavitation Experience:
  Cavitation has been minimum.

Significant or Unique Problems:
  None.

List of Licenses Required:
  None reported.

ENVIRONMENTAL FEATURES

Recreation:
  Picnic and day use area (walk in only) at the Balsam Meadow Forebay Dam. Snow play area is located in the vicinity of the dam.

Fish and Wildlife:

Social:

BALSAM MEADOW HYDROELECTRIC PROJECT

PROJECT DESCRIPTION

Edison's Balsam Meadow Hydroelectric Project is located in the Sierra Nevadas approximately 55 miles northeast of Fresno, California. The water for the project originates well above 9,000 feet elevation as snowmelt is captured and stored in higher elevation reservoirs. The water is then diverted through existing tunnels into Edison's Huntington Lake. From Huntington Lake (elevation 6,950) the water is diverted into Shaver Lake (elevation 5,370) through an existing tunnel (Tunnel 7) which was constructed in the 1920s. A new diversion tunnel connected to Tunnel 7 diverts water into a new reservoir formed by a rockfill dam at Balsam Meadow (elevation 6,670).

This forebay has approximately 1,700 acre-feet of useable storage to power a 200 MW turbine/generator. The facility will operate primarily as a peaking unit and will produce 200 million Kwh per year based on a normal water year. The majority of this energy will be produced during the late spring and summer months when "runoff" is the greatest.

The plant operation will provide for 9 hours of generation with a 14-hour pump back cycle to refill the forebay on the days when the system demand requires the extra energy. The unit has a nameplate capacity of 200 MW and is powered by a 268,000 hp reversible pump-turbine operating at 400 RPM. The maximum capacity is more than 230 MW at a maximum static head of 1,335 feet. The power is generated at 13.8 kV and stepped up to 220 kV in the underground powerhouse. A sodium hexafloride (SF_6) insulated bus duct is used to bring the power up to a surface switchyard through a 20-foot diameter vertical elevator shaft. The energy is then transmitted along a 4.5 mile transmission route to where it connects into Edison's existing Big Creek transmission system.

The principal civil engineering features are described below:

PRINCIPAL CIVIL ENGINEERING FEATURES

Tunnel/Outlet Works. A new 16-foot diameter horseshoe-shaped diversion tunnel connects into the existing tunnel. This tunnel terminates at the Balsam Meadow forebay for a total length of 5,873 feet. An outlet structure was constructed at the end of the diversion tunnel where it discharges into Balsam Meadow forebay. A roller wheel slide gate installed at the outlet works is used to isolate the reservoir from the existing tunnel system.

A new 10-foot square slide gate installed at the end of the existing Tunnel 7 is used to back up and divert the water into the new diversion tunnel. This gate is normally closed, while the diversion tunnel gate is open when diverting water from Huntington Lake. An 8-inch pipe has been installed around the slide gate to furnish water to North Fork Stevenson Creek to maintain the native fish population.

Balsam Meadow Forebay and Dam. The project water discharges into the Balsam Meadow forebay which is formed by a concrete-faced, rockfill dam constructed across the mouth of Balsam Meadow. The dam, which is 1,325 feet long and 126 feet high at its maximum section, forms a 1,680 acre-foot reservoir. With an active capacity of 1,548 acre-feet, the unit can operate for an 8-hour period in the generation mode with no inflow from Tunnel 7. When the pumping portion of the project is installed, the project can complete a generation and pump back mode in a 24-hour period.

Power Tunnel. The power tunnel and penstock begin at the southwesterly end of the Balsam Meadow forebay and terminate at the powerhouse. This pressurized tunnel and penstock transports water from the forebay to the turbine to produce the 200 MW of capacity. The original design for the tunnel system consisted of an 18-foot upper power tunnel (horseshoe-shaped cross section 2,857 feet in length), a 30-foot diameter surge tank, a vertical tunnel (constructed by a raise-bore method, 13.5 feet in diameter with a height of 1,048 feet), a lower tunnel (consisting of an 18-foot horseshoe cross section on a 4 percent slope for 967 feet) and a penstock approximately 320 feet in length which transitions from the horseshoe section to a 10-foot diameter circular steel-lined section. The penstock reduces down to approximately 7 feet where it terminates and connects to the turbine shut-off valve.

During final engineering, the upper power tunnel was steepened to 9.3 percent slope to reduce the possibility of water column separation at the upper elbow. This revision resulted in the elimination of the upper surge chamber. This condition was confirmed by the final hydraulic transient study. During the excavation of the powerhouse chamber and tunnels in the vicinity of the powerhouse, numerous unexpected open joints were encountered. After an extensive geotechnical study and hydrofracture tests were performed, it was determined that the rock joints which crossed the lower power tunnel, vertical shaft, and tailrace tunnel could result in considerable leakage into the powerhouse from the high pressure water in the power tunnel. Accordingly, a decision was made to steel line all of the lower power tunnel and the lower 440 feet of the vertical shaft to a point where the static water pressure equaled the minimum principle in-situ stress of the rock. It was also decided to concrete line approximately 425 feet of the tailrace tunnel immediately downstream from the powerhouse.

Powerhouse. The powerhouse cavern was excavated in competent granite and consists of an arched roof cavern which is 188 feet long, 80 feet wide, and 147 feet high at the maximum dimension. The cavern, which was excavated from the top down, contained 64,000 cubic yards of excavated rock.

The powerhouse is divided into four main levels: Basement floor at elevation 5,225 feet, turbine floor at elevation 5,242 feet, generator floor elevation 5,255 feet, and main floor at elevation 5,272 feet. The 220 kV transformer was set in a covered vault in the powerhouse at elevation 5,235 feet.

A 450-ton bridge crane was installed on concrete crane beams 2.7 feet wide by 9 feet deep, for the full length of the cavern. The crane beams are anchored to the rock wall with 30-foot long post-tensioned, grouted rock bolts.

Tailrace Tunnel. The tailrace tunnel, which is 7,543 feet in length, is an 18-foot diameter unlined tunnel having a horseshoe shape and a concrete-paved invert. The tailrace water discharges into Shaver Lake about 30 feet below the normal maximum water level. A gated outlet structure has been constructed at the end of the tunnel where it terminates in Shaver Lake. Two 9 feet by 17 feet bulkhead gates with 20-inch bypass systems are installed in the structure and, when closed, the tunnel can be dewatered. These bulkhead gates will be opened and closed only under balanced head conditions.

A 30-foot diameter surge chamber was constructed just downstream from the powerhouse. The chamber, constructed 55 feet to the east of tailrace tunnel, extends 276 feet in height above the tailrace invert. A side drift tunnel with a vertical orifice connects the surge chamber into the tunnel system. An 8-foot diameter vent shaft connects the top of the chamber to the ground surface.

PROJECT PLANNING AND DESIGN

The Balsam Meadow Hydroelectric Project was initially conceived in the early 20s as an addition to the Big Creek System. The project was revived in the early 70s by Edison Engineers at a time when a heavy emphasis was being placed on the development of renewable resource alternatives. Concentrated studies for the Balsam Meadow Hydroelectric Project actually began as early as 1973 with the issuance of a report entitled, "Potential Hydroelectric Development in the San Joaquin River Basin." It was during this time that the project was first envisioned as an underground pumped-storage facility. Over the next 15 years, the project evolved to its present configuration.

In the summer of 1978, geological field reconnaissance was initiated for the project. The commencement of civil/structural conceptual engineering and environmental studies followed in January 1979. Initial conceptual engineering activities involved a unit sizing study and an above versus below ground powerhouse evaluation.

In late 1979, preliminary engineering began with the layouts of tunnels, powerhouse, intake structures, and dam design in sufficient detail to support preparation of the FERC Application, Exhibit F "General Design Drawings." The Environmental Impact Report and the Exhibit F drawings were completed in early February 1980, and the FERC application was filed on February 29.

PROJECT CONSTRUCTION

The construction of the Balsam Meadow Project was divided into work packages to be bid consistent with contractor's capabilities and expertise as well as cost effectiveness. The two primary packages were the General Civil Contract and the Mechanical, Electrical, and Instrumentation (MEI) Contract.

Whenever possible, lump sum or fixed unit price bidding was used. Where the work could not be adequately defined because of unknown conditions time and material type contracting was utilized.

Edison supplied the major and long lead time equipment and materials such as the turbine/generator, switchgear, powerhouse crane, elevator, and switchyard/transmission structures. The contractors supplied all construction equipment and materials, including tunnel supports, concrete, and temporary facilities such as shops and warehousing, air compressors and blowers, batch and aggregate plants, and waste-treatment facilities.

The general civil contract, which comprised about 40% of the total project cost, and included excavation of all tunnels and shafts and the powerhouse chamber was bid in mid-1983 with notice to proceed on November 1, 1983.

Tunnels were driven by conventional drill-and-blast methods utilizing two Atlas-Copco hydraulic drill jumbos. One drill jumbo supported two hydraulic drills; the other supported three. An overall average progress of 36 lineal feet per day was achieved, working three 8-hour shifts, 5 days per week. Both drill jumbos, operating side-by-side, were used to first complete the Access Tunnel and related construction tunnels. They operated separately, using a swing heading technique, to complete all remaining tunnels ("swing heading" consisted of drilling one face while loading another and, at times, shooting and mucking at a third face).

Excavation started for the upper power tunnel and the diversion tunnel utilizing the swing heading technique until the Diversion Tunnel was advanced in the upstream direction to a point beyond the "saddle area", at which time priority was directed to excavating the Upper Power Tunnel. After both the Upper Power Tunnel and Diversion Tunnel were completed, the final tunnel construction was accomplished in the Tailrace Tunnel utilizing two headings. One heading was advanced downstream from the powerhouse side as the other was advanced in the upstream direction from the Shaver Lake side. All tunnel excavations were completed by December 1985.

Excavation of the vertical shafts, was by means of raised-bore techniques followed by slashing.

Powerhouse chamber excavation commenced on July 24, 1984 with removal of the crown section by way of a center drift from the crown access construction tunnel. Excavation of the initial 50 feet of crown consisted of multiple transverse drifts utilizing radial drill patterns to establish the crown's geometric shape. Following completion of the initial crown section, the balance of the crown was excavated utilizing multiple longitudinal drifts and horizontal drill patterns.

It was specified that rock bolts and shotcrete lining be applied to the crown perimeter surface within 8 hours after blasting, and with not more than two non-adjacent 300 square-foot areas of newly exposed rock being exposed to the air at any one time. However, because of the excellent quality of the rock encountered, the shotcreting requirements were relaxed which allowed the entire crown to be exposed prior to shotcret application.

Construction of the dam began in June 1984 and continued over three summers as planned, being shutdown due to snowfall during each winter. Following clearing and removal of native ground cover, weathered material was removed down to "foundation quality" rock for construction of the dam plinth or foundation. The rock embankment was then constructed utilizing material obtained from a quarry area located to the west of the dam. Incorporated into the "quarry" area is a spillway channel designed to accommodate emergency flows. The embankment was constructed using material classified and graded by size and placed in zones. Following completion of the embankment, a reinforced concrete slab was constructed on the upstream face of the dam using a slip forming operation. Finally, a concrete parapet wall was constructed on top of the dam.

The major equipment including the turbine, generator, and the main transformer arrived at Fresno during the August/September 1986 period and was transported to the jobsite. The generator stator and rotor were shipped in sections since their size and weights, when fully assembled, would be impossible to transport. The main transformer, weighing 165 tons, was the largest load ever transported over Highway 168 to the Shaver Lake area and was delivered successfully with no problems.

ENVIRONMENTAL IMPACTS

From the outset, environmental protection was established as a high priority. In 1979, Edison initiated consultations with the various Federal, State and local agencies exercising regulatory environmental oversight with respect to the Balsam Meadow Project. This process was completed in 1983 and resulted in formal agreements with these agencies. Collectively, these agreements defined the environmental protection, mitigation and recreational requirements for the project. They included such topics as: land use management, protection of soil, timber and water resources, public safety, fire protection, preservation of cultural resources, upgrading of local fisheries, establishment of public recreation facilities, enhancement of wildlife habitat and full restoration of areas impacted by construction.

In establishing and fulfilling all of these agreements, Edison not only maintained and preserved the local environment, but also made substantial improvements to benefit both the general public and the native wildlife. Early in the project, Edison established the position of Onsite Regulatory Liason Representative to monitor and coordinate environmental control and mitigation activities.

Specific examples of Edison's environmental activities related to the project include:

- Although the creation of the project forebay resulted in the flooding of approximately 60 acres of this native meadow, it did create a small lake which was further developed as a day-use picnic area for the public to enjoy. In addition, approximately 200 acres of surrounding timber/shrub habitat were improved to benefit wildlife.

- Rock was placed in terraces in both Shaver Lake and Stevenson Creek to provide habitat for native bass and trout populations and improve these fisheries.

- Picnic facilities, parking lots and restrooms were constructed in various locations along Shaver Lake and local snow ski areas.

- After first removing and stockpiling the topsoil from the badly eroded Stevenson Meadow, Edison layered the rock excavated from the tunnels and powerhouse into terraces, redistributed the topsoil and revegetated the meadow with trees, shrubs and grasses.

- A comprehensive program was initiated to identify and preserve any archaeological and historical sites or artifacts found. Several items were found and preserved through Edison's involvement with local Native American tribal organizations.

- Since the project utilizes an underground powerhouse and tunnels, there is virtually no visual impact other than a small switchyard and 4.5 miles of new transmission line.

The switchyard building was modeled after an alpine ski chalet and screened with trees and the visible transmission towers were painted green.

Edison planned, executed and continues to maintain a commitment to the environment. A fund has been established to provide for the monitoring and maintenance of these facilities in the future. The public and governmental agencies alike have favorably approved Edison's extensive environmental enhancements completed with the Balsam Meadow Project. The Regional Manager of the California Department of Fish & Game stated that "Edison was concerned about protecting the habitat and the environment up here ... we believe people will point to this project to show others what can be done for wildlife habitat."

This project description was provided by Dick Mascolo by exerpting the most pertinent sections of a report submitted to ASCE in nomination for the 1989 Outstanding Civil Engineering Achievement Award.

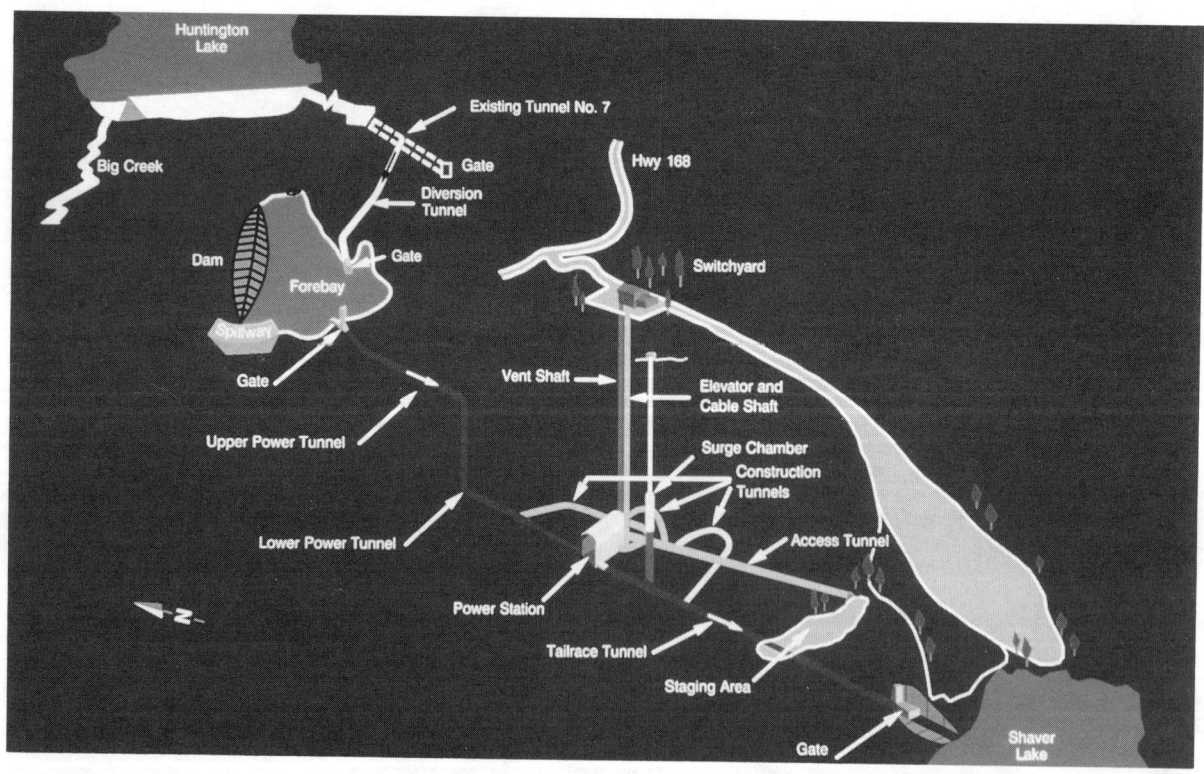

Balsam Meadow Pumped Storage Plant

BIBLIOGRAPHY

1. "Balsam Meadow Hydro Project Beats Time and Holds Budget," by David Martin, Tunnels and Tunnelling, March 1985.

2. "The Powerhouse Within," Dupont Magazine, June 1985.

3. "Dambuilders Inherit Cost Cap," Engineering News Record, August 15, 1985.

4. "The Balsam Meadow Hydroelectric Project Completes the Big Creek System," Prime Moyer Control, Woodward Governor Company, October 1987.

Plant Name: **BATH COUNTY**

Plant location:
  Warm Springs, VA
  Bath County

Rated capacity	2,100 MW
Average static head	1,179 ft
Plant efficiency	80.30 %
Stored energy	23,700 MWh
Number of units	6

Construction time: 8 years, 8 months
Construction cost: $639 per kW
Price level: 1988
First commercial power: December 1985
FERC project number: 2716

Owner: Virginia Power Company and Allegheny Generating Company
P.O. Box 26666
Richmond, Virginia 23261

Designers:
  Harza Engineering Company - All Civil Work
  Allis Chalmers - Pump Turbines
  Siemens-Allis - Motor/Generators

Plant Manager/Superintendent:
  Plant Manager
  Virginia Electric And Power Company
  Route A Box 280
  Warm Springs, Virginia 24484-9717
  (703) 279-3111

River or water source: Lower Reservoir - Back Creek, Upper Reservoir - Little Back Creek

	UPPER RESERVOIR	LOWER RESERVOIR
DAM		
Type	Zoned fill	Zoned fill
Height (ft)	460	135
Crest length (ft)	2,400	2,300
Volume (yd³)	17,003,000	4,002,000
RESERVOIR		
Type	On stream	On stream
Surface area (acres)	265	555
Usable power storage (acre-ft)	22,500	22,500
Power pool fluctuation (ft)	105.0	60.0
Operating levels		
Maximum (ft)	3,320.0	2,118.0
Minimum (ft)	3,215.0	2,058.0
Drainage area (miles²)	2.4	73.4
Seepage (ft³/s)	2.000	1.000
SPILLWAY		
Design flood		
Return period (years)	Overpump	PMF
Flow (ft³/s)	18,000	60,000
Capacity (ft³/s)	18,000	60,000
Type	Fuse plug	Chute
Gates		
Number	None	2
Type		Radial
Width (ft)		32.00
Height (ft)		30.00
OUTLET WORKS		
Discharge capacity (ft³/s)		5,700
Number of water passages	None	2
Dimensions of water passages		
Height (ft)		9.00
Width (ft)		6.00
Diameter (ft)		
Type of gates		Slide
Number of gates	None	2

Bath County - Page 1 (English)

Plant Name: **BATH COUNTY**

Plant location
  Warm Springs, VA
  Bath County
	Owner: Virginia Power Company and Allegheny Generating Company
	        P.O. Box 26666
	        Richmond, Virginia 23261

Rated capacity   2,100 MW
Average static head   359.4 m
Plant efficiency   80.30 %
Stored energy   23,700 MWh
Number of units   6
	Designers:
	    Harza Engineering Company - All Civil Work
	    Allis Chalmers - Pump Turbines
	    Siemens-Allis - Motor/Generators

Construction time: 8 years, 8 months
Construction cost: $639 per kW
Price level: 1988
First commercial power: December 1985
FERC project number: 2716
	Plant Manager/Superintendent:
	    Plant Manager
	    Virginia Electric And Power Company
	    Route A Box 280
	    Warm Springs, Virginia 24484-9717
	    (703) 279-3111

River or water source: Lower Reservoir - Back Creek, Upper Reservoir - Little Back Creek

	UPPER RESERVOIR	LOWER RESERVOIR
DAM		
Type	Zoned fill	Zoned fill
Height (m)	140.2	41.1
Crest length (m)	731.5	701.0
Volume (m³)	13,000,000	3,060,000
RESERVOIR		
Type	On stream	On stream
Surface area (Mm²)	1.07	2.25
Usable power storage (Mm³)	27.754	27.754
Power pool fluctuation (m)	32.00	18.29
Operating levels		
Maximum (m)	1,011.94	645.57
Minimum (m)	979.93	627.28
Drainage area (Mm²)	6.216	190.106
Seepage (m³/s)	0.0566	0.0283
SPILLWAY		
Design flood		
Return period (years)	Overpump	PMF
Flow (m³/s)	510	1,699
Capacity (m³/s)	510	1,699
Type	Fuse plug	Chute
Gates		
Number	None	2
Type		Radial
Width (m)		9.754
Height (m)		9.144
OUTLET WORKS		
Discharge capacity (m³/s)		161
Number of water passages	None	2
Dimensions of water passages		
Height (m)		2.743
Width (m)		1.829
Diameter (m)		
Type of gates		Slide
Number of gates	None	2

Bath County - Page 1 (Metric)

BATH COUNTY

INTAKES	UPPER INTAKE	LOWER INTAKE
Number	3	6
Type	Horizontal	Draft tube
Design discharge (ft³/s)	9,200	5,400
Gross inlet area (ft²)	4,408	1,420
(at trash racks)		
Bar racks		
spacing (in)	9.50	9.75
shape	Rectangular	Rectangular
depth/thickness (in)	5.00 / 0.63	4.00 / 0.75
diameter (in)		
Emergency gates		
number	6	None
height/width (ft)	28.67 / 12.75	
type	Bulkhead	
Service gates		
number	6	12
height/width (ft)	28.67 / 12.75	19.50 / 14.17
type	Fixed wheel	Slide gate
Bulkhead/stop logs (Y or N)	N	N
number of units serviced		
Hoists		
number	6	1
capacity (tons)	156	30
type	Single acting hydraulic cylinder	Gantry crane

WATER PASSAGES	Upper Tunnel	Shaft	Lower Tunnel	Surge Tanks Upper	Surge Tanks Lower	Penstocks	Tailrace Tunnel
Number	3	3	3	3		6	
Diameter (ft)	28.5	28.5	28.5	44.0		18.0	
Length (ft)	3,618	990	3,459	329		1,050	
Maximum velocity (ft/s)	16.9	16.9	16.9			21.2	
Concrete lining thickness (in)	18.00	24.00	18.00	33.00		24.00	
Total length of concrete sections (ft)	3,618	990	3,459	329		121	
Steel liner Thickness							
Minimum (in)						1.00	
Maximum (in)						2.00	
Material grade						See notes	
Total length of steel-lined sections (ft)						1,029	

Notes:
Penstocks: Grade 1,2: A516 3,4: A517 5,6: A537   Lengths: 1,2: 273m (896 ft) 3,4: 314m (1029 ft) 5,6: 383m (1257 ft).

Bath County - Page 2 (English)

BATH COUNTY

INTAKES	UPPER INTAKE	LOWER INTAKE
Number	3	6
Type	Horizontal	Draft tube
Design discharge (m³/s)	260.5	152.9
Gross inlet area (m²) (at trash racks)	409.5	131.9
Bar Racks:		
spacing (mm)	241	248
shape	Rectangular	Rectangular
depth/thickness (mm)	127 / 16	102 / 19
diameter (mm)		
Emergency gates		
number	6	None
height/width (m)	8.738 / 3.886	
type	Bulkhead	
Service gates		
number	6	12
height/width (m)	8.738 / 3.886	5.944 / 4.319
type	Fixed wheel	Slide gate
Bulkhead/stop logs (Y or N)	N	N
number of units serviced		
Hoists		
number	6	1
capacity (Mg)	142	27
type	Single acting hydraulic cylinder	Gantry crane

WATER PASSAGES	Upper Tunnel	Shaft	Lower Tunnel	Surge Tanks Upper	Surge Tanks Lower	Penstocks	Tailrace Tunnel
Number	3	3	3	3		6	
Diameter (m)	8.69	8.69	8.69	13.41		5.49	
Length (m)	1,102.8	301.8	1,054.3	100.4		320.0	
Maximum velocity (m/s)	5.16	5.16	5.16			6.47	
Concrete lining thickness (m)	0.457	0.610	0.457	0.838		0.610	
Total length of concrete sections (m)	1,102.8	301.8	1,054.3	100.4		36.8	
Steel liner Thickness							
Minimum (mm)						25	
Maximum (mm)						51	
Material grade						See notes	
Total length of steel-lined sections (m)						313.7	

Notes:
Penstocks: Grade 1,2: A516 3,4: A517 5,6: A537 Lengths: 1,2: 273m (896 ft) 3,4: 314m (1029 ft) 5,6: 383m (1257 ft).

Bath County - Page 2 (Metric)

BATH COUNTY

POWERHOUSE and RELATED FEATURES

Powerhouse Structure
Type: Conventional surface, indoor
Length: 502 ft   Width: 157 ft   Height: 194 ft

Guard Valves
Number: 6   Diameter: 9.5 ft
Type: Spherical

Transformers
Number: 9 single phase
Ratings: 300 - FOA each
Voltages: (kV) 500 / 20.5

Generator
Rating generating (MVA): 389.0   Rating pumping (MVA): 348.0
Insulation type: Class F1
Starting method: Direct water-cooled conductor design
Starting equipment:

Runners
Material: Stainless steel
Minimum unit submergence: 65.0 ft
WR^2: 148,300,000 (lbf x ft²)
Manufacturer: Allis Chalmers
Model test by: Allis Chalmers

	Reversible Runners	Reversible Motor/Generator			
Number	6				6
Diameter (ft)	20.90	Rotor	7.90	Stator	10.20
rpm synchronous	257.0				257.0
rpm overspeed	415.0				415.0
Type	Francis				Modified umbrella

Information on Runners

Condition:	Gross Head (ft) Generating	Gross Head (ft) Pumping	Capacity (MW) Generating	Capacity (MW) Pumping	Discharge (ft³/s) Generating	Discharge (ft³/s) Pumping	Turbine/Pump Eff.(%) Generating	Turbine/Pump Eff.(%) Pumping
Maximum head & maximum power	1,260	1,260	508	370	5,400	3,160	88.6	91.6
Minimum head & maximum power	1,080	1,080	395	420	4,900	4,000	89.6	92.3

Note: Data in the above table are based on field tests.

Condition:	Net Head (ft) Generating	Net Head (ft) Pumping	Capacity (MW) Generating	Capacity (MW) Pumping	Discharge (ft³/s) Generating	Discharge (ft³/s) Pumping	Turbine/Pump Eff.(%) Generating	Turbine/Pump Eff.(%) Pumping
Rated head @ best gate	1,080	1,100	380	420	4,700	4,200	90.7	92.7

Note: Data in the above table are based on field tests.

BATH COUNTY

POWERHOUSE and RELATED FEATURES

Powerhouse Structure
Type: Conventional surface, indoor
Length: 153.0 m   Width: 47.9 m   Height: 59.1 m

Guard Valves
Number: 6   Diameter: 2.90 m
Type: Spherical

Transformers
Number: 9 single phase
Ratings: 300 - FOA each
Voltages: (kV) 500 / 20.5

Generator
Rating generating (MVA): 389.0   Rating pumping (MVA): 348.0
Insulation type: Class F1
Starting method: Direct water-cooled conductor design
Starting equipment:

Runners
Material: Stainless steel
Minimum unit submergence: 19.81 m
WR^2: 61,300,000 (Newtons x m^2)
Manufacturer: Allis Chalmers
Model test by: Allis Chalmers

	Reversible Runners	Reversible Motor/Generator			
Number	6		6		
Diameter m	6.38	Rotor 2.409	Stator	3.109	
rpm synchronous	257.0			257.0	
rpm overspeed	415.0			415.0	
Type	Francis		Modified umbrella		

Information on Runners

Condition:	Gross Head (m)		Capacity (MW)		Discharge (m^3/s)		Turbine/Pump Eff.(%)	
	Generating	Pumping	Generating	Pumping	Generating	Pumping	Generating	Pumping
Maximum head & maximum power	384.0	384.0	508	370	152.9	89.5	88.6	91.6
Minimum head & maximum power	329.2	329.2	395	420	138.8	113.3	89.6	92.3

Note: Data in the above table are based on field tests.

Condition:	Net Head (m)		Capacity (MW)		Discharge (m^3/s)		Turbine/Pump Eff.(%)	
	Generating	Pumping	Generating	Pumping	Generating	Pumping	Generating	Pumping
Rated head @ best gate	329.2	335.3	380	420	133.1	118.9	90.7	92.7

Note: Data in the above table are based on field tests.

BATH COUNTY

Plant Data:

Average GWh generating per year:	3,308
Average GWh pumping per year:	4,115
Starting time from standstill (s):	180
Changeover time pumping to generating (min):	10
Planned/scheduled time between major overhauls (years):	15
Outage time required per unit during major overhauls (weeks):	22
Representative plant availability (%):	95.0
Representative planned outages (weeks per year):	1

Miscelleneous Notes:

The plant rated capacity is 2,100 MW at minimum head. Average capacity is 2,400 MW.

Penstocks average 320 meters (1,050 ft) in length but have different lengths of concrete and steel sections, as follows.

	Concrete-lined sections	Steel-lined sections
Penstock #1 & #2 -	5.97 m (19.6 ft)	273.1 m (896.1 ft)
Penstock #2 & #3 -	5.97 m (19.6 ft)	313.7 m (1029.1 ft)
Penstock #3 & #6 -	36.79 m (120.7 ft)	383.2 m (1257.1 ft)

The construction period of 8 years and 8 months included a 2.5-year shutdown period with limited powerhouse construction.

Planned outages are reported as 1 week per year, but every third year the planned outage is 3 weeks.

Changeover times are 10 minutes from pumping to generating but 20 minutes from generating to pumping.

Starting time is 3 minutes from standstill in the generating mode but 13 minutes in the pumping mode.

The upper intake rack bars have diagonal bracing that is 127 mm (5 in.) by 19 mm (0.75 in.).

The lower intake rack bars have diagonal bracing that is 1.63 m (64 in.) by 25.4 mm (1 in.).

Cavitation Experience:

Cavitation has ben very minor - two spots per vane on unit 1.

Significant or Unique Problems:

During initial filling of the power tunnels, unexpected high leak rates were observed, as were unexpected high pressures near the penstock liners. These high pressures caused the collapse of a 13.7 m (45 ft) section of the steel liner.

An extensive program of consolidation grouting was developed, and approximately 100,000 sacks of cement were injected into over 20,000

BATH COUNTY

drilled holes. The repair of the buckled steel liner was accomplised in parallel with the grouting program. The results of the grouting program were successfull, and unit operations were started in October 1985.

List of Licenses Required:
FERC.

ENVIRONMENTAL FEATURES

Recreation:
Downstream from the lower reservoir, two recreation ponds, campgrounds, and hiking trails have been provided. Nonpowered boating, a fishing pier, and a swimming beach are also present. Picnic areas, playground areas, camping areas, and one hunting area are available.

Fish and Wildlife:
Fish stocked in recreation pools attract anglers. A number of material waste pile areas and excavation slopes provide open grazing areas for deer.

Social:
The project provides improved flood control for downstream communities. It also provides, improvement in minimum flows in Little Back Creek and Back Creek.

Bath County, a 2100 MW development in the USA

by A. Zagars and J.M. Hagood, Jr
Vice President and Project Director* and Director of Hydroelectric Engineering**

With an installed capacity of 2100 MW, the Bath County pumped-storage plant currently under construction by the Virginia Electric and Power Company, will be the largest capacity development of its kind. General information about the project, which is sited about 13 km north of Mountain Grove, is given, as well as a description of the main features.

THE BATH COUNTY pumped-storage project will have an installed capacity of 2100 MW, the largest capacity of pumped-storage installations either already in operation or presently under construction. By midsummer of 1983 its first three units with 1050 MW aggregate capacity will commence operation making a substantial contribution to the security of Vepco's power system which is estimated to grow by that time to about 13 300 MW capacity with an annual net generation approaching 60 000 GWh. The other three units are scheduled for completion by summer of 1984 with the total installation then reaching the designed plant capacity of 2100 MW.

The total expenditures for the generation and transmission facilities including power storage and conveyance structures, will amount to about $785 million, including interest during construction and price escalation.

The principal purpose of the project is for peaking duties. Because of its location of two natural streams, the Commonwealth of Virginia required 3200 acre-ft ($3.95 \times 10^6 m^3$) of conservation storage to provide for minimum releases of 15 ft^3/s (0.424 m^3/s) during low flow periods, and a moderate flood control storage pool of 2500 acre-ft ($3.1 \times 10^6 m^3$) to minimize flood damages downstream. Except for these requirements, the scheme is intended to function as a pure pumped-storage project.

The major works will be constructed in the valleys of Back Creek (lower reservoir) and Little Back Creek

---
* Harza Engineering Company, Chicago, Illinois, USA.
** Virginia Electric and Power Company, Richmond, Virginia, USA.

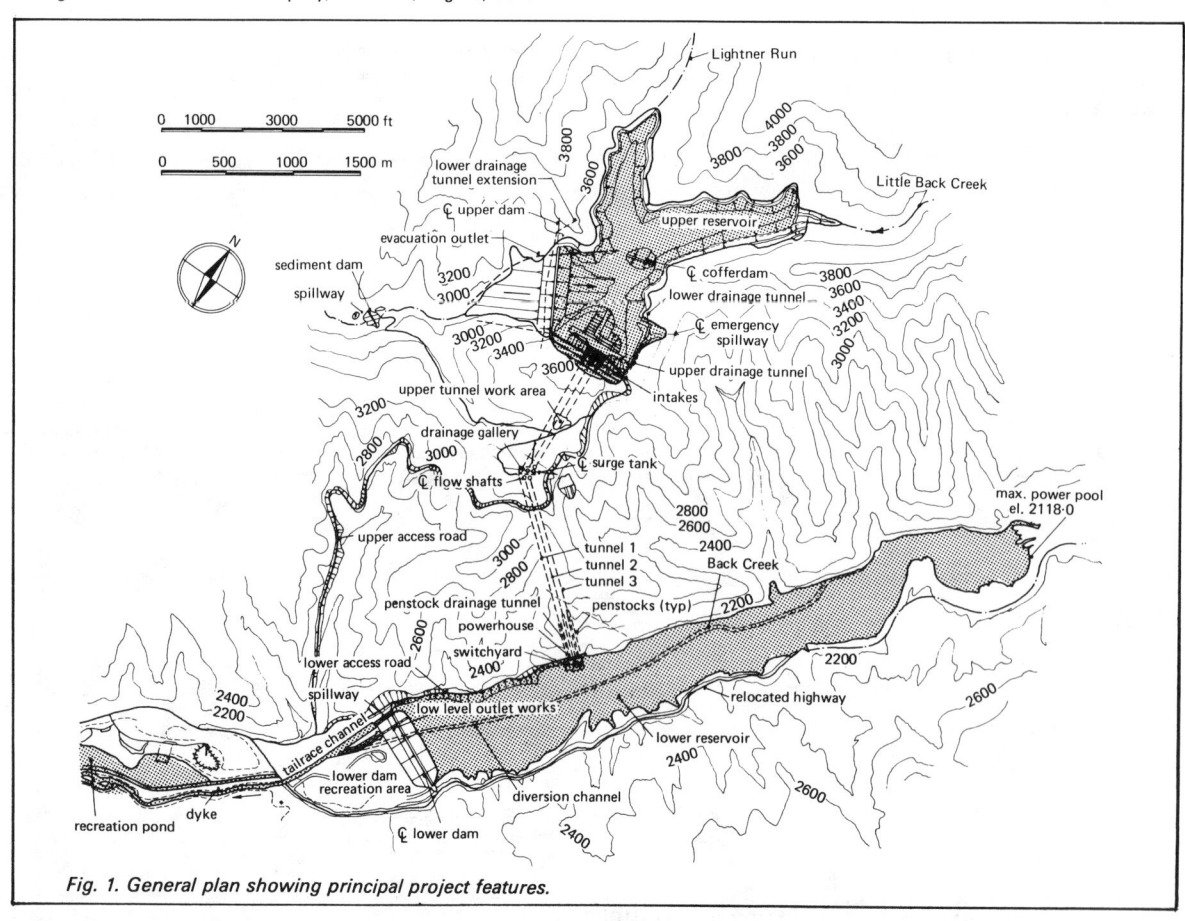

Fig. 1. General plan showing principal project features.

Reprinted, with permission, from <u>International Water Power and Dam Construction</u>, October 1977.

Reservoir details—Bath County		
	Lower Reservoir	Upper Reservoir
Area, acres (m² in brackets) (at full power pool)	555 (2.24×10⁶)	265 (1.07×10⁶)
Storage, acre-ft (m³×10⁶ in brackets)		
Inactive	2300 (2.84)	13 000 (16)
Conservation	3200 (3.95)	—
Power	22 500 (27.75)	22 500 (27.75)
Flood control	2500 (3.08)	—
Total controlled storage	30 500 (37.62)	35 500 (43.8)

(upper reservoir), which have an altitude differential of approximately 1000 ft (304 m).

One of its principal features will be a conventional surface powerhouse housing six 350 MW units. It will be located at the western rim of the lower reservoir and connected with the upper reservoir via an 8400 ft-(2560 m)-long water conduit system. The upper reservoir will be formed by constructing a 470 ft (143 m)-high (measured from original ground) zoned earth and rockfill dam, which, when constructed, will be the highest (measured from stripped ground) dam in the USA east of the Mississippi river. Conventional type power intakes, a diversion tunnel converted to permanent outlet works and an emergency spillway for the release of possible over-pumping flows, will comprise the other upper reservoir features. The lower reservoir structures will consist of a 135 ft (41 m)-high zoned earth and rockfill dam, a controlled chute type spillway and a twin-conduit low level outlet works passing under the lower dam.

The operating heads will vary between 1075 ft and 1280 ft (327 and 390 m).

Two 500 kV overland transmission lines, one 35 miles (56k m) and the other 50 miles (80 km) long, will connect the project with Vepco's existing 500 kV transmission network.

Studies

Appraisal studies of potential site developments were conducted during the late 1960's. Feasibility studies of a potential site, in the general vicinity of the present site, were conducted during the period of 1970-1971.

Conceptual and project optimization studies, including preparation of Federal Power Commission (FPC) License Application documents, began in March 1972. Engineering studies were facilitated by contracting with the engineer and the General Contractor simultaneously to ensure a complete flow of information between them from the beginning. Results of the feasibility studies had indicated a plant of 1500 MW capacity. Subsequent conceptual, optimization and load forecast studies uprated this to 2100 MW, as presently licensed for construction. The initial engineering schedule called for optimization of the project and preparation of FPC License documents by October 1972 and commencement of construction by January 1975.

Financial considerations, revaluation of load growth in Vepco's service area and an increase in the licensing lead time, resulted in rescheduling of construction for 1977, with the first three units going on line by the summer of 1983 and the other three by 1984.

Exploration efforts

Prefeasibility exploration, consisting of geologic mapping, core drilling, seismic refraction surveys, borehole geophysical tests and test pitting were conducted in 1971. These efforts were expanded during the period of 1972-1973 primarily to confirm the feasibility of the upper reservoir, and thus the feasibility of the pumped-storage scheme itself, and to obtain required information on geologic conditions for the various project features. Exploration was also carried out to locate sources for suitable construction materials. Approximately 60 000 linear ft (18 300 m) of core drilling—49 000 ft (15 000 m) for foundation exploration and design purposes, 11 000 ft (3350 m) for material exploration—was performed to obtain the information needed. The quantity for foundation drilling included exploratory holes drilled for several project alternatives investigated before the final concept was adopted.

An exploratory adit of approximately 900 ft length (274 m), which later will serve as one of the two permanent drainage tunnels above the six penstocks to control external pressures, was excavated in the spring of 1973. The purpose of the exploratory tunnel was to obtain a better understanding of geologic conditions over the penstocks than could be ascertained by the drill holes, to conduct tests for determination of penstock design parameters and to evaluate tunnelling conditions. The test programme for obtaining of design parameters consisted of tunnel convergence measurements, deformation measurements with multiple position borehole extensometers, plate bearing, borehole over coring and rock bolt pull-out tests. Shotcrete was applied in designated areas to compare protected and unprotected tunnel faces against exposure to atmosphere.

The far end of the tunnel was over-excavated to a 34.5 ft (10.5 m)—diameter semicircle, identical with the upper half of an excavated power tunnel. The roof was painted white and the invert covered with a plastic sheet in the over-excavated area to observe the amount and rate of rock fallout. After four years of exposure, a few small pieces of rock have separated from the excavated soffit face. Most of the paint has pealed off; however, it still can be traced over the rock surface.

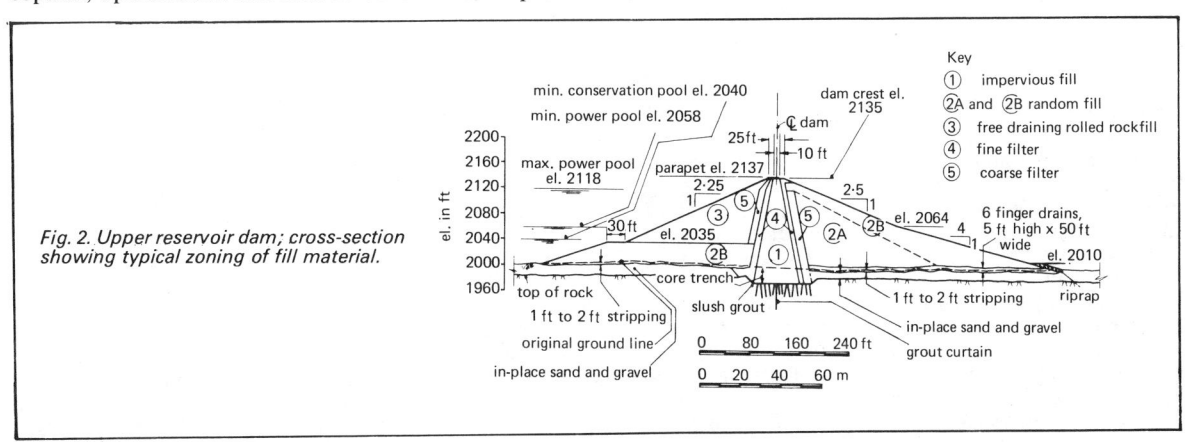

Fig. 2. Upper reservoir dam; cross-section showing typical zoning of fill material.

Key
1. impervious fill
2A and 2B random fill
3. free draining rolled rockfill
4. fine filter
5. coarse filter

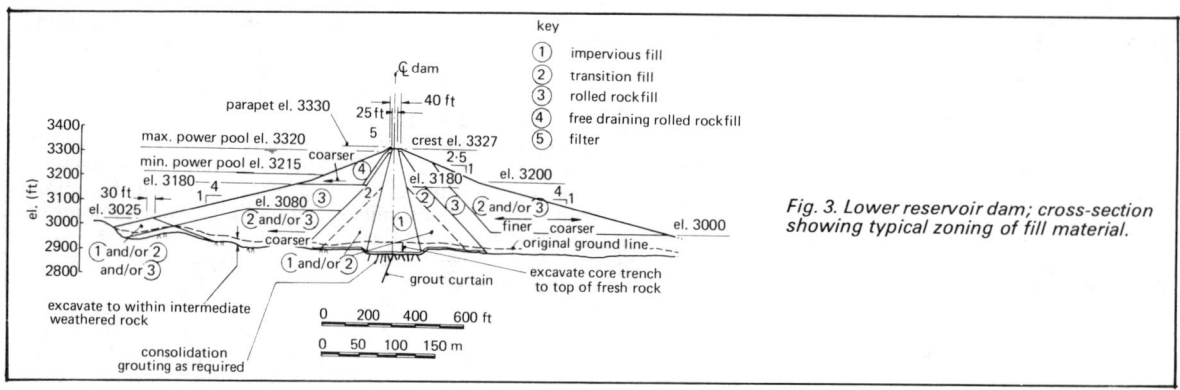

Fig. 3. Lower reservoir dam; cross-section showing typical zoning of fill material.

Project geology

All Project features will be situated in sedimentary rock formations of the Devonian age.

The lower reservoir, located in the Back Creek valley and contained within the Millboro/Brallier formation, is flanked by two large, dense rock masses which will provide an effective barrier against seepage. The Millboro/Brallier formation consists of thinly interbedded siltstone, sandstone and claystone which possess adequate strength to support safely the principal structures of the lower reservoir—the dam, powerhouse and spillway. Medium hard, dense, thin-bedded siltstone is the predominant rock of the formation. Siliceously cemented, hard and rather thinly bedded sandstone, and dense, medium soft, thinly bedded claystone form the interbeds of the formation. The beds dip steeply in a westerly direction over the wide valley floor; however, they become severely folded, forming sharp synclines and anticlines along the west bank of the valley. Joints are normal to bedding planes and closely spaced. They are closed or filled with calcite at reasonably shallow depths. Some joints, however, were found open at greater depth as indicated by high water takes.

The Back Creek valley floor is covered with alluvial material with an average thickness of about 10 ft (3 m). Weathering in the rock of the valley floor reaches to about 15 ft (4.5 m) depth. The colluvial cover amounts to about 30 ft (9 m) along the left bank of the valley and to a rather thin 2 ft (0.6 m) cover on the opposite right (west) bank of the valley. Rock weathering is extended about 40 ft (12 m) below the surface at the left abutment of the lower dam and up to approximately 50 ft (15 m) in depth at the right abutment of the dam. The exploratory adit exposed severely folded, weathered and faulted rocks through a 350 ft (107 m)-wide zone of the rock mass forming the right bank of the valley. West of this zone the rock is fresh, thinly bedded and dipping gently (about 8 to 10°) in north-westerly direction.

The contact between the underlying Millboro/Brallier formation and the overlying Chemung formation intersects the vertical flow shafts at approximately their midheight. Thus, the lower part of the water conduit system (lower power tunnels and the penstocks) will lie in the Millboro/Brallier formation. The upper power tunnels, surge tanks and intakes will be contained in the Chemung formation, consisting of dense interbedded sandstone, siltstone and shale beds. The Chemung formation is about 2000 ft (600 m) thick, is relatively impermeable and ensures watertightness for the upper reservoir which will lie in this formation.

Three distinctly identified marker beds, traced by drilling over a wide area between the lower and upper reservoirs, confirm continuity in the stratigraphy leading to the conclusion that no major geologic faults that could affect the water tightness of the reservoirs or the design of the project features exist in the area. The most common discontinuities in the rock masses are the joints and bedding planes. Slickensides, healed brecciated zones and gouge have been observed in extracted cores and in the exploratory adit as discontinuities in the bedding planes.

Weathering of the rock in the upper reservoir area varies approximately between 50 to 100 ft (15 m to 30 m) in depth.

Laboratory tests on core samples of the Chemung formation indicate average density of about 2643 kg/m³.; permeability to air about 3.5 m 10^{-10} m³/s; unconfined compressive strength of all tests average about 12 600 lb/in²; minimum is 3 560 lb/in² for interbedded rocks. Minimum rock strengths for sandstone, siltstone and claystone/shale amounted to about 9500, 5700, and 5000 lb/in² respectively. Maximum strength varies between 14 000 and 31 300 lb/in².

Triaxial tests yielded minimum/maximum values of 1800/5000, 800/3100 and 400/2000 lb/in² cohesion for sandstone, siltstone and claystone/shale, respectively. The corresponding results for friction angles were 43/55.5, 33/49 and 28/65 degrees, respectively. Young's modulus determined from dynamic field tests varied between 5.7×10^6 lb/in² for claystone/shale and 7.6×10^6 lb/in² for sandstone. Uniaxial dynamic laboratory tests yielded 4.8×10^6 lb/in² and 4.4×10^6 lb/in² for claystone/shale and sandstone, respectively.

Reservoirs

The lower reservoir will be formed by constructing a 135 ft (41 m)-high and 2400 ft (731 m)-long zoned earth and rockfill dam containing about 4×10^6 yd³ (3×10^6 m³) of fill material across Back Creek valley. An upper reservoir will be created by constructing a 470 ft (143 m)-high, 2400 ft (731 m)-long zoned earth and rockfill dam across the valley of Little Back Creek. This dam will contain approximately 24×10^6 yd³ (18×10^6 m³) of material. The pertinent data of the reservoirs is shown in the table.

The maximum power storage amounts to approximately 23.7 GWh of stored energy. Normal daily operations will, however, release somewhat less than this amount, because the demand for maximum energy during any one day will occur only under extreme conditions.

The lower reservoir will be approximately 0.4 miles (640 m) wide and 3.3 miles (5.28 km) long. Its maximum water level fluctuation during a daily cycle will be 60 ft (18 m) within the power storage zone. Except for clearing of wooded areas, the reservoir will require no special treatment to contain the water or for slope stabilization purposes.

The upper reservoir will be located just east of the border with West Virginia and will be contained by the upper reservoir dam, the rock mass of the Allegheny Mountain on the west and by a ridge of Little Mountain on the east, surrounded and underlain by the dense

Chemung formation. The reservoir will extend into Little Back Creek and Lightner Run valleys. It will be subjected to a maximum drawdown of 105 ft (32 m). However, the drawdown during the normal daily operating cycle will be somewhat less. To prevent major slides in the reservoir, the overburden material in the drawdown zone will be removed and for construction of the upper dam; its principal uses will be in the impervious core and transition zones.

Because of the dip of about 8° of the bedding planes into the reservoir along the east rim and the low friction resistance along the bedding planes, drainage galleries with drain holes will be provided in the rim to ensure stability in the rock mass under drawdown conditions. The galleries will drain by gravity into the creek downstream of the dam.

In the east rim area, the slopes in the drawdown zone will be protected against surficial sloughing and degradation of the exposed rock, and the resulting possible under-cutting of the slopes above the pool level, by a 12 ft (3.7 m)-thick rockfill layer. This type of protection offers the necessary free-draining characteristics, and, because of the thickness, lends itself well for placement with conventional construction equipment on the steep slopes.

Lower reservoir dam
The lower reservoir dam will be constructed of material from various sources. The excavation material from the power tunnels and the powerhouse will be placed in the upstream shell below the operating drawdown level and in the downstream shell; free draining quarrry material will be used for the upstream shell in the drawdown area and filters; clayey material from the valley floor will go into the impervious core. Originally, the easily obtainable alluvial valley gravels were intended for the upstream zones. However, a relatively high content of fine (minus 200) fraction and presence of dispersed clayey lenses make this material unsuitable for the zones requiring free-draining material without expensive processing.

The shells will be placed on top of the alluvial valley gravels after stripping the organic soil and a layer of silty material of about 2 ft (0.6 m) thickness. A core trench will form the foundation for the impervious core. The foundation will be consolidated and slush-grouted before placement of the core. A staged grout curtain will provide for a positive cut-off along the valley floor and the abutments. Several 50 ft (15 m)-wide free-draining zones (finger drains) will provide for drainage of the downstream filter zones.

The crest of the dam will be set 2 ft (0.6 m) above the probable maximum flood level. A 2 ft (0.6 m)-high rockfill parapet will be provided as an additional protection against wave action.

The downstream toe will be riprapped between the spillway and the outlet works to protect against turbulent flow effects resulting from operation of the stilling basin during high flows.

The dam will be instrumented with pressure cells and slope indicators to monitor internal movements.

Upper reservoir dam
This structure will also be a zoned earth and rockfill dam. Its height was determined on the basis of the studies conducted for project optimization and of the maximum acceptable water storage level that could be accommodated without encroaching upon the safety of the east rim of the reservoir. Except for the free draining rockfill in the drawdown zone of the upstream shell, the construction material will come from the required excavation in reservoir rim, from the dam foundation, forebay channel, and intake area.

The Zone 1, impervious fill, will be obtained from processed material excavated from the upper 5 ft to 8 ft (1.5 m to 2.4 m) of soil, rock fragments and deeply weathered rock under the top soil in the reservoir rim, dam foundation and forebay area. The Zone 2, transition fill material will come from the same areas excavated from material underlying the Zone 1 material. Zone 3 material will consist of fresh rock from the required excavation for the forebay and intakes. The free draining rockfill material for Zone 4 will be obtained from a sandstone quarry in the surge tank area.

Filters (Zone 5) will consist of processed quarry material. A 3 ft (1 m)-high rockfill parapet will provide 5 ft (1.5 m) of freeboard with respect to a reservoir level resulting from overpumping with all six units for about 1.5 h duration.

Stripping will be carried out to intermediately weathered rock under the shells and to fresh rock in the core area. The core trench and the transition zone foundations will be treated with slush grout or pneumatically applied mortar to prevent erosion of material from seepage through open joints. Also the shell foundations may require slush grout treatment to seal open joints, cracks, and so on.

Since the project is located in a general area that in the past has been subject to earthquakes of low to moderate intensity, the dam will be designed to account for earthquake effects. A network of monitors will be installed to record microseismic activity for the purpose of determining whether the reservoirs cause any seismic activity.

Power conduits
Three 28.5 ft (8.6 m)-diameter power tunnels will be excavated in two levels with connecting vertical flow shafts. The lengths of the upper tunnels vary between 3560 and 3660 ft (1085 m and 1115 m), the shafts will drop approximately 990 ft (301 m) to connect with the lower tunnels whose lengths vary between 3190 and 3640 ft (972 m and 1109 m). The tunnels will be inclined at a slope of 2 per cent which originally was based on proposed construction with rail mounted equipment. Tyre mounted equipment is under consideration now, but the grade has been left unchanged. Steeper grades to reduce the shaft length, or to eliminate them entirely, were considered, but it was concluded that the added construction efforts for working on steep slopes and the related increase in construction time, considering the relatively long conduits, would outweigh the savings gained in excavation and concrete quantities, including some negligible amount of gain in hydraulic efficiency. The concrete lining will be 24 in (60 cm) thick and is provided primarily for hydraulic reasons. The thickness was selected such that it would provide the necessary minimum working space between the forms and the excavated rock face. Blasting during excavation will be controlled to reduce the depth of rock disturbance. Nominal support will be 1 in (2.5 cm)-diameter, 10 ft (3 m)-deep epoxy grouted rock anchor bolts. It is expected that shotcrete requirements, because of the good rock quality and in-situ stress conditions, will be minimal. The concrete lining will be unreinforced, except in the elbows and zones of weak rock. It will be placed with the continuous sloping pour method, ie, with no vertical construction joints and no waterstops.

Shrinkage cracking of the concrete will be accepted and will not be considered critical because of the tight, dense rock formation surrounding the tunnels. Also, it is expected that external pressure build-up will be relieved through these cracks, although the lining would be capable of resisting the full groundwater pressures in ring compression.

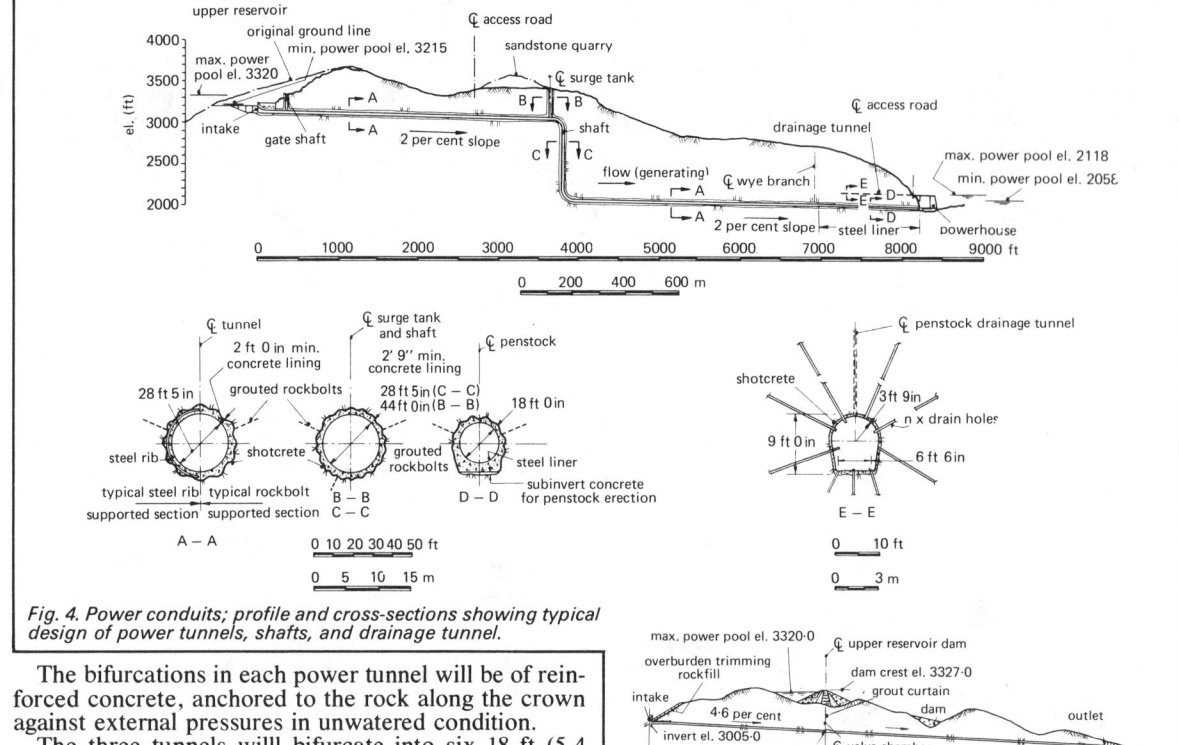

Fig. 4. Power conduits; profile and cross-sections showing typical design of power tunnels, shafts, and drainage tunnel.

The bifurcations in each power tunnel will be of reinforced concrete, anchored to the rock along the crown against external pressures in unwatered condition.

The three tunnels will bifurcate into six 18 ft (5.4 m)-diameter penstocks, their lengths varying between 860 ft and 1210 ft (262 m and 368 m). They will be steel lined from a point where the ground level above the penstocks is 40 per cent of the static head plus pressure rise on one unit. The adjoining concrete lining will be reinforced for a length of approximately 100 ft (30 m) upstream of the steel liners.

The design of the steel penstock liners will depend on rock assistance in the upstream zones where the rock will be allowed to resist about 70 per cent of the internal pressures as determined on the basis of elastic interaction between the steel liner, concrete embedment and the surrounding rock. The downstream zones of about 350 ft (106 m) length, where the rock is weak, as uncovered in the exploratory adit, were designed with no rock assistance with a factor of safety of 3 for most of the length, and a safety factor of 2.5 at the upstream end of that zone. The midsection was transitioned between the zones designed with and without rock assistance. The steel for the upstream zone will be 1 in (2.54 cm)-thick of ASTM A516 Grade 70 material. A 2 in (5 cm)-thick ASTM A517, Grade F steel pate will be used for the downstream zones. The thicknesses of the middle zones will change in $1/8$ in (3 mm) increments for the ASTM A517 Grade F material used also in these zones.

Two drainage tunnels will be provided about 150 ft (45 m) above the penstocks to limit external groundwater pressures to that head differential. Drainholes will be drilled from the drainage tunnels to intercept groundwater from above the drainage tunnels and possible seepage from the concrete lined penstocks and power tunnels upsteam from the steel lined sections.

Three 44 ft (13.4 m)-diameter surge tanks, one for each power tunnel, will be provided to cope with transient conditions during load rejection and acceptance cycles. The tanks will be constructed immediately upstream of the vertical flow shafts. Their location is governed mainly by the preference to locate them mostly underground for structural and aesthetic reasons because of their appreciable height of approximately 350 ft (106 m).

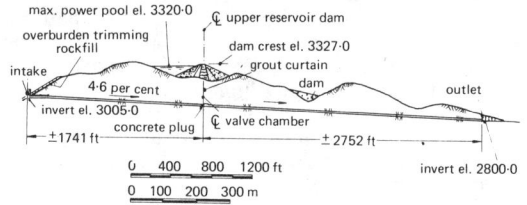

Fig. 5. Upper reservoir outlet works conduit; a 36 in-diameter (0.9 m) pipe will be installed in 6.5 ft×9 ft (1.98 m×2.7 m) horseshoe shape diversion tunnel and valved in valve chamber to provide for minimum flow releases and allow reservoir evacuation if necessary.

They will be concrete lined and reinforced in the zones of overburden and weathered rock. Their interior will be lined with a rubber-base coated membrane to contain seepage through small cracks in the concrete lining.

A 7 ft×9 ft (2.1 m×2.7 m) drainage gallery system around each tank, and surrounding all three tanks, will be provided near the bottom of the tanks to collect seepage that might develop through possible larger cracks in the walls. Holes will be drilled from the drainage gallery roof into the surrounding rock mass to drain seepage water into the galleries. This provision will prevent saturation of the rock near the exposed valley face and possible blowouts under saturated conditions.

Intakes

The tunnel controls will be located at the upper reservoir. Conventional type inlet structures will be constructed immediately upstream of the tunnel portals. They will be flared for gradual flow transitions, with particular concern for the flow conditions during the pumping mode. Intermediate walls will be provided to obtain satisfactory dispersion of the flow and also for structural reasons. The gross velocities through the trashracks during the pumping mode will be limited to 2 ft/s (0.6 m/s).

The trashracks will remain in place during both the pumping and generating cycles. They will be secured in position by removable wedges to control their movement under flow conditions in either direction and to expedite their removal for maintenance. Steel stoplogs can be placed in the trashrack slots to allow the inlet structure to be dewatered. The trashracks and the stoplogs will be handled with a mobile hoist.

Service and emergency closure gates will be located in an excavated shaft downstream of the inlet structure. This gate shaft location was selected for economic reasons since a tower type structure, designed for seismic effects, would have been more expensive.

The service gates, 12.75 ft × 28.5 ft (3.8 m × 8.7 m) will be of the fixed-wheel design and operated by hydraulic cylinder type hoists located on the intake deck. Sliding type bulkheads will be provided to enable unwatering of the tunnels in case of any malfunction of the service gates. They will be operated under balanced conditions only, by a mobile crane from the deck.

The service gates will be provided with fail-safe electrical controls to prevent their accidental lowering and closure of the tunnels during the pumping mode to prevent in pumping against the gates.

Powerhouse

The powerhouse will be of the conventional surface indoor type, housing six 65 ft (19.8 m) wide unit bays and an erection bay. It will abut the west bank of the lower reservoir at a location approximately one mile upstream of the dam. The shortest possible route for the power conduits, and topographically suitable location for the surge tanks were the governing factors that determined the powerhouse location.

During the early stages of the conceptual studies, an underground type powerhouse and also shaft (or silo) types were considered along with conventional ones. The topographical conditions and the multi-unit arrangement did not lend themselves well to the shaft type concept. Also, the then very short schedule for project licensing did not leave sufficient time for the necessary underground exploration and testing to confirm the technical feasibility of an underground powerhouse cavern. This conclusion was reached mainly on the basis of considerations for the deep setting of such a cavern in the type of sedimentary rock encountered.

The present shape of the powerhouse is an extrapolated out-growth from a similar structure developed for a 1500 MW plant, with six 250 MW units studied for the initial concept of the project. Its height of about 200 ft (61 m) is dictated by the minimum pump-turbine submergence requirement of 65 ft (20 m) the power storage depth of 60 ft (18 m), and an additional submergence of 18 ft (5.4 m) for conservation storage requirements. Also, a 5 ft (1.5 m) allowance for flood control storage, leaving 4 ft (1.2 m) freeboard with the powerhouse roof, is provided. Safety against flotation is the principal stability consideration.

The interior of the powerhouse will consists of a 65 ft (20 m)-wide and 490 ft (149 m)-long generator and erection bay hall flanked on the upstream side by a 35ft (10.6m)-wide service bay.

An enclosed unloading bay structure, with a 200 t bridge crane, will be constructed over the erection bay and unit bay 1 instead of individual hatches, with an outdoor gantry crane as frequently designed for semi-indoor type structures. This arrangement was selected to enable the erection of the six units under all weather conditions. Thus, most of the equipment will arrive at the powerhouse roof level el. 2127 to be unloaded with the 200 t overhead crane and will normally be lowered immediately to the erection bay floor at el. 2016.5. Some of the equipment may be stored temporarily in the unloading bay at the powerhouse roof level.

Two high-speed 40 t cranes will serve for installation of the lighter parts while a 700 t crane will handle the heavy parts of the pump-turbines and generator-motors. The provision of the two additional lighter cranes are for the purpose of expediting erection. Since the vertical space of the structure occupied by the cranes was dictated by the governing reservoir levels, the additional cost of the cranes and their accessories will be quickly offset by the savings gained from shortening the erection time by just a few days.

The steel draft tube liners, the spiral case and the stayrings will be moved into the powerhouse through a blockout in the erection bay end wall and lowered into

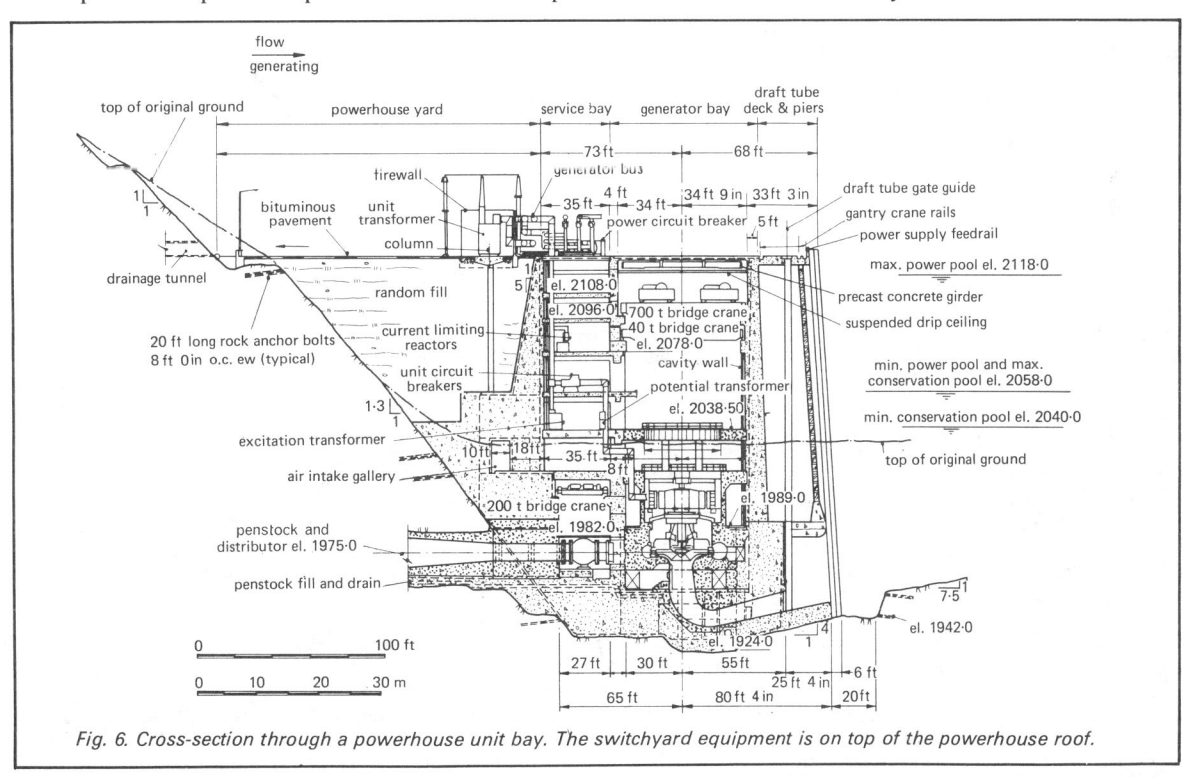

Fig. 6. Cross-section through a powerhouse unit bay. The switchyard equipment is on top of the powerhouse roof.

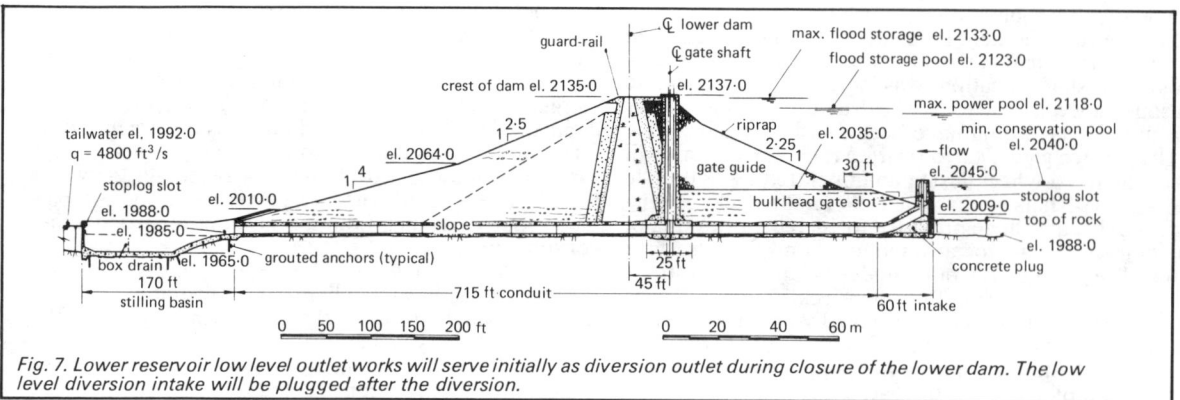

Fig. 7. Lower reservoir low level outlet works will serve initially as diversion outlet during closure of the lower dam. The low level diversion intake will be plugged after the diversion.

unit bays by a construction crane operating on a steel construction trestle erected along the centre of the unit bays. Because of the design of the pump-turbines, with thrust bearings mounted on the turbine headcover, the physical height of the units became appreciably lower than originally conceived with conventional thrust bearing arrangements. Thus, the top slab in the generator bay serves principally as a structural diaphragm for the exterior walls. The entire powerhouse structure will be of cast-in-place concrete except for the roof over the generating and erection bays which will be constructed with post-tensioned girders and with cast-in-place slab supported during the pouring of concrete by steel roof deck panels.

Powerhouse equipment

The powerhouse will have six reversible Francis-type pump-turbine generator-motor units, having a nominal rated generating capacity of 350 MW each, amounting to a total installed capacity of 2100 MW.

Each pump-turbine will be furnished with a 10 ft (3 m)-diameter steel spiral case inlet pipe, steel spiral case and stayring, movable wicket gates and a steel liner for the suction-draft tube wrapping around the elbow. The unlined portion of the draft tube will have one pier. The pump-turbine impeller-runner and wicket gates will be made of cast stainless steel. The guaranteed turbine output will be 380 000 kW at the normal minimum generating net head of 1080 ft (329 m). An output of 457 000 kW will be reached at the normal maximum net head of 1260 ft (384 m).

Each unit as a motor has a rated shaft output of 348 000 kW at 0.9 power factor, 20.5 kV, 60 Hz, three-phase. The rated electrical output as a generator will be 389 000 kVA at 0.9 power factor, 20.5 kV, 60 Hz, three-phase, with winding temperature rise limited to 60°C for both ratings. The generator-motors will also be capable of continuously operating at 115 per cent of the rated capacity in either mode with the winding temperature rise exceeding 60°C. An operating speed of 257.1 rev/min was selected for both the generating and pumping modes.

The guaranteed pump discharge at the normal maximum pump head of 1270 ft (387 m) is 3000 ft^3/s (85 m^3/s) per unit; at the normal minimum head of 1100 ft (335 m) it is 4100 ft^3/s (116 m^3/s) per unit.

Discharges in the generating mode will be 4600 ft^3/s (130 m^3/s) and 4500 ft^3/s (127 m^3/s) at the maximum net head of 1292 ft (393 m) and minimum head of 1077 ft (328 m), respectively.

Each unit will be funished with two oil-pressure operated double-acting servomotors for operation of the wicket gate mechanism. Modern electric-hydraulic speed governing systems with three term "PID" solid state type speed and acceleration sensing will be used as unit controls. The servomotor time will be adjustable between 10 and 60 s and will allow independent dual rate timing in both the opening and closing direction.

A minimum allowable servomotor closing time was specified to limit pressure head rise to approximately 37 per cent of the maximum static head on the units under pressure rise resulting from rejection of rated load on two units. Speed rise will not exceed 45 per cent under the same conditions with the specified minimum flywheel effect in the generator-motor. Each unit will be protected with a 9 ft 6 in (2.9 m)-diameter spherical type guard valve which will be closed when the units are idle to reduce high pressures on the wicket gates. The valves will also provide for emergency closure of the flow to the units. The valves will be operated by a high pressure oil system which is controlled automatically, but may be controlled manually if the need arises.

The excitation for the generator-motors will be static, potential-source, rectifier type with excitation power transformer, silicon power thyristors, and necessary accessories. The excitation supply transformer will be connected directly to the generator-motor main leads. A solid state-type voltage regulator will be furnished for each machine.

The units will be started in the pumping mode by either the synchronous starting or variable-frequency converter method, the former being limited for starting only 5 of the 6 units. The variable-frequency converter will be static-type with thyristor rectifiers, reactors, and thyristor inverters. This starting equipment will generate variable frequency current in the stator windings with resulting torque on the rotor of the unit selected for starting. Generator-motor bus, switchgear with accessories, and station service equipment will be installed in the service bay along the upstream side of the powerhouse. Three banks of single phase transformers rated 300 000 kVA will be located in the yard along the powerhouse roof, el. 2127. Each transformer will have two low-voltage windings and one high voltage winding. Two generator-motors will be connected to each transformer bank, one to each of the low voltage windings. The high voltage windings of the transformers will be connected to a 500-kV SF$_6$ gas insulated switchgear located on the powerhouse roof. The switchgear switching arrangement will be a four-breaker ring bus with connections to two transmission lines connecting into Vepco's existing 500 kV system at the Lexington and Valley substations, about 35 and 50 miles (56 km and 80 km) from the project, respectively.

A 25 t travelling-type draft tube gate gantry crane will be rail-mounted at the powerhouse roof deck level to service the draft tube gates and, if necessary, also the trashracks. The gates will be of the sliding bulkhead type and will be lowered and raised under balanced conditions. Six sets of gates will be provided to allow for closure of all draft tubes during simultaneous filling of the lower reservoir and of completion of the remaining work in the powerhouse during that period. The trashracks will remain in place during the generating mode. A 6 in (15 cm) bar spacing can be employed, because it is not expected the presence of the reservoir will enhance the presently rather poor aquatic habitat for an all-year-round fish population. Maximum gross velocity of 3.5 ft/s (0.9 m/s) and 2.8 ft/s (0.85 m/s) during generating and pumping mode, respectively, was used as the basis for sizing the trashrack openings.

Low level outlet works
The lower stages of the natural inflows of Back Creek, up to a maximum discharge of approximately 5000 ft³/s (140 m³/s), will be released through two 6 ft × 9 ft (1.8 m × 2.7 m) concrete conduits passing under the lower dam and terminating in a stilling basin. Originally, the conduit inlet will be set at the level of the creek bed to enable diversion of the Back Creek through the conduits during the final construction stage of the lower dam. After completion of the dam, the diversion inlet opening will be plugged. The inlet will be protected with reinforced concrete trash-beams to prevent possible clogging of the conduits with large size, water-logged timber entering the reservoir. The gates will be installed in a concrete gate shaft constructed just upstream of the impervious core of the dam, passing through the upstream shell of the dam. The shaft is sized to provide each conduit with one hydraulically operated vertical lift slide gate (regulating gate) and one vertical lift fixed wheel gate (emergency closure gate). The regulating gates will operate with openings not smaller than 1 ft (0.3 m), to avoid operation with small gate openings that could cause cavitation along the conduit invert. This gate opening will limit the minimum discharge through both conduits to approximately 300 ft³/s (8.49 m³/s), which is approximately three times the value of the mean annual flow. One jet-flow type gate, with a 22 in (0.5 m)-diameter orifice, will be installed in each regulating gate to regulate flows of less than 300 ft³/s (8.49 m³/s). The provision of these jet flow gates will minimize the need for frequent operation of the main regulating gates and make it unnecessary to operate the gates with openings less than 1 ft (0.3 m).

Minimum flow release requirements of 10 to 13 ft³/s (0.283 to 0.368 m³/s), a license requirement for low flow augmentation during periods of drought, will be discharged through "constant discharge facilities" provided in each conduit to avoid operation of the jet flow gates with very small openings over prolonged periods. These facilities will consist of a 12 in (0.3 m)-diameter conduit system with built-in water operated diaphragm type pressure reducing valve, a butterfly valve and an orifice header with multiple orifices. Irrespective of water fluctuations in the reservoir, a constant head will be maintained by the pressure reducing valve. Each facility will operate only in the completely open position.

Hydraulic cylinder hoist systems will operate the regulating, emergency, and jet flow gates. The butterfly valves of the constant discharge facilities will be operated with a motor operator or a handwheel mounted below the deck level in the gate shaft.

The hoist controls will be housed in a hydraulic control module next to the gate shaft at the deck level. It will be connected with a reservoir level sensing system and a remote control panel in the powerhouse.

Vertical lift, sliding type inlet bulkheads will be provided for closure of the inlet openings at time of inspection. The bulkheads will be dogged below the operating deck and normally will be submerged.

Spillway
Back Creek flows in excess of 5000 ft³/s (460 m³/s) will be discharged over a controlled, chute type spillway. Two gated openings with 32 ft wide × 30 ft high (9.7 m × 9.14 m) radial gates will be capable of discharging approximately 60 000 ft³/s (1700 m³/s) of the total routed 65 000 ft³/s (1840 m³/s) outflow of the probable maximum flood. The spillway approach channel will be formed by a moderate cut into the toe of the right bank of the valley and by appropriately curved retaining walls retaining the upstream shell of the lower dam.

The spillway structure, consisting of an ogee section with piers and an adjoining chute structure with stilling basin, will be constructed in the right abutment of the lower dam. The hydraulic jump type stilling basin will connect to the chute downstream of the toe of the dam and provide for exit of flows into an excavated tailrace channel connecting to Back Creek.

Hydraulic model tests were conducted at the Georgia Institute of Technology (Atlanta, Georgia) primarily to determine the design of the stilling basin that would ensure flow conditions causing least erosion to the dam and the abutments during flood stages. Also, approach conditions were observed and water level and pressure measurements were taken along the ogee and chute, and in the stilling basin. Tests were conducted for various stilling basin lengths, with and without baffle blocks, and for two settings of the stilling basin floor. The tests covered flow stages between 10 000 and 60 000 (283 m³/s and 1700 m³/s) (spillway discharge during probable maximum flood). The normal design flow was specified as the full gate discharge of 38 000 ft³/s (1100 m³/s) with reservoir level at the top of the gates (top of power pool plus 5 ft (1.5 m) of required flood storage).

A relatively short stilling basin design with baffle blocks indicated acceptable flow conditions not only for the design flood but also for the outflow of the probable

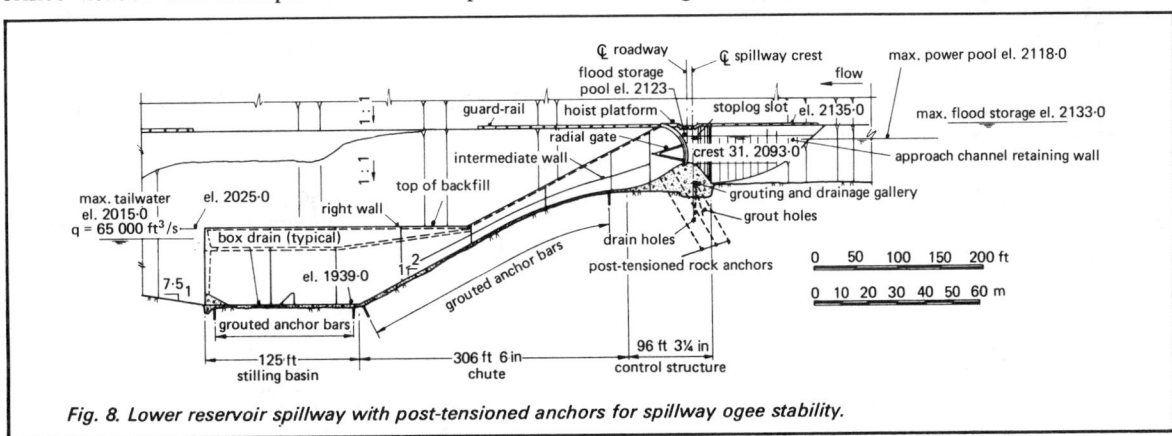

Fig. 8. Lower reservoir spillway with post-tensioned anchors for spillway ogee stability.

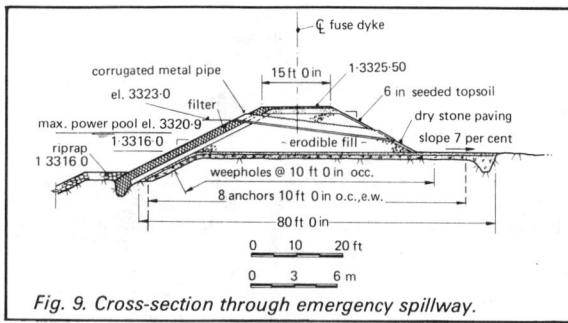
Fig. 9. Cross-section through emergency spillway.

maximum flood. The latter flow would be swept out without the baffle blocks for the same size of stilling basin.

Geologically, the entire spillway structure will be located in a very complex bedrock structure marked by intensely folded beds with series of small synclines and anticlines. Weathering is 20 to 75 ft (6 m to 23 m) deep and the rock is severely jointed and fractured. Slickensided along the bedding planes with a frictional resistance angle of only 17°, the foundation rock poses stability problems for the ogee structure. Several 500 t post-tensioned high strength anchors will be required to obtain the necessary resistance against sliding. The excavation slope along the right chute and stilling basin walls, and the foundation stability under the structure are controlled by the steeply (approximately 45°) dipping beds into the structure and the low frictional resistance along these beds. The beds fold upward under the structure and then dip gently into the left side of the structure. Rock bolts and post tensioned anchors will be used where stability problems will be encountered.

The 32 ft×30 ft (11 m×10.4 m) crest gates will be of radial design, trunnion-mounted on the pier faces. They will be operated by wire rope cables attached to motor-driven hoisting equipment. Emergency power will be supplied by a diesel type emergency generator located at the powerhouse. Appropriate slots will be provided in the piers to receive steel stoplogs in case the gates have to be raised because of maintenance.

Emergency spillway

The Federal Power Commission license requires that provisions for emergency releases in case of overpumping be incorporated in the design of the upper reservoir. A fuse dyke of erodible material will be built across a deepened natural saddle in the east rim of the reservoir. The spillway will be uncontrolled. An impervous membrane constructed of dense hydraulic grade asphalt will be placed on the upstream face of the fuse plug. Erosion of the 260-ft long and 10 ft-high (79 m × 3 m) dyke will be triggered when the reservoir level reaches el. 3323, and begins discharging through the 12 in (0.3 m)-diameter half-round, open-bottom corrugated metal pipes (erosion trigger pipes) penetrating the fuse dyke. The depth of erosion will be controlled by constructing the fuse dyke on a concrete slab. In case of overpumping, the overflows would discharge into a natural valley (Saw Pit Hollow) and empty into the lower reservoir. The trigger pipes will be set at an elevation that will allow for overpumping with all six units, at a rate of 18 000 ft³/s (509.4 m³/s) for about 1.5 h.

It is expected that fail-safe controls provided for the units will insure that the emergency spillway will never be triggered to operate.

An emergency spillway of the type described above has been triggered for operation once at the Seneca pumped-storage plant in Pennsylvania and functioned as expected.

Upper reservoir low level outlet works

License regulations require that at least 2 ft³/s (0.05 m³/s) be released from the upper reservoir. A low level outlet works to provide for this release and also to enable reservoir evacuation, in case the need arises, will be installed in the 4490 ft long (1368 m) and 6 ft×9 ft (1.8 m×2.74 m) horseshoe type diversion tunnel after completion of the diversion.

The evacuation conduit will consist of a 36 in (0.9 m)-diameter steel pipe, embedded in concrete between the inlet and the valve chamber, and supported on saddles in the downstream portion of the tunnel. Two 30 in-diameter (0.76 m) rotary plug-valves, installed in the valve chamber, will control the evacuation flows. The minimum flow releases will be released via a 6 in (15 cm) valved conduit bypassing the two plug valves. An energy dissipation structure will be built downstream of the tunnel to serve during diversion and in case the evacuation conduit is operated.

Recreation facilities

Recreation facilities, consisting of ponds for boating and swimming, camping and picnic grounds, and hiking trails, will be provided downstream of the lower dam. The lower reservoir area is unsuitable for this purpose because of the nearly 60 ft (18 m)-high daily fluctuations in the water levels. The land- and water-based recreation facilities will cover an area of approximately 400 acre (1.6×10^6 m) . It is expected that between fifty and sixty thousand visitors may visit the site yearly.

Present project status

At the time of writing this paper, the contractor was well along with preliminary construction activities. Clearing of the lower reservoir area and excavation in the diversion channel for the lower reservoir structures were under way. Also, excavation had progressed in the lower access road and powerhouse area, and relocation of the highway around the lower reservoir had begun. The wastewater treatment plant was under construction.

Currently, three main equipment contracts, (see below) for the pump-turbines and guard valves, generator motors, and the penstocks have been let. Other separate supply contracts for hydromechanical and electrical equipment will follow during 1977 and 1978.

The main contractor, acting as owner's agent, will perform most construction and equipment installation work. The penstocks will be erected by the supplier and all underground work will be executed by a sub-contractor. Contracts for the installation of the pump-turbines and generator-motors have not been awarded. All quality control aspects will be Agent's responsibility whereas the Owner, with the Engineer assisting, will perform the quality assurance aspects.

Board of consultants

The Federal Power Commission licence conditions require that three or more consultants review the design, specifications and construction of the project. The present Board members are: Mr. W. E. Johnson (Structures); Dr. R. B. Peck (Foundations, Embankments); Mr. W. V. Conn (Geology); and Mr. G. J. Vencill (Equipment). Messrs. Johnson and Peck served also as Board members during the preliminary period and have contributed greatly to various design and construction related matters. □

Main contractors/manufacturers;
Daniel Construction Company, Greenville, South Carolina;
Allis Chalmers, York, Pennsylvania;
Allis Chalmers, Milwaukee, Wisconsin;
Chicago Bridge & Iron Company, Oak Brook, Illinois;
Gates & Fox, Loomis, California.

PROJECT PLAN

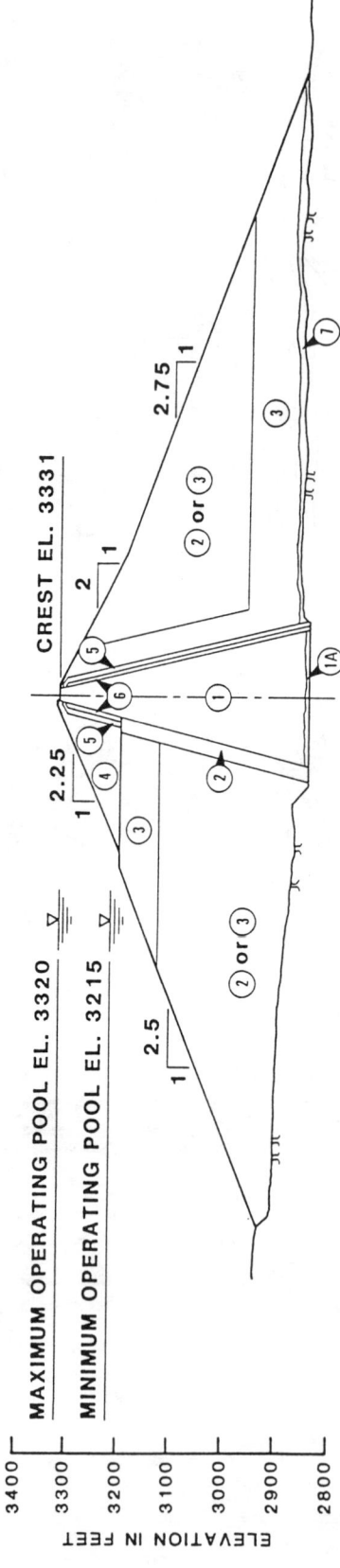

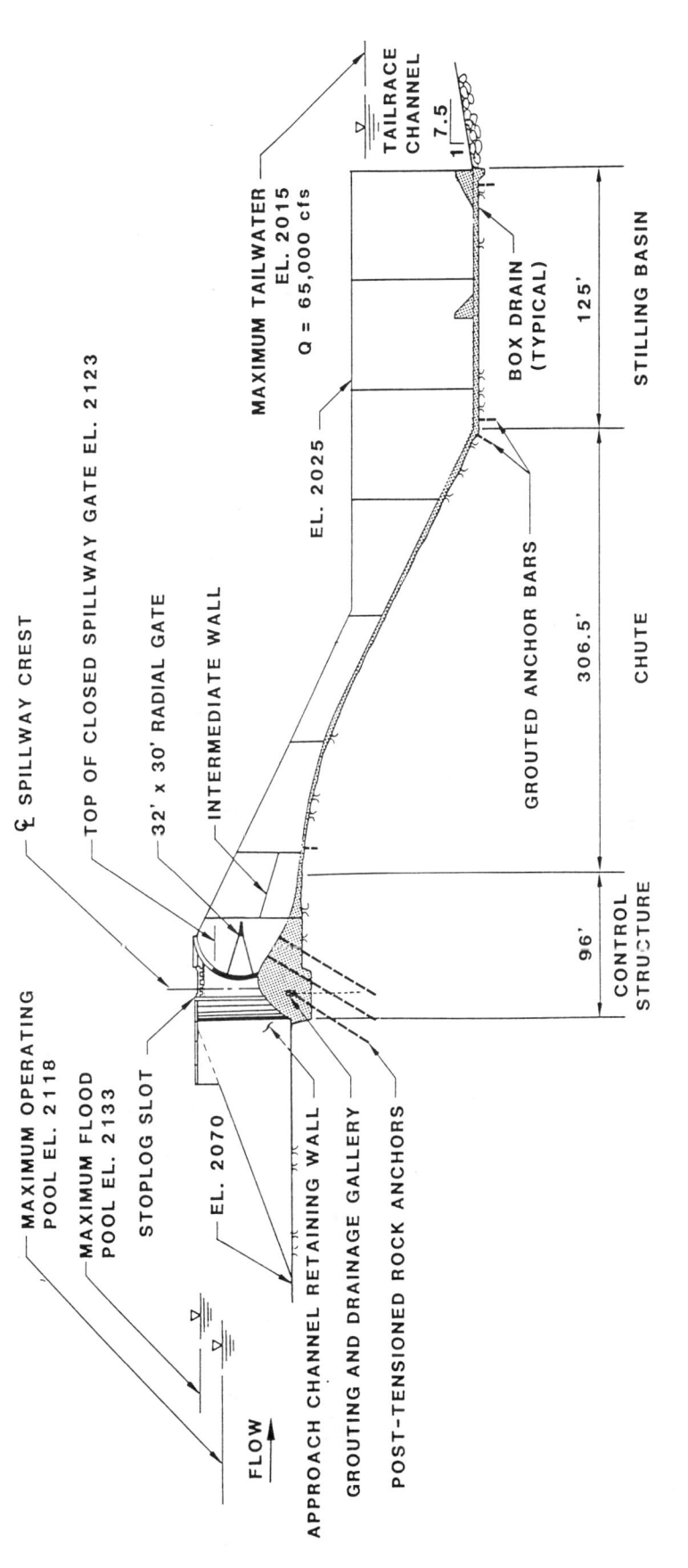

SPILLWAY

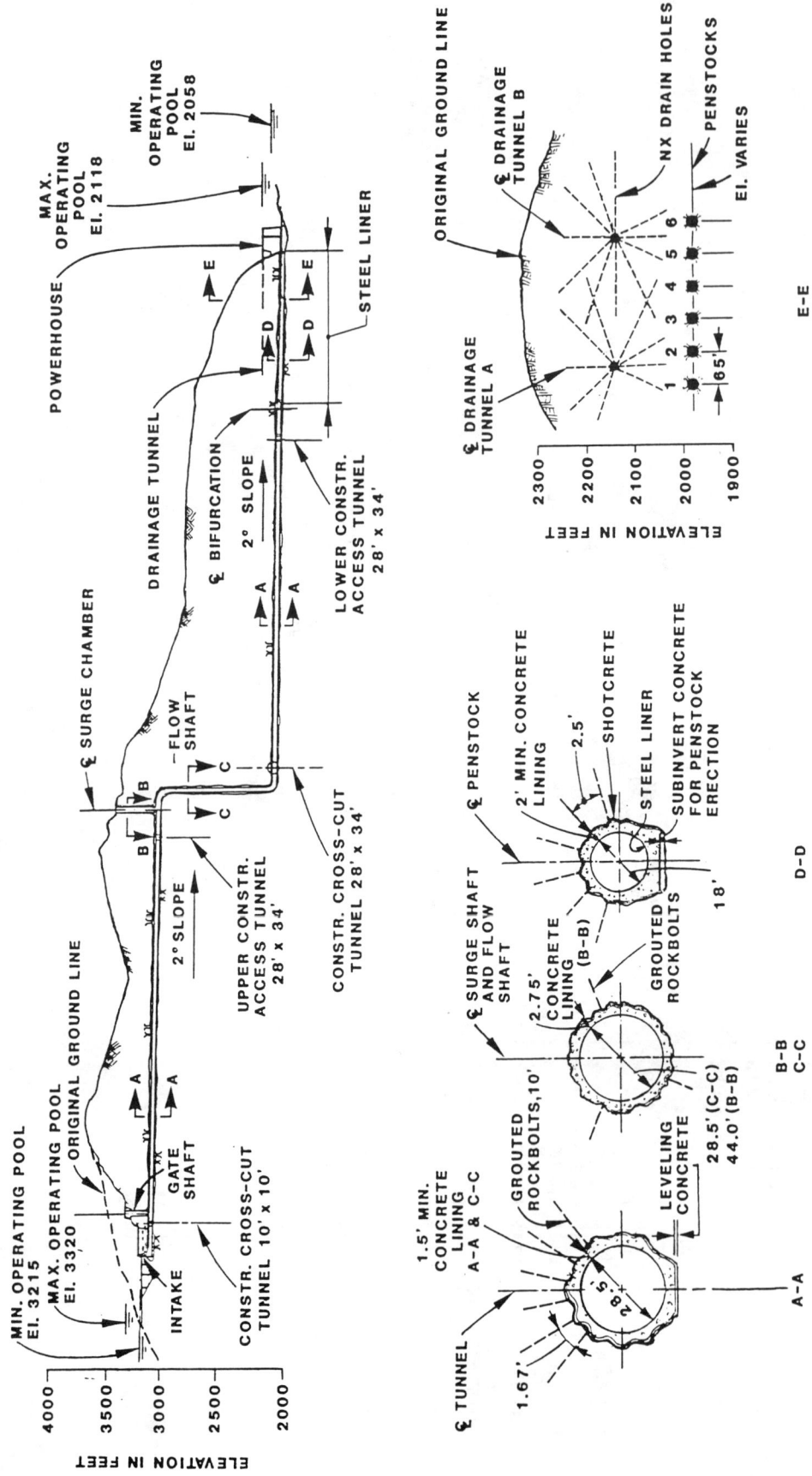

POWER CONDUITS

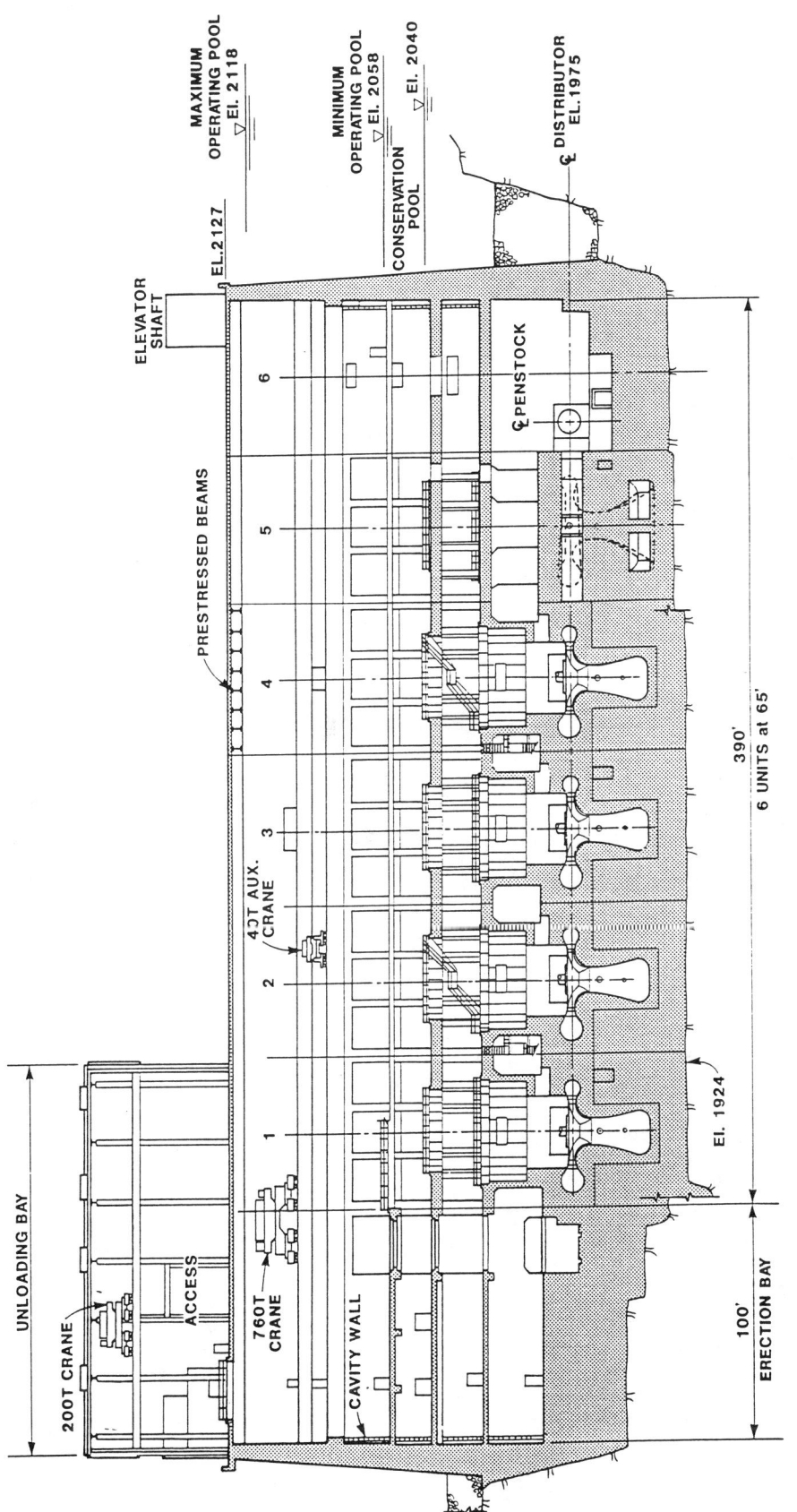

POWERHOUSE

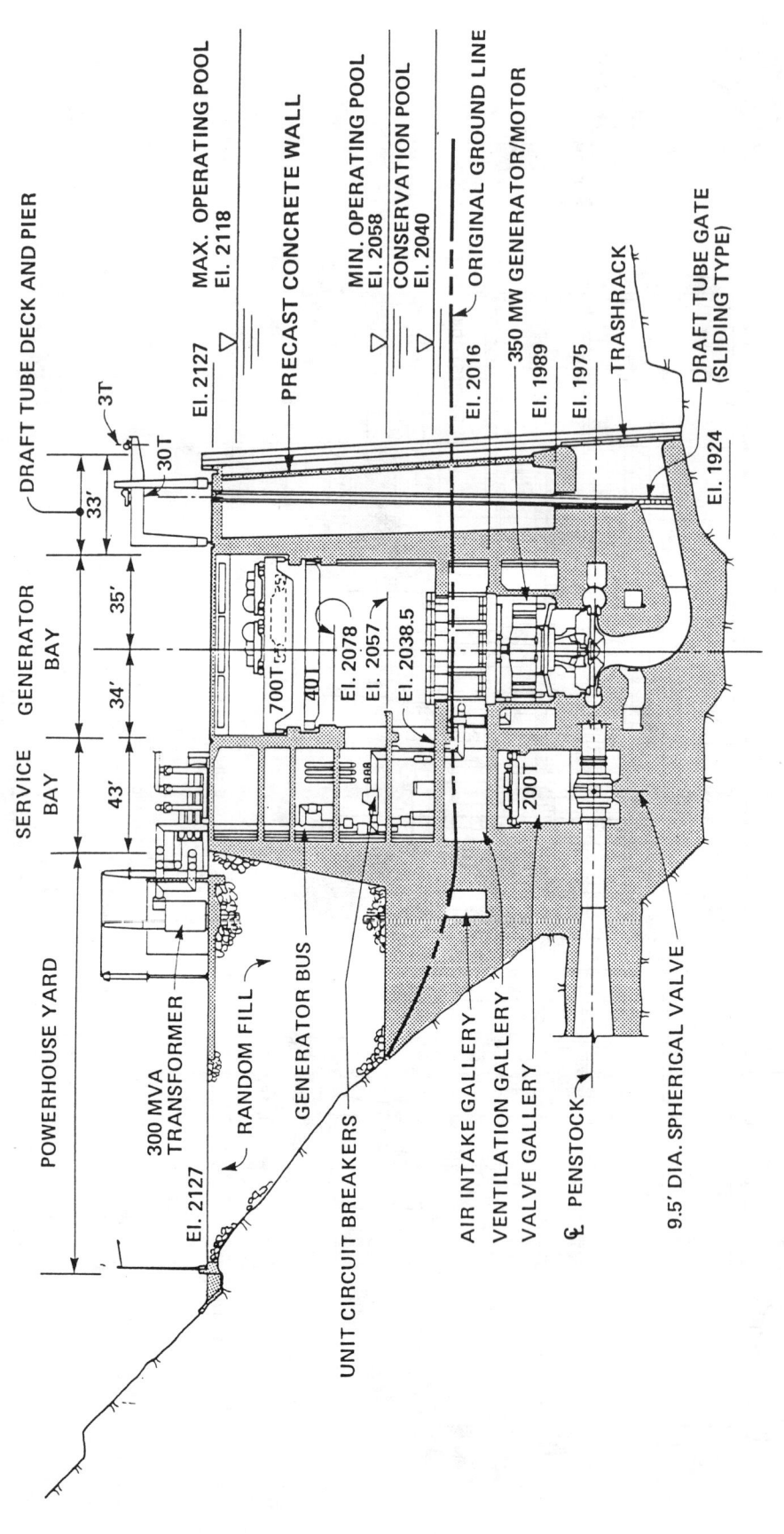

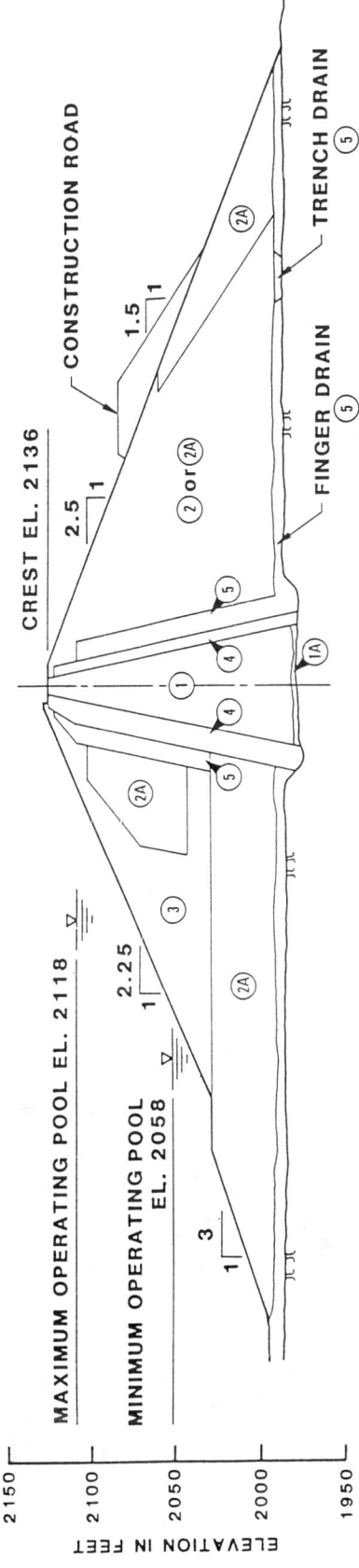

LOWER RESERVOIR DAM CROSS SECTION

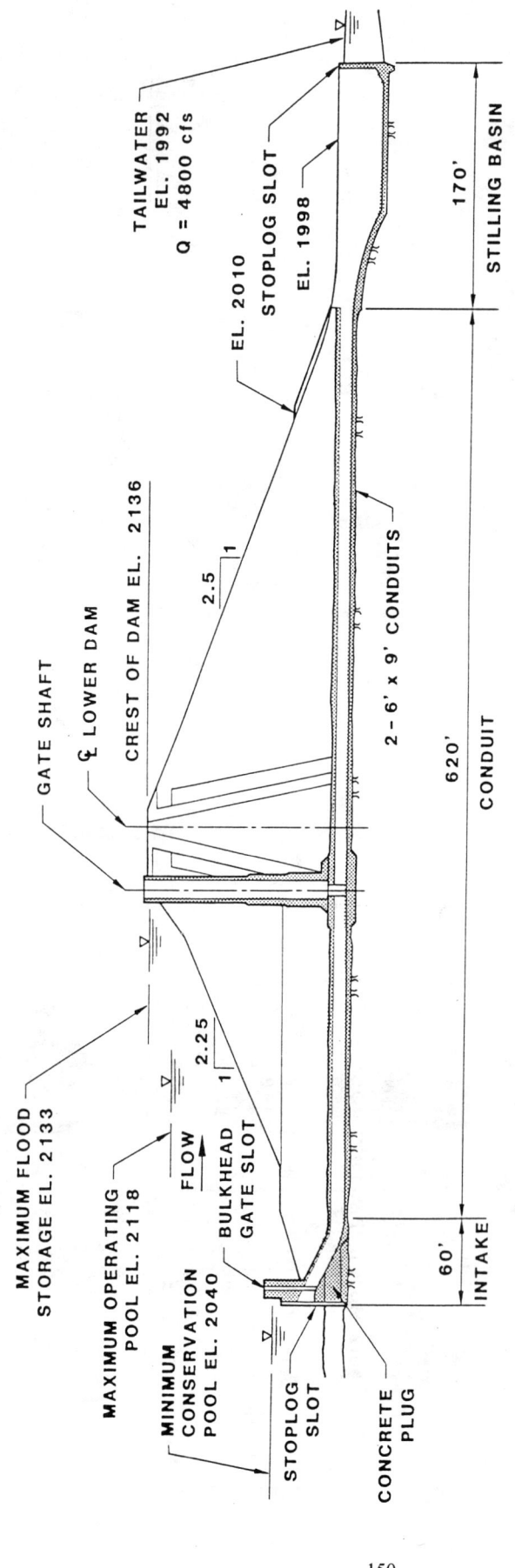

LOW LEVEL OUTLET

Upper Reservoir

Lower Reservoir

Bath County Pumped Storage Plant

Powerhouse Interior

Bath County Pumped Storage Plant

LIST OF BATH COUNTY PAPERS

1. "Bath County Pumped Storage Project Tunnel System-Evaluation of Remedial Measures," K. L. Wong, R. G. Oechsel, D. T. Waffle, ISRM Symposium for Rock Mechanics and Power Plants, September 1988, Madrid, Spain.

2. "High Pressure Grouting of the Bath County Tunnel System," W. J. Bogdovitz, N. B. Holst, WATERPOWER '87, August 1987, Portland, Oregon.

3. "Bath County Tunnel System/Remedial Treatment Program," D. E. Kleiner, R. G. Oechsel, K. L. Wong, N. B. Holst, A. H. Stukey, WATERPOWER '87, August 1987, Portland, Oregon.

4. "Bath County: The World's Largest Pumped-Storage Plant Commissioned," R. Sullenberger, Water Power & Dam Construction, February 1986.

5. "Instrutable Rock Tests Power Tunnel Grouters," Engineering News Record, May 1, 1986.

6. "The Bath County Pumped Storage Project," C. W. Pennington, John C. Clark, WATERPOWER '85, September 1985, Las Vegas, Nevada.

7. "Bath County Pumped Storage," Civil Engineering, July 1985.

8. "Slipforming the Bath County Shafts," W. D. Leech, P. S. Buskirk, Rapid Excavation and Tunneling Conference, June 1985, New York, N.Y.

9. "Bath County Pumped Storage Project Update," C. W. Pennington, D. J. Duck, American Power Conference, April 1985, Chicago, IL.

10. "Two Dams, Six Turbines and an Underwater Powerhouse Add Up to 2100 MW of Power at Virginia's Bath County Pumped Storage Project, the World's Largest," A. Danilevsky, Civil Engineering, November 1984.

11. "Bath County Development and Concept, Underground Works and Release Structures," Third Annual USCOLD Lecture, Stanton, Va., May 1983, J. M. Hagood, A. Zagars, K. L. Zerneke, D. E. Kleiner.

LIST OF BATH COUNTY PAPERS (cont.)

12. "Some Design Factors for the Bath County Trashracks," W. S. Hamilton, J. P. Tullis, Water Power & Dam Construction, August 1982.

13. "Bath County Pumped Storage Project (Virginia, USA) Underground Features," ISRM Symposium, Aachen, Germany, May 1982, A. Zagars.

14. "Electrical Features of the Bath County Pumped Storage Plant," R. J. Fostiak, W. L. Thompson, IEEE, 1982.

15. "Variable Frequency Starting Equipment for Pumped Storage Project," H. M. Henriksen, H. Brueckl, IEEE, 1982.

16. "Good Rock Cuts Time on Pumped Storage Job," Engineering News Record, December 7, 1978.

17. "Bath County Pumped Storage Project - Six 350 MW Pump/Turbines," H. L. Engleman, G. R. Frey, T. F. Armbruster, American Power Conference, April 1978, Chicago.

18. "Bath County, a 2100 MW Development in the USA," A. Zagars, J. M. Hagood, Water Power & Dam Construction, October 1977.

19. "World's Largest Pumped Storage - Bath County," USCOLD News, March 1977.

20. "Economic Analysis of Virginia Electric and Power Company's Bath County Pumped Storage Project," J. M. Hagood, H. H. Chen, ASCE Power Division Specialty Conference, University of Coolorado, Boulder, Colorado, August 12-14, 1974.

Plant Name: **BEAR SWAMP**

Plant location:
  Rowe, MA
  Berkshire County
	Owner: New England Power Company
	         25 Research Drive
	         Westborough, Massachusetts 01581

Rated capacity 600 MW Average static head 770 ft Plant efficiency 69.90 % Stored energy 3,690 MWh Number of units 2	Designers:

Construction time: 3 years
Construction cost: $200 per kW
Price level: 1974
First commercial power: January 1974
FERC project number: 2669
	Plant Manager/Superintendent:
	
	   Bear Swamp Pumped Storage Project
	   P.O. Box
	   Readsboro, Vermont 05350
	   ( ) -

River or water source: Deerfield River, a tributary to the Connecticut River	

DAM	UPPER RESERVOIR	LOWER RESERVOIR
Type	Rockfill central core	Earth and rockfill
Height (ft)	155	130
Crest length (ft)		900
Volume (yd³)	124,000	567,000
RESERVOIR		
Type	Constructed	On stream
Surface area (acres)	1,120	1,440
Usable power storage (acre-ft)	5,260	5,900
Power pool fluctuation (ft)	50.0	40.0
Operating levels		
Maximum (ft)	1,600.0	870.0
Minimum (ft)	1,550.0	830.0
Drainage area (miles²)	1,250.0	1,250.0
Seepage (ft³/s)		
SPILLWAY		
Design flood		
Return period (years)	Overpump	
Flow (ft³/s)	6,700	73,900
Capacity (ft³/s)		
Type	Overflow weir	Gated
Gates		
Number	None	2
Type		Tainter
Width (ft)		36.00
Height (ft)		40.00
OUTLET WORKS		
Discharge capacity (ft³/s)		
Number of water passages	None	None
Dimensions of water passages		
Height (ft)		19.00
Width (ft)		10.54
Diameter (ft)		
Type of gates		Slide
Number of gates	None	2

Bear Swamp - Page 1 (English)

Plant Name: **BEAR SWAMP**

Plant location
Rowe, MA
Berkshire County

Rated capacity	600 MW
Average static head	234.7 m
Plant efficiency	69.90 %
Stored energy	3,690 MWh
Number of units	2

Construction time: 3 years
Construction cost: $200 per kW
Price level: 1974
First commercial power: January 1974
FERC project number: 2669

Owner: New England Power Company
25 Research Drive
Westborough, Massachusetts 01581

Designers:

Plant Manager/Superintendent:

Bear Swamp Pumped Storage Project
P.O. Box
Readsboro, Vermont 05350
( ) -

River or water source: Deerfield River, a tributary to the Connecticut River

	UPPER RESERVOIR	LOWER RESERVOIR
DAM		
Type	Rockfill central core	Earth and rockfill
Height (m)	47.2	39.6
Crest length (m)		274.3
Volume (m³)	94,600	433,503
RESERVOIR		
Type	Constructed	On stream
Surface area (Mm²)	4.53	5.83
Usable power storage (Mm³)	6.488	7.278
Power pool fluctuation (m)	15.24	12.19
Operating levels		
Maximum (m)	487.68	265.18
Minimum (m)	472.44	252.98
Drainage area (Mm²)	3,237.503	3,237.503
Seepage (m³/s)		
SPILLWAY		
Design flood		
Return period (years)	Overpump	
Flow (m³/s)	190	2,093
Capacity (m³/s)		
Type	Overflow weir	Gated
Gates		
Number	None	2
Type		Tainter
Width (m)		10.973
Height (m)		12.192
OUTLET WORKS		
Discharge capacity (m³/s)		
Number of water passages	None	None
Dimensions of water passages		
Height (m)		5.791
Width (m)		3.212
Diameter (m)		
Type of gates		Slide
Number of gates	None	2

BEAR SWAMP

INTAKES	UPPER INTAKE	LOWER INTAKE
Number	1	2
Type	Glory hole	
Design discharge (ft³/s)	10,600	10,600
Gross inlet area (ft²) (at trash racks)	1,965	600
Bar racks		
spacing (in)		6.00
shape		
depth/thickness (in)		2.50 / 0.94
diameter (in)		
Emergency gates		
number	None	None
height/width (ft)		
type		
Service gates		
number	None	2
height/width (ft)		16.00 / 20.00
type		Slide
Bulkhead/stop logs (Y or N)		
number of units serviced		
Hoists		
number	None	1
capacity (tons)		30
type		

WATER PASSAGES	Upper Tunnel	Shaft	Lower Tunnel	Surge Tanks Upper	Surge Tanks Lower	Penstocks	Tailrace Tunnel
Number		1	1			2	2
Diameter (ft)		25.0	25.0			17.5	22.0
Length (ft)		749	410			175	500
Maximum velocity (ft/s)		5.4	5.4			5.5	3.5
Concrete lining thickness (in)		12.00	12.00				6.00
Total length of concrete sections (ft)							
Steel liner Thickness							
Minimum (in)						1.44	
Maximum (in)						1.50	
Material grade						ASTM A516 Grade 70	
Total length of steel-lined sections (ft)						175	

Notes:
Tailrace Tunnel: Horseshoe shape 6.71 m (22 ft) by 8.83 m (29 ft).

BEAR SWAMP

INTAKES	UPPER INTAKE	LOWER INTAKE
Number	1	2
Type	Glory hole	
Design discharge (m³/s)	300.2	300.2
Gross inlet area (m²)	182.6	55.7
(at trash racks)		
Bar Racks:		
spacing (mm)		152
shape		
depth/thickness (mm)		64 / 24
diameter (mm)		
Emergency gates		
number	None	None
height/width (m)		
type		
Service gates		
number	None	2
height/width (m)		4.877 / 6.096
type		Slide
Bulkhead/stop logs (Y or N)		
number of units serviced		
Hoists		
number	None	1
capacity (Mg)		27
type		

WATER PASSAGES	Upper Tunnel	Shaft	Lower Tunnel	Surge Tanks Upper	Surge Tanks Lower	Penstocks	Tailrace Tunnel
Number		1	1			2	2
Diameter (m)		7.62	7.62			5.33	6.71
Length (m)		228.3	125.0			53.3	152.4
Maximum							
velocity (m/s)		1.65	1.65			1.68	1.07
Concrete lining							
thickness (m)		0.305	0.305				0.152
Total length of concrete							
sections (m)							
Steel liner							
Thickness							
Minimum (mm)						37	
Maximum (mm)						38	
Material grade						ASTM A516 Grade 70	
Total length of steel-							
lined sections (m)						53.3	

Notes:
Tailrace Tunnel: Horseshoe shape 6.71 m (22 ft) by 8.83 m (29 ft).

BEAR SWAMP

POWERHOUSE and RELATED FEATURES

Powerhouse Structure
Type: Underground Rock Cavern
Length: 227 ft   Width: 79 ft   Height: 152 ft

Guard Valves
Number: 2   Diameter: 11.0 ft
Type: Spherical

Transformers
Number: 2
Ratings: 333
Voltages: (kV) 230 / 13.8

Generator
Rating generating (MVA): 333.0   Rating pumping (MVA): 309.0
Insulation type:
Starting method: Pony motor
Starting equipment: Motor 29,000 hp 12.3 kv

Runners
Material: Steel
Minimum unit submergence: 70.0 ft
WR^2:
Manufacturer: Hitachi
Model test by: Hitachi

	Reversible Runners	Reversible Motor/Generator	
Number	2		2
Diameter (ft)	19.20	Rotor Stator	
rpm synchronous	225.0		225.0
rpm overspeed			
Type	Francis		

Information on Runners

Condition:	Gross Head (ft) Generating	Gross Head (ft) Pumping	Capacity (MW) Generating	Capacity (MW) Pumping	Discharge (ft³/s) Generating	Discharge (ft³/s) Pumping	Turbine/Pump Eff.(%) Generating	Turbine/Pump Eff.(%) Pumping
Maximum head & maximum power	770		320		5,310			
Minimum head & maximum power	680					4,430		

Condition:	Net Head (ft) Generating	Net Head (ft) Pumping	Capacity (MW) Generating	Capacity (MW) Pumping	Discharge (ft³/s) Generating	Discharge (ft³/s) Pumping	Turbine/Pump Eff.(%) Generating	Turbine/Pump Eff.(%) Pumping
Rated head @ best gate	720	685	298			4,430	92.5	91.5

Bear Swamp - Page 3 (English)

BEAR SWAMP

POWERHOUSE and RELATED FEATURES

Powerhouse Structure
Type: Underground Rock Cavern
Length: 69.2 m   Width: 24.1 m   Height: 46.3 m

Guard Valves
Number: 2   Diameter: 3.35 m
Type: Spherical

Transformers
Number: 2
Ratings: 333
Voltages: (kV) 230 / 13.8

Generator
Rating generating (MVA): 333.0   Rating pumping (MVA): 309.0
Insulation type:
Starting method: Pony motor
Starting equipment: Motor 29,000 hp 12.3 kv

Runners
Material: Steel
Minimum unit submergence: 21.34 m
WR^2:
Manufacturer: Hitachi
Model test by: Hitachi

	Reversible Runners	Reversible Motor/Generator		
Number	2			2
Diameter m	5.84	Rotor	Stator	
rpm synchronous	225.0			225.0
rpm overspeed				
Type	Francis			

Information on Runners

Condition:	Gross Head (m)		Capacity (MW)		Discharge (m³/s)		Turbine/Pump Eff.(%)	
	Generating	Pumping	Generating	Pumping	Generating	Pumping	Generating	Pumping
Maximum head & maximum power	234.7		320		150.4			
Minimum head & maximum power	207.3					125.4		

Condition:	Net Head (m)		Capacity (MW)		Discharge (m³/s)		Turbine/Pump Eff.(%)	
	Generating	Pumping	Generating	Pumping	Generating	Pumping	Generating	Pumping
Rated head @ best gate	219.5	208.8	298			125.4	92.5	91.5

Bear Swamp - Page 3 (Metric)

BEAR SWAMP

Plant Data:
    Average GWh generating per year: 524
    Average GWh pumping per year: 716
    Starting time from standstill (s):
    Changeover time pumping to generating (min):   15
    Planned/scheduled time
        between major overhauls (years):   11
    Outage time required per unit
        during major overhauls (weeks):
    Representative plant availability (%):   90.3
    Representative planned outages (weeks per year):

Miscelleneous Notes:
    The cost of this project includes a 25-MW conventional hydropower plant in the tail pond.

    We have encountered very few operational difficulties and no major outages have resulted.

Cavitation Experience:

Significant or Unique Problems:

List of Licenses Required:

ENVIRONMENTAL FEATURES

Recreation:

Fish and Wildlife:

Social:

Bear Swamp pumped-storage plant will start up this summer

By Kuo-Hua Yang[*], FASCE, and Denton E. Nichols[**], MASCE

The authors of this article describe the overall design of the civil and mechanical features of this 600MW pumped-storage station, which incorporates a 10MW conventional station and is scheduled to produce first power in the coming summer

THE BEAR SWAMP pumped-storage project is located on the Deerfield River in the north-western Massachusetts towns of Florida and Rowe. The upper reservoir is situated in a natural basin known as Bear Swamp, 750ft above and 1600ft south of the Deerfield River.

Two main dykes and two saddle dykes contain the upper reservoir, being constructed of compacted rockfill shells with impervious glacial-till central cores, and filter-zone transitions between the core and the rockfill shells.

The north dyke has a maximum height of 140ft and the southern one of only 120ft (Fig. 1).

An emergency spillway is located east of the north dyke for maximum safety in case of over-pumping, despite the

---
[*] Project Engineer, Chas T. Main Inc, Boston, Massachusetts, USA.
[**] Senior Engineer, New England Power Service Co, Westboro, Massachusetts, USA.

fact that instrumentation and interlocks are provided to prevent this occurring. The crest of the spillway is set at el. 1602, two feet above the maximum upper reservoir level, and the spillway is excavated simply out of bedrock and designed as a broadcrested weir.

The lower reservoir is created by an earth dam, which is a 125ft-high earthfill embankment with an upstream rockfill shell. Adjoining the dam on the right bank is a non-overflow intake structure for a 10MW power plant, and abutting the intake structure is a gated concrete spillway with two Tainter gates 40ft high by 36ft wide. The reservoir has a usable power storage of 4900acre-ft, with a 40ft operating range (Fig. 2).

A reinforced-concrete diversion conduit is constructed in the right bank of the river to permit construction of the lower dam and related structures.

The hydraulic system between the upper and lower reser-

Fig. 1. An aerial view of the upper-reservoir area—a natural basin that gives the project its name

Reprinted, with permission, from <u>International Water Power and Dam Construction</u>, May 1974.

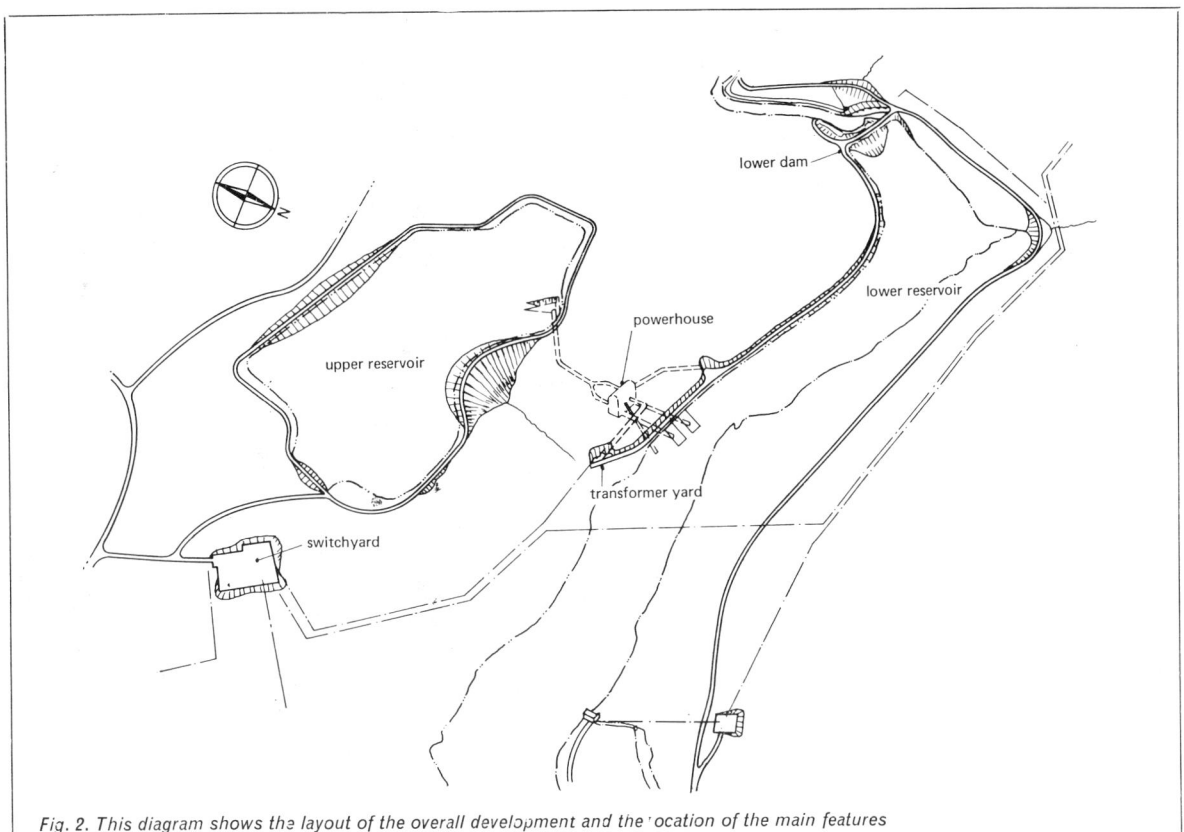
Fig. 2. This diagram shows the layout of the overall development and the location of the main features

voirs consists of an intake structure, a vertical concrete-lined pressure shaft, a concrete tunnel bifurcating into two steel-lined penstock tunnels and two shotcrete-lined tailrace tunnels. An underground powerhouse contains two vertical reversible pump-turbines, each rated at 400 000hp, with motor-generators rated at 300MW each.

A spherical valve 11ft in diameter is connected to the inlet section of each pump-turbine.

Since the project's lower reservoir will inundate an existing 15MW run-of-river hydro station, a new facility will be built half a mile upstream on the north side of the river. At the lower reservoir dam the new 10MW station will pass river flows and feed the four hydroelectric stations existing downstream.

Both new stations will be fully automated and will operate as run-of-river plants with some modification to fit the Bear Swamp operating cycle.

Intake
The intake location has been chosen to make maximum use of a natural pocket in the terrain with a steep rock back slope. A 1-to-50 scale model of the intake was built with the entire upper reservoir configuration for hydraulic model testing, so the approach velocities to the intake were represented correctly.

The purpose of the model testing was to determine the dimensions of the intake channel and the extent and configuration of training walls around the intake channel, so as to eliminate as much vortex action as possible within the 50ft operating range during the generation cycle.

Since the entire upper reservoir was cleared and grubbed, it was decided to eliminate the intake trashrack structure because of frequent trashrack structure failures at many pumped-storage plants.

The intake side walls are excavated vertically in rock and are lined with reinforced concrete. All rock faces are reinforced with rock bolts (Fig. 3).

Hydraulic system
The 25ft-diameter vertical shaft was originally designed as a steeply-inclined tunnel. Although the combined length of vertical shaft plus horizontal tunnel is greater than the length of an inclined tunnel, the contractor performing the excavation strongly recommended adoption of the vertical shaft in order to achieve not only speed in construction but also economy.

At the bottom of this vertical shaft a 90° elbow joins the main power tunnel, which runs at a grade of 7·5% into a bifurcation from which two 17·5ft-diameter tunnels continue to the 11ft-diameter spherical valves upstream of the pump-turbine units. About 175ft of the penstock tunnel are steel lined (Fig. 4).

The liner is ASTM 516-70 high-strength steel with a yield strength of 75 000 lb/in². It is designed for maximum hydraulic pressure, including water hammer to 50% of the yield point, and will be 100% X-rayed.

The rock excavation at the welded sections is enlarged to facilitate the field weld and X-ray examination. Although provision for drainage in the back-up concrete is provided, the liner was checked for external pressure (these drain pipes are extended into the powerhouse through gutters to the sump pit).

Several transient studies were performed in order to determine the best method of reducing water-hammer pressures. This can be achieved with either a time delay or variable-speed governor.

The former method involves delaying the gate closure until the turbine reaches runaway speed, while the latter means closing the wicket gates at a faster speed at the beginning of closure when the rate of reduction in turbine flow due to overspeed is moderate, and then at a slower speed of gate closure when the reduction of turbine flow accelerates.

The studies showed that to achieve the shortest possible governor time without significant increase in water hammer, the variable speed governor should be adopted, and this was also the turbine-manufacturer's recommendation. The intent is to minimize the duration of vibration after the unit reaches runaway speed.

The power shaft and tunnel will be concrete lined for hydraulic reasons, and the bifurcation section will be in reinforced concrete. As a safety precaution to protect the runners while pumping, the tailrace tunnels will be shotcreted along the arch and side walls to prevent any potential rock falls.

During excavation of the tunnels, two soft seams were encountered. Each is parallel to the steeply-dipping foliation of the rock and strikes approximately parallel to the longitudinal axis of the powerhouse.

The first seam is about 50ft downstream of the north wall and the second about 200ft upstream of the powerhouse south wall. The north seam, consisting primarily of soft silt and clay gouge from ½in to 16in wide, is severely weathered and open to water percolation.

This seam intersects the access tunnel, the cable tunnel and both tailrace tunnels, but it was the considered opinion of the engineers, geologists and the board of consultants that this seam would not influence the stability or integrity of the powerhouse cavern.

However, as a safety precaution, the rock on either side of the seam was reinforced with rock bolts in each of the tunnels, and in addition, steel straps and shotcrete were added to the tailrace tunnels in the areas of the seam.

Fig. 3. Construction in progress on the inlet (pumping-outlet) to the upper reservoir. The structure, which was located to make maximum use of a natural pocket in the terrain, will not incorporate a trashrack

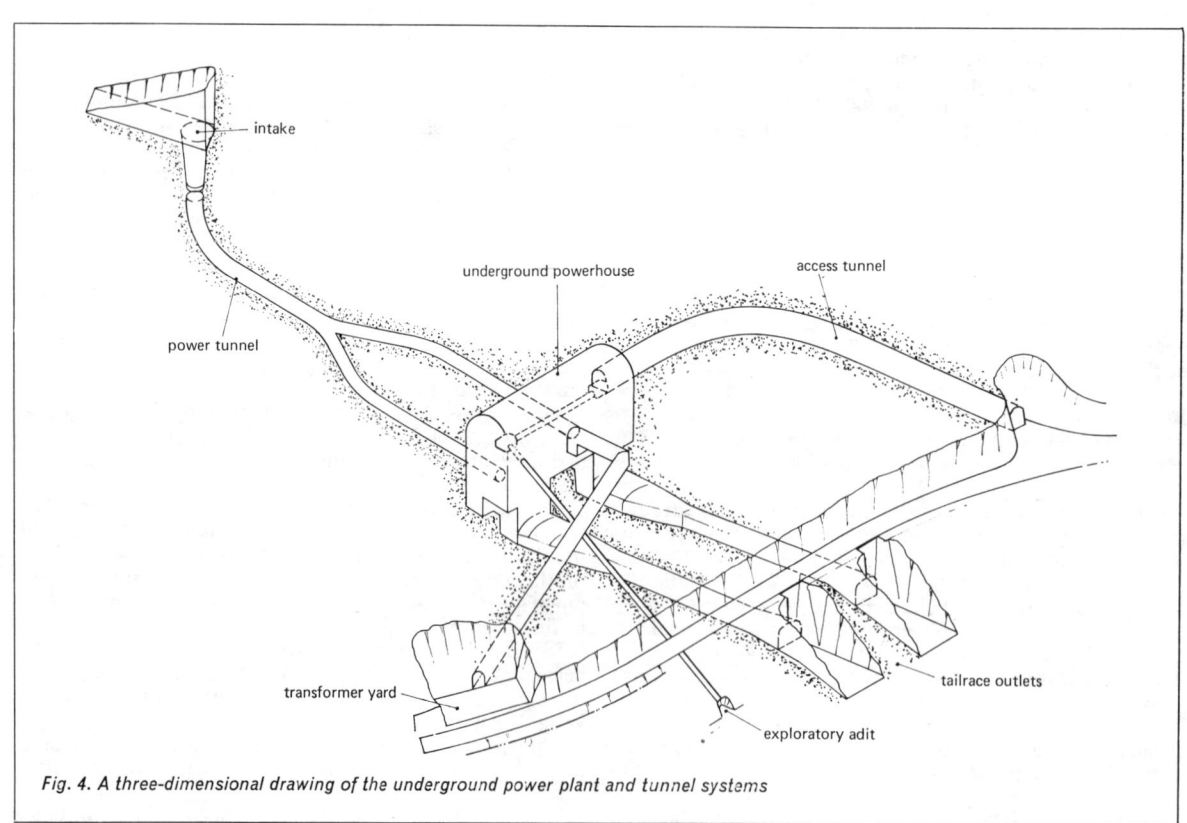

Fig. 4. A three-dimensional drawing of the underground power plant and tunnel systems

Water Power May 1974

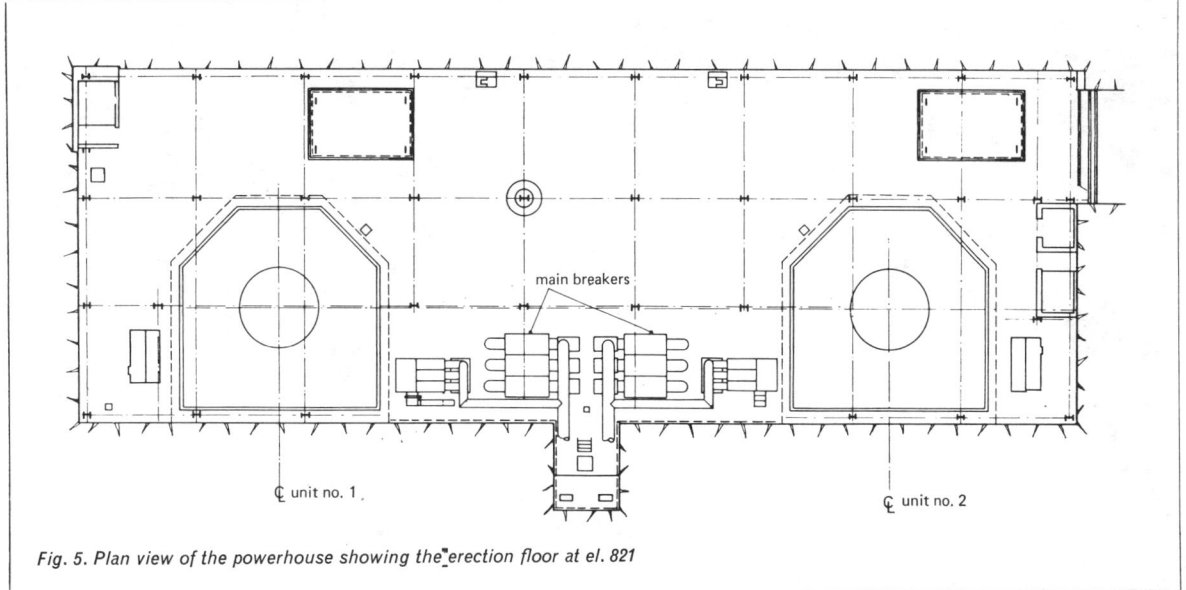

Fig. 5. Plan view of the powerhouse showing the erection floor at el. 821

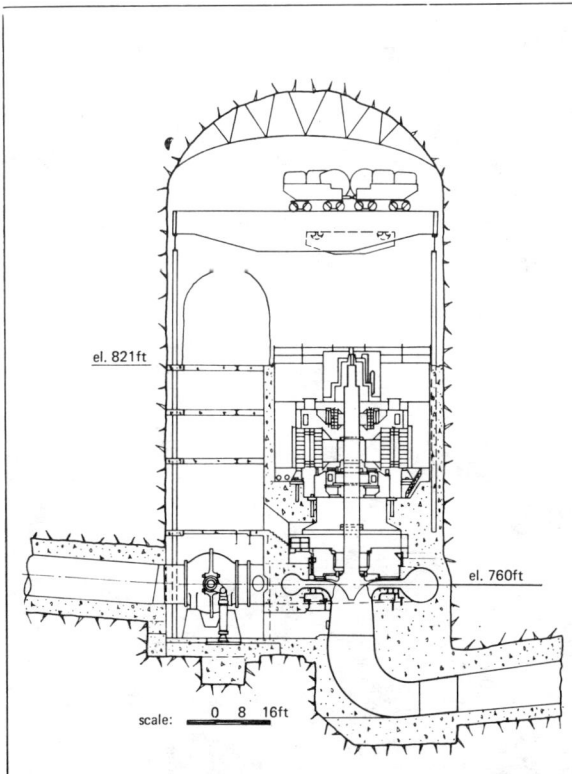

Fig. 6. Powerhouse cross-section including one of the two reversible machines and its spherical shut-off valve

Measuring gauges were installed straddling the seams at several locations in order to monitor any possible rock movement at these locations.

The south seam was encountered during excavation of the penstock tunnels. In general, this seam is narrow and unweathered with sound rock on both sides. No special treatment other than grouting was required during construction.

Powerhouse design

In the original concept of Bear Swamp, a surface powerhouse was proposed, but as studies related to geology, plant location and waterways became more detailed and design progressed, it became apparent that an underground powerhouse was more suitable for this project.

One of the primary reasons for this conclusion was the deep submergence of the pump-turbine units required by each of the manufacturers of whom inquiry was made.

Because of this deep submergence, the surface power plant would be subjected to enormous uplift pressures due to the lower reservoir water levels. This uplift pressure would require an extremely heavy concrete structure to counterbalance the buoyant forces, and a cost comparison indicated potential savings of about $2·5 million by locating the powerhouse underground.

The rock in the vicinity of the powerhouse is primarily a light green garnet-bearing, chlorite mica-schist with well-developed schistosity, containing occasional quartz seams. From the project exploration and other available geological information, it was not expected that underground construction would encounter any extraordinary engineering problems.

The proximity of the upper and lower reservoirs, which results in short power and tailrace tunnels, was also a factor favouring the selection of an underground design.

This configuration has the added advantage of preservation of the natural environment and the possibility of a year-round construction programme. On the basis of these considerations, deep submergence and economics, it was concluded that the powerhouse definitely belonged underground.

Several studies were made of the plant arrangement. With only two units it was more advantageous to place the erection area between these units (Figs. 5 and 6), and

Fig. 7. The portal structure which connects the lower storage with the twin shotcrete-lined tailrace tunnels

this type of layout also provides a simple and centralized location for electrical and mechanical equipment.

From the structural point of view, it allows a greater distance between tailrace tunnels thus ensuring the integrity of the rock pillar between them, and it also allows an arrangement wherein the tunnel for the main cables leaves the powerhouse at a central location.

With the powerhouse located a very short distance from the lower reservoir, and a difference of nearly 60ft in elevation between the station erection floor and the plant access road, it was impossible to locate the access tunnel so as to enter the powerhouse between the units in order to have direct access to the erection floor.

The tunnel thus runs into the powerhouse through the west end wall so a reasonable slope and curvature for transport of heavy equipment is maintained (a tunnel slope of 7·5% was selected to accommodate construction equipment as well as plant transport requirements).

The short length of the tailraces combined with the separation of the units by an erection bay dictated the layout with separate tailrace tunnels for each unit (Fig. 7).

The complex bifurcation, which would have been required by a common tunnel, was thus eliminated, and the short length and the low water-velocity in the tunnel preclude the need for a downstream surge chamber.

Without a surge chamber it seemed unreasonable to excavate space merely to accommodate the operating mechanism for draft tube gates; therefore, they are placed outside at the tailrace tunnel exit to the lower reservoir. This location does have the minor disadvantage of requiring a relatively long unwatering time for the tailrace tunnels.

Separate investigations were made to determine the optimum location for the main transformers. Cost comparisons indicated that the added structural cost to accommodate underground transformers exceeded the additional cost for the longer isolated phase bus ducts required to place the transformers outdoors.

Fig. 8. Inside the power cavern, showing machinery being installed

Fig. 9. Looking towards the end of the cavern containing unit No. 1 at erection-floor level

The possibility of placing the transformers above ground directly over the power plant was investigated; however, economics dictated that the transformer yard should be located near the lower reservoir just outside of the main cable tunnel.

This location had the advantage of allowing the HT lines to join an existing transmission right-of-way leading to the main switchyard adjacent to the upper reservoir.

The main cable tunnel leads from the powerhouse midway between the units and slopes upwards to the portal area. It is 16ft wide by 25ft high, and the isolated phase bus ducts for each unit are supported by a steel frame on each side wall. This tunnel will also serve as an emergency exit, and convey fresh air into the cavern for ventilation.

As in most underground powerhouses, it was desirable to erect the crane girders as early as possible in order to utilize the travelling crane during the early stages of construction. For this reason, a complete structural steel frame was designed to support the crane girder, making erection of the 606t crane independent of the concrete placement sequence (Figs. 8 and 9).

With the erection bay located at the centre of the powerhouse, an unloading area was needed in front of the access tunnel, and the floor area above the No. 2 spherical-valve chamber was designed for this purpose.

Rock-mechanics investigation

Since geological information defining the rock within the powerhouse cavern was limited to drill-hole cores 400ft away, an exploratory adit was driven into the proposed powerhouse chamber below the arch spring line and along the full length centreline of the cavern, prior to construction.

The 6ft-wide by 8ft-high adit had enlarged chambers at each end to permit in-situ stress measurements and rock-bolt testing. Rock stress measurements by the overcoring method were made to verify the proposed orientation of the powerhouse axis.

The in-situ stress results indicated that the principal stress within the rock is parallel to the longitudinal axis of the powerhouse. This confirmed the original orientation of the cavern.

Rock bolts were inserted in the test chamber to determine the best type of rock anchorage and the duration of time that the installed load would be retained. As a result of these tests, the type of anchorage, bolt strength and spacing could be recommended for construction.

These tests also defined the time that the bolt load would be retained prior to grouting.

The arch portion of the cavern is reinforced by 1in-diameter rock bolts, 20ft long, spaced at 5ft centres. This rock bolt pattern was extended 15ft below the arch spring line.

As the rock foliation dips about 65° to the upstream side, the downstream north wall is stabilized by rock bolts 20ft long, at about 8ft centres, installed approximately perpendicular to the foliation.

The bifurcation section of the power tunnel and the transition section of the tailrace tunnels are reinforced by 10-15ft-long rock bolts, spaced at approximately 5ft centres.

All the rock bolts are hollow groutable bolts, except in those areas where they are required only for construction safety and are to be covered eventually by structural concrete lining, in which case the bolts are solid and ungrouted.

In order to measure the rock movement in the cavern, eight extensometers were installed. There are three at the centreline of each unit, one at the arch crown, one at each side of the spring line, and at each end wall a single extensometer is installed in the direction of the principal stress.

The extensometers are the slope-indicator type, having

six positions, with the leads extended to a central read-out station. The heads and the cables are protected from rock-blasting damage by steel plates, channels and shotcrete cover.

These instruments were installed immediately after completion of the arch excavation and records available at the time of writing indicate no significant rock movement.

In order to provide an alternate means of obtaining a cross check on the sophisticated instrumentation, a series of measuring pins were installed along the powerhouse walls at various elevations.

The movement of the walls is being measured directly by using a nickel-steel tape and tape extensometer attached between opposite points. As the excavation progressed deeper, the direct measurement of some of these points became extremely difficult, and they were then monitored by means of triangulation techniques along a precise base-line.

Since the rock within the cavern is relatively sound, no lining of the arch had originally been contemplated, but during construction the excavation contractor elected to place a 4in shotcrete lining on the arch for safety purposes. The performance of the shotcrete to date has been excellent.

Prior to placing this lining, weep holes were provided at seepage points and suspected potential seepage areas in order to relieve any back-pressure buildup.

The final design calls for a drip roof to be installed as a second line of defence against possible water dripping on the equipment. This ceiling consists of corrugated "galbestos" sheets suspended from the roof rock-bolts by galvanized turnbuckle rods and a channel support system.

Mechanical and electrical equipment

The two reversible, single runner, vertical-shaft Francis pump-turbines are rated at 400 000hp at 720ft net head as turbines; at 685ft minimum total dynamic head, the guaranteed pump discharge is 4430cusecs; the synchronous speed in both directions is 225rev/min.

The centreline of the distributor is 70ft below the minimum tailwater elevation in order to obtain the submergence required by the pumping characteristics. The machine's guaranteed peak efficiencies are 92·5% as turbines and 91·5% as pumps.

The cast steel one-piece runners are 230in overall diameter with the area subject to cavitation welded with an ⅛in-thick stainless-steel overlay.

Welded high tensile-strength steel plate is used throughout the construction of the 11ft inlet diameter spiral cases, and all welds are 100% radiographed. Before concrete embedment, each casing was pressure tested at 650 lb/in², and during embedding normal operating pressure was maintained in the casing.

The electric-hydraulic governors are capable of closing the wicket gates from full open position in 10s, and two rate limiters in the closing direction prevent excessive pressure in the spiral case during generating load-rejection and flow reversal during pumping-load rejection.

The 11ft-diameter spherical valves are oil-pressure operated and provide a 30s closing capability to shut off the water to the pump-turbine in the event of failure of the gate-operating mechanism.

Both upstream and downstream seals are operated by water pressure: the valve is normally operated under balanced water pressure through a by-pass line.

The valve body is rigidly attached to the penstock line and any longitudinal movement is provided for by the valve feet sliding on lubricated foundation plates, together with a slip joint between the valve companion flange and the spiral case extension.

Each of the two generator/motors is rated 333MVA, 13·8kV, 0·90 power factor, 80°C temperature-rise as a

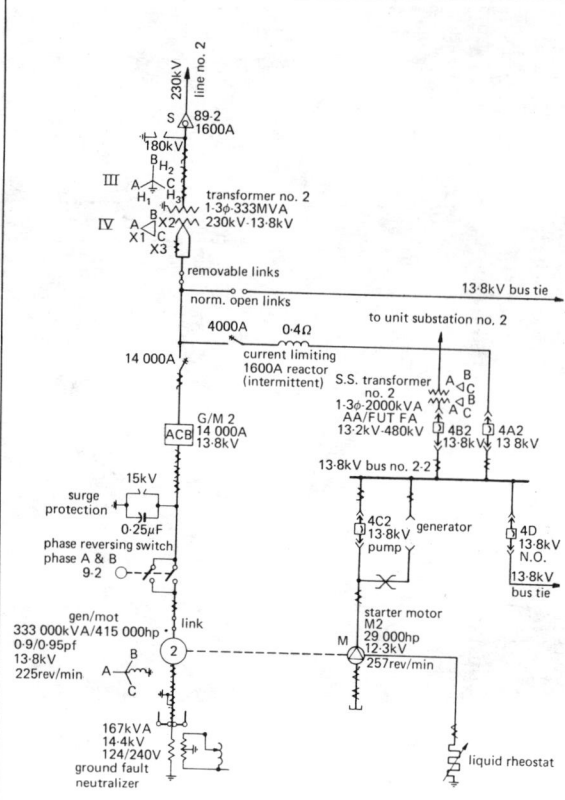

Fig. 10. Single-line diagram showing the layout of the main electrical installations

generator; and 415 000hp, 0·95 power factor, as a motor.

A reversible pony starting motor of the wound-rotor induction type is directly connected to the main shaft, and is rated at 12·3kV, 29 000hp at 257rev/min. By means of a liquid rheostat the motor can be started at rated torque without exceeding the rated current in the stator.

Two 3-phase, 333MVA main transformers are installed at the main cable tunnel portal. The main cables connecting the generator-motor in the underground power plant and the main transformer at the tunnel portal are isolated phase bus ducts rated at 13·8kV, 14 000A, with forced-air cooling.

The bus conductor and enclosure consists of aluminium sections welded together, and the three bus ducts for each unit are supported on a steel frame anchored at the side walls of the tunnel.

The ancillaries include the cooling equipment with air-to-water heat exchangers, phase-reversal switches, isolating switches and surge-protection equipment. Each set of the main cables has single-pole, air-blast, circuit breakers, rated at 14 000A, 3500MVA (Fig. 10).

Control of the Bear Swamp generator-motor units is centralized in the station control room, and can be remotely-controlled via microwave by a supervisory control system from the Deerfield River hydro power station upstream.

The project is being built by New England Power Co, and is scheduled for startup during the summer of 1974.

---

* Part of the paper was presented by the same authors at the ASCE National Structural Engineering meeting, Cleveland, Ohio, USA, in April 1972.

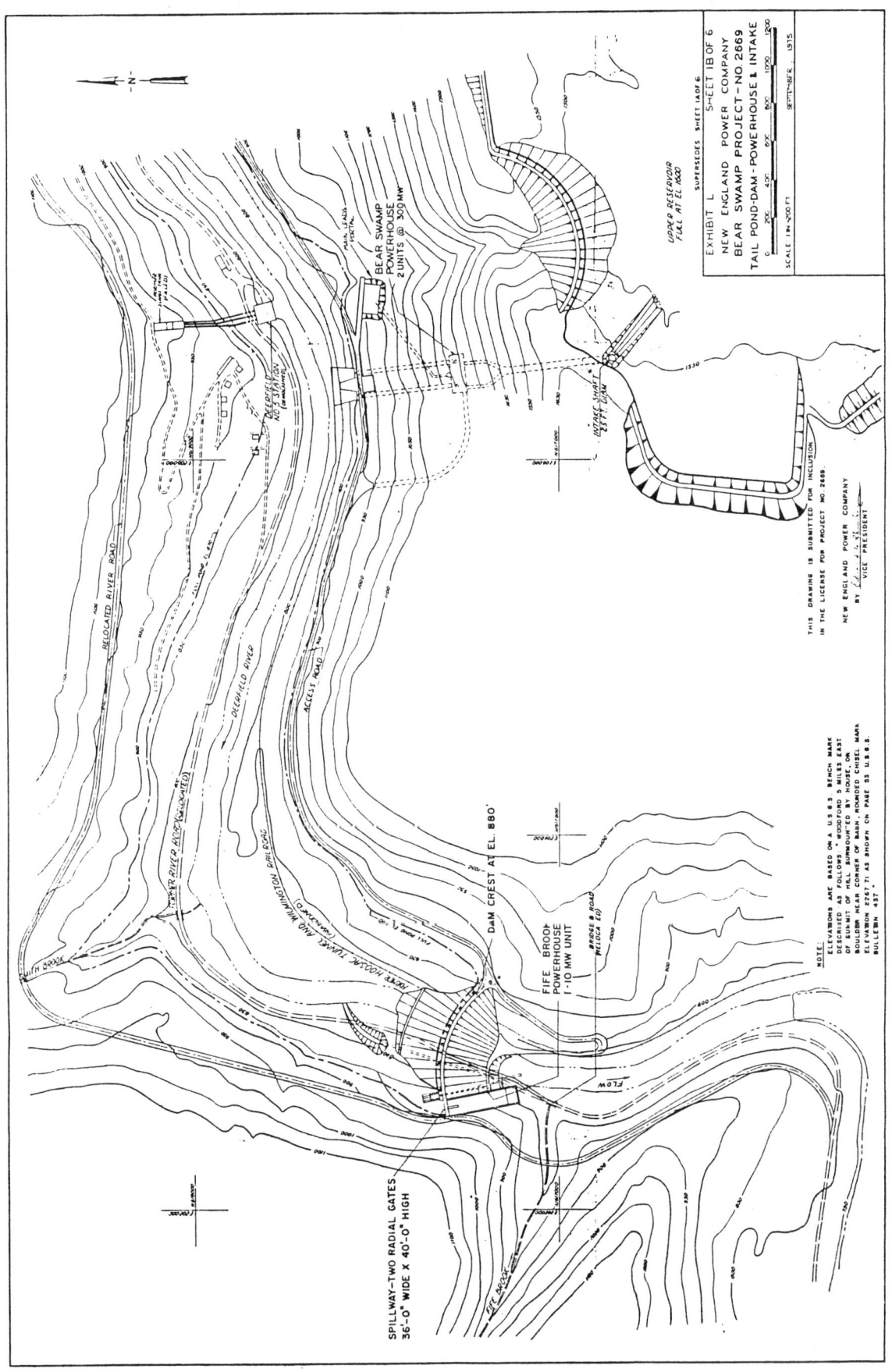

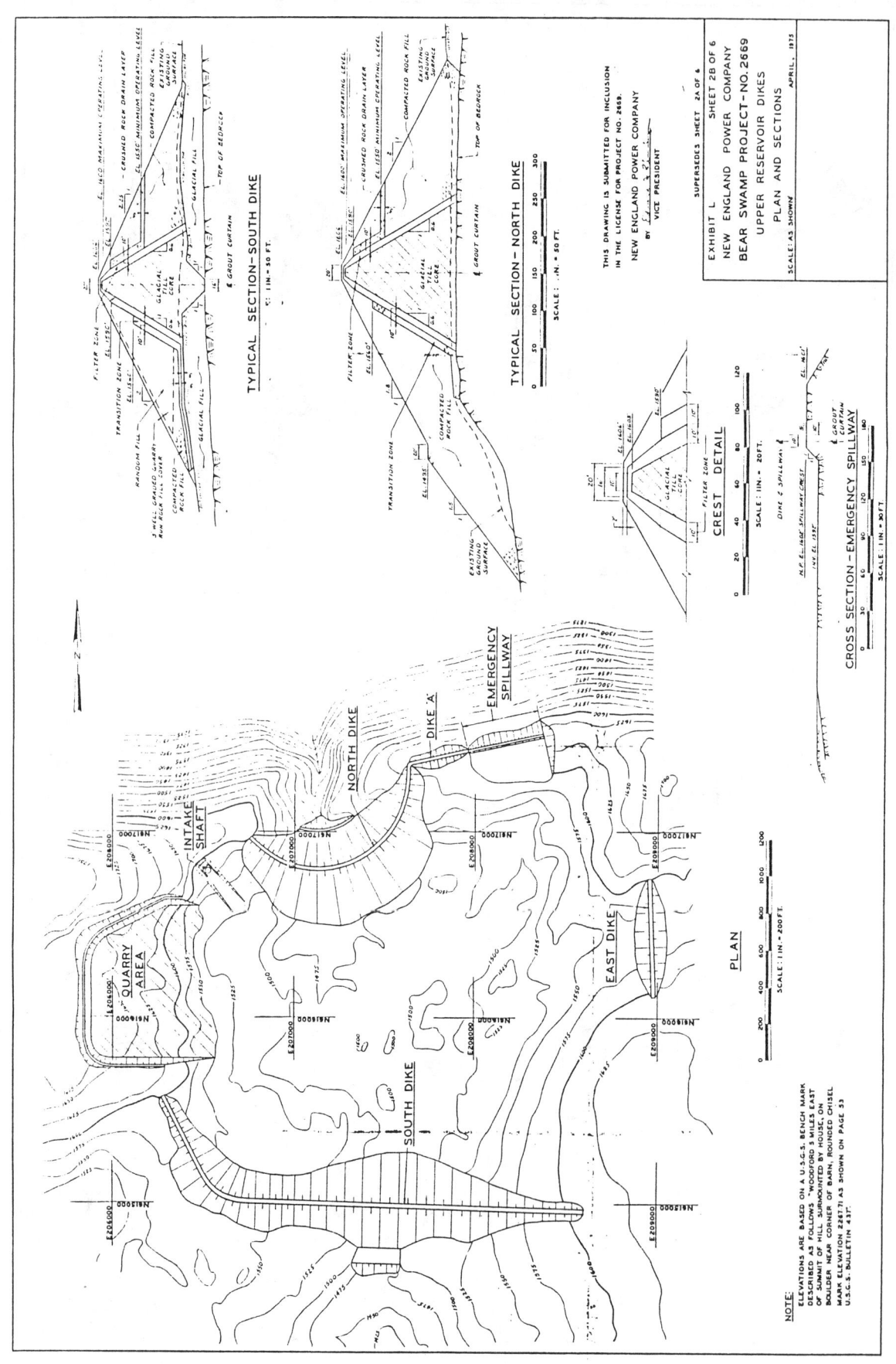

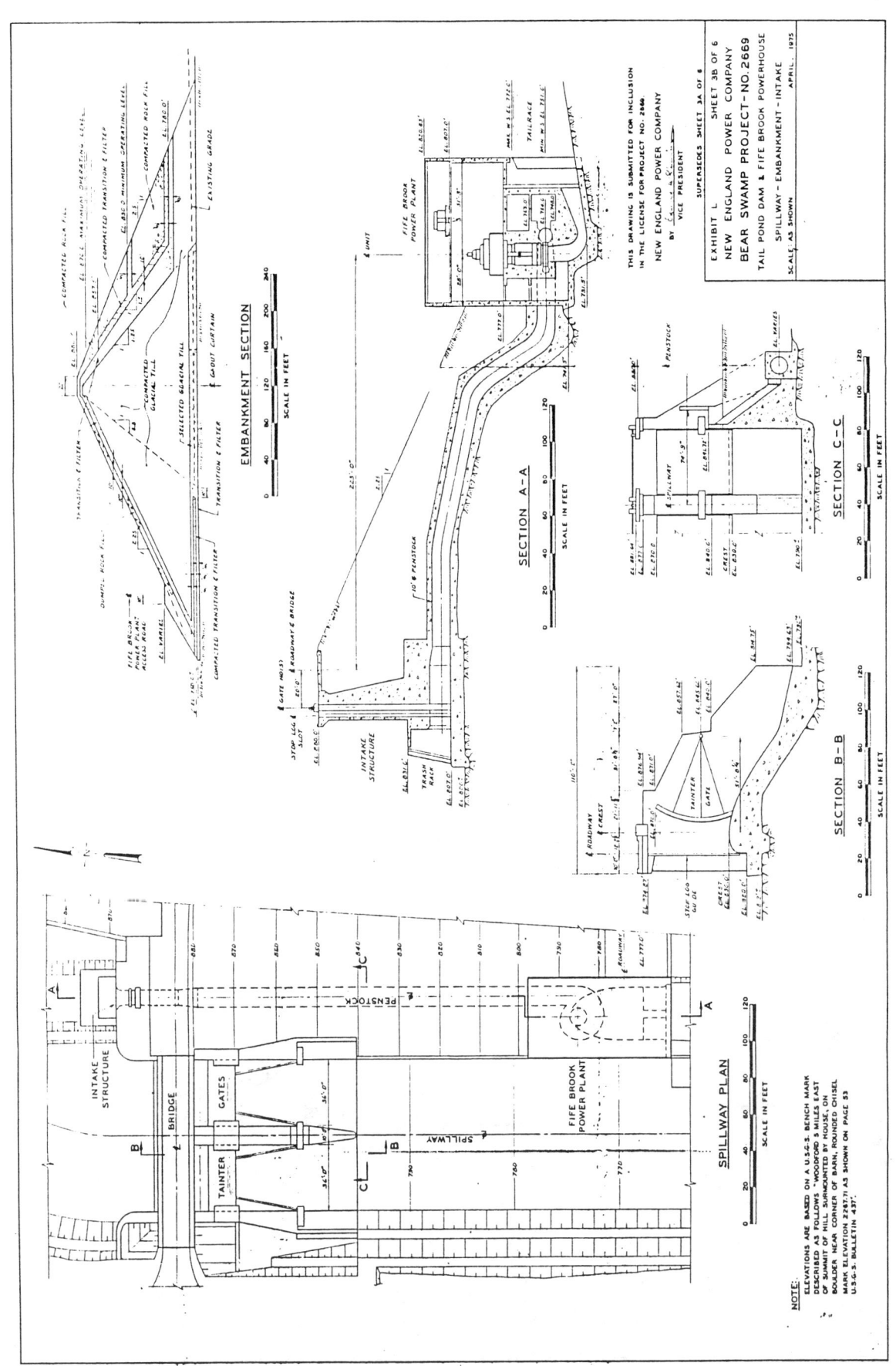

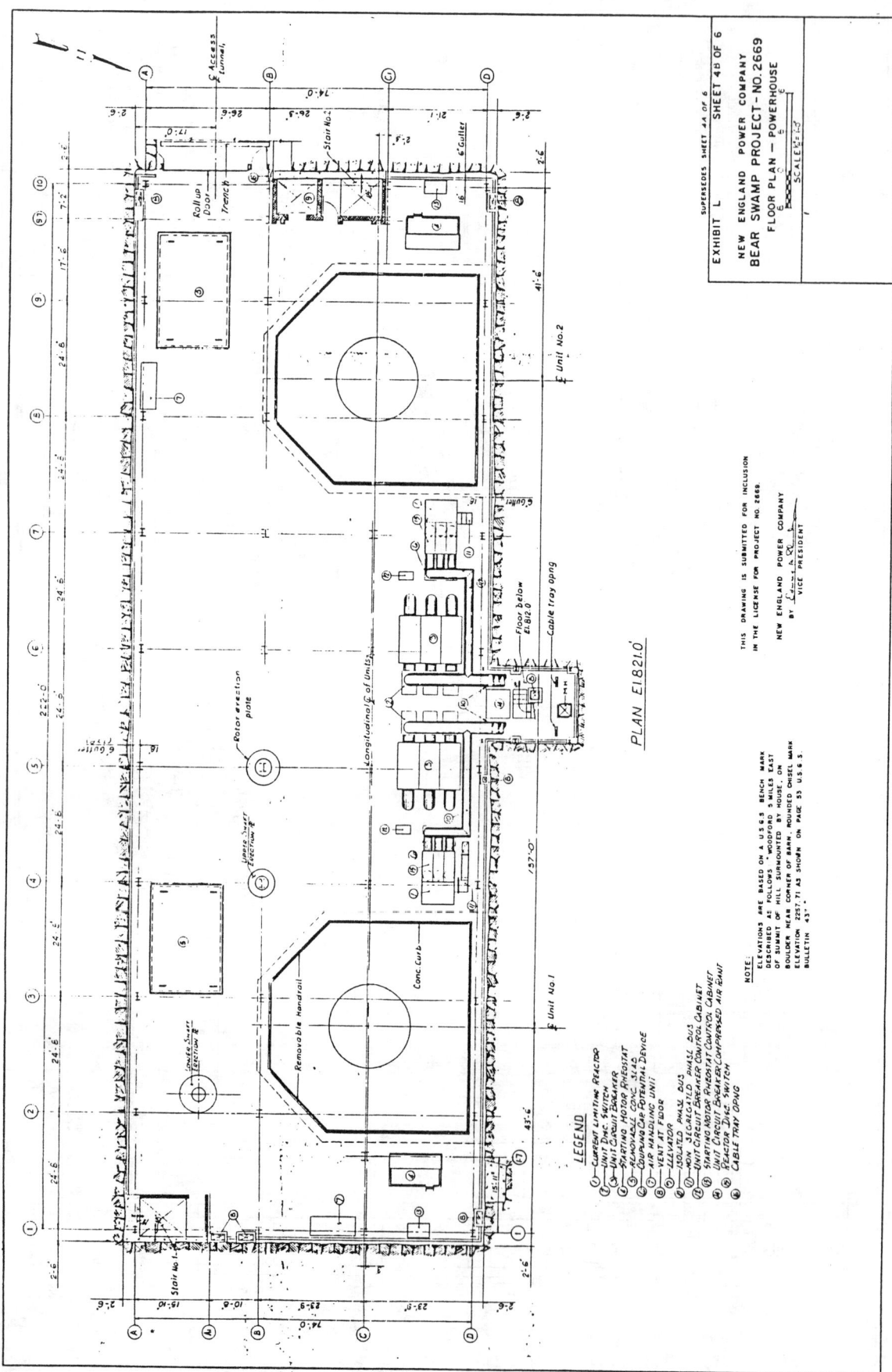

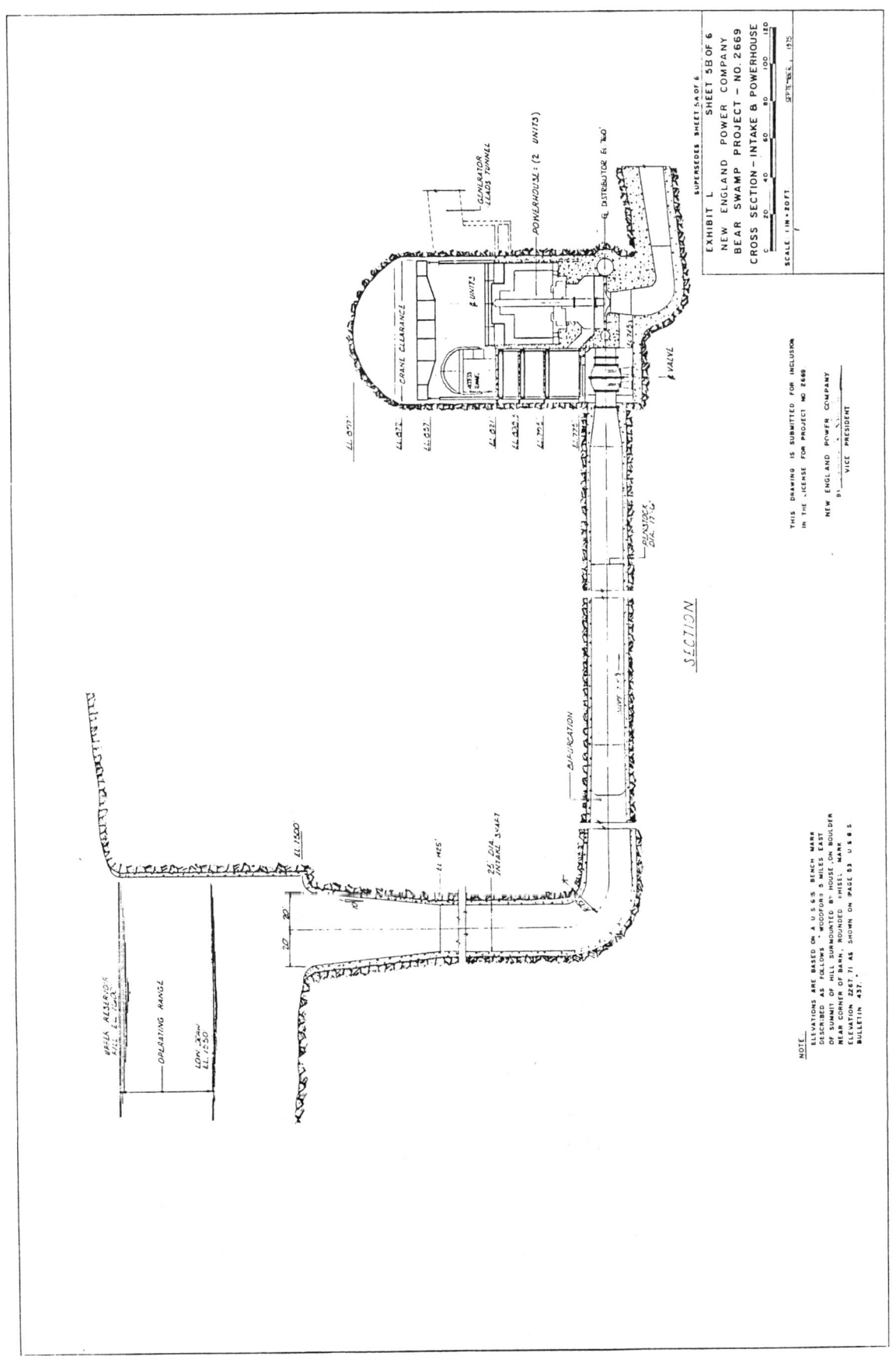

BIBLIOGRAPHY OF PAPERS
PRESENTED ON BEAR SWAMP PUMPED STORAGE PROJECT

Bear Swamp Pumped Storage Project

    Denton E. Nichols

        Presented to: ASCE National Water Resource
                           Engineering Meeting
                           Phoenix, Arizona, 1/11-15/71
                           Preprint - 1324

Design Features of Bear Swamp Project

    Denton E. Nichols
    K.H. Yang

        Presented to: ASCE National Structural
                           Engineering Meeting
                           Cleveland, Ohio, 4/24-28/72
                           Preprint - 1687

Environmental Design of the Bear Swamp Project

    Robert W. Kwratkowski
    Louis D. Pierce

        Presented to: ASCE Annual and National Environmental
                           Engineering Meeting
                           Houston, Texas, 10/16-22/72
                           Preprint - 1814

Engineering the Bear Swamp Project

    Robert W. Kwratkowski
    David R. Campbell

        Presented to: Boston Society of Civil Engineers,
                           Section, ASCE
                           Boston, Massachusetts, 10/23/74

Plant Name: **BLENHEIM-GILBOA PUMPED STORAGE PROJECT**

Plant location:
 Blenheim and Gilboa, NY
 Schoharie County

Rated capacity 1,000 MW
Average static head 1,100 ft
Plant efficiency 75.00 %
Stored energy 12,000 MWh
Number of units 4

Construction time: 4 years, 5 months
Construction cost: $149 per kW
Price level: 1973
First commercial power: July 1973
FERC project number: 2685

River or water source: Schoharie Creek

Owner: New York Power Authority
 123 Main Street
 White Plains, New York 10601

Designers:
 Uhl, Hall and Rich, Division of Chas. T. Main, Inc.
 Prudential Center, Boston Mass. 02199

Plant Manager/Superintendent:
 James J. McCarthy Jr.
 P.O. Box F
 Grand George, New York 12434

 (607) 588-6061

	UPPER RESERVOIR	LOWER RESERVOIR
DAM		
Type	Earth & rockfill	Earth and rockfill
Height (ft)	162	165
Crest length (ft)	11,900	18,000
Volume (yd³)	5,875,000	1,159,000
RESERVOIR		
Type	Constructed	Constructed
Surface area (acres)	390	430
Usable power storage (acre-ft)	15,000	15,000
Power pool fluctuation (ft)	48.0	40.0
Operating levels		
Maximum (ft)	2,003.0	900.0
Minimum (ft)	1,955.0	860.0
Drainage area (miles²)	0.6	40.0
Seepage (ft³/s)		
SPILLWAY		
Design flood		
Return period (years)		
Flow (ft³/s)	10,000	150,000
Capacity (ft³/s)		
Type	Emergency spill channel	Ogee, concrete gravity
Gates		
Number	None	3
Type		Radial
Width (ft)		38.00
Height (ft)		46.50
OUTLET WORKS		
Discharge capacity (ft³/s)		
Number of water passages	None	2
Dimensions of water passages		
Height (ft)		
Width (ft)		
Diameter (ft)		3.00
Type of gates		Howell-Bunger valve
Number of gates	None	1

Blenheim-Gilboa Pumped Storage Project - Page 1 (English)

Plant Name: BLENHEIM-GILBOA PUMPED STORAGE PROJECT

Plant location
 Blenheim and Gilboa, NY
 Schoharie County

Rated capacity   1,000 MW
Average static head   335.3 m
Plant efficiency   75.00 %
Stored energy   12,000 MWh
Number of units   4

Construction time: 4 years, 5 months
Construction cost: $149 per kW
Price level: 1973
First commercial power: July 1973
FERC project number: 2685

River or water source: Schoharie Creek

Owner: New York Power Authority
 123 Main Street
 White Plains, New York 10601

Designers:
 Uhl, Hall and Rich, Division of Chas. T. Main, Inc.
 Prudential Center, Boston Mass. 02199

Plant Manager/Superintendent:
 James J. McCarthy Jr.
 P.O. Box F
 Grand George, New York   12434

 (607) 588-6061

DAM	UPPER RESERVOIR	LOWER RESERVOIR
Type	Earth & rockfill	Earth and rockfill
Height (m)	49.4	50.3
Crest length (m)	3,627.1	5,486.4
Volume (m³)	4,491,762	886,120
RESERVOIR		
Type	Constructed	Constructed
Surface area (Mm²)	1.58	1.74
Usable power storage (Mm³)	18.502	18.502
Power pool fluctuation (m)	14.63	12.19
Operating levels		
Maximum (m)	610.51	274.32
Minimum (m)	595.88	262.13
Drainage area (Mm²)	1.450	103.600
Seepage (m³/s)		
SPILLWAY		
Design flood		
Return period (years)		
Flow (m³/s)	283	4,248
Capacity (m³/s)		
Type	Emergency spill channel	Ogee, concrete gravity
Gates		
Number	None	3
Type		Radial
Width (m)		11.582
Height (m)		14.173
OUTLET WORKS		
Discharge capacity (m³/s)		
Number of water passages	None	2
Dimensions of water passages		
Height (m)		
Width (m)		
Diameter (m)		0.914
Type of gates		Howell-Bunger valve
Number of gates	None	1

Blenheim-Gilboa Pumped Storage Project - Page 1 (Metric)

BLENHEIM-GILBOA PUMPED STORAGE PROJECT

INTAKES	UPPER INTAKE	LOWER INTAKE
Number	1	4
Type	Ungated submerged circular weir (morning glory)	Draft tube
Design discharge (ft³/s)	14,880	11,360
Gross inlet area (ft²) (at trash racks)	3,390	1,350
Bar racks		
spacing (in)	6.00	6.00
shape		
depth/thickness (in)	2.00 / 0.75	1.25 / 0.75
diameter (in)		
Emergency gates		
number	None	None
height/width (ft)		
type		
Service gates		
number	None	4
height/width (ft)		12.50 / 13.50
type		Bulkhead
Bulkhead/stop logs (Y or N)	N	Y
number of units serviced		4
Hoists		
number	None	1
capacity (tons)		30
type		Gantry crane

WATER PASSAGES	Upper Tunnel	Shaft	Lower Tunnel	Surge Tanks Upper	Surge Tanks Lower	Penstocks	Tailrace Tunnel
Number		1	1			4	
Diameter (ft)		20.0	28.0			12.0	
Length (ft)		1,035	900			1,960	
Maximum velocity (ft/s)		24.0	24.0			33.0	
Concrete lining thickness (in)		24.00	24.00			18.00	
Total length of concrete sections (ft)		1,035	900				
Steel liner Thickness							
Minimum (in)						1.13	
Maximum (in)						1.31	
Material grade						A537, A517	
Total length of steel-lined sections (ft)						1,960	

Notes:

Blenheim-Gilboa Pumped Storage Project - Page 2 (English)

BLENHEIM-GILBOA PUMPED STORAGE PROJECT

INTAKES	UPPER INTAKE	LOWER INTAKE
Number	1	4
Type	Ungated submerged circular weir (morning glory)	Draft tube
Design discharge (m³/s)	421.4	321.7
Gross inlet area (m²) (at trash racks)	314.9	125.4
Bar Racks:		
spacing (mm)	152	152
shape		
depth/thickness (mm)	51 / 19	32 / 19
diameter (mm)		
Emergency gates		
number	None	None
height/width (m)		
type		
Service gates		
number	None	4
height/width (m)		3.810 / 4.115
type		Bulkhead
Bulkhead/stop logs (Y or N)	N	Y
number of units serviced		4
Hoists		
number	None	1
capacity (Mg)		27
type		Gantry crane

WATER PASSAGES	Upper Tunnel	Shaft	Lower Tunnel	Surge Tanks Upper	Surge Tanks Lower	Penstocks	Tailrace Tunnel
Number		1	1			4	
Diameter (m)		8.53	8.53			3.66	
Length (m)		315.5	274.3			597.4	
Maximum velocity (m/s)		7.32	7.32			10.06	
Concrete lining thickness (m)		0.610	0.610			0.457	
Total length of concrete sections (m)		315.5	274.3				
Steel liner Thickness							
Minimum (mm)						29	
Maximum (mm)						33	
Material grade						A537, A517	
Total length of steel-lined sections (m)						597.4	

Notes:

Blenheim-Gilboa Pumped Storage Project - Page 2 (Metric)

BLENHEIM-GILBOA PUMPED STORAGE PROJECT

POWERHOUSE and RELATED FEATURES

Powerhouse Structure
Type: Semi-Outdoor
Length: 526 ft   Width: 172 ft   Height: 132 ft

Guard Valves
Number: 4   Diameter: 9.2 ft
Type: Spherical

Transformers
Number: 5 (includes 1 spare)
Ratings: 285
Voltages: (kV) 345 / 16.2

Generator
Rating generating (MVA): 278.0   Rating pumping (MVA): 286.0
Insulation type:
Starting method: Reduced voltage - water depression
Starting equipment: Air depression system; 22,000 hp starting motors

Runners
Material:
Minimum unit submergence: 50.0 ft
WR^2:
Manufacturer: Hitachi
Model test by: Hitachi

	Reversible Runners	Reversible Motor/Generator		
Number	4			4
Diameter (ft)	19.90	Rotor	Stator	
rpm synchronous	257.0			257.0
rpm overspeed				
Type	Francis			Umbrella type

Information on Runners

Condition:	Gross Head (ft) Generating	Gross Head (ft) Pumping	Capacity (MW) Generating	Capacity (MW) Pumping	Discharge (ft³/s) Generating	Discharge (ft³/s) Pumping	Turbine/Pump Eff.(%) Generating	Turbine/Pump Eff.(%) Pumping
Maximum head & maximum power	1,143	1,173	270	264	3,120	2,300	92.0	84.0
Minimum head & maximum power	1,097	1,127	240	290	2,954	2,750	90.0	86.0

Note: Data in the above table are based on field tests.

Condition:	Net Head (ft) Generating	Net Head (ft) Pumping	Capacity (MW) Generating	Capacity (MW) Pumping	Discharge (ft³/s) Generating	Discharge (ft³/s) Pumping	Turbine/Pump Eff.(%) Generating	Turbine/Pump Eff.(%) Pumping
Rated head @ best gate	1,110	1,173	270	264	3,120	2,300	92.0	84.0

Note: Data in the above table are based on field tests.

BLENHEIM-GILBOA PUMPED STORAGE PROJECT

POWERHOUSE and RELATED FEATURES

Powerhouse Structure
Type: Semi-Outdoor
Length: 160.3 m   Width: 52.4 m   Height: 40.2 m

Guard Valves
Number: 4   Diameter: 2.82 m
Type: Spherical

Transformers
Number: 5 (includes 1 spare)
Ratings: 285
Voltages: (kV) 345 / 16.2

Generator
Rating generating (MVA): 278.0   Rating pumping (MVA): 286.0
Insulation type:
Starting method: Reduced voltage - water depression
Starting equipment: Air depression system; 22,000 hp starting motors

Runners
Material:
Minimum unit submergence: 15.24 m
WR^2:
Manufacturer: Hitachi
Model test by: Hitachi

	Reversible Runners	Reversible Motor/Generator		
Number	4			4
Diameter m	6.06	Rotor	Stator	
rpm synchronous	257.0			257.0
rpm overspeed				
Type	Francis			Umbrella type

Information on Runners

Condition:	Gross Head (m)		Capacity (MW)		Discharge (m³/s)		Turbine/Pump Eff.(%)	
	Generating	Pumping	Generating	Pumping	Generating	Pumping	Generating	Pumping
Maximum head & maximum power	348.4	357.5	270	264	88.3	65.1	92.0	84.0
Minimum head & maximum power	334.4	343.5	240	290	83.7	77.9	90.0	86.0

Note: Data in the above table are based on field tests.

Condition:	Net Head (m)		Capacity (MW)		Discharge (m³/s)		Turbine/Pump Eff.(%)	
	Generating	Pumping	Generating	Pumping	Generating	Pumping	Generating	Pumping
Rated head @ best gate	338.3	357.5	270	264	88.3	65.1	92.0	84.0

Note: Data in the above table are based on field tests.

Blenheim-Gilboa Pumped Storage Project - Page 3 (Metric)

BLENHEIM-GILBOA PUMPED STORAGE PROJECT

Plant Data:
Average GWh generating per year:	1,400
Average GWh pumping per year:	2,100
Starting time from standstill (s):	240
Changeover time pumping to generating (min):	15
Planned/scheduled time between major overhauls (years):	10
Outage time required per unit during major overhauls (weeks):	12
Representative plant availability (%):	95.0
Representative planned outages (weeks per year):	3

Miscelleneous Notes:

This plant has been built on Schoharie Creek, which is impounded by Gilboa Dam a few miles upstream to form Schoharie Reservoir, a New York City water supply reservoir. The drainage area of Schoharie Creek (at the Blenheim-Gilboa lower reservoir dam) below Gilboa Dam is 103.6 Mm² (40 mi²). However, when Gilboa Dam is spilling, the Blenheim-Gilboa plant recieves runoff from an additional 813 Mm² (314 mi²) area.

Turnaround times:
Generating to pumping   - 20 min
Pumping to generating   - 15 min
Mode changes   - 650 per machine per year

The five transformers include one which is spare.

Two service gates are required to close one draft tube, so the four service gates can be used to allow service to two draft tubes (at the most) at one time.

The four penstocks have an average length of 597.4 m (1960 ft).

The steel liners are as follows:
Thickness		
Maximum	33.3 mm (1.3125 in.)	34.9 mm (1.375 in.)
Minimum	28.6 mm (1.125 in.)	25.4 mm (1.0 in.)
Material	A537	A517
Length	353.6 m (1160 ft)	182.9 m (600 ft).

The discharge and the inlet area of the lower intake are combined values for all four lower intake units.

Cavitation Experience:

Cavitation is repaired once every 2 years when the units are out for routine maintenence.

Significant or Unique Problems:

List of Licenses Required:
FERC.

BLENHEIM-GILBOA PUMPED STORAGE PROJECT

ENVIRONMENTAL FEATURES

Recreation:
An admission-free visitor's center, housed in a restored 19th century barn, is part of an educational-historical-conservation complex created by the Power Authority and is operated as a museum by the Schoharie County Historical Society.

Fish and Wildlife:
The Power Authority built the 2.63 Mm² (650 acre) Mine Kill State Park in connection with the power project, acquired land for a wildlife management program in the project area, nd provided a 64 Km² (16 acre) fish pond about 25.9 Km (10 mi) northwest of the project near Summit.

Social:
The historic Lansing Manor, Visitor's Bureau exhibits, and a weather station are located near the plant. The conservation, recreation, and cultural facilities earned the Power Authority the Department of Interior's highest outdoor-recreation award of 1976.

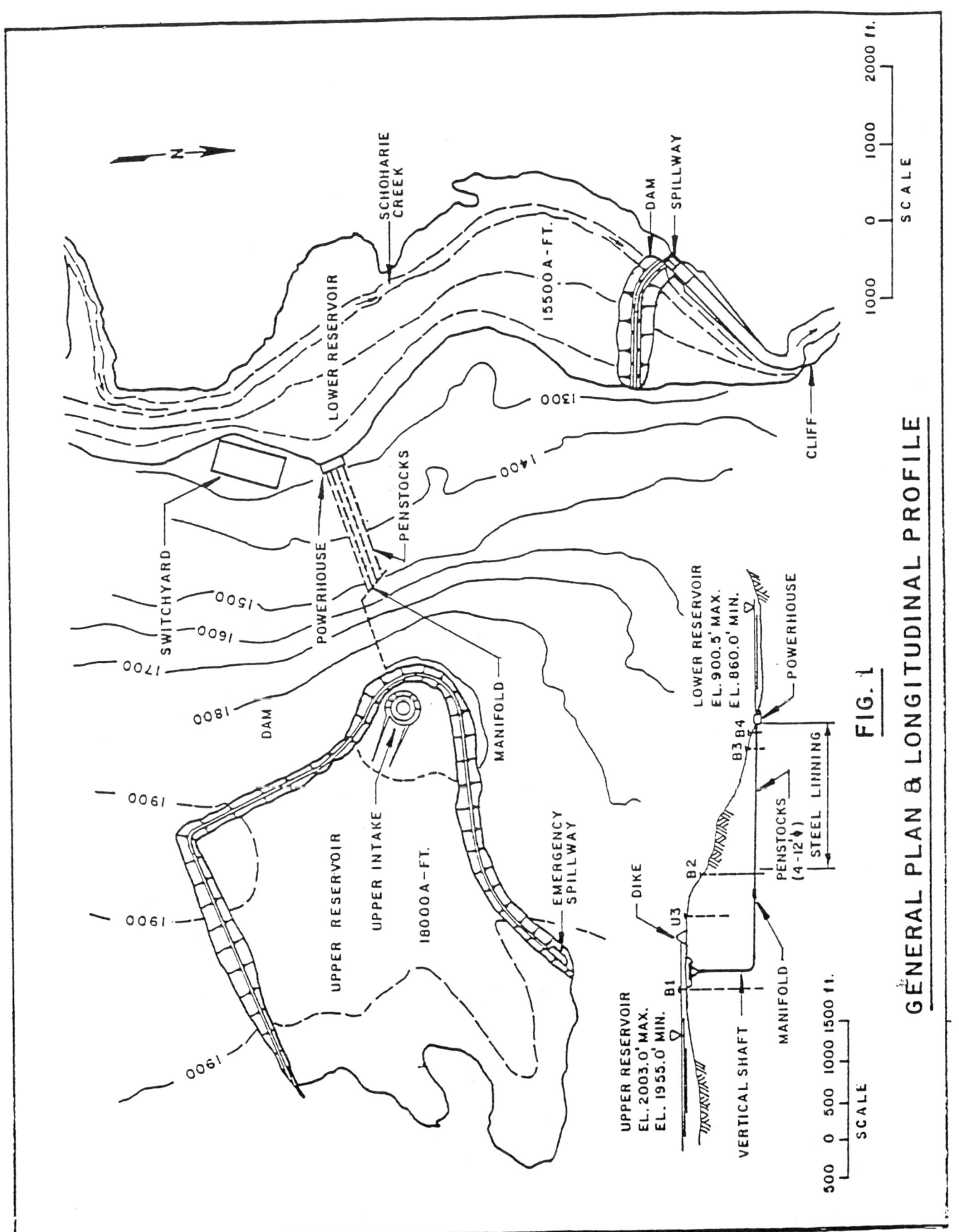

FIG. 1

GENERAL PLAN & LONGITUDINAL PROFILE

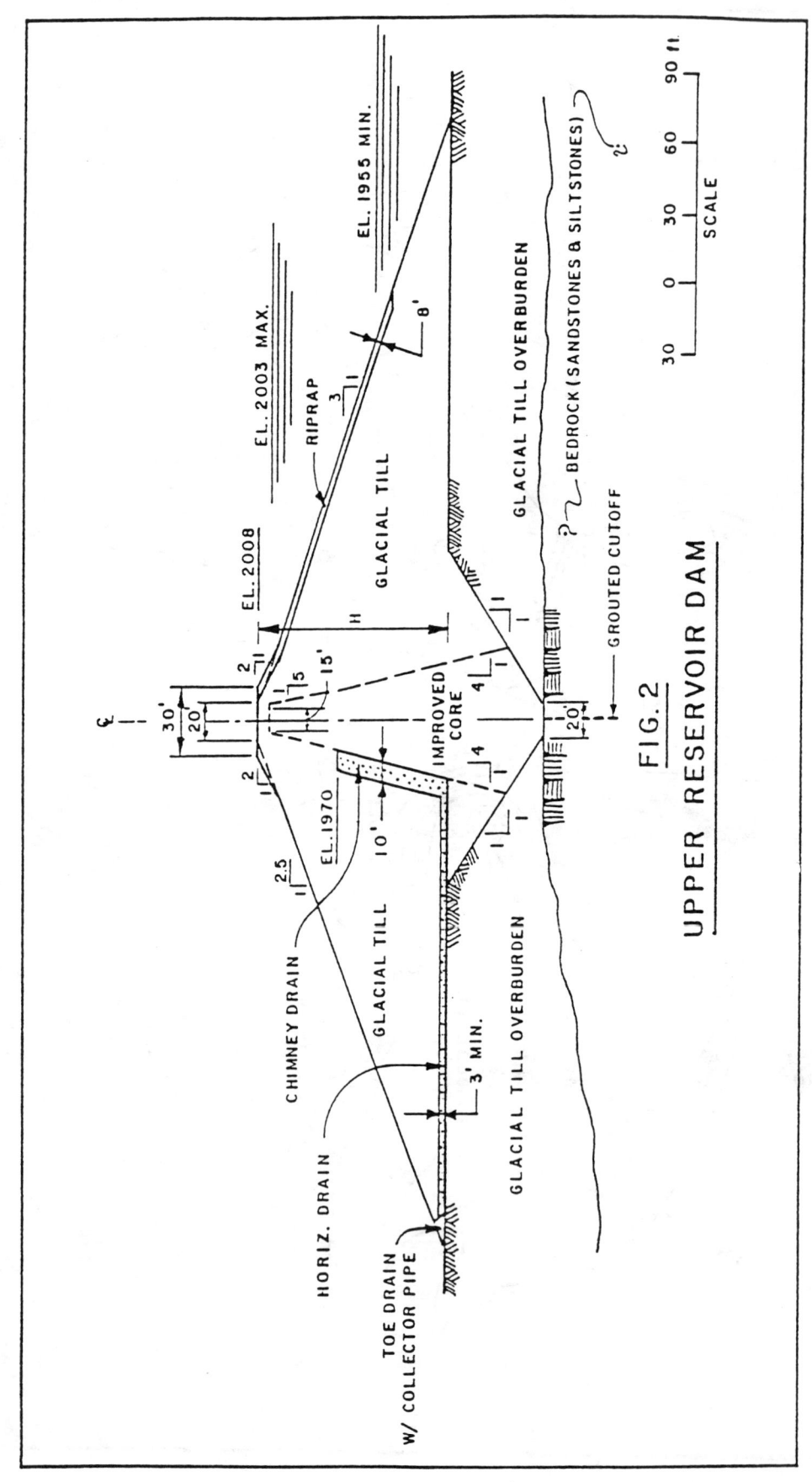

FIG. 2
UPPER RESERVOIR DAM

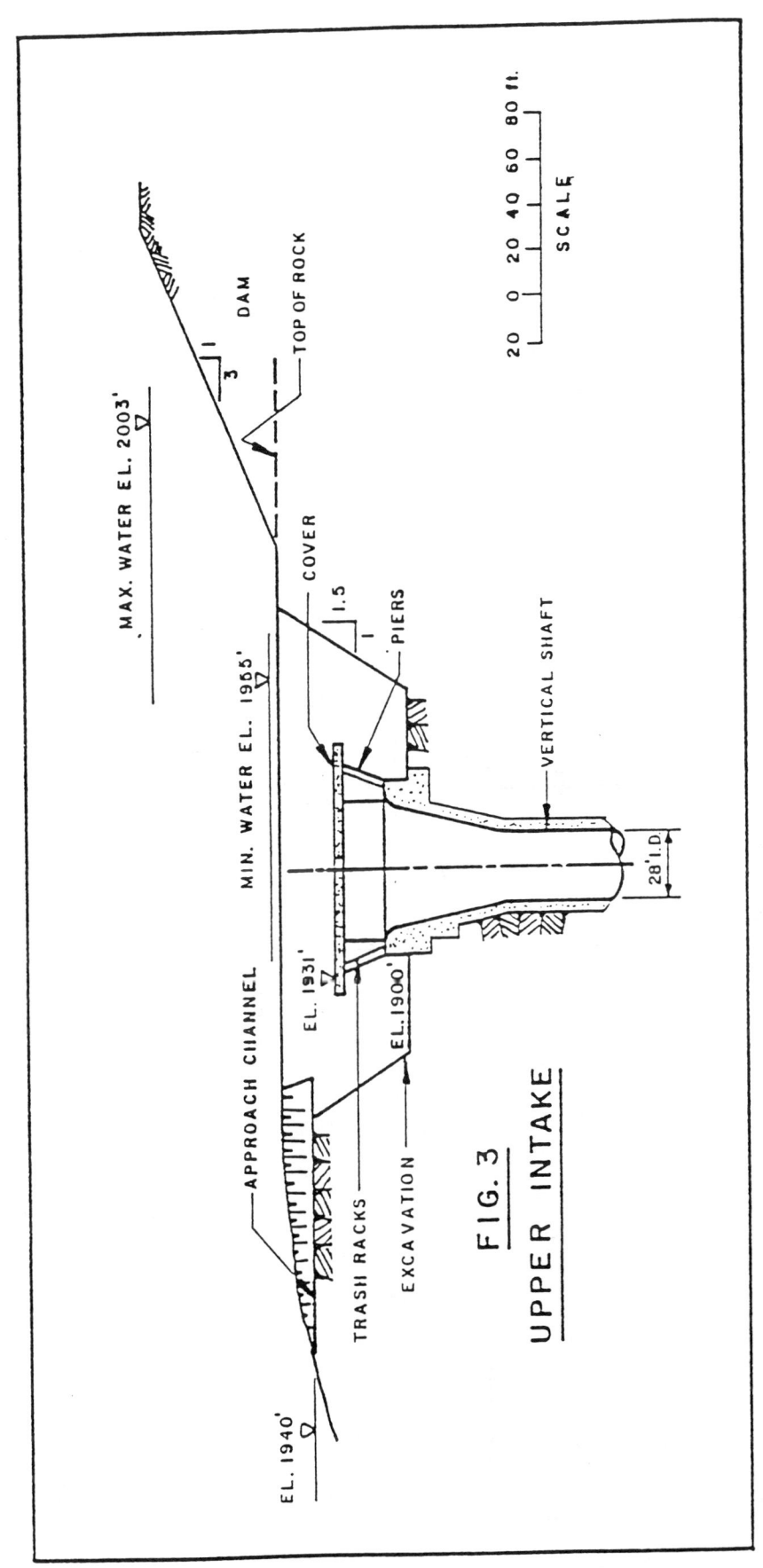

FIG. 3
UPPER INTAKE

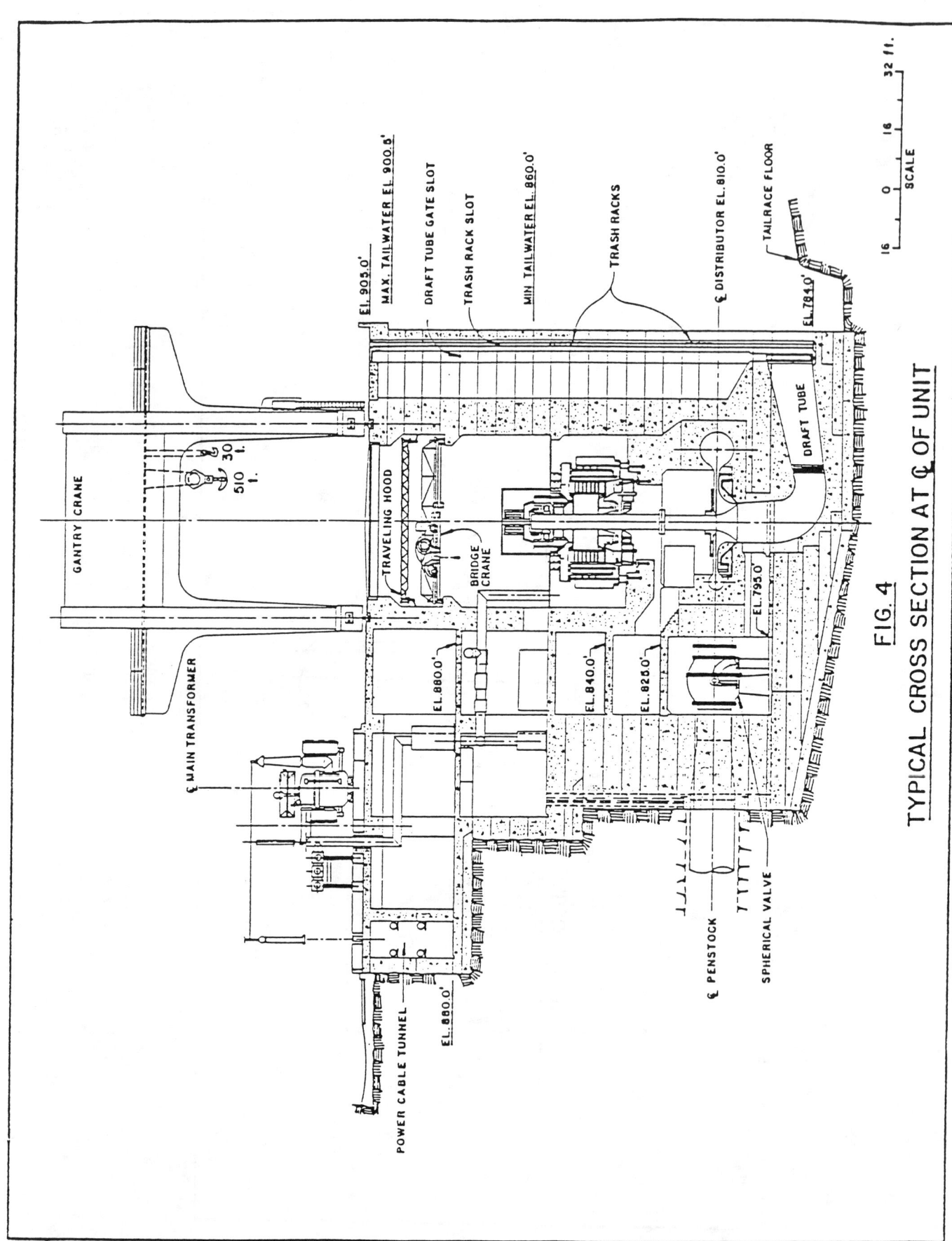

FIG. 4 TYPICAL CROSS SECTION AT ℄ OF UNIT

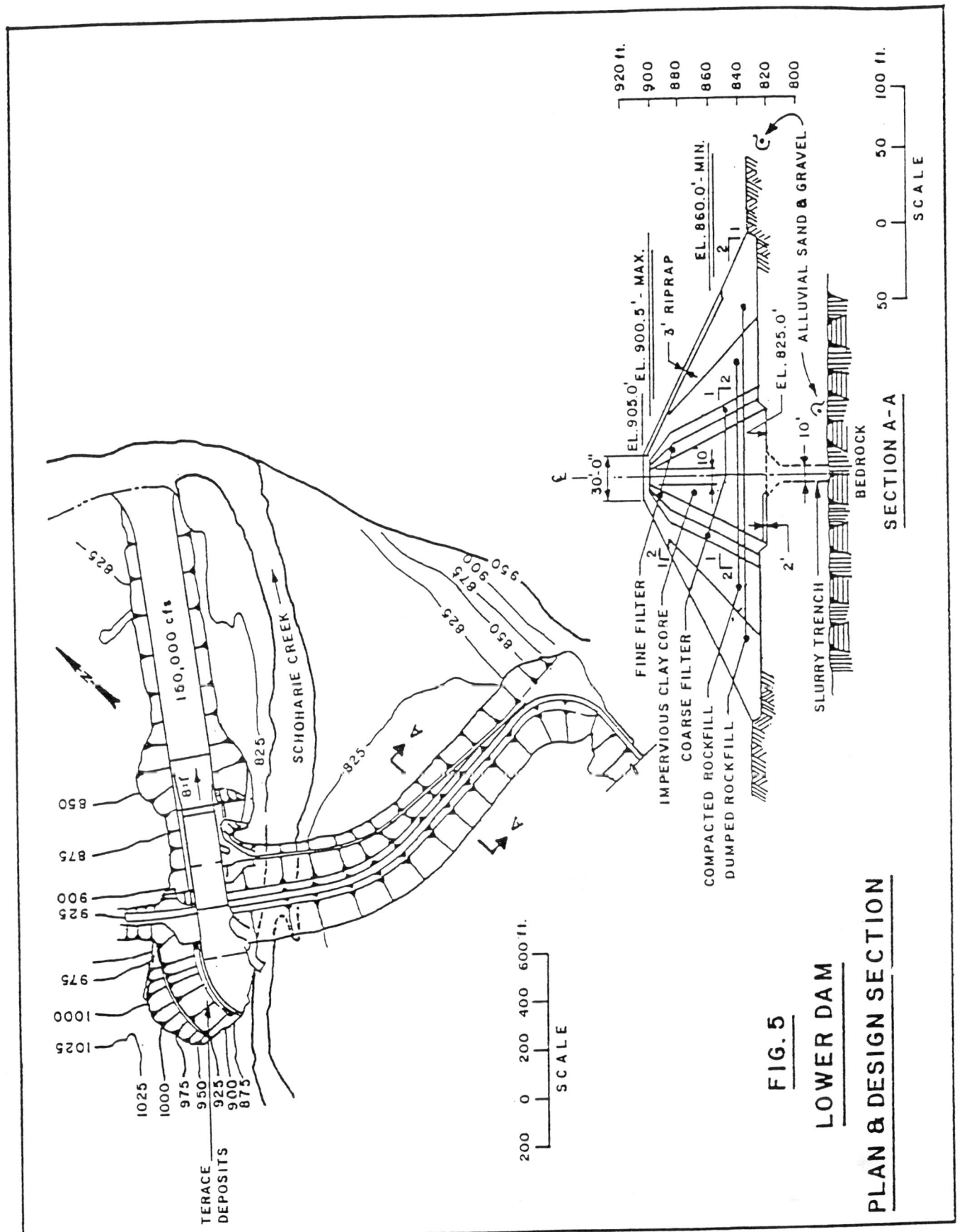

FIG. 5
LOWER DAM
PLAN & DESIGN SECTION

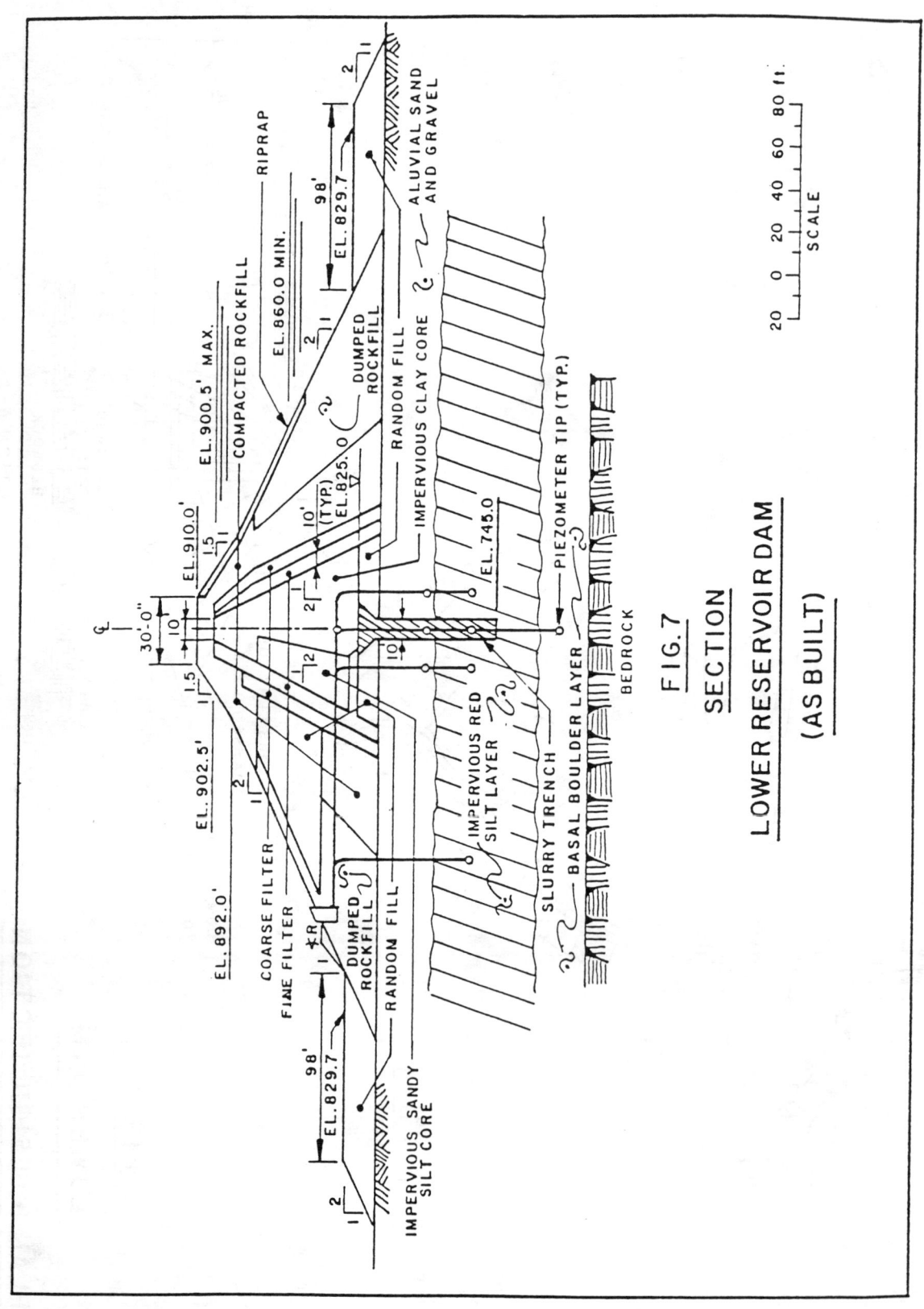

FIG. 7
SECTION
LOWER RESERVOIR DAM
(AS BUILT)

Aerial View

Powerhouse

Blenheim-Gilboa Pumped Storage Plant

Blenheim-Gilboa Pumped Storage Project
Bibliography

The following papers have been written on the Blenheim-Gilboa Project:

1. Wolman, Lee M.G., H. Kenneth Haase, Carlos Poggio and George F. Head, System Planning and Feasibility, Proceedings of the American Power Conference, 1976, Vol. 38, page 997.

2. Vasilescu, Mircea S., Richard V. Sausa, Fred H. Hartz, Richard S. Coleman and Alton P. Davis, Design and Construction, Proceedings of the American Power Conference, 1976, Vol. 38, page 1006.

3. Collyer, Jack M., Frank L. Brown, Francis M. Fullerton and James W. Conover, Environmental Considerations, Proceedings of the American Power Conference, 1976, Vol. 38, page 1025.

4. Lipsky, Charles I. and Edward T. Voelker, Startup and Operating Experience, Proceedings of the American Power Conference, 1976, Vol. 38, page 1032.

5. Data - Statistics, Blenheim-Gilboa Pumped Storage Project, Power Authority of the State of New York (pamphlet, no date).

Plant Name: **CABIN CREEK**

Plant location:
 Georgetown, CO
 Clear Creek County

Rated capacity	300 MW
Average static head	1,160 ft
Plant efficiency	67.00 %
Stored energy	1,450 MWh
Number of units	2

Construction time: 2 years, 11 months
Construction cost: $105 per kW
Price level: 1967
First commercial power: March 1967
FERC project number: 2351

River or water source: South Clear Creek

Owner: Public Service Company of Colorado
 5900 E. 39th Avenue
 Denver, Colorado 80207

Designers:
 Stone and Webster Engineering Corp.

Plant Manager/Superintendent:
 Charles R. McClain
 6276 County Road 381
 Georgetown, Colorado 80444

 (303) 569-3238

	UPPER RESERVOIR	LOWER RESERVOIR
DAM		
Type	Concrete faced, rockfill	Compacted earth and rockfill, impervious core
Height (ft)	210	95
Crest length (ft)	1,460	1,150
Volume (yd³)	1,194,000	1,043,000
RESERVOIR		
Type	On stream, constructed	Constructed
Surface area (acres)	23	52
Usable power storage (acre-ft)	1,400	1,850
Power pool fluctuation (ft)	91.0	42.0
Operating levels		
Maximum (ft)	11,196.0	10,012.0
Minimum (ft)	11,105.0	9,970.0
Drainage area (miles²)	1.0	14.0
Seepage (ft³/s)	1.000	0.550
SPILLWAY		
Design flood		
Return period (years)		10,000 yrs
Flow (ft³/s)		13,600
Capacity (ft³/s)	50	13,600
Type	Overflow weir, ungated	Ogee crest, 44.2 m (145 ft) wide + tunnel spillway.
Gates		
Number	None	None
Type		
Width (ft)		
Height (ft)		
OUTLET WORKS		
Discharge capacity (ft³/s)		
Number of water passages	None	2
Dimensions of water passages		
Height (ft)		
Width (ft)		
Diameter (ft)		6.00
Type of gates		Manual
Number of gates		3

Plant Name: **CABIN CREEK**

Plant location Georgetown, CO Clear Creek County	Owner: Public Service Company of Colorado 5900 E. 39th Avenue Denver, Colorado 80207
Rated capacity 300 MW Average static head 353.4 m Plant efficiency 67.00 % Stored energy 1,450 MWh Number of units 2	Designers: Stone and Webster Engineering Corp.
Construction time: 2 years, 11 months Construction cost: $105 per kW Price level: 1967 First commercial power: March 1967 FERC project number: 2351 River or water source: South Clear Creek	Plant Manager/Superintendent: Charles R. McClain 6276 County Road 381 Georgetown, Colorado 80444 (303) 569-3238

	UPPER RESERVOIR	LOWER RESERVOIR
DAM		
Type	Concrete faced, rockfill	Compacted earth and rockfill, impervious core
Height (m)	64.0	29.0
Crest length (m)	445.0	350.5
Volume (m³)	912,879	797,431
RESERVOIR		
Type	On stream, constructed	Constructed
Surface area (Mm²)	0.09	0.21
Usable power storage (Mm³)	1.727	2.282
Power pool fluctuation (m)	27.74	12.80
Operating levels		
Maximum (m)	3,412.54	3,051.66
Minimum (m)	3,384.80	3,038.86
Drainage area (Mm²)	2.590	36.260
Seepage (m³/s)	0.0283	0.0156
SPILLWAY		
Design flood		
Return period (years)		10,000 yrs
Flow (m³/s)		385
Capacity (m³/s)	1	385
Type	Overflow weir, ungated	Ogee crest, 44.2 m (145 ft) wide + tunnel spillway.
Gates		
Number	None	None
Type		
Width (m)		
Height (m)		
OUTLET WORKS		
Discharge capacity (m³/s)		
Number of water passages	None	2
Dimensions of water passages		
Height (m)		
Width (m)		
Diameter (m)		1.829
Type of gates		Manual
Number of gates		3

Cabin Creek - Page 1 (Metric)

CABIN CREEK

INTAKES	UPPER INTAKE	LOWER INTAKE
Number	1	2
Type	Vertical shaft with cover	Draft tube type
Design discharge (ft³/s)	4,400	
Gross inlet area (ft²)	5,736	
(at trash racks)		
Bar racks		
spacing (in)	4.00	
shape	Rectangular	
depth/thickness (in)	4.50 / 0.44	
diameter (in)		
Emergency gates		
number	None	None
height/width (ft)		
type		
Service gates		
number	None	None
height/width (ft)		
type		
Bulkhead/stop logs (Y or N)	N	
number of units serviced		
Hoists		
number	None	2
capacity (tons)		90
type		Gantry (90 ton), bridge (15 ton)

WATER PASSAGES	Upper Tunnel	Shaft	Lower Tunnel	Surge Tanks Upper	Surge Tanks Lower	Penstocks	Tailrace Tunnel
Number		1	1			2	
Diameter (ft)		15.0	12.0			0.5	
Length (ft)		1,019	3,093			188	
Maximum velocity (ft/s)		25.0	38.9			38.8	
Concrete lining thickness (in)		14.40	12.00				
Total length of concrete sections (ft)		1,019	3,093				
Steel liner Thickness							
Minimum (in)			0.75				
Maximum (in)			1.13				
Material grade			A517				
Total length of steel-lined sections (ft)			1,530				

Notes:
Penstocks: Diameters are 2.59 m (8.5 ft) and 1.68 m (5.5 ft). Length includes steel bifurcation.

Cabin Creek - Page 2 (English)

CABIN CREEK

INTAKES	UPPER INTAKE	LOWER INTAKE
Number	1	2
Type	Vertical shaft with cover	Draft tube type
Design discharge (m³/s)	124.6	
Gross inlet area (m²) (at trash racks)	532.9	
Bar Racks:		
spacing (mm)	102	
shape	Rectangular	
depth/thickness (mm)	114 / 11	
diameter (mm)		
Emergency gates		
number	None	None
height/width (m)		
type		
Service gates		
number	None	None
height/width (m)		
type		
Bulkhead/stop logs (Y or N)	N	
number of units serviced		
Hoists		
number	None	2
capacity (Mg)		82
type		Gantry (90 ton), bridge (15 ton)

WATER PASSAGES	Upper Tunnel	Shaft	Lower Tunnel	Surge Tanks Upper	Surge Tanks Lower	Penstocks	Tailrace Tunnel
Number		1	1			2	
Diameter (m)		4.57	3.66			2.59	
Length (m)		310.6	942.7			57.3	
Maximum velocity (m/s)		7.62	11.86			11.83	
Concrete lining thickness (m)		0.366	0.305				
Total length of concrete sections (m)		310.6	942.7				
Steel liner Thickness							
Minimum (mm)			19				
Maximum (mm)			29				
Material grade			A517				
Total length of steel-lined sections (m)			466.3				

<u>Notes:</u>
Penstocks: Diameters are 2.59 m (8.5 ft) and 1.68 m (5.5 ft). Length includes steel bifurcation.

Cabin Creek - Page 2 (Metric)

CABIN CREEK

POWERHOUSE and RELATED FEATURES

Powerhouse Structure
Type: Semi-outdoor
Length: 145 ft   Width: 99 ft   Height: 100 ft

Guard Valves
Number: 2   Diameter: 5.5 ft
Type: Spherical

Transformers
Number: 2 sets of 3, single phase
Ratings: 39/52/65 OA/FA/FOA
Voltages: (kV) 230 / 13.8, 1 auto trans w/ 11.5

Generator
Rating generating (MVA): 166.7   Rating pumping (MVA): 191.7
Insulation type:
Starting method: Depress water level, start units spinning, open valves and gates
Starting equipment: 11.5 kv, 14,000 horsepower, wound rotor motors

Runners
Material: Stainless steel welded overlay 3.2 mm (1/8-in.) thick
Minimum unit submergence: 35.0 ft
WR^2: 36,100,000 (lbf x ft²)
Manufacturer: Allis Chalmers
Model test by: Allis Chalmers

	Reversible Runners	Reversible Motor/Generator		
Number	2			2
Diameter (ft)	13.80	Rotor	18.23	Stator 24.92
rpm synchronous	360.0			360.0
rpm overspeed	485.0			485.0
Type	Francis	Synchronous, Umbrella		

Information on Runners

Condition:	Gross Head (ft) Generating	Gross Head (ft) Pumping	Capacity (MW) Generating	Capacity (MW) Pumping	Discharge (ft³/s) Generating	Discharge (ft³/s) Pumping	Turbine/Pump Eff.(%) Generating	Turbine/Pump Eff.(%) Pumping
Maximum head & maximum power	1,226	1,226	170	117		860	87.5	82.9
Minimum head & maximum power	1,095	1,095	142	130		1,140	85.3	90.4

Note: Data in the above table are based on design data.

Condition:	Net Head (ft) Generating	Net Head (ft) Pumping	Capacity (MW) Generating	Capacity (MW) Pumping	Discharge (ft³/s) Generating	Discharge (ft³/s) Pumping	Turbine/Pump Eff.(%) Generating	Turbine/Pump Eff.(%) Pumping
Rated head @ best gate	1,190	1,230	160			840		

Cabin Creek - Page 3 (English)

CABIN CREEK

POWERHOUSE and RELATED FEATURES

Powerhouse Structure
Type: Semi-outdoor
Length: 44.2 m   Width: 30.2 m   Height: 30.5 m

Guard Valves
Number: 2   Diameter: 1.68 m
Type: Spherical

Transformers
Number: 2 sets of 3, single phase
Ratings: 39/52/65 OA/FA/FOA
Voltages: (kV) 230 / 13.8, 1 auto trans w/ 11.5

Generator
Rating generating (MVA): 166.7   Rating pumping (MVA): 191.7
Insulation type:
Starting method: Depress water level, start units spinning, open valves and gates
Starting equipment: 11.5 kv, 14,000 horsepower, wound rotor motors

Runners
Material: Stainless steel welded overlay 3.2 mm (1/8-in.) thick
Minimum unit submergence: 10.67 m
WR^2: 14,922,000 (Newtons x m^2)
Manufacturer: Allis Chalmers
Model test by: Allis Chalmers

	Reversible Runners	Reversible Motor/Generator		
Number	2			2
Diameter m	4.19	Rotor 5.556	Stator	7.595
rpm synchronous	360.0			360.0
rpm overspeed	485.0			485.0
Type	Francis	Synchronous, Umbrella		

Information on Runners

Condition:	Gross Head (m) Generating	Gross Head (m) Pumping	Capacity (MW) Generating	Capacity (MW) Pumping	Discharge (m³/s) Generating	Discharge (m³/s) Pumping	Turbine/Pump Eff.(%) Generating	Turbine/Pump Eff.(%) Pumping
Maximum head & maximum power	373.7	373.7	170	117		24.4	87.5	82.9
Minimum head & maximum power	333.8	333.8	142	130		32.3	85.3	90.4

Note: Data in the above table are based on design data.

Condition:	Net Head (m) Generating	Net Head (m) Pumping	Capacity (MW) Generating	Capacity (MW) Pumping	Discharge (m³/s) Generating	Discharge (m³/s) Pumping	Turbine/Pump Eff.(%) Generating	Turbine/Pump Eff.(%) Pumping
Rated head @ best gate	362.7	374.9	160			23.8		

Cabin Creek - Page 3 (Metric)

CABIN CREEK

Plant Data:
Average GWh generating per year:	75
Average GWh pumping per year:	116
Starting time from standstill (s):	390
Changeover time pumping to generating (min):	60
Planned/scheduled time between major overhauls (years):	18
Outage time required per unit during major overhauls (weeks):	
Representative plant availability (%):	95.0
Representative planned outages (weeks per year):	3

Miscelleneous Notes:
Turnaround times:
  generating to pumping - 1500 seconds
  pumping to generating - 1500 seconds
  mode changes   - 740 per machine per year

Seepage from the upper reservoir varies from 0.014 (0.5 ft³/s) to 0.034 m³/s (1.2 ft³/s).

Seepage from the lower reservoir varies from 0.008 (0.3 ft³/s) to 0.017 m³/s (0.6 ft³/s).

A 44.2 m (145 ft) ogee crest spillway currently (1990) under construction for the lower resrvoir will bring the total capacity of the tunnel spillway and the overflow spillway to 385 m³/s (13,600 ft³/s).

The 57.3 m (188 ft) length of the penstocks includes the steel bifurcation.

The generator/motor rating in MVA is 166.7 at 60 degrees celsius and 191.7 at 80 degrees celsius.

Cavitation Experience:
Cavitation has occurred to the runner, draft tube, and wicket gate bushings.

Significant or Unique Problems:

List of Licenses Required:
FERC.

ENVIRONMENTAL FEATURES

Recreation:
There is no recreation on the Cabin Creek site. Fishing from the bank is permitted on Clear Lake immediately downstream from the lower reservoir. No boating is permitted on Clear Lake.

Fish and Wildlife:

CABIN CREEK

Goats and sheep can be seen on the slopes near the lower reservoir. An abundance of marmots are on the site. Elk can be seen above the upper reservoir.

Social:
None.

Figures 1-10 are from an IEEE Power Group paper entitled:

<u>The Design and Construction</u>
<u>of</u>
<u>Cabin Creek Pumped-Storage Hydroelectric Project</u>
by Lawrence M. Robertson.

© 1966 IEEE. Reprinted, with permission, from Proceedings of the 1966 ASME/IEEE Joint Power Conference; Denver, Colorado; pp. 1-23.

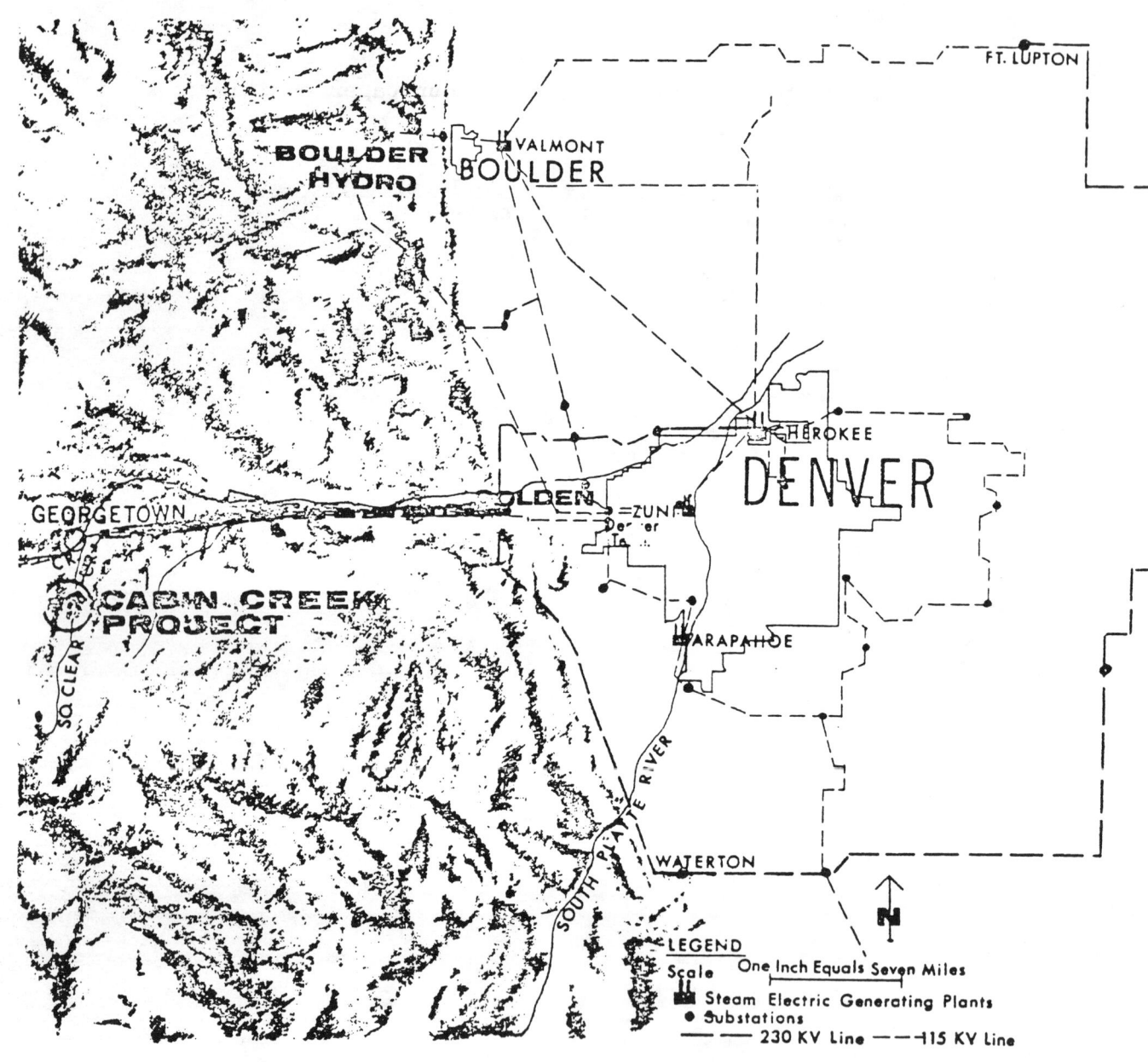

Fig. 1. Map of location of Cabin Creek Project
(Figure 2 removed)

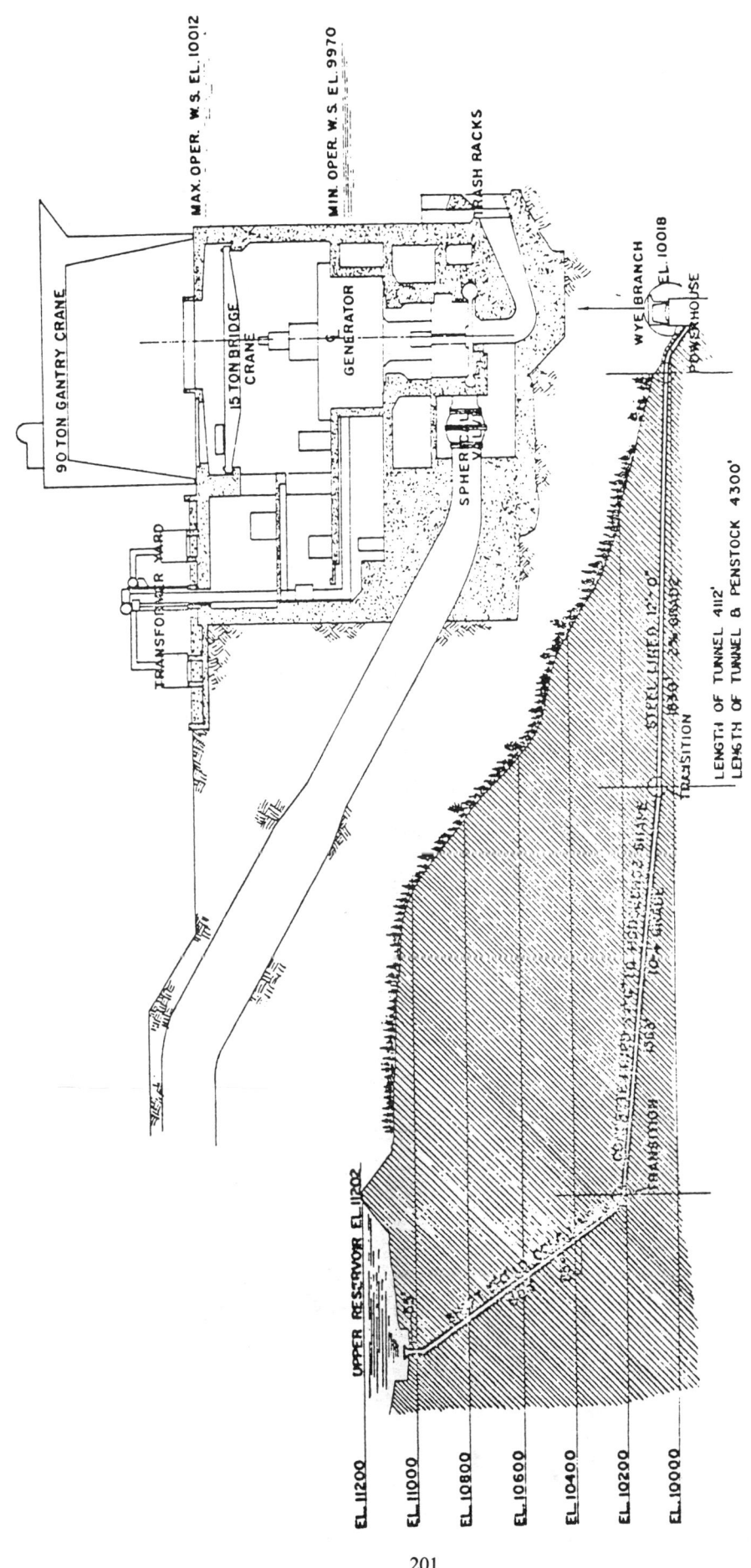

Fig. 3. Section of plant, profile of tunnel and reservoirs. (Figures 4&5 removed)

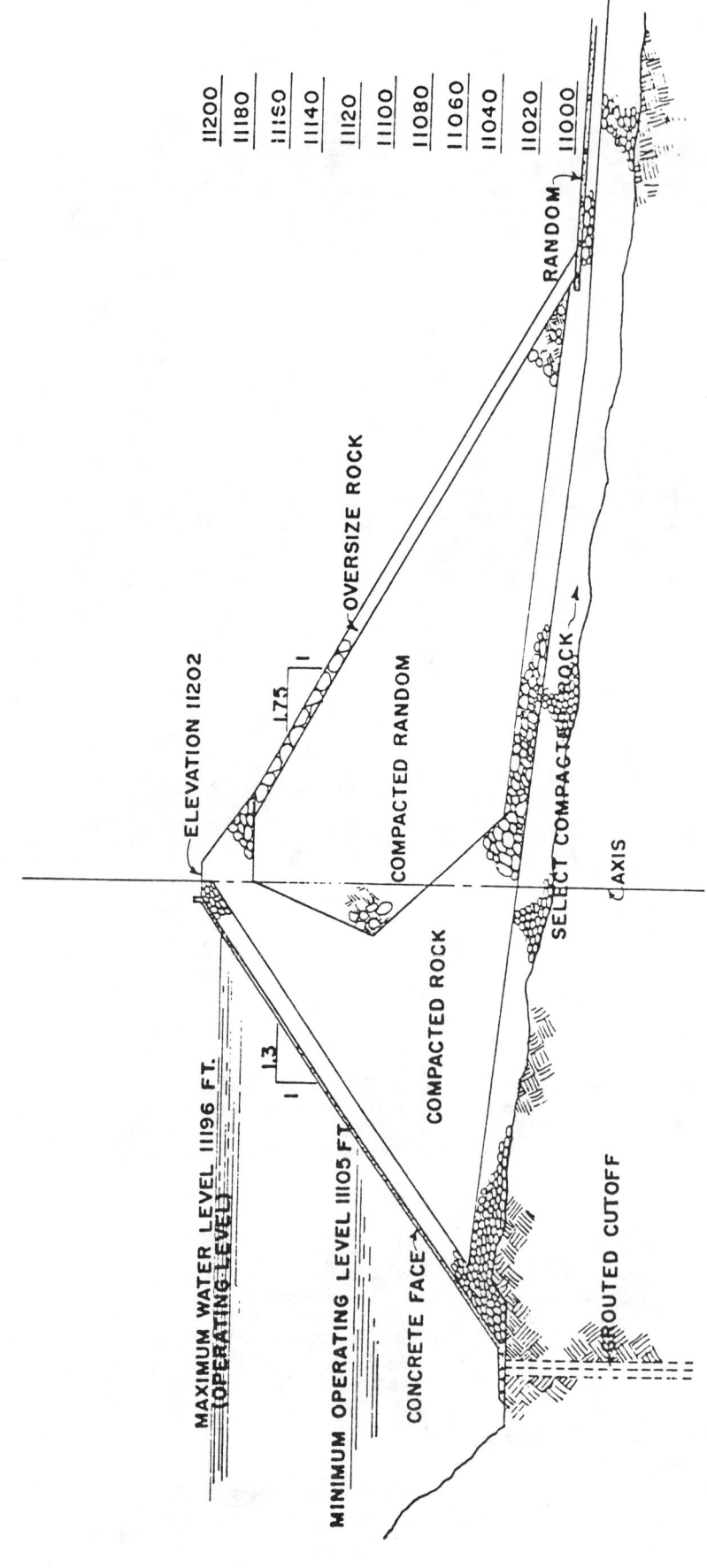

Fig. 6. Cross-section of upper dam.

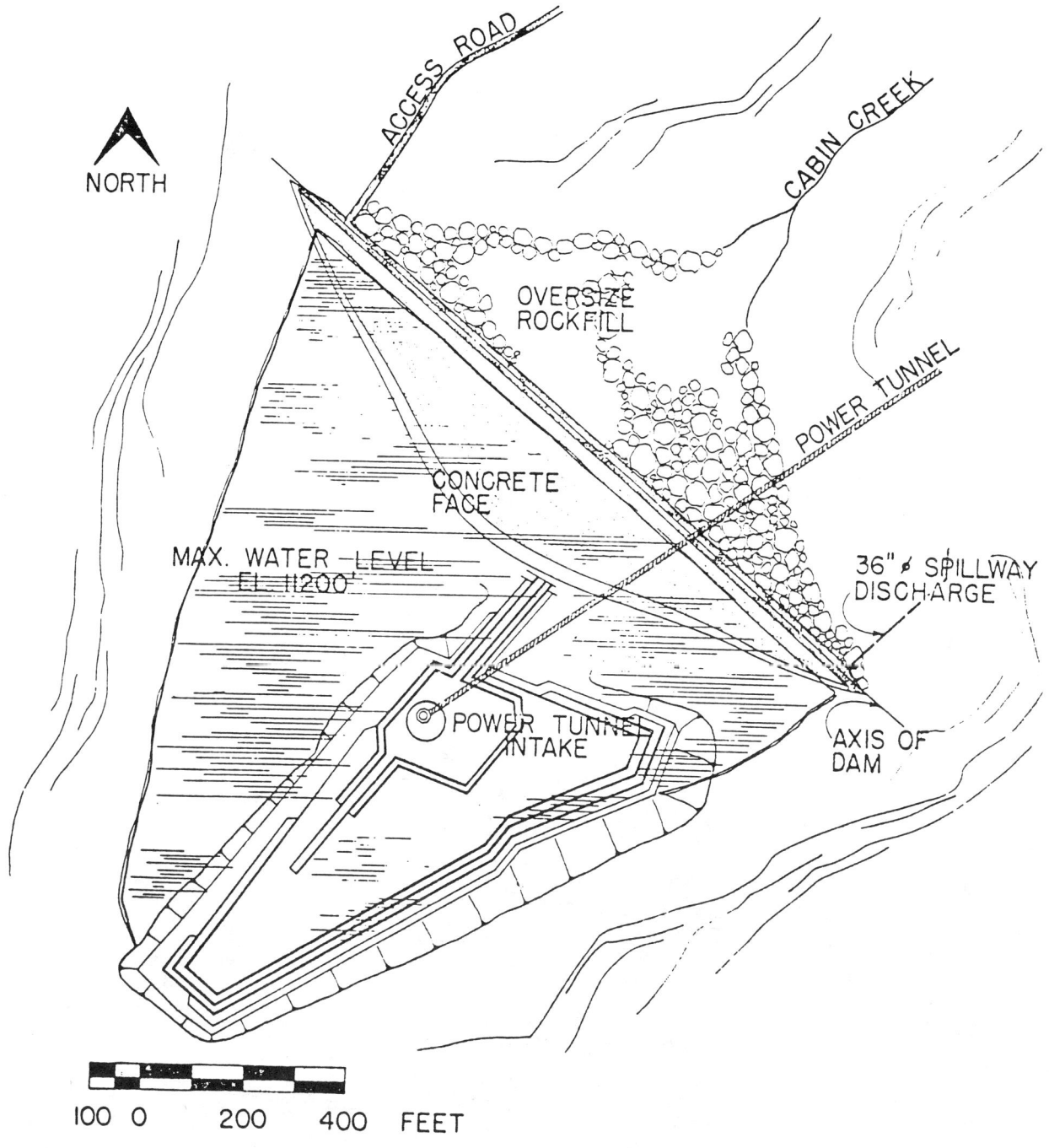

Fig. 7. Plan of upper reservoir, dam, tunnel intake and power tunnel.

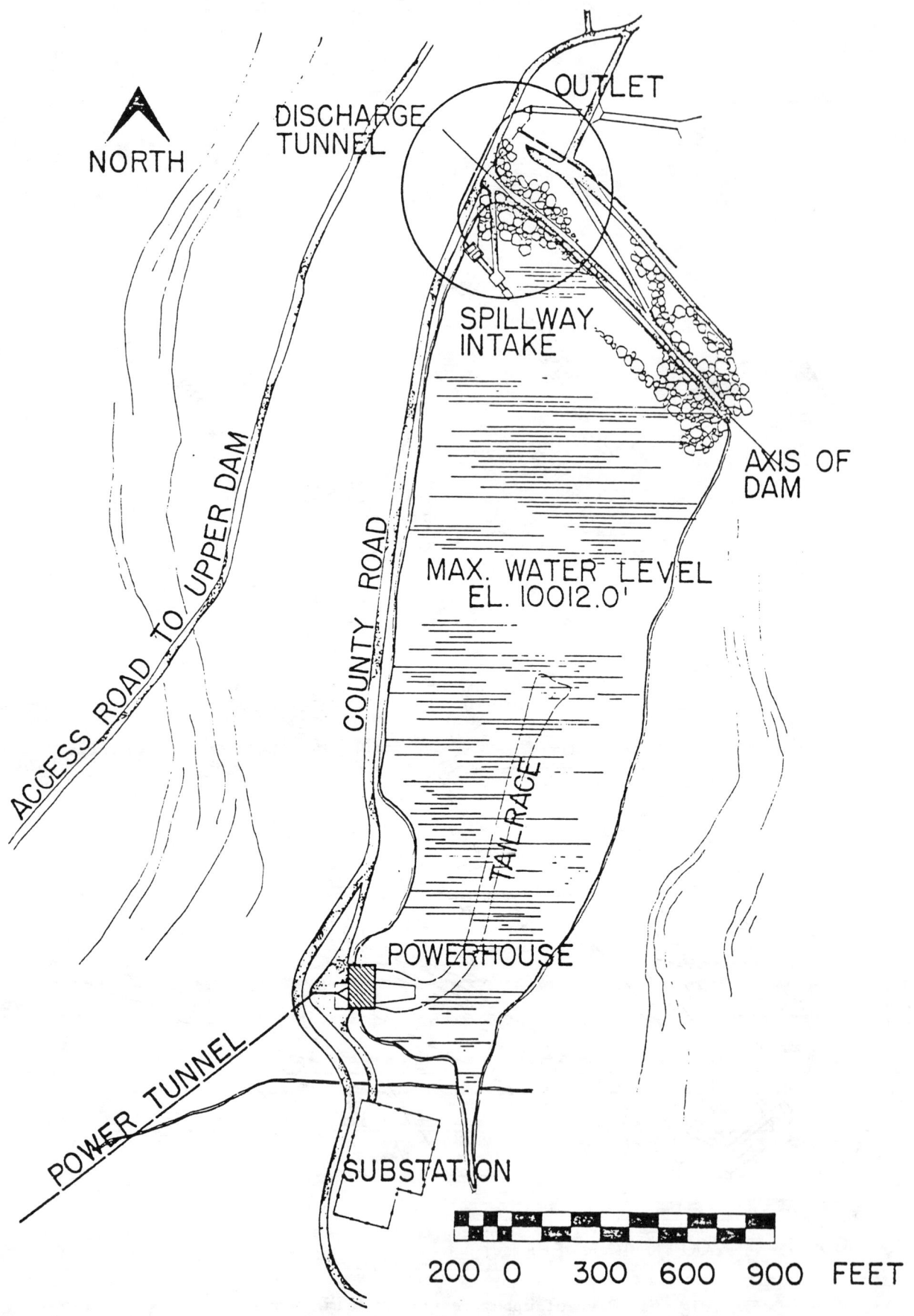

Fig. 8. Plan of lower reservoir, dam, powerhouse, substation

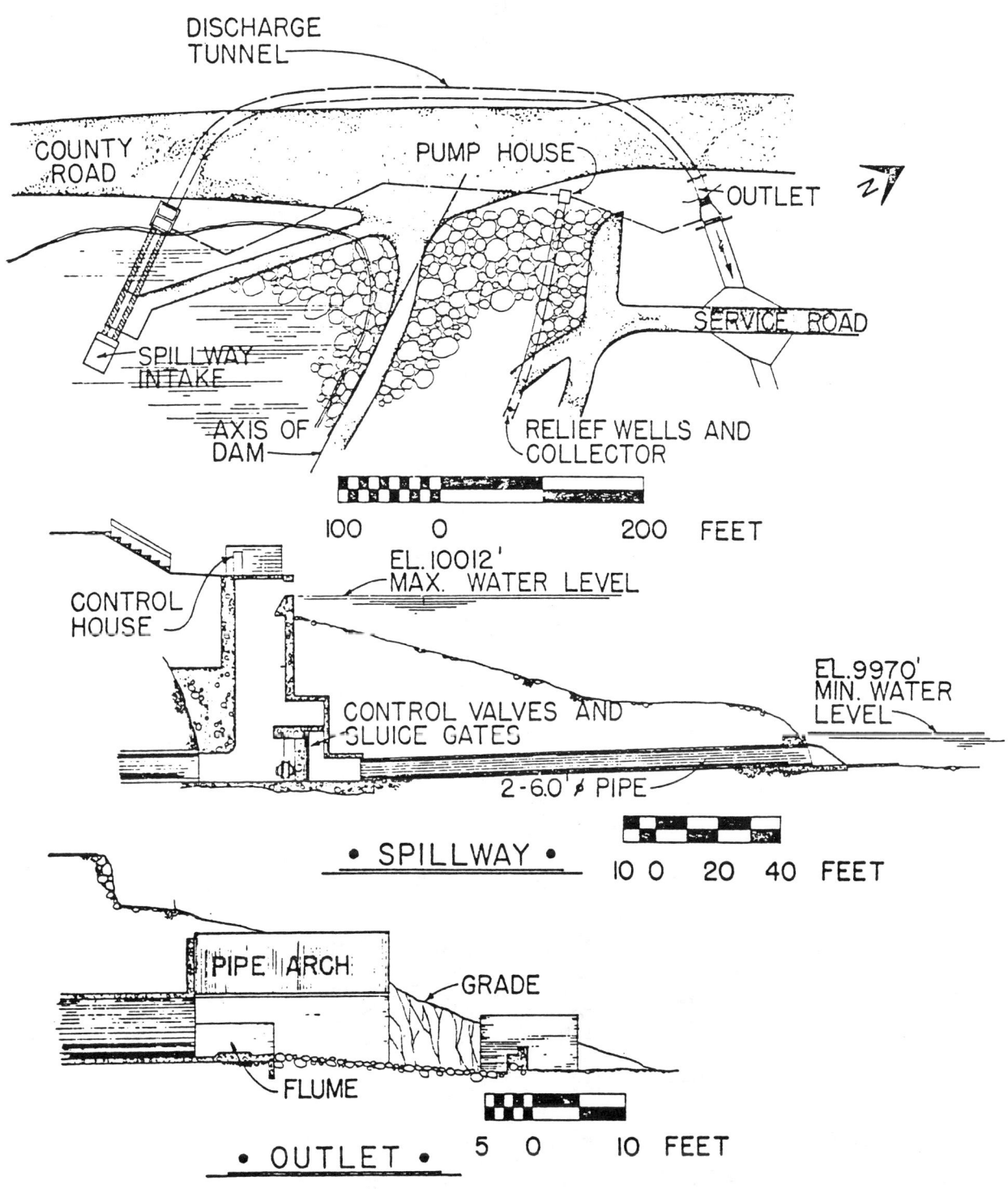

Fig. 9. Details of lower reservoir spillway, diversion tunnel, control valves, sluice gates and relief

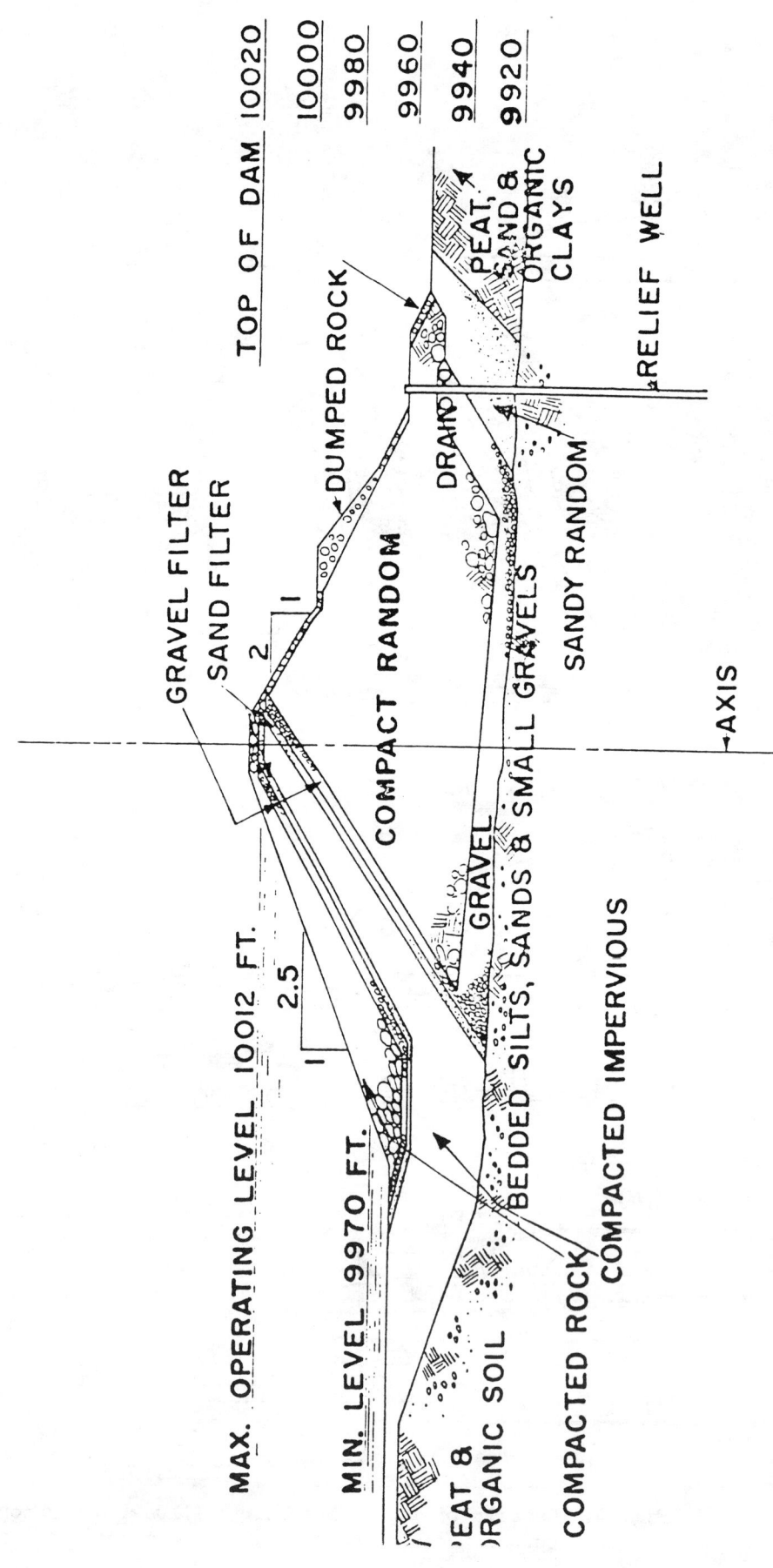

Fig. 10. Cross-section of lower dam.

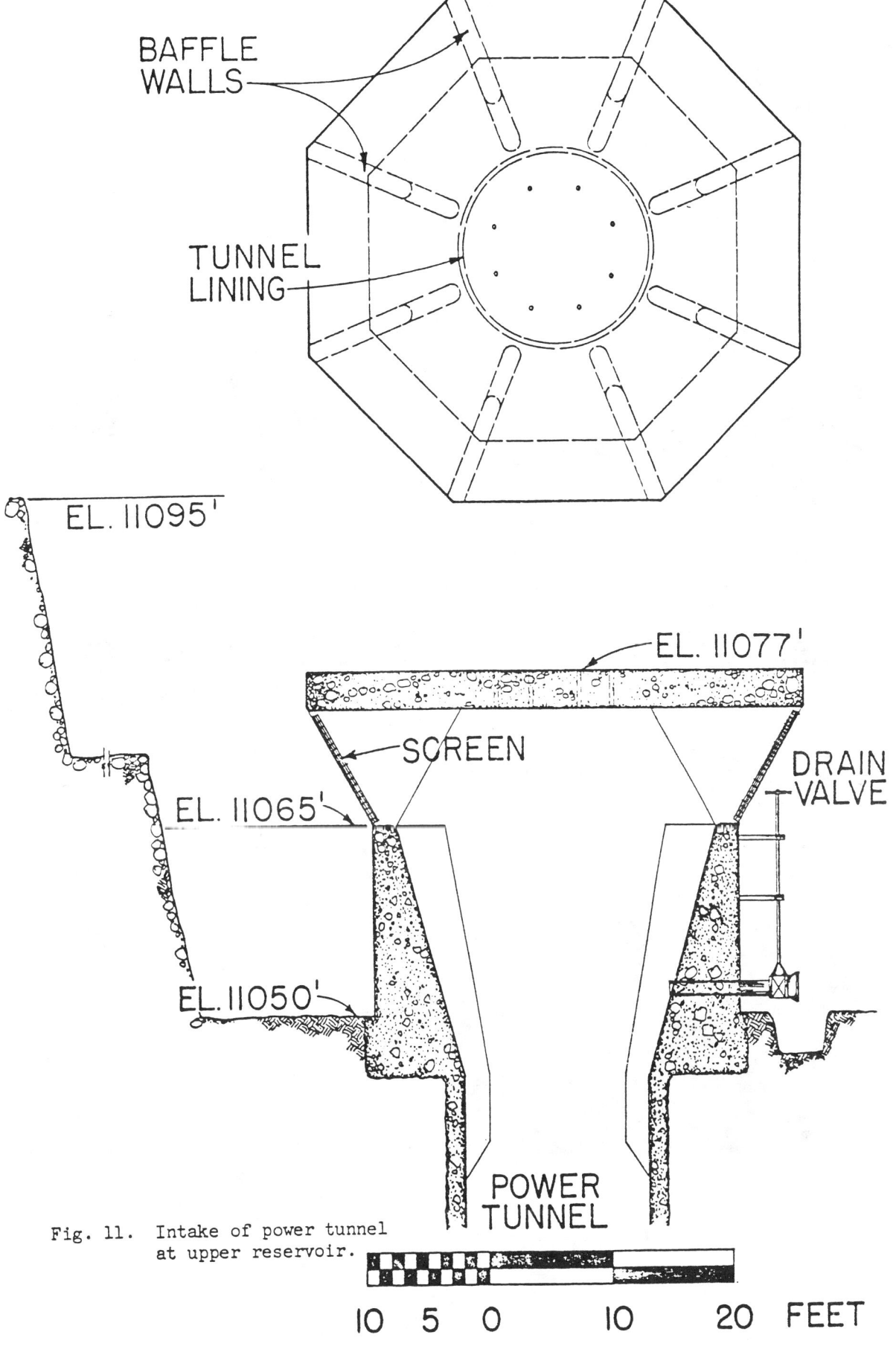

Fig. 11. Intake of power tunnel at upper reservoir.

Aerial View

Cabin Creek Pumped Storage Plant

LIST OF PROJECT ARTICLES

1. The Design and Construction of Cabin Creek Pumped-Storage Hydroelectric Project, L.M. Robertson, IEEE Paper No. 31 PP 66-547, June 26, 1967.

2. Cabin Creek Pumped Storage Hydroelectric Project, H.W. Hight, Proceedings of the ASCE, PO 1, January 1971.

3. Recent Experience At Cabin Creek Pumped Storage Hydroelectric Project, F.W. Eastom, Proceedings of the American Power Conference, Volume 35, 1973.

4. Electrical Design and Testing of Cabin Creek Pumped-Storage Hydroelectric Project, L.M. Robertson, F.W. Eastom, F.L. Brennan, L.S. Burt, C.W. Hubbard, and J.F. Fuller, IEEE Paper No. 31 TP 67-487.

5. Operation and Maintenance of Cabin Creek Pumped-Storage Hydroelectric Project, L.M. Robertson and F.W. Eastom, IEEE Conference Paper No. C72 613-8, July 1972.

6. The Cabin Creek Pumped Storage Hydroelectric Project-Performance and Field Testing, L.M. Robertson, F.W. Eastom, B. Eyden, R. Dunaiski, L.S. Burt, C.W. Hubbard, J. Howe, and P. Johrde, IEEE, ASME joint meeting, Denver, Colorado, Jan. 12, 1968.

3913r/el

Plant Name: **CARTERS**

Plant location:
Chatsworth, GA
Murray County

Owner: Mobile District, Corps of Engineers
P.O. Box 2288
Mobile, Alabama 36628-0001

Rated capacity	258 MW
Average static head	345 ft
Plant efficiency	70.00 %
Stored energy	1,889 MWh
Number of units	2

Designers:
Mobile District, Corps of Engineers

Construction time: 13 years
Construction cost: $271 per kW
Price level: 1988
First commercial power: January 1976
FERC project number: None

Plant Manager/Superintendent:
Plant Manager
Route 3
Box 3358
Chatworth, Georgia 30705-9312
(404) 334-2640

River or water source: Coosawatee River

	UPPER RESERVOIR	LOWER RESERVOIR
DAM		
Type	Rockfill	Concrete
Height (ft)	445	35
Crest length (ft)	2,053	208
Volume (yd^3)	15,000,000	766,000
RESERVOIR		
Type	Constructed	Constructed
Surface area (acres)	3,220	870
Usable power storage (acre-ft)	134,900	17,210
Power pool fluctuation (ft)	52.0	15.0
Operating levels		
Maximum (ft)	1,072.0	707.0
Minimum (ft)	1,022.0	665.5
Drainage area (miles^2)	376.0	530.0
Seepage (ft^3/s)		
SPILLWAY		
Design flood		
Return period (years)	PMF	
Flow (ft^3/s)	194,200	197,800
Capacity (ft^3/s)	194,200	197,800
Type	Concrete gravity	
Gates		
Number	5	4
Type	Tainter	Tainter
Width (ft)	42.00	42.00
Height (ft)	36.50	36.50
OUTLET WORKS		
Discharge capacity (ft^3/s)	100,000	
Number of water passages	1	None
Dimensions of water passages		
Height (ft)		
Width (ft)		
Diameter (ft)	16.50	
Type of gates	Low level sluice	
Number of gates	4	

Plant Name: **CARTERS**

Plant location	Owner: Mobile District, Corps of Engineers
Chatsworth, GA	P.O. Box 2288
Murray County	Mobile, Alabama 36628-0001

Rated capacity	258 MW	
Average static head	105.2 m	Designers:
Plant efficiency	70.00 %	Mobile District, Corps of Engineers
Stored energy	1,889 MWh	
Number of units	2	

Construction time: 13 years
Construction cost: $271 per kW
Price level: 1988
First commercial power: January 1976
FERC project number: None

Plant Manager/Superintendent:
Plant Manager
Route 3
Box 3358
Chatworth, Georgia 30705-9312
(404) 334-2640

River or water source: Coosawatee River

	UPPER RESERVOIR	LOWER RESERVOIR
DAM		
Type	Rockfill	Concrete
Height (m)	135.6	10.7
Crest length (m)	625.8	63.4
Volume (m³)	11,468,328	585,649
RESERVOIR		
Type	Constructed	Constructed
Surface area (Mm²)	13.03	3.52
Usable power storage (Mm³)	166.398	21.228
Power pool fluctuation (m)	15.85	4.57
Operating levels		
Maximum (m)	326.75	215.49
Minimum (m)	311.51	202.84
Drainage area (Mm²)	973.841	1,372.701
Seepage (m³/s)		
SPILLWAY		
Design flood		
Return period (years)	PMF	
Flow (m³/s)	5,499	5,601
Capacity (m³/s)	5,499	5,601
Type	Concrete gravity	
Gates		
Number	5	4
Type	Tainter	Tainter
Width (m)	12.802	12.802
Height (m)	11.125	11.125
OUTLET WORKS		
Discharge capacity (m³/s)	2,832	
Number of water passages	1	None
Dimensions of water passages		
Height (m)		
Width (m)		
Diameter (m)	5.029	
Type of gates	Low level sluice	
Number of gates	4	

Carters - Page 1 (Metric)

CARTERS

INTAKES	UPPER INTAKE	LOWER INTAKE
Number	2	
Type	Concrete	
Design discharge (ft³/s)	21,600	
Gross inlet area (ft²)	7,064	
(at trash racks)		
Bar racks		
spacing (in)		
shape		
depth/thickness (in)		
diameter (in)		
Emergency gates		
number	None	
height/width (ft)		
type		
Service gates		
number	4	
height/width (ft)	20.50 / 14.00	
type	Tractor	
Bulkhead/stop logs (Y or N)	Y	
number of units serviced	2	
Hoists		
number	2	
capacity (tons)	26	
type	Fixed	

WATER PASSAGES	Upper Tunnel	Shaft	Lower Tunnel	Surge Tanks Upper	Surge Tanks Lower	Penstocks	Tailrace Tunnel
Number						2	
Diameter (ft)						18.0	
Length (ft)						838	
Maximum velocity (ft/s)						21.2	
Concrete lining thickness (in)						30.00	
Total length of concrete sections (ft)							
Steel liner							
Thickness							
Minimum (in)						0.63	
Maximum (in)						1.88	
Material grade						ASTM A516 Grade 60	
Total length of steel-lined sections (ft)						833	

Notes:
Penstocks: Four penstocks total, two reported for reversible units.

CARTERS

INTAKES	UPPER INTAKE	LOWER INTAKE
Number	2	
Type	Concrete	
Design discharge (m³/s)	611.6	
Gross inlet area (m²) (at trash racks)	656.2	
Bar Racks:		
spacing (mm)		
shape		
depth/thickness (mm)		
diameter (mm)		
Emergency gates		
number	None	
height/width (m)		
type		
Service gates		
number	4	
height/width (m)	6.248 / 4.267	
type	Tractor	
Bulkhead/stop logs (Y or N)	Y	
number of units serviced	2	
Hoists		
number	2	
capacity (Mg)	24	
type	Fixed	

WATER PASSAGES	Upper Tunnel	Shaft	Lower Tunnel	Surge Tanks Upper	Surge Tanks Lower	Penstocks	Tailrace Tunnel
Number						2	
Diameter (m)						5.49	
Length (m)						255.4	
Maximum velocity (m/s)						6.46	
Concrete lining thickness (m)						0.762	
Total length of concrete sections (m)							
Steel liner Thickness							
Minimum (mm)						16	
Maximum (mm)						48	
Material grade						ASTM A516 Grade 60	
Total length of steel-lined sections (m)						253.9	

Notes:
Penstocks: Four penstocks total, two reported for reversible units.

CARTERS

POWERHOUSE and RELATED FEATURES

Powerhouse Structure
Type: Surface
Length: 361 ft   Width: 69 ft   Height: 54 ft

Guard Valves
Number: None   Diameter: ft
Type:

Transformers
Number:
Ratings: 176.9
Voltages: (kV) 230 / 13.2

Generator
Rating generating (MVA): 132.0   Rating pumping (MVA): 138.0
Insulation type:
Starting method: Cross line
Starting equipment:

Runners
Material: Steel
Minimum unit submergence: 16.5 ft
WR^2:
Manufacturer: Allis Chalmers
Model test by: Allis Chalmers

	Reversible Runners	Reversible Motor/Generator		
Number	2			2
Diameter (ft)	20.70	Rotor	Stator	
rpm synchronous	150.0			150.0
rpm overspeed				163.6
Type	Francis			

Information on Runners

Condition:	Gross Head (ft)		Capacity (MW)		Discharge (ft³/s)		Turbine/Pump Eff.(%)	
	Generating	Pumping	Generating	Pumping	Generating	Pumping	Generating	Pumping
Maximum head & maximum power	406	406						
Minimum head & maximum power	315	315						

Condition:	Net Head (ft)		Capacity (MW)		Discharge (ft³/s)		Turbine/Pump Eff.(%)	
	Generating	Pumping	Generating	Pumping	Generating	Pumping	Generating	Pumping
Rated head @ best gate	345	347	129	110	5,200	4,435		

Note: Data in the above table are based on design data.

CARTERS

POWERHOUSE and RELATED FEATURES

Powerhouse Structure
Type: Surface
Length: 110.2 m   Width: 21.0 m   Height: 16.4 m

Guard Valves
Number: None   Diameter: m
Type:

Transformers
Number:
Ratings: 176.9
Voltages: (kV) 230 / 13.2

Generator
Rating generating (MVA): 132.0   Rating pumping (MVA): 138.0
Insulation type:
Starting method: Cross line
Starting equipment:

Runners
Material: Steel
Minimum unit submergence: 5.03 m
WR^2:
Manufacturer: Allis Chalmers
Model test by: Allis Chalmers

	Reversible Runners	Reversible Motor/Generator		
Number	2			2
Diameter m	6.30	Rotor	Stator	
rpm synchronous	150.0			150.0
rpm overspeed				163.6
Type	Francis			

Information on Runners

Condition:	Gross Head (m)		Capacity (MW)		Discharge (m³/s)		Turbine/Pump Eff.(%)	
	Generating	Pumping	Generating	Pumping	Generating	Pumping	Generating	Pumping
Maximum head & maximum power	123.7	123.7						
Minimum head & maximum power	96.0	96.0						

Condition:	Net Head (m)		Capacity (MW)		Discharge (m³/s)		Turbine/Pump Eff.(%)	
	Generating	Pumping	Generating	Pumping	Generating	Pumping	Generating	Pumping
Rated head @ best gate	105.2	105.8	129	110	147.3	125.6		

Note: Data in the above table are based on design data.

CARTERS

Plant Data:
   Average GWh generating per year:
   Average GWh pumping per year:
   Starting time from standstill (s):
   Changeover time pumping to generating (min):
   Planned/scheduled time
      between major overhauls (years):   13
   Outage time required per unit
      during major overhauls (weeks):   39
   Representative plant availability (%):   89.0
   Representative planned outages (weeks per year):   7

Miscelleneous Notes:
   None.

Cavitation Experience:
   Pumping units needed major cavitation repair after 15 years.

Significant or Unique Problems:
   At net head of approximately 116 m (380 ft), penstock sudden pressure can be exceeded on loss of load (load rejection) at gate openings greater than 72% on unit No. 3.

List of Licenses Required:
   None -- direct congressional authorization.

ENVIRONMENTAL FEATURES

Recreation:
   Significant recreational opportunities were developed as part of the project. These included facilities for boating, hiking, camping, fishing, swimming, picnicking, hunting, bird watching, and a visitor center. Visitation exceeds 1 million people per year.

Fish and Wildlife:
   The project does not have any fish and wildlife issues.

Social:
   Positive socioeconomic impact has occurred. Tourism has increased, new residential developments have been started, and new time share houseboats are present at the marina. Land use has changed from only rural with timber production, to now including housing. Historic DeSoto camp area was inundated. The reservoir will be used for water supply by the city of Chatsworth.

Journal of the
POWER DIVISION
Proceedings of the American Society of Civil Engineers

DESIGN OF CARTERS PUMPED STORAGE PROJECT[a]

By Walter W. Burdin,[1] M. ASCE

INTRODUCTION

The Carters Project is located on the Coosawattee River, one of the headwaters of the Coosa River, near Carters, in Murray County, Georgia (Fig. 1). It consists of a high, rockfill main dam with a powerhouse and appurtenant works on the right bank, an emergency spillway on the left bank, and a low, earth-and-rockfill reregulation dam about 2 miles downstream of the main dam. The main dam is about 27 miles above the mouth of the Coosawattee and about 75 miles north of Atlanta, Georgia. The reregulation dam and reservoir, the main dam, and a portion of the upper reservoir will be in Murray County, with the majority of the main reservoir in Gilmer County, Georgia (Fig. 2). Average annual power production is estimated at 406,200,000 kwhr. Of this amount 230,200,000 kwhr is attributable to the inclusion of pump-turbine units.

Below the dam site the river enters a rolling valley, which was a huge granary before the turn of the century. Barges traveled the river between Carters and Rome, bringing up the fertilizer and taking down oats. According to the Atlanta Journal, this region gave our language the expression "more than Carter's got oats." Farrish Carter bought this land from the Indians about 1830 and the family still owns most of it. Mr. Carter, who was then reputed to be the richest man in Georgia, built his summer home about 5 miles from the river. It still stands.

So far as can be determined, the first proposal for comprehensive development of the Coosawattee River was in the report on the Alabama-Coosa Branch of the Mobile River System submitted by the Chief of Engineers on

---

Note.—Discussion open until November 1, 1970. To extend the closing date one month, a written request must be filed with the Executive Secretary, ASCE. This paper is part of the copyrighted Journal of the Power Division, Proceedings of the American Society of Civil Engineers, Vol. 96, No. PO3, June, 1970. Manuscript was submitted for review for possible publication on September 25, 1969.

[a] Presented at the October 13-17, 1969, ASCE Annual and Environmental Engineering Meeting held at Chicago, Ill.

[1] Hydraulic Engr., Design Branch, Engrg. Div., U.S. Army Engr. District, Mobile, Ala.

November 14, 1934, in accordance with the River and Harbor Act of 1927. This report analyzed three possible sites on the Coosawattee River.

The River and Harbor Act of 1945 provided broad authority for the development of the entire Alabama-Coosa System for navigation, flood control, power, and other purposes. The Carters Project is a feature of the overall

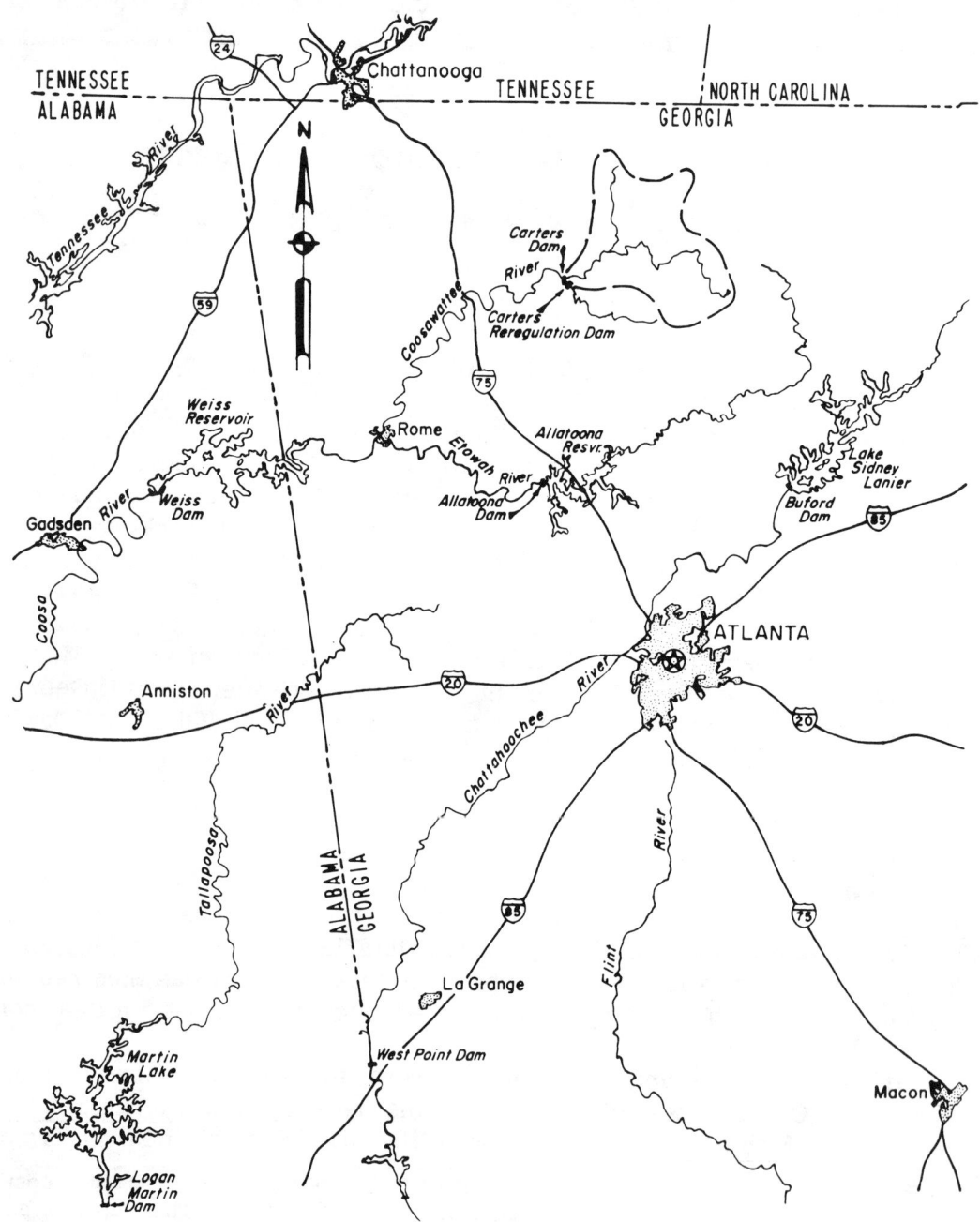

FIG. 1.—LOCATION MAP

plan for the development of this system. The current work on the Alabama River is carried on under this same authority.

A study of the Oostanaula River and its tributaries made in 1957 indicated that a flood control-power project on the Coosawattee tributary was feasible.

Preliminary study in 1959 indicated further study was warranted, and in 1961 a Site Selection Report was submitted which resulted in the choice of the present site.

Construction officially began in April, 1962 with a contract for an access

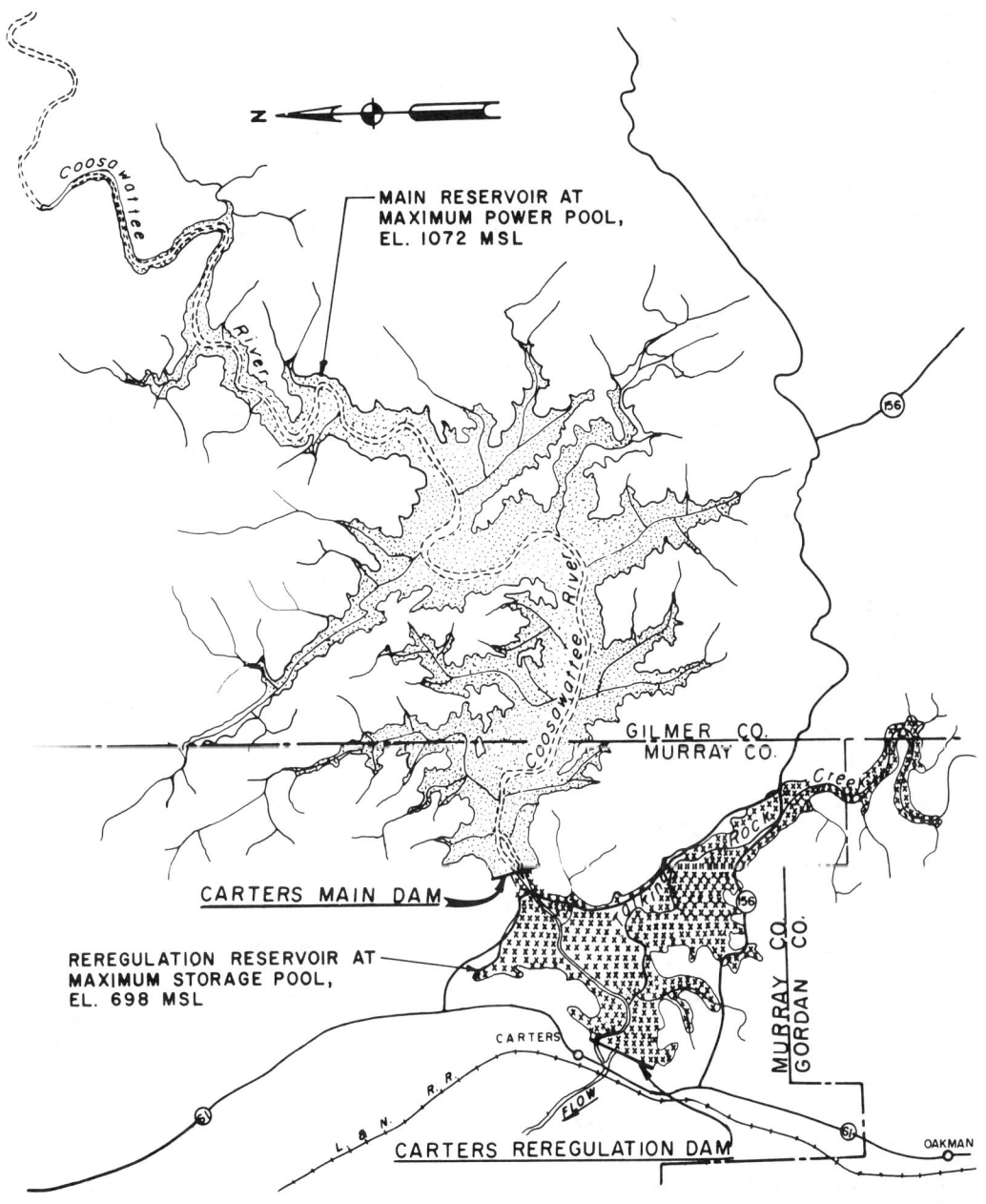

FIG. 2.—RESERVOIR MAP

road, site clearing and test fills to provide design data for the main dam. The presently scheduled completion date is January, 1974.

GEOLOGY

The Carters Project is located at the irregular escarpment which separates the Piedmont province from the Apalachian Valley province. The main dam is about 1/2 mile upstream of this escarpment in a 600-ft deep gorge. Specifically, the main dam and reservoir are in the Dahlonega Plateau subdivision of the Piedmont province. This region is characterized by rugged, mountainous terrain. One of the major thrust faults of the United States, the Cartersville fault, is located along the boundary escarpment. The escarpment is the result of this fault and of differential erosion between the harder crystalline rocks of the Piedmont and the softer sedimentary rocks of the Apalachian Valley. The reregulation dam will be built within the Apalachian Valley province and about 1.8 miles downstream from the main dam (Fig. 3). Broad valley lands with occasional north-trending ridges typify this province.

The Ocoee series of rocks underlies the main dam site and extends eastward through the reservoir area. At the project site this formation consists

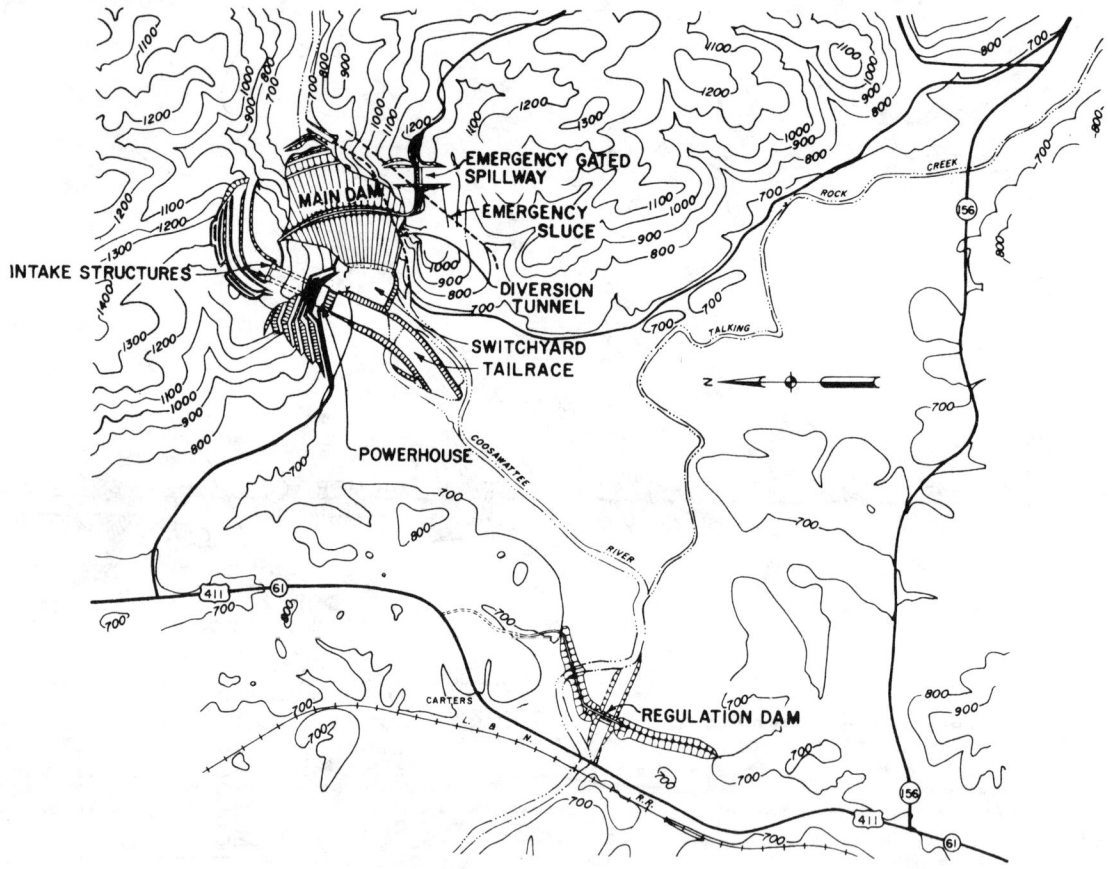

FIG. 3.—SITE PLAN

mainly of quartzite, argillite, and phyllite with quartzite predominating. The quartzites are generally massive rocks with compressive strength ranging to about 27,000 psi. When shot, they break into blocky shapes with abrasive faces. The term argillite is used to define a group of rocks whose properties lie between the quartzites and the phyllites. These rocks are hard and resemble slate, except that they do not cleave into thin slabs but quarry as semislabby to massive rock. The phyllites are moderately hard rocks that have a characteristic wavy foliation and a silky sheen on fracture surfaces. With a compressive strength of about 10,000 psi, they are not as strong as the other rocks, but their strength far exceeds the foundation requirements for this dam.

The Conasauga formation, consisting of shale with a few limestone beds, underlies the Apalachian Valley province in the reregulation dam area. The topography is controlled to a large extent by the geologic structure. The lower, nearly flat, valley areas are developed upon the shale, and the higher plateaus and north-trending ridges are generally capped by harder limestone which has resisted erosion. The spillway for the reregulation dam will be founded on the highest portion of a sound moderately hard shale formation (Fig. 3).

HYDROLOGY

The Coosawattee River basin has a temperate climate with long, warm summers and short, usually mild winters. Subfreezing temperatures occur frequently but are of short duration and subzero temperatures are rare. The frost-free season averages 200 days a year and the mean annual temperature is about 60°.

The watershed lies in a region with abundant precipitation, which is well-distributed throughout the year. Average annual rainfall is 60 in., and about 49% of this appears as runoff. Light snow is not unusual in the winter, but contributes little to the total precipitation. Intense flood-producing storms, usually of the frontal type, occur mostly in the winter and spring. The storms that occur in the summer and fall are usually thunderstorms with high intensities over smaller areas.

The steep slopes of the mountains and channels produce flashy runoff characteristics. Heavy local storms coupled with these characteristics have produced flash floods which endangered lives in the area on a number of occasions. It is not uncommon for the Coosawattee River to rise 10 ft in a few hours. The wide valleys and flat slopes produce entirely different characteristics in the region downstream of the project.

RESERVOIR REGULATION AND DEVELOPMENT

The main dam will form a reservoir with an area of 3,224 acres at maximum power pool El. 1072. At this level the reservoir will have a length of about 11 miles and a shoreline of 55 miles (Fig. 2). During the dryer fall months, the pool may be drawn down as much as 20 ft to maintain power production, but the minimum power pool at El. 1022 will be reached only in the event of a severe drought.

The project will be operated to provide the maximum practicable flood control benefits. Generally speaking, all floods coming down the Coosawattee will be impounded in the main reservoir until the downstream crest has passed Resaca, a town on the Oostanaula River about 30 miles downstream from Carters. Minor releases to meet minimum power commitments will be permitted, but will be stored in the reregulation reservoir until the flood is over. As an additional benefit, it is anticipated that regulated releases for normal operation will mitigate pollution in the Coosa River at Rome, Georgia.

The water of the Coosawattee is now of good quality. The Federal Water Pollution Control Agency is presently making a preimpoundment water quality survey, under contract, for use as a benchmark. Although experience is limited, present information indicates that pumping has a beneficial effect on both the temperature and the oxygen content of the water. It is also anticipated that impoundment in the relatively broad and shallow reregulation pool and the turbulence in the stilling basin below the reregulation dam will have further beneficial effects. For these reasons no special water quality features will be built initially. However, water quality will be monitored closely, particularly during the early years of operation, and measures will be taken later if necessary.

It is estimated that 1,750,000 people live within a 75-mile driving distance of the Carters Reservoir. Because of its location in the Blue Ridge Mountains, the lake will be one of the most beautiful in Georgia and will provide opportunities for activities such as swimming, boating, water skiing, fishing, hiking, picnicking and camping. It is anticipated that total attendance will reach 1,250,000 people annually. To prepare for this influx, seven public use areas covering 1,610 acres will be developed and 14 miles of access roads will be built. Water supply and sanitary facilities will be provided for the camping areas. One or more areas will be leased to concessionaires for marina and resort development. The remaining government-owned land around the reservoir will be open for hiking and primitive camping.

PROJECT DESCRIPTION

General.—As previously noted, the main dam is in a deep gorge about 1/2 mile upstream of the escarpment (Fig. 3). The emergency spillway will be in a channel cut through the top of Horn Mountain, which forms the left bank of the gorge. The diversion tunnel was bored through the base of this mountain around the dam, and the emergency sluice will be built in another tunnel to be driven south of the diversion tunnel. Over on the top of Bell Mountain, the right bank, a huge bench was cut from the peak at El. 1370 to El. 1112.3 to provide fill for the dam and to accommodate the headrace, which is cut even deeper into the mountain. The powerhouse will be in a bench cut into the foot of Bell Mountain. About 1.8 miles downstream from the main dam, the reregulation dam will be built to smooth out daily power surges and serve as a reservoir for the pump turbines.

River Diversion.—River flow was diverted around the dam through a 23-ft, D-shaped (or straight-legged horseshoe) tunnel excavated 2,407 ft through the left abutment. This tunnel is unlined but the upstream portal has a reinforced concrete structure with slots for the closure stoplogs. To increase hydraulic efficiency, this structure was given a constant-radius bellmouth. For reservoir filling the stoplogs will be placed and the tunnel will be closed with a concrete plug under the dam centerline.

The diversion tunnel and upstream cofferdam were designed to control the third largest flood of record, which had a maximum discharge of 18,400 cfs. Studies indicated that the cofferdam could be built to El. 760, which would contain the design flood with 1.5 ft of freeboard, within the 4-to-5-month period prior to the usual flood season. (Design freeboard was 3 ft, but the remainder was to be added after start of the flood season.)

On October 4, 1964, with the cofferdam crest at El. 698, heavy rains upstream caused the river to rise 4 ft above the temporary crest. The flood subsided within 24 hr and caused little damage. Apparently, only a minor amount of fines was washed out of the rockfill and the contractor had a few days of pumping and cleanup before resuming normal operations.

The upstream cofferdam was designed as an integral feature of the main dam, and was incorporated as construction progressed. The downstream cofferdam was simply fill-dumped into the old river channel. It was used for access during first-stage construction and later covered by the switchyard fill.

Rock-Fill Dam.—The main dam is a rolled-rock structure with an impervious core. The crest of the dam is at El. 1112.3, which was set by a 5-ft freeboard requirement above the spillway design flood peak elevation. The total height of the dam above its foundation is 440 ft, which makes it the highest rock dam east of the Mississippi. The length along the crest is about 1800 ft

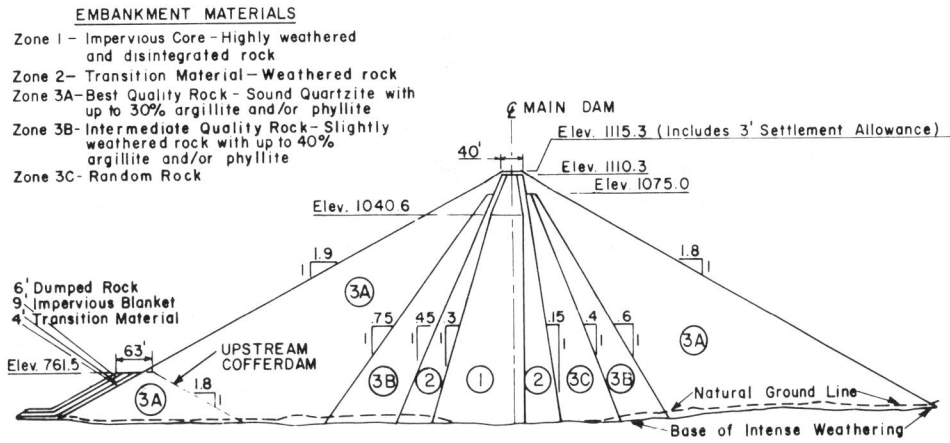

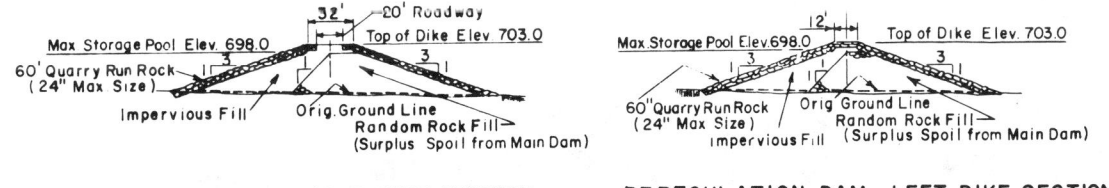

FIG. 4.—TYPICAL DAM SECTIONS

and its total volume is about 15,000,000 cu yd, of which 12,000,000 cu yd is rock. The main dam was topped-out on November 13, 1969.

The final design section of the dam is shown in Fig. 4. The core is an impervious material which classifies as a brown inorganic plastic silt with gravel, while the other zones are rock of varying degrees of hardness and weathering. Most of the dam material was obtained from the headrace, spillway, and powerhouse excavations. The rock in these areas is primarily quartzite, with some phyllite and argillite. Two types of instrument systems are being embedded in the main dam, an Idel position measurement system to determine vertical and horizontal displacement and a Glotzl system to determine the vertical and horizontal soil pressure at selected points within the dam and pore-water pressure in the impervious core.

During preliminary design it was assumed that the valley foundation would be relatively smooth. Inspection of the stripped foundation disclosed a rough

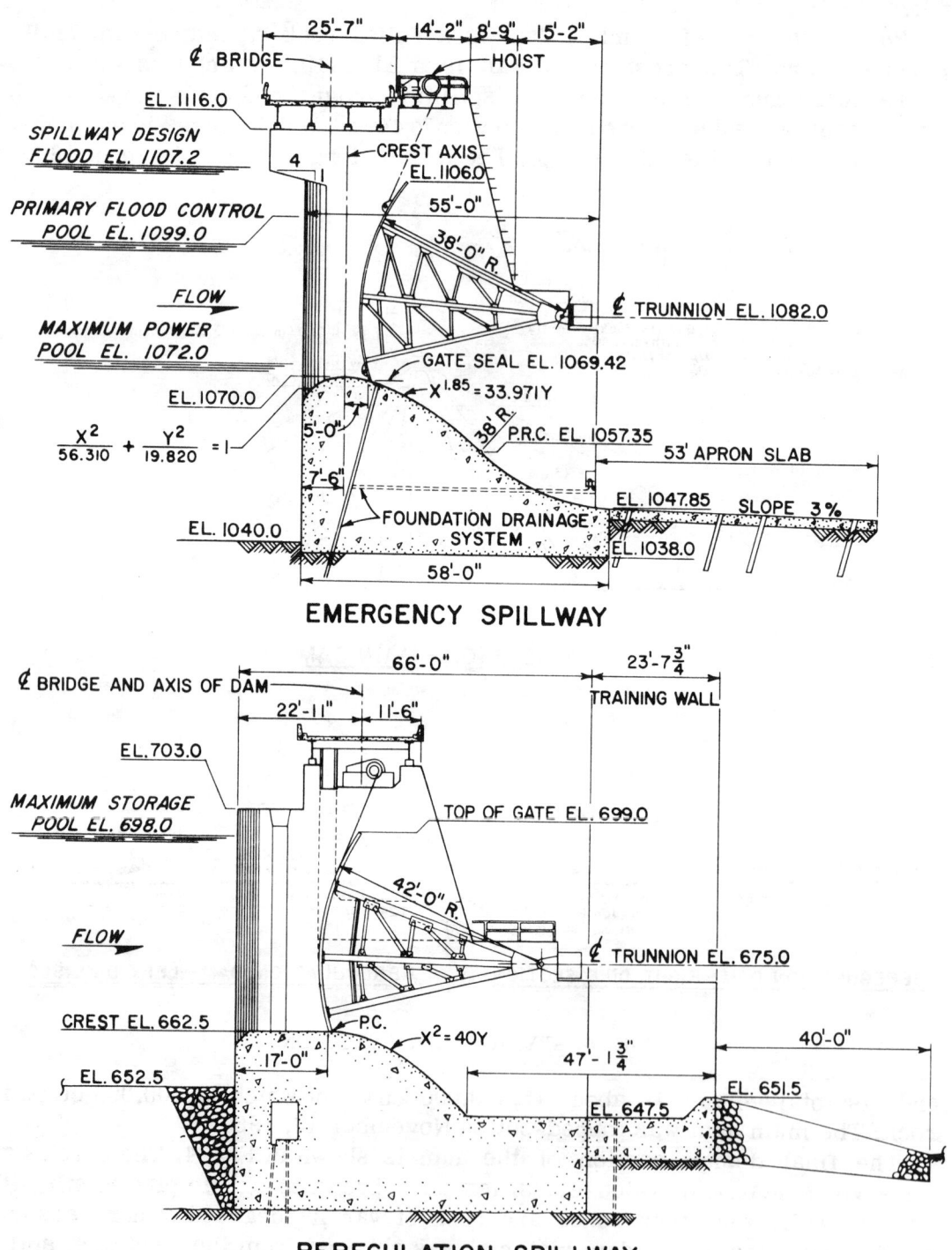

FIG. 5.—SPILLWAY SECTIONS

surface with lateral ridges that gave a higher friction factor. Using a conservative estimate of the new factor, new stability computations were made which allowed steepening the upstream dam slope to 1 on 1.9 and the downstream to 1 on 1.8 for a saving of nearly 1,300,000 cu yd of rockfill.

Emergency Spillway.—The emergency spillway is located on the left, or south, bank about 500 ft east of the dam. (See Fig. 3.) The control structure will be a concrete gravity-type ogee weir shaped to current Corps standards (Fig. 5). The crest will be at El. 1070, 2 ft below the level of the maximum power pool. Flow over the crest will be regulated by five tainter gates, 42 ft long and 36 ft high. These gates will have individual electric hoists. This spillway will handle, with adequate safety factors, the peak spillway design flood discharge of 199,200 cfs. (The spillway design flood is the probable maximum flood for this location. This flood, in turn, is based on the runoff from the hypothetical probable maximum precipitation.)

The spillway channel is one of the quarry areas and, consequently, the shape of the upstream portion has varied with the quantity estimates for the main dam. It is presently being held firm at the minimum section which meets our hydraulic requirements. This is a trapezoidal channel with the bottom the width of the crest, and level at El. 1040 with 1 on 10 sides slopes.

This spillway was originally planned as a service spillway which would be used annually for flood regulation. As the capacity of the powerplant was increased, it was decided that all floods possible would be regulated through the turbines. This decision made the spillway an emergency spillway, which would be required only for floods greatly exceeding the 100-yr flood magnitude. Most of the service-type features, which included a flip-bucket downstream of the crest and extensive clearing and cascades in the discharge valley, were accordingly eliminated.

Emergency Sluice.—A low-level sluice through Horn Mountain, which forms the left abutment, will permit lowering the pool below the crest of the emergency spillway or the turbine intakes if this necessity ever arises. This sluice will have an average discharge capacity of 10,000 cfs at the volumetric midpoint pool elevation of 950 msl.

The final decision to include an emergency sluice was made after construction was well under way and it was no longer feasible to construct a sluice in the diversion tunnel. A new 21-ft, D-shaped tunnel 2,900 ft long will be driven south of the diversion tunnel (Fig. 3). This tunnel will be unlined except for the midportion where there will be a 116-ft long concrete plug with dual gate passages 10.5 ft high and 5 ft wide followed by about 200 ft of impact paving for the floor and the walls below the springline. A 5-ft diam vertical shaft from the surface in the vicinity of the emergency spillway will provide access to the gate machinery. The gates will be remotely operated from a station near the spillway. There will be a reinforced concrete intake structure with stoplog slots at the tunnel entrance.

Headrace.—The headrace is a deep channel cut through the rock around the right end of the dam. It is about 1,600 ft long from the point where the excavation begins to its termination at the intake structures. This headrace has two major purposes: First, it minimizes the length of the penstocks, and, second, it serves as the major quarry area for the main dam. It became apparent rather early that the quarry requirements would result in a channel which was oversized insofar as the hydraulic requirements were concerned. Some of the major dimensions have therefore tended to fluctuate a bit with the estimates of the dam fill requirements.

The final shape of the headrace is shown on Fig. 3. The 300-ft wide section at the intake structures had to be held since it was controlled by the spacing of the units in the powerhouse. Excavation was already down to the wide berm of El. 1112.3 when the need for adjustment became apparent so this elevation was held firm. The bottom was held at El. 979, about 1 ft below the intake inverts. The left bank was held firm to simplify layout work and excavation quantities were adjusted by changing the right bank. The final width of the channel from the entrance to the beginning of the intake transitioning is 130 ft.

The vertical rock faces which the intake structures will abut have been excavated using presplitting methods to obtain faces of undisturbed rock within a tolerance of plus-or-minus 6 in. Rock bolts 30 ft long will be grouted on about a 10-ft grid pattern into these vertical rock faces. Other sound rock faces in the headrace have a tolerance of plus-or-minus 12 in.

Penstocks and Intake Towers.—Each penstock will have its own reinforced concrete intake tower (Fig. 6). These will be all alike except for minor variations in towers 3 and 4, to allow for flow reversal during pumping. Each tower will be 31 ft wide and 51 ft deep at the base, and will have a height of 138.5 ft above the base. There will be a reinforced concrete transition, which will be entirely inside the rock, between each intake structure and penstock. Each passage will be closed by a 14-ft by 20.5-ft upstream-sealing tractor gate. Each gate will have an electric hoist, with an auxiliary hydraulic lowering device for emergency closure, on the deck at El. 1112.5.

There will be a work bay at El. 1080 in each tower for servicing the head gates. Inspection access for the gates and penstocks will be provided by a portable electric manlift, which will operate in a recess immediately downstream of the gate on the centerline of the passage. One set of steel stoplogs will be provided, to be installed where needed with a road crane. Trash racks at each intake will be removable with a road crane.

Four steel-lined penstocks are being constructed between the intake structures and the powerhouse. Each penstock will have an inside diameter of 18 ft (as determined by an economic study) and an overall length of about 835 ft. The penstocks were designed to be as alike as possible, but there are necessary differences between those for the pump-turbine units and those for the conventional units. Penstocks 3 and 4, for the pump-turbines, will be 9 ft deeper at the powerhouse end. They will also have a long diffuser-type transition between the penstock section and the 13.5-ft diam entrance-discharge section for the pump-turbines instead of the more abrupt convergence between the penstock section and the 16-ft diam entrance section on the conventional turbines. Also, the liner plate thickness will be a little greater through most of the length of penstocks 3 and 4 because of greater waterhammer pressures produced by the pump-turbines.

Steel for the linings will conform to ASTM Standard A 516-67, Grade 60. Normally, a higher-strength steel would be required for a project with a penstock design head (including waterhammer) as great as Carters'. However, the tunnels are covered with an adequate depth of excellent rock, so the steel design assumed the rock was capable of resisting 37.5 % of the internal penstock pressure except for 100 ft at the lower end where rock cover was inadequate. Also, about half of the length of the liner was designed for minimum thickness rather than internal water pressure. For these reasons the selected steel competed economically with the higher-strength steels.

The penstock liners will be erected within 23-ft diam tunnels. After the

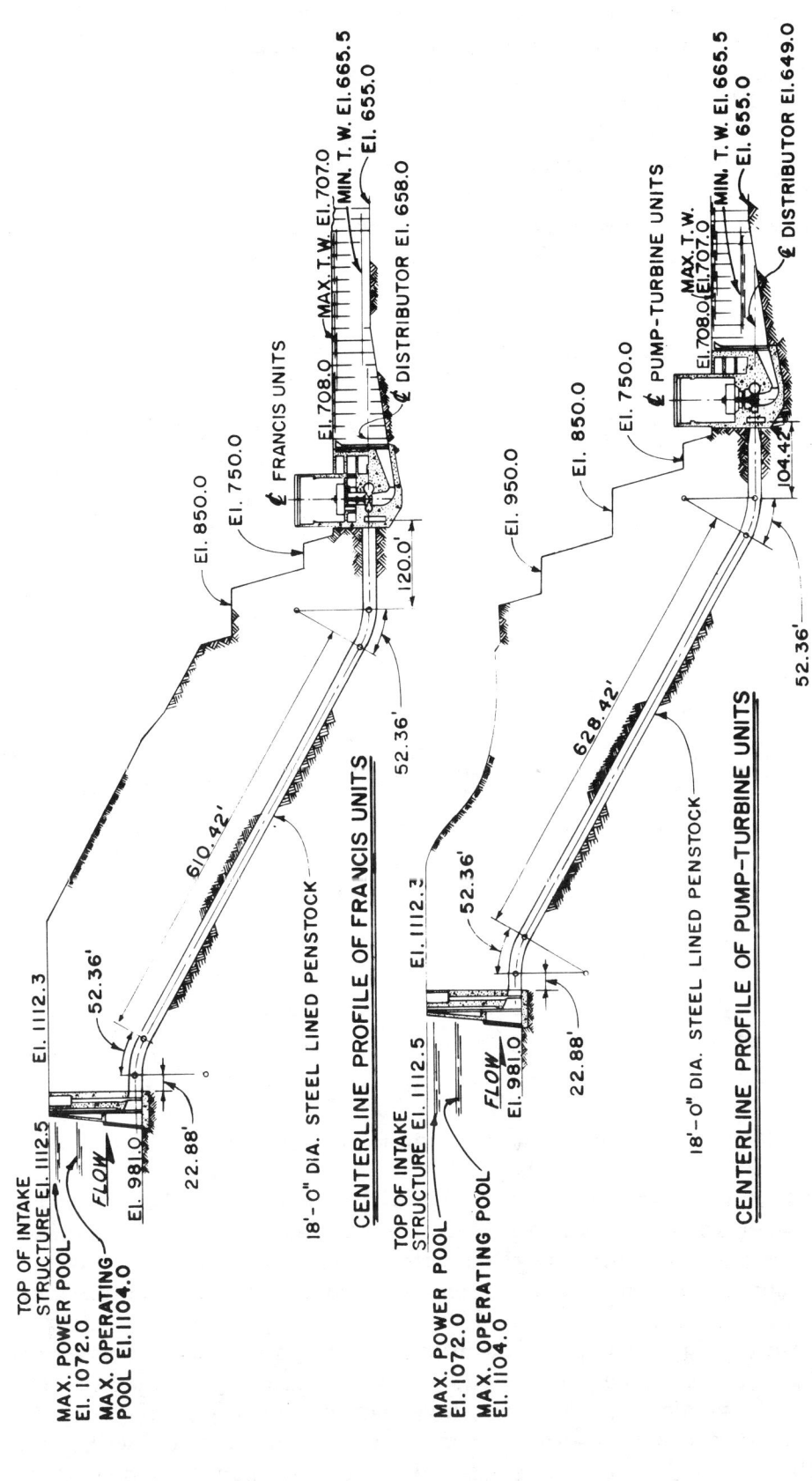

FIG. 6.—POWERPLANT SECTION

steel liners are in place, the space between the liners and the rock will be filled with concrete, after the concrete has set, contact grouting will be used to fill all voids between concrete and rock and concrete and liner.

The penstock alignment adopted is the most economical of the several profiles studied. Since the cost of the steel and concrete in the liners represents about 70 % of the total penstock cost, the alignment adopted is that with the shortest overall length. The 30° tunnel slope of the selected alignment increased the tunnel construction cost because it was difficult to muck; however, the overall cost is still the lowest.

Powerhouse.—The powerhouse structure will be 390 ft long and 115 ft wide. It will house four hydrogenerator units, which are described in greater detail

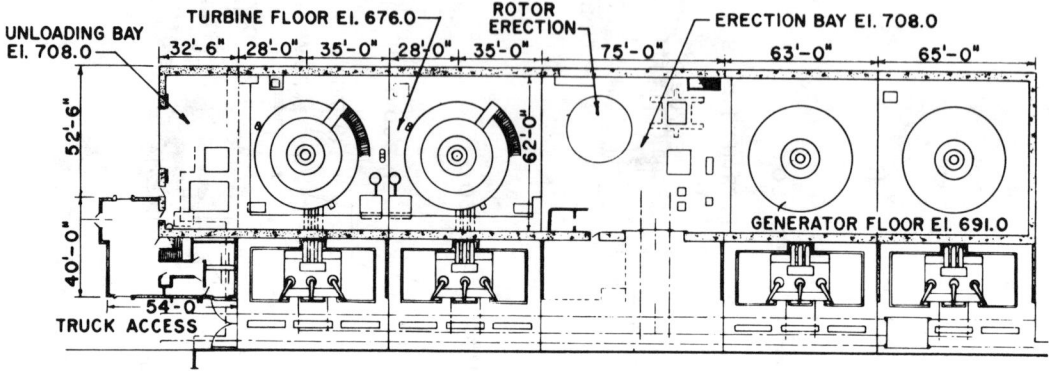

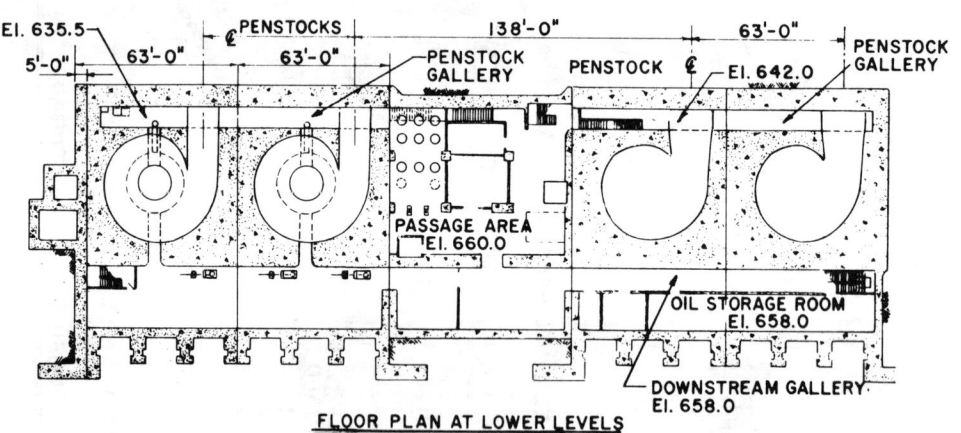

FIG. 7.—POWERHOUSE FLOOR PLANS

in the following paragraphs, and will include an erection bay, an unloading bay, and an entrance wing. The exterior architecture will be divided into two horizontal zones. The lower portion will have an exposed aggregate surface while the upper portion will be formed concrete, using plywood forms, with V-groove rustication.

The erection bay will be in the middle of the powerhouse between units 2 and 3 (Fig. 7). This came about because it was originally planned to build the powerplant in two stages. The first stage was to include only the two conventional units (units 1 and 2), with their penstocks and intakes, and with a powerhouse which included only these units and an erection bay. The pump-turbines (units 3 and 4) with all their appurtenances would be installed later in the second stage. When the decision was made to install all units initially, the first stage powerhouse was well advanced in design. For reasons of economy the major portion of this work was retained.

The conventional units, with the distributor centerline at El. 658, will be served with a generator floor at El. 691 and a turbine floor at El. 676 (Fig. 8). However, the pump-turbine distributor center line will be at El. 649, and, in addition, it is considered desirable to have the reversible units as close-

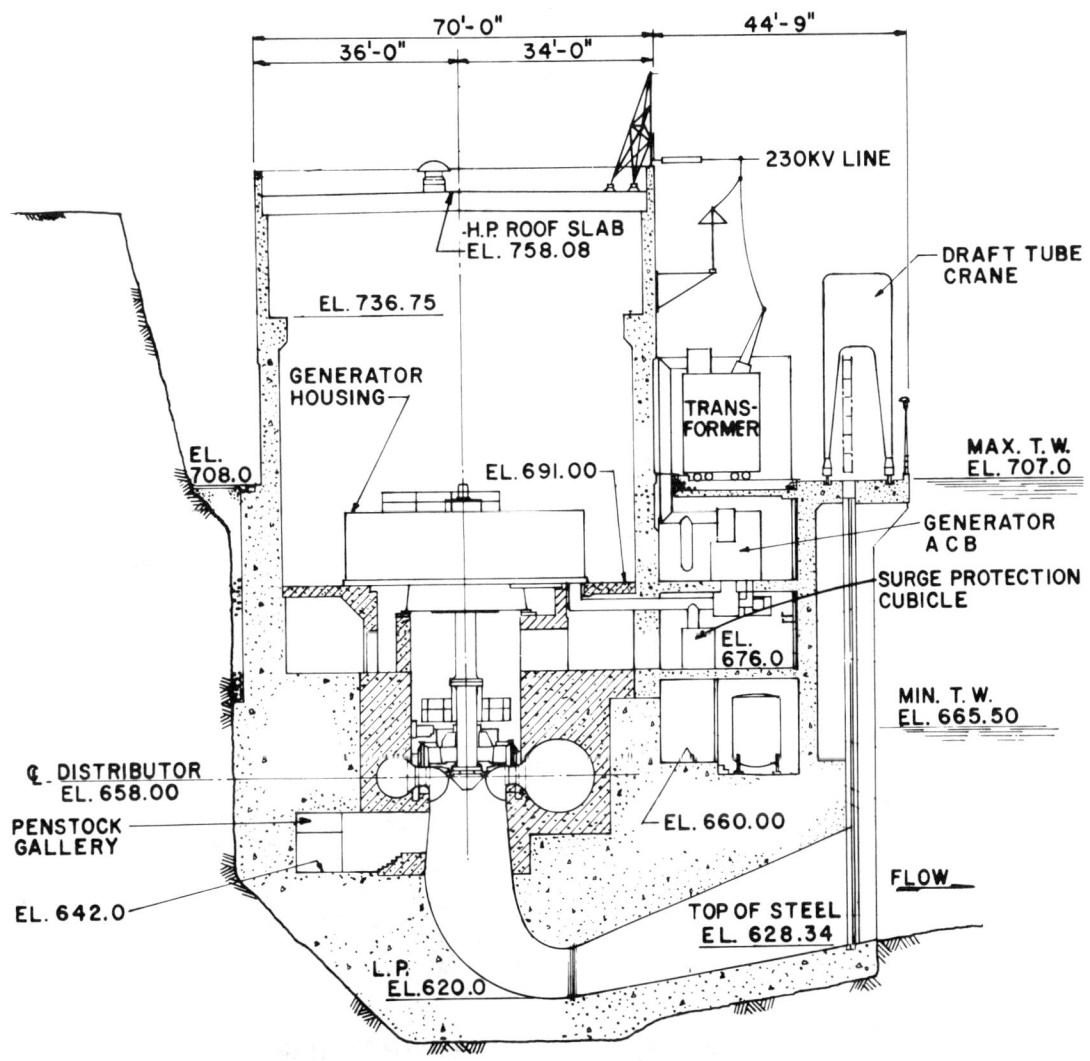

FIG. 8.—TRANSVERSE SECTION—FRANCIS UNITS

coupled as possible for stiffness. It will therefore be possible to place the generator floor for the reversible units at El. 676, the same elevation as the turbine floor for the conventional units (Fig. 9). There will be no floor at El. 691 over units 3 and 4. Also, there will be no turbine floor, as such, but access to the pump-turbines will be provided by passages from the service bay floor at El. 660.

All units will be installed and maintained with a 350-ton bridge crane. The crane will have two trolleys, each of which will have a 175-ton sister hook and a 25-ton auxiliary hook. Maximum capacity will be developed with a lifting beam.

The service bay will be located on the downstream side of the structure below the draft tube deck and will house sewage treatment, water treatment, oil storage, and other services. An extensive collector drain system along the upstream wall of the powerhouse will be provided to reduce the differential head to 10 ft.

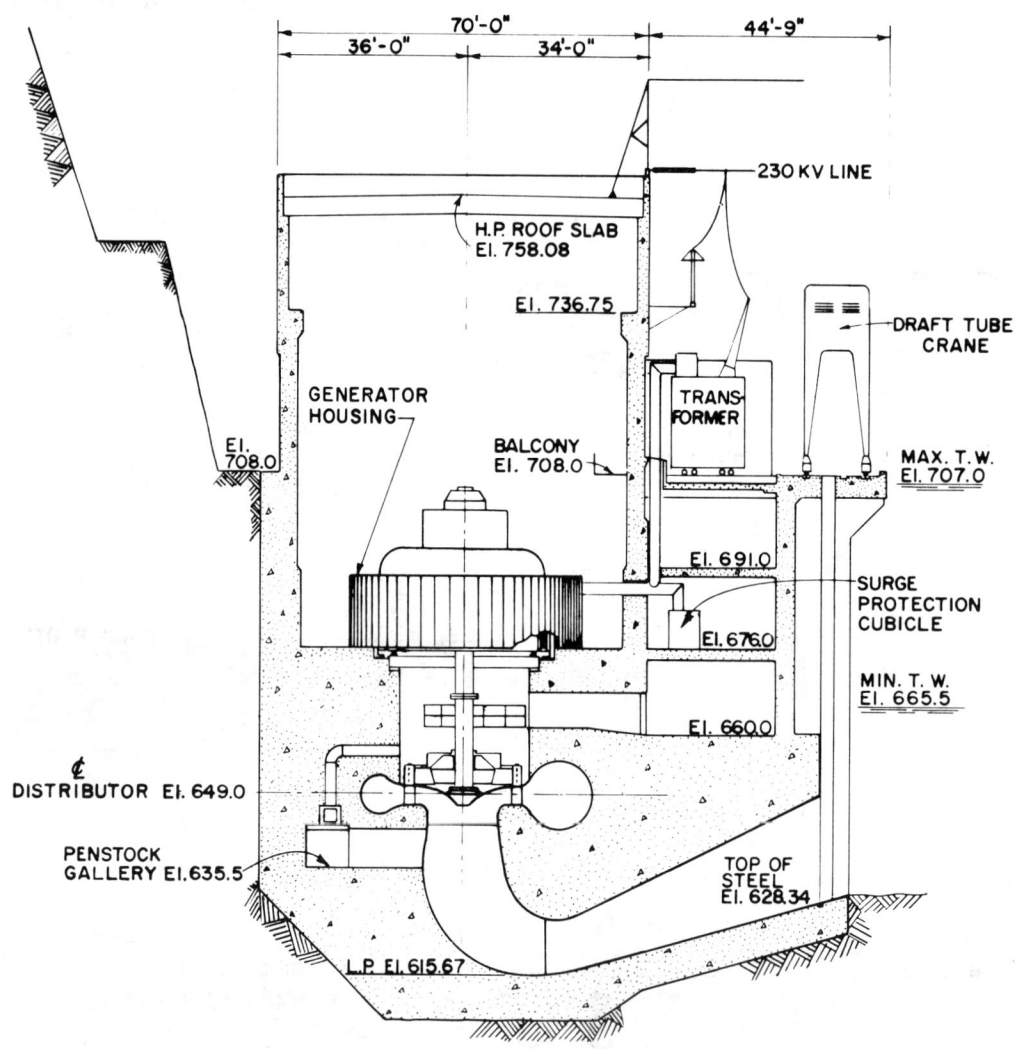

FIG. 9.—TRANSVERSE SECTION—PUMP TURBINE UNITS

Tailrace.—The tailrace will have a bottom width of 317 ft, which is the out-to-out dimension from the unit 1 to unit 4 draft tube piers, plus 1 ft for excavation tolerance. This width was chosen during early planning, when it was proposed to construct the powerhouse in two stages as it simplified the cofferdamming. The draft tubes will discharge from the powerhouse at an invert El. of 629, and the tailrace will slope upward at a 1 on 6 slope from the draft tubes to a bottom elevation of 655 and then hold this elevation for the remainder of its length. The bottom elevation of 655 will keep tailrace velocities acceptably low at the beginning of generation when the reregulation pool is at its minimum level, El. 665.5.

Reregulation Dam.—The reregulation dam, consisting of a concrete gated spillway and composite rock-and-earth dikes, will be located approximately 1.8 miles downstream of the main dam. This site is below the confluence of the Coosawattee River and its major tributary in this vicinity, Talking Rock Creek. This necessitates designing the spillway to handle floods from Talking Rock Creek, but no alternate site provided adequate storage. The spillway will be a concrete gravity-type structure 208 ft long with four 42-ft wide tainter gates (Fig. 5). The gates will be raised and lowered by individual electric hoists located on top of the piers. Automatic gate control is planned to maintain steady flows downstream of the project while the level of the reregulation pool fluctuates during generation and pumping. Pool fluctuation will be about 15 ft with a daily operating cycle, somewhat more for a weekly cycle, and can reach a maximum of 30 ft with a severe flood on Talking Rock Creek. The two center gates will be controllable in 0.05-ft increments for the first foot of movement to provide accurate release of minimum flows.

A concrete non-overflow wall will be provided at each end of the spillway to transition to the embankment. The dikes will have a combined length of about 2,900 ft to span the flood plain about 700 ft upstream of U.S. Highway No. 411. The top of the dikes will be at El. 703 to provide 5 ft of freeboard above the maximum storage pool (Fig. 4). As the pool happens to be the same as the standard project flood, the dikes will probably never be overtopped. The downstream half, approximately, of each dike will be random rock fill which is surplus from the main dam. The upstream portion will be impervious fill. Both dike faces will be riprapped with 60-in. quarry run stone.

POWER UNITS—CONVENTIONAL AND REVERSIBLE

Units 1 and 2 are rather conventional hydroelectric generators (Fig. 8). The turbines are of the Francis type rated at 172,000 hp each while operating at 163.6 rpm under a net head of 345 ft. The runner has a diameter of about 154 in. at the center line of the distributor. This center line will be installed at El. 658. The turbines will have no guard valve between the penstock and the scroll case, but will depend entirely upon rubber seals on the wicket gates to prevent leakage when the units are not operating. These turbines are being supplied by Newport News Shipbuilding and Dry Dock Company.

The generators are rated at 131,579 kva each at 0.95 power factor and 13.8 kv between phases. The exciters are larger than usual to provide for pump-unit startup. While nominally rated at 375 volts, they will put out 500 volts at 80 % speed and maintain the output voltage proportional to speed for a period of three minutes during pump-unit startup. These generators are being supplied by Allis-Chalmers.

Units 3 and 4 are Francis-type, reversible single-stage pump-turbine generator-motor units (Fig. 9). As turbines the units are rated at 173,000 hp at 345 ft net head. As pumps they are rated at 4435 cfs at 347 ft total head. Operating speed is 150 rpm in both modes. Runner diameter is about 246 in. at the distributor centerline. Due to their size, these runners are being constructed in sections and will be finally assembled on-site. Distributor center line will be at El. 649, 9 ft lower than that of the conventional units. They will also have rubber seals on the wicket gates. These pump-turbines will be supplied by Allis-Chalmers.

The generator-motors will have the same rating as generators as the conventional units. As motors they will have a rated output of 185,000 hp. Specifications for the generator-motors were issued in September 1968, but the bid opening was postponed until model testing of the pump-turbines could be completed to assure the best match of characteristics. These tests are now complete and it is presently planned to award this contract in July, 1970.

As pump-turbine units are inherently rougher than conventional units, each pump-turbine and generator-motor will be as close coupled as possible. The generator-motor will have two guide bearings, and the thrust bearing will be located above the motor. Since the pump-turbines will be set lower than the conventional turbines, it will be possible to place the generator floor for units 3 and 4 at the same elevation as the turbine floor for units 1 and 2.

Either of the conventional units can be used to start the reversible units for the pumping cycle. Assuming all units are shut down, the starting cycle, very briefly, will be as follows:

1. The starting unit will be automatically started to operate isolated from the system at a fixed gate position at about 80 % speed at no load and zero voltage.

2. In the required sequence, the auxiliaries, bearing oil pumps, cooling water pumps, etc., will start on the unit to be started; high pressure oil will be supplied to the generator-motor thrust bearing; the impeller-runners will be unwatered; the generator-motor brakes will be released; and the exciter field rheostats and voltage regulators on both units will be automatically run to the proper settings.

3. The two units will be connected through the 230 kv auxiliary bus and excitation applied to the starting generator. Constant exciter field current will be maintained automatically on the driving generator to produce initially 150 % of rated motor current for the generator-motor.

4. The generator-motor will start as an induction motor on its amortisseur winding. As it accelerates the starting generator speed will fall. When the motor reaches about 95 % of the starting generator speed (about 40 % of synchronous speed), excitation will be applied to the motor field and the two units will pull into synchronism.

5. When the units have settled into synchronous operation, the reduced frequency induction motor start will change to a synchronous motor start. The wicket gates for the starting unit will open automatically under governor control to accelerate both units to rated speed.

6. At 95 % of rated speed the pump-turbine casing will be rewatered to prime the pump-turbine and load the generator-motor for pumping at rated speed against the closed wicket gates.

7. The automatic synchronizing gear will take over, match both units to the system, and close the breaker connecting both units to the main 239 kv bus and the system.

8. Load will be transferred to the system and the unit will shut down automatically.

9. The pump-turbine wicket gates will open to attain the desired pumping rate.

This analysis is not intended to imply that all problems connected with the pump-turbines have been solved. As an alternative to step 6, the pump-turbine can be rewatered after the unit is connected to the system. It is tentatively planned to blow down the tailwater after the wicket gates are closed for shutdown.

COSTS AND BENEFITS

As of July 1969, cost estimates on current price levels indicate a total cost of $92.2 million for the Carters Project. The major contracts for this

TABLE 1.—CARTERS PROJECT MAJOR CONTRACTS

Item (1)	Contractor (2)	Amount, in dollars (3)
(a) Construction Contracts		
Access road, site clearing	Ledbetter Bros., Inc.	144,495
Diversion tunnel portals	Ledbetter Bros., Inc.	177,176
Diversion tunnel	Cowin & Co., Inc.	601,264
Main dam, Phase I	Roy Ryan Sons Co., Inc.	1,679,854
Main dam, Phase II	Clement Bros., Co.	18,126,734
Access road and saddle dikes	Phillips & Jordan, Inc.	1,015,710
Warehouses	Engineered Bldgs., Inc.	115,700
Penstock tunnels	W. L. Hailey & Co., Inc.	1,090,600
Penstock intakes and liners	Al Johnson Construction Co.	7,859,743
Emergency spillway	—	1,897,200[b]
Low-level sluice	—	2,530,000[b]
Reregulation dam	—	4,356,000[b]
Powerhouse[a]	—	12,109,000[b]
Reservoir clearing	—	1,672,000[b]
(b) Supply Contracts		
Francis turbines (2)	Newport News Shipbuilding & Dry Dock Co.	1,641,630
Generators (2)	Allis-Chalmers	2,534,395
Pump-turbines (2)	Allis-Chalmers	2,525,185
Generator—motors (2)	—	3,027,000[b]
Transformers	—	1,738,000[b]
Governors	—	330,000[b]
Switchgear	—	2,333,000[b]
Powerhouse crane	—	1,016,000[b]
Draft tube crane	—	180,000[b]
Switchyard steel and equipment	—	2,437,000[b]
Intake trash racks	—	150,000[b]

[a] Design of powerhouse: first stage by Tippetts-Abbott-McCarthy-Stratton; second stage by Mobile District, Corps of Engineers.
[b] Estimated amount.

project, both actual and scheduled, are listed in Table 1. If the present schedule is maintained, the project will be essentially complete in January, 1974.

The total annual charges for the project are estimated at $4,628,000. The annual monetary benefits are estimated at $9,776,000, of which $8,611,000 is for power and the balance is attributed to flood control, recreation, and area redevelopment. These figures will yield a benefit/cost ratio of 2.1.

The Carters Project was given a high priority for development when early studies indicated it would have a higher b/c ratio than others on the system. The previously given data, which reflect current costs, substantiate this decision.

CONCLUSIONS

The topography and geology of the region made the Carters site an attractive location for a dam. The proximity of the Atlanta load center made a high capacity peaking plant desirable, while the relatively small annual flow of the Coosawattee River dictated a pumping plant to attain this capacity. This project will add 500 Mw of needed peaking capacity to the network.

As a multiple-use project, Carters Dam will also provide significant flood protection downstream and will add substantially to the recreational opportunities in North Georgia.

APPENDIX.—BIBLIOGRAPHY

"Basic Hydrology," Carters Dam Design Memoranda, Mobile District, Corps of Engineers, No. 1, Nov., 1961.

"Emergency Gated Spillway," Carters Dam Design Memoranda, Mobile District, Corps of Engineers, No. 9, Feb., 1965.

"Gates Spillway for Reregulation Dam," Carters Dam Design Memoranda, Mobile District, Corps of Engineers, No. 15, Aug., 1966.

"General Design," Carters Dam Design Memoranda, Mobile District, Corps of Engineers, No. 5, July, 1963, and supplement Sept., 1964.

"Intake Structures for Powerhouse," Carters Dam Design Memoranda, Mobile District, Corps of Engineers, No. 17, Oct., 1966.

"Penstocks," Carters Dam Design Memoranda, Mobile District, Corps of Engineers, No. 14, Supplement, June, 1968.

"Powerhouse Penstocks," Carters Dam Design Memoranda, Mobile District, Corps of Engineers, No. 14, Dec., 1965.

Robeson, F. A., and Crisp, R. L., Jr., "Rockfill Design—Carters Dam," *Journal of the Construction Division,* ASCE, Vol. 92, No. CO3, Proc. Paper 4906, September, 1966, pp. 51–74.

Spieth, J. T., "Carters Dam—Georgia, Reversible Pump—Turbine Units Nos. 3 & 4," Powerhouse Supervisors Conference 1967, Mobile, Alabama, Oct. 17–19, 1967.

"Carters Dam; Spot of Beauty in the Wilderness," *Atlanta Journal,* Feb. 22, 1961.

"Radio Signals Help Design Dam and Warn of Movement," *Engineering News Record,* Aug. 25, 1966.

"Carters Dam Brings a New Tunneling Angle," *Engineering News Record,* Feb. 6, 1969.

Aerial View of Dam and Powerhouse

Carters Pumped Storage Plant

BIBLIOGRAPHY

1. "Basic Hydrology," Carters Dam Design Memoranda, Mobile District, Corps of Engineers, No. 1, November 1961.

2. "Emergency Gated Spillway," Carters Dam Design Memoranda, Mobile District, Corps of Engineers, No. 9, February 1965.

3. "Gates Spillway for Reregulation Dam," Carters Dam Design Memoranda, Mobile District, Corps of Engineers, No. 15, August 1966.

4. "General Design," Carters Dam Design Memoranda, Mobile District, Corps of Engineers, No. 5, July 1963, and supplement September 1964.

5. "Intake Structures for Powerhouse," Carters Dam Design Memoranda, Mobile District, Corps of Engineers, No. 17, October 1966.

6. "Penstocks," Carters Dam Design Memoranda, Mobile District, Corps of Engineers, No. 14, Supplement, June 1968.

7. "Powerhouse Penstocks," Carters Dam Design Memoranda, Mobile District, Corps of Engineers, No. 14, December 1965.

8. Robeson, F.A. and Crisp, R.L., Jr., "Rockfill Design - Carters Dam," Journal of the Construction Division, American Society of Civil Engineers, Vol. 92, No. CO3, Proc. Paper 4906, September 1966, pages 51-74.

9. Spieth, J.T., "Carters Dam - Georgia, Reversible Pump - Turbine Units Nos. 3 and 4, "Powerhouse Supervisors Conference 1967, Mobile, Alabama, October 17-19, 1967.

10. "Carters Dam; Spot of Beauty in the Wilderness," Atlanta Journal, February 22, 1961.

11. "Radio Signals Help Design Dam and Warn of Movement," Engineering News Record, August 25, 1966.

12. "Carters Dam Brings a New Tunneling Angle," Engineering News Record, February 6, 1969.

13. "Design of Carters Pumped Storage Project, Journal of the Power Division, Proceedings of the American Society of Civil Engineers, June 1970.

Plant Name: **CASTAIC POWER PROJECT**

Plant location: Castaic, CA Los Angeles County	Owner: Los Angeles Department of Water and Power 111 North Hope Street Los Angeles, California 90012-2694
Rated capacity 1,250 MW Average static head 1,070 ft Plant efficiency 67.00 % Stored energy 12,000 MWh Number of units 6	Designers:
Construction time: 3 years Construction cost: $596 per kW Price level: 1988 First commercial power: January 1973 FERC project number: 2426	Plant Manager/Superintendent: Mary Beck 37700 Templin Highway Castaic, California 91301 (805) 257-3573
River or water source: West Branch California Aqueduct	

	UPPER RESERVOIR	LOWER RESERVOIR
DAM		
Type	Earth and rockfill	Earth and rockfill
Height (ft)	4,000	170
Crest length (ft)	1,090	1,960
Volume (yd³)	6,860,000	5,903,000
RESERVOIR		
Type	Constructed, Pyramid Reservoir	Constructed, Castaic Reservoir
Surface area (acres)	1,357	493
Usable power storage (acre-ft)	20,000	20,000
Power pool fluctuation (ft)	18.0	60.0
Operating levels		
Maximum (ft)	2,578.0	1,540.0
Minimum (ft)	2,340.0	1,480.0
Drainage area (miles²)	293.0	76.0
Seepage (ft³/s)		
SPILLWAY		
Design flood		
Return period (years)	1,000 yrs	1,000 yrs
Flow (ft³/s)		
Capacity (ft³/s)	150,000	28,750
Type	Broad crested weir	Broad crested weir
Gates		
Number	1	None
Type		
Width (ft)		
Height (ft)		
OUTLET WORKS		
Discharge capacity (ft³/s)	80,000	8,000
Number of water passages	8	1
Dimensions of water passages		
Height (ft)	12.00	
Width (ft)	8.00	
Diameter (ft)		21.00
Type of gates	Sluice gates	Outlet tower with sluice gates
Number of gates	8	8

Castaic Power Project - Page 1 (English)

Plant Name: **CASTAIC POWER PROJECT**

Plant location
Castaic, CA
Los Angeles County

Rated capacity   1,250 MW
Average static head   326.0 m
Plant efficiency   67.00 %
Stored energy   12,000 MWh
Number of units   6

Construction time: 3 years
Construction cost: $596 per kW
Price level: 1988
First commercial power: January 1973
FERC project number: 2426

River or water source: West Branch California Aqueduct

Owner: Los Angeles Department of Water and Power
111 North Hope Street
Los Angeles, California 90012-2694

Designers:

Plant Manager/Superintendent:
Mary Beck
37700 Templin Highway
Castaic, California   91301

(805) 257-3573

	UPPER RESERVOIR	LOWER RESERVOIR
DAM		
Type	Earth and rockfill	Earth and rockfill
Height (m)	1,219.2	51.8
Crest length (m)	332.2	597.4
Volume (m³)	5,244,849	4,513,552
RESERVOIR		
Type	Constructed, Pyramid Reservoir	Constructed, Castaic Reservoir
Surface area (Mm²)	5.49	2.00
Usable power storage (Mm³)	24.670	24.670
Power pool fluctuation (m)	5.49	18.29
Operating levels		
Maximum (m)	785.77	469.39
Minimum (m)	713.23	451.10
Drainage area (Mm²)	758.871	196.840
Seepage (m³/s)		
SPILLWAY		
Design flood		
Return period (years)	1,000 yrs	1,000 yrs
Flow (m³/s)		
Capacity (m³/s)	4,248	814
Type	Broad crested weir	Broad crested weir
Gates		
Number	1	None
Type		
Width (m)		
Height (m)		
OUTLET WORKS		
Discharge capacity (m³/s)	2,265	227
Number of water passages	8	1
Dimensions of water passages		
Height (m)	3.658	
Width (m)	2.438	
Diameter (m)		6.401
Type of gates	Sluice gates	Outlet tower with sluice gates
Number of gates	8	8

Castaic Power Project - Page 1 (Metric)

CASTAIC POWER PROJECT

INTAKES	UPPER INTAKE	LOWER INTAKE
Number	1	2
Type	Reinforced concrete gravity structure, multiple compartmented	West Branch of California Aqueduct and Castiac Creek.
Design discharge (ft³/s)	18,600	
Gross inlet area (ft²) (at trash racks)	352	
Bar racks		
spacing (in)		
shape		
depth/thickness (in)		
diameter (in)		
Emergency gates		
number	1	None
height/width (ft)	32.00 / 25.00	
type	Flat-leaf coaster type	
Service gates		
number	None	None
height/width (ft)		
type		
Bulkhead/stop logs (Y or N)		
number of units serviced		
Hoists		
number	None	1
capacity (tons)		30
type		Gantry crane

WATER PASSAGES	Upper Tunnel	Shaft	Lower Tunnel	Surge Tanks Upper	Surge Tanks Lower	Penstocks	Tailrace Tunnel
Number	1			1		6	
Diameter (ft)	30.0			120.0		12.5	
Length (ft)	37,775			476		3,500	
Maximum velocity (ft/s)	26.0						
Concrete lining thickness (in)							
Total length of concrete sections (ft)	1,350						
Steel liner Thickness							
Minimum (in)						0.88	
Maximum (in)						1.88	
Material grade						ASTM A537 Grade A	
Total length of steel-lined sections (ft)				476		3,500	

Notes:

Castaic Power Project - Page 2 (English)

CASTAIC POWER PROJECT

INTAKES	UPPER INTAKE	LOWER INTAKE
Number	1	2
Type	Reinforced concrete gravity structure, multiple compartmented	West Branch of California Aqueduct and Castiac Creek.
Design discharge (m³/s)	526.7	
Gross inlet area (m²) (at trash racks)	32.7	
Bar Racks:		
spacing (mm)		
shape		
depth/thickness (mm)		
diameter (mm)		
Emergency gates		
number	1	None
height/width (m)	9.754 / 7.620	
type	Flat-leaf coaster type	
Service gates		
number	None	None
height/width (m)		
type		
Bulkhead/stop logs (Y or N)		
number of units serviced		
Hoists		
number	None	1
capacity (Mg)		27
type		Gantry crane

WATER PASSAGES	Upper Tunnel	Shaft	Lower Tunnel	Surge Tanks Upper	Surge Tanks Lower	Penstocks	Tailrace Tunnel
Number	1			1		6	
Diameter (m)	9.14			36.58		3.81	
Length (m)	11,513.8			145.1		1,066.8	
Maximum velocity (m/s)	7.93						
Concrete lining thickness (m)							
Total length of concrete sections (m)	411.5						
Steel liner Thickness							
Minimum (mm)						22	
Maximum (mm)						48	
Material grade						ASTM A537 Grade A	
Total length of steel-lined sections (m)				145.1		1,066.8	

Notes:

Castaic Power Project - Page 2 (Metric)

CASTAIC POWER PROJECT

POWERHOUSE and RELATED FEATURES

Powerhouse Structure
Type: 2/3 Underground
Length: 600 ft   Width: 187 ft   Height: 190 ft

Guard Valves
Number: 6   Diameter: 8.7 ft
Type: Spherical

Transformers
Number: 7
Ratings: 6 each 290   1 each 70
Voltages: (kV) 230 / 18   1 at 11 /230

Generator
Rating generating (MVA): 250.0   Rating pumping (MVA): 227.0
Insulation type:
Starting method: Solid state
Starting equipment: Direct start with pelton wheel or back-to-back start

Runners
Material: Solid cast stainless steel 13% chromium, 5% nickel
Minimum unit submergence: 50.0 ft
WR^2:
Manufacturer: Hitachi
Model test by: Hitachi

	Reversible Runners	Reversible Motor/Generator		
Number	6			7
Diameter (ft)	19.20	Rotor	Stator	
rpm synchronous	257.0			257.0
rpm overspeed				
Type	Modified Francis		6-Sync	1-Sync

Information on Runners

	Gross Head (ft)		Capacity (MW)		Discharge (ft³/s)		Turbine/Pump Eff.(%)	
Condition:	Generating	Pumping	Generating	Pumping	Generating	Pumping	Generating	Pumping
Maximum head & maximum power	1,098	1,098	261		3,500			
Minimum head & maximum power	800			261		3,200		

	Net Head (ft)		Capacity (MW)		Discharge (ft³/s)		Turbine/Pump Eff.(%)	
Condition:	Generating	Pumping	Generating	Pumping	Generating	Pumping	Generating	Pumping
Rated head @ best gate	1,078	1,065	255	227	3,500	2,250		

Castaic Power Project - Page 3 (English)

CASTAIC POWER PROJECT

POWERHOUSE and RELATED FEATURES

Powerhouse Structure
Type: 2/3 Underground
Length: 182.9 m   Width: 57.0 m   Height: 57.9 m

Guard Valves
Number: 6   Diameter: 2.64 m
Type: Spherical

Transformers
Number: 7
Ratings: 6 each 290   1 each 70
Voltages: (kV) 230 / 18   1 at 11 /230

Generator
Rating generating (MVA): 250.0   Rating pumping (MVA): 227.0
Insulation type:
Starting method: Solid state
Starting equipment: Direct start with pelton wheel or back-to-back start

Runners
Material: Solid cast stainless steel 13% chromium, 5% nickel
Minimum unit submergence: 15.24 m
WR^2:
Manufacturer: Hitachi
Model test by: Hitachi

	Reversible Runners	Reversible Motor/Generator		
Number	6			7
Diameter m	5.87	Rotor	Stator	
rpm synchronous	257.0			257.0
rpm overspeed				
Type	Modified Francis		6-Sync	1-Sync

Information on Runners

Condition:	Gross Head (m)		Capacity (MW)		Discharge (m³/s)		Turbine/Pump Eff.(%)	
	Generating	Pumping	Generating	Pumping	Generating	Pumping	Generating	Pumping
Maximum head & maximum power	334.7	334.7	261		99.1			
Minimum head & maximum power	243.8			261		90.6		

Condition:	Net Head (m)		Capacity (MW)		Discharge (m³/s)		Turbine/Pump Eff.(%)	
	Generating	Pumping	Generating	Pumping	Generating	Pumping	Generating	Pumping
Rated head @ best gate	328.6	324.6	255	227	99.1	63.7		

Castaic Power Project - Page 3 (Metric)

CASTAIC POWER PROJECT

Plant Data:

Average GWh generating per year:	
Average GWh pumping per year:	
Starting time from standstill (s):	
Changeover time pumping to generating (min):	
Planned/scheduled time between major overhauls (years):	6
Outage time required per unit during major overhauls (weeks):	14
Representative plant availability (%):	92.0
Representative planned outages (weeks per year):	8

Miscelleneous Notes:

The powerhouse contains seven units total, six of which are reversible. The tables present data on the six reversible units, except that the plant capacity is for all seven units.

The outlet works for the upper reservoir has eight gates. Six are 3.66 m (12 ft) high and two are 2.743 m (9 ft) high. All gates are 2.44 m (8 ft) wide.

The upper reservoir includes one gate-controlled and one uncontrolled spillway. The gated spillway capacity is 623 m³/s (22,000 ft³/s). The combined capacity of both spillways is 4,248 m³/s (150,000 ft³/s).

The lower reservoir has one morning glory-type spillway and one ogee-crested emergency spillway. The emergency spillway has a capacity of 481 m³/s (17,000 ft³/s) and the combined capacity of both spillways is 014.0 m³/s (28,747 ft³/s).

Cavitation Experience:

Significant or Unique Problems:

List of Licenses Required:
FERC, State DWR, DSOD.

ENVIRONMENTAL FEATURES

Recreation:
None

Fish and Wildlife:

Social:

Castaic pumped-storage project in California

By Carl D. Haase*

An article covering civil-engineering design and construction, as well as the electrical and mechanical equipment, being incorporated in the 1250MW pumped-storage development at Castaic in California. Besides making use of the 1060ft-head for peaking purposes, the scheme will include a 50MW Pelton set to develop the flow in the west branch of the California State Aqueduct

THE CASTAIC POWER PROJECT is a pumped-storage hydroelectric development designed to provide peak-load power from the falling water on the west branch of the California State Aqueduct. The project, which is located approximately 25 miles north of the Los Angeles city limits, is a co-operative venture between the Department of Water and Power of the City of Los Angeles, California (DWP), and the Department of Water Resources of the State of California (DWR).

The California State Aqueduct's headwaters are located at Oroville dam north of Sacramento. From there, water will flow both by gravity and by pumping to Southern California. After being pumped through the Tehachapi Mountains to the Tehachapi after-bay, the aqueduct will divide into the east and west branches to serve Southern California.

The east branch will traverse the Antelope Valley, cross the San Bernardino Mountains, and terminate at Perris reservoir, just west of Riverside. The ultimate average annual flow of 2283cusecs in the west branch will be lifted from the Tehachapi after-bay by the Oso pumping plant to Quail Lake and will then flow by open channel and pipeline into the Pyramid reservoir (see Fig. 1).

Pyramid reservoir will form the upper storage reservoir for the Castaic power project.

From Pyramid reservoir, the water will travel through the 7½ mile-long Angeles tunnel, through the Castaic power project (including the pumping forebay reservoir) and into Castaic reservoir which is the terminal reservoir for the west branch of the aqueduct. Pyramid reservoir, the Angeles tunnel and the Castaic power project comprise the co-operative venture between DWP and DWR (see Fig. 2).

DWR originally planned to develop the Castaic power plant as a base-load generating facility. A continuous flow of water would have been utilized to develop a maximum of 160MW; however, this plant was redesigned to a 232MW plant which could be partly utilized for peaking. Under the co-operative venture, DWP will be required to furnish DWR with comparable capacity and energy.

The Castaic power project will utilize a 1060ft drop in elevation from the Pyramid reservoir to the Castaic reservoir, 90% of which will be available for generation of power at full-plant output. Since this project is not a pure pumped-storage project (it will utilize aqueduct flow and Pyramid reservoir), the economic sizing of the plant involved evaluating only the cost of enlarging

---
* Civil-Engineering Associate, Civil-Engineering Design Subsection, Power Design and Construction Division, Los Angeles Department of Water and Power, Los Angeles, California, 90054, USA.

Fig. 1. Map showing the location of the Castaic pumped-storage project

Pyramid reservoir, the Angeles tunnel, its intake, and the surge chamber, and constructing the powerhouses, tailrace, and forebay dam.

Pyramid was enlarged to provide the guaranteed allotment of 10 000 acre-ft of storage although additional storage is available for use during an emergency.

The project was, therefore, sized for a total capacity of 1250MW to be provided by six 200MW reversible pump-turbine/generator-motor units (units 1-6) and a 50MW impulse unit (No. 7).

The Unit 7 power plant, which is constructed on a separate site approximately 1000ft from the main powerhouse, was required to pass the initial aqueduct flow prior to the completion of the first 200MW unit. The Unit 7 power plant can be utilized to start any of the main 200MW units in the pumping mode and is required for starting the last of the main units in the pumping mode when five units are already pumping.

According to the present schedule (June, 1971), the aqueduct will deliver 500cusecs starting in October of this year. This water will pass through the completed Unit 7 power plant until the first of six main units is put on the line in April, 1973. The second main unit is scheduled for completion in January, 1974, and the

Reprinted, with permission, from International Water Power and Dam Construction, September 1971.

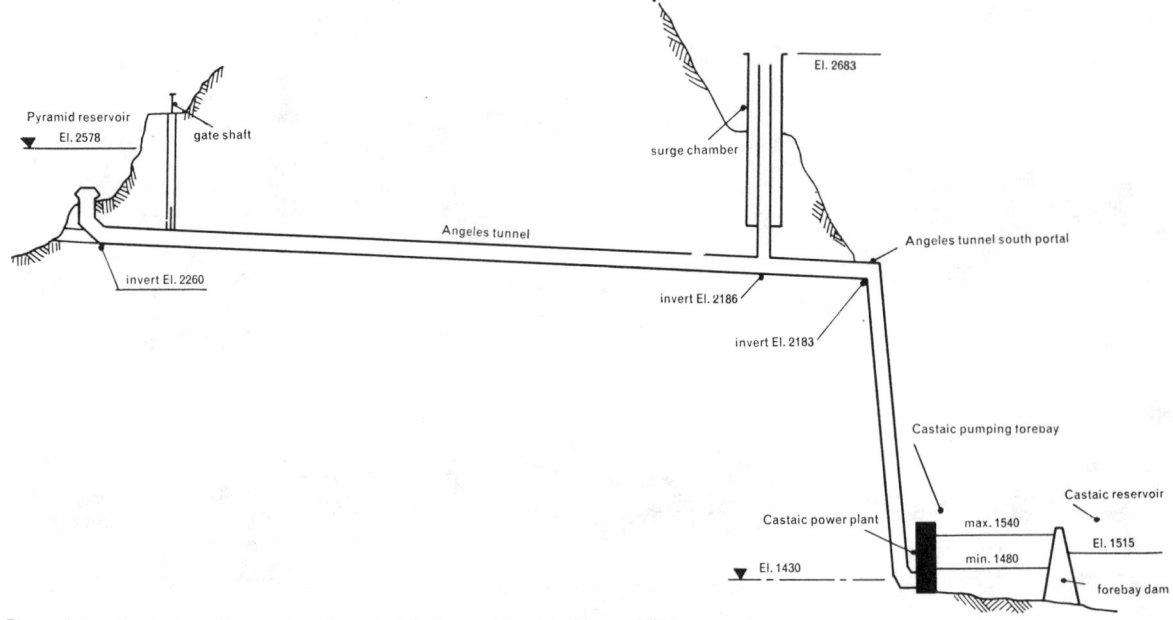

Fig. 2. A longitudinal profile showing the salient features of the development

remaining units will be completed one year apart beginning with Unit 3 in July, 1975.

The design for the enlarged surge chamber and power facilities (including the forebay dam) was performed by DWP engineers.

Angeles tunnel and surge chamber

The Angeles tunnel was increased in size from the 17ft-diameter tunnel originally designed by DWR, which had a capacity of 3000cusecs, to the 30ft diameter tunnel, now nearing completion, which is sufficient to pass 18 600 cusecs.

Due to the length of the Angeles tunnel, care was taken to minimize the head losses. Each incremental foot of head loss is estimated to cost the project approximately $400 000 in generation benefits over the assumed 70-year life of the project. Therefore, the Angeles tunnel specification provides for the removal by grinding of projections in the line of flow as well as the use of steel forms and steel trowel finishing of the concrete invert to minimize friction head loss in the tunnel.

The height of the differential-type surge chamber (see Fig. 3) was controlled by the hydraulic grade line for pumping, while the diameter was controlled by full-load rejection by all six main units, with a water velocity in the tunnel of approximately 24ft/s. The internal riser has an inside diameter of 28ft; the main tank is 393ft-high and has an interior diameter of 120ft, with the upper 158ft extending above ground.

The surge chamber allows full-load acceptance on all six main generating units, providing quick pickup capability to improve system reliability, and will accept the surge resulting from full-load rejection without spilling.

Because the active San Andreas fault (California's best-known earthquake fault) is only 9·5 miles away, the seismic loadings for design are quite large. It was assumed a ground acceleration of 0·5g would occur during the pumping cycle, when the water surface in

◀ *An artist's impression of the completed Castaic scheme*

Water Power September 1971

Fig. 3. A cross-section through the differential-type surge chamber

the surge tank would be 25ft below the top of the surge chamber. However, a value of 0·3g ground acceleration was used for the full-load rejection conditions, when the water surface in the main tank is at the top of the internal riser.

This reduction was made because of the very remote

likelihood of a seismic disturbance coinciding exactly in time with full-load rejection. Assuming that full-load rejection could be caused by a seismic disturbance, the maximum elevation of the water in the surge chamber would occur 240 seconds later, at which time the main shock would have subsided.

The bid price for construction of the surge chamber was approximately $13·4 million, which included excavation of 142 000yd³, use of 16·5 million pounds of No. 18-S reinforcing steel bars (2¼in-diameter), and use of 11 million pounds of steel liner plate for the tank.

The construction of the Angeles tunnel and the surge chamber is being performed under a contract administered by DWR. The low bid was based on construction methods for excavation which involved blasting only the upper half of the Angeles tunnel, supporting the upper half with steel sets supported on concrete footings and excavating the lower portion with conventional earth moving equipment.

The surge chamber was excavated by first drilling a 6ft-diameter vertical shaft into the crown of the Angeles tunnel. The remaining material was excavated by ripping, pushed into the vertical shaft and hauled out of the Angeles tunnel south portal by conventional earth-moving equipment.

Manifold and penstock

At the Angeles tunnel south portal, the 30ft-diameter tunnel section divides into six penstocks for the main units and one smaller penstock for Unit 7. The six-main-penstock scheme was decided upon after considering a trifurcation feeding three main penstocks, each with a bifurcation just upstream of the powerhouse.

The 30ft-diameter Angeles tunnel divides into the six main penstocks through a manifold designed by Escher-Wyss Ltd., Switzerland. This patented design is less expensive to construct than the trifurcation and bifurcations and will create less than one foot of head loss at rated flow compared to an estimated five to eight feet of loss in the alternative scheme.

The manifold is designed for 600ft maximum head and will be the largest structure of this type. A scale model of the manifold was built and tested by Escher-Wyss to predict the stress and flow characteristics of the prototype.

The steel used for the manifold shell and the penstocks conforms to a modified ASTM A537 Grade A Specification. The internal reinforcing ribs for the manifold are forged vanadium nitrogen-treated steel.

The external diameter of the main penstocks varies from 13ft 6in at the manifold to 11ft 6in at the powerhouse, and plate thicknesses vary from $\frac{7}{8}$in to $1\frac{7}{8}$in. The main penstocks have a vertical drop of 767ft in a horizontal distance of approximately 2260ft. Penstock sections are coupled where the penstocks are above grade and welded where they are buried.

The penstocks for the Unit 7 power plant branches from an extension of the Angeles tunnel liner. This penstock varies in diameter from 7ft at the 'wye' branch to 6ft at the Unit 7 power plant, and the penstock plate thicknesses vary from $\frac{7}{16}$in to $\frac{7}{8}$in.

Each of the penstocks has a butterfly valve at its upper end where it connects to either the manifold or 'wye' branch. These valves may be closed from the main powerhouse or at each valve, and will close automatically if the flow rate increases above a predetermined amount.

The penstocks for Units 1 and 2, which at this location are 12ft 6in in diameter. Unit 1 penstock in the foreground is undergoing pressure tests

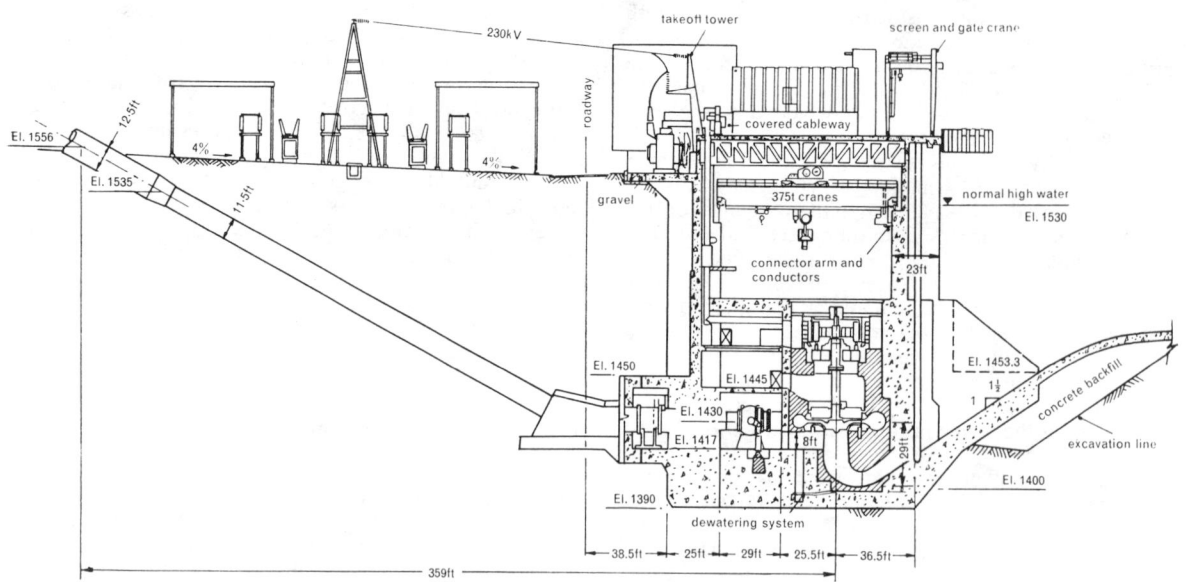

Fig. 4. A cross-section through the main powerhouse which is being constructed in an open excavation

The butterfly valves for Units 1 to 6 are $145\frac{5}{8}$in internal diameter and the butterfly valve for Unit 7 is 86·6in internal diameter.

Main powerhouse

Early geological investigations showed that the possible powerhouse sites were not suitable for an underground powerhouse; the make-up of the soil was primarily siltstone, with some sandstone lenses, dipping at approximately 15°. Therefore, the main powerhouse and Unit 7 power plant are being built in an open excavation.

The pump-turbine distributor centre-line was set at El. 1430ft, providing a minimum turbine submergence of 50ft. The final location of the main powerhouse was based on a comparison of the cost of excavating for the main powerhouse and the additional length of tailrace channel with the cost-saving for shorter penstocks.

The main powerhouse is 500ft long by 100ft wide by 150ft high, with the roof at El. 1561ft. This allows for six main units with centre-line spacings of 65ft, and a 100ft by 100ft service bay located at the east end of the structure.

The construction of the powerhouse and tailbay required that a hole 900ft wide by 800ft long at the top and 235ft deep be excavated. Except for the tailbay, the area around the main powerhouse is being backfilled with approximately 550 000yd³ of material. The finished grade will be at El. 1544ft, and the filled area on the penstock side of the main powerhouse will be used for the switchyard.

The combination of the pumping forebay high-water elevation of 1540ft and backfilling the main powerhouse from El. 1440ft to 1544ft presented a major design problem. The siltstone backfill material obtained from the powerhouse excavation, when saturated, has little or no cohesion value. Under the ground acceleration of 0·4g, used for the design earthquake factor, saturated siltstone would impose unacceptably high loadings on the main powerhouse.

It was decided that a subsurface drainage system capable of preventing saturation of the backfill material would be installed because no acceptable backfill material other than siltstone was available at the site.

The subsurface drainage system consists of a sand-like drainage envelope placed upon the exposed excavated surfaces on the penstock side of the main powerhouse followed by siltstone backfill. At each end of the powerhouse, a vertical curtain zone of the same sand-like material will also be installed.

The entire drainage envelope will slope to the rear of the powerhouse where any seepage water will be collected in a perforated pipe collection system. This collection system drains into a sump located adjacent to the main powerhouse from which the water will be pumped into the tailbay.

In June, 1968, the site grading contract was awarded; this $2·9 million contract involved the excavation of approximately 3·1 million yd³ of earth for the main powerhouse and 950 000yd³ of earth for the penstock ridges.

Upon completion of this contract, DWP's crews began construction of the main powerhouse (see Fig. 4), the Unit 7 power plant, the penstock piers and anchors, and the manifold foundation. The structures, when completed, will contain an estimated 20 000t of reinforcing steel (mostly No. 18-S and No. 11 bars).

All of the reinforcing steel is being fabricated by Bethlehem Steel Corporation, which also has the responsibility of placing over 15 000t of reinforcing steel for the main powerhouse. The remaining quantity is being placed by DWP's crews.

Approximately 250 000yd³ of concrete will be used, of which the greater part will be in place by early 1972.

Cranes and elevators

The machine floor of the main powerhouse will be serviced by two 375t-capacity bridge cranes. Each crane has a 25t auxiliary hook and 10t monorail hoist. The two cranes can be coupled together, and by the use of a lifting beam a single lift in excess of 700t can be made. This operation will be necessary when installing the rotors for the generator-motors.

Construction work on the main powerhouse excavation began in January, 1970

A bridge crane of 125t-capacity will move material from a drive-through area on the penstock side of the main powerhouse to the interior service bay. A gantry crane with a 30t-capacity main hook and 3t-capacity auxiliary hook runs on the roof of the main powerhouse along a 460ft crane runway to service the draft tube gates and trashrack.

A freight elevator (lift) of 10t-capacity will operate between an outside entrance to the main powerhouse at El. 1544ft and the interior service bay at El. 1486ft. Two passenger elevators of 2t-capacity each will also be installed in the main powerhouse.

Pump turbines

The main powerhouse will contain six 257rev/min, reversible pump-turbines manufactured by Hitachi, Japan. Runner submergence will vary from 50ft to 110ft. The maximum net generating head at the pump-turbine distributor centre-line will be approximately 1075ft, and the maximum total pumping head will be approximately 1140ft. The maximum input for pumping is 350 000hp.

The impeller runners are solid cast stainless-steel with a diameter of 19ft 3in. The stainless steel for Unit 1 is an 18% chromium, 8% nickel alloy while the stainless steel for Units 2 to 6 is a 13% chromium, 5% nickel alloy. The governors for all six pump-turbines are of the electric-hydraulic type.

The scroll cases and draft tubes are made of welded plate steel. Spherical-type shut-off valves with an internal diameter of 104in are located at each pump-turbine inlet.

On the tailbay side of the main powerhouse, the lowering of vertical gates over the draft tube outlets will permit dewatering of the draft tubes, the gates being operated by the 30t gantry crane located on the roof.

Trashrack screens will be lowered in front of each draft tube opening during the pumping operation and raised during the generation cycle. The screens are operated by electrically-driven hoists located on the main powerhouse roof.

Electrical equipment

The generator-motors, being manufactured by ASEA, Inc, Sweden, are designed with a generating rating of 250MVA at 60°C rise and 287MVA at 80°C rise (power factor = 0·85 over-excited) and motor ratings of 310 000hp at 60°C rise and 357 000hp at 80°C rise (power factor = 0·94 over-excited) when operating to drive the pumps.

Each generator-motor will be connected to the 230kV

The picture shows the separate Unit 7 power-plant and switchyard area at the time work was nearing completion

switchrack through a 270MVA power transformer. When in full operation, all energy flowing through the switchrack will be transported to or from bulk power-receiving stations in DWP's system by four 230kV transmission circuits.

The generator-motor units will be vertical shaft synchronous, modified umbrella-type machines with two guide bearings, a high-pressure oil lift thrust bearing and a thyristor-type static excitation system. The thrust bearing will support a static load in excess of 600t. The generator-motors will be set in a pit that has a 45ft diameter and is 23ft deep.

Isolated phase-bus will connect the generator-motor terminals to the low-voltage side of the 270MVA main transformer. The phase-reversal switch located in the isolated phase-bus will provide the means whereby a generator-motor can be connected to the system as either a generator or motor.

Any main unit can be started in the pumping mode by either the synchronous or semi-synchronous methods. In order to provide for synchronous back-to-back starting by use of a main unit, a double bus, 230kV switchrack is provided. One bus is designated as the operating bus while the other one is designated as the starting bus.

The procedure for synchronous starting by use of Unit 7 is as follows:
(1) both Unit 7 and the pump-motor unit will be brought to a standstill, and then connected to the 230kV starting bus;
(2) the main unit phase-reversal switch will be placed in a pump mode and the runner cavity of the pump-turbine will be dewatered by means of compressed air;
(3) excitation will be applied to both machines;
(4) both machines will be brought up to speed synchronously by opening the jets of the Unit 7 turbine;
(5) with both units up to speed, they will be synchronized to the system, and the generator-motor power circuit-breakers closed to the 230kV operating bus;
(6) the generator-motor will be disconnected from the Unit 7 bus;
(7) the pump-motor will then be loaded by first releasing air pressure and allowing water to rise in the pump-turbine impeller space, and then by opening the wicket gates.

The Castaic power plant will be designed for operation from a central control room and will have provision for ultimate operation by remote supervisory control. The central control room will house all the control equipment necessary to operate the six main units and Unit 7.

Tailrace channel

The main powerhouse was located at the upper end of the pumping forebay to minimize the Angeles tunnel and penstock costs. An excavated channel was therefore required in the canyon bottom to convey water back to the units during pump back.

The excavated channel, called the tailrace channel, is basically a trapezoidal-shaped channel with side slopes of 3:1 in the streambed alluvium and 1:1 in the siltstone bedrock. The tailrace channel bottom width is 68ft and its length is approximately one mile. The excavated depth varies according to the streambed profile and diminishes to zero at the down-canyon terminus. The invert varies from El. 1474ft at the tailbay to 1470ft at the down-canyon terminus.

The tailrace channel incorporates several innovative features including 3000 lineal feet of slurry trench through 40ft of alluvium forming a groundwater cutoff between the tailrace channel and the adjacent storm bypass channel in order to stabilize the channel's side slopes. In addition, soil cement is being used for both lining the channel and for buttressing the side slopes where adversely dipping bedding planes or highly fractured bedrock are encountered.

The tailrace channel is being lined only in those portions where highly erodible siltstone bedrock is exposed. The soil-cement mix, consisting of Portland cement and streambed sands and gravels used in the channel invert, was designed to have a permeability high enough to eliminate the need to provide a separate subdrainage system.

Tailbay

The tailbay configuration, including the curved transition into the tailrace channel, was modelled in a hydraulic model study and vortices found to occur noted during pumping.

When pumping, particularly at lower forebay water-surface elevations, the flow from the tailrace channel did not diverge sufficiently and continued directly towards the main powerhouse.

This caused flow to be directed laterally along the face of the main powerhouse forming eddies at each of the main powerhouse counterfort and trashrack support piers. Some modification of the tailbay-tailrace transition was made but appreciable lengthening of the curve configuration to improve flow divergence was not possible because of space limitations.

Therefore, a combination of vertical skirts and aprons spanning the main powerhouse piers was added at lower elevations to reduce the tendencies to form eddies and possibly-damaging vortices.

Forebay dam

The pumping forebay is formed behind a forebay dam which is located in the upper portion of the Castaic Canyon arm of Castaic reservoir. Since the Castaic reservoir could be seasonally drawn down to elevations which are more than 150ft below the streambed elevation at the main powerhouse site, it is necessary to construct an embankment which retains water for pump back during low Castaic reservoir conditions.

The forebay dam is located approximately 2½ miles downstream from the main powerhouse. It is a zoned earth and rockfill embankment 170ft high with an approximate crest length of 2000ft.

The dam, when completed in June, 1974, will provide for 34 000 acre-ft total storage at El. 1540ft, the maximum water level in the pumping forebay. Twelve thousand acre-ft of this is below El. 1480ft, the minimum operating level, and is considered dead storage.

The embankment, because of its location in Castaic reservoir and because of severe drawdown conditions, posed unusual design problems. The upstream slope was found to require a free-draining rockfill zone in order to satisfy stability criteria of seismic loading during the rapid daily drawdown which could be expected during pumped-storage operation.

The downstream face is subject to wave action and requires armouring over almost the entire height and (because of the possibility of the drawing down of Castaic reservoir) down to well below the toe of the embankment.

The materials being utilized in the construction of the forebay dam naturally make maximum use of on-site material. The impervious core will be of siltstone material similar to that used to backfill the main powerhouse. This material was obtained from the main powerhouse excavation and stockpiled under the site grading contract.

The shell materials will be obtained from stream-bed sands and gravels, with boulders and cobbles for the rockfill zone. The downstream slope protection will be of soil cement made from the mixing of cement with the stream-bed sands and gravels.

Soil-cement slope protection was preferred to rip-rap since suitable quantities of rock for rip-rap are not available at the site nor are there adequate quarry sites available within economic haul distances.

Outlet works

The design of the outlet works for the forebay dam was

The main powerhouse area seen from the tailbay side. Draft tube intakes for Unit 6 are shown in the foreground and the surge chamber (under construction) can just be seen on the skyline. Buried sections of the penstocks for Units 1 and 2 are in place on the right

A view of the main powerhouse construction seen from the penstock side. The service bay is shown in the far left of the picture, Unit 6 in the far right, and the penstocks for Units 1 and 2 in the foreground

governed by topographic and geologic constraints that required a cut-and-cover outlet conduit located at the right abutment of the dam. The size of the outlet conduit was determined by using a time-weighted average of generating head lost due to the elevation of the pumping forebay during a planned weekly operational cycle.

The size study primarily considered full conduit head losses under the conditions when Castaic reservoir will be high enough to influence the ability of the conduit to discharge the required aqueduct flows. The optimum outlet conduit size determined was a 21ft-diameter circular conduit.

At the pumping forebay end of the outlet conduit is a gated vertical outlet tower located within the upstream slope of the dam. A free-standing outlet tower was selected because it met the requirements for operational flexibility throughout the wide range of water surface elevations of the pumping forebay and Castaic reservoir. Adequate flexibility could not be met by a single-gate control works.

The free-standing outlet tower allows for the placement of eight sluice-type control gates relatively high in the pumping forebay, a location where cavitation when Castaic reservoir is extremely low need not be considered.

The outlet tower also serves as a service spillway incorporating a circular overflow crest which discharges into the outlet tower. The flow from the service spillway is common with the flow through the gate inlets.

The outlet works are designed to pass service spillway flows sufficient to provide protection against a flood with a probable frequency of recurrence of one-in-100-years.

The configuration of the outlet works presented some unusual design problems. Expulsion of entrained air during submerged operation and hydrodynamic transients during the transition from open channel flow to full-pipe flow were the most difficult of these.

Model studies indicated that the entrainment of air could not be reduced significantly. By setting the scope of the outlet conduit at zero to negate buoyancy forces and by installing, just upstream from the stilling basin portion of the outlet works, a "slotted cover air diffuser section" in order to disperse air slugs, it was found that entrained air could be passed satisfactorily.

Pressure build-up and blow-back due to hydraulic transients were eliminated by providing an exceptionally large amount of ventilation at the tower.

The stilling basin at the terminus of the structure is patterned after a conventional US Bureau of Reclamation design. The sizing was determined by the free discharge operating conditions, and modifications to the wall heights and end sill were made to accommodate operations under partially and totally submerged conditions.

The Castaic power project, with its maximum peaking capacity of 1250MW when completed in 1978, will approximately triple DWP's present hydroelectric generating capacity. Since it will have the ability to respond quickly to system demands and is located relatively close to the city limits, it will greatly contribute to DWP's system reliability.

In order to meet future system demands, estimated at 6262MW in 1980 and 13 714MW in 2000, DWP has recently entered into an agreement with DWR for a pilot study of a second pump-storage project on the west branch of the California aqueduct.

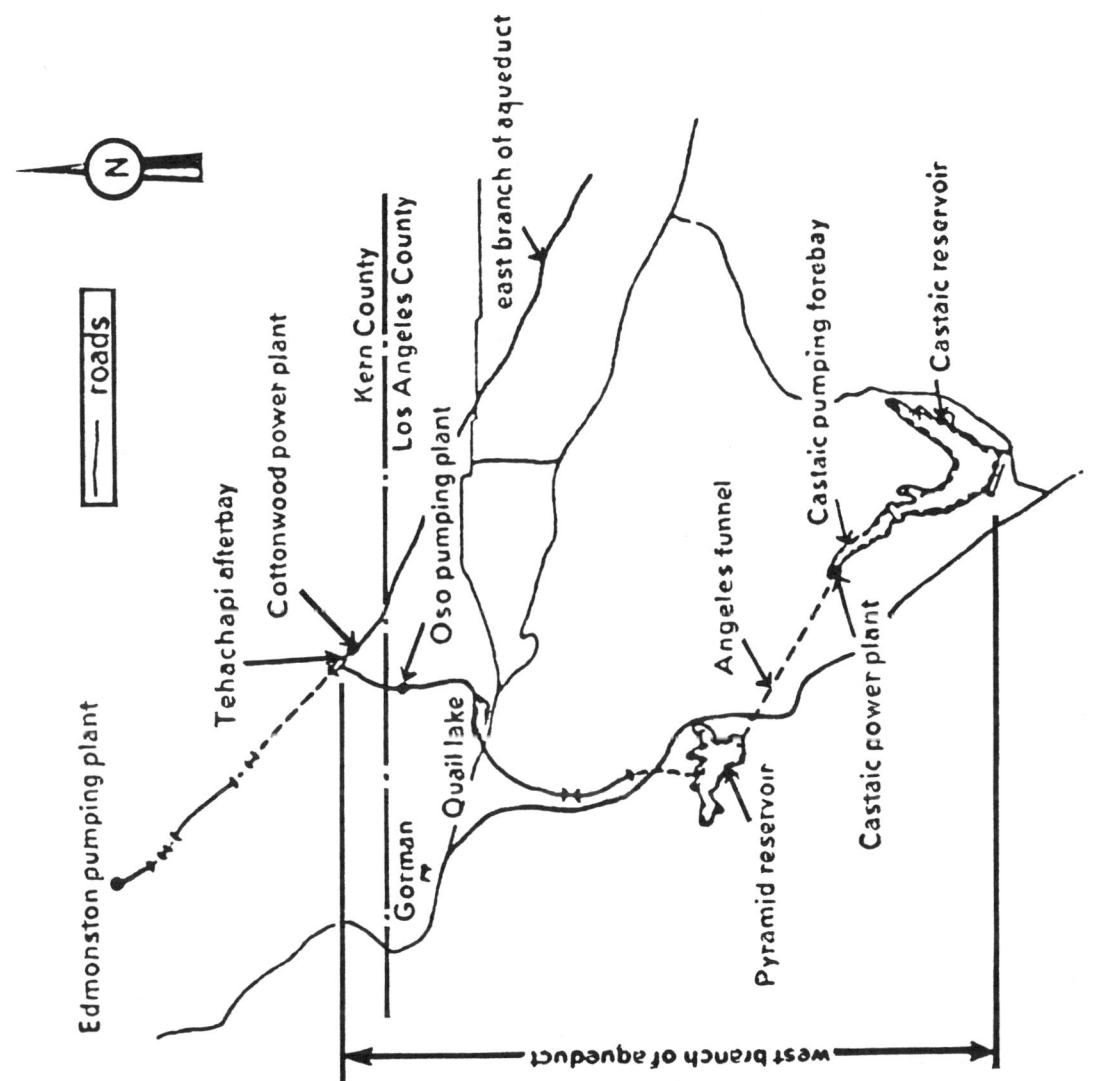

Map showing the location of the Castaic pumped-storage project

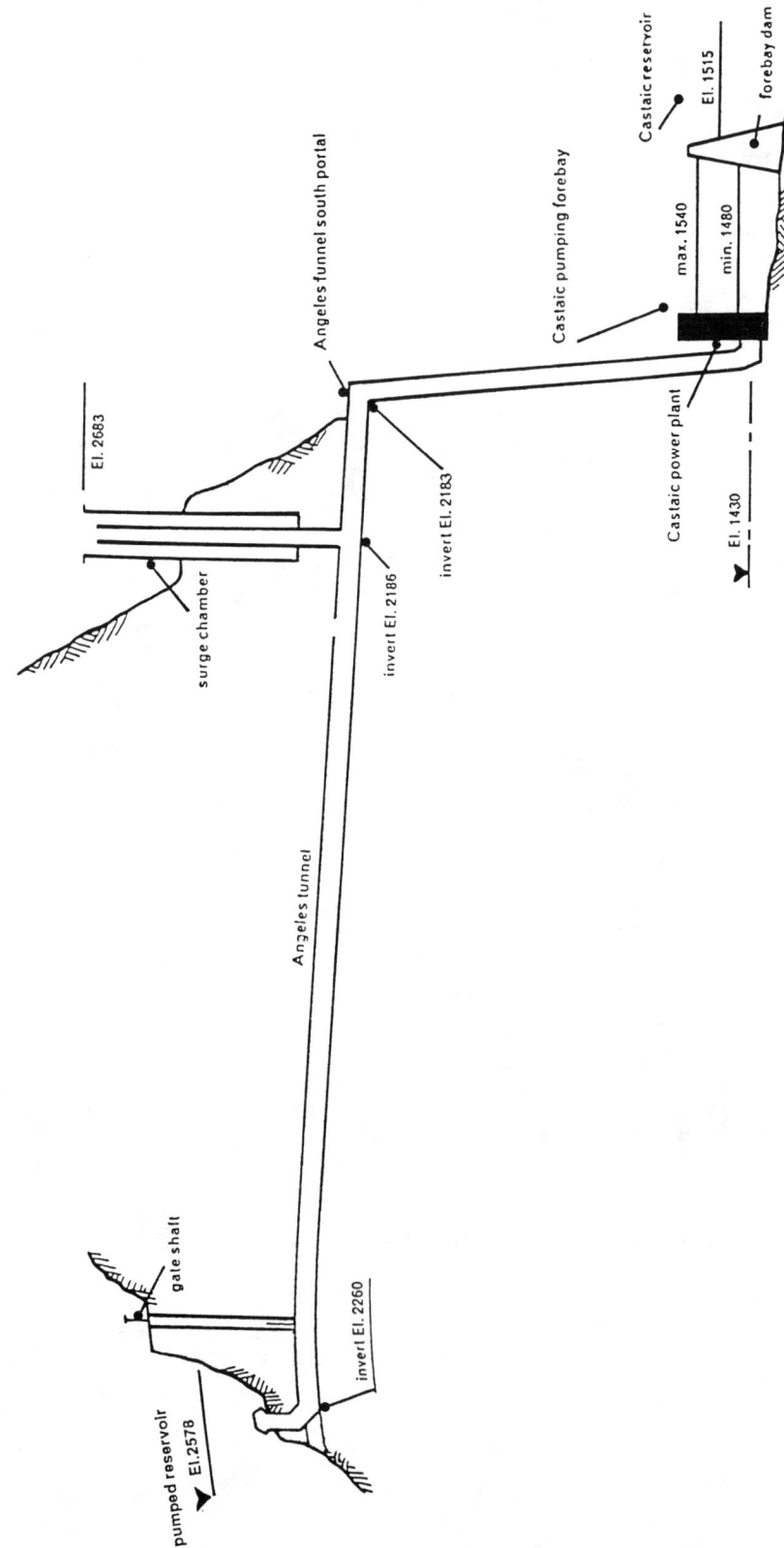

A longitudinal profile showing the salient features of the development

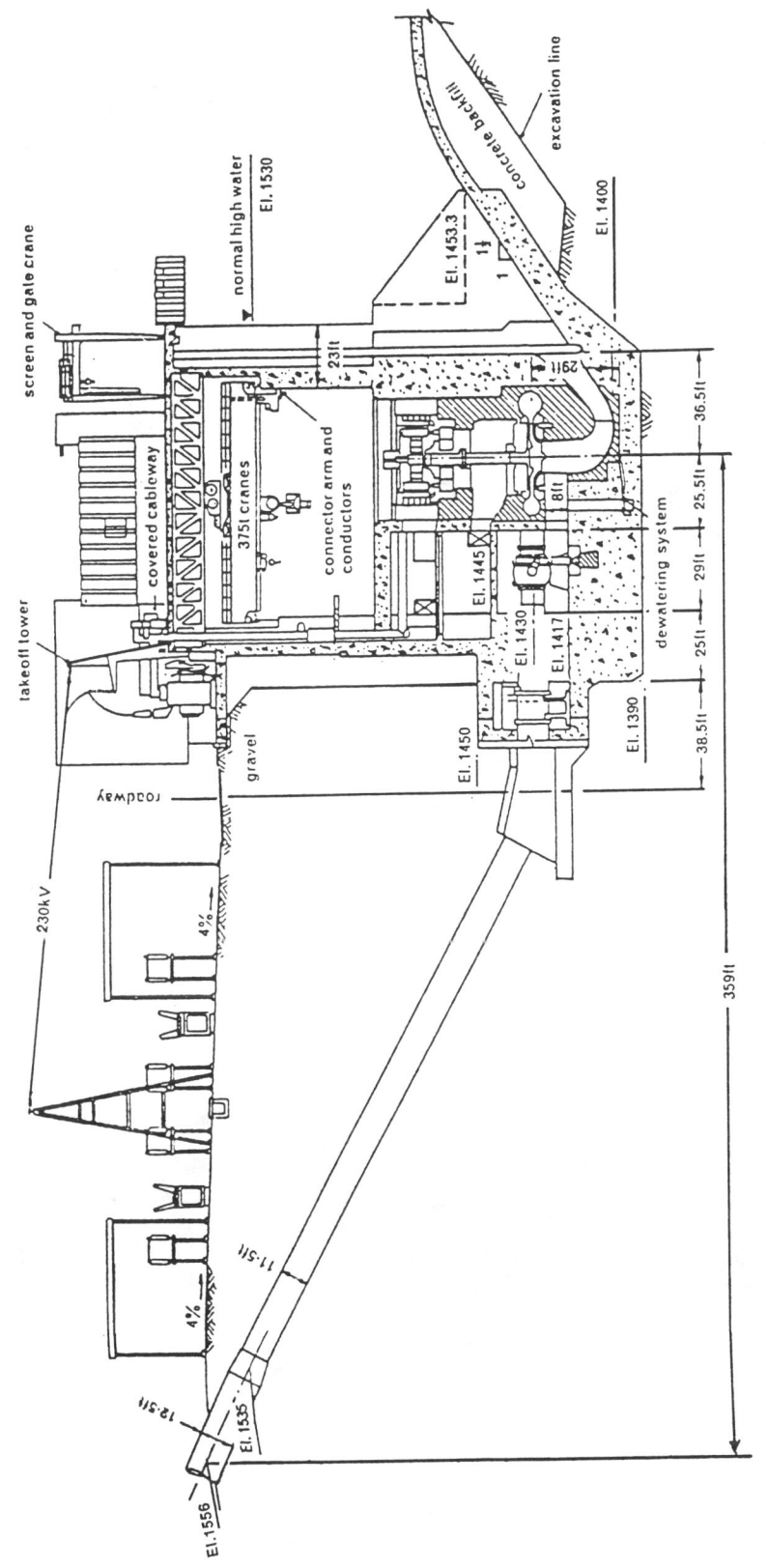

A cross-section through the main powerhouse which is being constructed in an open excavation

Aerial View of Outlet Works

Powerhouse

Castaic Pumped Storage Plant

List of Reports
Castaic Power Project

1. "Castaic Pumped-Storage Project in California", WaterPower magazine, September 1971.

2. "Inspection and Review of Project Facilities, Castaic Power Plant Complex - Devil Canyon Power Plant Complex", Federal Energy Regulatory Commission Project No. 2426 for the State of California and the Los Angeles Department of Water and Power, by Karl V. Taylor, March 1981.

3. "Inspection and Review of Project Facilities, Castaic Power Plant Complex - Devil Canyon Power Plant Complex", Federal Energy Regulatory Commission Project No. 2426 for the State of California and the Los Angeles Department of Water and Power, by Joe Sciandrone, March 1985.

4. "Third Safety Inspection Report - Elderberry Facilities of the Castaic Pumped Storage Development", Federal Energy Regulatory Commission California Aqueduct Project No. 2426 for the State of California and the Los Angeles Department of Water and Power, by Woodward-Clyde Consultants, April 1990.

5. "Performance Report for Elderberry Forebay Dam and Castaic Power Plant - January 1985 to December 1989", by the City of Los Angeles Department of Water and Power Power Design and Construction Division.

6. "Seismic Stability Analysis of Elderberry Dam", Federal Energy Regulatory Commission California Aqueduct Project No. 2426 for the Los Angeles Department of Water and Power, by Woodward-Clyde Consultants, March 1990.

Plant Name: **CLARENCE CANNON**

Plant location:
  Center, MO
  Ralls County

Rated capacity   31 MW
Average static head   75 ft
Plant efficiency   %
Stored energy   265 MWh
Number of units   1

Construction time:
Construction cost:
Price level:
First commercial power: November 1984
FERC project number: None

Owner: St. Louis District, Corps of Engineers
       Box 20 B
       Monrae City, Missouri 63456

Designers:
  St. Louis District, Corps of Engineers
  Omaha District, Corps of Engineers

Plant Manager/Superintendent:
  W. Dale Russel
  Box 20 B
  Monrae City, Missouri 63456

  (314) 565-2618

River or water source: Salt River reregulating dam (located 9.5 miles downstream of the main dam)

	UPPER RESERVOIR	LOWER RESERVOIR
DAM		
Type	Compacted earth, Clarence Cannon Dam	Rolled earth
Height (ft)	106	31
Crest length (ft)	1,690	1,550
Volume (yd³)	2,170,000	
RESERVOIR		
Type	Constructed, Mark Twain Lake	Constructed, reregulating dam
Surface area (acres)	38,400	1,200
Usable power storage (acre-ft)	355,000	4,960
Power pool fluctuation (ft)	71.0	7.0
Operating levels		
Maximum (ft)	638.0	528.0
Minimum (ft)	567.0	521.0
Drainage area (miles²)	2,318.0	29.0
Seepage (ft³/s)	1.766	1.766
SPILLWAY		
Design flood		
Return period (years)	PMP	
Flow (ft³/s)	276,500	
Capacity (ft³/s)	276,500	31,500
Type	Concrete ogee, gated	Concrete, gravity type
Gates		
Number	4	2
Type	Tainter	Tainter
Width (ft)	50.00	30.00
Height (ft)	39.00	31.00
OUTLET WORKS		
Discharge capacity (ft³/s)	10,500	50
Number of water passages	2	1
Dimensions of water passages		
Height (ft)		
Width (ft)		
Diameter (ft)	26.00	3.00
Type of gates	Powerhouse	Sluice, for minimum flows
Number of gates	4	1

Clarence Cannon - Page 1 (English)

Plant Name: **CLARENCE CANNON**

Plant location Center, MO Ralls County	Owner: St. Louis District, Corps of Engineers Box 20 B Monroe City, Missouri 63456
Rated capacity 31 MW Average static head 22.9 m Plant efficiency % Stored energy 265 MWh Number of units 1	Designers: St. Louis District, Corps of Engineers Omaha District, Corps of Engineers
Construction time: Construction cost: Price level: First commercial power: November 1984 FERC project number: None	Plant Manager/Superintendent: W. Dale Russel Box 20 B Monroe City, Missouri 63456 (314) 565-2618
River or water source: Salt River reregulating dam (located 9.5 miles downstream of the main dam)	

DAM	UPPER RESERVOIR	LOWER RESERVOIR
Type	Compacted earth, Clarence Cannon Dam	Rolled earth
Height (m)	32.3	9.4
Crest length (m)	515.1	472.4
Volume (m³)	1,659,085	
RESERVOIR		
Type	Constructed, Mark Twain Lake	Constructed, reregulating dam
Surface area (Mm²)	155.40	4.86
Usable power storage (Mm³)	437.889	6.118
Power pool fluctuation (m)	21.64	2.13
Operating levels		
Maximum (m)	194.46	160.93
Minimum (m)	172.82	158.80
Drainage area (Mm²)	6,003.626	75.110
Seepage (m³/s)	0.0500	0.0500
SPILLWAY		
Design flood		
Return period (years)	PMP	
Flow (m³/s)	7,830	
Capacity (m³/s)	7,830	892
Type	Concrete ogee, gated	Concrete, gravity type
Gates		
Number	4	2
Type	Tainter	Tainter
Width (m)	15.240	9.144
Height (m)	11.887	9.449
OUTLET WORKS		
Discharge capacity (m³/s)	297	1
Number of water passages	2	1
Dimensions of water passages		
Height (m)		
Width (m)		
Diameter (m)	7.925	0.914
Type of gates	Powerhouse	Sluice, for minimum flows
Number of gates	4	1

Clarence Cannon - Page 1 (Metric)

CLARENCE CANNON

INTAKES	UPPER INTAKE	LOWER INTAKE
Number	1	1
Type	Flaired, integral with powerhouse	Draft tube
Design discharge (ft³/s)	5,500	6,600
Gross inlet area (ft²) (at trash racks)	1,728	1,620
Bar racks		
spacing (in)	8.00	8.00
shape	Rectangular	Rectangular
depth/thickness (in)	3.00 / 0.69	3.00 / 0.69
diameter (in)		
Emergency gates		
number	None	None
height/width (ft)		
type		
Service gates		
number	1	1
height/width (ft)	37.25 / 17.50	69.00 / 19.50
type	Vertical, slide type - 3 bulkheads	Vertical lift, slide type - 3 bulkheads
Bulkhead/stop logs (Y or N)	Y	Y
number of units serviced	1	1
Hoists		
number	4	1
capacity (tons)	30	30
type	Electric cable	Draft tube Gantry crane

WATER PASSAGES	Upper Tunnel	Shaft	Lower Tunnel	Surge Tanks Upper	Lower	Penstocks	Tailrace Tunnel
Number						1	
Diameter (ft)						26.0	
Length (ft)						70	
Maximum velocity (ft/s)						11.3	
Concrete lining thickness (in)							
Total length of concrete sections (ft)							
Steel liner							
Thickness							
Minimum (in)							
Maximum (in)							
Material grade							
Total length of steel-lined sections (ft)						70	

Notes:

Clarence Cannon - Page 2 (English)

CLARENCE CANNON

INTAKES	UPPER INTAKE	LOWER INTAKE
Number	1	1
Type	Flaired, integral with powerhouse	Draft tube
Design discharge (m³/s)	155.7	186.9
Gross inlet area (m²) (at trash racks)	160.5	150.5
Bar Racks:		
spacing (mm)	203	203
shape	Rectangular	Rectangular
depth/thickness (mm)	76 / 17	76 / 17
diameter (mm)		
Emergency gates		
number	None	None
height/width (m)		
type		
Service gates		
number	1	1
height/width (m)	11.354 / 5.334	21.031 / 5.944
type	Vertical, slide type - 3 bulkheads	Vertical lift, slide type - 3 bulkheads
Bulkhead/stop logs (Y or N)	Y	Y
number of units serviced	1	1
Hoists		
number	4	1
capacity (Mg)	27	27
type	Electric cable	Draft tube Gantry crane

WATER PASSAGES	Upper Tunnel	Shaft	Lower Tunnel	Surge Tanks Upper	Surge Tanks Lower	Penstocks	Tailrace Tunnel
Number						1	
Diameter (m)						7.93	
Length (m)						21.3	
Maximum velocity (m/s)						3.44	
Concrete lining thickness (m)							
Total length of concrete sections (m)							
Steel liner Thickness							
Minimum (mm)							
Maximum (mm)							
Material grade							
Total length of steel-lined sections (m)						21.3	

Notes:

Clarence Cannon - Page 2 (Metric)

CLARENCE CANNON

POWERHOUSE and RELATED FEATURES

Powerhouse Structure
Type: Surface
Length: 225 ft   Width: 150 ft   Height: 150 ft

Guard Valves
Number: None   Diameter: ft
Type:

Transformers
Number:
Ratings:
Voltages: (kV) 69 / 13.2 3 phase, 2 winding

Generator
Rating generating (MVA): 32.6   Rating pumping (MVA): 32.5
Insulation type:
Starting method:
Starting equipment:

Runners
Material:
Minimum unit submergence: 6.0 ft
WR^2: 69,180,000 (lbf x ft²)
Manufacturer: Allis Chalmers
Model test by: Allis Chalmers

	Reversible Runners	Reversible Motor/Generator	
Number	1		1
Diameter (ft)	22.20	Rotor	Stator
rpm synchronous	75.0		75.0
rpm overspeed			
Type	Francis		

Information on Runners

Condition:	Gross Head (ft)		Capacity (MW)		Discharge (ft³/s)		Turbine/Pump Eff.(%)	
	Generating	Pumping	Generating	Pumping	Generating	Pumping	Generating	Pumping
Maximum head & maximum power	117	87	30	32	6,450	3,400	84.0	
Minimum head & maximum power	59	59	20	34	4,900	5,500	87.0	

Note: Data in the above table are based on design data.

Condition:	Net Head (ft)		Capacity (MW)		Discharge (ft³/s)		Turbine/Pump Eff.(%)	
	Generating	Pumping	Generating	Pumping	Generating	Pumping	Generating	Pumping
Rated head @ best gate	75	75	26	34	4,750	4,500	90.0	

Note: Data in the above table are based on design data.

Clarence Cannon - Page 3 (English)

CLARENCE CANNON

POWERHOUSE and RELATED FEATURES

Powerhouse Structure
Type: Surface
Length: 68.6 m Width: 45.7 m Height: 45.7 m

Guard Valves
Number: None Diameter: m
Type:

Transformers
Number:
Ratings:
Voltages: (kV) 69 / 13.2 3 phase, 2 winding

Generator
Rating generating (MVA): 32.6 Rating pumping (MVA): 32.5
Insulation type:
Starting method:
Starting equipment:

Runners
Material:
Minimum unit submergence: 1.83 m
WR^2: 28,595,000 (Newtons x m^2)
Manufacturer: Allis Chalmers
Model test by: Allis Chalmers

	Reversible Runners	Reversible Motor/Generator		
Number	1			1
Diameter m	6.77	Rotor	Stator	
rpm synchronous	75.0			75.0
rpm overspeed				
Type	Francis			

Information on Runners

Condition:	Gross Head (m)		Capacity (MW)		Discharge (m³/s)		Turbine/Pump Eff.(%)	
	Generating	Pumping	Generating	Pumping	Generating	Pumping	Generating	Pumping
Maximum head & maximum power	35.7	26.5	30	32	182.6	96.3	84.0	
Minimum head & maximum power	18.0	18.0	20	34	138.8	155.7	87.0	

Note: Data in the above table are based on design data.

Condition:	Net Head (m)		Capacity (MW)		Discharge (m³/s)		Turbine/Pump Eff.(%)	
	Generating	Pumping	Generating	Pumping	Generating	Pumping	Generating	Pumping
Rated head @ best gate	22.9	22.9	26	34	134.5	127.4	90.0	

Note: Data in the above table are based on design data.

Clarence Cannon - Page 3 (Metric)

CLARENCE CANNON

Plant Data:
Average GWh generating per year:	76
Average GWh pumping per year:	
Starting time from standstill (s):	105
Changeover time pumping to generating (min):	10
Planned/scheduled time between major overhauls (years):	15
Outage time required per unit during major overhauls (weeks):	20
Representative plant availability (%):	90.0
Representative planned outages (weeks per year):	3

Miscelleneous Notes:
None.

Cavitation Experience:
Cavitation has occurred to blades and draft tubes.

Significant or Unique Problems:

List of Licenses Required:
None -- direct congressional authorization.

ENVIRONMENTAL FEATURES

Recreation:
The re-regulating dam and Cannon dam offer fishing, boating, swimming, camping, waterskiing, hunting, wildlife habitat, and so forth.

Fish and Wildlife:
Improved water quality below the re-regulating dam, increased fishery and increased wildlife habitat have resulted.

Social:
Benefits include community water supply, flood control, navigation, and hydropower.

CLARENCE CANNON DAM & RESERVOIR
POWER PLANT

The authorized purposes of the project include flood control, hydro-electric power, water supply, fish and wildlife conservation, recreation, and incidental navigation.

The Clarence Cannon Dam and Mark Twain Lake (formerly the Joanna Reservoir) on the Salt River, Missouri, was authorized by the Flood Control Act of 23 October 1962.

The Salt River Basin lies in the northeastern portion of Missouri and embraces all or parts of Adair, Audrain, Boone, Callaway, Knox, Macon, Marion, Monroe, Pike, Ralls, Randolph, Schuyler, and Shelby Counties. The total drainage area of the basin is approximately 2,920 square miles. Many smaller streams enter the river above the damsite and the resulting ravines and valleys form a very irregular shoreline of approximately 285 miles at joint-use pool elevation. The approximate water depth at the damsite is 66 feet above the flood plain at the normal recreation pool level, elevation 606.0.

The project is located principally in Monroe and Ralls Counties in northeastern Missouri. The main damsite is located at mile 63.0 on the Salt River and situated approximately 12 miles southeast of Monroe City, in Ralls County, Missouri. The re-regulation dam is located 9.5 miles downstream from the main damsite. The project area is served on the north by U.S. Highway 24 and on the south by State Highway 154. State Highway 107 and State Highway J run north-east through the project.

a. <u>Main Dam</u>. The main dam consists of a compacted earth embankment, a grated concrete spillway, and a concrete hydroelectric power intake section. The spillway is located adjacent to the right abutment with the hydroelectric power intake section adjoining immediately to the left. The overall length of the dam is approximately 1,690 feet including the spillway and power intake sections. State Highway J crosses the dam.

b. <u>Spillway</u>. Four 50-foot wide by 39-foot tainter gates control the spillway releases. The overall length of the spillway, including the non-overflow sections and the power intake section, is 857 feet. The overflow section has a gross length between abutment piers of 230 feet, which includes the four 50-foot wide tainter gates and three 10-foot thick intermediate piers. The established spillway crest is elevation 600.0. The top of the tainter gates in the closed position is at elevation 639.0, one foot above the designed flood control pool.

c. <u>Temperature Control Structures</u>. In order to eliminate cold water releases detrimental to the downstream warm water fishery, a fixed crest weir has been located 70 feet upstream of the power intakes. This weir has been designed with a crest elevation of 575.0 which has been indicated as satisfactory by the U.S. Fish and Wildlife Service. However this structure raises the intake temperature therefore decreasing the cooling abilities of the units.

d. <u>Re-Regulation Dam</u>. The re-regulation dam is located 9.5 miles downstream of the main dam. This dam creates a pool to store power releases for re-regulation of downstream flow and for pumpback. Approximately 800 surface acres of water is available at maximum normal pool, elevation 530.0. The dam consists of compacted earth embankment and a gated concrete spillway having a crest width of 60 feet controlled by two 30-foot wide by 31-foot high tainter gates. The combined length of the spillway and embankment is 3,150 feet at its crest elevation 530.0 A 36-inch telescoping valve exiting through the control house into the stilling basin for passing the minimum downstream water release requirement of 50 c.f.s. and maintaining water quality.

e. <u>Units</u>. One conventional vertical shaft, single-runner, adjustable-blade, propeller type Kaplan turbine and one vertical shaft, single-runner, Modified Francis type pump-turbine are installed downstream of the intake gates in a welded plate-steel spiral case. The spiral case is embedded in the concrete substructure of the powerhouse and the turbine is designed and constructed so that all removable parts, including the impeller-runner, shaft, guide bearing, guide bearing support, head cover, gate mechanism, and wicket gates can be removed from above. The Kaplan Turbine is designated as Unit No. 1, and the Modified Francis Turbine is designated as Unit No. 2. The turbines are numbered from left to right when looking downstream. The conventional turbine has a single Kaplan-type adjustable blade runner bolted to a vertical shaft which turns the generator. The runner is driven in counterclockwise rotation (viewed from above) by water which enters through the spiral case and wicket gates and discharged through an elbow-type draft tube. The vertical pump-turbine is similar in construction to a conventional Francis hydraulic turbine. Performance, however, is different as the pump-turbine is capable of pumping water from the tailrace to the upper reservoir. During the generating cycle, the impeller-runner is driven in a counterclockwise rotation (viewed from above) by water from the spiral case, discharging through the elbow-type draft tube to the tailrace. When pumping, the impeller-runner is driven in a clockwise rotation (viewed from above) by the generator-motor. Under action of the impeller-runner, water flows from the tailrace to the impeller-runner and discharges out the spiral case through the intake and to the upper reservoir.

Intake gate operating mechanisms are outside of the water passages and are accessible for adjustment and repair. All removable parts of the turbine will pass through the generator stator.

A "Woodward" oil-pressure, cabinet-actuator-type governor, with an electrically-driven speed sensitive element, controls the speed of the units during generation and will maintain correct position of the wicket gates when pumping.

Pertinent Data

Type	Kaplan Turbine	Modified Francis Pump-Turbine
Manufactured by	Allis-Chalmers	Allis-Chalmers
Rated power generating, hp	42,900	42,900
Rated Speed, generating, r/min	128.6	75
Rated Speed, pumping, r/min	NA	75
Design Head,	75	75
Dynamic Head, pumping, ft	NA	75
Rated Discharge, generating, c.f.s.	5,500	5,515
Rated Discharge, pumping, c.f.s.	NA	4,500
Runner max. dia., inches	185	266.38
Runner throat., inches	185	231
Weight of rotating parts of turbine, pounds	150,500	334,000
Main shaft dia., inches	28	34
Servometer cylinder dia., inches	25.25	32

Pertinent Data

Type	Kaplan Turbine	Modified Francis Pump-Turbine
Servometer piston stroke, inches	18.56	26.185
Rate Power Pumping Max		46,800 Hp.

The spiral case is embedded in the powerhouse concrete. The centerline of the spiral case and stay ring is at elevation 521.0. The spiral case is fabricated of welded plate steel and designed for a hydrostatic pressure of 64 psi. The inside diameter of the Kaplan Turbine spiral case inlet and the pump-turbine spiral case inlet is 26 feet. The stay ring is made of cast steel and provides passage and guidance for the flow of water to the turbine runner. The stay ring also serves to support the weight of the turbine internal parts and structure above the spiral case, and to resist the forces caused by internal water pressure in the spiral case. The discharge ring is cast steel and is permanently set in the concrete structure. The discharge ring forms the foundation base for the stay ring and extends below the bottom of the runner. The upper base of the ring is machined to form a seat to which the bottom ring is bolted. The lower end of the discharge ring provides a seat for supporting the runner and shaft when they are disconnected from the generator shaft. The bottom ring is of cast steel construction, and is bolted and doweled to the discharge ring. The lower stem of each wicket gate is supported in the bottom ring by a bronze-lined guide bearing. The head cover is cast steel and spans the top of the runner and wicket gates, which supports the turbine guide bearing, turbine shaft packing box, upper stem of the wicket gates, and the gate operating mechanism. It forms the upper portion of the distributor and is bolted and doweled to the stay ring with corrosion-resistant steel bolts and nuts. Removable and renewable facing plate of chrome-nickel steel are fastened to the head cover forming the water surface immediately above the sicket gates. The upper stationary wearing ring is bolted to the head cover. Twenty wicket gates are provided to control the flow of water to the Kaplan Turbine, and twenty-eight wicket gates are provided to control the flow of water to the pump-turbine. In the case of the pump-turbine, they control the flow in either direction. Each gate is equipped with a lever which is linked in a common operating ring. The operating ring is positioned by two servo-motor which are actuated by oil from the governor. The wicket gates are of the balanced type with integral stems. The stems are drilled their entire length to supply grease to the lower stem bushings. Each gate has three bearings, one in the bottom ring and the other two in the head cover. Packing boxes, packed with asbestos-teflon packing, prevent water leakage around the gate stems where they extend through the head cover. The weight of the gate is carried by a bronze bearing below the gate arm. A bronze thrust collar is provided to prevent upward movement of the gate. The shear pin joining the gate lever to the shear lever is designed to be the weakest element in the gate mechanism. If an obstruction is lodged between two gates and the gate mechanism is moved, the shear pin will break. The failure of any shear pin will activate an alarm. A restraining device which upon shear pin failure prevents uncontrolled swinging of the wicket gates. The gate operating ring is a circular steel plate weldment made in one piece and connected through connecting rods to the servomotors. The large, movable ring transfers the motion and force impacted to it by the servomotor pistons, to the gate links and levers and then to all of the wicket gates simultaneously. The ring moves on bronze bearing segments which are attached to the head cover and are grooved for admittance of grease. The servomotors arc oil pressure operated, double acting cylinders that are hydraulically connected to the turbine governor by high pressure piping. The servometers operate on a maximum oil pressure of 350 psi. The servomotors are capable, under maximum operating conditions, of moving the turbine gates to a full opening or closing stroke in 5 seconds. Each servomotor is equipped with an adjustable spring-loaded cushioning valve which will cushion the gate mechanism at the end of each closing stroke.

Unit #1 runner is of the adjustable 6- blade propeller type, with blades automatically operated by the governor through the medium of oil pressure admitted to a servomotor by pipes passing through the hollow shaft from the top of the generator. The hub and blades are made of cast steel. The areas subjected to cavitation are covered with chrome-nickel-steel inlay. A cone is attached to the hub of the runner to guide the water as it leaves the runner. A drain valve is provided at the bottom of the cone for draining oil from the hub.

The pump-turbine, Unit #2, impeller-runner is a composite of a cast steel crown, band and six buckets fabricated into an integral unit by welding. The areas subject to cavitation are covered with a welded overlay of chrome-nickel steel. A cone is attached to the underside of the hub to guide the water as it leaves the runner. Wearing ring surfaces are machined on the outside of the runner band and crown. The turbine shaft is made of forged steel and is coupled to the runner with fitted bolts. A removable corrosion-resistant steel sleeve is provided on the shaft where it passes through the packing box in the head cover. The turbine guide bearing consists of two shell half sections lined with babbit which is grooved for oil circulation. The O.D. of the shell has a machined taper to fit the machined tapered bore in the bearing housing. The bearing housing is bolted and doweled to the head cover.

The bearings are lubricated by a pressure system. The pumps are positive displacement, rotary geared pumps, each having a capacity of 4.09 gpm at 45 psig (pump turbine) and 5.7 gpm at 75 psig (Kanlan). The main pump, which normally circulates the oil, is driven by an ac electric motor. The other pump, a standby, is driven by a dc electric motor. Each pump takes oil from the sump in the packing box. Each pump discharges oil through an illuminated sight flow indicator and check valve. The oil flows past a thermometer, through a strainer and flow switch to the bearing. The flow switch provides annunciation on low oil flow and allows the unit to be started only after normal oil flow is established. A flow switch connected to the ac pump discharge line starts the dc pump and activates an alarm when the flow drops to 2.14 gpm. The dc pump will continue to operate until the flow from the ac pump reaches normal. One oil pressure gage is provided on the gate board in the oil pump alcove and one oil pressure gage is provided on the governor actuator cabinet. An oil reservoir is formed in the top of the bearing shell to supply oil to the bearing in event the main oil supply is interrupted. Overflow from this reservoir drains to the oil sump. Oil from the bottom of the bearing is deflected into an oil catcher by an oil flinger and then drained to the oil sump. The oil sump is provided with a sight glass oil level indicator on the side of the sump and an oil level transmitter for the remote oil level indicator on the gage board in the oil pump alcove. An oil level switch is provided on the sump for high and low level alarm.

The packing box is installed below the turbine bearing where the main shaft passes through the head cover and forms a seal between the shaft and head cover assembly. The cast steel packing box housing is formed in halves to facilitate assembly and dismantling and is bolted and doweled to the head cover. The packing box is packed with four rings of packing which are held in compression by an adjustable packing gland. The adjusting bolts holding the packing gland to the housing should be tightened only enough to prevent excess leakage. A lantern ring separates the packing so that water can be admitted for cooling and lubrication. There are two rings of packing above and two rings of packing below the lantern ring. A flow switch, flow control valve and sight flow indicator are installed in the water supply line to the packing box. The globe valve in the branch line that supplies water to the top of the packing box in the pump-turbine should be adjusted for a trickle of water to keep the packing saturated.

Removable and renewable wearing rings are provided in the head cover and discharge ring opposite the impeller-runner sealing surfaces in the pump-turbine. Dissimilar characteristics of the wearing rings and runner sealing surface prevent seizing in case of accidental rubbing due to the close running clearance between the parts.

All moving parts, joints and sliding connections throughout the turbine operate on bronze bushings lubricated by a Farval Centralized Lubricating System. A motor operated grease pump, controlled by a timer, automatically supplies grease through two lines to the metering valves. A measured quantity of grease is supplied through individual metering valves to each lubricated point. Timer controller solenoid operated shut-off valves installed in the system allow the intermediate and lower wicket gate stem bearings, which are subject to water washing, to be lubricated more frequently than the other bearings.

The elbow type draft tube is lined with a steel draft tube liner to the low point of the draft tube. A 24 in. X 36 in. access mandoor in the draft tube liner at El. 508.5 provides entrance into the unwatered draft tube for inspection and maintenance. A test cock is provided below the mandoor to determine if the water level in the draft tube is below the sill of the door. The draft tube liner and anchorage is designed for full hydrostatic pressure on the outside due to tailwater at elevation 530.0 and a vacuum on the inside of the liner. A circumferential chamber at the top of each draft tube liner is provided for admission of air if required.

A 8-inch automatic spring loaded air valve in the Kaplan turbine is provided to relieve the vaccuum below the head cover. The valve is designed to break the vacuum upon rejection of load. The pump turbine is provided a connection for a future 8-inch air admission line and is provided with a 4-inch vent line.

Connections for a water depressing system are provided to allow the unit be operated as a synchronous condenser when the wicket gates are closed and tailwater depressed. A 10-inch compressed air line is provided to pass compressed air through the head cover to a space above the runner. Pipe connections are provided in the draft tube liner for water level controls to maintain the water level below the runner after the initial depression.

A 10-inch connection for a water depressing system is provided to allow the unit to be operated as a synchronous condenser and to reduce the power demand for pump start-up. Pipe connections are provided in the draft tube liner for water level controls to maintain the water level below the impeller-runner after the initial depression. During the pump prive phase, after sunchronizing, the depression air is vented through a line to the atmosphere by the 4-inch compressed air exhaust valve.

The equalizing system consists of two 12-inch main equalizing pipes to equalize pressure above and below the impeller-runner and prevent excessive pressure under the head cover.

The runner band drain system provides for drainage of the water surrounding the runner band into the draft tube to minimize load during pump start and synchronous condensing operation. The drain valve is located in the alcove in the draft tube access passage. The valve is actuated by an air cylinder.

Table 1
Clarence Cannon Dam and Reservoir - pertinent data

<u>General</u>

 Project purposes Flood control, power, fish and wildlife conservation, and incidental navigation (Miss. River).

<u>Location of dam</u>
 Stream Salt River, Missouri
 Main damsite (river mile) 63.0
 Re-regulation damsite (river mile) 53.5
 County Ralls
 Nearest Town Monroe City

<u>Location of reservoir</u>
 River miles 63.0 to 112.0
 Counties Ralls, Monroe, and Shelby

<u>Drainage area (square miles)</u>
 Salt River Basin 2,920
 Above main dam 2,318
 Between dam 29

Reservoir

Inactive storage pool
    Top elevation, ft. m.s.l.   567.2
    Area, acres   2,900

Joint-use pool
    Top elevation, ft. m.s.l.   606.0
    Area, acres   18,600
    Average depth of water at
        damsite, ft.   66
    Shoreline miles   285

    Regulated outflow
        Main dam, maximum c.f.s.   11,400
        Re-regulation dam, minimum c.f.s.   50

Flood control pool (lower zone)
    Top elevation, ft. m.s.l.   624.8
    Area, acres   28,600

    Regulated outflow
        Mississippi River not in flood
        Main dam, maximum c.f.s.   12,000
        Re-regulation dam, maximum c.f.s.   12,000
        Mississippi River in flood
        Main dam, maximum c.f.s.   8,870
        Re-regulation dam, maximum c.f.s.   2,171

Flood control pool (upper zone)
    Top elevation, ft. m.s.l.   638.0
    Area, Acres   38,400

Induced surcharge pool
    Top elevation, ft. m.s.l.   642.0
    Area, acres   42,200
    Maximum outflow, c.f.s.   217,000

Surcharge pool (total)
    Top elevation, ft. m.s.l.   648.0
    Area, acres   48,700
    Maximum outflow, c.f.s.   276,500

Freeboard
    Top elevation, feet   653.0
    Area, acres   53,200

Main Dam elevation, ft. m.s.l.   653.0
    Height above stream bed, ft.   138.0
    Length of crest, ft.   1,690
    Spillway (gate controlled)
        Crest elevation, ft. m.s.l.   600.0
        Crest width (gross) ft.   230.0
        Number of gates   4
        Gate width, ft.   50
        Gate height, ft.   39

Tailwater elevations
    12,000 c.f.s., ft. m.s.l.   533.0
    276,500 c.f.s., ft. m.s.l.   577.0

Re-Regulation Pool
    Top elevation, ft. m.s.l.   530.0
    Height above streambed, ft.   31.0
    Length of crest, ft.   3,150

Spillway (gate controlled)
    Crest elevation, ft. m.s.l.   499.0
    Crest width (gross) ft.   60
    Number of gates   2
    Gate width, ft.   30
    Gate height, ft.   31

Pool elevations (Regulated)
    Maximum ft. m.s.l.   528.0
    Minimum ft. m.s.l.   521.0

Sluice
    Diameter, inches   36

Tailwater elevation
    12,000 c.f.s., ft. m.s.l.   517.0
    50 c.f.s., ft. m.s.l.   500.0

Power
- Installation (nameplate), KW — 58,000
- Number of units — 2
  - Conventional unit KW — 27,000
  - Reversible unit KW — 31,000

Turbine releases
- At design head (75 feet) and rated capacity
- Conventional unit, c.f.s. — 4,750
- Reversible unit, c.f.s. — 4,900
- Maximum release c.f.s. — 10,600
- Average net head (pool elev. 590), ft. — 73
- Drawdown (elev. 606 to 590), ft. — 16

Project Acreage. Total project acreage for the main reservoir area is estimated at 61,955 acres. An additional 1,704 acres is required to maintain an area below the main dam for the regulation pool.

This project description was provided by Dennis Fenske and is taken from a Corps of Engineers report on the project.

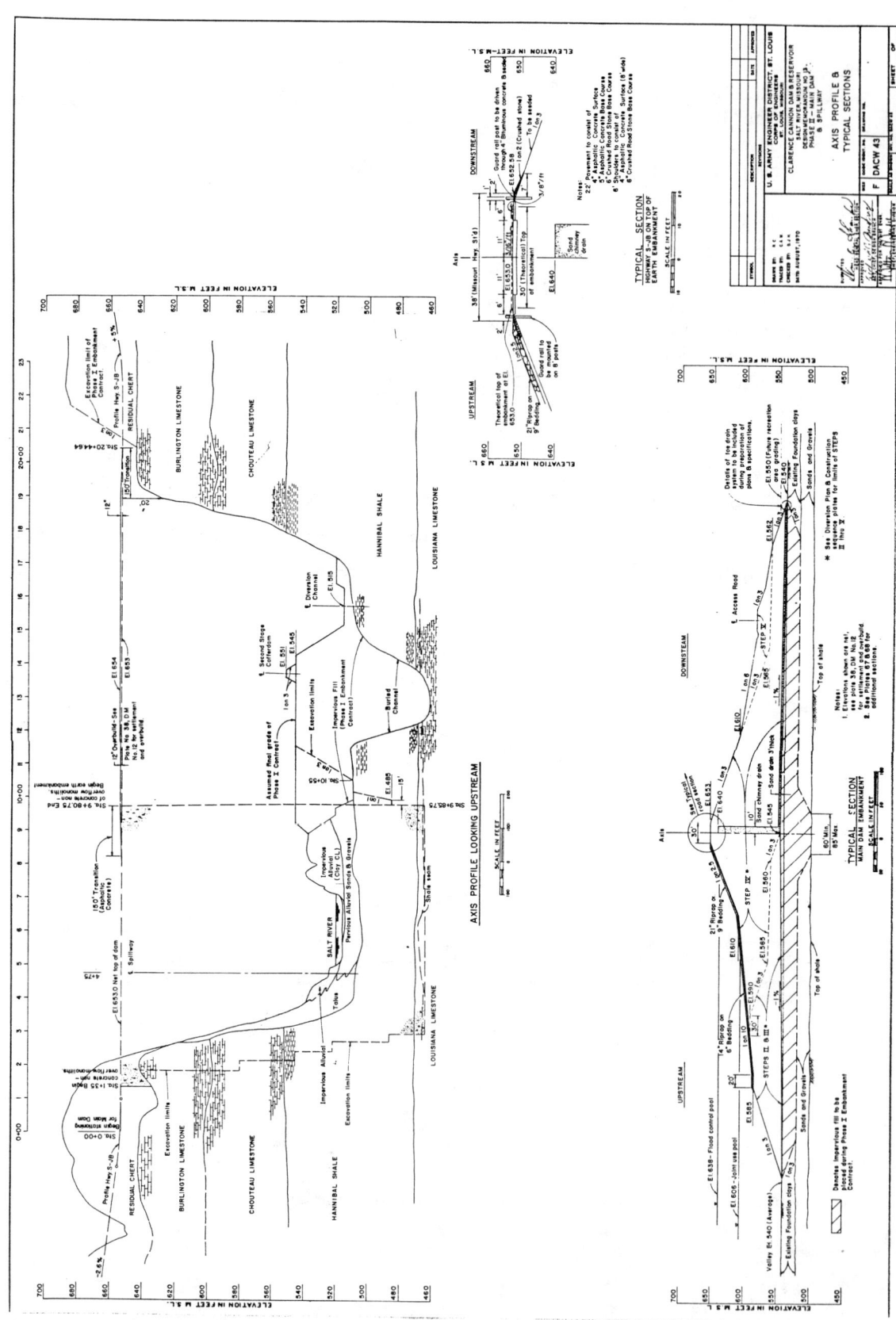

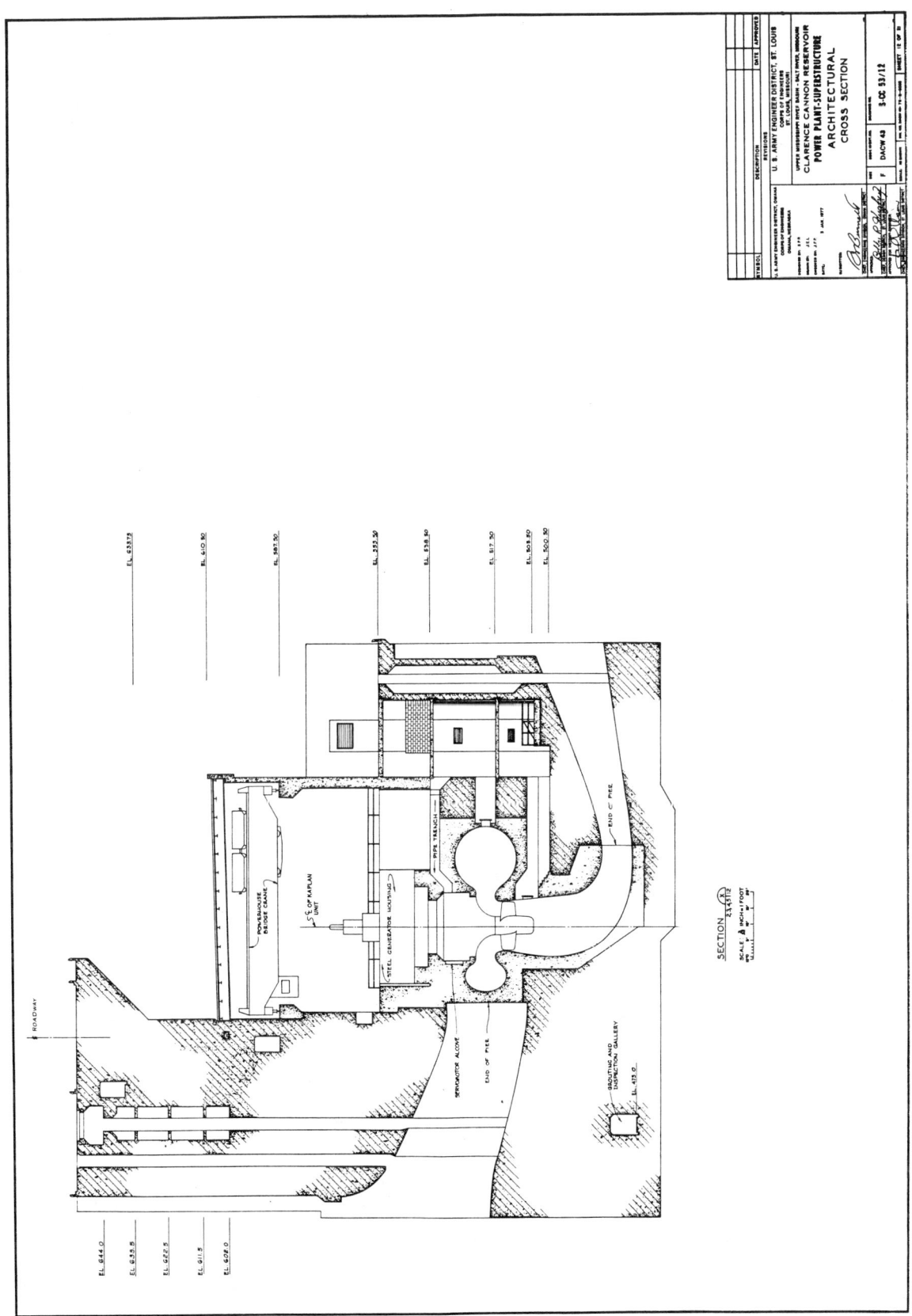

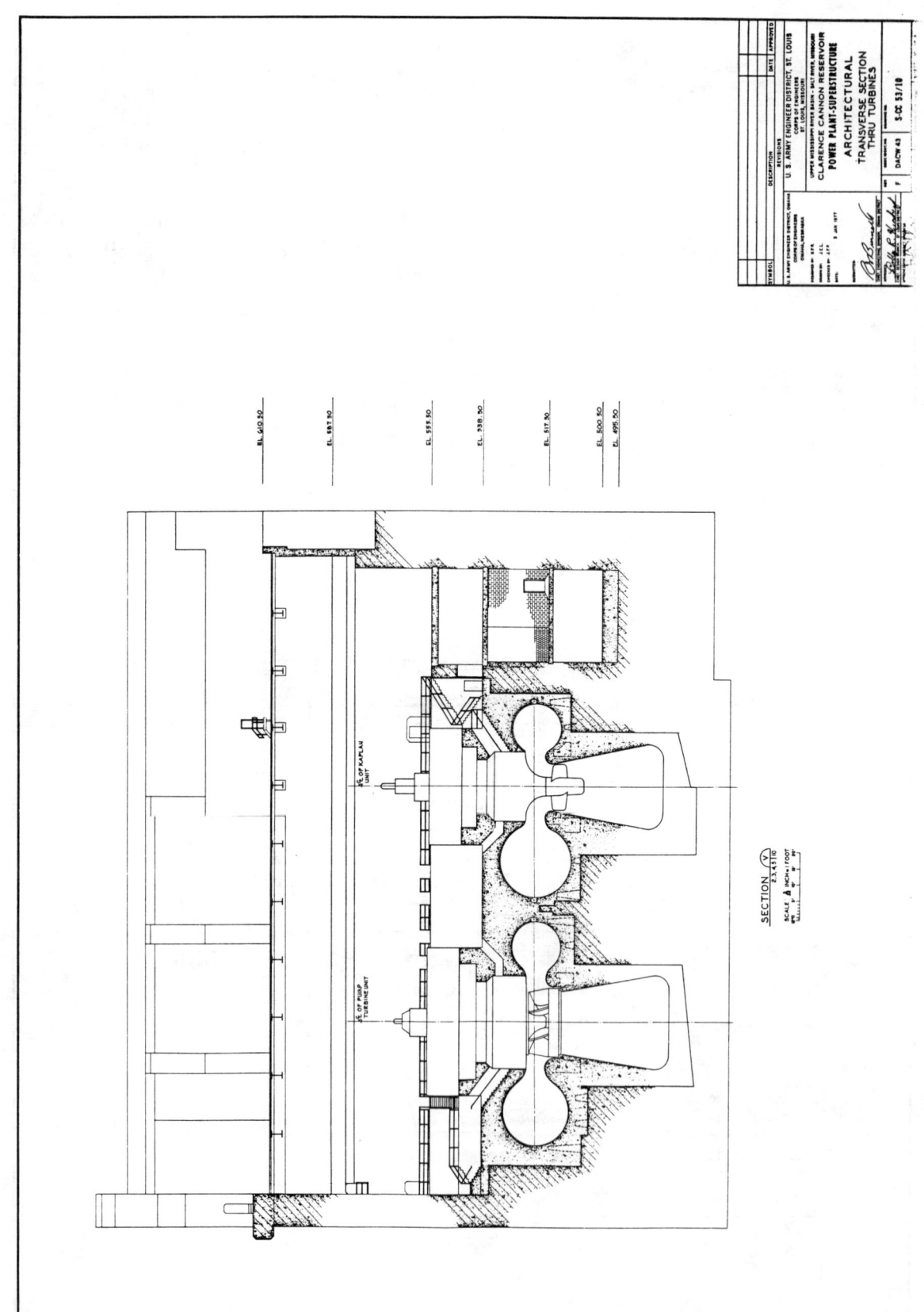

Power Plant and Mark Twain Lake

Reregulating Dam

Clarence Cannon Pumped Storage Plant

Plant Name: **DE GRAY DAM AND RESERVOIR**

Plant location:
  Arkadelphia, AK
  Clark County
	Owner: Vicksburgh District
	  Corps of Engineers
	  P.O. Box 60
	  Vicksburg, Mississippi 39180-0060

Rated capacity   28 MW
Average static head 171 ft
Plant efficiency   76.00 %
Stored energy   191 MWh
Number of units   1
	Designers:
	  Vicksburgh District, Corps of Engineers
	  Omaha District, Corps of Engineers

Construction time: 7 years, 10 months
Construction cost: $1,059 per kW
Price level: 1972
First commercial power: December 1971
FERC project number: None
	Plant Manager/Superintendent:
	  James Colwell
	  Blakely Mountain Powerhouse
	  P.O. Box 4
	  Mountain Pine, Arkansas 71956
	  (501) 767-2401

River or water source: Reregulating Dam on the Caddo River	

	UPPER RESERVOIR	LOWER RESERVOIR
DAM		
Type	Rolled earthfill, De Gray Dam	Rolled earthfill
Height (ft)	243	40
Crest length (ft)	3,400	1,071
Volume (yd^3)	6,831,000	56,000
RESERVOIR		
Type	Constructed, De Grey Lake	Constructed, reregulating dam
Surface area (acres)	13,400	430
Usable power storage (acre-ft)	393,200	1,600
Power pool fluctuation (ft)	41.0	4.0
Operating levels		
Maximum (ft)	402.0	221.0
Minimum (ft)	367.0	217.0
Drainage area (miles^2)	453.0	27.0
Seepage (ft^3/s)	1.716	0.172
SPILLWAY		
Design flood		
Return period (years)	>500 yrs.	SPF
Flow (ft^3/s)	67,000	45,000
Capacity (ft^3/s)	67,000	45,000
Type	Rock saddle, ungated	Ogee, ungated 94.5 m (310 ft) crest length
Gates		
Number	None	None
Type		
Width (ft)		
Height (ft)		
OUTLET WORKS		
Discharge capacity (ft^3/s)	6,000	4,400
Number of water passages	1	5
Dimensions of water passages		
Height (ft)	6.00	9.00
Width (ft)	11.50	5.00
Diameter (ft)		
Type of gates	Vertical lift	Sluice
Number of gates	1	5

Plant Name: **DE GRAY DAM AND RESERVOIR**

Plant location
Arkadelphia, AK
Clark County

Rated capacity	28 MW
Average static head	52.1 m
Plant efficiency	76.00 %
Stored energy	191 MWh
Number of units	1

Construction time: 7 years, 10 months
Construction cost: $1,059 per kW
Price level: 1972
First commercial power: December 1971
FERC project number: None

Owner: Vicksburgh District
Corps of Engineers
P.O. Box 60
Vicksburg, Mississippi 39180-0060

Designers:
Vicksburgh District, Corps of Engineers
Omaha District, Corps of Engineers

Plant Manager/Superintendent:
James Colwell
Blakely Mountain Powerhouse
P.O. Box 4
Mountain Pine, Arkansas 71956
(501) 767-2401

River or water source: Reregulating Dam on the Caddo River

	UPPER RESERVOIR	LOWER RESERVOIR
DAM		
Type	Rolled earthfill, De Gray Dam	Rolled earthfill
Height (m)	74.1	12.2
Crest length (m)	1,036.3	326.4
Volume (m³)	5,222,677	42,815
RESERVOIR		
Type	Constructed, De Grey Lake	Constructed, reregulating dam
Surface area (Mm²)	54.23	1.74
Usable power storage (Mm³)	485.008	1.974
Power pool fluctuation (m)	12.50	1.22
Operating levels		
Maximum (m)	122.53	67.36
Minimum (m)	111.86	66.14
Drainage area (Mm²)	1,173.271	69.930
Seepage (m³/s)	0.0486	0.0049
SPILLWAY		
Design flood		
Return period (years)	>500 yrs.	SPF
Flow (m³/s)	1,897	1,274
Capacity (m³/s)	1,897	1,274
Type	Rock saddle, ungated	Ogee, ungated 94.5 m (310 ft) crest length
Gates		
Number	None	None
Type		
Width (m)		
Height (m)		
OUTLET WORKS		
Discharge capacity (m³/s)	170	125
Number of water passages	1	5
Dimensions of water passages		
Height (m)	1.829	2.743
Width (m)	3.505	1.524
Diameter (m)		
Type of gates	Vertical lift	Sluice
Number of gates	1	5

DE GRAY DAM AND RESERVOIR

INTAKES	UPPER INTAKE	LOWER INTAKE
Number	1	1
Type	Vertical, octagonal tower, with selective temperature control	Draft tube
Design discharge (ft³/s)	28,000	21,995
Gross inlet area (ft²) (at trash racks)	1,890	240
Bar racks		
spacing (in)	6.00	
shape	Rectangular	
depth/thickness (in)	3.00 / 0.63	
diameter (in)		
Emergency gates		
number	4	None
height/width (ft)	23.50 / 22.50	
type	Trashrack, bulkhead	
Service gates		
number	1	2
height/width (ft)		16.92 / 17.67
type	Cylinder, 8.83 m (29 ft) diameter	Vertical lift, sliding
Bulkhead/stop logs (Y or N)	Y	Y
number of units serviced	2	2
Hoists		
number	3	1
capacity (tons)	17	10
type	Screw	Gantry crane - 2 drum

WATER PASSAGES	Upper Tunnel	Shaft	Lower Tunnel	Surge Tanks Upper	Surge Tanks Lower	Penstocks	Tailrace Tunnel
Number		1	1			1	
Diameter (ft)		29.0	29.0			13.2	
Length (ft)		118	1,258			194	
Maximum velocity (ft/s)		15.4	15.4			20.5	
Concrete lining thickness (in)		27.00	30.00				
Total length of concrete sections (ft)		118	423				
Steel liner Thickness							
Minimum (in)			0.88			0.50	
Maximum (in)			1.13			0.81	
Material grade			ASTM 201 Grade B			ASTM 210 Grade B	
Total length of steel-lined sections (ft)			835			194	

Notes:

De Gray Dam And Reservoir - Page 2 (English)

DE GRAY DAM AND RESERVOIR

INTAKES	UPPER INTAKE	LOWER INTAKE
Number	1	1
Type	Vertical, octagonal tower, with selective temperature control	Draft tube
Design discharge (m³/s)	792.9	622.8
Gross inlet area (m²) (at trash racks)	175.6	22.3
Bar Racks:		
spacing (mm)	152	
shape	Rectangular	
depth/thickness (mm)	76 / 16	
diameter (mm)		
Emergency gates		
number	4	None
height/width (m)	7.163 / 6.858	
type	Trashrack, bulkhead	
Service gates		
number	1	2
height/width (m)		5.156 / 5.385
type	Cylinder, 8.83 m (29 ft) diameter	Vertical lift, sliding
Bulkhead/stop logs (Y or N)	Y	Y
number of units serviced	2	2
Hoists		
number	3	1
capacity (Mg)	16	9
type	Screw	Gantry crane - 2 drum

WATER PASSAGES	Upper Tunnel	Shaft	Lower Tunnel	Surge Tanks Upper	Surge Tanks Lower	Penstocks	Tailrace Tunnel
Number		1	1			1	
Diameter (m)		8.84	8.84			4.01	
Length (m)		36.0	383.4			59.1	
Maximum velocity (m/s)		4.69	4.69			6.25	
Concrete lining thickness (m)		0.686	0.762				
Total length of concrete sections (m)		36.0	128.9				
Steel liner Thickness							
Minimum (mm)			22			13	
Maximum (mm)			29			21	
Material grade			ASTM 201 Grade B			ASTM 210 Grade B	
Total length of steel-lined sections (m)			254.5			59.1	

Notes:

De Gray Dam And Reservoir - Page 2 (Metric)

DE GRAY DAM AND RESERVOIR

POWERHOUSE and RELATED FEATURES

Powerhouse Structure
Type: Surface
Length: 150 ft   Width: 70 ft   Height: 110 ft

Guard Valves
Number: 1   Diameter: 14.0 ft
Type: Butterfly

Transformers
Number: 1
Ratings: 30/40
Voltages: (kV) 115 / 13.2 3 Phase

Generator
Rating generating (MVA): 31.1   Rating pumping (MVA):
Insulation type: Class B
Starting method: Start pump/turbine as pump from conventional unit generator
Starting equipment:

Runners
Material:
Minimum unit submergence: 0.0 ft
WR^2: 30,800,000 (lbf x ft²)
Manufacturer: Newport News Shipbuilding and Drydock Co.
Model test by: Newport New Shipbuilding and Drydock Co.

	Reversible Runners	Reversible Motor/Generator			
Number	1				1
Diameter (ft)	16.20	Rotor	24.61	Stator	30.92
rpm synchronous	128.5				128.5
rpm overspeed	182.0				182.0
Type	Francis				

Information on Runners

Condition:	Gross Head (ft)		Capacity (MW)		Discharge (ft³/s)		Turbine/Pump Eff.(%)	
	Generating	Pumping	Generating	Pumping	Generating	Pumping	Generating	Pumping
Maximum head & maximum power	206	206	33	32	3,000	1,750	90.5	88.0
Minimum head & maximum power	146	146	27	30	2,580	2,030	84.0	88.0

Note: Data in the above table are based on design data.

Condition:	Net Head (ft)		Capacity (MW)		Discharge (ft³/s)		Turbine/Pump Eff.(%)	
	Generating	Pumping	Generating	Pumping	Generating	Pumping	Generating	Pumping
Rated head @ best gate	171	175	29	31	2,150	1,850	93.0	88.0

Note: Data in the above table are based on design data.

DE GRAY DAM AND RESERVOIR

POWERHOUSE and RELATED FEATURES

Powerhouse Structure
Type: Surface
Length: 45.7 m   Width: 21.3 m   Height: 33.5 m

Guard Valves
Number: 1   Diameter: 4.27 m
Type: Butterfly

Transformers
Number: 1
Ratings: 30/40
Voltages: (kV) 115 / 13.2   3 Phase

Generator
Rating generating (MVA): 31.1   Rating pumping (MVA):
Insulation type: Class B
Starting method: Start pump/turbine as pump from conventional unit generator
Starting equipment:

Runners
Material:
Minimum unit submergence: m
WR^2: 12,731,000 (Newtons x m^2)
Manufacturer: Newport News Shipbuilding and Drydock Co.
Model test by: Newport New Shipbuilding and Drydock Co.

	Reversible Runners	Reversible Motor/Generator		
Number	1			1
Diameter m	4.94	Rotor 7.501	Stator	9.424
rpm synchronous	128.5			128.5
rpm overspeed	182.0			182.0
Type	Francis			

Information on Runners

Condition:	Gross Head (m) Generating	Gross Head (m) Pumping	Capacity (MW) Generating	Capacity (MW) Pumping	Discharge (m^3/s) Generating	Discharge (m^3/s) Pumping	Turbine/Pump Eff.(%) Generating	Turbine/Pump Eff.(%) Pumping
Maximum head & maximum power	62.8	62.8	33	32	85.0	49.6	90.5	88.0
Minimum head & maximum power	44.5	44.5	27	30	73.1	57.5	84.0	88.0

Note: Data in the above table are based on design data.

Condition:	Net Head (m) Generating	Net Head (m) Pumping	Capacity (MW) Generating	Capacity (MW) Pumping	Discharge (m^3/s) Generating	Discharge (m^3/s) Pumping	Turbine/Pump Eff.(%) Generating	Turbine/Pump Eff.(%) Pumping
Rated head @ best gate	52.1	53.3	29	31	60.9	52.4	93.0	88.0

Note: Data in the above table are based on design data.

De Gray Dam And Reservoir - Page 3 (Metric)

DE GRAY DAM AND RESERVOIR

Plant Data:

Average GWh generating per year:	91
Average GWh pumping per year:	1
Starting time from standstill (s):	240
Changeover time pumping to generating (min):	15
Planned/scheduled time between major overhauls (years):	
Outage time required per unit during major overhauls (weeks):	
Representative plant availability (%):	98.0
Representative planned outages (weeks per year):	2

Miscelleneous Notes:

De Gray Dam and Reservoir includes two units, one conventional and one reversible unit. The information in the tables is for the reversible unit.

The upper dam includes, in addition to the main embankment reported in the tables, a rolled earthfill dike 30.5 m (100 ft) high, and 4.0 Km (2.5 miles) long, and contains a volume of 4,681,000 m³ (6,122,000 yd³).

Under normal operating conditions unit submergence is 1.5 m (5 ft). A shutdown alarm sounds for a condition of zero submergence.

An 8.8 m (29 ft) reinforced concrete tunnel that is 111.6 m (366.1 ft) in length connects the intake structure and vertical shaft to the upper reservoir. The tunnel is only used during emergency conditions to drain the upper reservoir.

Cavitation Experience:

Cavitation has affected 2 ft² per bucket, 7 buckets inlain with stainless steel.

Significant or Unique Problems:

A major fault was encountered in the left abutment toe. This was excavated and backfilled with impervious clay-gravel. A three line grout curtain was installed. Due to the discovery of two smaller faults located near the downstream end of tunnel and powerhouse penstock area, rock bolts were used upstream from the powerhouse wall.

List of Licenses Required:

None -- direct congressional authorization.

ENVIRONMENTAL FEATURES

Recreation:

There are 21 recreation sites, including 6 Class A camp areas, 19 boat-launching ramps, and 10 swimming beaches. Visitation in 1990 was 9,252,000 visitor hours.

DE GRAY DAM AND RESERVOIR

Fish and Wildlife:
There are over 121 Mm² (30,000 acres) of land and water for fisheries, forestry, and wildlife management. Natural resources are indigenous to this region of Arkansas. There is a 546 Km² (1,350 acre) waterfowl refuge near the re-regulation pool. Bald eagles are winter residents.

Social:
A total of 1.726 Mm³ (1,400 acre-feet) in the re-regulation pool is maintained for continuous discharge downstream for water supply, fish and wildlife purposes and pollution abatement.

Page 4 (Continued)

DEGRAY PUMPED STORAGE
DESCRIPTION OF PROJECT FEATURES

Project Authorization. Construction of the DeGray Project was authorized (as a multiple-purpose project) by the River and Harbor Act of 1950, approved 17 May 1950, as recommended by the Chief of Engineers Senate Document Number 117, Eighty-first Congress, first session.

Purpose. Flood control, hydroelectric power, water supply, recreation, pollution abatement.

Initial Work Started. July 1962, on access road. Power production commenced in November 1971.

General Description of Project. DeGray Lake Project is located in southwestern Arkansas, on the Caddo River, a tributary of the Oauchita River. The main dam is located approximately 7 miles northwest of the town of Arkadelphia. The project consists of the following pertinent features which are described in detail on the following pages.

- An earthfill main dam approximately 3,400 feet long, 243 feet high above the stream bed, with a crest elevation of 453 feet, m.s.l.

- An earthfill saddle dike approximately 13,200 feet long, a maximum height of 98 feet and a crest elevation of 453 feet, m.s.l.

- One main tunnel 29 feet in diameter, approximately 1,700 feet long is used for flood control release and power production.

- Two branch tunnels which serve as power taps for operation of the turbines. One tunnel is 15 feet 9 inches in diameter and 100 feet long; the other tunnel is 13 feet 2 inches in diameter and approximately 194 feet long.

- An intake structure with access bridge approximately 400 feet upstream of the dam axis. The intake structure is 96 feet wide, 112 feet high, with a 29-foot diameter vertical shaft extending to the main tunnel.

- An unlines broad-crested type spillway located approximately 4,000 feet east of the dam. The spillway has a crest at elevation 423 feet, m.s.l., with a width of 200 feet.

- A powerplant which initially has a 40,000 kW conventional Francis turbine and a 28,000 kW reversible Francis pump-turbine. Provisions have been made to add an additional 40,000 kW conventional Francis turbine when justified in the future.

- A reregulating dam located approximately 3 miles downstream of the main dam for pumped storage use. It is an earthfill structure with a concrete spillway section. The earthfill portions are a total of 1,070 feet long with a crest elevation of 235 feet, m.s.l. The spillway section is 310 feet wide with a crest elevation of 221 feet, m.s.l. Flow through the spillway section is regulated by five 5-foot by 9-foot gated sluices.

TUNNELS AND INTAKE STRUCTURE

Main Tunnel. The main tunnel is excavated through sandstone and shale in the west abutment. This tunnel extends from the diversion intake to the outlet works and is 29 feet in diameter, approximately 1,700 feet long, and has an embedded steel liner from the axis of the dam embankment to the power taps. The remainder of the tunnel is lined with reinforced concrete.

This tunnel served for diversion of the river during construction of the dam embankment and is now used for flood control releases and power. Flood control releases are controlled by a vertical lift slide gate located at the downstream end of the main tunnel.

Construction of the main tunnel was started in January 1964 and completed in June 1966.

Branch Tunnels. There are two branch tunnels, which serve as power taps for operation of the turbines. These tunnels extend from the main tunnel to the powerhouse and are lined with steel plate for their entire length. One power tap has an inside diameter of 15'9" and is approximately 194 feet long. The branch tunnels were completed in July 1968.

Intake Structure. The intake structure is located on the west abutment approximately 400 feet upstream from the axis of the dam. This is a reinforced concrete structure 96 feet wide and 112 feet high. Water enters through four sides of the structure and is discharged down a 19-foot diameter vertical shaft into the main tunnel. Closure is effected by a cylinder gate located inside the structure. The temperature of the water released into the Caddo River is controlled through the use of baffle gates and bulkheads which may be placed to provide for intake of water from three different elevations. A semi-gantry crane is located on the deck of the intake structure for handling the bulkheads and baffle gates and for operation of the trashrack rake.

An access bridge connects the intake structure deck to the road on the dam. This bridge consists of prestressed concrete box girders supported by three reinforced concrete piers.

Construction of the intake structure and bridge was started in June 1966 and completed in July 1968.

Low Level Outlet Facilities. Provision has been made to unwater the lake in case of emergency. These facilities consist of a prestressed concrete vertical lift slide gate (12'4" x 12'4") located in the main tunnel just upstream of the intake shaft/tunnel intersection. Cables are provided from the gate to the deck of the intake structure and the intake semi-gantry crane will be used to lift the gate. A reinforced concrete orifice has been constructed at the upstream end of the main tunnel to reduce the opening to be spanned by the emergency bulkhead and provide more favorable flow conditions. The emergency bulkhead was used for tunnel closure during installation of the gate and will be used for emergency closure in case the gate needs repairs.

The low level outlet facilities were constructed under the powerplant contract.

EMBANKMENT

Dam. The main dam is of earthfill construction, 243 feet high above the streambed and 3,400 feet long, with a crest elevation of 453 feet, m.s.l. Approximately 6,831,000 cubic yards of material were used in construction of the embankment. Both the upstream and downstream slopes are protected by riprap.

Construction was started on the dam in June 1965 and was completed in October 1969.

Dike. The dike is of earthfill construction and is located along the drainage divide between the Caddo River and Ouachita River. It extends northward from the high sandstone ridge northeast of the damsite for approximately 13,200 feet. The crest of the dike is at elevation 453.0 feet, m.s.l., and its maximum height is approximately 98 feet. The dike slope on the lakeside is protected by riprap and contains approximately 6,122,000 cubic yards of material.

Construction on the dike was started in June 1965 and completed in November 1969.

SPILLWAY

The spillway, located approximately 4,000 feet east of the dam, is an uncontrolled, unlined, broad-crested type with a crest elevation of 423.0 feet, m.s.l. It consists of an approach channel, a flat control section, and an outlet channel. The control section is 200 feet wide and approximately 250 feet long. It is located in sandstone sufficiently resistant to scour to afford a satisfactory control. The inlet and outlet channels slope away from the control section at 0.5 percent to provide drainage. The excavated channel section has 2V on 1H side slopes up to the top of rock and 1V on 1.5H side slopes above the top of rock. A 20-foot berm is provided at the top of rock to increase stability of the side slopes. A 4-foot high chain link type fence extends around both sides of the spillway cut.

Construction of the spillway was started in July 1965 and completed in May 1967.

POWERPLANT

The powerhouse is located riverward of the outlet works and provisions are made for the addition of a third unit at a future date without disturbing the initial units' operation. The switchyard is located on a berm on the left side of the powerhouse and provides room for additional switchyard equipment in the future.

The present switchyard provides two 115 kV lines and three bays for future use, one for the third unit, and two for additional lines. The transformer yard is located adjacent to the right downstream corner of the powerhouse and also provides space for a second transformer when required. Location of the future third unit, a conventional generating type, will be at the downstream end of the present 29-foot diameter tunnel.

The powerplant contains a conventional Francis turbine on the right or landward side, and a reversible Francis pump-turbine on the left or riverward side. The conventional Francis turbine has a rated horsepower of 63,600 at a rated head of 171 feet with a speed of 150 r.p.m. The pump-turbine unit, in the generating cycle, has a rated horsepower of 44,500 at the rated head of 171 feet and at a speed of 128.5 r.p.m. In the pumping cycle, the pump-turbine unit has a rated capacity of 1,900 c.f.s. at a rated dynamic head of 170 feet and a speed of 128.5 r.p.m.

The conventional generator is rated at 42,105 kVA with a .95 power factor. The pump-turbine unit, operating as a generator, is rated at 29,474 kVA with a .95 power factor.

The DeGray Powerplant is unmanned and controlled from Blakely Mountain. No visitors' facilities have been provided for the powerhouse, although sidewalks, steps and parking facilities for public fishing have been provided on the exterior.

The contract for construction of the powerplant was awarded in July 1968 and completed in January 1972.

REREGULATING DAM

The reregulating dam is located on the Caddo River, approximately 3 miles downstream from the main dam. It is an earthfill structure with a concrete spillway section. The spillway lies in the present water course of the river. The earthfill portions extend westward from the spillway approximately 190 feet, abutting the right valley wall and eastward approximately 880 feet across a wide alluvial shelf.

The spillway consists of a 310-foot weir section with a crest at elevation 221.0, a stilling basin, and upstream and downstream walls. Flow is regulated by five 5-foot by 9-foot gated sluices. All components of the spillway rest on a rock foundation.

The five weir sluiceways are operated to provide continuous discharges downstream of the regulating dam for water supply, fish and wildlife purposes and pollution abatement. The sluiceway gates can be operated locally or remotely from the DeGray or Blakely Mountain Powerhouses. The design contemplated a pool between spillway crest elevation (221.0) and a minimum pool (209.0) with a total storage capacity of 2,000 acre feet. The pump back and water supply portion of the pool between elevations 221.0 and 217.0 would contain 1,600 acre feet, and the remainder of the pool (1,400 acre feet) between elevations 217 and 209 would provide storage for downstream releases during periods on non-generation. Drawdown to elevation 209 would occur only during most adverse operation conditions.

Initial continuous releases from the reregulating pool are 152 c.f.s. through the sluiceways with excess water, if any, allowed to go over the spillway. The ultimate continuous releases will be 387 c.f.s.

Bids were opened on 22 July 1969, and the contract was awarded on 5 August 1969 and completed in November 1972.

LAKE

Construction of this project creates a lake which controls rainfall runoff from a drainage area of 453 square miles, has a maximum storage capacity of 882,000 acre-feet, and maximum surface area of 17,000 acres at flood control pool.

The power and water supply pool, is at 408 feet, m.s.l., and has a total storage capacity of 393,200 acre-feet and a surface area of 13,400 acres.

All lake clearing has been completed including tree topping operations which were finished in August 1970.

Initial filling of the lake began in August 1969.

This article was provided by Luther Newton, and is taken from a Corps of Engineers report.

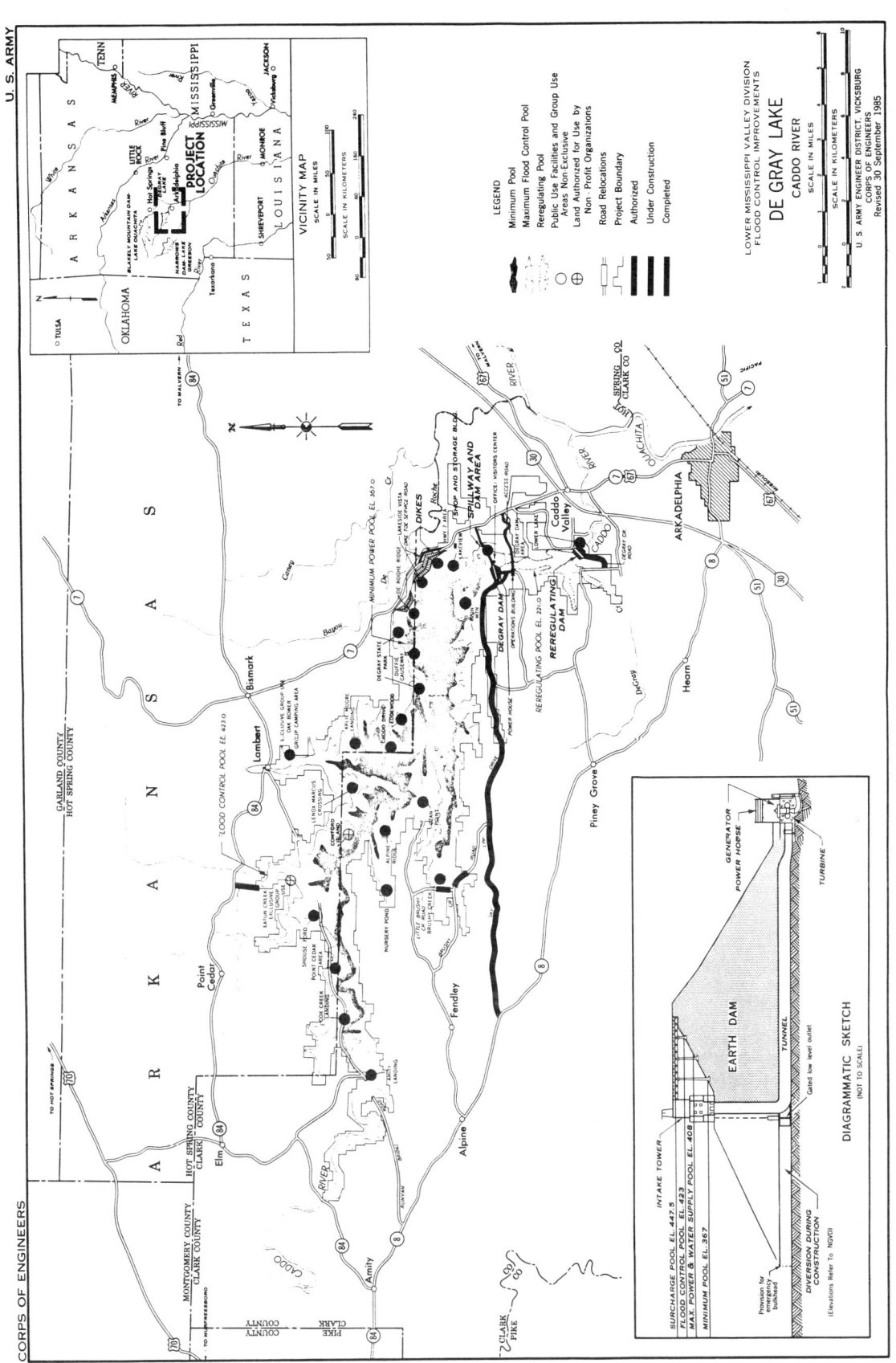

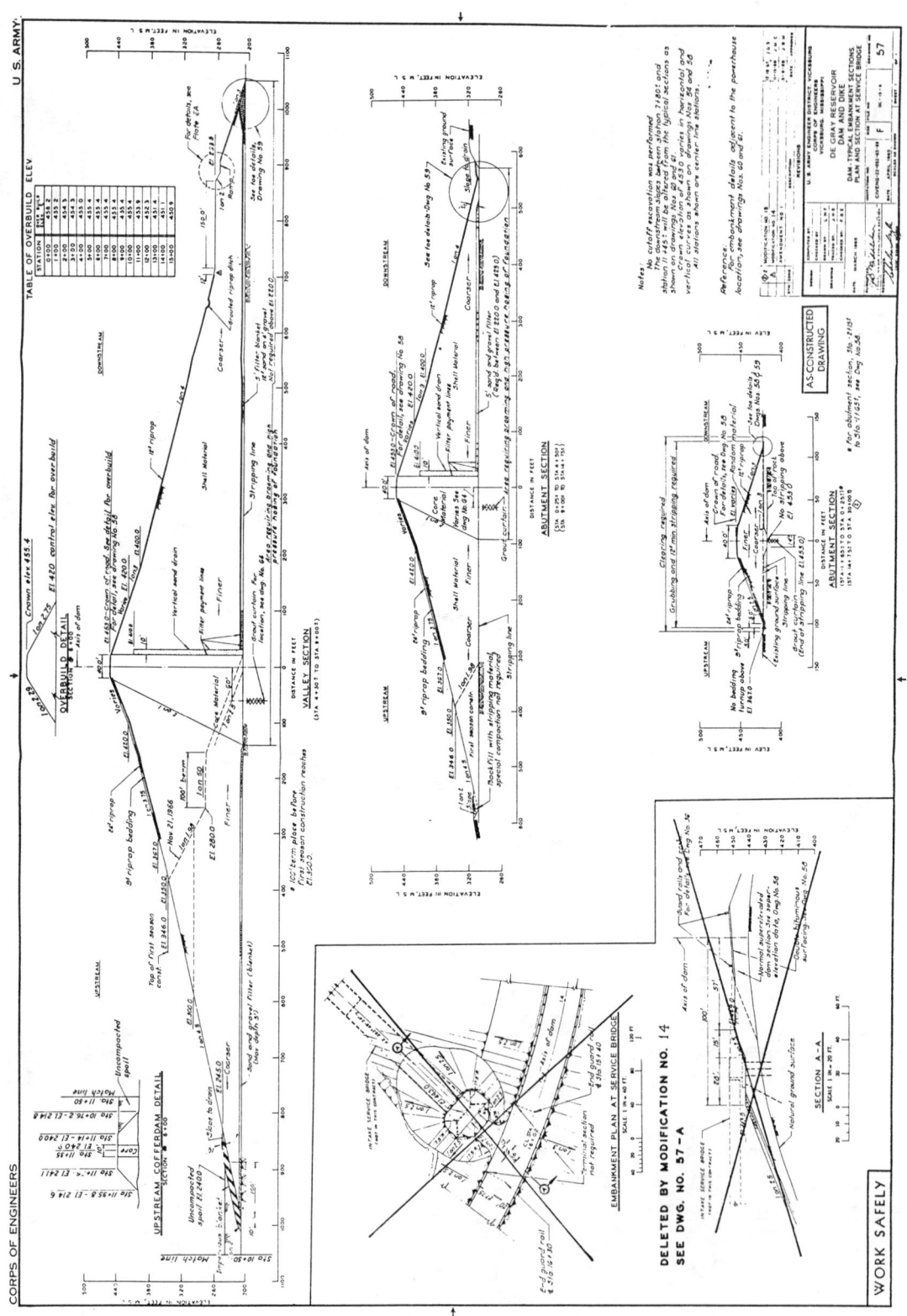

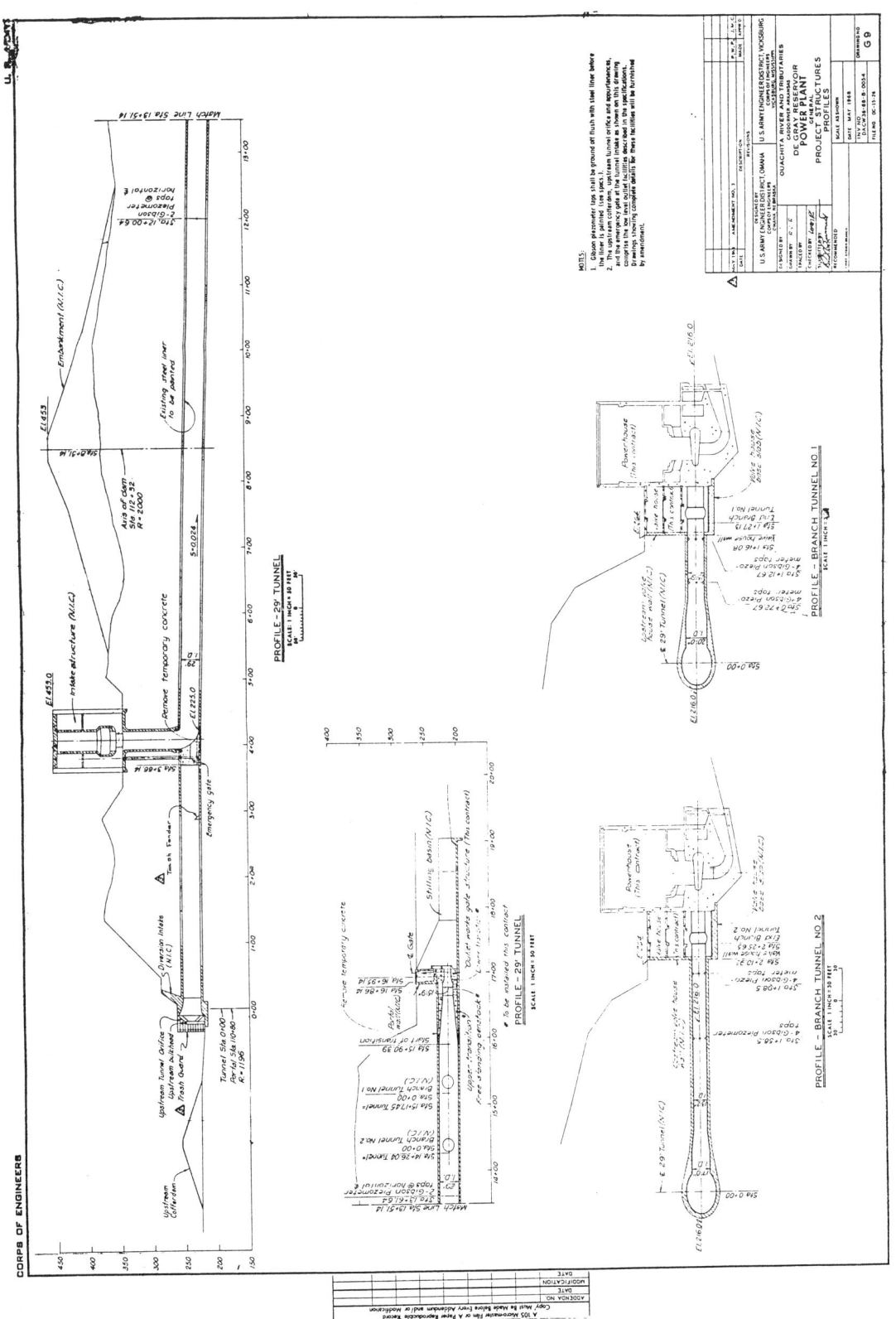

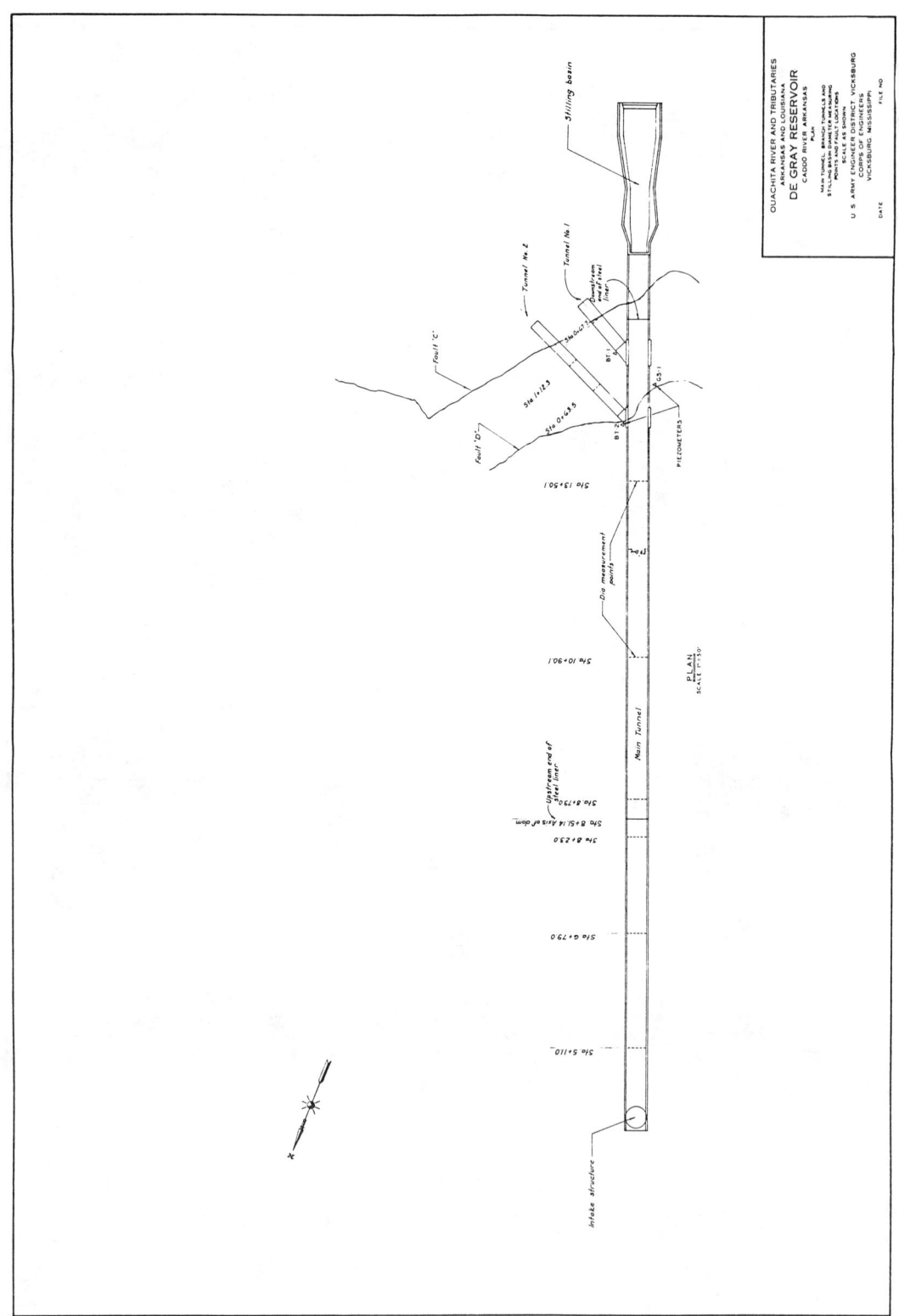

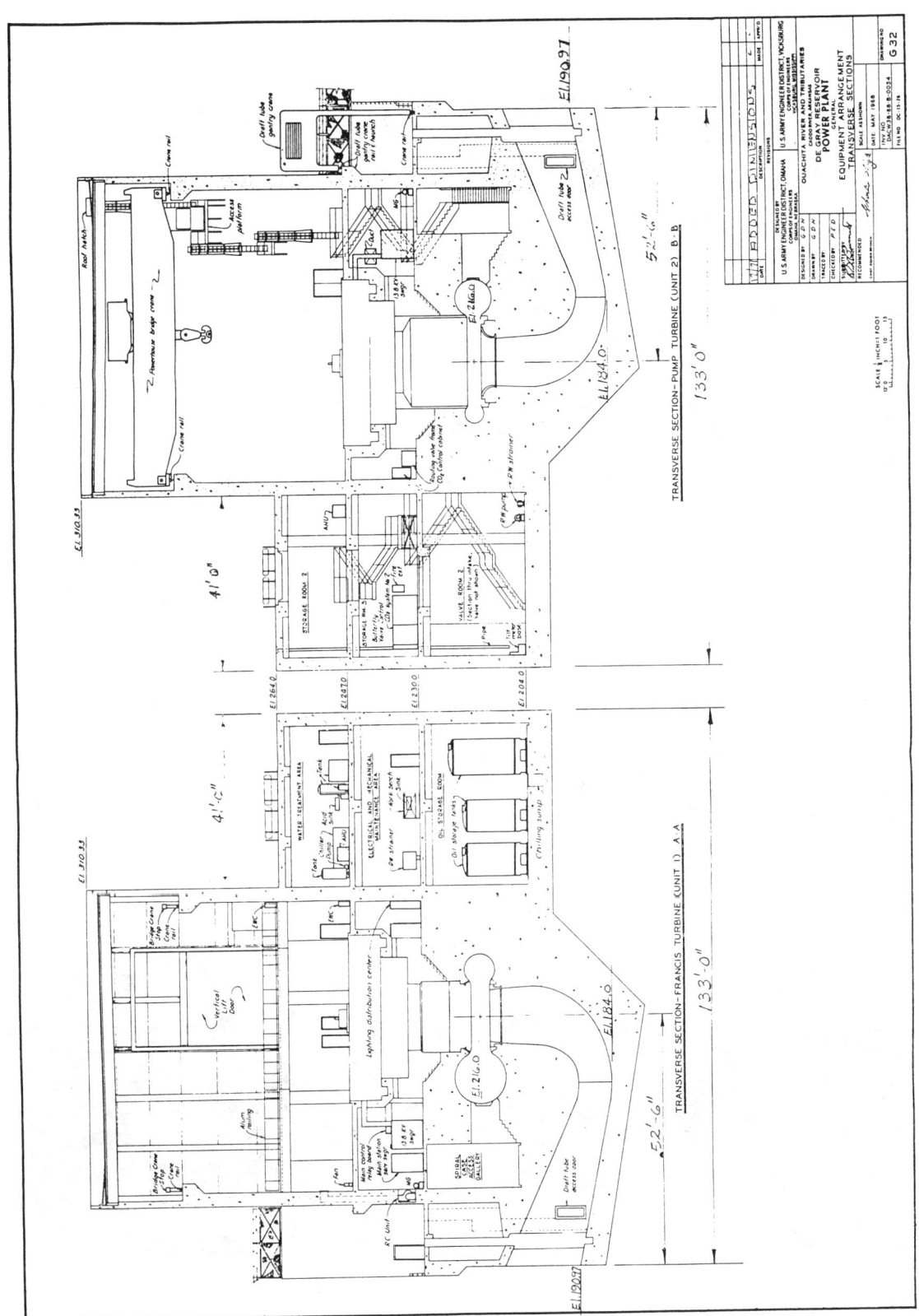

DeGray Pumped Storage Plant

VIEW OF WESTERN END OF DEGRAY DAM. INTAKE AND ACCESS BRIDGE ON LAKESIDE AND POWERHOUSE AT DOWNSTREAM TOE.

VIEW LOOKING SOUTHWEST AT DEGRAY RE-REGULATING DAM.

DeGray Pumped Storage Plant

Plant Name: **EDWARD HYATT POWER PLANT**

Plant location:
Oroville, CA
Butte County

Rated capacity 293 MW
Average static head 615 ft
Plant efficiency   %
Stored energy   MWh
Number of units   3

Construction time: 6 years, 1 month
Construction cost: $102 per kW
Price level: 1967
First commercial power: October 1967
FERC project number: 2100

River or water source: Feather River

Owner: California Department of Water Resoources
1416 Ninth Street
Sacramento, California 95814

Designers:
State of California, Department of Water Resources
Division of Design and Construction

Plant Manager/Superintendent:
Rolland Williams
Chief, Oroville Field Division
460 Glen Drive
Oroville, California 95965
(916) 534-2413

	UPPER RESERVOIR	LOWER RESERVOIR
DAM		
Type	Zoned earthfill, Oroville Dam	Concrete gravity
Height (ft)	770	143
Crest length (ft)	6,920	1,300
Volume (yd³)	80,000,000	154,000
RESERVOIR		
Type	Constructed, Oroville Reservoir	Constructed
Surface area (acres)	15,800	323
Usable power storage (acre-ft)	2,685,000	13,000
Power pool fluctuation (ft)	170.0	4.0
Operating levels		
Maximum (ft)	900.0	225.0
Minimum (ft)	730.0	221.0
Drainage area (miles²)	3,611.0	0.1
Seepage (ft³/s)	0.089	0.007
SPILLWAY		
Design flood		
Return period (years)	450 yrs	
Flow (ft³/s)	440,000	
Capacity (ft³/s)	150,000	650,000
Type	Concrete channel	Gated ogee crest
Gates		
Number	8	14
Type	Radial	Radial
Width (ft)	17.58	40.00
Height (ft)	33.50	23.00
OUTLET WORKS		
Discharge capacity (ft³/s)	5,400	800
Number of water passages	2	3
Dimensions of water passages		
Height (ft)		
Width (ft)		
Diameter (ft)	6.00	5.00
Type of gates		
Number of gates	None	None

Plant Name: **EDWARD HYATT POWER PLANT**

Plant location Oroville, CA Butte County	Owner: California Department of Water Resoources 1416 Ninth Street Sacramento, California 95814
Rated capacity 293 MW Average static head 187.5 m Plant efficiency % Stored energy MWh Number of units 3	Designers: State of California, Department of Water Resources Division of Design and Construction
Construction time: 6 years, 1 month Construction cost: $102 per kW Price level: 1967 First commercial power: October 1967 FERC project number: 2100	Plant Manager/Superintendent: Rolland Williams Chief, Oroville Field Division 460 Glen Drive Oroville, California 95965 (916) 534-2413
River or water source: Feather River	

	UPPER RESERVOIR	LOWER RESERVOIR
DAM		
Type	Zoned earthfill, Oroville Dam	Concrete gravity
Height (m)	234.7	43.6
Crest length (m)	2,109.2	396.2
Volume (m³)	61,164,418	117,742
RESERVOIR		
Type	Constructed, Oroville Reservoir	Constructed
Surface area (Mm²)	63.94	1.31
Usable power storage (Mm³)	3,311.918	16.035
Power pool fluctuation (m)	51.82	1.22
Operating levels		
Maximum (m)	274.32	68.58
Minimum (m)	222.50	67.36
Drainage area (Mm²)	9,352.499	0.259
Seepage (m³/s)	0.0025	0.0002
SPILLWAY		
Design flood		
Return period (years)	450 yrs	
Flow (m³/s)	12,459	
Capacity (m³/s)	4,248	18,406
Type	Concrete channel	Gated ogee crest
Gates		
Number	8	14
Type	Radial	Radial
Width (m)	5.359	12.192
Height (m)	10.211	7.010
OUTLET WORKS		
Discharge capacity (m³/s)	153	23
Number of water passages	2	3
Dimensions of water passages		
Height (m)		
Width (m)		
Diameter (m)	1.829	1.524
Type of gates		
Number of gates	None	None

Edward Hyatt Power Plant - Page 1 (Metric)

EDWARD HYATT POWER PLANT

INTAKES	UPPER INTAKE	LOWER INTAKE
Number	2	
Type	Inclined screen	
Design discharge (ft³/s)	9,000	
Gross inlet area (ft²)	380	
(at trash racks)		
Bar racks		
spacing (in)	4.50	
shape	Elliptical	
depth/thickness (in)	3.50 / 0.75	
diameter (in)		
Emergency gates		
number	2	
height/width (ft)	32.75 / 22.50	
type	Hydraulic, inclined slide gate	
Service gates		
number	None	
height/width (ft)		
type		
Bulkhead/stop logs (Y or N)	N	
number of units serviced		
Hoists		
number	2	
capacity (tons)	50	
type	Gantry	

WATER PASSAGES	Upper Tunnel	Shaft	Lower Tunnel	Surge Tanks Upper	Surge Tanks Lower	Penstocks	Tailrace Tunnel
Number	2					6	2
Diameter (ft)	22.0					12.0	35.0
Length (ft)	385					196	
Maximum velocity (ft/s)						30.0	10.0
Concrete lining thickness (in)	18.00						
Total length of concrete sections (ft)	385						
Steel liner Thickness							
Minimum (in)							
Maximum (in)						1.50	
Material grade						ASTM A-441	
Total length of steel-lined sections (ft)						196	

Notes:
Upper Tunnel: Each tunnel feeds three penstocks.
Penstocks: Average length. Three for pump-turbines and three for conventional units.

Edward Hyatt Power Plant - Page 2 (English)

EDWARD HYATT POWER PLANT

INTAKES	UPPER INTAKE	LOWER INTAKE
Number	2	
Type	Inclined screen	
Design discharge (m³/s)	254.9	
Gross inlet area (m²) (at trash racks)	35.3	
Bar Racks:		
spacing (mm)	114	
shape	Elliptical	
depth/thickness (mm)	89 / 19	
diameter (mm)		
Emergency gates		
number	2	
height/width (m)	9.982 / 6.858	
type	Hydraulic, inclined slide gate	
Service gates		
number	None	
height/width (m)		
type		
Bulkhead/stop logs (Y or N)	N	
number of units serviced		
Hoists		
number	2	
capacity (Mg)	45	
type	Gantry	

WATER PASSAGES	Upper Tunnel	Shaft	Lower Tunnel	Surge Tanks Upper	Lower	Penstocks	Tailrace Tunnel
Number	2					6	2
Diameter (m)	6.71					3.66	10.67
Length (m)	117.3					59.7	
Maximum velocity (m/s)						9.14	3.05
Concrete lining thickness (m)	0.457						
Total length of concrete sections (m)	117.3						
Steel liner							
Thickness							
Minimum (mm)							
Maximum (mm)						38	
Material grade						ASTM A-441	
Total length of steel-lined sections (m)						59.7	

Notes:
Upper Tunnel: Each tunnel feeds three penstocks.
Penstocks: Average length. Three for pump-turbines and three for conventional units.

Edward Hyatt Power Plant - Page 2 (Metric)

EDWARD HYATT POWER PLANT

POWERHOUSE and RELATED FEATURES

Powerhouse Structure
Type: Underground
Length: 550 ft   Width: 69 ft   Height: 137 ft

Guard Valves
Number: 6   Diameter: 9.5 ft
Type: Ball valve

Transformers
Number: 6 Maloney
Ratings: 127
Voltages: (kV) 230 / 13.8

Generator
Rating generating (MVA): 123.0   Rating pumping (MVA): 129.0
Insulation type:
Starting method: Direct across the line start
Starting equipment:

Runners
Material: ASTM A-27 class 70-36
Minimum unit submergence: 30.0 ft
WR^2:
Manufacturer: Allis Chalmers
Model test by: Allis Chalmers

	Reversible Runners	Reversible Motor/Generator	
Number	3		3
Diameter (ft)	19.10	Rotor	Stator
rpm synchronous	189.5		189.5
rpm overspeed	284.0		284.0
Type	Francis		Westinghouse

Information on Runners

Condition:	Gross Head (ft)		Capacity (MW)		Discharge (ft³/s)		Turbine/Pump Eff.(%)	
	Generating	Pumping	Generating	Pumping	Generating	Pumping	Generating	Pumping
Maximum head & maximum power	675	660	123	133	2,500	2,200	91.6	92.0
Minimum head & maximum power	500	500	92	122	2,280	2,350	88.7	87.5

Note: Data in the above table are based on design & model data.

Condition:	Net Head (ft)		Capacity (MW)		Discharge (ft³/s)		Turbine/Pump Eff.(%)	
	Generating	Pumping	Generating	Pumping	Generating	Pumping	Generating	Pumping
Rated head @ best gate	615	592	98			1,870	91.6	

Edward Hyatt Power Plant - Page 3 (English)

EDWARD HYATT POWER PLANT

POWERHOUSE and RELATED FEATURES

Powerhouse Structure
Type: Underground
Length: 167.6 m   Width: 21.0 m   Height: 41.8 m

Guard Valves
Number: 6   Diameter: 2.90 m
Type: Ball valve

Transformers
Number: 6 Maloney
Ratings: 127
Voltages: (kV) 230 / 13.8

Generator
Rating generating (MVA): 123.0   Rating pumping (MVA): 129.0
Insulation type:
Starting method: Direct across the line start
Starting equipment:

Runners
Material: ASTM A-27 class 70-36
Minimum unit submergence: 9.14 m
WR^2:
Manufacturer: Allis Chalmers
Model test by: Allis Chalmers

	Reversible Runners	Reversible Motor/Generator	
Number	3		3
Diameter m	5.82	Rotor	Stator
rpm synchronous	189.5		189.5
rpm overspeed	284.0		284.0
Type	Francis		Westinghouse

Information on Runners

Condition:	Gross Head (m) Generating	Pumping	Capacity (MW) Generating	Pumping	Discharge (m³/s) Generating	Pumping	Turbine/Pump Eff.(%) Generating	Pumping
Maximum head & maximum power	205.7	201.2	123	133	70.8	62.3	91.6	92.0
Minimum head & maximum power	152.4	152.4	92	122	64.6	66.5	88.7	87.5

Note: Data in the above table are based on design & model data.

Condition:	Net Head (m) Generating	Pumping	Capacity (MW) Generating	Pumping	Discharge (m³/s) Generating	Pumping	Turbine/Pump Eff.(%) Generating	Pumping
Rated head @ best gate	187.5	180.4	98			53.0	91.6	

Edward Hyatt Power Plant - Page 3 (Metric)

EDWARD HYATT POWER PLANT

Plant Data:
```
Average GWh generating per year:
Average GWh pumping per year:
Starting time from standstill (s):                    300
Changeover time pumping to generating (min):           60
Planned/scheduled time
   between major overhauls (years):                     1
Outage time required per unit
   during major overhauls (weeks):                      2
Representative plant availability (%):               96.0
Representative planned outages (weeks per year):        2
```

Miscelleneous Notes:
The plant contains six units, three reversible and three conventional units. Each reversible unit is rated at 97.8 MW and each conventional unit is rated at 117 MW.

The reversible units increase the dependable capacity of the plant by about 75% and also increase the conservation water yield.

The units discharge into two 10.67 m (35-ft) diameter tailrace tunnels, used originally for diversion of the Feather River during construction of the dam. The two tunnels are permanantly plugged at the upper end.

The project incorporates an inclined intake structure built into the sloping hillsides of the reservoir near the upstream toe of the dam.

Cavitation Experience:

Significant or Unique Problems:
Debris has caused blockage of the intakes.

List of Licenses Required:

ENVIRONMENTAL FEATURES

Recreation:
In addition to boating, fishing, camping, waterskiing, swimming, and hunting, there are public boat ramps, picnic areas, campgrounds, and beaches available.

Fish and Wildlife:
King salmon, trout, bass, catfish, and panfish, in addition to deer, ducks, geese and turkeys are present.

Social:

Edward Hyatt Power Plant - Page 4

Journal of the POWER DIVISION

Proceedings of the American Society of Civil Engineers

EDWARD HYATT (OROVILLE) UNDERGROUND POWER PLANT[a]

By Alfred R. Golzé,[1] F. ASCE

INTRODUCTION

Oroville Dam and Reservoir and associated power facilities of the California State Water Project are located on the Feather River approximately 5 miles east of the city of Oroville at the northeastern edge of the Central Valley where the Cascade Range coming south meets the Sierra Nevada Range going north. Oroville Reservoir, with a storage capacity of 3,537,577 acre-ft, is the principal conservation unit of the State Water Project. Normal releases from the Reservoir are through the Oroville Underground Power Plant, now known as the Edward Hyatt Power Plant, into the natural channel of the Feather River downstream of Oroville Dam. The Feather River discharges into the Sacramento River which in turn flows into the Delta, formed by the confluence of the Sacramento River flowing south and the San Joaquin River flowing north, (see Fig. 1).

Oroville Dam (Fig. 2) currently the highest dam in the United States, was completed in 1967 and water storage began in November of that year. The reservoir filled for the first time in July, 1969. Power is included in the Oroville development for two major purposes. One is to use the value of the falling water from the reservoir as a means of earning revenue to assist in financing the cost of the water project, and the second is to aid in meeting the growing power loads of the northern part of the State of California.

A contract was negotiated with the Pacific Gas and Electric Company, the Southern California Edison Company, and the San Diego Gas and Electric Company for the purchase of all power produced at Oroville and its adjoining Thermalito Power Plant, which is operated in tandem with the Oroville plant.

Note.—Discussion open until August 1, 1971. To extend the closing date one month, a written request must be filed with the Executive Director, ASCE. This paper is part of the copyrighted Journal of the Power Division, Proceedings of the American Society of Civil Engineers, Vol. 97, No. PO2, March, 1971. Manuscript was submitted for review for possible publication on March 9, 1970.

[a] Presented at the January 29, 1970, ASCE National Meeting on Water Resources Engineering, held at Memphis, Tenn.

[1] Deputy Director, Dept. of Water Resources, The Resources Agency, State of California, Sacramento, Calif.

As constructed and operated, the power complex has a combined dependable capacity of 725 Mw and an average annual energy output of 2,758,000,000 kwhr.

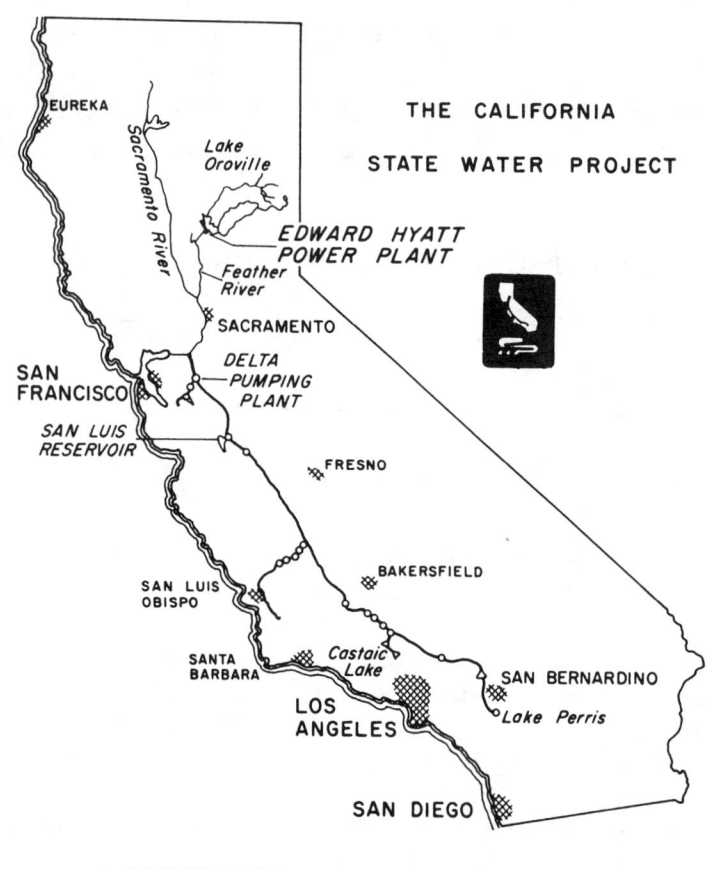

FIG. 1.—LOCATION MAP

FIG. 2.—OROVILLE DAM AND RESERVOIR

The revenues from the sale of this power exceed $16 million per yr and are sufficient to finance two revenue bond issues totaling $244,995,000. The income from the sale of these bonds finances the direct costs of the power facilities and an allocated part of the cost of Oroville Dam, the intake facilities to the powerplant and other joint facilities.

It should be noted at this point that the State Water Project is entirely self-liquidating with revenues from the sale of water and power being sufficient to pay off the entire cost of the Project with interest over a period of about 50 yr from its completion date.

For maximum return of revenue, pumped storage operation is required at Oroville. During offpeak hours, water is pumped back into the Oroville Reservoir for release during onpeak hours. For this purpose, three of the turbines and generators at Oroville out of the total of six, are reversible. They act as generating units during the hours of maximum load demand between 7:00 a.m. and 10:00 p.m. on weekdays, and as pumping units during hours of light load from 10:00 p.m. to 7:00 a.m. on weekdays and all day on Sundays and holidays. Studies indicate that an average of 556,000,000 kwhr of energy will be required annually for the offpeak pump-back operation at the Oroville Thermalito complex.

The cost of the energy consumed in pumping will be recovered many times over because the pump-back operation makes it possible to increase the dependable generating capacity of the power complex by about 75 %. This additional power, generated and marketed during peak hours, is of much higher value than the energy required for pumping during offpeak hours. The pump-back operation also increases the conservation water yield of the Project by about 100,000 acre-ft.

DESIGN CONSIDERATIONS

In deciding what type of power plant to construct at the Oroville site first consideration was given to a conventional surface plant located at the toe of the dam. An outdoor plant would have been restricted as to size, it would have adversely affected the appearance of the great dam rising behing it and it would have required penstocks of considerable length through the dam, or through one of the abutments, with the attendant problem of handling hydraulic transients. After consideration of various types of underground plants such as a spherical rock cavern with a circular power plant, or a rectangular rock cavern, it was decided to construct the latter type of plant with six generators arranged in-line. The study of the economics involved also considered the location of the main transformers both on the surface and in the cavern. The final design selected, resulted in a single rectangular rock cavern (see Fig. 3) of approximately 550 ft in length, 137 ft in height and 69 ft in width.

The cavern does contain all equipment including the large transformers and the penstock valves. Holding the rock excavation to a minimum required a compact installation of the essential equipment. The various units of heavy equipment can be removed for maintenance or repair without disturbing other major elements of equipment. In fact, this has already been accomplished during the initial year of operation of the plant.

Two circular penstock tunnels each of 22-ft diam leading from an inclined intake structure on the left abutment of the dam, convey the reservoir water into the underground installation (see Fig. 4). The entrance to these penstocks

FIG. 3.—NEARLY COMPLETED EXCAVATION OF ROCK CAVERN FOR OROVILLE POWER PLANT

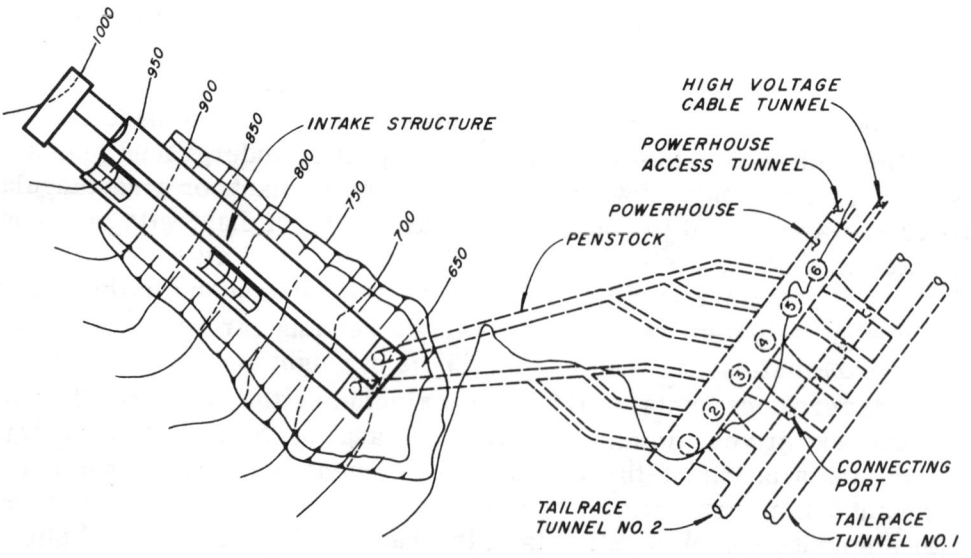

FIG. 4.—GENERAL PLAN INTAKE STRUCTURE AND POWERHOUSE

is at El. 613. During normal operation of the Oroville Reservoir, its water surface will fluctuate between El. 730 and El. 900. After passing through the turbines at El. 205, the water is discharged into two tailrace tunnels, each of which is 35 ft in diameter and are designed for a combined flow of 16,500 cfs during the generating cycle. The tailrace tunnels extending to the upstream toe of the dam were used originally to divert the main stream of the Feather River during construction of the dam. For this latter purpose, the capacity of the two tunnels was as high as 190,000 cfs. The two tunnels have been since permanently plugged at the upper end. One of the tunnels contains an outlet valve which permits maintaining riverflows during any periods when the powerplant may be shut down.

As shown in Figs. 5, 6(a) and 6(b), the underground facility includes a generator room, switchgear gallery, turbine floor, access and high voltage gal-

FIG. 5.—CUTAWAY MODEL OF OROVILLE POWER PLANT

leries, transformer vaults, valve pits, erection bay, and space for general station-operation requirements. To service the machine hall, two 200-ton overhead cranes are available.

Access to the machine hall is obtained from a tunnel constructed in the left abutment from the south side of the Oroville Dam about 1,500 ft into the machine hall. The access tunnel begins at El. 340 and terminates at El. 252, which is the elevation of the powerhouse generator floor. This requires a constant grade of 4° in the tunnel. The tunnel was used during construction of the power plant at its full cross-sectional width of 30.5 ft. After excavation of the machine hall was completed, the tunnel was concrete lined and then divided into two sections, one a passageway 15 ft in width by 23 ft in height and the other a compartment

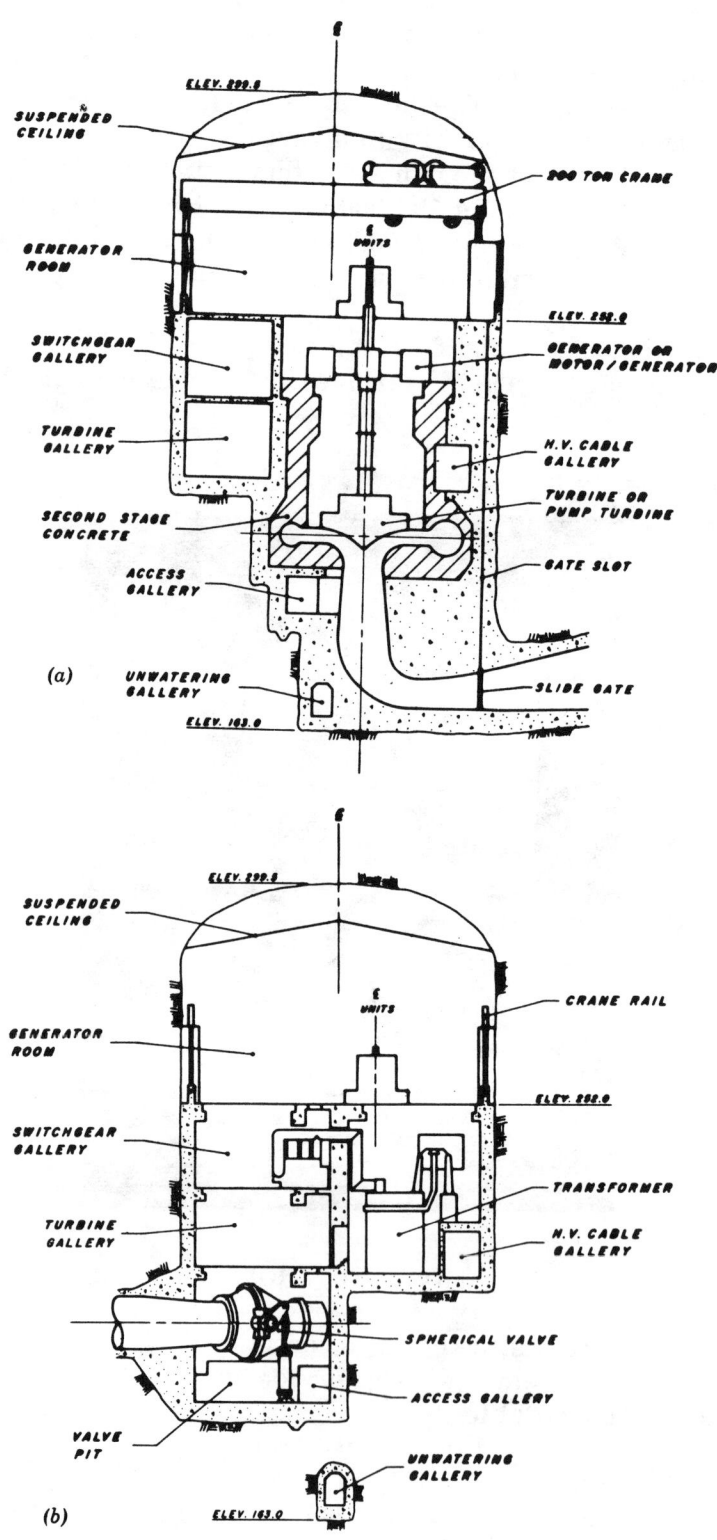

FIG. 6.—(a) SECTION THROUGH CENTER LINE UNIT; (b) SECTION THROUGH TRANSFORMER VAULT

8 ft in width by 23 ft in height in which are housed the control and high voltage cables.

GEOLOGY

The geologic setting for the Oroville Power Plant is in the foothills on the western slope of the Sierra Nevada Range which is a giant granite-cored tilted fault block. A series of tightly folded, steeply dipping metamorphic rocks overlies the granite core along its western and northwestern flanks. Oroville Power Plant was excavated in the upper paleozoic metavolcanic rock formation, one of several units within this metamorphic series. The rock formation is a very hard, dense and fine-grained rock.

A thorough and comprehensive program of geological exploration was executed before the final plant layout was adopted. The final exploration for underground excavation consisted of an exploration tunnel and core drilling. The exploration tunnel, 1,884 ft long, consisted of an adit, drift, and crosscuts. The main drift was located at elevation 250, approximately at the center of the machine hall excavation and at the generator floor level. The total length of all core holes amounted to more than a mile, 5,351 ft in length.

The purpose of this exploration was to evaluate the tectonic features of the rock, establish location of the plant, examine physical features of the joints and shear zones, and permit direct examination of rock conditions by visual inspection and field testing by first the designers and later the prospective contractors.

The six draft tube tunnels are approximately elliptical in shape at the junction with the powerhouse chamber, and circular at the junction with the tailrace tunnels. The horizontal and vertical axes of each ellipse are 36 ft and 20 ft, respectively, and the diameter of the circular section is 12 ft. The six penstock branch-line tunnels enter the powerhouse chamber at El. 205, and are 15 ft in diameter.

The roof of the powerhouse excavation was formed by a compound curve, with a rise of 19 ft and with a maximum width of 72 ft at the section through the draft tube gate slots. The problem facing the designers was to install exceptionally large equipment such as the generators and transformers and yet maintain economy and safety of excavation. For this reason, the roof span of the main cavern was kept to a minimum and the rock stabilization was achieved by rock bolting and grouting.

ROCK REINFORCEMENT AND DRAINAGE

The term rock reinforcement should first be distinguished from rock support. Conventional methods of supporting rock assume that a certain amount of rock acts on the support, such as a steel rib, as a dead load. The effectiveness of any support requires that an intimate contact be obtained between the rock and support system. This is usually accomplished by adequate lagging and blocking between the support system and the rock. As the rock becomes distressed it fails by loosening and load is transferred to the support system. In reality, then, the support system reacts in a passive manner.

Rock reinforcement by rock bolting provides a stress envelope which reinforces a mass of jointed bedded rock, or both, into a structural entity capable of providing its own support. Rock reinforcement requires bolts to which a positive force is applied as opposed to unstressed anchor bars or bolts used for other purposes. A system of stressed and grouted rock bolts installed in a regular pattern provides an active force to the rock mass and creates an envelope of compression which acts as a structural member. This structural member helps the rock to provide its own support, therefore, the term rock reinforcement.

For the Oroville Underground Power Plant, the rock reinforcement system consists of stressed and grouted rock bolts supplemented by chain link fabric, and guniting of the entire roof. As the underground cavern for Oroville Power Plant represents a large area of excavation, a general pattern of bolting was chosen for rock reinforcement. In addition, the specifications stated that rock reinforcement shown on drawings is typical only, and modifications directed by the engineer can include variations in the pattern, spacing and length, as conditions may require. Two basic patterns of bolting were designed to strengthen and to stabilize the rock. They are: (1) The powerhouse roof was bolted with 20-ft long, 1-in. diam, high-strength bolts spaced approximately 4 ft on centers; and (2) vertical rock faces in the machine hall received the same bolts except that the spacing was increased to approximately 6 ft on centers. Additional bolts were used at the junction of various tunnels with the powerhouse chamber and in the areas where special treatment of rock was required. This simplification of rock bolting has definite, practical advantages and it assured uniform and systematic progress of the work.

Immediately after blasting, the bolts were installed as close to the working face as possible. Early installation of rock bolts was essential to enhance the safety of rock excavation and to minimize relaxation of the so-called decompression zone at the excavated surface. Bolts were anchored in place by means of an expansion anchor, tensioned to a specified stress, packed and sealed at the rock face, and finally grouted.

The Oroville Power Plant excavation is subject to water seepage from all possible directions. The rock structure with grouted tight joints reduces the inflow of water to quantities that are easily controlled by installed drainage. Envelope grouting was performed by drilling from the powerhouse excavation chamber and it overlaps the grouting accomplished from the tailrace tunnel. The end walls of the chamber are protected in a similar manner.

Actual first-stage grouting of the underground excavation extends from a depth of 40 ft to a depth of 60 ft using pressures up to 100 psi, thus providing a transition to the high-pressure grout zone. The high-pressure zone is grouted with pressures up to 300 psi with a thickness of 30 ft and it constitutes the main barrier against the seeping water. Thus the water pressure will be exerted at a considerable distance from the actual face of the excavation where increased rock stresses will be safely absorbed by the rock structure.

Drain holes were drilled in the rock surrounding the powerhouse draft tubes, diversion tunnels and access tunnel after the envelope consolidations and contact grouting was completed. Provision has been made for future drains if they are required. The minimum diameter of the holes is not less than that produced by a standard NX size drill bit. The spacing of drill holes was approximately 25 ft and their length varied from 25 ft to 50 ft.

Any water finding its way through to the roof of the powerplant excavation is intercepted by the suspended ceiling and drains by gravity into the peripheral gap between the structure and the rock. The seepage from the drain holes around the power plant is collected by the pipe drainage system and is removed through a wet sump with automatically controlled pumps. An emergency standby unit is located in the downstream end of the plant. The amount of ground water intercepted with the reservoir full has been minimal and well within the capability of the drain system to handle.

TAILRACE TUNNELS

Use of the two large diversion tunnels as tailrace tunnels for the power plant following completion of Oroville Dam, required close attention to their hydraulic characteristics and capabilities. In order to be certain that the design was complete, the California Department of Water Resources contracted with the office of the Chief Engineer of the Bureau of Reclamation in 1962 to make a model

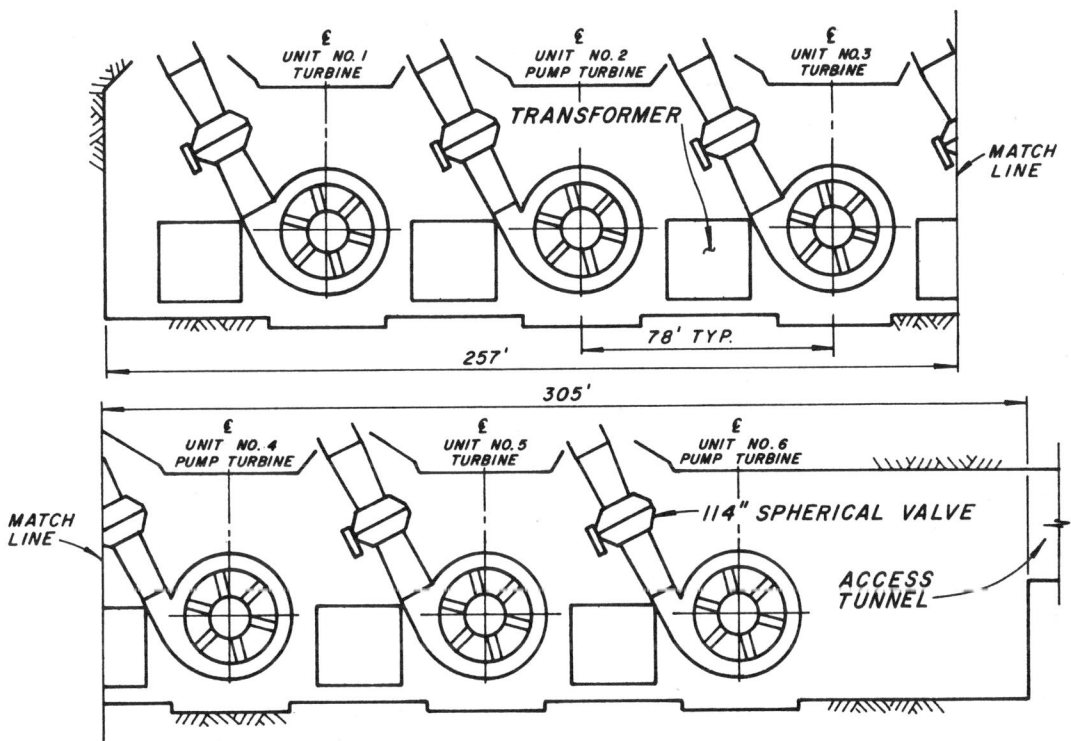

FIG. 7.—PLANT LAYOUT

study on the draft tube connections and tailrace surge problems arising from operation of the underground power plant.

The underground power plant contains six units which discharge into the downstream portions of the original diversion tunnels (Fig. 7). The odd number units, numbers 1, 3 and 5 are conventional Francis-type turbines and the even-number units, 2, 4 and 6 are reversible pump turbines. The draft tubes for units 1 and 2 discharge directly into tunnel 2 and the draft tubes of unit numbers 3, 4, 5 and 6 pass under tunnel 2 and connect with tunnel number 1.

Tailrace Tunnel 1 begins at the tunnel plug with the invert at El. 205.33 and slopes downward 3% for a distance of 1,400 ft. It then slopes upward to El. 182 at the submerged outlet portal. Thus, most of this tunnel is below the tailwater elevation and it operates as a pressure tunnel. At the upstream end, near the plug the top of the tunnel is above the tailwater and a free water surface exists.

Tunnel 2 is constructed with the invert horizontal and at El. 207.50 it flows about half-full at the normal tailwater El. of 225.0 and remains partly full at all other operating conditions. Large open ports connect Tunnel 2 with the draft tubes of units 3, 4, 5 and 6 that pass beneath it, thus Tunnel 2 acts as a surge chamber to receive water from or supply water to the draft tubes and Tunnel 1 during load changes on the system. The area of the port in each draft tube equals the cross-sectional area of the tube.

To provide atmospheric pressure on the free water surface of Tunnel 1 at all times, a vent interconnects the upstream end of both tunnels downstream of the plugs. When all power plant units are generating at their installed capacity at the design reservoir head of 620 ft, the discharge into the tailrace is about 13,250 cfs. However, a maximum discharge of 16,500 cfs can be obtained when all units are generating at full load under a 500-ft head. The latter discharge produces a flow velocity of 11 fps in each tunnel.

If the load on one or more generators is suddenly cut off, the wicket gates on the affected units will automatically close and stop the flow into the tailrace tunnels. An excessive draw-down of the water surface in the tunnels is prevented by drawing upon water stored between the plug and upstream unit in each tunnel and by an interchange of water from one tunnel to another through the surge ports into draft tubes numbers 3, 4, 5 and 6. During load acceptance, the wicket gates open and flows start through the turbines. These flows move through the draft tubes and into the tunnels to produce advancing wave fronts that travel toward the plugs and the downstream portals.

At the conclusion of the model testing by the Bureau of Reclamation it reported that its tests "showed that satisfactory conditions would prevail in the power plant tailrace system under even the most extreme load rejections or acceptances theoretically possible. This conclusion substantiates the one reached by the design engineers of the State of California through the analytical studies. The tailrace tunnel system as designed by the California engineers performs well and is satisfactory for prototype use."

INTAKE STRUCTURE

The design of the intake for the underground Oroville Power Plant received particular attention. It was at first concluded that a standing free tower with gates at various elevations would suffice. Further studies with full consideration of seismic loadings indicated that an inclined system of shutters feeding into a penstock system resting on the rock of the left abutment would be a more secure and economical design.

In order to maximize the downstream propagation of fishlife and also to protect the extensive rice crops downstream from the reservoir, a limited range of temperatures must be sustained in the reservoir releases. This is achieved by selective withdrawal from various levels within the reservoir because temperatures in the reservoir vary with the depth of water.

The power plant intake structure (Fig. 8) consists of two inclined chutes, rectangular in shape built into the sloping hillsides of the reservoir near the upstream toe of the dam. Each of the two intakes has a cross-sectional area of 1,100 sq ft with a continuous row of movable shutters on top. The elevation at which water will be admitted to the intake depending on the desired temperature, will be controlled by removal of the shutters at the proper level. The shutters, which are primarily steel plates, will be either placed or retrieved from each chute by means of a hoist located at the upper end of the structure. They may be completely removed and stored in storage space at the upper end of the intake structure. Stainless steel trash racks have been installed along the chutes covering an area of 54,000 sq ft. The United States Bureau of Rec-

FIG. 8.—INCLINED INTAKE STRUCTURE WITH STAINLESS STEEL TRASHRACKS

FIG. 9.—INSTALLATION OF GENERATORS

lamation tested the hydraulic characteristics of the intake structure by use of a model at its Denver laboratory.

Joining the lower end of each intake structure are 22-ft diam power penstocks. Each of the penstocks is designed as a concrete-lined tunnel terminating at a manifold having three 12-ft diam penstock branches leading directly to the turbine scroll cases in the powerplant. The manifold and the 12-ft branches are lined with 1-1/2-in. steel plate.

For normal maintenance or emergency closure of the penstock, a flat-leaf coaster-type gate has been provided at the base of each of the intake chutes. It rides down the chute on a continuous roller train at each side and is moved by means of a lifting beam and gantry housed at the top of the intake structure. This gate closes over the top of the penstock and when closed effectively dewaters the entire penstock to the powerplant. At the lower end of each penstock branch line there is an 114-in. diam spherical valve provided for each of the six units. These valves are used as part of the normal operation of the plant and are designed against full reservoir head.

EQUIPMENT

Major equipment installed in the plant (see Fig. 9) consists of three 117 Mw generating units driven by Francis turbines with three 97.8 Mw generating units

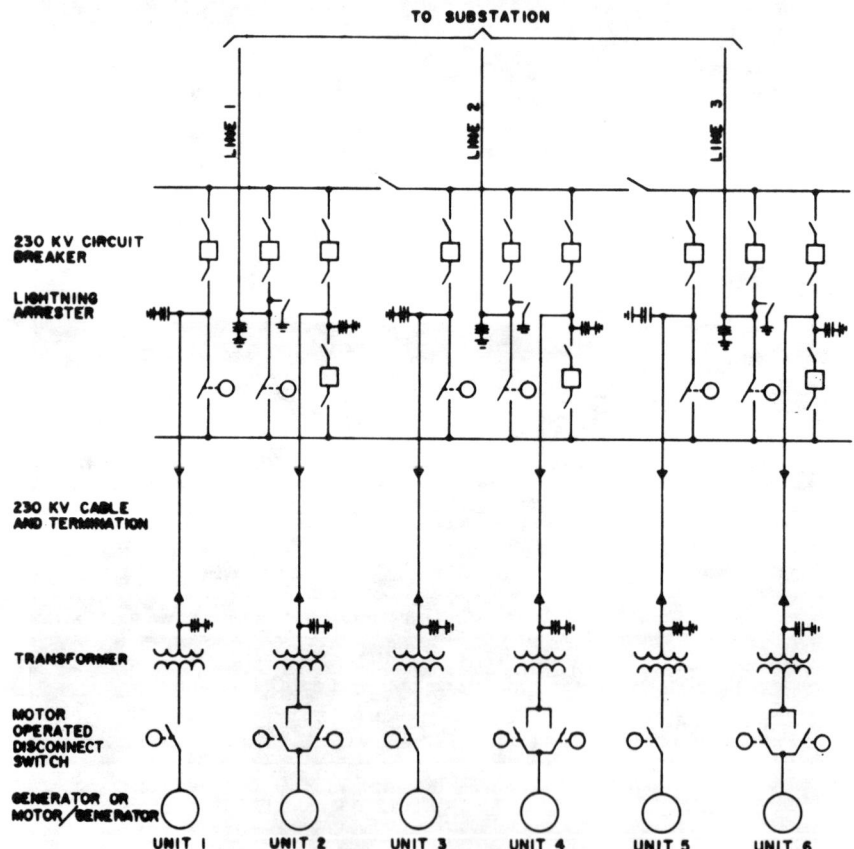

FIG. 10.—SINGLE LINE DIAGRAM

driven by reversible Francis pump-turbines. The equipment is designed for a rated output of 161,000 hp under a net head of 615 ft for the Francis turbines and a rated output of 120,000 hp under the same head for the pump-turbines. Each pump-turbine has a pump capacity of 1,870 cfs under a total dynamic head of 592 ft. The guaranteed efficiency of the Francis turbines is 94.0 % and that of the pump-turbines is 91.6 %, operating as a turbine.

The pump-turbine units are started in the pumping phase with the pump unwatered by using one of the local generators as a source of starting power. The generator is held at half speed and voltage, while the pump motor is brought up to half speed as an induction motor. The motor field is then applied to synchronize the two machines. Both machines will be brought slowly to full speed

FIG. 11.—SWITCHYARD AND CONTROL BUILDING

FIG. 12.—OROVILLE POWER PLANT IN OPERATION

and voltage under governor control. The motor will be synchronized to the running power source, the generator will be disconnected, and the air released from the pump to allow water to fill the pump. This will result in approximately half load on the motor. The wicket gates will then be opened to start pumping and fully load the motor.

The 117 Mw generators are designed to operate at 200 rpm, 60 cps, and the 173,000 hp-97.8 Mw motor generators are designed to operate at 189.5 rpm at 60 cps. The generators and motor-generators at the Oroville Underground Plant are the vertical-shaft, hydraulic-turbine driven, synchronous type with the thrust bearing above the rotor, two guide bearings, directly connected exciter, voltage regulator with magnetic amplifier, excitation cubicle, and a closed system of ventilation with surface air coolers, generation is at 12.5 kv and pumping will be at 12 kv.

Each generator is connected through disconnect switches to a three-phase transformer by means of an isolated phase bus (see Fig. 10). Each motor generator is connected through phase-reversing and disconnect switches to a three-phase transformer in a similar manner. The 230 kv transformer, rated at 127 Mva, is located between the generating units and is delta connected on the low-voltage side and solidly grounded wye on the high-voltage side. Power is transmitted to the switchyard (see Fig. 11) located above ground 2,000 ft away, by means of a 231 kv high-pressure oil-pipe-type cable carried through the access tunnel. Normal station-service power is supplied from the low-voltage side of two of the transformers.

ARCHITECTURAL TREATMENT

In designing the architectural appearance of the underground power plant, it was the intent of the architects to maintain an atmosphere built on brightness and color that would minimize the underground location and give an internal appearance equivalent to any surface generating station (Fig. 12). By the generous use of flourescent lighting and bright colors of red and turquoise which symbolize power and water, this objective has been substantially achieved. Red and turquoise is a symbolic color system used throughout all California Department of Water Resources structures, both surface and subsurface.

The main generating floor at El. 252 has been paved with terrazzo made with a polyester binder. This material is not only attractive but will stand considerable wear. Wall panels below the crane rails are of precast concrete which are surfaced with exposed quartz aggregates to create additional sparkle and life.

The enamelled metallic ceiling is suspended from the native rock and is merged into wall panels above the crane rails. At the downstream end of the plant, an observation platform is provided for visitors and a time vault has been constructed. Sealed in the time vault are duplicates of all the plans and specifications, papers and reports dealing not only with the construction of the Oroville complex, but the entire State Water Project at the time of sealing early in 1969.

SUMMARY AND CONCLUSIONS

The first contract for construction of the Oroville Power Plant was awarded to the McNamara-Fuller Company for excavation of the machine hall, the placing of the basic concrete and the construction of the intake chutes, penstocks, and draft tubes. This contract was awarded on May 31, 1963, and was completed on May 16, 1967. The turbines were manufactured by the Allis-Chalmers Company of York, Pennsylvania, and the generators were manufactured by the Westinghouse Electric Corporation of Pittsburgh. The first complete unit at Oroville went on the line on March 28, 1968, and the last of the six went on the line on February 20, 1969.

Following installation of all six units, additional testing was necessary before the power plant could be considered completed and the energy made available to the power companies without restriction as required by the contracts executed with them. All tests were completed to the satisfaction of the utilities by July 20, 1969, at which time the contract for use of the facilities and beginning of full payments to the State became effective. As previously mentioned herein, revenues from the full operation of the Oroville Power Plant and its Thermalito subsidiary total $16 million per yr. Oroville-Thermalito energy delivered to Pacific Gas and Electric's Table Mountain substation for the period February, 1968 through July, 1970 was about 5 billion kwhr.

In conclusion, it may be stated that the Oroville Underground Plant is meeting its functions of providing an income of cash to the State Water Project and is fulfilling the part designed for it to meet the continually growing power loads in California.

APPENDIX.—REFERENCES

1. Erwin, W. B., "Oroville Dam," *World Dams Today,* Japan Dam Association, Tokyo, 1967, p. 248.
2. "Exploration Data Oroville Powerplant," Butte Co., Calif. *Project Geology Report D-12,* Oct., 1962.
3. Gilbert, P. H., "Edward Hyatt Power Plant Instake Structure," *Journal of the Power Division,* ASCE, Vol. 95, No. PO1, Proc. Paper 6471, March, 1969, pp. 101–118.
4. Gilbert, P. H., and Smith, T. A., "Oroville Powerplant Instake Structures," *Technical Memorandum No. 16,* California Department of Water Resources, Sacramento, Calif., July, 1965.
5. Golzé, Alfred R., "California Water Project Power Plans and Demands," *Civil Engineering,* Vol. 35, No. 2, February, 1965, pp. 35–39.
6. Golzé, Alfred R., "Power from Oroville," *Water Power,* Mar., 1967.
7. "Hydraulic Model Studies of Oroville Dam Powerplant Intake Structure," *U.S. Bureau of Reclamation Hydraulics Branch Report No. HDY 509,* 1965.
8. "Hydraulic Model Studies of the Diversion Tunnels for Oroville Dam," *U.S. Bureau of Reclamation Hydraulics Branch Report No. HYD 502,* 1963.
9. Lanning, C. C., "Oroville Dam Diversion Tunnels," *Journal of the Power Division,* ASCE, Vol. 93, No. PO2, Proc. Paper 5506, October, 1967, pp. 51–65.
10. "Model Studies of the Draft Tube Connections and Surge Characteristics of the Tailrace Tunnels for Oroville Powerplant," *U.S. Bureau of Reclamation Hydraulics Branch Report No. HYD 507,* Apr., 1963.
11. O'Neill, A. L., "Rock Reinforcement in Underground Construction," *Technical Memorandum No. 22,* California Department of Water Resources, Sacramento, Calif., Feb., 1967.
12. Paul, K., Jr., "Stress Analysis of Oroville Powerplant Pump-Turbine Head Covers and Thermalito Powerplant Turbine Stay Rings by Finite Element Techniques," *Technical Memorandum No. 25,* California Department of Water Resources, Sacramento, Calif., July, 1967.
13. Pona, M., "Oroville Underground Powerplant," *Technical Memorandum No. 8,* California Department of Water Resources, Sacramento, Calif., Sept., 1964.
14. Schulz, W. G., Thayer, D. P., and Doody, J. J., "Oroville Underground Power Plant," *Seventh International Congress on Large Dams,* Vol. 2, Rome, 1961, o. 439.

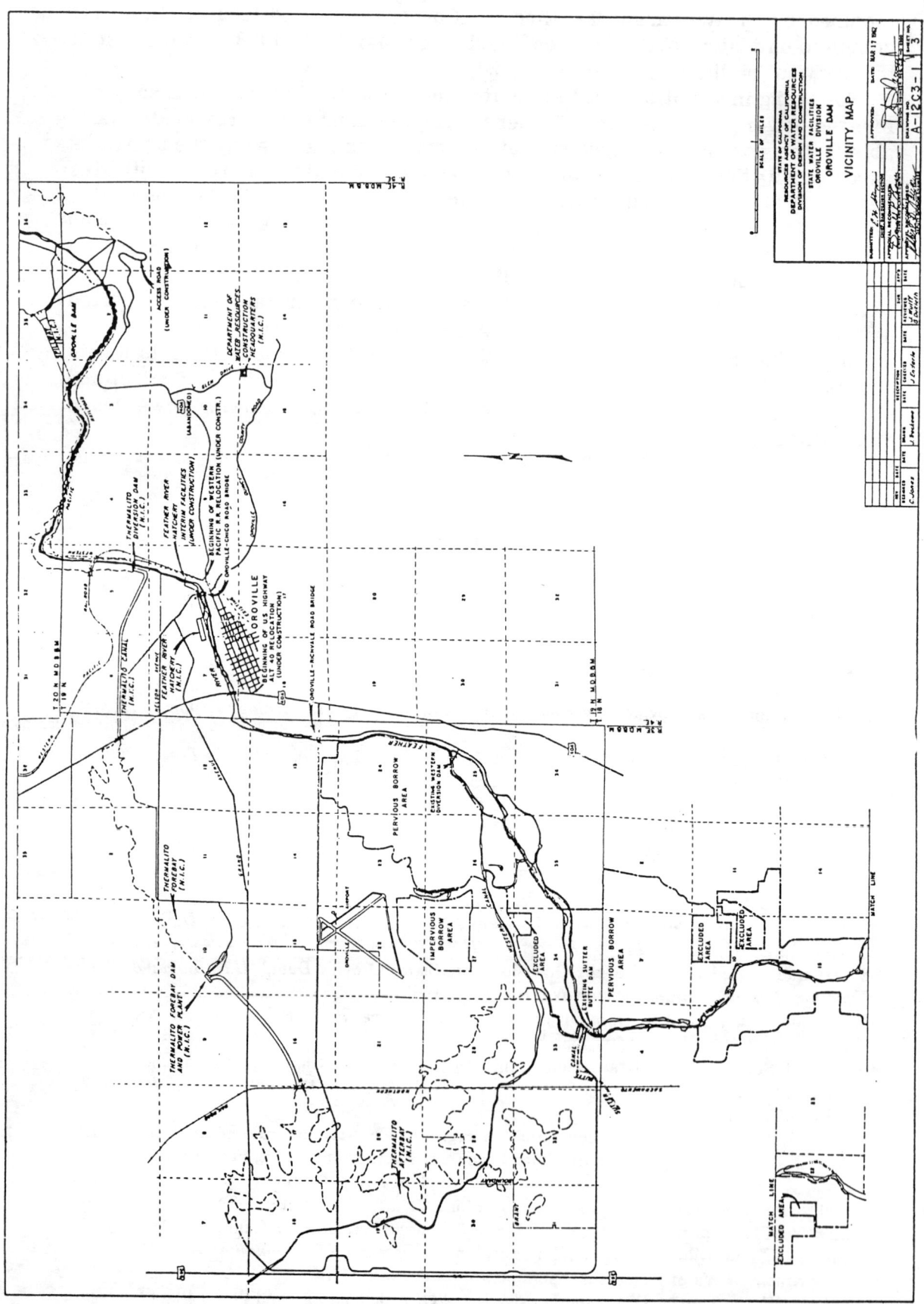

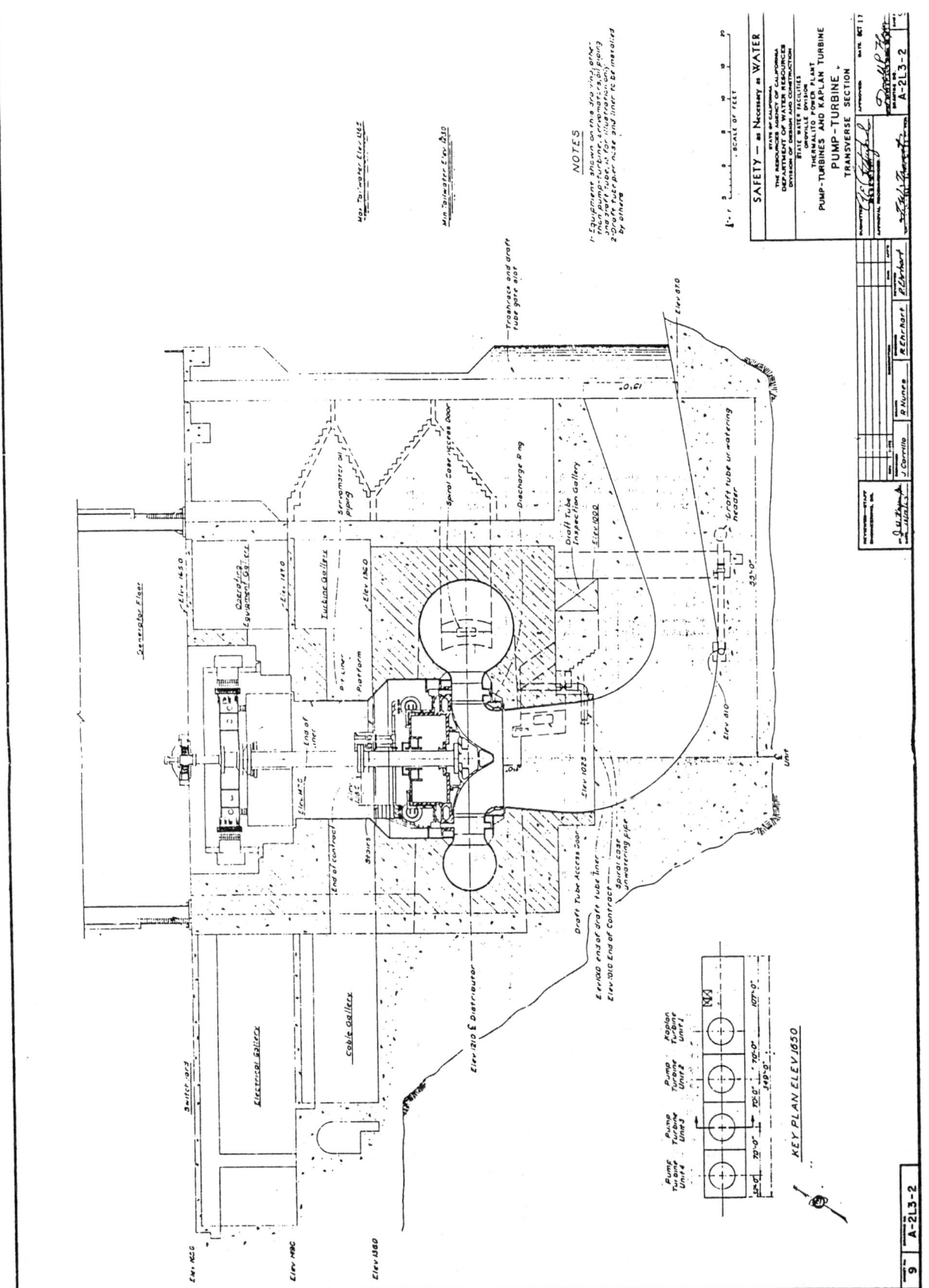

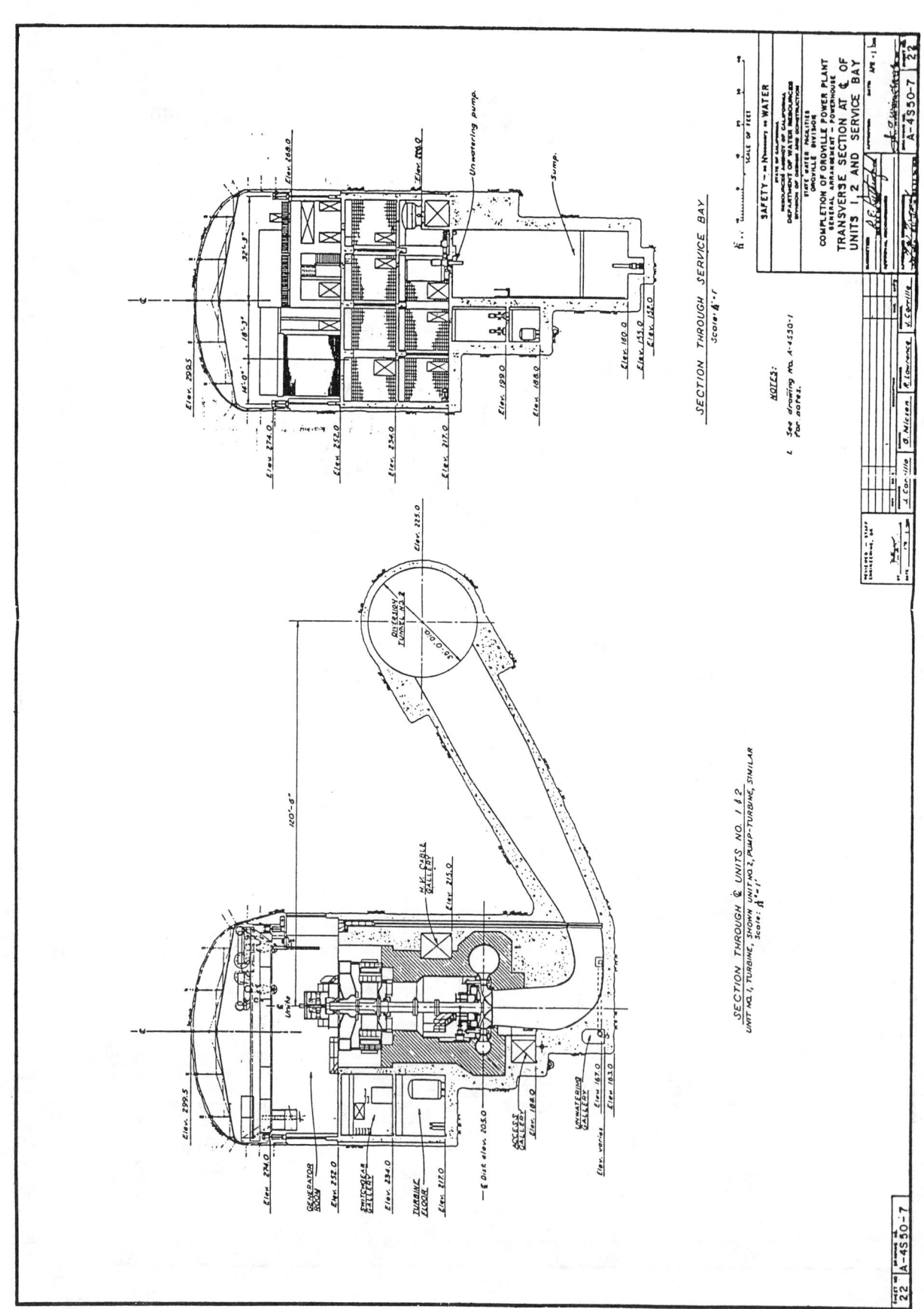

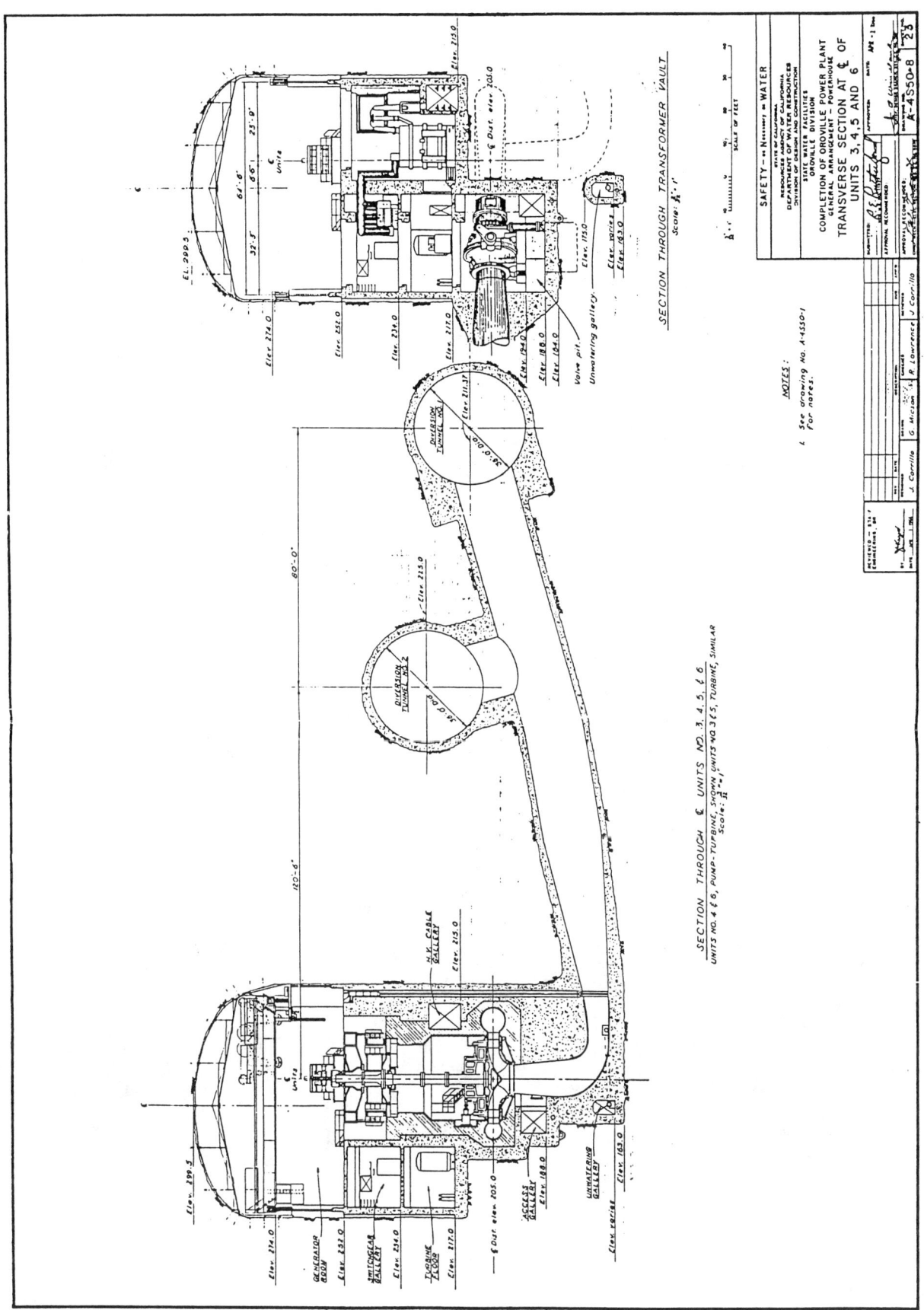

Aerial View

Oroville Dam - Powerhouse Below Left Abutment

Edward Hyatt Pumped Storage Plant

BIBLIOGRAPHY

1. Erwin, W.B., "Oroville Dam," World Dams Today, Japan Dam Association, Tokyo, 1967, page 248.

2. "Exploration Data Oroville Powerplant," Butte County, California, Project Geology Report D-12, October, 1962.

3. Gilbert, P.H., "Edward Hyatt Powerplant Intake Structure," Journal of the Power Division, American Society of Civil Engineers, Vol. 95, No. PO1, Proc. Paper 6471, March 1969, pages 101-118.

4. Gilbert, P.H., and Smith, T.A., "Oroville Powerplant Intake Structures," Technical Memorandum No. 16, California Department of Water Resources, Sacramento, California, July 1965.

5. Golze´, Alfred R., "California Water Project Power Plans and Demands," Civil Engineering, Vol. 35, No. 2, February 1965, pages 35-39.

6. Golze´, Alfred R., "Power from Oroville," Water Power, March 1967.

7. "Hydraulic Model Studies of Oroville Dam Powerplant Intake Structure," U.S. Bureau of Reclamation Hydraulics Branch Report No. HYD 509, 1965.

8. "Hydraulic Model Studies of the Diversion Tunnels for Oroville Dam," U.S. Bureau of Reclamation Hydraulics Branch Report No. HYD 502, 1963.

9. Lanning, C.C., "Oroville Dam Diversion Tunnels," Journal of the Power Division, American Society of Civil Engineers, Vol. 93, No. PO2, Proc. Paper 5506, October 1967, pages 51-65.

10. "Model Studies of the Draft Tube Connections and Surge Characteristics of the Tailrace Tunnels for Oroville Powerplant," U.S. Bureau of Reclamation Hydraulics Branch Report No. HYD 507, April 1963.

11. O'Neill, A.L., "Rock Reinforcement in Underground Construction," Technical Memorandum No. 22, California Department of Water Resources, Sacramento, California, February 1967.

12. Paul, K., Jr., "Stress Analysis of Oroville Powerplant Pump-Turbine Head Covers and Thermalito Powerplant Turbine Stay Rings by Finite Element Techniques," Technical Memorandum No. 25, California Department of Water Resources, Sacramento, California, July 1967.

13. Pona, M., "Oroville Underground Powerplant," Technical Memorandum No. 8, California Department of Water Resources, Sacramento, California, September 1964.

14. Schulz, W.G., Thayer, D.P., and Doody, J.J., "Oroville Underground Powerplant," Seventh International Congress on Large Dams, Vol. 2, Rome 1961, page 439.

15. Wachter, G.F., "Pumped Storage and Oroville Design and Initial Operation, Paper No. 70PP541-PWR, Institute of Electrical and Electronics Engineers.

16. Golze´, Alfred R., Journal of the Power Division, Proceedings of the American Society of Civil Engineers, March 1971.

Plant Name: **FAIRFIELD**

Plant location:
  Jenkinsville, SC
  Fairfield County
	Owner: South Carolina Electric & Gas Company
	       1426 Main Street
	       Columbia, South Carolina 29201

Rated capacity 512 MW Average static head 150 ft Plant efficiency 70.40 % Stored energy 4,096 MWh Number of units 8	Designers: Gibbs & Hill

Construction time: 4 years, 3 months
Construction cost: $388 per kW
Price level: 1978
First commercial power: June 1978
FERC project number: 1894
	Plant Manager/Superintendent:
	  S.E. Stockman
	  Fairfield Pumped Storage
	  P.O. Box 57
	  Jenkinsville, South Carolina 29065
	  (803) 345-4523

River or water source: Broad River	

	UPPER RESERVOIR	LOWER RESERVOIR
DAM		
Type	Random-filled	Concrete, Crest Gates
Height (ft)	180	46
Crest length (ft)	11,001	2,001
Volume (yd³)	9,996,000	62,000
RESERVOIR		
Type	Constructed, Monticello Reservoir	Existing, Parr Reservoir
Surface area (acres)	6,795	4,398
Usable power storage (acre-ft)	29,023	29,024
Power pool fluctuation (ft)	4.5	10.0
Operating levels		
Maximum (ft)	425.0	266.0
Minimum (ft)	420.5	256.0
Drainage area (miles²)	0.0	4.7
Seepage (ft³/s)	49.441	
SPILLWAY		
Design flood		
Return period (years)		
Flow (ft³/s)		228,000
Capacity (ft³/s)		228,000
Type	None	Overflow
Gates		
Number	None	10
Type		Bascule
Width (ft)		200.00
Height (ft)		8.99
OUTLET WORKS		
Discharge capacity (ft³/s)		
Number of water passages	None	None
Dimensions of water passages		
Height (ft)		
Width (ft)		
Diameter (ft)		
Type of gates		
Number of gates	None	None

Fairfield - Page 1 (English)

Plant Name: **FAIRFIELD**

Plant location
Jenkinsville, SC
Fairfield County

Rated capacity	512 MW
Average static head	45.7 m
Plant efficiency	70.40 %
Stored energy	4,096 MWh
Number of units	8

Construction time: 4 years, 3 months
Construction cost: $388 per kW
Price level: 1978
First commercial power: June 1978
FERC project number: 1894

River or water source: Broad River

Owner: South Carolina Electric & Gas Company
1426 Main Street
Columbia, South Carolina 29201

Designers:
Gibbs & Hill

Plant Manager/Superintendent:
S.E. Stockman
Fairfield Pumped Storage
P.O. Box 57
Jenkinsville, South Carolina 29065
(803) 345-4523

	UPPER RESERVOIR	LOWER RESERVOIR
DAM		
Type	Random-filled	Concrete, Crest Gates
Height (m)	54.9	14.0
Crest length (m)	3,353.0	610.0
Volume (m³)	7,642,563	47,500
RESERVOIR		
Type	Constructed, Monticello Reservoir	Existing, Parr Reservoir
Surface area (Mm²)	27.50	17.80
Usable power storage (Mm³)	35.800	35.800
Power pool fluctuation (m)	1.37	3.05
Operating levels		
Maximum (m)	129.54	81.08
Minimum (m)	128.17	78.03
Drainage area (Mm²)	0.039	12.173
Seepage (m³/s)	1.4000	
SPILLWAY		
Design flood		
Return period (years)		
Flow (m³/s)		6,456
Capacity (m³/s)		6,456
Type	None	Overflow
Gates		
Number	None	10
Type		Bascule
Width (m)		60.960
Height (m)		2.740
OUTLET WORKS		
Discharge capacity (m³/s)		
Number of water passages	None	None
Dimensions of water passages		
Height (m)		
Width (m)		
Diameter (m)		
Type of gates		
Number of gates	None	None

Fairfield - Page 1 (Metric)

FAIRFIELD

INTAKES	UPPER INTAKE	LOWER INTAKE
Number	4	1
Type	Gated intake structure	16 Draft tube openings
Design discharge (ft³/s)	12,001	3,000
Gross inlet area (ft²) (at trash racks)	1,501	567
Bar racks		
spacing (in)	7.00	7.00
shape	Rectangular	Rectangular
depth/thickness (in)	3.00 / 1.00	3.00 / 1.00
diameter (in)		
Emergency gates		
number	4	16
height/width (ft)	30.51 / 18.34	15.09 / 25.00
type	Roller	Slide
Service gates		
number	1	None
height/width (ft)	30.51 / 18.34	
type	Roller	
Bulkhead/stop logs (Y or N)	N	N
number of units serviced		
Hoists		
number	4	1
capacity (tons)	45	185
type	Electrically driven hoist rope	Gantry

WATER PASSAGES	Upper Tunnel	Shaft	Lower Tunnel	Surge Tanks Upper	Surge Tanks Lower	Penstocks	Tailrace Tunnel
Number	4					4	
Diameter (ft)	24.0					26.0	
Length (ft)	375					720	
Maximum velocity (ft/s)						22.3	
Concrete lining thickness (in)							
Total length of concrete sections (ft)	375						
Steel liner Thickness							
Minimum (in)						0.75	
Maximum (in)						0.88	
Material grade						ASTM A516 GRADE 60	
Total length of steel-lined sections (ft)						720	

Notes:
Upper Tunnel: Section transitions from 9.1m (30ft) by 15.2m (50ft) to a 7.3m (24ft) diameter.

Fairfield - Page 2 (English)

FAIRFIELD

INTAKES	UPPER INTAKE	LOWER INTAKE
Number	4	1
Type	Gated intake structure	16 Draft tube openings
Design discharge (m³/s)	339.8	85.0
Gross inlet area (m²) (at trash racks)	139.4	52.7
Bar Racks:		
spacing (mm)	178	178
shape	Rectangular	Rectangular
depth/thickness (mm)	76 / 25	76 / 25
diameter (mm)		
Emergency gates		
number	4	16
height/width (m)	9.300 / 5.590	4.600 / 7.620
type	Roller	Slide
Service gates		
number	1	None
height/width (m)	9.300 / 5.590	
type	Roller	
Bulkhead/stop logs (Y or N)	N	N
number of units serviced		
Hoists		
number	4	1
capacity (Mg)	41	168
type	Electrically driven hoist rope	Gantry

WATER PASSAGES	Upper Tunnel	Shaft	Lower Tunnel	Surge Tanks Upper	Surge Tanks Lower	Penstocks	Tailrace Tunnel
Number	4					4	
Diameter (m)	7.32					7.92	
Length (m)	114.5					219.5	
Maximum velocity (m/s)						6.80	
Concrete lining thickness (m)							
Total length of concrete sections (m)	114.5						
Steel liner Thickness							
Minimum (mm)						19	
Maximum (mm)						22	
Material grade						ASTM A516 GRADE 60	
Total length of steel-lined sections (m)						219.5	

Notes:
Upper Tunnel: Section transitions from 9.1m (30ft) by 15.2m (50ft) to a 7.3m (24ft) diameter.

Fairfield - Page 2 (Metric)

FAIRFIELD

POWERHOUSE and RELATED FEATURES

Powerhouse Structure
Type: Semi-outdoor
Length: 520 ft   Width: 150 ft   Height: 108 ft

Guard Valves
Number: None   Diameter: ft
Type:

Transformers
Number: 4
Ratings: 170
Voltages: (kV) 230 / 13.8

Generator
Rating generating (MVA): 71.0   Rating pumping (MVA): 77.2
Insulation type: Polyester-Class B
Starting method: Reduced voltage
Starting equipment: 6.9 kv motor starting breaker, 13.8 kv motor run breaker

Runners
Material: ASTM A-27-7040 cast/welded
Minimum unit submergence: 32.0 ft
WR^2: 4,900,000 (lbf x ft²)
Manufacturer: Allis Chalmers
Model test by: Allis Chalmers

	Reversible Runners	Reversible Motor/Generator
Number	8	8
Diameter (ft)	16.20	Rotor 23.88 Stator 29.63
rpm synchronous	150.0	150.0
rpm overspeed	241.0	241.0
Type	Francis	Av-Gen/Motor

Information on Runners

Condition:	Gross Head (ft) Generating	Pumping	Capacity (MW) Generating	Pumping	Discharge (ft³/s) Generating	Pumping	Turbine/Pump Eff.(%) Generating	Pumping
Maximum head & maximum power	169	169	75	77			88.0	92.1
Minimum head & maximum power	155	155	67	77			88.2	92.5

Condition:	Net Head (ft) Generating	Pumping	Capacity (MW) Generating	Pumping	Discharge (ft³/s) Generating	Pumping	Turbine/Pump Eff.(%) Generating	Pumping
Rated head @ best gate	162	162	62	77	6,003	4,803	90.4	92.3

Fairfield - Page 3 (English)

FAIRFIELD

POWERHOUSE and RELATED FEATURES

Powerhouse Structure
Type: Semi-outdoor
Length: 158.5 m   Width: 45.7 m   Height: 32.9 m

Guard Valves
Number: None   Diameter: m
Type:

Transformers
Number: 4
Ratings: 170
Voltages: (kV) 230 / 13.8

Generator
Rating generating (MVA): 71.0   Rating pumping (MVA): 77.2
Insulation type: Polyester-Class B
Starting method: Reduced voltage
Starting equipment: 6.9 kv motor starting breaker, 13.8 kv motor run breaker

Runners
Material: ASTM A-27-7040 cast/welded
Minimum unit submergence: 9.75 m
WR^2: 2,025,000 (Newtons x m^2)
Manufacturer: Allis Chalmers
Model test by: Allis Chalmers

	Reversible Runners	Reversible Motor/Generator		
Number	8			8
Diameter m	4.93	Rotor 7.280	Stator	9.030
rpm synchronous	150.0			150.0
rpm overspeed	241.0			241.0
Type	Francis			Av-Gen/Motor

Information on Runners

Condition:	Gross Head (m) Generating	Gross Head (m) Pumping	Capacity (MW) Generating	Capacity (MW) Pumping	Discharge (m³/s) Generating	Discharge (m³/s) Pumping	Turbine/Pump Eff.(%) Generating	Turbine/Pump Eff.(%) Pumping
Maximum head & maximum power	51.5	51.5	75	77			88.0	92.1
Minimum head & maximum power	47.1	47.1	67	77			88.2	92.5

Condition:	Net Head (m) Generating	Net Head (m) Pumping	Capacity (MW) Generating	Capacity (MW) Pumping	Discharge (m³/s) Generating	Discharge (m³/s) Pumping	Turbine/Pump Eff.(%) Generating	Turbine/Pump Eff.(%) Pumping
Rated head @ best gate	49.4	49.4	62	77	170.0	136.0	90.4	92.3

FAIRFIELD

Plant Data:

Average GWh generating per year:	431
Average GWh pumping per year:	615
Starting time from standstill (s):	135
Changeover time pumping to generating (min):	11
Planned/scheduled time between major overhauls (years):	14
Outage time required per unit during major overhauls (weeks):	14
Representative plant availability (%):	88.8
Representative planned outages (weeks per year):	3

Miscelleneous Notes:

This plant is designed for operation in conjunction with the Virgil Summer Nuclear Station. (The upper reservoir serves as the cooling pond for the nuclear power station.)

The first unit overhaul is planned for the fall of 1991, 13 years after the power-on-line date of June 1978.

The crest length of 3,353 m (11,000 ft) is for four dams.

The upper tunnels are an extenstion of the intake structure, which is described as follows:

The intake-outlet structure is located at the left abutment of the main dam. The structure consists of an approach channel formed by two concrete gravity walls and the actual reinforced-concrete, gated, intake structure, with four water passages that are approximately 9.14 m (30 ft) wide by 15.24 m (50 ft) high near the trash racks. The passages decrease in size as they near the gates and then make the transition to the 7.92 m (26 ft) diameter penstocks, also making a 30 degree bend.

The intake gates are structural steel, fixed-wheel gates that are approximately 9.14 m (30 ft) high by 6.1 m (20 ft) wide.

The outage time per unit for a major overhaul is estimated at 12 to 16 weeks per pair of units.

The reported plant availability of 88% is a 5-year average.

The spillway capacity of 6,400 m³/s (226,000 ft³/s) is based on the flood of record.

Cavitation Experience:

Cavitation has occurred, repair requires 120 pounds per 2000 hours operation.

Significant or Unique Problems:

Problems have occurred with the power transformer, switchgear, and wicket gate bushings.

List of Licenses Required:

FAIRFIELD

FERC.

ENVIRONMENTAL FEATURES

Recreation:
Facilities include a 1.21 Mm² (300-acre) recreational lake, five boat landings, two Fairfield County parks and a scenic overlook.

Fish and Wildlife:
Waterfowl impoundments on the Broad river and the Enoree river.

Social:
The local fire department has a facility for filling tankers from the upper reservoir.

Page 4 (Continued)

ENGINEERING FOUNDATION CONFERENCE

Converting Existing Hydro-Electric Dams
and Reservoirs into

PUMPED STORAGE

Facilities

Franklin Pierce College
Rindge, N.H.
August 18-23, 1974

CHAIRMEN:
Ellis L. Armstrong
Ted W. Mermel

Published by
American Society of Civil Engineers
345 East 47th Street
New York, N.Y. 10017

FAIRFIELD PUMPED STORAGE FACILITY

By: E. H. Crews, Jr., Vice President
Construction & Production Engineering
South Carolina Electric & Gas Co.

INTRODUCTION

The South Carolina Electric and Gas Company is an investor owned utility serving 24 of the 46 counties in South Carolina. The projected load growth of electrical energy for the periods of 1970-1985 conservatively indicated the need of base load and peaking power.

The fuel supplies indicated a strong look at nuclear plants for the base load. The efficient operation of a nuclear plant requires a high load factor to utilize full time the low cost fuel. The most desirable choice of energy for peaking power was hydro.

The company began the search for a site to combine these desired generation methods and ultimately decided on the Parr area, located approximately 26 miles northwest of Columbia, the capital city. (See Figure 1 page 3.) The area is rather remote and has a long history of power development. The local people are quite familiar with this power development.

The Parr Hydro Plant was completed in 1914 and consists of 6 units capable of 14.9 MW. The conventional 72.5 MW coal fired Parr Steam Plant was built adjacent to Parr Hydro being completed in 1929. Parr Steam Plant was subsequently updated in 1971 by the addition of gas turbines and waste heat boilers to furnish steam for continued operation of the Parr Steam Plant turbines. The small experimental pressurized water reactor (CVTR) was built in this immediate area in 1963. The CVTR was decommissioned in 1967.

The Parr Hydro plant has a 35' high dam which is 2000 feet long and contains a pond of 1,850 acres. Redevelopment will involve utilizing the existing 1,850 acre reservoir, enlarged by means of spillway crest gates, Figure 2 page 4, as the 29,000 acre feet lower pool of a 480 MW pumped storage project known as Fairfield Pumped Storage Facility (Figure 3 page 5). An entirely new 6800 acre upper pool (Monticello Impoundment) will be constructed and will, if authorized, serve as a cooling impoundment for a 900 MW pressurized water reactor of the Virgil C. Summer Nuclear Station. The entire complex is shown on Figure 4 page 6.

The existing Parr Dam that will be modified has an overall length of 2000 feet and is a concrete Ogee type overflow section with an average height of about 35 feet. West of the concrete section is an earth dike approximately 300 feet long. East of the concrete section is the existing Parr Hydro plant, which is about 300 feet long and has an installed capacity of 14.9 MW. East of the hydro plant is about 90 feet of concrete non-overflow section. Beyond this, further East, is a short earth-filled section. Raising of the earth dikes and non-overflow section is not required as their crest will still be 5 feet higher than the increased water level. The general layout of the existing Parr Hydro facility with the gates superimposed on the existing dam crest is shown on Figure 5 page 7.

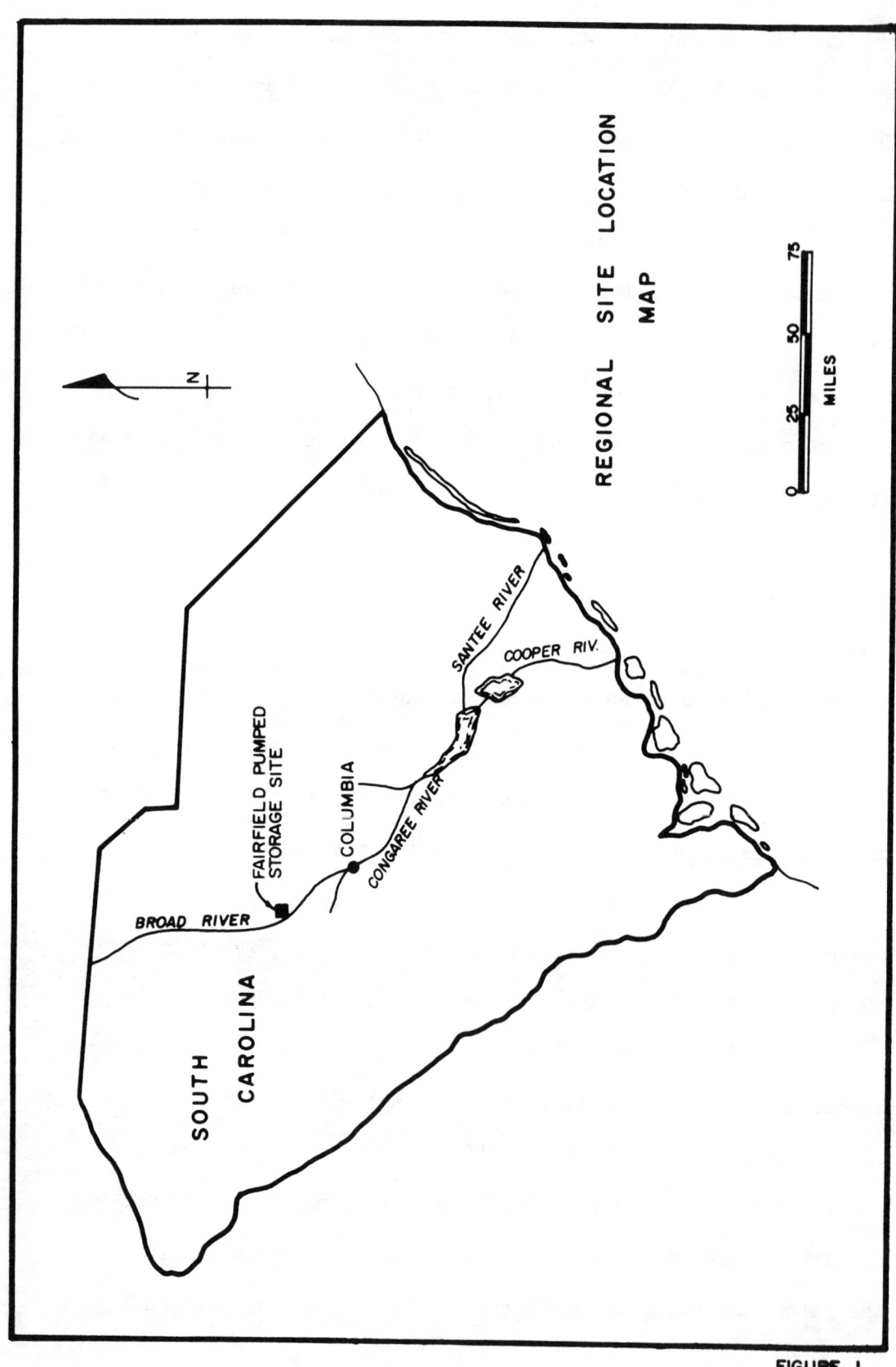

FIGURE 1

**PARR POWERHOUSE AND DAM
WITH BASCULE GATES**
FAIRFIELD PUMPED STORAGE FACILITY
South Carolina Electric & Gas Company

Figure 2

**FAIRFIELD PUMPED - STORAGE FACILITY
Parr, South Carolina
SOUTH CAROLINA ELECTRIC AND GAS CO.**

Figure 3

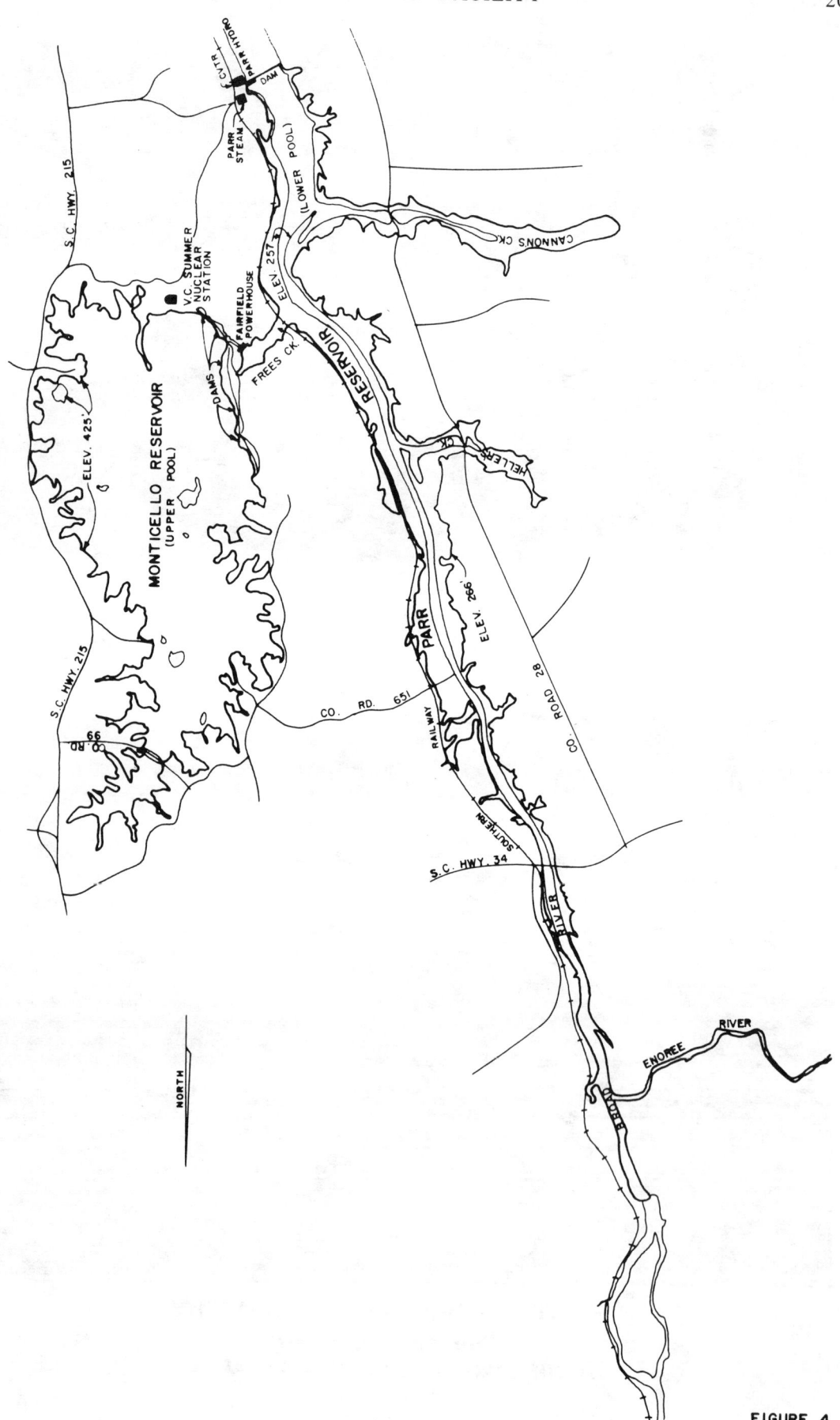

FIGURE 4

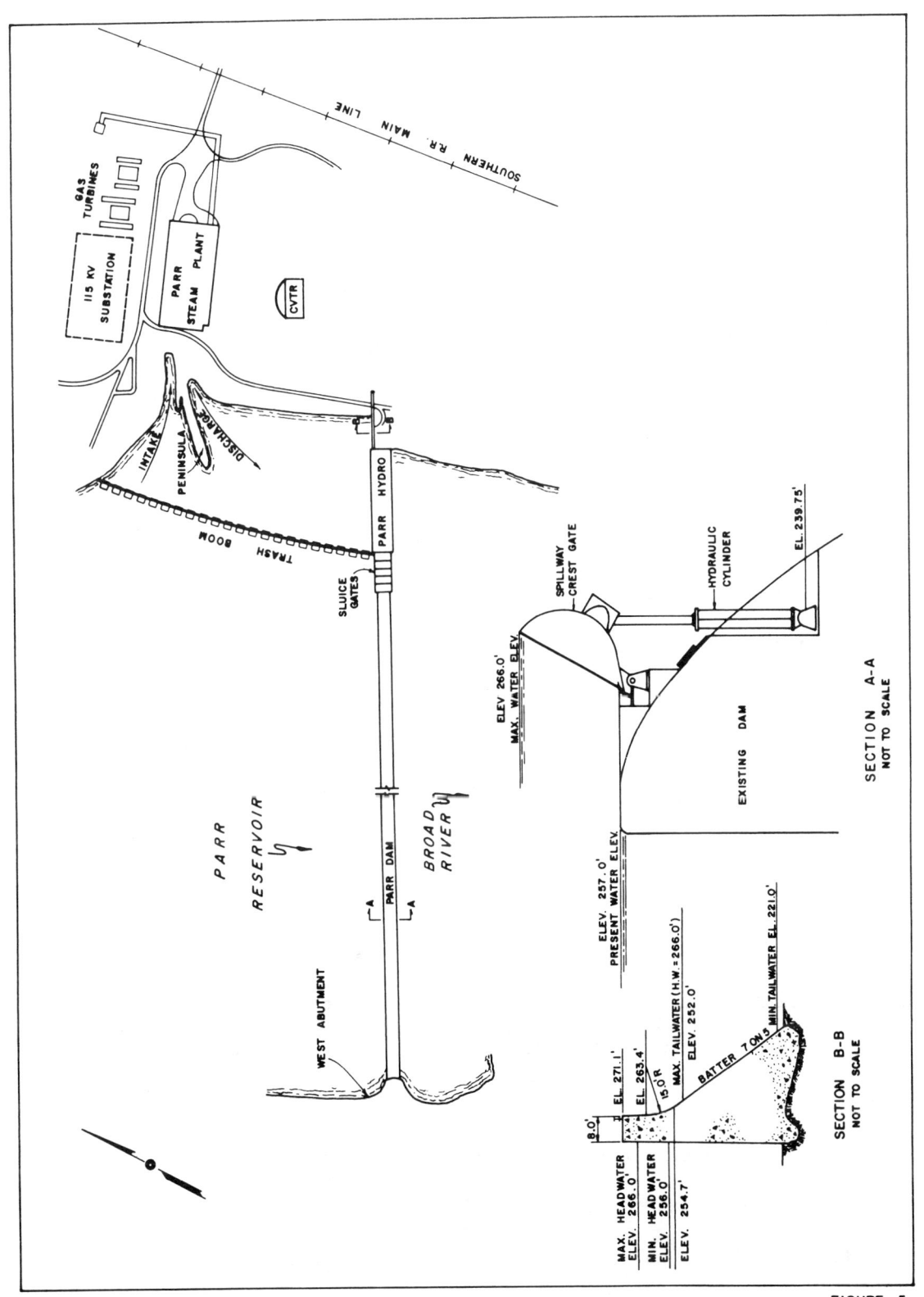

FIGURE 5

The Parr Dam was constructed in 1914 and though the structure is 60 years old, it is still in good, sound condition. No obvious damage to the concrete surface has been found by visual inspection. Some patchwork repairs exist on the surface, but these are surficial only.

Core drillings were performed on Parr Dam, and cores of concrete were obtained down to approximately 30 feet below the top of the Dam. The results of the concrete compression tests made on samples of the concrete cores, approximately 4 to 6 feet below the top, indicated compression strength from 1,300# to 3,400# psi which we regarded as quite satisfactory to withstand the increase in head.

The dam is founded on solid rock and no deterioration in the foundation condition is apparent. Core drilling into the dam foundation revealed a sound, compact rock condition requiring no special foundation treatment for the increase in hydraulic head. The bedding planes, and joints are sufficiently sealed to leakage, especially seeps and small flows. No weathering of the foundation rock was observed.

The stability analysis of the existing dam structure has been made with an increase in headwater to El. 266 and minimum tailwater at El. 221 with silt to El. 257. For earthquake forces, the safety factor was found to be 1.25 against sliding, and 1.58 against overturning at the new head. At the existing powerhouse, these factors will be 1.57 and 1.60 respectively. Thus, the structure was found to be safe for the proposed increase in headwater, and from the stability points of view, no strengthening of the dam is required.

Enlargement of any reservoir, by modification of an existing dam to raise the reservoir water level, requires consideration of the additional submergence of land and the effects of higher water levels on existing developments and other important structures. By raising the present normal water level elevation (lower pool) of Parr Reservoir by 9 feet, an additional area of approximately 2550 acres will be submerged each day. The vast majority of this land area consists of hardwood, pine and mixed pine forests. Fortunately the area has not developed, except for a 12.5 mile section of the Southern Railroad which runs parallel to the east bank of the existing reservoir and County Road 28 which runs parallel to the west bank of the existing reservoir. The railroad will require 5.5 miles of roadbed modification, including replacement of 1260 feet of timber trestle. County Road 28 will require raising the grade of 5600 feet of roadbed including rebuilding two bridges, each with an average length of 230 feet. No additional structures will be influenced by the higher water levels on the lower pool except the company's Parr Hydro Station and Parr Steam Plant. The aforementioned items are shown on Figure 4 page 6

Hydraulically operated crest gates, with automatic and manual controls, were chosen as the preferable alternative of effectively raising the existing dam height in contrast to permanently raising the concrete spillway section. These gates are being supplied and erected by the Dravo Corporation. The existing Parr Hydro Station will pass the average Broad River flow of 6000 cfs. During the times the river flow exceeds 6000 cfs, the crest gates can be automatically or manually operated, either lowered or raised, from a control board to be located in the existing Parr Hydro Station. Control of the gate spill position, when coordinated with the incoming water inventory, will greatly reduce backwater flooding compared to that which would result if the fixed crest of the existing dam were permanently raised. This arrangement, in addition to reducing the amount of new lower pool land required, will reduce the modification and relocation expenses of the two highway bridges, raising of the rail and greatly decrease the amount of fill and rip-rap on the railroad. Also, modifications to the existing Parr Hydro and Parr Steam Plant will be lessened considerably. The crest gates will provide the required increased storage volume of 29,000 acre feet which is equivalent to 8 continuous hours of generation by the pumped storage facility.

After the crest gate supplier was selected, hydraulic model tests were performed by Alden Research Laboratories, Holden, Massachusetts, to determine as accurately as possible the coefficient of discharge of the existing dams Ogee spillway and the coefficient of discharge of the existing dam modified with the spillway crest gates installed. The resultant coefficients were 3.85 and 3.00. Also, during the hydraulic tests at Alden, an uplift force was found to occur on a gate as increasing water flowed over the gate surface. To compensate for the combined uplift force, sixty (60) post tension anchors will be installed through the dam along its axis and anchored into bedrock to resist the resultant tension on the dam.

The planned modifications of the existing dam consists of installing 10 movable spillway crest gates. Each movable gate will be 198.4 feet long and 9 feet high and each will be operated by three (3) hydraulic cylinders. Concrete piers, each 4 feet wide, will be located between pairs of gates. The main purpose of the piers is to admit air under the gates when water is spilling over. Taking into account the reduction in the coefficient of discharge and the obstruction due to four piers, the overall reduction in the spillway capacity will be approximately 22%.

Subsequent results of backwater studies revealed that when experiencing a 25 year flood (163,000 cfs) the water elevation at the dam, with all currently planned spillway crest gates down, will be one foot higher with no contributing flow from the upper pool. This increased backwater effect becomes non existant about five miles upstream of the dam. For the remaining 13 mile length of the project's influence on the river, the profile is actually lower, due to improved river flow resulting from clearing. With full generation flow from the upper pool combined with the 25 year flood the water elevation at the dam with all gates down will be 3 foot higher. This resulting backwater effect becomes non existant about 7 miles upstream of the dam and similiarly a lower profile is realized for the remaining upstream influence of the project. All of the backwater profile data was developed by contract arrangements with the State Water Resources Department. The actual studies were conducted for the state by the USGS for different flows ranging from 6000 cfs (baseflow-Av. daily flow) to 242,000 cfs (100-year flood) for the following three conditions:

1. Baseline conditions with no modification of Parr Dam or the lower pool

2. Parr Dam modified with crest gates installed and down in horizontal position, lower pool cleared but the pumped storage

units not operating.

3. All conditions as in #2 but the pumped storage units in full generation mode, except in the case of a 50-year flood, generation reduced to one-half.

The effect of these conditions on the river was studied for a distance of 18 miles upstream from Parr Dam which is the extent of the projects influence on the Broad River. In this study zone the main tributaries of the Broad River were considered as also being in corresponding flood conditions.

We are now determining the need for, and the most practical means of, increasing the capacity of the planned spillway. The studies to determine the most practical means of increasing the capacity includes:

(1) Providing additional length of gated spillway (tainter or crest gates) on the west earth dike portion of the dam.

(2) Installing tainter gates in one of the 10 spillway crest gate positions.

Detailed field investigation consisting of topographical surveys and geological exploration on the west earth dike portion of the existing dam are in progress. These investigations will assist in determining the feasibility of providing additional length of gated spillway. If feasible the optimum length, height, and depth of the new section and the possible requirements of an approach and spill channel must then be determined. Feasibility studies are also in progress for substituting one of the crest gates with one or more tainter gates and determining the optimum dimensions of the tainter gates. If a tainter gate is used, we anticipate that the portion of the crest of the existing concrete dam will be cut as deep as possible to effect a considerable increase in discharge per foot of dam, however, no approach or spill channel modifications are required for this arrangement.

The results of the two previously mentioned methods of increasing spillway capacity will be integrated in a complete overall economic study of the projects impact, with and without increasing the spillway capacity, and will determine the degree of additional design modifications. This study will consist of analyzing conditions along the river, additional land or structures that may be endangered by higher than anticipated water levels, cost of land and flooding rights, cost of generation lost by possibly less than full pumped storage plant operation during flood conditions on the river, project schedule delay, and increased construction and material cost. At the conclusion of this economic analysis, the most feasible plan will be implemented.

Modifications of the existing Parr Hydro and Parr Steam facilities were previously mentioned. Compared to the total project these modifications are not extensive but had to be revealed early to assist design changes, scheduling and material procurement.

At the hydro plant a trash boom about 700' long protects the intake channel. It consists of approximately 35 rectangular hollow steel plate boxes, each about 20' long. These are strung between two cables which are anchored to the dam at one end and to a deadman at the shore end. The anchorage of the trash boom at the dam and the shore end will have to be redesigned to permit the trash boom to rise and fall 9 feet during pumping and generating modes of the Fairfield Pumped Storage Plant. The track level of the hydro intake gate hoist will have to be raised about one foot to clear the daily high water level. The main head gates, gate seals and trash racks will be inspected at a time when the Parr Hydro Plant is shut down. Minor repairs and replacement of portions of the gate seals is anticipated.

Within the hydro plant itself, new structural foundations will be provided for the hydraulic system and control panel required to operate the spillway crest gates. Minor alterations and repairs will be made in the hydro plant to accommodate the increased hydraulic head.

The hydraulic machinery at Parr Hydro will operate against an additional 9' head which represents an increase of 25%. It is, therefore, necessary to check the operating characteristics of the turbine to ascertain the effect of this 25% increase. Whatever modifications are made as a result of this increase in head will be of such a nature as not to interfere with the operation at its present head since the increased head will vary with each day from 0 - 25%. Additionally the governor operation and forces required to protect wicket gates will be evaluated.

The turbine-generators are approximately 60 years old. Records of turbine characteristics, if such records did exist, are currently not available and subsequent records of changes that may have been made to the turbine have not been found, so it becomes a difficult task to evaluate turbine performance at the new head condition. Investigations and analyses by our consultants indicated that there should not be any operating problems or characteristics that will require major equipment changes. Due to the lack of original design information, it will be necessary to wait until the turbines are run at the higher head to determine what adjustments and safe guards will be necessary to ensure continued safe reliable operations.

The raising of the water level in the Broad River will require relocation of assorted electrical conduits at the Parr Hydro Plant. The amount of work is considered minor in scope.

At the steam plant, a small earth berm separates the condenser circulating water intake and discharge channels. The purpose of the separation is to prevent recirculation of the heated discharge. With the new normal water level nine feet higher, the peninsula would be completely submerged each day. To maintain the desired separation of the channels we presently plan to drive sheet piling around the perimeter of this peninsula and add earth fill in the interior. The condenser circulating water intake and discharge tunnels themselves will require pressure grouting repair. Other repairs involve replacing concrete protection on steel beams and repair or replacement of a column footing which is moderately loaded. Similar minor alterations will be made in the lower elevations of the steam plant.

Borings are being taken at the lower pool end of the steam plant to determine soil permeability characteristics and water table level in order to estimate the effect of the normal 9'-0" increase in water level with regard to boiler room flooding. A simple protective system is anticipated.

Only minor electrical modification will be required at both existing steam and hydro facilities due to the change in water level. This mainly consists of raising various conduits to avoid innundation.

The perimeter of the lower pool and islands will be selectively cleared of trees and vegetation from the present water elevation (El. 257) up to the water profile developed by an incoming river flow of 36,000 cfs combining with a 48,000 cfs generating flow from the pumped storage facility.

Selective clearing will involve close examination of the developed backwater profile data and topographic maps of the area, then clearing only those areas that would cause an increase in the backwater profile. Selective clearing will, in contrast to clear cutting, result in preserving wooded areas infrequently covered with water, reduce initial cost and reduce future maintenance clearing. We anticipate approximately 50% reduction in the lower pool area clearing requirement. The cumulative area to be cleared is estimated at 3,200 acres.

The Environmental Reports for both the nuclear and pumped storage projects were completed to meet licensing authority needs. The company retained the services of Dames & Moore, Inc. for preparation of the reports including the majority of data collection relative to the project area.

The Alden Research Laboratory was retained to perform all of the thermal analyses necessary to clearly establish the resultant thermal influence of the heated circulating water discharge from the nuclear station on the upper and lower pools and the river downstream of the modified Parr Dam.

During the early stage of license preparation the company developed a very active speakers list and took advantage of all opportunities to explain the project to the citizens of the state and particularly to the citizens of the immediate area.

A citizens committee was established by company efforts with the purpose of bringing to the attention of the company any concerns of the local population. The meetings of the citizens committee were open for anyone to attend but active participation remained limited to the area representatives serving by company request as the committee. The committee representatives were selected by the company to achieve a representation of all the areas interest.

The committee made an inspection trip to Alden Laboratory and to a recently completed pumped storage project. At Alden the committee participated in discussions concerning the operation of the overall project and observed the physical model in operation. Observing the model repeatedly operating through daily cycles and studying maps of the project area provided each of the members with a very good understanding of the project's size and the land area effected.

During the committee's visit to the recently completed pumped storage facility they observed an operation which is comparable to the operation of our company's planned project, but more important, had the opportunity to discuss the area impact of the project with local residents. During these discussions, the experienced local residents gave a very unbiased story to our committee members of how the project affected their personal lives and the lives of the other people in the area. Recommendations were made to our committee and to our company representatives participating in the trip that itemized ways to ease the impact of the Fairfield Project.

Two interventions were filed against Fairfield Project. One intervention was based on environmental damage to the natural habitat and the other on the amount of land the company was requiring from an individual for the project. Both of the intervenors served on the citizens committee.

The company visited every local, county, state and Federal Agency whose possible interests would influence the licensing of the project and explained the need for the project and the details. All problems that resulted from these visits were resolved to the mutual satisfaction of all parties.

The application for license for the Pumped Storage Project was filed with the Federal Power Commission, July 26, 1972. The F.P.C. circulated the company's environmental report for comments September 27, 1972. The F.P.C. circulated its initial environmental report for comments September 7, 1973. The final F.P.C. environmental report was issued March 1974. These dates are expressly pointed out to bring to your attention that getting a license through the F.P.C. to a point of receiving the license to permit construction or being advised that the license request is denied is an extremely drawn out affair.

Engineering for the Fairfield Pumped Storage Facility began in 1971. Gibbs and Hill, Inc. are the Architect-Engineers for the major part of the Pumped Storage Project. Specifications for the pump/turbines - motor/generators were completed in September 1971, and these long lead items were committed to Allis-Chalmers in November 1972. Delay in receiving the project license has been costly and in this current period of inflation and long delivery lead times the cost in time and dollars is not fully known. A minimum of other long lead items have been committed to the fabricators. Structural and architectural design of the Fairfield Powerhouse and related facilities is nearly complete. Mechanical and electrical design is proceeding at a slower pace.

Daniel Construction Company has been retained to perform construction services for the facility. Licensing delays have also prevented Daniel from producing a firm construction schedule as well as starting their work. The inability to phase the engineering, procurement, and construction into a workable, consistant schedule grows constantly worse as months pass without license approval by the F.P.C. SCE&G is consequently in the unenviable position of attempting to complete the design (by providing the engineer with enough data to do so), and avoiding escalation and lead times on equipment that might not phase expeditiously into the construction schedule upon granting of the license.

Current expectations are that units 1 and 2 will begin pumping to completely fill the upper pool during the second half of 1976. This is necessary to obtain the water supply for initial flush and cleaning of the nuclear facility. Start up of the eighth pumped storage unit is planned to occur in the fall of 1977 and will provide SCE&G Co. with a total of 480 MW of pumped storage capacity.

At the present time the V. C. Summer Nuclear Station is under construction. Construction of the Fairfield Pumped Storage Facility has not begun.

In conclusion, the SCE&G Company is convinced that the present plans to utilize the off peak generation of the V. C. Summer Nuclear Station to provide the basic portion of pump back energy to the upper pool will meet the system load forecast requirements in the most economical manner and with a minimum of environmental impact as compared to any alternate methods available.

*This paper was presented August 18, 1974 and the Federal Power Commission issued a license for the Fairfield Pumped Storage Facility on August 28, 1974.

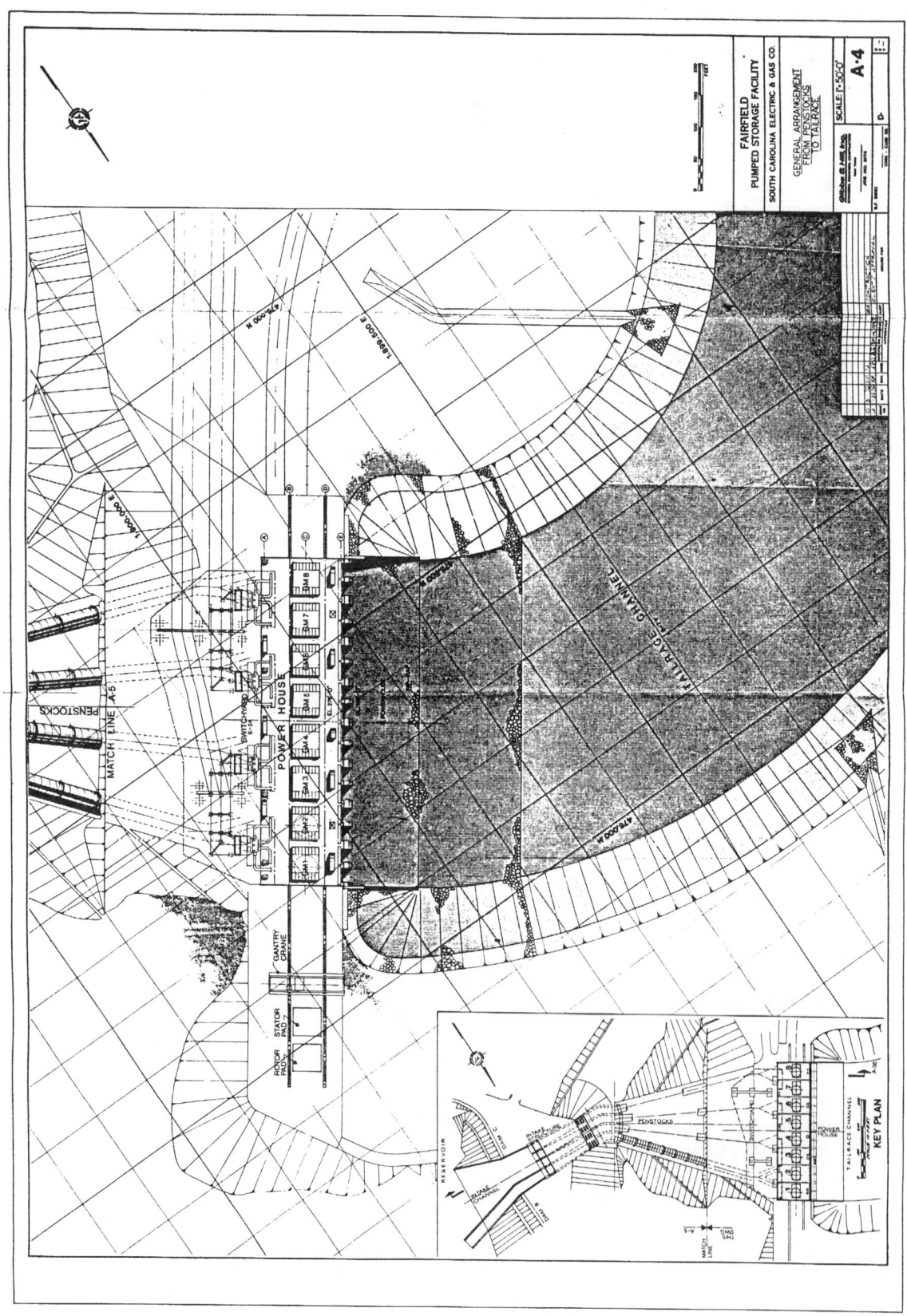

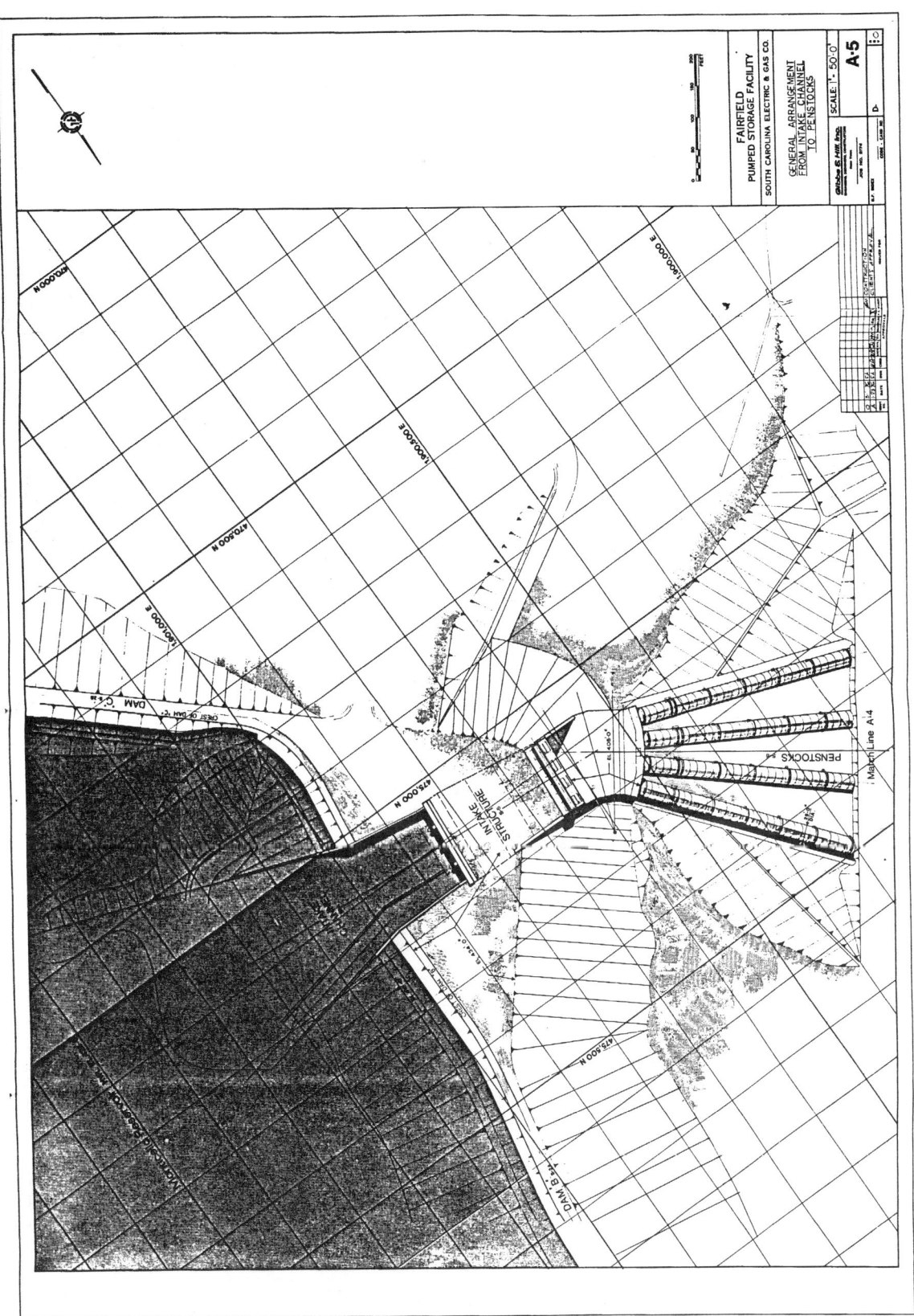

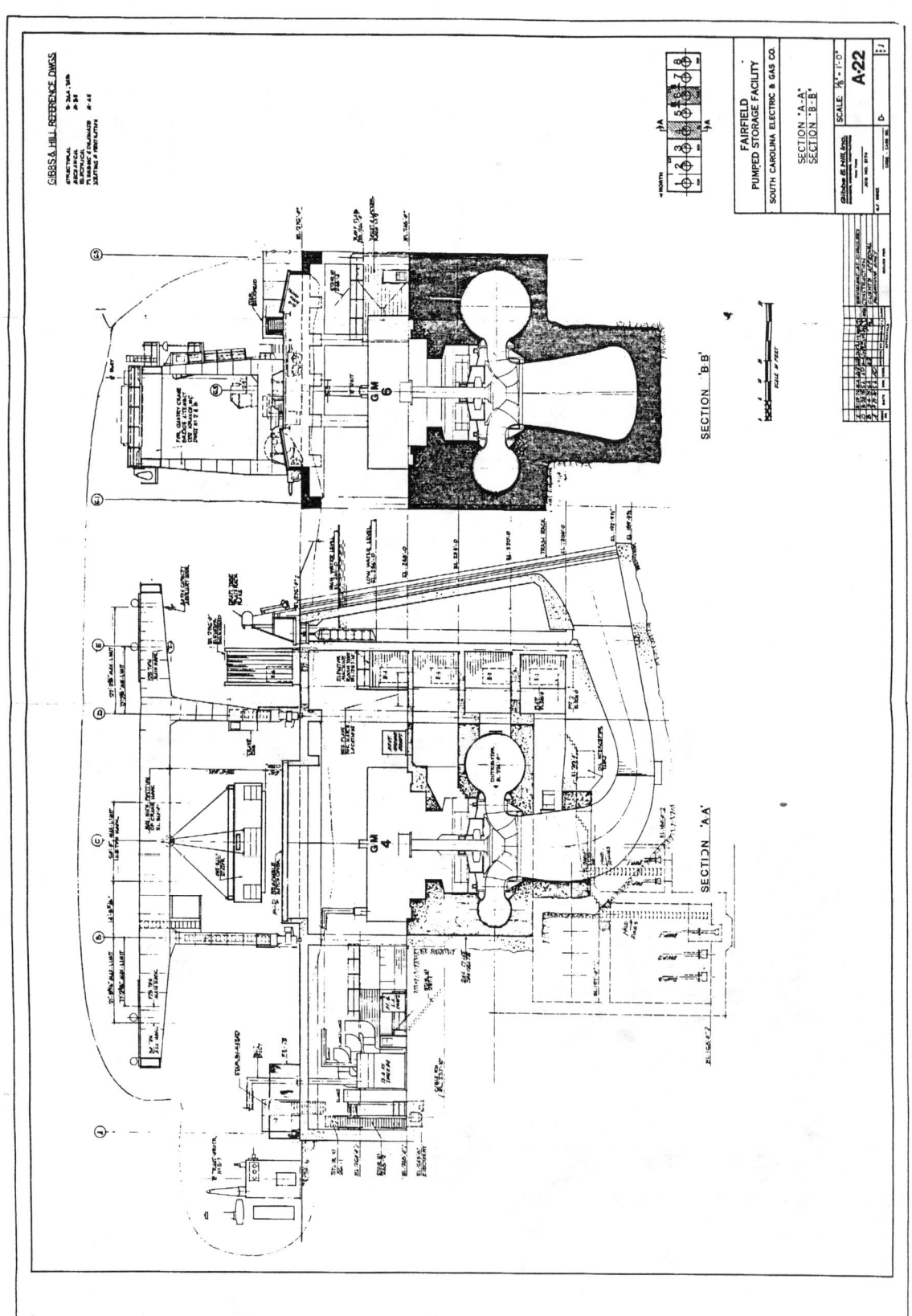

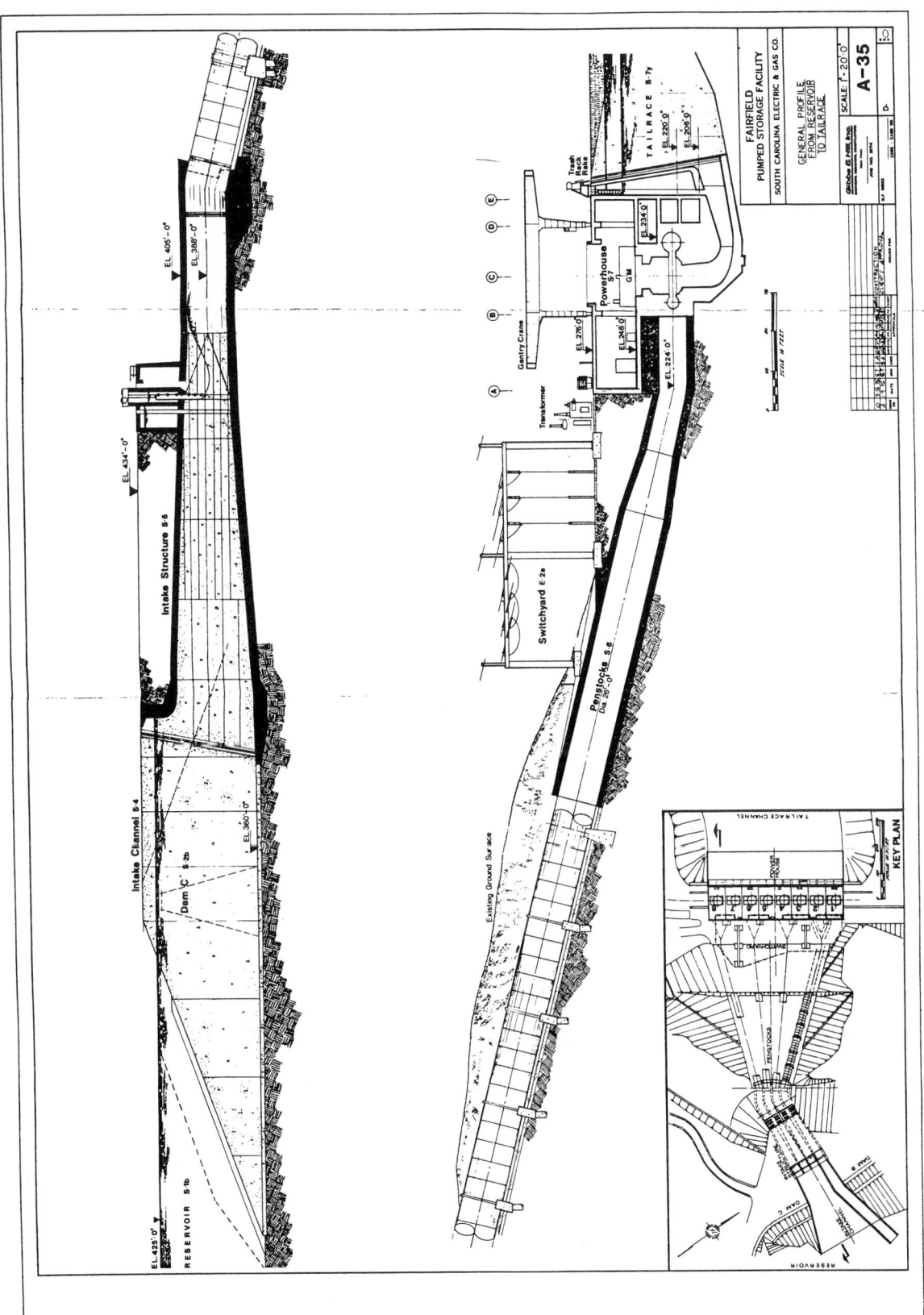

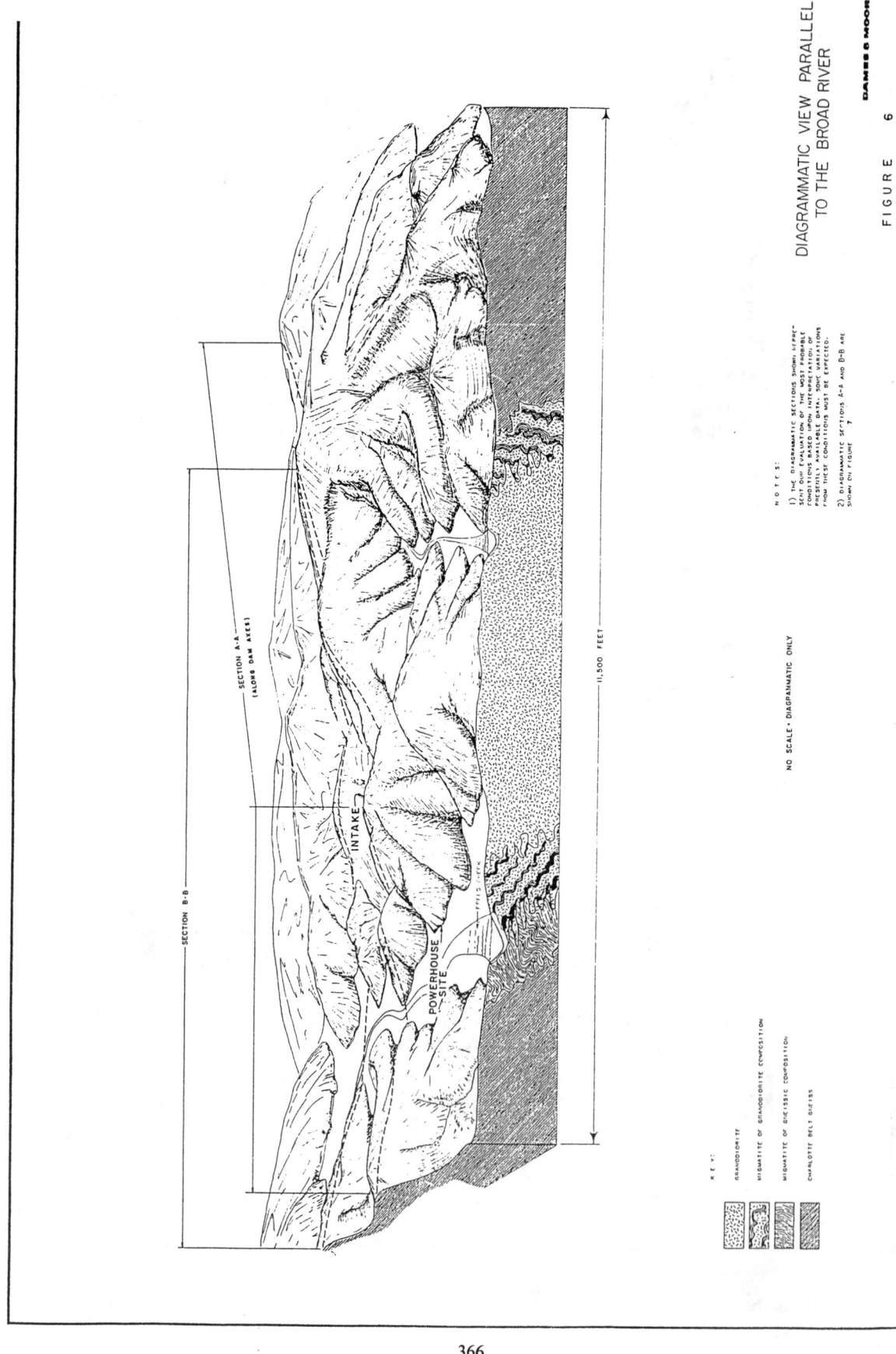

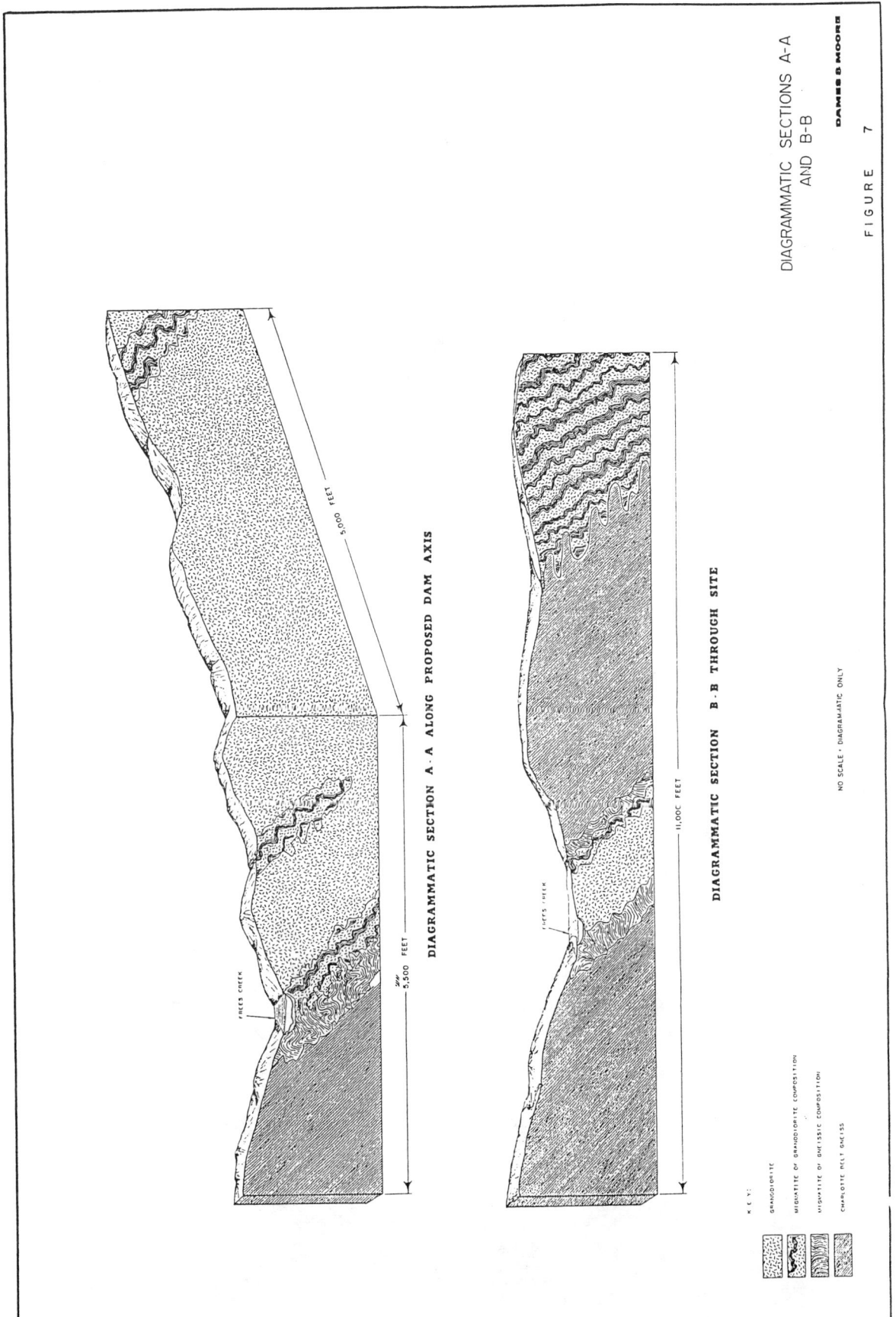

Powerhouse

Fairfield Pumped Storage Plant

LIST OF PROJECT ARTICLES PUBLISHED

TITLE: FAIRFIELD PUMPED STORAGE FACILITY

AUTHOR: Crews, E. H. Jr.

SOURCE: Converting Existing Hydro-Electric Dams and Reservoirs into Pumped Storage Facil, Eng Found Conf, Proc, Franklin Pierce Coll, Rindge, NH, Aug 18-23 1974 p197-214, Publ by ASCE, New York, NY 1975

PUBLICATION YEAR: 1974

TITLE: ELECTRICAL FEATURES OF FAIRFIELD PUMPED STORAGE PROJECT, SOUTH CAROLINA ELECTRIC AND GAS COMPANY

AUTHOR: Biro, P.

SOURCE: IEEE Power Eng Soc, Winter Meet, Prepr, New York, NY Feb 3-8 1980 Publ by IEEE (Cat n CH1523-4/80), New York, NY, 1980 Pap F80 249-3, 6 p

PUBLICATION YEAR: 1980

TITLE: ELECTRICAL FEATURES OF FAIRFIELD PUMPED STORAGE PROJECT, SOUTH CAROLINA ELECTRIC AND GAS COMPANY

AUTHOR: Biro, P

SOURCE: IEEE Transactions on Power Apparatus and Systems v PAS-99 n 5 Sep-Oct 1980 p 1752-1757

PUBLICATION YEAR: 1980

TITLE: DIGITAL DATA SYSTEM IMPROVES BEARING MONITORING
       Stockman, S.; Frick, N.

SOURCE: Power Eng. (USA) vol.92, no. 12, pp.:22-3

PUBLICATION YEAR: Dec. 1988

ADDITIONAL ARTICLES: Seven Papers in Proceedings of American Power Conference

PUBLICATION YEAR: 1982

Plant Name: **FLATIRON POWERPLANT**

Plant location:
  Loveland, CO
	Owner: Bureau of Reclamation
	         Attn: D-3000
	         P.O. Box 25007
	         Denver, Colorado 80225

Rated capacity 9 MW Average static head 290 ft Plant efficiency % Stored energy MWh Number of units 1	Designers: Bureau of Reclamation

Construction time: 2 years, 10 months
Construction cost:
Price level:
First commercial power: January 1954
FERC project number:
	Plant Manager/Superintendent:
	         Bureau of Reclamation
	         995 Wilson Avenue
	         Loveland, Colorado 80539

River or water source: Colorado and Big Thompson Rivers

DAM	UPPER RESERVOIR	LOWER RESERVOIR
Type	Zoned earthfill	Zoned earthfill
Height (ft)	214	86
Crest length (ft)	1,959	1,726
Volume (yd^3)	4,264,000	382,000
RESERVOIR		
Type	Constructed, Carter Lake	Constructed, Flatiron Afterbay
Surface area (acres)	1,142	47
Usable power storage (acre-ft)		440
Power pool fluctuation (ft)	135.0	10.8
Operating levels		
Maximum (ft)	5,760.0	5,472.8
Minimum (ft)	5,625.0	5,462.0
Drainage area (miles^2)	5.7	7.4
Seepage (ft^3/s)	2.700	
SPILLWAY		
Design flood		
Return period (years)		
Flow (ft^3/s)		23,630
Capacity (ft^3/s)		
Type	None	Concrete, ungated
Gates		
Number	None	None
Type	None	
Width (ft)		
Height (ft)		
OUTLET WORKS		
Discharge capacity (ft^3/s)	1,261	929
Number of water passages	1	2
Dimensions of water passages		
Height (ft)	3.00	6.75
Width (ft)	3.00	9.00
Diameter (ft)		
Type of gates	Slide	Radial
Number of gates	2	2

Flatiron Powerplant - Page 1 (English)

Plant Name: **FLATIRON POWERPLANT**

Plant location	Owner:	Bureau of Reclamation
Loveland, CO		Attn: D-3000
		P.O. Box 25007
		Denver, Colorado 80225
Rated capacity 9 MW		
Average static head 88.4 m	Designers:	
Plant efficiency %		Bureau of Reclamation
Stored energy MWh		
Number of units 1		
Construction time: 2 years, 10 months	Plant Manager/Superintendent:	
Construction cost:		Bureau of Reclamation
Price level:		995 Wilson Avenue
First commercial power: January 1954		Loveland, Colorado 80539
FERC project number:		

River or water source: Colorado and Big Thompson Rivers

	UPPER RESERVOIR	LOWER RESERVOIR
DAM		
Type	Zoned earthfill	Zoned earthfill
Height (m)	65.2	26.2
Crest length (m)	597.0	526.0
Volume (m³)	3,260,000	292,000
RESERVOIR		
Type	Constructed, Carter Lake	Constructed, Flatiron Afterbay
Surface area (Mm²)	4.62	0.19
Usable power storage (Mm³)		0.543
Power pool fluctuation (m)	41.15	3.29
Operating levels		
Maximum (m)	1,755.65	1,668.11
Minimum (m)	1,714.50	1,664.82
Drainage area (Mm²)	14.763	19.166
Seepage (m³/s)	0.0765	
SPILLWAY		
Design flood		
Return period (years)		
Flow (m³/s)		669
Capacity (m³/s)		
Type	None	Concrete, ungated
Gates		
Number	None	None
Type	None	
Width (m)		
Height (m)		
OUTLET WORKS		
Discharge capacity (m³/s)	36	26
Number of water passages	1	2
Dimensions of water passages		
Height (m)	0.914	2.057
Width (m)	0.914	2.743
Diameter (m)		
Type of gates	Slide	Radial
Number of gates	2	2

Flatiron Powerplant - Page 1 (Metric)

FLATIRON POWERPLANT

INTAKES	UPPER INTAKE	LOWER INTAKE
Number	1	
Type	Reinforced concrete	
Design discharge (ft³/s)	550	
Gross inlet area (ft²) (at trash racks)	162	
Bar racks		
spacing (in)	2.00	
shape		
depth/thickness (in)	0.50 / 0.50	
diameter (in)		
Emergency gates		
number	1	None
height/width (ft)	8.00 /	
type	Butterfly	
Service gates		
number	None	1
height/width (ft)		6.33 /
type		Butterfly
Bulkhead/stop logs (Y or N)	Y	
number of units serviced		1
Hoists		
number	None	1
capacity (tons)		100
type		Gantry

WATER PASSAGES	Upper Tunnel	Shaft	Lower Tunnel	Surge Tanks Upper	Surge Tanks Lower	Penstocks	Tailrace Tunnel
Number	1			1		1	
Diameter (ft)	10.0			50.0		8.0	
Length (ft)	6,912					480	
Maximum velocity (ft/s)							
Concrete lining thickness (in)	9.96						
Total length of concrete sections (ft)	6,912						
Steel liner Thickness							
Minimum (in)						0.38	
Maximum (in)							
Material grade						C	
Total length of steel-lined sections (ft)						480	

Notes:
Upper Surge Tank: Restricted-orifice type, underground

FLATIRON POWERPLANT

INTAKES	UPPER INTAKE	LOWER INTAKE
Number	1	
Type	Reinforced concrete	
Design discharge (m³/s)	15.6	
Gross inlet area (m²) (at trash racks)	15.1	
Bar Racks:		
spacing (mm)	51	
shape		
depth/thickness (mm)	13 / 13	
diameter (mm)		
Emergency gates		
number	1	None
height/width (m)	2.438 /	
type	Butterfly	
Service gates		
number	None	1
height/width (m)		1.929 /
type		Butterfly
Bulkhead/stop logs (Y or N)	Y	
number of units serviced		1
Hoists		
number	None	1
capacity (Mg)		91
type		Gantry

WATER PASSAGES	Upper Tunnel	Shaft	Lower Tunnel	Surge Tanks Upper	Surge Tanks Lower	Penstocks	Tailrace Tunnel
Number	1			1		1	
Diameter (m)	3.05			15.24		2.44	
Length (m)	2,106.8					146.3	
Maximum velocity (m/s)							
Concrete lining thickness (m)	0.254						
Total length of concrete sections (m)	2,106.8						
Steel liner Thickness							
Minimum (mm)						10	
Maximum (mm)							
Material grade						C	
Total length of steel-lined sections (m)						146.3	

Notes:
Upper Surge Tank: Restricted-orifice type, underground

Flatiron Powerplant - Page 2 (Metric)

FLATIRON POWERPLANT

POWERHOUSE and RELATED FEATURES

Powerhouse Structure
Type: Semi-outdoor (Gantry Not Showing)
Length: 56 ft   Width: 71 ft   Height: 49 ft

Guard Valves
Number: 1   Diameter: 6.3 ft
Type: Butterfly

Transformers
Number:
Ratings:
Voltages: (kV) 13.8 / 115

Generator
Rating generating (MVA): 8.5   Rating pumping (MVA): 8.0
Insulation type:
Starting method: Amortisseur bar
Starting equipment:

Runners
Material: Steel
Minimum unit submergence: 9.0 ft
WR^2:
Manufacturer: Allis Chalmers
Model test by: Allis Chalmers

	Reversible Runners	Reversible Motor/Generator	
Number	1		1
Diameter (ft)	8.40	Rotor	Stator
rpm synchronous	300.0		300.0
rpm overspeed	555.0		555.0
Type	Francis		

Information on Runners

Condition:	Gross Head (ft) Generating	Gross Head (ft) Pumping	Capacity (MW) Generating	Capacity (MW) Pumping	Discharge (ft³/s) Generating	Discharge (ft³/s) Pumping	Turbine/Pump Eff.(%) Generating	Turbine/Pump Eff.(%) Pumping
Maximum head & maximum power	298	298						
Minimum head & maximum power	153	153						

Condition:	Net Head (ft) Generating	Net Head (ft) Pumping	Capacity (MW) Generating	Capacity (MW) Pumping	Discharge (ft³/s) Generating	Discharge (ft³/s) Pumping	Turbine/Pump Eff.(%) Generating	Turbine/Pump Eff.(%) Pumping
Rated head @ best gate	290	240	9	10	470	370		

Note: Data in the above table are based on design data.

Flatiron Powerplant - Page 3 (English)

FLATIRON POWERPLANT

POWERHOUSE and RELATED FEATURES

Powerhouse Structure
Type: Semi-outdoor (Gantry Not Showing)
Length: 17.1 m   Width: 21.6 m   Height: 15.0 m

Guard Valves
Number: 1   Diameter: 1.93 m
Type: Butterfly

Transformers
Number:
Ratings:
Voltages: (kV) 13.8 / 115

Generator
Rating generating (MVA): 8.5   Rating pumping (MVA): 8.0
Insulation type:
Starting method: Amortisseur bar
Starting equipment:

Runners
Material: Steel
Minimum unit submergence: 2.74 m
WR^2:
Manufacturer: Allis Chalmers
Model test by: Allis Chalmers

	Reversible Runners	Reversible Motor/Generator	
		Rotor	Stator
Number	1		1
Diameter m	2.57		
rpm synchronous	300.0		300.0
rpm overspeed	555.0		555.0
Type	Francis		

Information on Runners

Condition:	Gross Head (m)		Capacity (MW)		Discharge (m³/s)		Turbine/Pump Eff.(%)	
	Generating	Pumping	Generating	Pumping	Generating	Pumping	Generating	Pumping
Maximum head & maximum power	90.8	90.8						
Minimum head & maximum power	46.6	46.6						

Condition:	Net Head (m)		Capacity (MW)		Discharge (m³/s)		Turbine/Pump Eff.(%)	
	Generating	Pumping	Generating	Pumping	Generating	Pumping	Generating	Pumping
Rated head @ best gate	88.4	73.2	9	10	13.3	10.5		

Note: Data in the above table are based on design data.

Flatiron Powerplant - Page 3 (Metric)

FLATIRON POWERPLANT

Plant Data:
  Average GWh generating per year:
  Average GWh pumping per year:
  Starting time from standstill (s):
  Changeover time pumping to generating (min):
  Planned/scheduled time
    between major overhauls (years):
  Outage time required per unit
    during major overhauls (weeks):   11
  Representative plant availability (%):
  Representative planned outages (weeks per year):

Miscelleneous Notes:
  Flatiron Powerplant is a feature of the Colorado-Big Thompsen Project, a transcontinental water project.

  Flatiron Powerplant is located approximately 12.1 Km (7.5 mi) west of Loveland, Colorado, and 6.4 Km (4 mi) south of the Big Thompson River.

  Flatiron Powerplant units No. 1 and No. 2 operate on regulated flows; the water falls 337.7 m (1,108 ft) from Pinewood Lake to Flatiron Afterbay for peaking purposes and irrigation demands.

  Unit No. 3 pumps water from Flatiron Afterbay to Carter Lake during periods of low power demand. Additional power for the system is developed by reversing the flow during peak power and/or irrigation demand periods.

  The main generators (units No. 1 and No. 2) are each driven at 514 rpm by a Francis-type turbine rated at 48,000 hp at 321.6 m (1,055 ft) of head and utilizing flows up to 27.2 m³/s (960 ft³/s). Each turbine spiral case is directly connected to a 1.12 m (44 in.) pressure-regulating valve.

  The pump-turbine unit (unit No. 3) is rated to pump 10.5 m³/s (370 ft³/s) at 73.1 m (240 ft) of head running at 300 rpm. As a turbine, it has a nameplate rating of 12,000 hp at a head of 88.4 m (290 ft) and utili zes flows up to 13.3 m³/s (470 ft³/s). The turbine speed is 257 rpm.

Cavitation Experience:
  Repair has been required on the stay vanes. A lot of cavitation has occurred around the butterfly valve housing.

Significant or Unique Problems:
  Wave problems during pumping restricts boating on the afterbay.

List of Licenses Required:
  None -- direct congressional authorization.

FLATIRON POWERPLANT

ENVIRONMENTAL FEATURES

Recreation:
  Recreation activities include day use and overnight camping.

Fish and Wildlife:

Social:
  The twin penstocks are considered landmarks by many of the residents of Loveland, Colorado.

Page 4 (Continued)

Design Features of the Flatiron Power and Pumping Plant

This paper* describes the design features of the generating units at the Flatiron Power and Pumping Plant of the Colorado-Big Thompson project. It includes a description of the 48,000-h.p. Francis-type turbines which operate under a maximum static head of 1,118 ft. and also a large 2-speed reversible pump-turbine motor-generator unit

By JOHN PARMAKIAN,
Head of the Technical Engineering Analysis Section, Design Division,
Bureau of Reclamation, U.S. Department of Interior.

THE Flatiron power and pumping plant is the largest and the terminal plant which utilises the power drop in the transmountain water-diversion system of the Colorado-Big Thompson Project. This completed project is a recent Bureau of Reclamation development which diverts water from the Colorado River on the western slope of the Continental Divide through the 13-mile-long Adams Tunnel to the eastern slope. It provides irrigation, hydro-electric power and other benefits for a large area northeast of Denver, Colorado. The profile diagram of the power-drop and irrigation features of the diverted water on the eastern slope is shown in Fig. 1. The design of this system is based on a water supply of 340,000 acre-ft., of which 310,000 acre-ft. are diverted from the western slope and 30,000 are obtained from the Big Thompson river. Water for the Flatiron power plant is supplied via the Flatiron penstocks and the Bald Mountain pressure tunnel from the Flatiron reservoir. This reservoir receives its discharge from the Pole Hill power plant. The afterbay of the Flatiron power plant supplies water to the Horsetooth feeder canal. Carter lake, which supplies water to the St. Vrain supply canal, is supplied from the Flatiron afterbay by the reversible pump-turbine in the Flatiron pumping plant. Provisions are also made for the generation of power at the pump-turbine when Carter lake water is released back through the plant to the Flatiron afterbay.

General Features

The Flatiron power and pumping plant structure is a semi-indoor installation. It consists of a reinforced-concrete structure which rests on siltstone, claystone, and sandstone. The roof is located high above the bed of the creek and is above any flash flood that is expected to occur in this locality. The main powerhouse crane of 100-ton capacity is mounted on the roof deck.

The combination plant contains two 48,000 h.p. Francis-type turbines and one reversible pump-turbine, motor-generator unit. The two main generating units have a combined hydraulic capacity of approximately 960 cusecs. and operate under a maximum static head of 1,118 ft. These units are designed to provide peak-load generation for the interconnected power system. The reversible pump-turbine unit has a rated pumping capacity of 370 cusecs. at a head of 240 ft. and is designed to supply Carter lake during off-peak hours. The unit may also be used for peak-load generation at any time.

Surge Tank

The Bald Mountain pressure tunnel conveys water from the Rattlesnake reservoir to the Flatiron penstocks. This 6,700-ft.-long tunnel is served by an inlet channel and trashrack structure in the reservoir and is provided with a 50 ft. diameter surge tank near the outlet portal of the tunnel. This simple, restricted-orifice type underground surge tank is designed to minimise the pressure surges induced by the sudden rejection of the flow at the turbines and to supplement the tunnel supply when the plant load is suddenly increased. It provides for contemplated fluctuations of 70 ft. between maximum downsurge and upsurge in the surge tank.

Penstocks

Two parallel, variable-diameter, 84-in., 78-in., and 72-in. welded and riveted, plate-steel penstocks are provided as waterways to carry water from the outlet of the Bald Mountain pressure tunnel to the main generating units in the Flatiron power plant. The penstocks are about 5,800 ft. in length and have a maximum shell thickness of $1\frac{5}{16}$ in. at lower end. For the upstream portion of each penstock comprising about 2,700 ft., an ASTM Designation A285 Grade C steel was used. For the remaining lower portion of each penstock a higher strength ASTM Designation A212 Grade B steel was used. All penstock sections fabricated from the A212 steel and from the A285 steel over 1 in. in thickness were thermally stress-relieved

*Contributed by the Hydraulic Division for presentation at the A.S.M.E. Diamond Jubilee Spring Meeting, Baltimore.

Reprinted, with permission, from *International Water Power and Dam Construction*, November 1955.

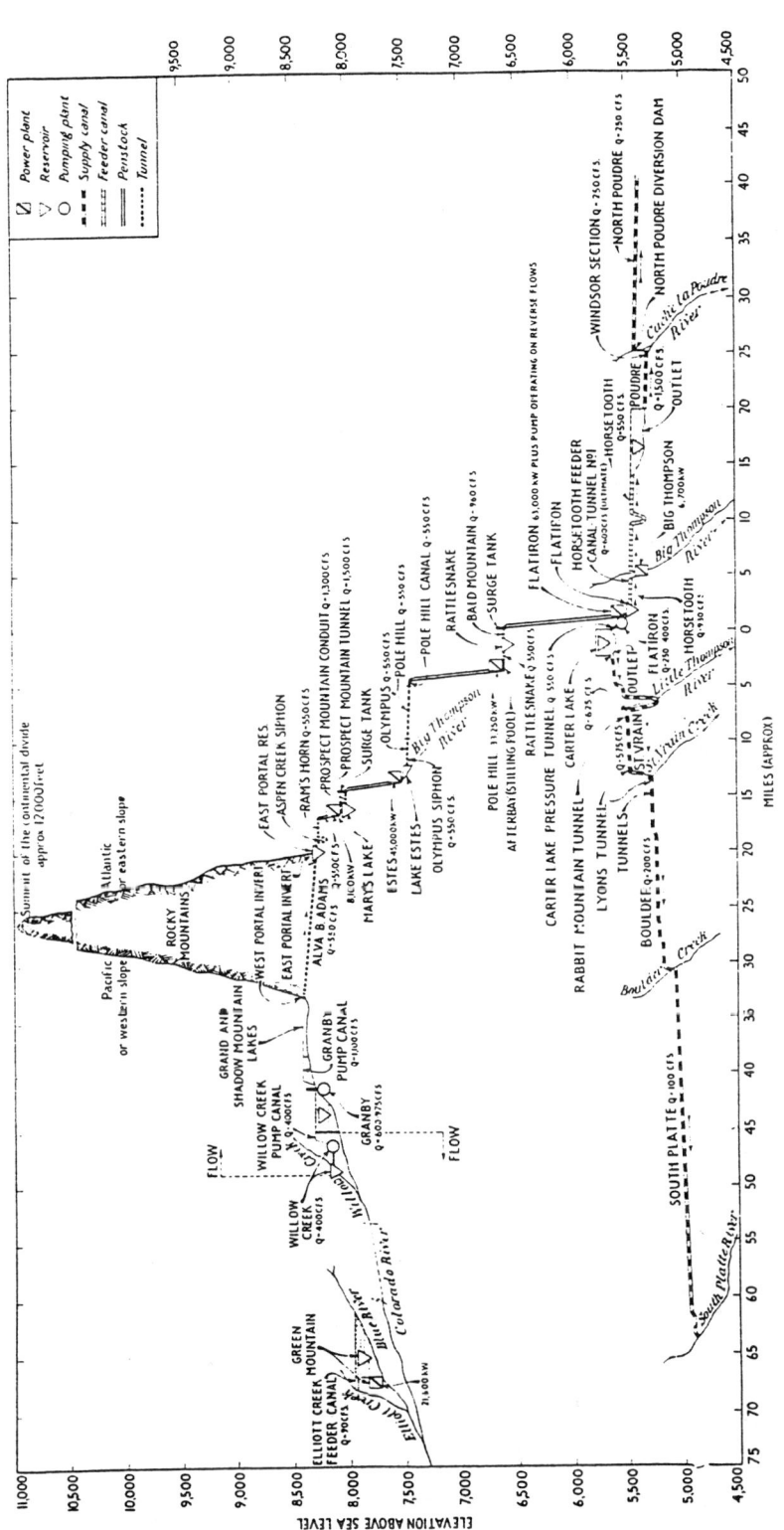

Fig. 1. Profile diagram of Colorado-Big Thompson project

after welding. Each individual erection section was then subjected to a hydrostatic pressure test in a specially designed pressure testing machine where a test pressure of 1¼ times the design pressure was applied. The field girth joints which were made from A285 steel were butt welded and those made from A212 steel were butt riveted, except the downstream end sections which were Dresser-coupled. For the most part the penstocks were installed above the ground with supports consisting of double-ring girders welded to the pipe. The penstocks were fabricated and installed by the Southwest Welding and Manufacturing Company.

Turbines

The main turbine installation for the plant consists of two vertical-shaft, Francis-type, 514 revs. per min. turbines, each having a capacity of 48,000 h.p. at full gate opening when operating under an effective head of 1,055 ft. The general arrangement through one of these units is shown in Fig. 2. Each turbine spiral case, pressure regulator valve body, and other parts subject to penstock water pressure, are designed for 600 lb. per sq. in., which is the maximum pressure, including waterhammer, that can be obtained with the pressure regulator functioning. They were shop tested at 900 lb. per sq. in., field tested at 600 lb. per sq. in., and embedded in concrete while under an internal pressure of 450 lb. per sq. in.

Each turbine is designed and constructed so that all removable parts may be removed from above by means of the main powerhouse crane, and can pass through the circular pit below the generator and through the generator stator. Provision is made in the design to permit the turbine shaft and runner to be raised or lowered as a unit for purposes of checking or inspecting the generator thrust bearing or disconnecting the main shaft coupling.

A critical shortage of nickel existed during the period of construction of the Flatiron turbines and for this reason aluminium bronze was substituted in most of the parts where stainless steel was originally called for in the specifications. This substitution included the runners which are cast of "Ampco" aluminium bronze. However, a spare interchangeable carbon-steel runner was also provided. The turbines and pressure regulator valves were designed and furnished by the Pelton Water Wheel Company.

Pressure Regulators

A pressure regulator of the internal-needle type is installed at each turbine to reduce the pressure rise in the penstock due to the rapid closure of the turbine wicket gates and to provide a water bypass

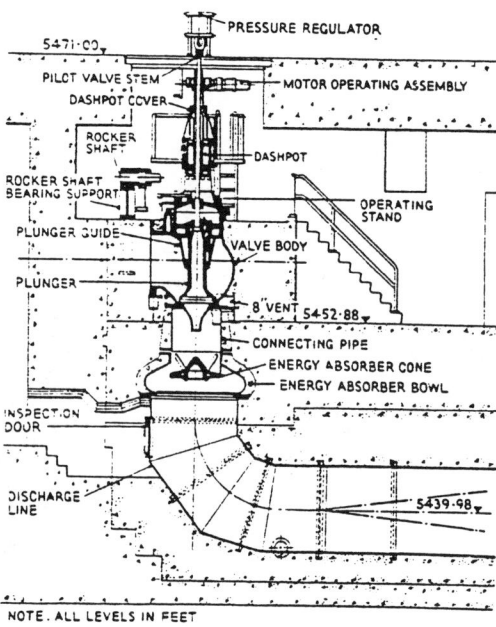

Fig. 3. Cross section through pressure regulator

during periods of shutdown. The water passage for the pressure regulator or relief valve for each unit is directly connected to the turbine spiral case and is designed to operate either as a synchronous bypass or as a water saving relief valve having a capacity equal to 100 per cent. of the flow capacity of the connected turbine. It is mechanically connected to the turbine wicket-gate mechanism and will stall the servomotor if the pressure regulator valve sticks or fails to open, thus ensuring a limitation on the water-hammer during rapid load rejections. The pressure regulator is normally set to operate as a water-saving relief valve which opens in step with the wicket-gate closure at rapid load rejection and then slowly recloses under the control of a dashpot. It does not open when the wicket gates slowly close during normal load changes. Each pressure regulator also can be operated as a continuous bypass when the turbine wicket gates are closed.

The energy of the jet leaving the pressure regulator-valve orifice is dissipated in the lower part of the pressure regulator shown in Fig. 3. The jet impinges on

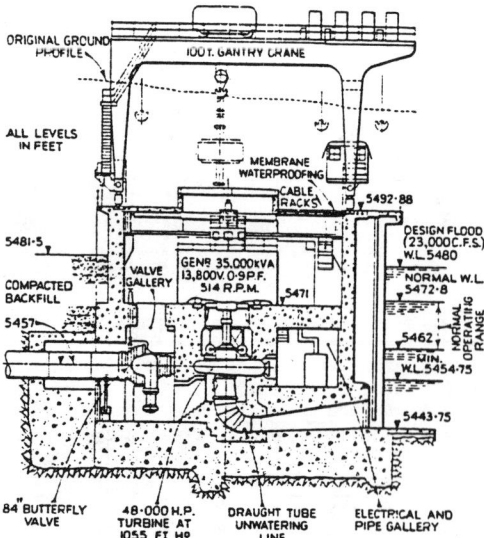

Fig. 2. Cross section through main generating unit

WATER POWER November 1955

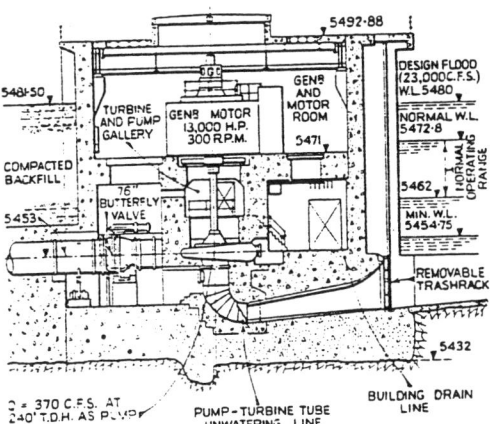

Fig. 4. Cross section through pump-turbine motor-generator unit

the steel energy-absorber cone and is doubled back on itself by a steel bowl surrounding the base of the cone. A steel-lined elbow draft tube conducts the flow to the tailrace.

Governors and Generators

The governors for regulating the speed of the turbines are of the oil-pressure, cabinet-actuator type, with electrically driven speed-responsive elements. The governor rating corresponding to the capacity of the turbine servomotors is 70,000 ft.-lb. The actuator main relay valves are adjusted to limit the rate of movement of the turbine gates between 5 and 15 sec. for a full gate opening or gate closing stroke. The governors were designed and furnished by the Woodward Governor Company.

The two main generators are 3-phase, 60-cycle, alternating-current generators with an output of 35,000 kVA each at the rated voltage of 13,800 V and at 90 per cent. power factor. Each generator has a self-aligning vertical equalising Kingsbury type thrust bearing and an upper guide bearing located above the rotor, and a lower guide bearing below the rotor. All parts of the generators and exciters are designed and constructed to withstand safely the stresses resulting from a runaway speed of 950 r.p.m. The generators were manufactured and installed by the Westinghouse Electric Corporation.

Pump Turbine

The pump turbine at the Flatiron power and pumping plant is of particular interest as it is one of the first large units of its type to be placed in service in the United States. The general arrangement through this unit is shown in Fig. 4. The pump rating is 370 cusecs. at 240 ft. total head. As a turbine-driven generator the unit is rated at 8,500 kVA. The combination unit is basically a pump with only 6 impeller vanes but it has fixed stay vanes like a turbine. It has no wicket gates or governor and therefore always operates as a turbine at "full gate." The butterfly valve adjacent to the unit is held in the wide open position when the unit is pumping or generating. Because of the dual type of operation, trashracks are provided at both headwater and tailwater. A 42-in. diameter submerged tube valve is installed in the conduit system to bypass the pump-turbine unit when the irrigation flow requirements exceed the water capacity of the unit as a turbine.

At this installation the head on the unit varies from 290 to 140 ft. for the turbine and 170 to 300 ft. for the pump. Model tests by the pump-turbine manufacturer for the particular impeller selected indicated that with these head ranges higher efficiencies were obtained by running the turbine at a lower speed than the pump. The adopted speeds were 300 revs. per min. when pumping and 257 when turbining.

The pump-turbine spiral case and other parts subject to penstock pressure are designed for 200 lb. per sq. in., which is the maximum pressure that can be obtained with the discharge valve closed and a runaway speed in the pumping direction of 340 revs. per min. The pump-turbine was designed and built by the Allis-Chalmers Manufacturing Company.

Motor Generator

The motor generator at the Flatiron power and pumping plant is designed for use either as a motor to drive the pump or as a turbine-driven generator. The unit is of the vertical-shaft, alternating-circuit, 3 phase, 60 cycle, 13,800 V, synchronous type and is directly connected to the pump turbine. The motor generator has a thrust bearing and an upper guide bearing located above the rotor, and a lower guide bearing located below the rotor. The entire rotor assembly of the motor generator, shaft, and pump-turbine runner is supported by a Kingsbury-type thrust bearing. This bearing is also arranged for pressure lubrication so that it can support the weight of the rotor assembly on a film of oil when starting in either direction to give a low starting torque. The unit is designed for operation at speeds of 300 and 257 revs. per min. either as a pump motor or as a turbine-driven generator. It is capable of operating as a 13,000 h.p. synchronous motor at a speed of 300 revs. per min., a 10,700 h.p. motor at a speed of 257 revs. per min., as a generator rated 8,500 kW at a speed of 257 revs. per min., and as a 10,000 kW generator at a speed of 300 revs. per min. These ratings are based on operation at 100 per cent. power factor. The unit can also be operated as a synchronous condenser.

The arrangement for obtaining the two-speed synchronous machine is by the use of two electrically independent windings on the stator, either one of which can be selected for use by selector switches and by a rotor winding consisting of 24 poles. For motor operation at 300 revs. per min. the 24-pole stator windings are used with the 24-pole rotor windings. For generator operation at a speed of 257 revs. per min., the 28-pole stator windings are used and an equivalent of 28 poles on the 24-pole rotor windings are obtained by a special switching apparatus which also reverses the polarity of some of the poles.

The unit is designed and constructed so that it is capable of operation at both speeds and in both directions, but as normal operation will be either as a motor at a speed of 300 revs. per min. or as a generator at a speed of 257 revs. per min., the switching arrangement is only designed to connect the windings for these two methods of operation. All parts of the

(Continued on page 428)

From page 416

motor-generator unit are designed and constructed to withstand a runaway speed of 555 revs. per min. in either direction. The motor generator was designed and built by the Allis-Chalmers Manufacturing Company.

Operation of Pump Turbine

To date the pump-turbine unit has been used most of the time as a pump because its primary function is to store water in the Carter Lake Reservoir. However, it is also used as a turbine whenever it is necessary to deliver the water stored in Carter lake to the Horsetooth feeder canal. As a pump the motor is started at full voltage with the impeller unwatered by means of compressed air to reduce the duration of the disturbances of the electrical system. After the unit is synchronised, the air is released and the discharge butterfly valve is opened. The unit is started as a turbine by partially opening the butterfly valve and using the bypass valve as a vernier to obtain synchronous speed. The main circuit breaker is then closed and the butterfly valve opened fully to load the unit. To stop the unit the butterfly valve is closed before opening the circuit breaker.

Initial operation of the reversible pump-turbine at the Flatiron plant has shown good performance both as a pump and as a turbine. Preliminary results of field tests using the salt velocity method indicate that all guaranteed efficiencies will be exceeded. Present indications are that reversible units of this type will find increasing applications in modern hydro-electric systems.

WATER POWER November 1955

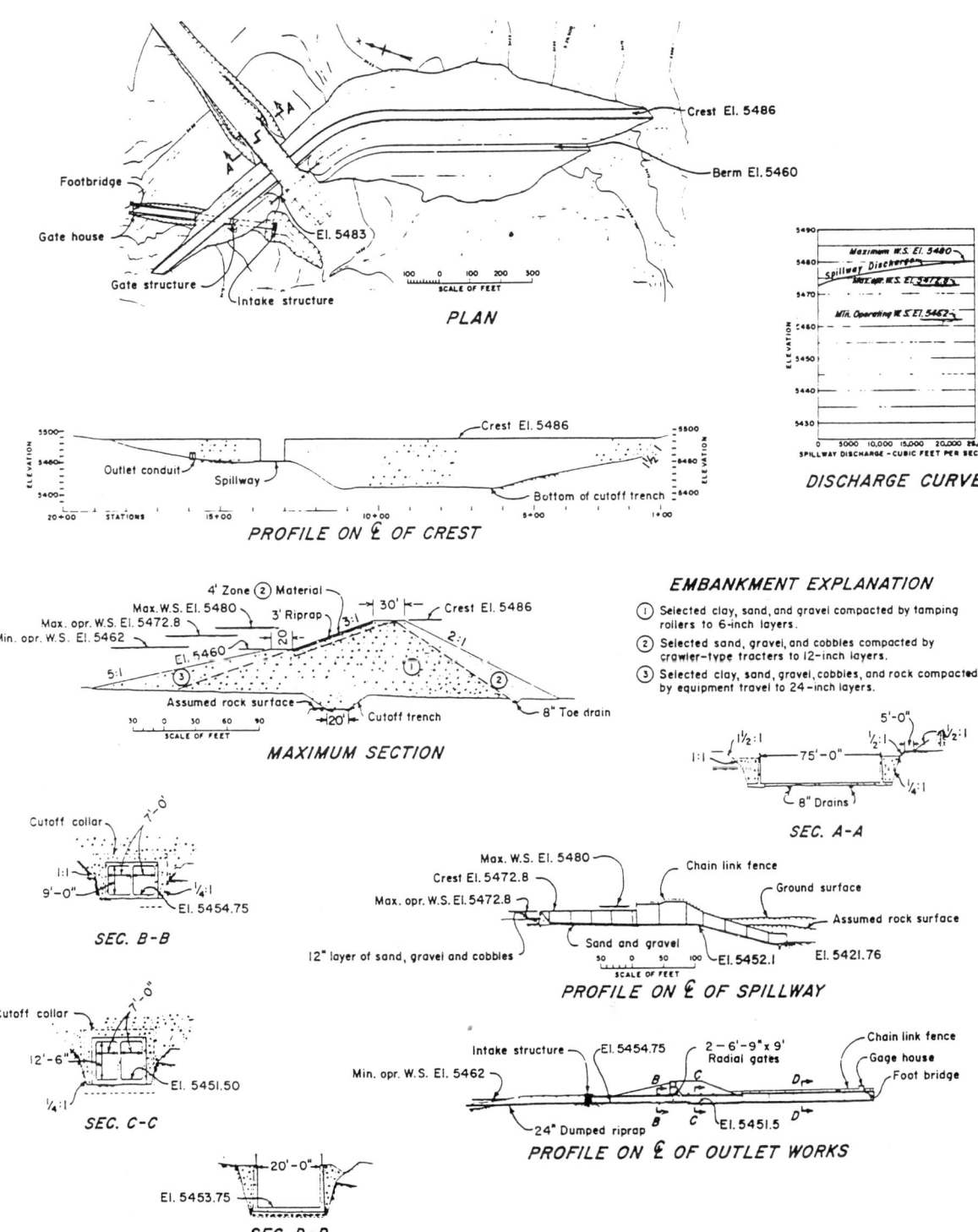

Flatiron Dam, Plan and Sections

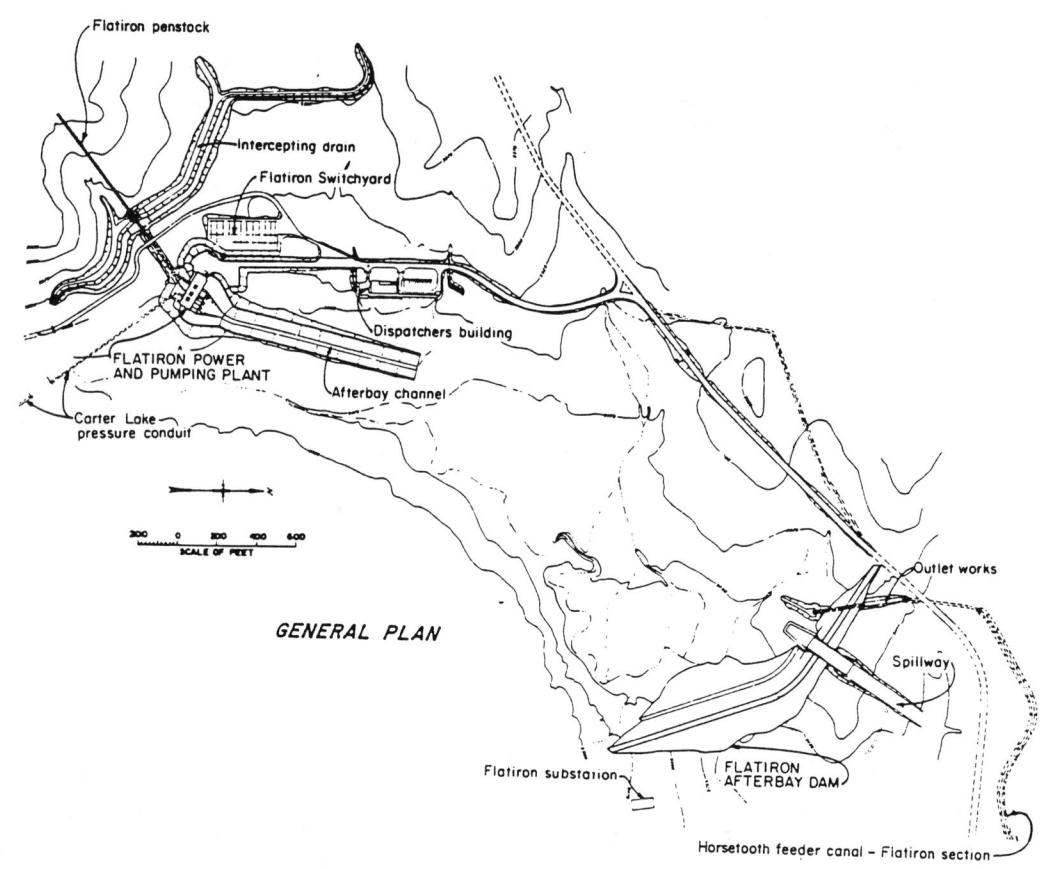

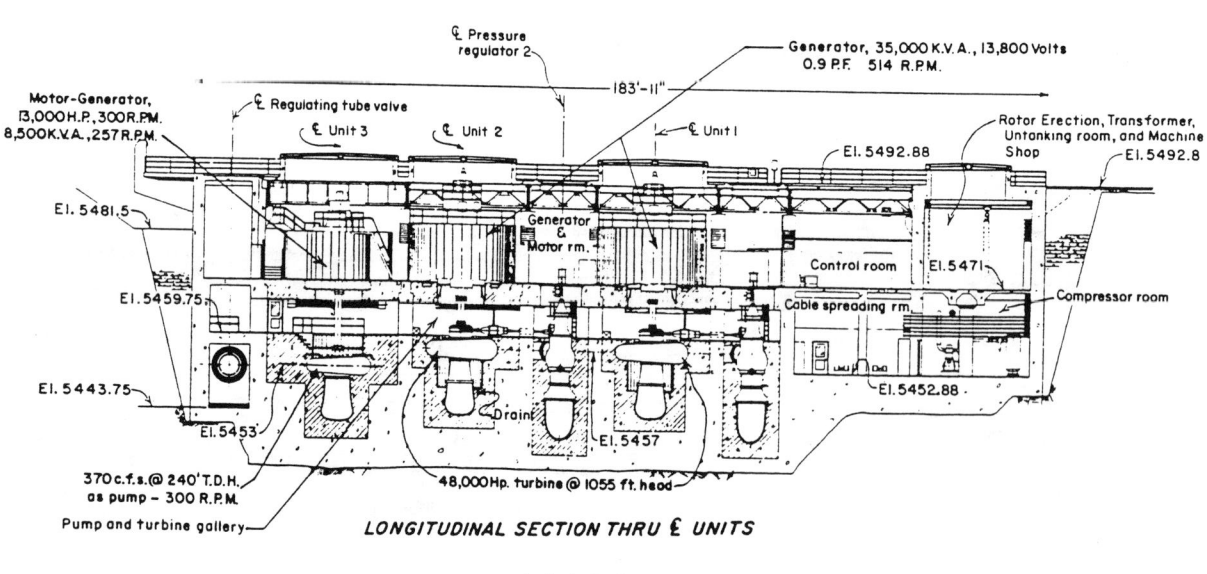

Flatiron Power and Pumping Plant, Plan and Section

BIBLIOGRAPHY

1. "Pumped Storage and Hydro Generation at Flatiron Powerplant," S.M. Denton and H.O. Britt, AEE1 Trans. Paper No. 55-164, October 21, 1954.

2. "Design Features of the Flatiron Power and Pumping Plant", John Parmakian, Head of Technical Engineering Analysis Section, Design Division, Bureau of Reclamation, U.S. Department of Interior, Water Power, November 1955.

Plant Name: **GIANELLI SAN LUIS PUMP-GENERATE PLANT**

Plant location: Los Banos, CA Merced County	Owner: Bureau of Reclamation Attn: D-3000 P.O. Box 25007 Denver, Colorado 80225
Rated capacity 400 MW Average static head 197 ft Plant efficiency 83.00 % Stored energy 509,707 MWh Number of units 8	Designers: Bureau of Reclamation, Denver, Colorado
Construction time: 3 years Construction cost: $342 per kW Price level: 1968 First commercial power: January 1968 FERC project number: Exempt	Plant Manager/Superintendent: Duane Knittel, Chief San Luis Field Division B.F. Sisk San Luis Dam 31770 W. Hwy 152 Santa Nella, California 95322 (209) 826-0718
River or water source: California Aqueduct / Delta Mendota Canal	

	UPPER RESERVOIR	LOWER RESERVOIR
DAM		
Type	Zoned earthfill (B.F. Sisk-San Luis)	Earth and rockfill (Forebay Dam - O'Neill)
Height (ft)	314	87
Crest length (ft)	18,600	14,301
Volume (yd³)	80,285,000	2,877,000
RESERVOIR		
Type	Constructed	Constructed - no inflow drainage area.
Surface area (acres)	13,000	2,250
Usable power storage (acre-ft)	1,909,219	
Power pool fluctuation (ft)	218.0	
Operating levels		
Maximum (ft)	544.0	225.0
Minimum (ft)	326.0	217.0
Drainage area (miles²)	84.7	
Seepage (ft³/s)	0.388	0.742
SPILLWAY		
Design flood		
Return period (years)	100 yrs	
Flow (ft³/s)	1,030	3,300
Capacity (ft³/s)	1,030	3,250
Type	Morning Glory 2.89 m (9.5 ft) diameter.	Morning Glory - 3.6 m (11.8 ft) diameter
Gates		
Number	None	None
Type		
Width (ft)		
Height (ft)		
OUTLET WORKS		
Discharge capacity (ft³/s)		
Number of water passages	None	None
Dimensions of water passages		
Height (ft)		
Width (ft)		
Diameter (ft)		
Type of gates		
Number of gates		None

Plant Name: **GIANELLI SAN LUIS PUMP-GENERATE PLANT**

Plant location	Owner: Bureau of Reclamation
Los Banos, CA	Attn: D-3000
Merced County	P.O. Box 25007
	Denver, Colorado 80225

Rated capacity 400 MW
Average static head 60.0 m
Plant efficiency 83.00 %
Stored energy 509,707 MWh
Number of units 8

Designers:
    Bureau of Reclamation, Denver, Colorado

Construction time: 3 years
Construction cost: $342 per kW
Price level: 1968
First commercial power: January 1968
FERC project number: Exempt

Plant Manager/Superintendent:
    Duane Knittel, Chief San Luis Field Division
    B.F. Sisk San Luis Dam
    31770 W. Hwy 152
    Santa Nella, California 95322
    (209) 826-0718

River or water source: California Aqueduct / Delta Mendota Canal

	UPPER RESERVOIR	LOWER RESERVOIR
DAM		
Type	Zoned earthfill (B.F. Sisk-San Luis)	Earth and rockfill (Forebay Dam - O'Neill)
Height (m)	95.7	26.5
Crest length (m)	5,669.3	4,359.0
Volume (m³)	61,382,622	2,200,000
RESERVOIR		
Type	Constructed	Constructed - no inflow drainage area.
Surface area (Mm²)	52.61	9.11
Usable power storage (Mm³)	2,355.000	
Power pool fluctuation (m)	66.45	
Operating levels		
Maximum (m)	165.81	68.58
Minimum (m)	99.36	66.14
Drainage area (Mm²)	219.373	
Seepage (m³/s)	0.0110	0.0210
SPILLWAY		
Design flood		
Return period (years)	100 yrs	
Flow (m³/s)	29	93
Capacity (m³/s)	29	92
Type	Morning Glory 2.89 m (9.5 ft) diameter.	Morning Glory - 3.6 m (11.8 ft) diameter
Gates		
Number	None	None
Type		
Width (m)		
Height (m)		
OUTLET WORKS		
Discharge capacity (m³/s)		
Number of water passages	None	None
Dimensions of water passages		
Height (m)		
Width (m)		
Diameter (m)		
Type of gates		
Number of gates		None

Gianelli San Luis Pump-Generate Plant - Page 1 (Metric)

GIANELLI SAN LUIS PUMP-GENERATE PLANT

INTAKES	UPPER INTAKE	LOWER INTAKE
Number	4	8
Type	Roller gate	Suction tube
Design discharge (ft³/s)	4,238	2,122
Gross inlet area (ft²) (at trash racks)	2,151	628
Bar racks		
spacing (in)	6.00	6.50
shape	Rectangular	Rectangular
depth/thickness (in)	2.00 / 0.63	2.00 / 0.63
diameter (in)		
Emergency gates		
number	4	8
height/width (ft)	22.01 / 17.49	15.75 / 12.80
type	Roller	Stop log
Service gates		
number	None	None
height/width (ft)		
type		
Bulkhead/stop logs (Y or N)	Y	Y
number of units serviced	2	8
Hoists		
number	1	1
capacity (tons)	50	15
type	Gantry	Gantry

WATER PASSAGES	Upper Tunnel	Shaft	Lower Tunnel	Surge Tanks Upper	Surge Tanks Lower	Penstocks	Tailrace Tunnel
Number						4	
Diameter (ft)						17.5	
Length (ft)						2,146	
Maximum velocity (ft/s)						17.7	
Concrete lining thickness (in)						43.32	
Total length of concrete sections (ft)						2,165	
Steel liner Thickness							
Minimum (in)						1.19	
Maximum (in)						1.31	
Material grade						A 201 B	
Total length of steel-lined sections (ft)						1,191	

Notes:
Penstocks: Penstock length varies from 648.0 m (2126 ft) to 659.9 m (2165 ft). See also page 4 notes.

GIANELLI SAN LUIS PUMP-GENERATE PLANT

INTAKES	UPPER INTAKE	LOWER INTAKE
Number	4	8
Type	Roller gate	Suction tube
Design discharge (m³/s)	120.0	60.1
Gross inlet area (m²) (at trash racks)	199.8	58.3
Bar Racks:		
spacing (mm)	152	165
shape	Rectangular	Rectangular
depth/thickness (mm)	51 / 16	51 / 16
diameter (mm)		
Emergency gates		
number	4	8
height/width (m)	6.710 / 5.330	4.800 / 3.900
type	Roller	Stop log
Service gates		
number	None	None
height/width (m)		
type		
Bulkhead/stop logs (Y or N)	Y	Y
number of units serviced	2	8
Hoists		
number	1	1
capacity (Mg)	45	14
type	Gantry	Gantry

WATER PASSAGES	Upper Tunnel	Shaft	Lower Tunnel	Surge Tanks Upper	Surge Tanks Lower	Penstocks	Tailrace Tunnel
Number						4	
Diameter (m)						5.33	
Length (m)						654.0	
Maximum velocity (m/s)						5.39	
Concrete lining thickness (m)						1.100	
Total length of concrete sections (m)						660.0	
Steel liner Thickness							
Minimum (mm)						30	
Maximum (mm)						33	
Material grade						A 201 B	
Total length of steel-lined sections (m)						363.0	

Notes:
Penstocks: Penstock length varies from 648.0 m (2126 ft) to 659.9 m (2165 ft). See also page 4 notes.

Gianelli San Luis Pump-Generate Plant - Page 2 (Metric)

GIANELLI SAN LUIS PUMP-GENERATE PLANT

POWERHOUSE and RELATED FEATURES

Powerhouse Structure
Type: Indoor
Length: 483 ft	Width: 155 ft	Height: 49 ft

Guard Valves
Number: 8	Diameter: 13.0 ft
Type: Butterfly

Transformers
Number: 4 pad mounted 3∅ delta connect low side
Ratings:
Voltages: (kV) 105 MVA forced oil and forced air cooler

Generator
Rating generating (MVA): 53.0	Rating pumping (MVA): 44.7
Insulation type: Asphatum and glass (Class B Insulation)
Starting method: Reduced voltage start
Starting equipment: none

Runners
Material: 18-8 stainless steel
Minimum unit submergence: 15.0 ft
WR^2: 4,900,000 (lbf x ft²)
Manufacturer: Hitachi
Model test by: Hitachi

	Reversible Runners	Reversible Motor/Generator			
Number	8				8
Diameter (ft)	17.70	Rotor	21.33	Stator	23.62
rpm synchronous	150.0				150.0
rpm overspeed	206.0				206.0
Type	Francis		2 Speed Syncronous		

Information on Runners

Condition:	Gross Head (ft) Generating	Gross Head (ft) Pumping	Capacity (MW) Generating	Capacity (MW) Pumping	Discharge (ft³/s) Generating	Discharge (ft³/s) Pumping	Turbine/Pump Eff.(%) Generating	Turbine/Pump Eff.(%) Pumping
Maximum head & maximum power	324	324	50	45	2,122	1,413	86.6	89.7
Minimum head & maximum power	117	69	5	34	871	2,200	55.9	73.0

Note: Data in the above table are based on field tests.

Condition:	Net Head (ft) Generating	Net Head (ft) Pumping	Capacity (MW) Generating	Capacity (MW) Pumping	Discharge (ft³/s) Generating	Discharge (ft³/s) Pumping	Turbine/Pump Eff.(%) Generating	Turbine/Pump Eff.(%) Pumping
Rated head @ best gate	304	341	45	48	2,031	1,487	86.1	89.7

Note: Data in the above table are based on field test.

Gianelli San Luis Pump-Generate Plant - Page 3 (English)

GIANELLI SAN LUIS PUMP-GENERATE PLANT

POWERHOUSE and RELATED FEATURES

Powerhouse Structure
Type: Indoor
Length: 147.2 m   Width: 47.3 m   Height: 14.8 m

Guard Valves
Number: 8   Diameter: 3.96 m
Type: Butterfly

Transformers
Number: 4 pad mounted 3∅ delta connect low side
Ratings:
Voltages: (kV) 105 MVA forced oil and forced air cooler

Generator
Rating generating (MVA): 53.0   Rating pumping (MVA): 44.7
Insulation type: Asphatum and glass (Class B Insulation)
Starting method: Reduced voltage start
Starting equipment: none

Runners
Material: 18-8 stainless steel
Minimum unit submergence: 4.57 m
WR^2: 2,025,000 (Newtons x m²)
Manufacturer: Hitachi
Model test by: Hitachi

	Reversible Runners	Reversible Motor/Generator			
Number	8				8
Diameter m	5.40	Rotor	6.500	Stator	7.200
rpm synchronous	150.0				150.0
rpm overspeed	206.0				206.0
Type	Francis			2 Speed Syncronous	

Information on Runners

Condition:	Gross Head (m)		Capacity (MW)		Discharge (m³/s)		Turbine/Pump Eff.(%)	
	Generating	Pumping	Generating	Pumping	Generating	Pumping	Generating	Pumping
Maximum head & maximum power	98.8	98.8	50	45	60.1	40.0	86.6	89.7
Minimum head & maximum power	35.7	21.0	5	34	24.7	62.3	55.9	73.0

Note: Data in the above table are based on field tests.

Condition:	Net Head (m)		Capacity (MW)		Discharge (m³/s)		Turbine/Pump Eff.(%)	
	Generating	Pumping	Generating	Pumping	Generating	Pumping	Generating	Pumping
Rated head @ best gate	92.7	103.9	45	48	57.5	42.1	86.1	89.7

Note: Data in the above table are based on field test.

Gianelli San Luis Pump-Generate Plant - Page 3 (Metric)

GIANELLI SAN LUIS PUMP-GENERATE PLANT

Plant Data:

Average GWh generating per year:	321
Average GWh pumping per year:	389
Starting time from standstill (s):	50
Changeover time pumping to generating (min):	20
Planned/scheduled time between major overhauls (years):	8
Outage time required per unit during major overhauls (weeks):	12
Representative plant availability (%):	86.0
Representative planned outages (weeks per year):	16

Miscelleneous Notes:

The Gianelli plant is a part of the Central Valley irrigation project and the California Water Plan.

The Gianelli Pump-Generate Plant is located at the downstream base of B.F. Sisk-San Luis Dam. The plant is a seasonal pumped storage plant. The lower reservoir is the O'Neill Forebay Dam.

In addition to the 50-ton gantry crane hoist in the upper reservoir, four roller-mounted hydraulic hoists, each capable of lifting 900,000 pounds, are included.

The concrete lining thickness in the four penstocks varies from 0.762 m (2.5 ft) to 1.1 m (3.6 ft). The total length of steel section varies from 357.9 m (1,174 ft) to 369.7 m (1213 ft).

The width of the powerhouse varies from 29.5 m (96.8 ft) to 47.3 m (155 ft).

The dam was built from 1963 to 1967, and the powerplant was built from 1965 to 1968.

Cavitation Experience:

Cavitation has occurred on the stay vanes, the butterfly valves, and on the penstock near the butterfly valves.

Significant or Unique Problems:

Problems have included breaking of the amortisseur straps, a hollow spot in the casting of the stainless steel impellers, breaking on wear rings. Low efficiency at high heads on the 150 rpm units, and vibration at low generating heads.

List of Licenses Required:

None.

ENVIRONMENTAL FEATURES

Recreation:

Recreation activities include powerboating and sailboating, wind surfing, swimming, picnicing, camping, biking, fishing, waterskiing,

GIANELLI SAN LUIS PUMP-GENERATE PLANT

hiking, horseback riding, nature study, bird watching and wild flower viewing.

Fish and Wildlife:
Fish and wildlife at the project include deer, squirrels, wild boar (feral pigs), eagles, rabbits, fish (shad, sucker, carp, balackfish, hitch, hardhead, catfish, perch, mosquito fish, crappie, bass, war mouth, sunfish, bluegill, salmon, flounder, sturgeon, spittail, and striped bass), pheasant, quail, pigeon, pelican, egret, heron, kit, red fox, and coyote.

Social:
Domestic water from the plant is supplied to the Departments of Forestry and Parks and Recreation. Improvements include 563 Km (350 miles) of roads, 40.5 Km² (10 acres) of parking lots, irrigation.

Page 4 (Continued)

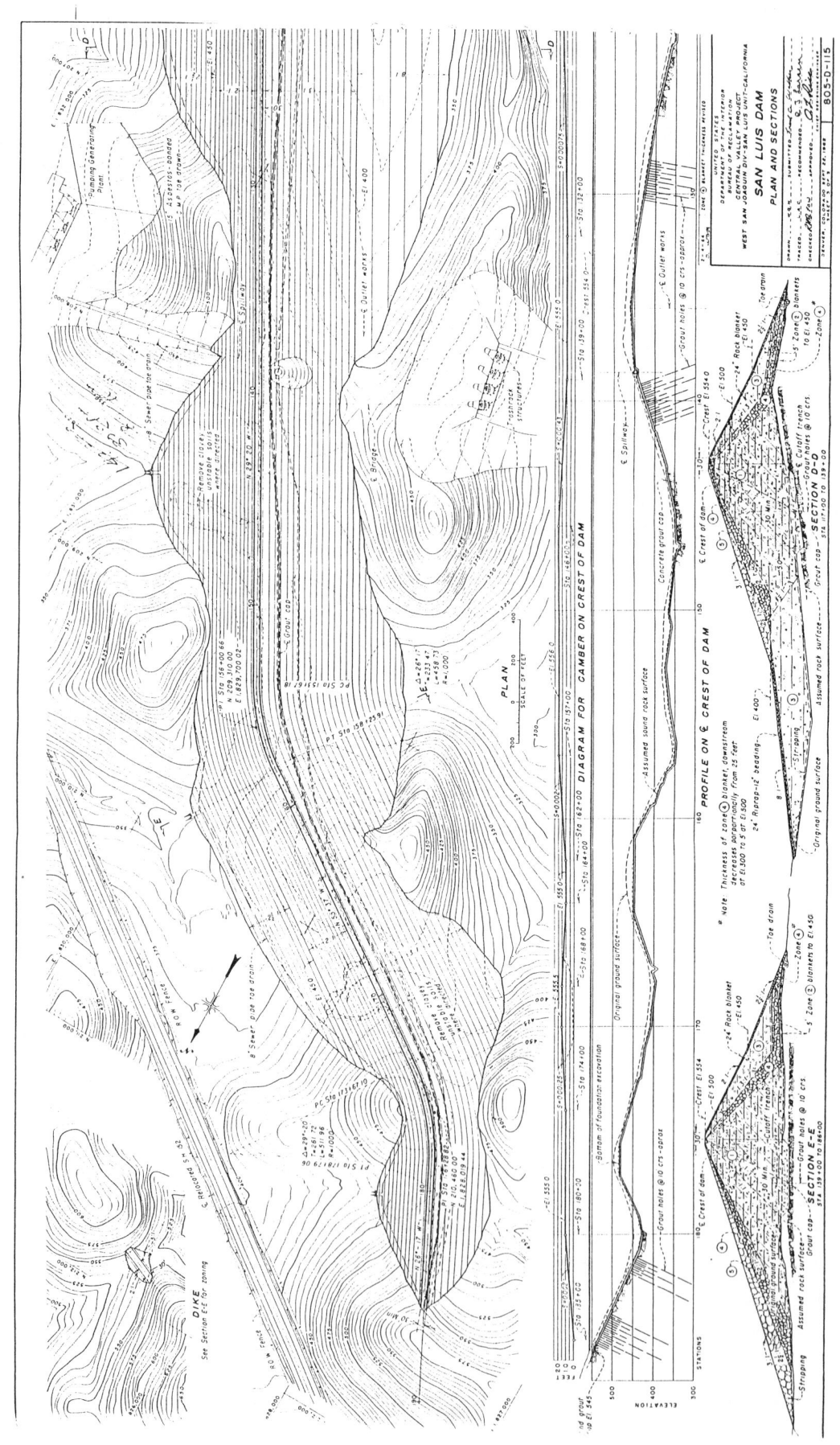

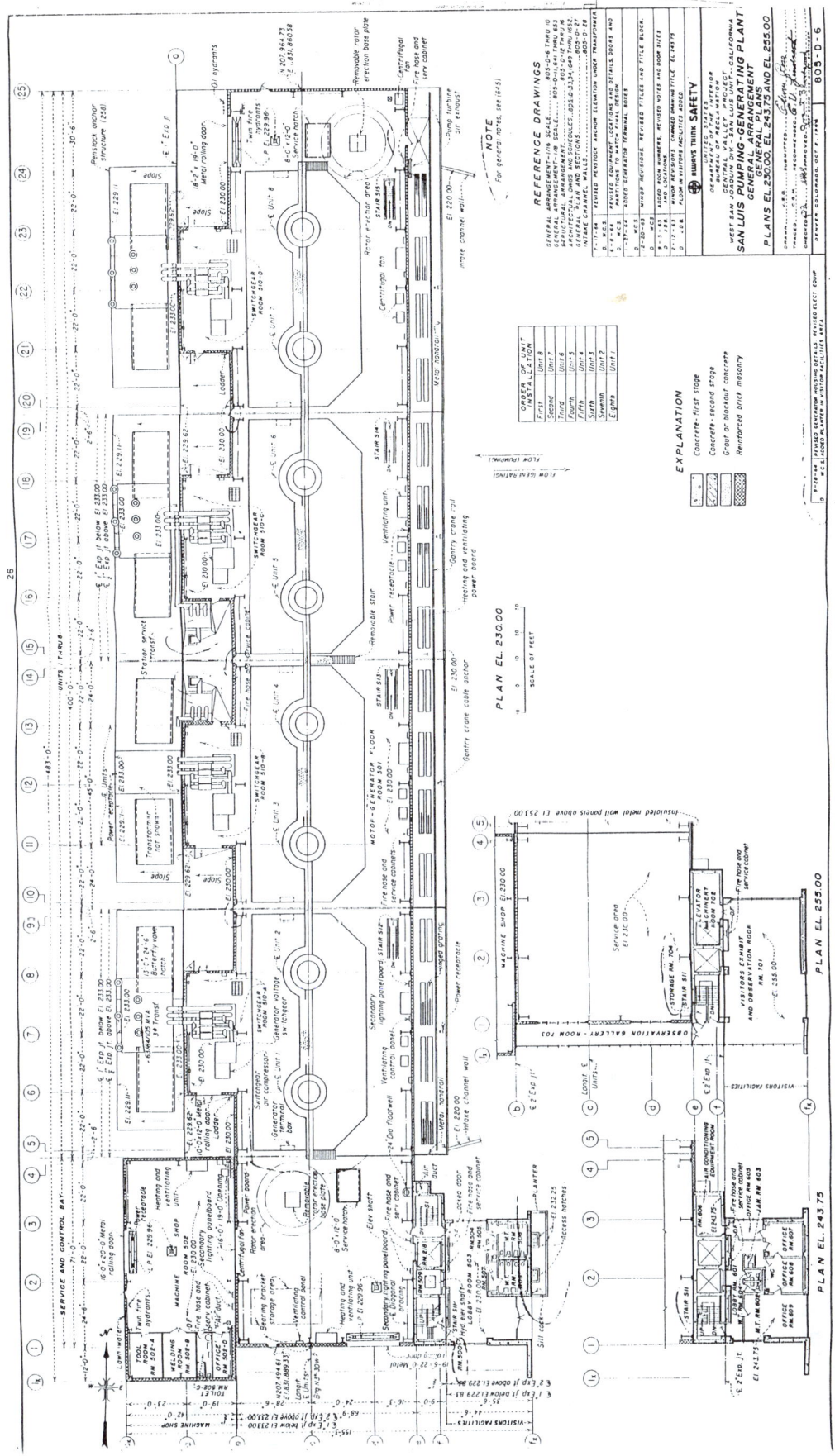

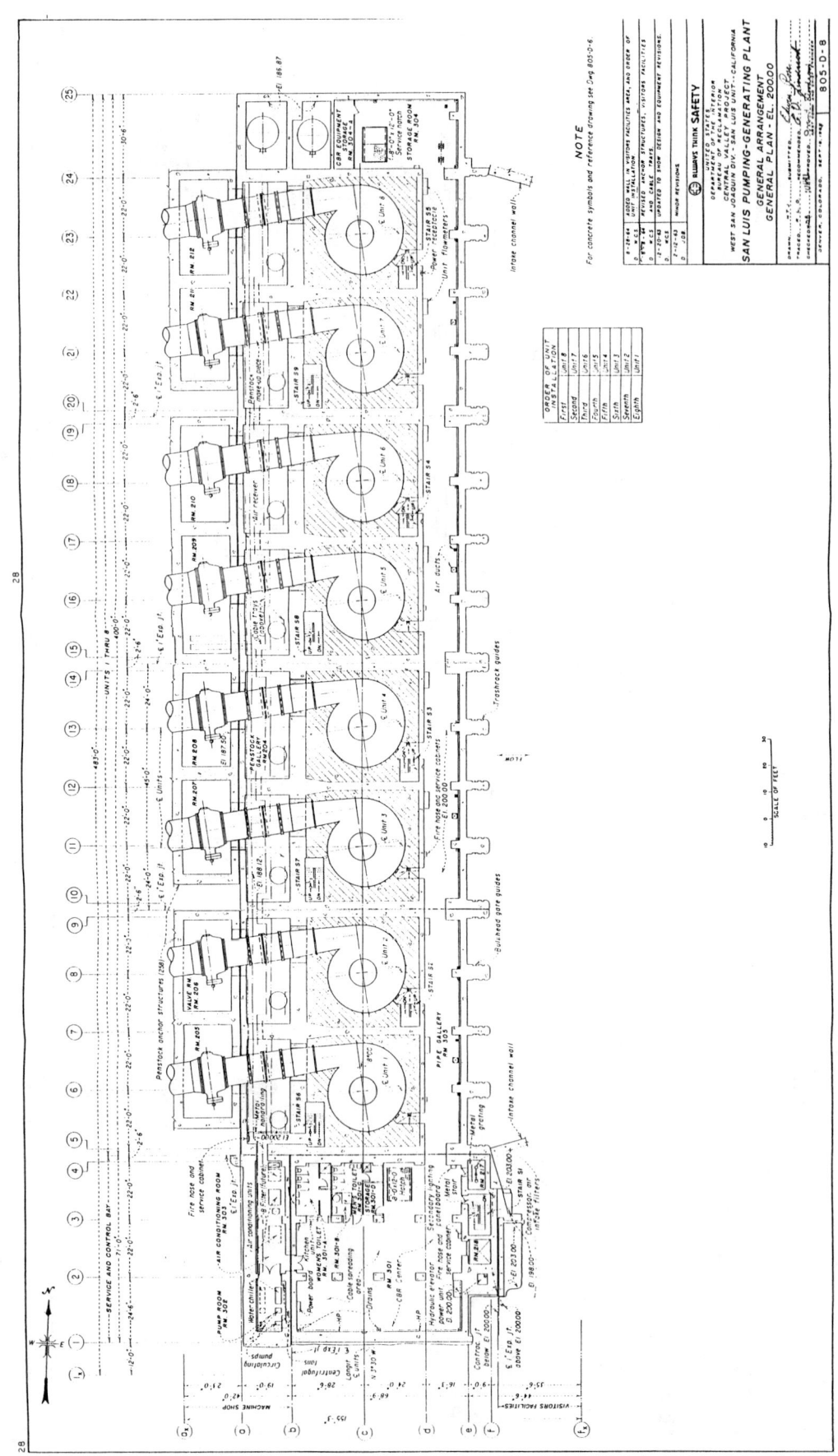

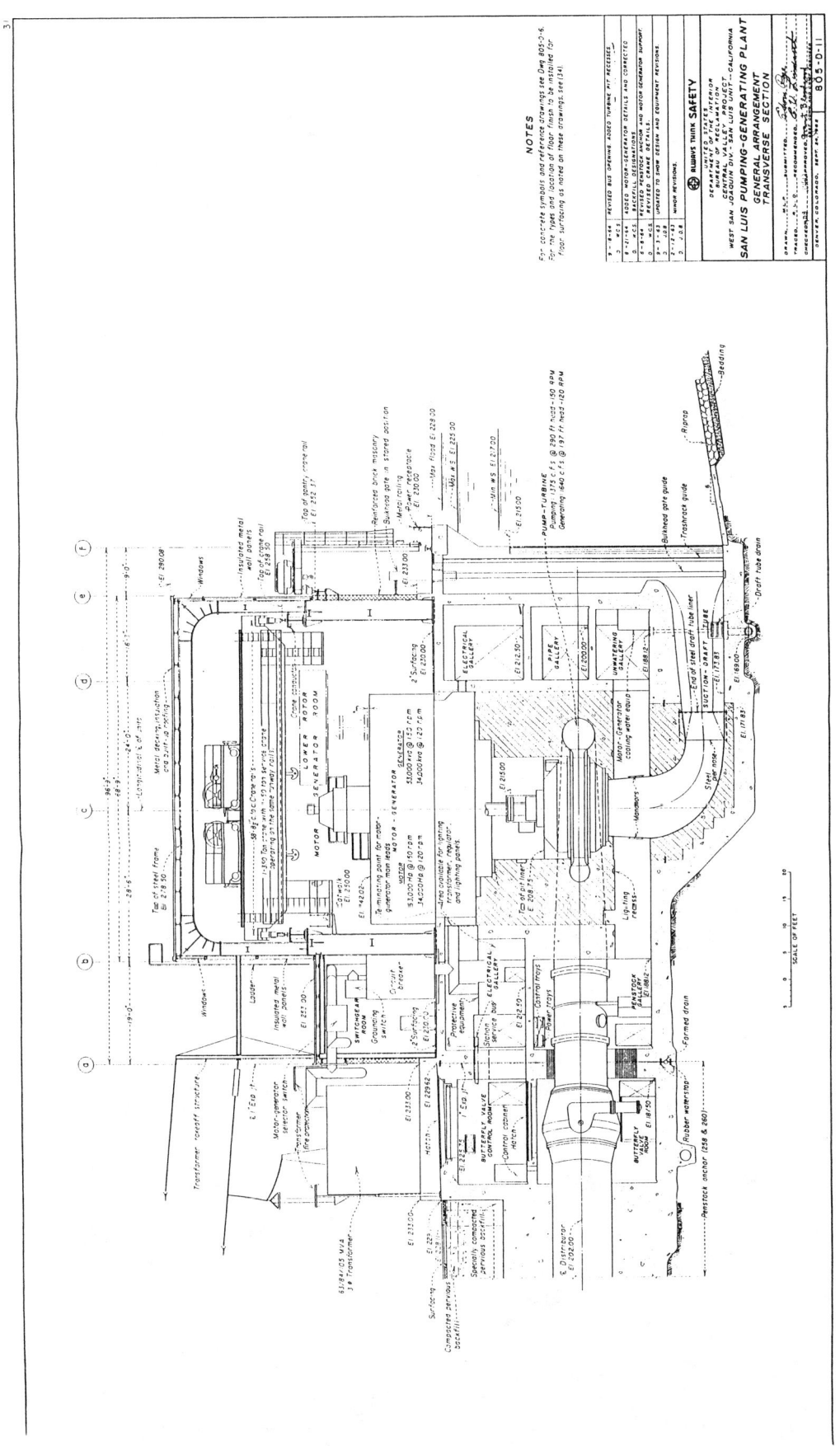

Aerial View of Reservoir (San Luis Dam)

Gianelli Pumped Storage Plant

BIBLIOGRAPHY

1. Goodman, D.L., "San Luis Pumped Storage Unit to be Among Nation's Largest," Power Engineering, 68:42-4, May 1964.

2. "San Luis Unit Technical Record of Design and Construction," Bureau of Reclamation, vols. 1-7 (117875), 1974.

3. "Performance Report No. 4, San Luis Dam, State of California," The Resources Agency, Department of Water Resources, Division of Operations and Maintenance, April 1984.

Plant Name: **GRAND COULEE**

Plant location:
  Grand Coulee, WA
  Grant County
		Owner: Bureau of Reclamation
		       Attn: D-3000
		       P.O. Box 25007
		       Denver, Colorado 80225

Rated capacity	314 MW	
Average static head	265 ft	Designers:
Plant efficiency	%	Bureau of Reclamation
Stored energy	11,115 MWh	
Number of units	6	
Construction time:		Plant Manager/Superintendent:
Construction cost: $1 per kW		Bureau of Reclamation
Price level: 1980		P.O. Box 620
First commercial power: October 1973		Grand Coulee, Washington 99133
FERC project number:		
		(509) 633-9300
River or water source: Columbia River		

DAM	UPPER RESERVOIR	LOWER RESERVOIR
Type	Two zoned earthfill, North Dam	Concrete gravity, Grand Coulee Dam
Height (ft)	145	550
Crest length (ft)	1,170	5,223
Volume (yd³)	1,659,000	11,977,000
RESERVOIR		
Type	Constructed, Feeder Canal from Banks Lake	On stream, Franklin D. Roosevelt Lake
Surface area (acres)	27,000	82,304
Usable power storage (acre-ft)	715,000	5,185,400
Power pool fluctuation (ft)	6.0	82.0
Operating levels		
Maximum (ft)	1,571.0	1,290.0
Minimum (ft)	1,557.0	1,208.0
Drainage area (miles²)	0.0	74,100.0
Seepage (ft³/s)		
SPILLWAY		
Design flood		
Return period (years)		
Flow (ft³/s)		1,000,000
Capacity (ft³/s)		1,000,000
Type	None	Overflow chute
Gates		
Number	None	11
Type		Drum
Width (ft)		135.00
Height (ft)		30.00
OUTLET WORKS		
Discharge capacity (ft³/s)	13,200	225,000
Number of water passages	1	40
Dimensions of water passages		
Height (ft)	12.00	
Width (ft)	18.00	
Diameter (ft)		8.50
Type of gates	Radial gates	
Number of gates	1	40

Plant Name: **GRAND COULEE**

```
Plant location                              Owner: Bureau of Reclamation
  Grand Coulee, WA                                 Attn: D-3000
  Grant County                                     P.O. Box 25007
                                                   Denver, Colorado  80225
Rated capacity            314 MW
Average static head       80.8 m            Designers:
Plant efficiency             %                     Bureau of Reclamation
Stored energy         11,115 MWh
Number of units              6

Construction time:                          Plant Manager/Superintendent:
Construction cost: $1 per kW                       Bureau of Reclamation
Price level: 1980                                  P.O. Box 620
First commercial power: October 1973               Grand Coulee, Washington     99133
FERC project number:
                                                   (509) 633-9300
River or water source: Columbia River
```

DAM	UPPER RESERVOIR	LOWER RESERVOIR
Type	Two zoned earthfill, North Dam	Concrete gravity, Grand Coulee Dam
Height (m)	44.2	167.6
Crest length (m)	356.6	1,592.0
Volume (m³)	1,268,049	9,157,350
RESERVOIR		
Type	Constructed, Feeder Canal from Banks Lake	On stream, Franklin D. Roosevelt Lake
Surface area (Mm²)	109.27	333.08
Usable power storage (Mm³)	881.945	6,396.134
Power pool fluctuation (m)	1.83	24.99
Operating levels		
Maximum (m)	478.84	393.19
Minimum (m)	474.57	368.20
Drainage area (Mm²)		191,919.192
Seepage (m³/s)		
SPILLWAY		
Design flood		
Return period (years)		
Flow (m³/s)		28,317
Capacity (m³/s)		28,317
Type	None	Overflow chute
Gates		
Number	None	11
Type		Drum
Width (m)		41.148
Height (m)		9.144
OUTLET WORKS		
Discharge capacity (m³/s)	374	6,371
Number of water passages	1	40
Dimensions of water passages		
Height (m)	3.658	
Width (m)	5.486	
Diameter (m)		2.591
Type of gates	Radial gates	
Number of gates	1	40

GRAND COULEE

INTAKES	UPPER INTAKE	LOWER INTAKE
Number	2	2
Type	Canal	Concrete
Design discharge (ft³/s)	1,950	1,950
Gross inlet area (ft²) (at trash racks)	288	1,380
Bar racks		
spacing (in)	6.25	6.25
shape		
depth/thickness (in)	3.00 / 0.75	2.50 / 0.63
diameter (in)		
Emergency gates		
number	1	2
height/width (ft)		20.11 / 13.00
type	76.2 cm (30 in) spool valve	Reverse flow wheel
Service gates		
number	None	None
height/width (ft)		
type		
Bulkhead/stop logs (Y or N)	N	
number of units serviced		
Hoists		
number	None	1
capacity (tons)		
type		Hydraulic

WATER PASSAGES	Upper Tunnel	Shaft	Lower Tunnel	Surge Tanks Upper	Surge Tanks Lower	Penstocks	Tailrace Tunnel
Number	6	6				6	6
Diameter (ft)	18.0	18.0				12.0	14.0
Length (ft)	220	226				181	89
Maximum velocity (ft/s)	9.4	9.4				21.2	15.5
Concrete lining thickness (in)	36.00						
Total length of concrete sections (ft)							
Steel liner Thickness							
Minimum (in)	0.50	0.50				1.19	
Maximum (in)	0.50	1.19				1.19	
Material grade	A285	A285				A285	
Total length of steel-lined sections (ft)	115	226				181	

Notes:
Upper Tunnel: Upper tunnel length of steel-lined section varies.

GRAND COULEE

INTAKES	UPPER INTAKE	LOWER INTAKE
Number	2	2
Type	Canal	Concrete
Design discharge (m³/s)	55.2	55.2
Gross inlet area (m²) (at trash racks)	26.8	128.2
Bar Racks:		
spacing (mm)	159	159
shape		
depth/thickness (mm)	76 / 19	64 / 16
diameter (mm)		
Emergency gates		
number	1	2
height/width (m)		6.129 / 3.962
type	76.2 cm (30 in) spool valve	Reverse flow wheel
Service gates		
number	None	None
height/width (m)		
type		
Bulkhead/stop logs (Y or N)	N	
number of units serviced		
Hoists		
number	None	1
capacity (Mg)		
type		Hydraulic

WATER PASSAGES	Upper Tunnel	Shaft	Lower Tunnel	Surge Tanks Upper	Surge Tanks Lower	Penstocks	Tailrace Tunnel
Number	6	6				6	6
Diameter (m)	5.49	5.49				3.66	4.27
Length (m)	67.1	68.9				55.2	27.0
Maximum velocity (m/s)	2.87	2.87				6.46	4.73
Concrete lining thickness (m)	0.914						
Total length of concrete sections (m)							
Steel liner Thickness							
Minimum (mm)	13	13				30	
Maximum (mm)	13	30				30	
Material grade	A285	A285				A285	
Total length of steel-lined sections (m)	35.1	68.9				55.2	

Notes:
Upper Tunnel: Upper tunnel length of steel-lined section varies.

Grand Coulee - Page 2 (Metric)

GRAND COULEE

POWERHOUSE and RELATED FEATURES

Powerhouse Structure
Type: Surface
Length: 607 ft Width: 110 ft Height: 140 ft

Guard Valves
Number: None Diameter: ft
Type:

Transformers
Number: 2
Ratings: 1-175 AND 1-185
Voltages: (kV) 13.6 / 230

Generator
Rating generating (MVA): 50.0 Rating pumping (MVA):
Insulation type:
Starting method:
Starting equipment: Line starting

Runners
Material: Stainless steel
Minimum unit submergence: 17.0 ft
WR²:
Manufacturer: Nohab, Toshiba
Model test by: Nohab, Toshiba

	Reversible Runners	Reversible Motor/Generator	
		Rotor	Stator
Number	6		6
Diameter (ft)	14.30		
rpm synchronous	200.0		200.0
rpm overspeed			
Type	Francis		Vertical Shaft

Information on Runners

Condition:	Gross Head (ft)		Capacity (MW)		Discharge (ft³/s)		Turbine/Pump Eff.(%)	
	Generating	Pumping	Generating	Pumping	Generating	Pumping	Generating	Pumping
Maximum head & maximum power	363	363	67	46	2,575	1,431	92.0	89.0
Minimum head & maximum power	267	267	41	52	2,150	1,980	89.0	91.6

Note: Data in the above table are based on model tests.

Condition:	Net Head (ft)		Capacity (MW)		Discharge (ft³/s)		Turbine/Pump Eff.(%)	
	Generating	Pumping	Generating	Pumping	Generating	Pumping	Generating	Pumping
Rated head @ best gate	280	292	45	51	2,200	1,813	91.5	91.6

Note: Data in the above table are based on model tests.

GRAND COULEE

POWERHOUSE and RELATED FEATURES

Powerhouse Structure
Type: Surface
Length: 185.0 m   Width: 33.7 m   Height: 42.7 m

Guard Valves
Number: None   Diameter: m
Type:

Transformers
Number: 2
Ratings: 1-175 AND 1-185
Voltages: (kV) 13.6 / 230

Generator
Rating generating (MVA): 50.0   Rating pumping (MVA):
Insulation type:
Starting method:
Starting equipment: Line starting

Runners
Material: Stainless steel
Minimum unit submergence: 5.18 m
WR^2:
Manufacturer: Nohab, Toshiba
Model test by: Nohab, Toshiba

	Reversible Runners	Reversible Motor/Generator	
Number	6		6
Diameter m	4.37	Rotor	Stator
rpm synchronous	200.0		200.0
rpm overspeed			
Type	Francis		Vertical Shaft

Information on Runners

Condition:	Gross Head (m)		Capacity (MW)		Discharge (m³/s)		Turbine/Pump Eff.(%)	
	Generating	Pumping	Generating	Pumping	Generating	Pumping	Generating	Pumping
Maximum head & maximum power	110.6	110.6	67	46	72.9	40.5	92.0	89.0
Minimum head & maximum power	81.4	81.4	41	52	60.9	56.1	89.0	91.6

Note: Data in the above table are based on model tests.

Condition:	Net Head (m)		Capacity (MW)		Discharge (m³/s)		Turbine/Pump Eff.(%)	
	Generating	Pumping	Generating	Pumping	Generating	Pumping	Generating	Pumping
Rated head @ best gate	85.3	89.0	45	51	62.3	51.3	91.5	91.6

Note: Data in the above table are based on model tests.

GRAND COULEE

Plant Data:
   Average GWh generating per year:
   Average GWh pumping per year:
   Starting time from standstill (s):
   Changeover time pumping to generating (min):
   Planned/scheduled time
     between major overhauls (years):
   Outage time required per unit
     during major overhauls (weeks):
   Representative plant availability (%):
   Representative planned outages (weeks per year):   3

Miscelleneous Notes:
   Of the six reversible units, four are rated at 70,000 hp pumping and 2 are rated at 67,500 hp pumping. All are rated at 67,500 hp generating. The units are 436.6 cm (171 7/8 in.) and 428.6 cm (168 3/4 in.) in diameter.

   The Grand Coulee Pump-Generating Plant also has six pumping units, each rated at 67,500 hp when generating. The power plant is one of four powerhouses at the Grand Coulee Dam. Three of the powerhouses contain conventional hydropower facilities. The lower reservoir to the pumped storage units is the Franklin D. Roosevelt Lake, the lake formed by Grand Coulee Dam.

   The pumped storage units are located in the left abutment. The upper reservoir for the pumped storage units is Banks Lake. Water from Banks Lake is fed to the pumped storage units through a feeder canal.

   The upper reservoir is actually formed by two embankments. The tables report the quantities for the main embankment. A second embankment has a total volume of 1,126,000 m³ (1,473,000 yd³) of material. The second embankment does not include a spillway or outlet works.

Cavitation Experience:

Significant or Unique Problems:

List of Licenses Required:
   None -- direct congressional authorization.

ENVIRONMENTAL FEATURES

Recreation:
   All types of recreation occurs including swimming, boating, hiking, camping, and fishing. The area has about 2 million visitors per year including 100,000 campers, 50,000 boat launches, 485,000 visitor-center visitors. Archeologists look for artifacts during low water.

Fish and Wildlife:

GRAND COULEE

Fish enhancement includes two net pens. Additionally, plans are being negotiated to enhance fish and wildlife programs.

Social:
Relatively few people visited this area prior to dam construction. Grand Coulee and the associated irrigation project created two large reservoirs and four permanent towns whose primary industry is project maintenance and recreation.

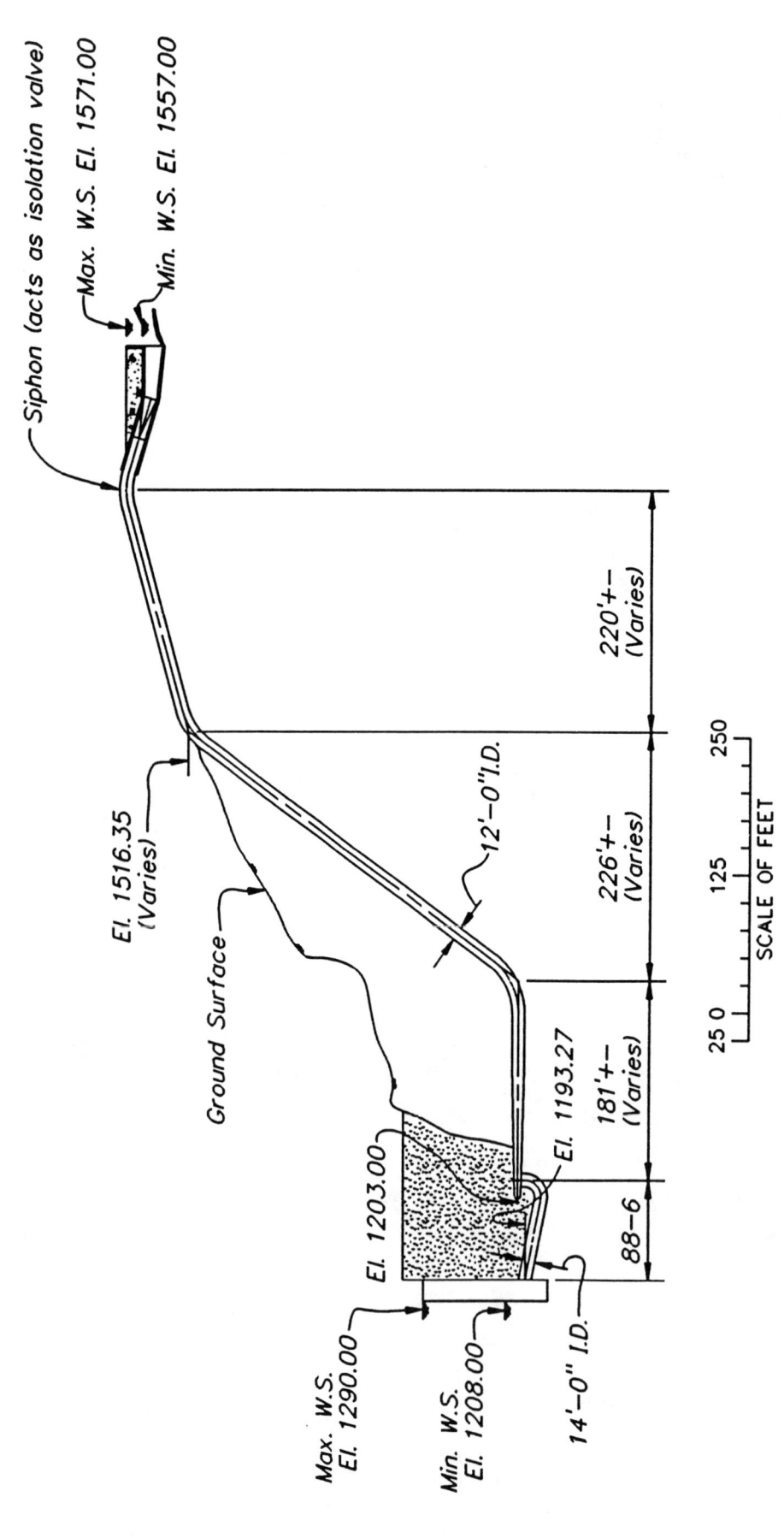

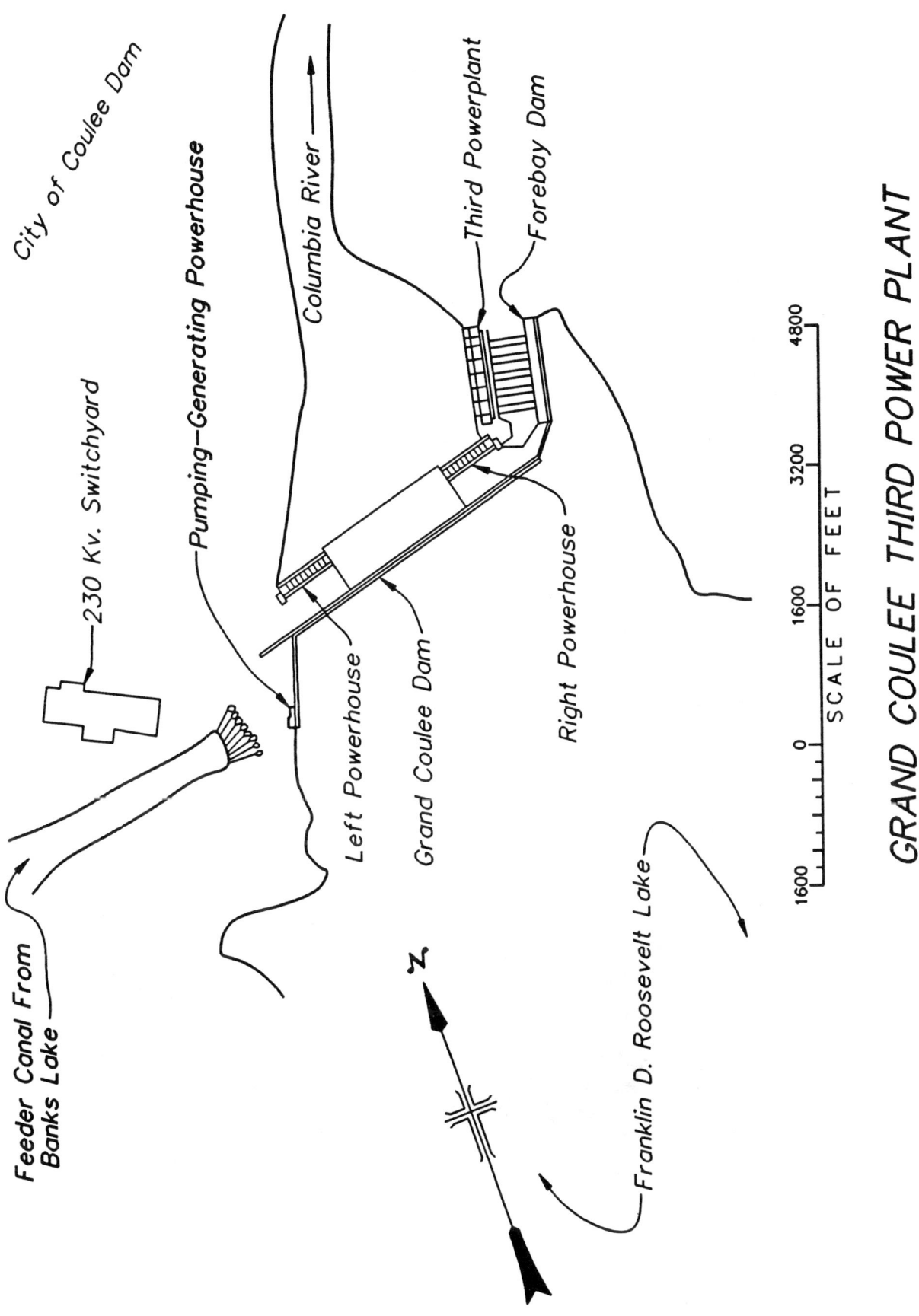

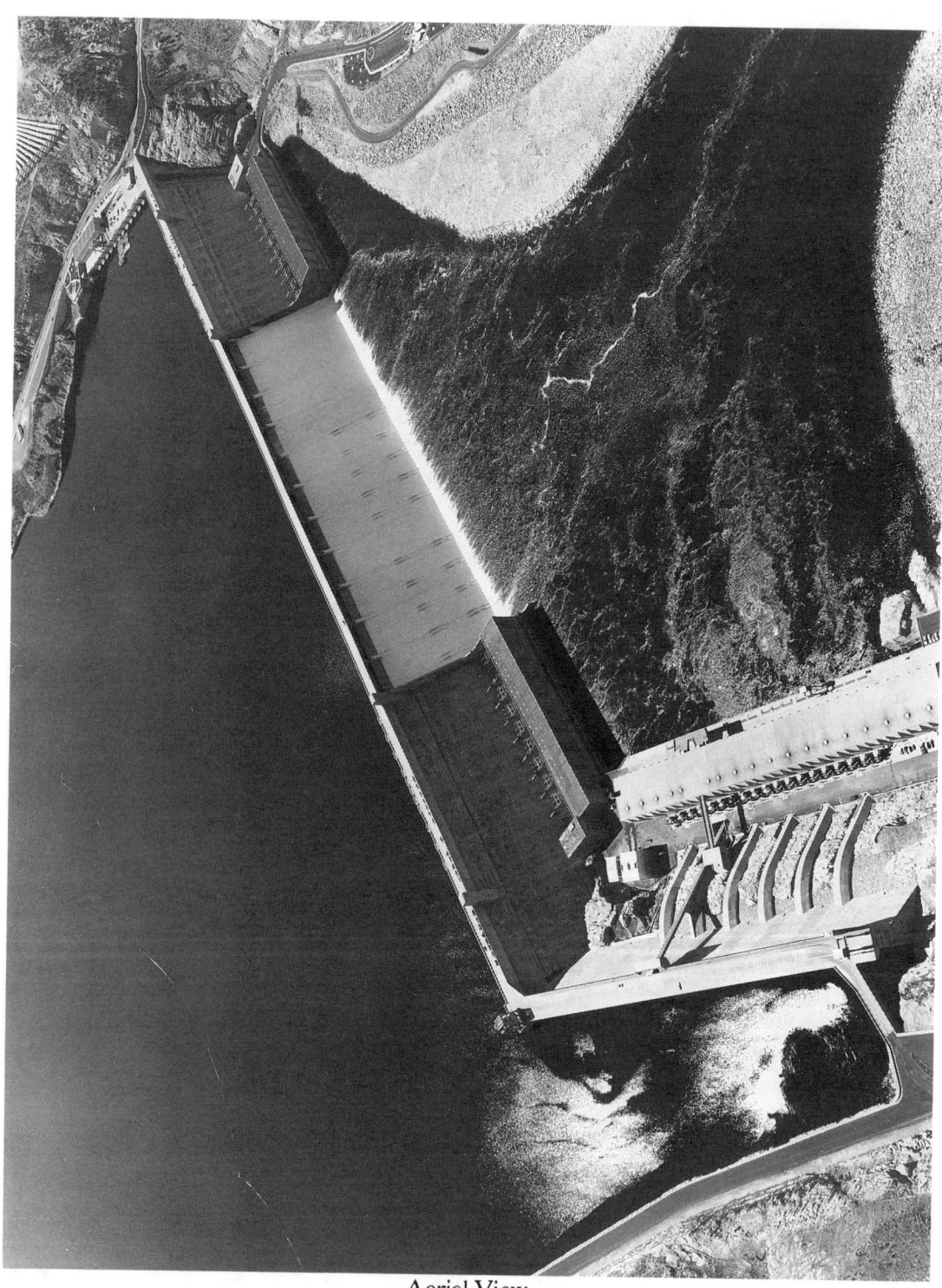

Aerial View

Grand Coulee Pumped Storage Plant

Grand Coulee Dam

Grand Coulee Pumped Storage Plant

August 22, 1990

Articles About Grand Coulee Dam

4/01/34, Grand Coulee Dam and Power Plant on Columbia River, A. Gilbert Darwin, Western Construction News, pages 2-16. Historical Data-Vol.1

2/01/37, The Biggest Thing On Earth, Richard L. Neuberger, Harper's Magazine, pages 247-258. Historical Data-Vol.1

3/01/37, The Columbia Basin - Grand Coulee Project, Spokane C of C, Brochure, 40 pages. Historical Data-Vol.1

1/01/38, A Curb-side Story of Grand Coulee Dam and a Last Frontier, Fred M. Weil, Brochure, 24 pages. Historical Data-Vol.1

1/16/38, Biggest Job on Earth (Progress Edition), Author unknown, The Spokesman-Review, 24 pages. Historical Data-Vol.1

1/01/40, Grand Coulee Dam - The Eighth Wonder of the World, Author unknown, Brochure, 32 pages. Historical Data-Vol.1

1/02/40, Grand Coulee Dam - The Biggest Man-made Structure of All Time, Author unknown, Brochure, 32 pages. Historical Data-Vol. 1

1/01/51, The Columbia Basin Project, Bureau of Reclamation, Brochure, 4 pages. Historical Data-Vol.1

6/01/58, A Brief Chronological History of the Construction of Grand Coulee Dam, Alvin F. Darland, Report, 5 pages. Historical Data-Vol.1

9/13/61, Notes on Some Features of the Design, Construction, Operation..., A.F. Darland, Report, 27 pages. Historical Data-Vol.1

11/15/66, Effects of Dam Construction on Temperatures of Columbia River, Robert T. Jaske, Journal AWWA, pages 935-942. Historical Data-Vol.1

3/01/67, Proposed Third Power Plant at Grand Coulee Dam, Harold G. Arthur, ASCE Journal-Power Division, pages 15-29. Historical Data-Vol.1

7/10/67, 12 600-Mw Units Planned for Grand Coulee No. 3, B.P. Bellport, Electrical World, 4 pages. Historical Data-Vol.1

1/01/68, 820,000-Horsepower Turbines for Grand Coulee Third Powerplant, B.P. Bellport, Proc. of American Power Conference, pages 837-843. Historical Data-Vol.1

5/01/68, 600-Mw Units for Grand Coulee, Carlos Bates, Power Engineering, pages 58-59. Historical Data-Vol.1

7/18/68, Coulee's Powerplant Will Really Be Grand, Author unknown, Engineering News-Record, pages 34-41. Historical Data-Vol.1

11/01/68, Transporting Equipment for Grand Coulee Third Plant,
Harold G. Author, ASCE Journal of Power Division, pages 125-141.
Historical Data-Vol.1

11/12/68, Enlarging the World's Eighth Wonder, Author unknown,
Dupont Magazine, pages 2-6. Historical Data-Vol.1

3/01/69, Job 119 - Grand Coulee Dam, Author unknown, The Green Trumpet,
page 9. Historical Data-Vol.1

5/01/69, An Even Bigger Eighth Wonder, Author unknown, Philnews, pages 5-8.
Historical Data-Vol.1

5/08/69, Excavation Upstages Blasting at Grand Coulee, Author unknown,
Engineering News-Record, pages 34-36. Historical Data-Vol.1

6/01/69, Grand Coulee Forebay, Author unknown, The Green Trumpet, page 3.
Historical Data-Vol.1

8/05/69, Photogrametry Saves on Power Plant Survey Costs, Ervin E. Miller,
Electrical World, pages 71-71. Historical Data-Vol.1

9/19/69, BuRec Readies Grand Coulee to Supply Power for the Next Century,
Author unknown, Pacific Builder & Engineer, pages 54-59.
Historical Data-Vol.1

9/20/69, R. Granger - From Hoover Dam to Coulee, Author unknown, Pacific
Builder & Engineer, page 59. Historical Data-Vol.1

6/01/70, $112,525,612 Contract Signed for Construction - World's Largest Power
Complex, Author unknown, Vinnell, unknown, pages 10-13.
Historical Data-Vol.2

10/01/70, Concrete Placement Work Due to Start This Month, Author unknown,
Northwest Public Power Bulltn, page 13. Historical Data-Vol.1

12/01/70, First Concrete Placed at Grand Coulee Third Powerplant, Author
unknown, Vinnell, pages 10-13. Historical Data-Vol.2

12/04/70, Grand Coulee Powerhouse Special and Columbia Basin Project History,
Author unknown, Pacific Builder & Engineer, 30 pages. Historical Data-Vol.2

4/30/71, Bidders Need Sharp Pencil, Author unknown, North Star, pages 6-7.
Historical Data-Vol.2

7/30/71, Coulee Project is a Challenge, Author unknown, North Star, page 13.
Historical Data-Vol.2

8/01/71, Blasting Mass-Concrete at Grand Coulee Dam, J.R. Granger, Civil
Engineering-ASCE, pages 28-31. Historical Data-Vol.2

1/01/72, More Power to the Grand Coulee Dam!, David Foraker, Big, pages 2-7.
Historical Data-Vol.2

5/01/72, Pumped Storage Potential in the Pacific Northwest, Author unknown, Western Construction, page 23. Historical Data-Vol.2

8/24/72, Coulee Dam's Third Powerplant is a Study in Superlatives, Author unknown, Engineering News-Record, pages 28-29. Historical Data-Vol.2

10/01/72, Contractor Maintains Tight Schedule on Third Power Plant, Ralph Monson, Contractors & Engineers Mag., pages 30-31. Historical Data-Vol.2

10/01/73, Goodall and the History of the Grand Coulee Dam, Joe H. Moss, Goodall News, 4 pages. Historical Data-Vol.2

12/01/73, Penstock Construction at Grand Coulee Dam, Wayne H. John, Civil Engineering-ASCE, pages 52-54. Historical Data-Vol.2

12/02/73, Use of Dollies Enables Haul of Big Shaft Over Dam Spillway Structure, Author unknown, Transportation Engineer, pages 18-19. Historical Data-Vol.2

1/01/74, Winter Work at Grand Coulee, Author unknown, North Star, page 5. Historical Data-Vol.2

2/01/85, Higher Interest on Hydro Debt Threatens Region, Tina Nelson, Northwest Public Power Bulltn., pages 8-11. Historical Data-Vol.2

Plant Name: **HELMS**

Plant location:
  Shaver Lake, CA
  Fresno County

Rated capacity	1,206 MW
Average static head	1,608 ft
Plant efficiency	74.00 %
Stored energy	184,000 MWh
Number of units	3

Construction time: 8 years, 1 month
Construction cost: $590 per kW
Price level: 1981
First commercial power: June 1984
FERC project number: 2735

River or water source: North Fork Kings River

Owner: Pacific Gas & Electric Company
  77 Beale Street
  San Francisco, California 94536

Designers:
  PG&E'S Engineering Department

Plant Manager/Superintendent:
  Bobby Mooneyhem
  57800 McKinley Grove Road
  Shaver Lake, California 93664

  (209) 488-7896

	UPPER RESERVOIR	LOWER RESERVOIR
DAM		
Type	Rockfill concrete faced, Courtright Dam	Rockfill concrete faced
Height (ft)	315	260
Crest length (ft)	863	3,330
Volume (yd³)	1,560,000	3,700,000
RESERVOIR		
Type	Existing, constructed, Courtright Lake	Existing, constructed, Lake Wishon
Surface area (acres)	1,646	1,166
Usable power storage (acre-ft)	119,174	89,989
Power pool fluctuation (ft)	164.0	109.9
Operating levels		
Maximum (ft)	8,182.4	6,548.6
Minimum (ft)	8,018.4	6,440.3
Drainage area (miles²)	39.3	177.2
Seepage (ft³/s)	3.531	14.832
SPILLWAY		
Design flood		
Return period (years)		
Flow (ft³/s)	12,010	30,020
Capacity (ft³/s)	19,990	56,010
Type	Overflow ogee	Overflow, ogee
Gates		
Number	None	6
Type		Radial
Width (ft)		40.03
Height (ft)		11.52
OUTLET WORKS		
Discharge capacity (ft³/s)	4,767	3,800
Number of water passages	1	1
Dimensions of water passages		
Height (ft)		
Width (ft)		
Diameter (ft)	6.99	6.99
Type of gates	Howell-Bunger	Howell-Bunger Valve
Number of gates	1	1

Helms - Page 1 (English)

Plant Name: **HELMS**

Plant location
Shaver Lake, CA
Fresno County

Owner: Pacific Gas & Electric Company
77 Beale Street
San Francisco, California 94536

Rated capacity	1,206 MW
Average static head	490.0 m
Plant efficiency	74.00 %
Stored energy	184,000 MWh
Number of units	3

Designers:
PG&E'S Engineering Department

Construction time: 8 years, 1 month
Construction cost: $590 per kW
Price level: 1981
First commercial power: June 1984
FERC project number: 2735

Plant Manager/Superintendent:
Bobby Mooneyhem
57800 McKinley Grove Road
Shaver Lake, California 93664

(209) 488-7896

River or water source: North Fork Kings River

	UPPER RESERVOIR	LOWER RESERVOIR
DAM		
Type	Rockfill concrete faced, Courtright Dam	Rockfill concrete faced
Height (m)	96.0	79.2
Crest length (m)	263.0	1,015.0
Volume (m³)	1,192,620	2,828,650
RESERVOIR		
Type	Existing, constructed, Courtright Lake	Existing, constructed, Lake Wishon
Surface area (Mm²)	6.66	4.72
Usable power storage (Mm³)	147.000	111.000
Power pool fluctuation (m)	50.00	33.50
Operating levels		
Maximum (m)	2,494.00	1,996.00
Minimum (m)	2,444.00	1,963.00
Drainage area (Mm²)	101.800	458.950
Seepage (m³/s)	0.1000	0.4200
SPILLWAY		
Design flood		
Return period (years)		
Flow (m³/s)	340	850
Capacity (m³/s)	566	1,586
Type	Overflow ogee	Overflow, ogee
Gates		
Number	None	6
Type		Radial
Width (m)		12.200
Height (m)		3.510
OUTLET WORKS		
Discharge capacity (m³/s)	135	108
Number of water passages	1	1
Dimensions of water passages		
Height (m)		
Width (m)		
Diameter (m)	2.130	2.130
Type of gates	Howell-Bunger	Howell-Bunger Valve
Number of gates	1	1

Helms - Page 1 (Metric)

HELMS

INTAKES	UPPER INTAKE	LOWER INTAKE
Number	1	1
Type	Submerged, horizontal	Submerged, horizontal
Design discharge (ft³/s)	7,487	9,888
Gross inlet area (ft²) (at trash racks)	4,596	3,671
Bar racks		
spacing (in)	8.50	8.50
shape	Circular	Circular
depth/thickness (in)		
diameter (in)	1.50	1.50
Emergency gates		
number	1	None
height/width (ft)	32.15 / 21.33	
type	Fixed wheel	
Service gates		
number	2	2
height/width (ft)	26.90 / 11.81	15.42 / 33.79
type	Slide	Slide
Bulkhead/stop logs (Y or N)	N	N
number of units serviced		
Hoists		
number	2	2
capacity (tons)	23	37
type	Hydraulic cylinder	Hydraulic cylinder

WATER PASSAGES	Upper Tunnel	Shaft	Lower Tunnel	Surge Tanks Upper	Surge Tanks Lower	Penstocks	Tailrace Tunnel
Number	1	1	1	1	1	3	3
Diameter (ft)	27.0	27.0	27.0	60.4	44.0	15.4	15.5
Length (ft)	13,215	1,745	4,068	571	440	505	270
Maximum velocity (ft/s)	17.3	17.3	17.3			38.1	17.4
Concrete lining thickness (in)	11.76	26.76	11.76	18.12	18.12	26.76	15.00
Total length of concrete sections (ft)	13,215	1,745	4,068	571	440	505	270
Steel liner Thickness							
Minimum (in)	0.77					1.62	1.00
Maximum (in)	1.25					3.00	1.00
Material grade	A537 C1					A537 C1, A516 G70	A516 G70
Total length of steel-lined sections (ft)	1,709					505	70

Notes:
Shaft: Shaft slope is 55 degrees.
Upper Surge Tank: Diameter varies from 14.3 m (47 ft) to 18.3 m (60 ft).
Penstocks: Diameter varies from 4.57 m (15 ft) to 2.74 m (9 ft).
Tailrace Tunnel: Lengths of tailrace tunnels and concrete liners are averages.

HELMS

INTAKES	UPPER INTAKE	LOWER INTAKE
Number	1	1
Type	Submerged, horizontal	Submerged, horizontal
Design discharge (m³/s)	212.0	280.0
Gross inlet area (m²) (at trash racks)	427.0	341.0
Bar Racks:		
spacing (mm)	216	216
shape	Circular	Circular
depth/thickness (mm)		
diameter (mm)	38	38
Emergency gates		
number	1	None
height/width (m)	9.800 / 6.500	
type	Fixed wheel	
Service gates		
number	2	2
height/width (m)	8.200 / 3.600	4.700 / 10.300
type	Slide	Slide
Bulkhead/stop logs (Y or N)	N	N
number of units serviced		
Hoists		
number	2	2
capacity (Mg)	21	34
type	Hydraulic cylinder	Hydraulic cylinder

WATER PASSAGES	Upper Tunnel	Shaft	Lower Tunnel	Surge Tanks Upper	Surge Tanks Lower	Penstocks	Tailrace Tunnel
Number	1	1	1	1	1	3	3
Diameter (m)	8.23	8.23	8.23	18.40	13.40	4.70	4.72
Length (m)	4,028.0	532.0	1,240.0	174.0	134.0	154.0	82.3
Maximum velocity (m/s)	5.27	5.27	5.27			11.60	5.30
Concrete lining thickness (m)	0.300	0.680	0.300	0.460	0.460	0.680	0.380
Total length of concrete sections (m)	4,028.0	532.0	1,240.0	174.0	134.0	154.0	82.3
Steel liner Thickness							
Minimum (mm)	20					41	25
Maximum (mm)	32					76	25
Material grade	A537 C1					A537 C1, A516 G70	A516 G70
Total length of steel-lined sections (m)	521.0					154.0	21.3

Notes:
Shaft: Shaft slope is 55 degrees.
Upper Surge Tank: Diameter varies from 14.3 m (47 ft) to 18.3 m (60 ft).
Penstocks: Diameter varies from 4.57 m (15 ft) to 2.74 m (9 ft).
Tailrace Tunnel: Lengths of tailrace tunnels and concrete liners are averages.

HELMS

POWERHOUSE and RELATED FEATURES

Powerhouse Structure
Type: Underground cavern
Length: 336 ft   Width: 85 ft   Height: 144 ft

Guard Valves
Number: 3   Diameter: 8.0 ft
Type: Spherical

Transformers
Number: 9 + 1 spare, single phase
Ratings: 130 @ 55 degrees C
Voltages: (kV) 230 / 18

Generator
Rating generating (MVA): 390.0   Rating pumping (MVA): 320.0
Insulation type: Class B
Starting method: Starting motor
Starting equipment: Starting motor

Runners
Material: HT60 cast stainless steel
Minimum unit submergence: 200.1 ft
WR^2: 61,900,000 (lbf x ft²)
Manufacturer: Hitachi
Model test by: Hitachi

	Reversible Runners	Reversible Motor/Generator
Number	3	3
Diameter (ft)	16.80	Rotor 20.08 Stator
rpm synchronous	360.0	360.0
rpm overspeed	532.0	532.0
Type	Francis	Water cooled stator

Information on Runners

Condition:	Gross Head (ft)		Capacity (MW)		Discharge (ft³/s)		Turbine/Pump Eff.(%)	
	Generating	Pumping	Generating	Pumping	Generating	Pumping	Generating	Pumping
Maximum head & maximum power	1,745	1,942	407	279	3,069	1,670	91.7	91.0
Minimum head & maximum power	1,470	1,778	335	356	3,150	2,599	89.8	91.5

Note: Data in the above table are based on model tests.

Condition:	Net Head (ft)		Capacity (MW)		Discharge (ft³/s)		Turbine/Pump Eff.(%)	
	Generating	Pumping	Generating	Pumping	Generating	Pumping	Generating	Pumping
Rated head @ best gate	1,624	1,499	358	350	2,818	2,511	91.8	91.9

Note: Data in the above table are based on model tests.

HELMS

POWERHOUSE and RELATED FEATURES

Powerhouse Structure
Type: Underground cavern
Length: 102.4 m   Width: 25.9 m   Height: 43.9 m

Guard Valves
Number: 3   Diameter: 2.44 m
Type: Spherical

Transformers
Number: 9 + 1 spare, single phase
Ratings: 130 @ 55 degrees C
Voltages: (kV) 230 / 18

Generator
Rating generating (MVA): 390.0   Rating pumping (MVA): 320.0
Insulation type: Class B
Starting method: Starting motor
Starting equipment: Starting motor

Runners
Material: HT60 cast stainless steel
Minimum unit submergence: 61.00 m
WR^2: 25,586,000 (Newtons x m^2)
Manufacturer: Hitachi
Model test by: Hitachi

	Reversible Runners	Reversible Motor/Generator
Number	3	3
Diameter m	5.13	Rotor 6.120 Stator
rpm synchronous	360.0	360.0
rpm overspeed	532.0	532.0
Type	Francis	Water cooled stator

Information on Runners

Condition:	Gross Head (m)		Capacity (MW)		Discharge (m³/s)		Turbine/Pump Eff.(%)	
	Generating	Pumping	Generating	Pumping	Generating	Pumping	Generating	Pumping
Maximum head & maximum power	532.0	592.0	407	279	86.9	47.3	91.7	91.0
Minimum head & maximum power	448.0	542.0	335	356	89.2	73.6	89.8	91.5

Note: Data in the above table are based on model tests.

Condition:	Net Head (m)		Capacity (MW)		Discharge (m³/s)		Turbine/Pump Eff.(%)	
	Generating	Pumping	Generating	Pumping	Generating	Pumping	Generating	Pumping
Rated head @ best gate	495.0	457.0	358	350	79.8	71.1	91.8	91.9

Note: Data in the above table are based on model tests.

HELMS

Plant Data:
Average GWh generating per year:	584
Average GWh pumping per year:	772
Starting time from standstill (s):	120
Changeover time pumping to generating (min):	20
Planned/scheduled time between major overhauls (years):	3
Outage time required per unit during major overhauls (weeks):	8
Representative plant availability (%):	97.0
Representative planned outages (weeks per year):	4

Miscelleneous Notes:
The diameter of the upper penstock varies from 14.3 m (46.9 ft) to 18.4 m (60.4 ft).

The diameter of the penstocks varies from 4.7 m (15.4 ft) to 2.4 m (7.9 ft).

Cavitation Experience:
Very little.

Significant or Unique Problems:
Sept. 1982, the 6.7 m diameter by 42.7 m long steel pipe at Lost Canyon failed shortly after being pressurized, causing approximately 11 months delay.

Generator/motor roto dynamic problems required shaft and bearing modifications resulting in 16 month delay.

List of Licenses Required:
FERC, USFS Special Use Permit.

ENVIRONMENTAL FEATURES

Recreation:
Boat launching & docking, camping, fishing access & picnic areas.

Fish and Wildlife:
Plant fish to maintain pre-project catch rate. Wildlife enhancement in selected areas.

Social:
Rebuilt and paved USFS and county roads. Yearly snow removal on 33 miles of USFS and county roads.

HELMS PUMPED STORAGE PROJECT
SELECTION AND ALTERNATIVES

J. A. Davis, Helms Project Manager, M. ASCE
A. G. Strassburger, Consultant and
Former Helms Project Manager, F. ASCE

Pacific Gas and Electric Company
77 Beale Street
San Francisco, CA 94106

INTRODUCTION

As in any large project there is an extensive amount of early planning and studies. Many alternatives have to be examined before the final layout is made. This was true with the Helms Pumped Storage Project. Many of the company's engineers who did the early studies, some of whom have retired or moved on to other assignments, have contributed greatly towards the present reality of a working power plant, which provides electrical peaking power for PGandE's system.

The company's engineers who designed and managed the construction of Wishon and Courtright dams in the 1950's had a vision that someday a power plant (possibly a pumped storage plant) would be built between the reservoirs. The site was ideal, having two large capacity reservoirs (125,000 ac. ft. each) with a high (1,635 ft.) differential elevation and a modest distance apart (about 4 miles). Early studies that were conducted as the project started to develop consisted of:

- Size of plant
- Number and size of units
- Location of water conduits
- Location of powerhouse
- Type of intake-discharge structures at Courtright
- Lost Canyon Crossing

As a result of these early studies the final layout was a very efficient, concise and economic design. The layout met all of the engineering and construction criteria.

SIZE OF PLANT

One of the first decisions that had to be made was the size of the plant. A pumped storage plant is used mainly for two purposes; 1) to provide capacity to meet the peak demands of the electrical power system and 2) to provide standby power (spinning reserve) if elsewhere in the system generation is lost or a major transmission line fails. Over the years PGandE's vast hydro system had provided adequate power for the peaks, but with the increased use of air conditioning on hot days in Northern California it was projected that we would need a large plant to provide for future peaks. It was also decided that the project should provide spinning reserve of a capacity to match some of the larger thermal units in the company's system, such as Pittsburg 7 (750 MW) and the Diablo Canyon nuclear units (1100 MW each). Therefore, the company's generation planners were looking for capacity in the neighborhood of 700 to 1200 MW.

NUMBER AND SIZE OF UNITS

In determining the size of the units, major manufacturers of reversible pump-turbines throughout the world were contacted for advice regarding limitations on size, rotating speed and economy of

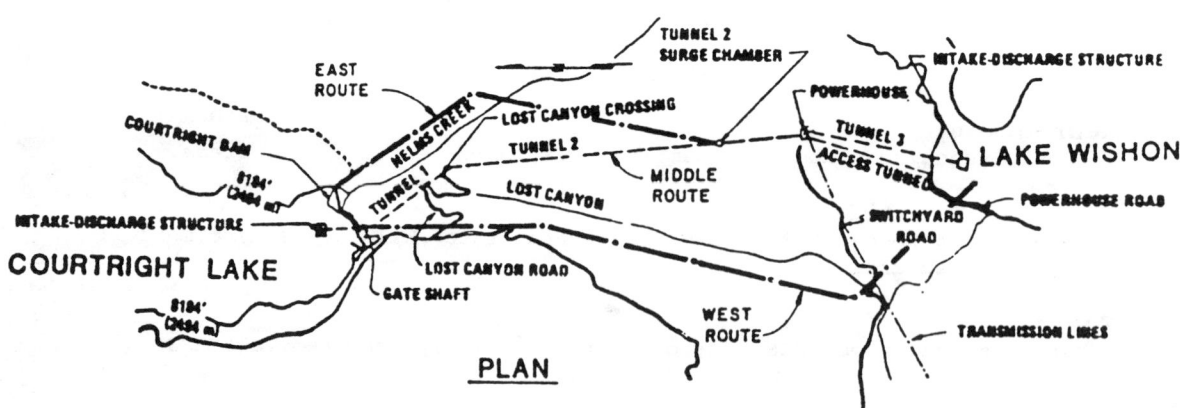

HELMS PUMPED STORAGE PROJECT

scale of units for normal static head of approximately 1,625 feet. Units in Japan and elsewhere were being made in the 1,500 foot range but with somewhat smaller capacity. The company determined that a reversible pump-turbine somewhere around 350 MW capacity, rotating at 360 rpm would be "state-of-the-art," but feasible.

Studies were then made for a two-unit (700 MW), a three-unit (1050 MW) and a four-unit (1400 MW) plant. Estimated quantities of excavation, concrete, cost of the units, number of tunnels, and construction schedules were considered. The result of this study was that three units (1050 MW) were the most economical and met the range of total capacity that the company planners had set as a criteria. (Recent start-up tests show that the units operate very well at 115% of rated capacity. This provides a total plant capacity of about 1200 MW.)

LOCATION OF WATER CONDUITS

Three locations for the water conduits were studied: the east, middle and west routes. (The routes are shown on the Plan.) The east route had the best rock coming out of the left abutment of Courtright Dam but required a difficult crossing of Helms Creek. The west route came out of the right abutment in poorer rock but required no canyon crossing. The middle rock came out the right abutment, crossed over Lost Canyon and under a narrow ridge. The middle route was selected because it was the most direct and the rock was considered as competent as the west route.

LOCATION OF POWERHOUSE

There were a number of reasons why the powerhouse is situated at its present location. Each one of the reasons will be discussed.

Submergence

It is a requirement that reversible pump-turbines have a positive pressure on the runner during the pumping operation. This pressure prevents cavitation on the runner. The various pump-turbine manufacturers that were contacted had advised us that to pump against a head of up to 1,950 feet (maximum Courtright El. 8,184 less runner El. 6,224) would require a minimum submergence of 200 feet. Since Lake Wishon could fluctuate in excess of 100 feet over a year cycle, the pump runner would have to be located 310 feet below Wishon's maximum level. A powerhouse at Lake Wishon's edge, 310 feet deep, would be very difficult to construct, requiring a shaft or other underground structure subject to hydrostatic pressures on the order of 310 feet.

Surge Chamber

There was only one good physical location between the two lakes and somewhere near the tunnel alignment for the upper tunnel surge chamber. This was on Lost Peak. Lost Peak is high enough above Courtright's maximum water surface to provide for surge conditions. To keep the hydraulic water hammer pressures to an acceptable value at the powerhouse it was prudent to locate the powerhouse as close as possible to the surge chamber. Therefore, it was desirable to locate the powerhouse back into the mountain away from Lake Wishon's edge.

Location of Switchyard

It was determined that there was no suitable site around Lake Wishon for a switchyard that was flat enough and would not impact on the visual quality of the lake. Therefore, it was decided to locate the switchyard above the powerhouse with the electrical power leads brought up a vertical shaft. The switchyard was located on the last flat spot before a steep grade up to the surge chamber.

Access Tunnel Grade

By locating the powerhouse approximately three-quarters of a mile back into the mountain the access tunnel could be driven at a grade favorable for construction of the tunnel with rubber tired vehicles. That is about 8% grade. Rail equipment could not be used.

Competent Geology

By being deeper in the mountainside it was felt that the rock would not have been subjected to weathering. (Later geological exploration by core drilling showed the rock to be competent.)

Based on the above parameters the powerhouse chamber was located where it was.

TYPE OF INTAKE-DISCHARGE STRUCTURE AT COURTRIGHT

In order to provide as much useable capacity for conventional generation as possible, the intake-discharge structure had to be located on the bottom of Courtright Reservoir. This would also minimize fish attraction to the intake and preclude ice damage. The initial plan conceived a "morning glory" type intake with a vertical shaft a horizontall tunnel to Lost Canyon. Model tests were run on the structure to determine if vortexing would occur during generating and defusing of the discharge during pumping without water surface disturbance for low reservoir levels. The structure had to have bulkhead gates that would close and allow inspection of Tunnel 1. The model showed that to prevent the bore effect during discharge the diameter of the structure had to be very large. The larger the structure became the more difficult it was to design it to withstand the water pressure whenever Tunnel 1 was drained. Therefore, a more conventional, horizontal type intake-discharge structure was designed and model tested. Tunnel 1 is horizontal to the edge of the lake and then slopes modestly down to Lost Canyon.

LOST CANYON CROSSING

Two alternatives for the layout of the crossing were considered; a surface location or cut and cover. The only good reasons for the latter were aesthetic. A number of governmental agencies and environmental groups opposed having an exposed structure at Lost Canyon that would eliminate an existing pond in the area. The advantages of surface construction were numerous. It would: 1) allow simpler driving of tunnels from two headings and disposal of the spoil in the canyon, thereby greatly shortening the construction schedules, 2) provide a location for an emergency shut off valve (the possibility of installing a butterfly type shut off valve was investigated but was abandoned in favor of the wheel gate in Tunnel 1), 3) provide access into tunnels 1 and 2 for future inspection and maintenance, and 4) reduce the height of Tunnel 2 Surge Chamber. The alternative of going under Lost Canyon was ruled out and the design and rehabilitation of Lost Canyon was made aesthetically pleasant.

SUMMARY

Only the major civil studies and alternatives have been addressed here. There were numerous alternatives for many of the individual items noted above. All of the early studies proved to be very beneficial and worth the effort. The result was basically a well developed layout and an efficient hydro pumped storage plant.

HELMS PUMPED STORAGE PROJECT
THE FACILITIES

D. Zayakov, Supervising Civil Engineer, F. ASCE
J. A. Davis, Helms Project Manager, M. ASCE
A. G. Strassburger, Consultant, F. ASCE
Pacific Gas and Electric Company
77 Beale Street
San Francisco, CA 94106

INTRODUCTION

The Helms Pumped Storage Project with its 1,744 foot maximum head ranks among the world's largest high-head projects with reversible pump-turbines. Additionally, because of a large amount of snow melt draining into the upper reservoirs, Helms can be operated as a conventional hydro project.

The project is located approximately 70 miles northeast of Fresno in California's Sierra Nevada mountains. Two existing Pacific Gas and Electric reservoirs, Courtright Lake and Wishon Lake, were utilized in the development of the project. The large capacity of these reservoirs and the high unused head between them were important factors in making the Helms Project economic. The project's three units can generate more than 1125 MW. Low cost off-peak energy is used for the pumping of the water from Lake Wishon to Courtright Lake.

DESCRIPTION OF THE MAIN PROJECT COMPONENTS

Intake-Discharge Structures

The Intake-Discharge Structures in both reservoirs consist of submerged gate structures, housing two bulkhead gates each, and trashrack structures.

The Courtright gates are 12 feet wide by 27 feet high; the Wishon gates are 15.5 feet wide by 34 feet high. Both are hydraulically operated with submerged hydraulic cylinders and can be closed and opened only under balanced head conditions. A reinforced concrete transition assures smooth water flow into and out of the tunnels.

The trashrack structures were designed and hydraulically model tested to ensure uniform and eddy-free velocity distribution through the trashracks. Velocities are low enough to not attract fish into the conduits.

Tunnel 1, Wheel Gate Shaft and House

Emergency upstream closure of the flow from Courtright Lake is provided by a 32 foot high by a 21 foot wide fixed wheel gate. The gate is located within a gate shaft about 1,000 feet downstream from the Courtright Intake-Discharge Structure.

The concrete-lined gate shaft is about 210 feet high and surfaces near the south end of the Courtright Lake. The hydraulic cylinder and controls for the wheel gate are located in a surface gate house. Because of the long distance (±220 feet) between the gate and the operator located above in a house, a series of rigid stems coupled together, are provided to lower and raise the gate. The gate is equipped with upstream inflatable seals.

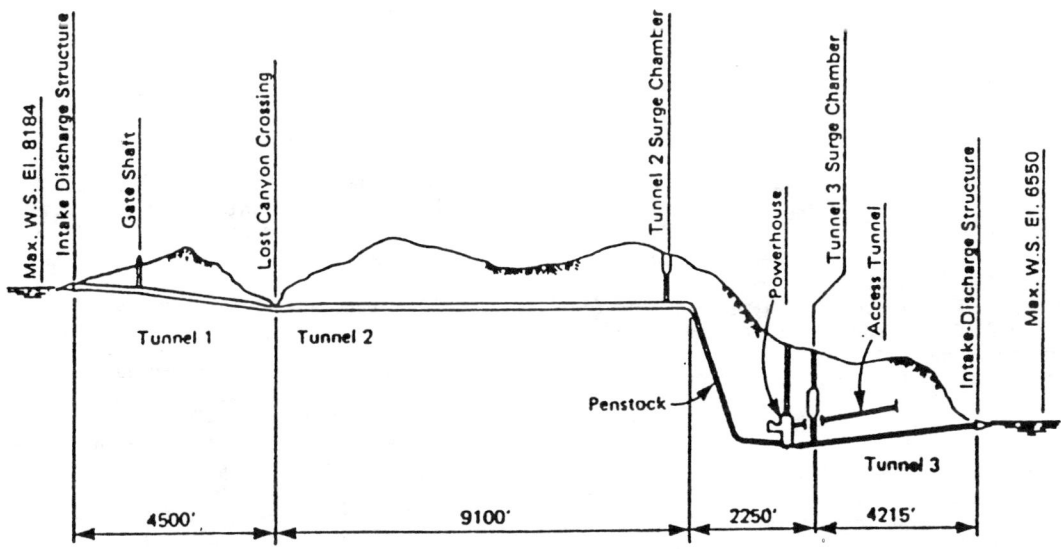

Schematic section through the main project features

Two bypass valves (6" and 16" diameter) are used to fill the tunnel with water before opening the gate.

The gate is designed to be closed from the powerhouse by remote control as well as automatically when under excess tunnel flow conditions.

The gate house has a subground chamber, equipped with a dogging device, which allows the gate to be raised for inspection and repairs. The gate house provides space for: the hydraulic cylinder and control system, standby power system, communication equipment, as well as emergency housing for snow survey crews.

The Tunnels, Penstock and Surge Chambers

All Helms underground facilities, including the tunnels, are situated in relatively competent granites and granodiorites.

The tunnels are in three categories: water conduits (Tunnels 1, 2 and 3), access, and construction tunnels.

Tunnel 1 is 27 feet in diameter and concrete lined. It begins at the Courtright intake-discharge structures, the length is 4,250 feet, and terminates in Lost Canyon.

To cross Lost Canyon and connect Tunnels 1 and 2, a 22 foot diameter steel conduit was constructed. The conduit is supported on concrete saddles, and has Dresser couplings between the pipe sections. Because of the relatively high water pressure in the tunnels and less rock cover near the Lost Canyon portals of Tunnels 1 and 2, steel lining was installed in both tunnels. This lining extended 900 feet into Tunnel 1 and 720 feet into Tunnel 2.

Tunnel 2 is also 27 feet diameter concrete lined, and the length is 8,980 feet. Tunnel 2 Surge Chamber is located at the south end of the tunnel. The surge chamber is of the orifice type. It extends for 560 feet from the tunnel to the surface. It is 47 feet in diameter except for the upper 60 feet where it widens to 60 feet diameter. The entire shaft is concrete lined, with a wire strand safety barrier over the top opening.

The penstock consists of a 1,750 foot long, 27 foot diameter inclined shaft and a horizontal portion, including the bifurcation of 1,140 feet. The penstock is concrete lined with 500 feet of the bifurcation for each unit steel lined.

Tunnel 3 connects the turbine draft tubes with Lake Wishon. It is a 27 foot diameter, concrete lined, 3,730 foot long tunnel. The downstream surge chamber is located about 650 feet from the powerhouse. This chamber was necessary because of the relatively long Tunnel 3. It extends for 970 feet from the tunnel to the surface, the upper 530 feet being a 10 foot diameter unlined shaft, with the middle 300 feet, 44 feet in diameter and the lower 130 feet, 27 feet in diameter being concrete lined. At the surface the chamber has a wire strand safety barrier similar to the upstream surge chamber. To our knowledge, the Helms conduit system is unique in that it combines pump-turbines with two surge chambers.

The Powerhouse Access Tunnel starts at the upper end of Lake Wishon. It is a 3,730 foot long, 30 foot diameter unlined horseshoe section. It slopes at 8.2% toward the powerhouse and is asphalt paved.

Construction Tunnels

A number of construction tunnels were provided to facilitate underground construction. One provides access from the Powerhouse Access Tunnel to the crowns of the Powerhouse and Transformer Chambers. A second tunnel extends from one end of the Powerhouse Chamber to the horizontal penstock and was used for access to the base of the inclined shaft and to the penstock manifold. A concrete plug is provided to seal off this tunnel to penstock pressure. A portion of the tunnel houses permanent powerhouse equipment such as air and oil storage tanks. The third tunnel provides access to Tunnel 3 and also has a plug resisting lower reservoir (Wishon) pressure.

Another access tunnel served as a construction bypass around the powerhouse and directly connected the Powerhouse Access to the penstock. A fifth access (or adit) was added late in construction to provide additional access to Tunnel 2. This is located downstream of Lost Canyon and permitted simultaneous installation of Lost Canyon Crossing and tunnel liners along the concreting of the Incline Shaft. This was necessitated to expedite the construction schedule.

Powerhouse and Transformers Chambers

The Powerhouse Chamber excavation is 83 feet wide, 125 feet high and 336 feet long. The walls and roof are reinforced with rockbolts. In addition, the roof is shotcreted and a false ceiling was installed to prevent water dripping over the equipment in the power plant. The walls are concreted below the generator floor only, leaving exposed rock above it. Two 270 ton cranes, one low-speed and one high-speed, were installed to service the heavy equipment.

Three reversible units generate a total of 1125 MW. Each Francis type reversible pump-turbines has a rated capacity of 480,000 hp, with a head of 1,560 feet and a design flow of 3,000 cfs. The rating as a pump is 455,000 hp, with a flow of 2,400 cfs and a head of 1,500 feet. The static head on the units varies between 1,470 feet and 1,744 feet. Each generator-motor is rated as 350 MW generator and rotates at 360 rpm. In the pumping mode the motor is rated at 345 MW. A spherical valve upstream of each turbine acts as a guard valve. Testing has proven each unit capable of producing in excess of 400 MW, making these the world's largest high-head pump-turbine units.

The Transformer Chamber is located upstream of the powerhouse and is connected with it by three short tunnels. The chamber is 270 feet long, 44 feet high and 31 feet wide. As in the powerhouse, the walls and roof are reinforced with rockbolts. Three single-phase transformers are provided for each unit. One additional single phase transformer is available as spare. The transformers are connected to the switchyard above ground via 230 KV oil filled cables installed in the 1,100 foot vertical shaft. The shaft is equipped with stairs and an elevator, and can be used as an emergency exit.

The powerhouse units connect with Tunnel 3 through the three draft tube extensions and a manifold. Each unit's draft tube is steel lined and equipped with a 10 foot by 15.5 foot tailrace gate. These gates are vertically hinged and have inflatable seals. The gates are hydraulically operated with controls in the powerhouse and can open and close only under balanced head.

Helms major structures are all located underground. Only the valve house near Courtright, about 200 feet of surface conduit and the switchyard are above ground. This reduces Helms environmental impact to a minimum.

HELMS PUMPED STORAGE PROJECT
DESIGN THROUGH CONSTRUCTION

A. G. Strassburger, Consultant and
Former Helms Project Manager, F. ASCE
D. W. Moller, Civil Engineer, M. ASCE
J. A. Davis, Helms Project Manager, M. ASCE
R. V. Farley, Helms Project Superintendent, M. ASCE

Pacific Gas and Electric Company
77 Beale Street
San Francisco, CA 94106

INTRODUCTION

The Helms Pumped Storage Project is a large, complex facility incorporating many state-of-the-art design and construction challenges. Field conditions encountered were not always anticipated and design decisions made earlier were often challenged and at times changed. High hydraulic pressures in large water conduits, exceeding in-situ rock strengths and resulting in hydro-fracturing of rock, called for some unique design and analytical methods. The pump-turbines are believed to be the world's largest high-head (500 meter class) reversible machines. The generator/motors, operating at 360 rpm, and capable of generating 400 MW each are at the leading edge of technology for generators. These machines required special design and erection attention.

Extremely difficult weather conditions during four of the six winter construction seasons, bringing deep snows and low temperatures far beyond normal, caused many secondary construction problems. A geologic shear zone located immediately adjacent to a 600 ft. deep surge tank required decisions which called for either; a complete structural redesign, relocation, or expensive adaptation of construction methods to in-situ conditions. Another shear zone near the high pressure penstock plug required extensive high pressure grouting with especially fine cements. The significance of the zone was elevated by the high hydrostatic pressure and expected hydrofracturing of rock. An unprecedented inclined shaft, about 1,800 feet long, and in broken rock required some unusual excavation and concreting techniques. These and other very challenging problems are discussed.

GEOTECHNICAL AND WATER PRESSURE CONSIDERATIONS

The project is located in the Sierra Nevada mountains between 8,200 and 6,500 feet in elevation. Although essentially in granitics a number of rock classifications within that general category were present. The two existing rockfill dams, and the Haas underground hydroelectric project, all built in the 1950's provided some insight into the detailed geology to be expected. Extensive surface geologic mapping of the project was done along with exploratory drilling, logging and evaluation. Special studies made in the powerhouse area included determination of the in-situ rock strength, modulus of elasticity and principle stresses. Strength and stress measurements were accomplished by hydrofracturing techniques, which were only in their development stages at that time. Minimum confined strength findings indicated that these were at or below the hydrostatic pressures under which the high pressure penstock would be operating.

Steel lining of the entire reach of conduit subject to this phenomenon was believed to be impractical and too expensive. Therefore, impacts of this hydrofracturing upon the concrete lined conduits, the Powerhouse and Transformer Chambers, and the penstock plug were considered in detail. The hydrofracturing was concluded to be a condition which had to be accepted and structures had to be designed accordingly.

Cement grouting at the head of the 500 ft. steel lining adjacent to the Powerhouse was used to reduce potential leakage and to provide reduction of pressure at the Powerhouse walls. Drainage and pressure monitoring devices were installed in the vicinity of the Powerhouse.

A number of significant shear zones penetrated the rock formation in the vicinity of the major excavations including one which passed in close proximity to the permanent concrete plug located in the Penstock Access Tunnel.

Upon filling the pressure tunnel in August 1982, leakage quantities and pressures, immediately downstream of the plug, were somewhat higher than desired. Flows approached 2 cfs, however, this was not an alarming quantity. Of more concern was the fact that one piezometer indicated pressures of about 80% of the upstream pressure and responded rapidly. There was concern that with time, and with the consideration that hydraulic transients might become too high. Thus, a secondary high pressure grouting program was initiated.

This program included additional consolidation grouting behind the concrete liner and plug in addition to high pressure staged grouting of the shear zone. Although chemical grouting was considered, and initially planned, the engineering contractor, in conjunction with the owner, chose instead to use ultra fine cement for the plugging medium. Some of the cement, available domestically, had a Blaine fineness of about 5500. This was used for initial stages. A finer cement with a fineness of about 8800 was used for the deeper and higher pressure work. Of additional interest in this procedure was the fact that access to the work was restricted to a 30" diameter opening through the concrete plug. The program was at least partially successful, particularly in reducing downstream pressures and in increasing response time. This program is discussed in more detail by Moller, et al.*

Another area of geotechnical interest which involved significant study related to rock response characteristics of the Powerhouse and Transformer Chambers. This work is reported in more detail by Sweeney, et al.** This study resulted in the decision to re-orient the penstock manifold. This was done by flopping the main leg of the manifold so that its long leg crossed the direction of the primary joint planes instead of being parallel. The studies did not specifically result in any significant layout changes of the Powerhouse or Transformer Chambers. Deformations of the rock walls during and after excavation were in close agreement with projections. The studies however pointed out that "slabbing" and "rock popping" of excavated walls could be expected. Thus when they occurred, they were of no surprise and could be remedied quickly.

The studies also gave confidence to the design concepts and details of the Powerhouse crane support system. It has been a practice in most underground powerhouses to essentially complete all excavation, then to install steel columns for rail supports, along with necessary concrete and struts to rock. At Helms we chose to use reinforced concrete beam rail supports. These were anchored to rock with post-tensioned rock bolts. This system permits much earlier installation of a construction, or the permanent, cranes. Although significant rock overbreak on one side increased the beam size, it is believed the system used was the correct choice. This system was also recently used with success at PGandE's Kerckhoff 2 Powerhouse.

HYDRAULIC STRUCTURES AND TRANSIENTS

Project layout alternatives are discussed in an accompanying paper by Davis. In addition, several design/layout alternatives were included in the bid package submitted to potential contractors. It was intended that the successful contractor should be allowed to participate in some of these design decision in order to minimize construction costs.

---

*D. W. Moller, H. L. Minch and J. P. Welsh, "Ultrafine Cement Pressure Grouting to Control Ground Water in Fractured Granitic Rock, Helms Pumped Storage Project," ACI Temp. Ref. No. Ms 5703(s).

**N. F. Sweeney and H. J. Hovland, "A Large Underground Opening Measured Versus Predicted Rock Deformations," Proceedings of Rapid Excavation and Tunnelling Conference, San Francisco, 1981, AIME/ASCE.

One of the alternatives related to the size of the main water tunnels. PGandE conducted extensive parametric studies related to size of the tunnels for both lined and unlined conditions. It was concluded that maximum velocities for an unlined tunnel should be limited to 8 to 8.5 fps, based on previous experience which indicated a tendency for rock to be loosened and moved along the invert for higher velocities. In this case of reversed flows, that tendency could be increased. This limiting velocity resulted in a 37 ft. modified horseshoe (vertical sides) section.

This tunnel was rather large, however, and it was feared that economy of scale might be offset by costs of additional supports. It was felt a smaller concrete lined tunnel might be more economic. Thus, PGandE designed an alternate concrete lined section with hydraulic characteristics equivalent to the unlined section. This resulted in a 27 ft. diameter circular section with maximum velocities of about 16-1/2 fps. All bid proposals showed the concrete section to be the most economic.

One of the most innovative and state-of-the-art design challenges encountered early in the design phase was the analysis of hydraulic transients in the total hydraulic conduit. Helms provided a combination of structures and machinery which to our knowledge has never been encountered elsewhere. Certainly hydraulic transient studies have been performed many times on long tunnels, for high heads, for restricted orifice surge tanks, for reversible pump-turbines with their unique four quandrant characteristics and probably for two surge tank systems. An analytical program using the method of characteristics and combining all of these features was developed in-house by PGandE engineers. Subsequent operations and testing of the machinery and conduits has thoroughly verified the predicted results.

To supplement these hydraulic transient analyses, it was believed prudent to verify the theoretical orifice coefficents used in the two surge tanks and to verify the total hydraulic performance of the structural complex including the surge tanks, a rock/sand trap and the elbow at the head of the inclined shaft. Hydraulic model studies were performed. Orifice coefficient studies and rock trap performance studies were conducted separately. These were then combined in a study incorporating all facilities to verify overall performance characteristics. Results of these studies had a welcome but unexpected benefit later on during construction and directly influenced some major construction decisions when some unfavorable geologic features were encountered adjacent to the Tunnel 2 Surge Tank.

Predesign exploratory work at the Surge Tank indicated that generally good rock should be encountered during excavation and the lining was designed accordingly. During excavation of Tunnel 2, an 80 ft. wide shear zone, requiring steel supports, was encountered immediately adjacent to the surge tank which at this time had been excavated to a 10 ft. diameter reamed hole. Further exploration and studies indicated that the shear zone, inclined very steeply, probably extended to the surface about 500 ft. above. It could likely be intercepted during the enlargement of the surge tank excavation. This would result in the potential release of a long steeply inclined wedge, with subsequent loading upon the surge tank lining. This called for a study of various options. These included: use of construction supports, as necessary, with a redesigned concrete lining and a larger excavation to accommodate the thicker lining; moving the surge tank, although already partially excavated, to a new location; retain the reamed hole excavation and move the remaining excavation off-center.

Numerous factors affected the decision. These included; construction costs and schedule, the winter season and heavy snow which impacted additional exploratory work; time and expense of a new design; additional geologic uncertainties; topography; and the level of the hydraulic gradient at the elbow of the inclined shaft. Another important factor was the knowledge which had been gained during the previously mentioned hydraulic model studies. These indicated that there were distinct advantages to not changing the overall configuration by moving the surge tank. It was thus left where it was, full permanent ground support was incorporated in the "temporary" excavation supports, and the original design was not changed. This is a good example of the benefits of close cooperation between the designers and constructors.

Among the other interesting hydraulic structures are the three draft tube gates. Because of the need to deeply submerge the runners of the reversible turbine-pumps, the Powerhouse Chamber is located deep underground, with the turbine approximately 300 feet below a full lower reservoir. A common draft tube dewatering arrangement is to provide separate wheel or bulkhead gates for each unit with a common hoist system. This also requires extensive excavation and construction of a separate gate chamber. In the case of Helms, and its deep submergence, this type of scheme would have been impractical or, at best, very expensive. The gate arrangement chosen consists of a vertically hinged hydraulic cylinder operated draft tube gate at each unit. Although controls and seats required some modifications and adjustments, the gates work well.

CONSTRUCTION PROBLEMS

Many of the construction problems encountered are already addressed elsewhere by the lead author.* We wish here to expand somewhat on the rationale

---

*"High-Head Underground Power Project Presents Design/Construction challenges," Proceedings, Rapid Excavation and Tunnelling Conference, San Francisco, 1981, AIME/ASCE.

for the incline, vs. a vertical shaft, to address concreting of the shaft and to revise some earlier conclusions.

Because of the difficulties of assessing all of the pros and cons of vertical vs. inclined shafts, PGandE, in submitting the bid documents to bidders, chose to ask for pricing on both a vertical shaft and one inclined to $55°$ from the horizontal. All bidders submitted unit prices such that, on an evaluated basis, the incline was the economic choice. A later estimate by the selected contractor again projected the incline shaft to be cheaper. Actual construction then resulted in a number of problems and delays, both during excavation and again in concreting.

Difficulties during concreting evolved primarily because of the extremely long length and high head under which concrete had to be placed. A placing method, similar to that used at Churchill Falls was chosen. This basically involved delivering concrete to a hopper at the upper end of the shaft and releasing it through a pipe to the forms. Concrete quality control and consistency, flow control and velocity control are all extremely critical. The length of drop here exceeded Churchill Falls by about 400 feet. This incremental length apparently was important. The contractor never was able to drop concrete beyond about 1,400 feet. The lower lining concrete was finally pumped up from the bottom, but at the expense of affecting the schedule.

Moving of forms was also difficult in the lower reaches. During excavation the 12" pilot bore meandered somewhat. This meandering was followed by the raise-bore machine, and then by the drill jumbo during final slashing. The form jumbo then also had to follow these meanders, making it difficult to move and match forms.

In the 1981 paper the lead author stated, "The writer, given the opportunity again, would... be more favorably inclined toward a vertical shaft...." That opinion was written while construction was still underway. Subsequent thereto, work was completed and the author has had an opportunity to reflect on all aspects, information, risks and results. The author now believes that each situation must be examined on its own merits and conditions, and that a deliberate bias toward a vertical shaft is not appropriate.

CONCLUSION

The Helms Project was declared commercial on June 30, 1984 and is now operating.

Its advantages are far greater than anticipated. It is being operated much more than planned and its economics are being proven positive in spite of the cost increases. The electric system dispatchers love it and we're proud of it.

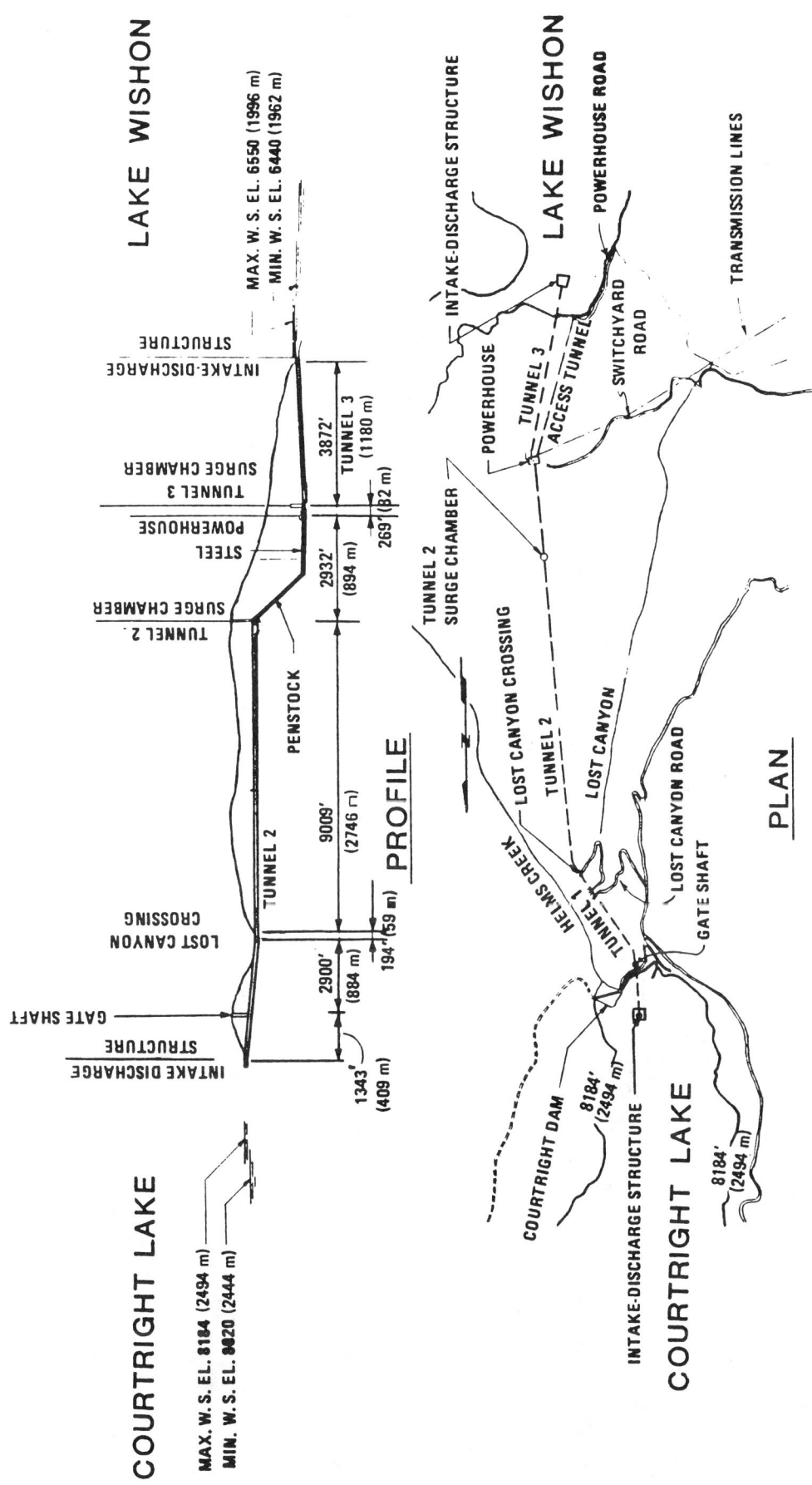

Figure 1 – HELMS PUMPED STORAGE PROJECT

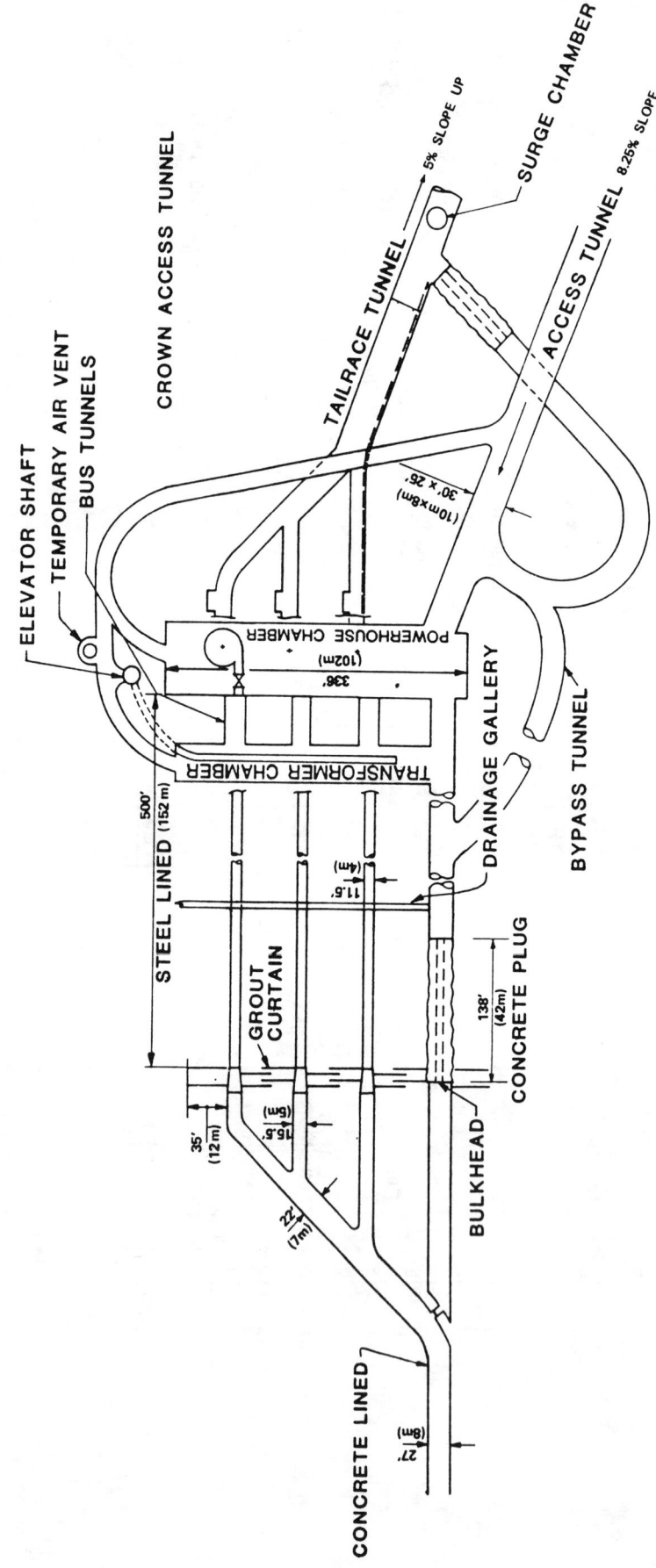

EXCAVATION POWERHOUSE COMPLEX

FIGURE 2

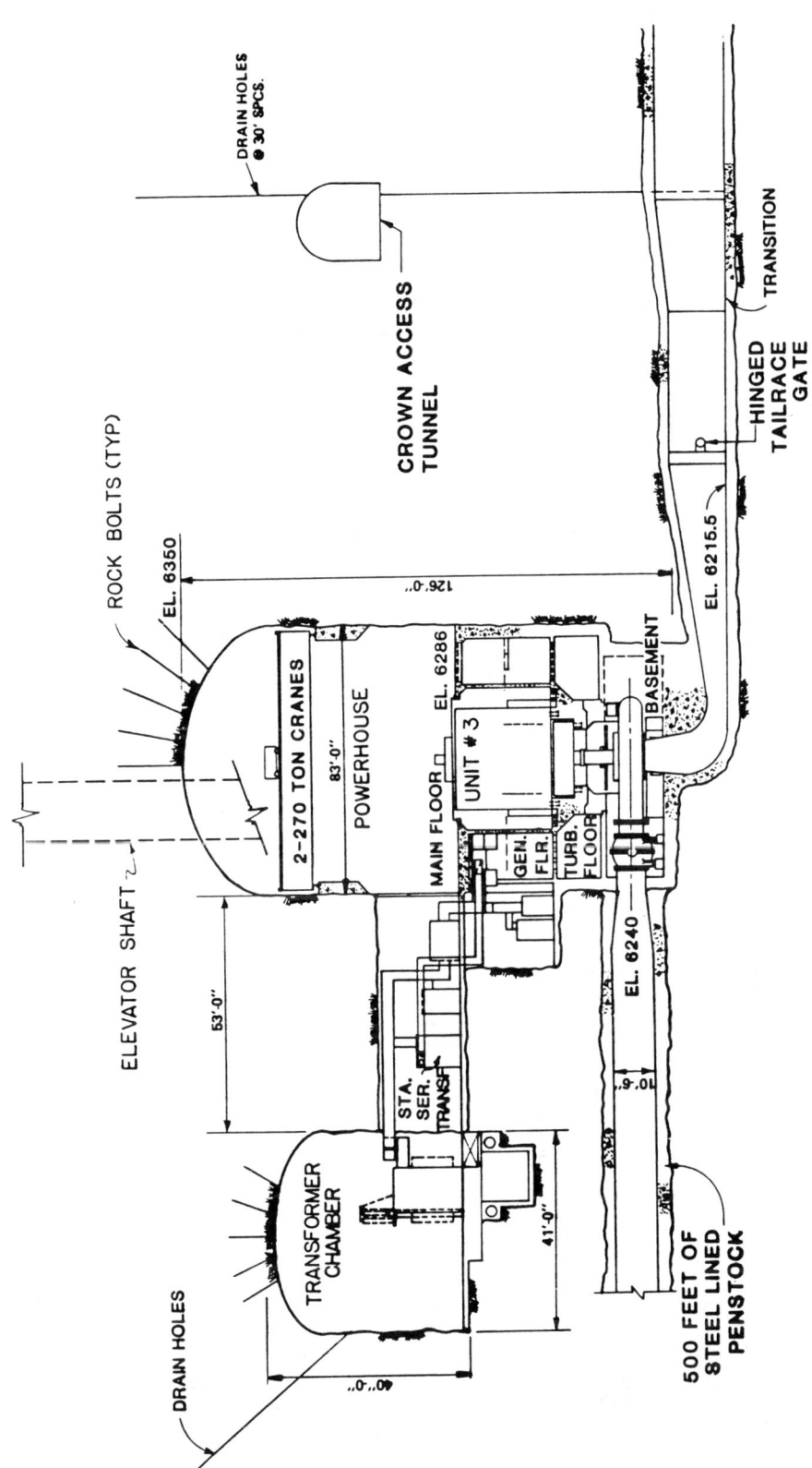

POWERHOUSE TRANSVERSE SECTION

FIGURE 3

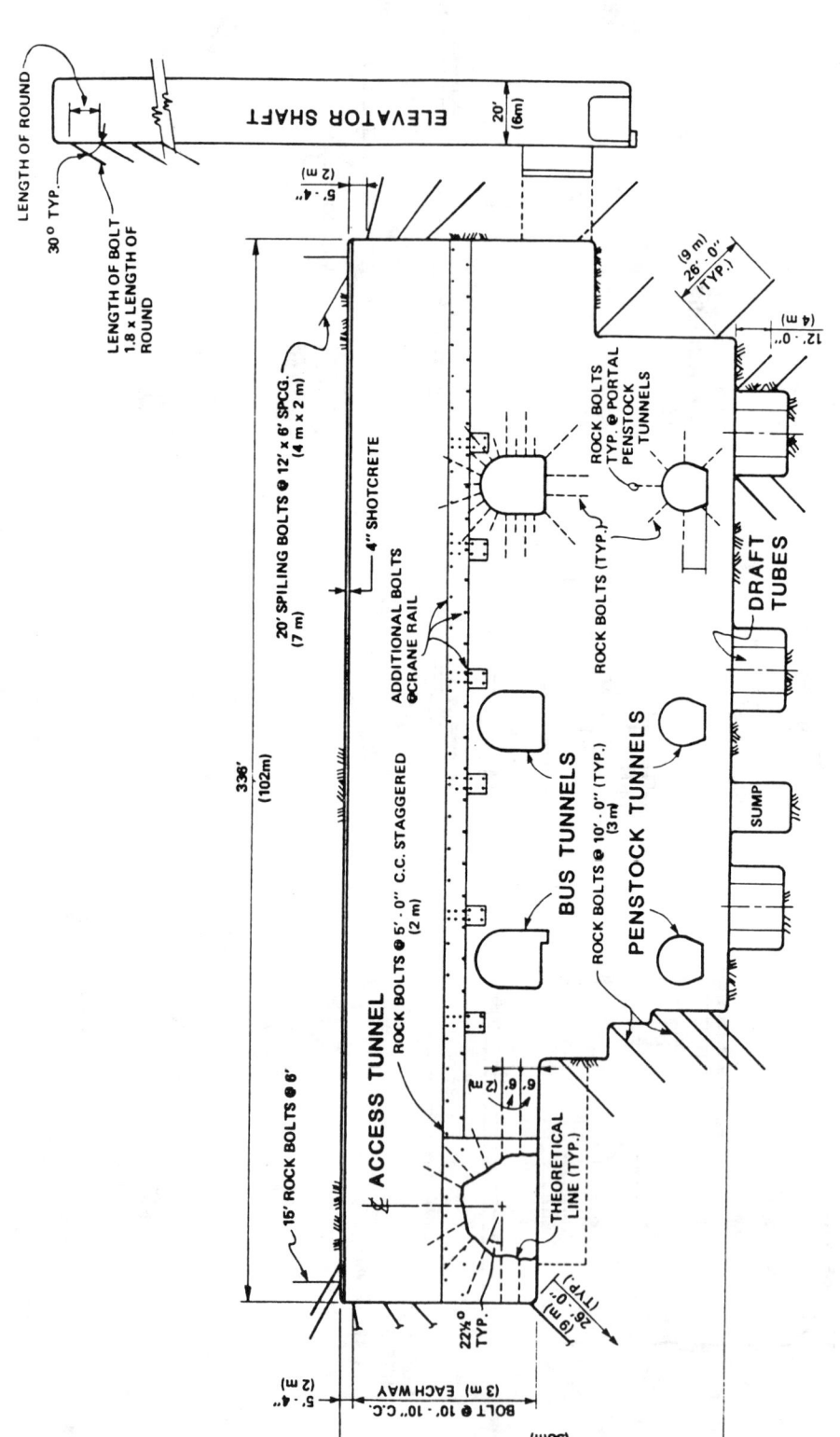

POWERHOUSE LONGITUDINAL SECTION
(EXCAVATION AND ROCK BOLTING)

FIGURE 4

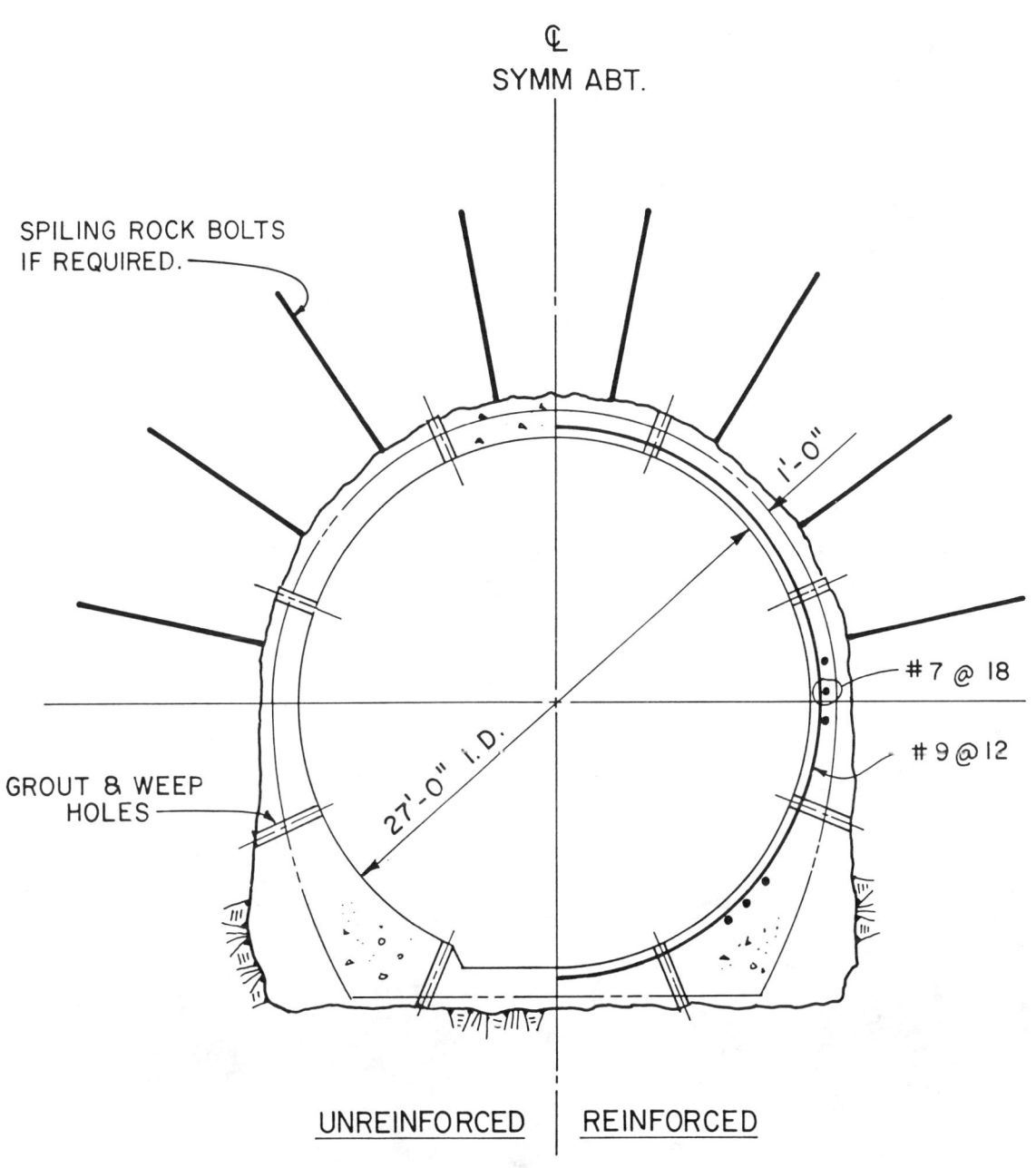

FIGURE 5

Underground Powerhouse with Reversible Units

Helms Pumped Storage Plant

Publications Related to the Helms Project

1) Tor L. Brekke, Arthur G. Strassburger, Jerry A. Davis, "Application of Some Novel Techniques to Establish Design Criteria for an Underground Pumped Storage Project," unpublished outline and notes presented at ASCE/SEAONC. New Concepts in Engineering, Oakland, California, April 1975.

2) B. C. Haimson, "Design of Underground Powerhouse and the Importance of Stress Measurements," Proceedings, 16th U.S. Symposium on Rock Mechanics, Minneapolis, Minnesota, September 1975.

3) D. F. Willoughby and H. J. Hovland, "Finite Element Analysis of Stages of Excavation of Helms Underground Powerhouse," Proceedings, 19th U.S. Symposium on Rock Mechanics, University of Nevada, Reno, Nevada, May 1978.

4) A. G. Strassburger, R. R. Friedrichs, J. A. Davis, "California Pumped-Storage Project Advances State of the Art in High-Head Tunneling," Tunneling Technology Newsletter, U.S. National Committee on Tunneling Technology, June 1978.

5) W. J. Tomei, "Environmental Effects of PG&E's Helms Project," ASCE, Portland, Oregon, April 14, 1980.

6) N. F. Sweeney, J. A. Davis, A. G. Strassburger, "A Practical/Economic Rock Mechanics Program," Proceedings, 21st U.S. Rock Mechanics Symposium, Salt Lake City, Utah, May 1980.

7) Norman F. Sweeney and H. John Hovland, "A Large Underground Opening-Measured Versus Predicted Rock Deformations," Proceedings of the Rapid Excavation and Tunneling Conference, AIME/ASCE, San Francisco, California, 1981.

8) Arthur G. Strassburger, "High-Head Underground Power Project Presents Design/Construction Challenges," Ibid.

9) Arthur G. Strassburger, "Helms Pumped Storage Project," U.S. Committee on Large Dams Newsletter, Issue #68, July 1982.

10) A. G. Strassburger, "Hydro-Fracturing at Helms," Water Power and Dam Construction, October 1982.

11) David W. Moller, Henry L. Minch and Joseph P. Welsh, "Ultrafine Cement Pressure Grouting to Control Ground Water in Fractured Granitic Rock at Helms Pumped Storage Project," American Concrete Institute, Kansas City, Missouri, September 1983.

12) J. A. Davis and A. G. Strassburger, "Helms Pumped Storage Project - Selection and Alternatives," Proceedings, Energy '84, ASCE, Pasadena, California, August 1984.

13) D. Zayakov, J. A. Davis and A. G. Strassburger, "Helms Pumped Storage Project - The Facilities," Ibid.

14) A. G. Strassburger, D. W. Moller and J. A. Davis, "Helms Pumped Storage Project - Design Through Construction," Ibid.

15) Anderson, S. F., The Helms Underground Storage Project Shaft Development, Proceedings of the Rapid Excavation Tunneling Conference, San Francisco, Califonria, May 1981.

Plant Name: **HIWASSEE-UNIT 2**

Plant location:
  Murphy, NC
  Cherokee County
	Owner: Tennessee Valley Authority
	  818 Power Building
	  Chattanooga, Tennessee 37401

Rated capacity 68 MW Average static head 190 ft Plant efficiency 76.90 % Stored energy 289,200 MWh Number of units 1	Designers:

Construction time: 1 year, 5 months
Construction cost: $93 per kW
Price level: 1955
First commercial power: May 1956
FERC project number: None
	Plant Manager/Superintendent:
	  D.M. Keith
	  Route 4
	  Murphy, North Carolina 28906
	
	  (704) 644-5121

River or water source: Hiwassee River at river mile 75.8	

DAM	UPPER RESERVOIR	LOWER RESERVOIR
Type	Concrete, Hiwassee Dam	Concrete
Height (ft)	307	150
Crest length (ft)	1,375	1,309
Volume (yd³)	7,200,000	4,055,000
RESERVOIR		
Type	On stream	On stream
Surface area (acres)	5,856	2,142
Usable power storage (acre-ft)	421,568	8,837
Power pool fluctuation (ft)	73.8	8.0
Operating levels		
Maximum (ft)	1,528.2	1,276.6
Minimum (ft)	1,450.5	1,272.3
Drainage area (miles²)	968.3	49.8
Seepage (ft³/s)		
SPILLWAY		
Design flood		
Return period (years)		
Flow (ft³/s)	129,990	156,590
Capacity (ft³/s)	129,990	156,590
Type	Concrete	Concrete
Gates		
Number	7	10
Type	Radial	Radial
Width (ft)	31.99	31.99
Height (ft)	23.00	23.00
OUTLET WORKS		
Discharge capacity (ft³/s)	22,990	
Number of water passages	4	None
Dimensions of water passages		
Height (ft)		
Width (ft)		
Diameter (ft)	8.50	
Type of gates	Ring seal	
Number of gates	4	None

Plant Name: **HIWASSEE-UNIT 2**

```
Plant location                              Owner:  Tennessee Valley Authority
   Murphy, NC                                       818 Power Building
   Cherokee County                                  Chattanooga, Tennessee  37401

Rated capacity              68 MW
Average static head         57.9 m
Plant efficiency            76.90 %           Designers:
Stored energy               289,200 MWh
Number of units             1

Construction time:  1 year, 5 months          Plant Manager/Superintendent:
Construction cost:  $93 per kW                   D.M. Keith
Price level:  1955                               Route 4
First commercial power:  May 1956                Murphy, North Carolina    28906
FERC project number:  None

                                                 (704) 644-5121

River or water source:  Hiwassee River at river mile 75.8
```

DAM	UPPER RESERVOIR	LOWER RESERVOIR
Type	Concrete, Hiwassee Dam	Concrete
Height (m)	93.6	45.7
Crest length (m)	419.0	399.0
Volume (m³)	5,505,000	3,100,000
RESERVOIR		
Type	On stream	On stream
Surface area (Mm²)	23.70	8.67
Usable power storage (Mm³)	520.000	10.900
Power pool fluctuation (m)	22.50	2.44
Operating levels		
Maximum (m)	465.80	389.10
Minimum (m)	442.10	387.80
Drainage area (Mm²)	2,508.000	129.000
Seepage (m³/s)		
SPILLWAY		
Design flood		
Return period (years)		
Flow (m³/s)	3,681	4,434
Capacity (m³/s)	3,681	4,434
Type	Concrete	Concrete
Gates		
Number	7	10
Type	Radial	Radial
Width (m)	9.750	9.750
Height (m)	7.010	7.010
OUTLET WORKS		
Discharge capacity (m³/s)	651	
Number of water passages	4	None
Dimensions of water passages		
Height (m)		
Width (m)		
Diameter (m)	2.590	
Type of gates	Ring seal	
Number of gates	4	None

Hiwassee-Unit 2 - Page 1 (Metric)

HIWASSEE-UNIT 2

INTAKES	UPPER INTAKE	LOWER INTAKE
Number	1	1
Type	Semicircular tower	Concrete draft tube
Design discharge (ft³/s)	3,700	4,300
Gross inlet area (ft²)	3,050	1,641
(at trash racks)		
Bar racks		
spacing (in)	6.00	6.00
shape	Rectangular	
depth/thickness (in)	6.00 / 1.00	3.00 / 0.63
diameter (in)		
Emergency gates		
number	2	None
height/width (ft)	26.00 / 19.00	
type	Roller train lift gate	
Service gates		
number	None	3
height/width (ft)		13.00 / 15.33
type		Slide
Bulkhead/stop logs(Y or N)		
number of units serviced		
Hoists		
number	2	1
capacity (tons)	60	25
type	Fixed hoist	Auxiliary

WATER PASSAGES	Upper Tunnel	Shaft	Lower Tunnel	Surge Tanks Upper	Surge Tanks Lower	Penstocks	Tailrace Tunnel
Number						1	
Diameter (ft)						18.0	
Length (ft)						190	
Maximum velocity (ft/s)						19.3	
Concrete lining thickness (in)							
Total length of concrete sections (ft)							
Steel liner Thickness							
Minimum (in)						0.75	
Maximum (in)						1.38	
Material grade						A10-34 mild steel	
Total length of steel-lined sections (ft)						190	

Notes:

Hiwassee-Unit 2 - Page 2 (English)

HIWASSEE-UNIT 2

INTAKES	UPPER INTAKE	LOWER INTAKE
Number	1	1
Type	Semicircular tower	Concrete draft tube
Design discharge (m³/s)	104.8	121.8
Gross inlet area (m²) (at trash racks)	283.4	152.5
Bar Racks:		
spacing (mm)	152	152
shape	Rectangular	
depth/thickness (mm)	152 / 25	76 / 16
diameter (mm)		
Emergency gates		
number	2	None
height/width (m)	7.925 / 5.791	
type	Roller train lift gate	
Service gates		
number	None	3
height/width (m)		3.962 / 4.674
type		Slide
Bulkhead/stop logs (Y or N)		
number of units serviced		
Hoists		
number	2	1
capacity (Mg)	54	23
type	Fixed hoist	Auxiliary

WATER PASSAGES	Upper Tunnel	Shaft	Lower Tunnel	Surge Tanks Upper	Lower	Penstocks	Tailrace Tunnel
Number						1	
Diameter (m)						5.49	
Length (m)						57.8	
Maximum velocity (m/s)						5.88	
Concrete lining thickness (m)							
Total length of concrete sections (m)							
Steel liner Thickness							
Minimum (mm)						19	
Maximum (mm)						35	
Material grade						A10-34 mild steel	
Total length of steel-lined sections (m)						57.8	

Notes:

HIWASSEE-UNIT 2

POWERHOUSE and RELATED FEATURES

Powerhouse Structure
Type: Outdoor
Length: 190 ft　　　　　Width: 90 ft　　　　　Height: 76 ft

Guard Valves
Number: None　　　　　Diameter: ft
Type:

Transformers
Number: 1
Ratings: 6.6
Voltages: (kV) 13.2 / 6.6

Generator
Rating generating (MVA): 59.7　　　　Rating pumping (MVA): 76.0
Insulation type:
Starting method: Direct control at powerhouse
Starting equipment:

Runners
Material: Cast steel
Minimum unit submergence: 1.0 ft
WR^2: 102,400,000 (lbf x ft²)
Manufacturer: Allis Chalmers
Model test by: Allis Chalmers

	Reversible Runners	Reversible Motor/Generator		
			Rotor	Stator
Number	1			1
Diameter (ft)	15.20			
rpm synchronous	105.9			105.9
rpm overspeed	161.0			121.0
Type	Francis		Enclosed vertical water-cooled	

Information on Runners

Condition:	Gross Head (ft)		Capacity (MW)		Discharge (ft³/s)		Turbine/Pump Eff.(%)	
	Generating	Pumping	Generating	Pumping	Generating	Pumping	Generating	Pumping
Maximum head & maximum power	255							
Minimum head & maximum power	173							

Condition:	Net Head (ft)		Capacity (MW)		Discharge (ft³/s)		Turbine/Pump Eff.(%)	
	Generating	Pumping	Generating	Pumping	Generating	Pumping	Generating	Pumping
Rated head @ best gate	190	205	60	76	4,180	3,900		

Hiwassee-Unit 2 - Page 3 (English)

HIWASSEE-UNIT 2

POWERHOUSE and RELATED FEATURES

Powerhouse Structure
Type: Outdoor
Length: 57.9 m   Width: 27.3 m   Height: 23.2 m

Guard Valves
Number: None   Diameter: m
Type:

Transformers
Number: 1
Ratings: 6.6
Voltages: (kV) 13.2 / 6.6

Generator
Rating generating (MVA): 59.7   Rating pumping (MVA): 76.0
Insulation type:
Starting method: Direct control at powerhouse
Starting equipment:

Runners
Material: Cast steel
Minimum unit submergence: 0.31 m
WR^2: 42,327,000 (Newtons x m^2)
Manufacturer: Allis Chalmers
Model test by: Allis Chalmers

	Reversible Runners	Reversible Motor/Generator		
Number	1			1
Diameter m	4.62	Rotor	Stator	
rpm synchronous	105.9			105.9
rpm overspeed	161.0			121.0
Type	Francis	Enclosed vertical water-cooled		

Information on Runners

Condition:	Gross Head (m)		Capacity (MW)		Discharge (m³/s)		Turbine/Pump Eff.(%)	
	Generating	Pumping	Generating	Pumping	Generating	Pumping	Generating	Pumping
Maximum head & maximum power	77.7							
Minimum head & maximum power	52.7							

Condition:	Net Head (m)		Capacity (MW)		Discharge (m³/s)		Turbine/Pump Eff.(%)	
	Generating	Pumping	Generating	Pumping	Generating	Pumping	Generating	Pumping
Rated head @ best gate	57.9	62.5	60	76	118.4	110.4		

Hiwassee-Unit 2 - Page 3 (Metric)

HIWASSEE-UNIT 2

Plant Data:
```
Average GWh generating per year:                    90
Average GWh pumping per year:                       58
Starting time from standstill (s):                 110
Changeover time pumping to generating (min):        20
Planned/scheduled time
   between major overhauls (years):
Outage time required per unit
   during major overhauls (weeks):
Representative plant availability (%):            94.0
Representative planned outages (weeks per year):     1
```

Miscelleneous Notes:

This hydropower plant has two units for generating but only one of the units is a pump-turbine.

The stored energy of 289,000 MWh is inclusive of normal hydropower generating operation. Generally, the water pumped in during a day must be generated out before any more generation takes place. We are not allowed to alter the lower pool's operation significantly.

The upper reservoir outlet works includes an emergency gate in addition to the four gates listed in the tables.

Cavitation Experience:

Cavitation has occurred to large areas on low pressure side.

Siqnificant or Unique Problems:

Thirteen main field pole connections failed between 1973 and 1991 (lack of flexibility). Cage winding connections failed in 1986.

Trashrack damaged during initial startup.

List of Licenses Required:

ENVIRONMENTAL FEATURES

Recreation:

Fish and Wildlife:

Social:

TECHNICAL DATA

GENERAL

Purpose	Flood control, power and recreation
Location—State/Nearest City	North Carolina/Murphy
River	Hiwassee
Drainage Basin Area	968 mi^2 (2,507 km^2)
Max. Recorded Flow	38,848 cfs (1,100 m^3/s)
Date of Completion	May 1940

RESERVOIR

Total Storage Capacity	434,000 ac-ft (535 Hm^3)
Active Storage Capacity	306,000 ac-ft (377 Hm^3)
Surface Area	6,230 acres (25 km^2)

DAM

Type	Straight, gravity-type, overflow dam
Height Above Foundation	307 ft (94 m)
Length of Crest	1,027 ft (313 m) nonoverflow
Volume of Dam	793,000 cy (606,328 m^3)

SPILLWAY

Type	Gated overflow crest, open channel chute; stilling basin with weir
Crest Length	260 ft (79 m)
Max. Discharge Capacity	130,000 cfs (3,680 m^3/s)

RIVER OUTLETS

Description	Four low-level conduits through dam, 102-in (2.6-m) diameter
Max. Discharge Capacity	23,000 cfs (651 m^3/s)

POWER FACILITIES

Power Conduit Description	18-ft (5.5-m) dia., 217.5 ft (66.3 m) and 189.8 ft (57.9 m) long
Power Plant Capacity	134,665 kW
Number and Type of Units	Unit 1—Vertical-shaft Francis; Unit 2—pump turbine
Rated Head	Unit 1—190 ft (57.9 m)
	Unit 2—Turbine 208 ft (63.4 m)/pump 205 ft (62.5 m)

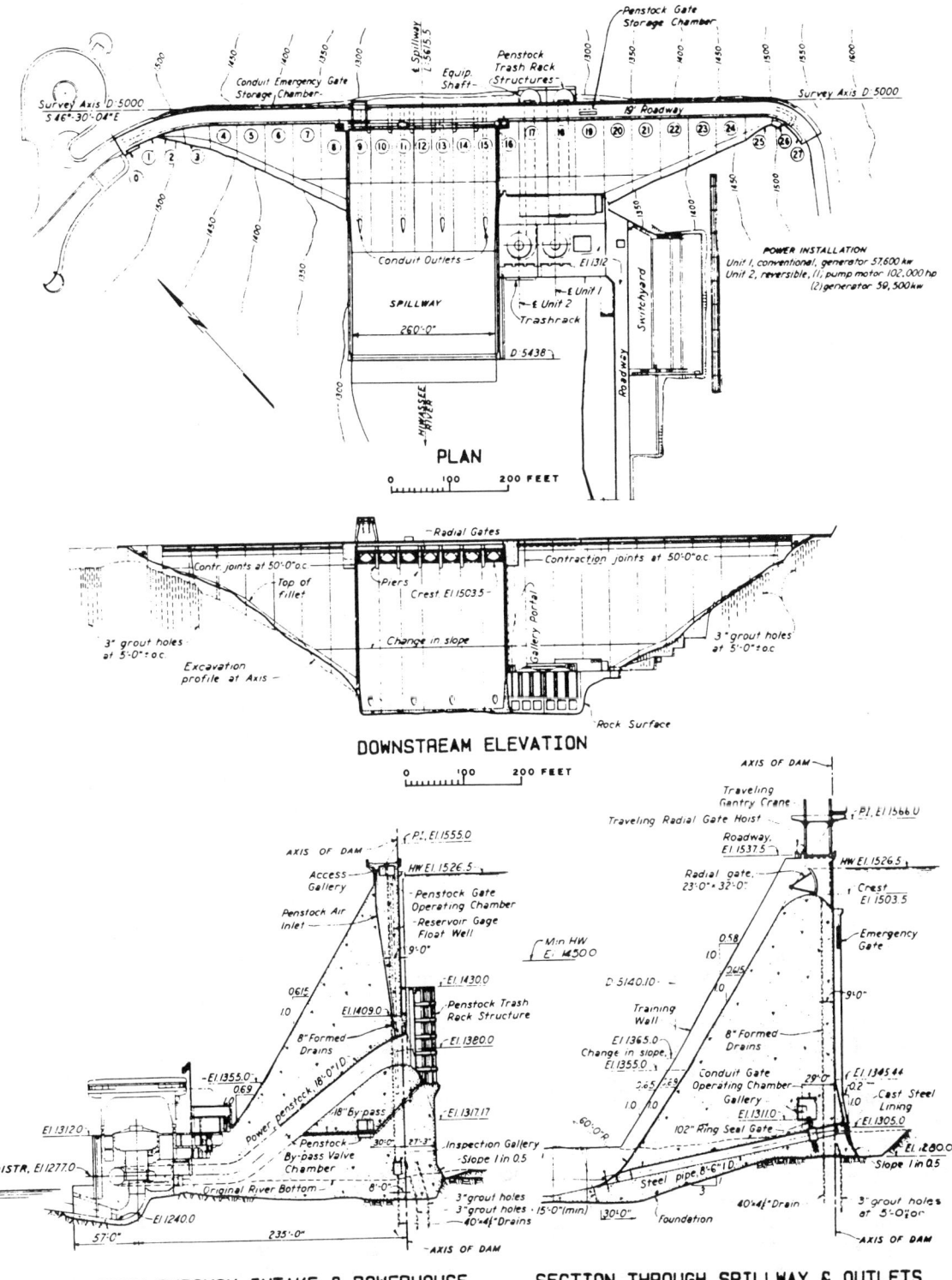

Location—The Hiwassee Dam is a straight concrete gravity dam located about 20 miles west of Murphy, North Carolina.

General Description—Hiwassee Dam rises 307 feet above the deepest bedrock and spans a deep gorge at Fowler Bend on the Hiwassee River. This 793,000-cy concrete structure is the key structure in the $110 million (1938 dollars for five dams) Hiwassee Valley Project, owned and operated by the Tennessee Valley Authority (TVA). The project develops the full potential in hydroelectric power and conservation of water of the Hiwassee River.

Background—Planning for the project began in 1934, construction started in 1936, and all construction was completed in 1943. The other principal functional units that comprise the Hiwassee Valley Project are Appalachia, Chatuge, Nottely, and Ocoee No. 3 Dams.

Purpose—The primary functions of the Hiwassee Valley Project include:

- Flood control
- Hydroelectric power generation
- Water conservation
- Recreation

SITE CONDITIONS

Geology—The damsite is located in the Wehutty Formation of the Great Smokey Group. The rock types that constitute the majority of the foundation are gneisses, mica quartzite, and schists. Overburden at the dam consisted of residual material and weathered rock which ranged from 0 to 35 feet in thickness, but averaged less than 7 feet.

Extensive geological investigations and subsurface drilling were used for evaluation of the geological characteristics of the site and the physical properties of the foundation rock. Generally the rock was of excellent quality and chemically stable, with compressive strengths averaging 1,500 to 35,000 psi.

Geologic Structures—The damsite is located on the southeast limb of a large overturned anticline with the rock strata striking from N45°E to N60°E and the dip ranging from 50° to 80° to the southeast.

Faults are numerous in the Hiwassee area in general, but relatively few were encountered in the excavation at the damsite. A few strike faults of southeast dip and indeterminable displacement were encountered in the river section of the foundation. These faults were perfectly tight features which showed no evidence of movement subsequent to the Appalachian orogeny.

Joints were extensively developed in the upper several feet of bedrock in the Hiwassee area and consisted of strike joints, dip joints, and sheet joints. Sheet joints in the river section formed badly decayed rock seams which resulted in the more serious foundation problems. These seams were either excavated or grouted to form a stable foundation.

Seismicity—The damsite is located in the Southern Appalachian tectonic province. The original design of the dam did not include seismic loading. The concrete structure is similar to other dams that have been analyzed and judged adequate to resist earthquake forces equal to or greater than a Maximum Credible Earthquake with a peak acceleration of 0.15 g; consequently, no siesmic analysis was performed.

Hydrology—The Hiwassee River system drains an area of 2,700 square miles. The drainage area begins on the northwest slope of the Blue Ridge Mountains in northern Georgia, at elevations ranging from 2,500 to 4,500 feet above sea level, and empties into the Tennessee River 35 miles northeast of Chattanooga, Tennessee. The watershed area receives an average annual rainfall of 58 inches.

The average annual system discharge at Hiwassee Dam is 2,200 cfs, corresponding to runoff from the basin of 30.6 inches.

Current national and TVA practice for computing design floods for dams since the 1960s is to determine project design flow using the hydrometeorological approach. For dams such as Hiwassee, whose failure would cause a large loss of life and significant economic losses, the dam would be designed to pass the probable maximum flood (PMF).

The PMF for Hiwassee would be a storm producing 28.9 inches of rainfall in three days. This storm would be preceded by a three-day storm producing 8.7 inches of rainfall, which would end three days prior to the start of the main storm.

DESCRIPTION OF MAIN FEATURES

Nonoverflow Dam—The nonoverflow dam is a straight gravity structure with a total crest length of 1,027 linear feet. It is divided into two parts, one on the right (north) side of the spillway and the other on the left (south) side of the powerhouse intake. The nonoverflow sections are built on the steep slopes of the rock abutments and are simple gravity sections with heavy fillets along the toe and heel. A concrete cutoff wall was built into the spur ridge, forming a continuation of the north nonoverflow section.

The typical cross section for the nonoverflow dam was selected with the basic considerations to keep the resultant from overturning forces within the middle third and limiting the maximum stresses at critical locations.

The stresses in the dam were obtained for three loading conditions.

Load Case I: Dead load only

Load Case II: With headwater at El. 1,532, corresponding to PMF

Load Case III: With headwater up to El. 1,537, top of roadway on dam

Main Spillway—The concrete structure containing the spillway is a gravity dam 602 feet long, forming an integral part of the dam. The spillway is centered about 70 feet north, or toward the right abutment from the midpoint of the dam, and in approximately the same relation to the centerline of the original riverbed. Flow over the spillway is confined between training walls and guided to the apron and energy dissipating pool at the toe of the spillway. The spillway consists of seven 32-ft wide ogee overfall sections at El. 1,503.5, which is approximately 235 feet above the original riverbed.

The discharge over the spillway is controlled by seven radial gates 32 feet wide by 23 feet high. The main spillway capacity is 130,000 cfs with the reservoir at its maximum elevation of 1,532 feet. The top of the concrete was set 5.5 feet above maximum headwater elevation.

Sluiceways—The primary function of the sluiceways is to provide a means for drawing the upper pool down to levels below spillway crest El. 1,503.5. There are four 102-in diameter ring seal gates located directly below the spillway gates. The centerline of the inlet to the sluiceways is located at El. 1,305. The discharge capacity of the sluices is 23,000 cfs with a headwater elevation of 1,532 feet. The discharge velocity under the above conditions is about 90 feet per second.

Power Facilities—The power facilities comprise the steel-lined conduits through the nonoverflow dam, the powerhouse located on the toe of the dam, a tailrace channel, and the switchyard. All waterways and structures were built to accommodate two units.

The first unit was installed in 1940 during initial construction. The waterwheel is a vertical shaft Francis turbine rated at 80,000 hp at a 190-ft head. The second unit, added in 1956, is a reversible pump turbine unit. The waterwheel is a vertical shaft Francis turbine rated at 102,000 hp at a 208-ft head. When operating as a pump, the reversible unit is rated at 1,750,000 gpm against a head of 205 feet. At this head, the pump will require a motor output of 102,000 hp.

CONSTRUCTION

Diversion—The construction of the dam across the riverbed was by means of three cofferdam stages. The cofferdams were constructed to be overtopped, if necessary, by the river at flood stage. The cofferdams were designed only high enough to take care of normal high water. The rocky bed of the river favored rock-filled timber crib construction.

Foundation Treatment—The foundation treatment consisted principally of the four following procedures:

1. Development of a watertight curtain just downstream from the axis of the dam by drilling and grouting groups of wagon-drill and diamond-drill holes, using various depths of drilling and various group pressures to suit groups of holes. The spacing of these holes was generally 2.5 or 5 feet.

2. General blanket or consolidation grouting over the entire foundation, with holes drilled at sufficient intervals in both directions as to give adequate consolidation.

3. Drilling of 36-in core holes at some points to inspect foundation rock or to open, clean, and grout weathered seams in the rock.

4. Development of a complete system of foundation drains by drilling a large number of holes just downstream from the dam axis along the full length of the dam, and draining them to tailwater by a system of wood-box drains and a sump pump.

Materials and Methods—All coarse and fine aggregates were manufactured from rock (graywacke) found in the area of the dam. The plant was located at a quarry 2,500 feet downstream of the dam. The rock was hauled from the quarry in Hug rear-dump, boulder-type quarry trucks to feed the primary crusher. The rock from the primary crusher was stockpiled and then reclaimed when needed. The aggregate was conveyed from the stockpile to other crushers as required and ended up at the concrete mixing plant.

The batching and mixing plant, located on the right bank of the river, was equipped with three 3-cy mixers with a designed output of 180 cubic yards per hour. The concrete was transported from the mixing plant to the forms by transfer trains and cableway. The trains hauled the concrete from the mixing plant to a transfer trestle where the mix was dumped into buckets and then moved by cableway to the point of deposit in the forms.

Dam Construction—Concrete placing started April 20, 1938, and at completion on January 31, 1940 a total of 793,000 cy of concrete was placed in the structure.

Project Cost and Financing—The original total cost for the Hiwassee Project was $16,844,000, but with the addition of Unit 2 in 1956, the total cost was increased to $23,200,000. Funding was provided by appropriations from the U.S. Congress.

PERFORMANCE

Instrumentation and Monitored Behavior—For the purpose of studying the behavior of the dam during and after construction, an extensive program of physical measurements and observations was initiated during construction. The dam was instrumented with numerous measuring devices consisting of strain and stress meters, resistance thermometers, joint meters, uplift and pore pressure cells, and plumb lines. The plan involved measurement of internal temperatures, strains and stresses, deflection of the structure, uplift pressures on the base and in the concrete pores, opening or closing of the contraction joints, and the extent of cracking.

The uplift data indicates that fluctuations in intensities at the individual pressure cells have been consistent for many years. Uplift readings are still well within design values. Recent problems caused by concrete growth, due to alkali-aggregate reaction, have warranted installation of crack gauges and surveys to monitor movement of the spillway piers. Plumb lines in the central spillway portion of the dam have shown an upstream movement over the years due to the concrete growth reaction.

Incidents and Repairs—The overall condition of the concrete in the dam is good, with the exception of cracking in several of the curved right and left nonoverflow blocks. This cracking, first observed in 1964, led to a petrographic study of concrete cores from the dam which detected the presence of an alkali-silica concrete growth reaction. The observed cracks in the nonoverflow blocks are $1/16$-inch to $1/2$-inch wide and for the most part appear to follow horizontal construction joints in a step-like fashion. Gauges have been installed to monitor these cracks and generally show an opening trend. Another apparent result of the concrete growth problem has been the incidence of spillway gate binding. Overcoring has been done periodically at this project to evaluate the magnitude and changes in stresses. No repairs have been required to date except trimming the skin plate of the spillway gates.

ACKNOWLEDGEMENTS

This project is owned and operated by the Tennessee Valley Authority, which was responsible for all phases of engineering and construction.

Hiwassee Dam

Hiwassee-Unit #2 Pumped Storage Plant

Plant Name: **HORSE MESA #4**

Plant location:
 Tortilla Flat, AZ
 Maricopa County

Rated capacity	97 MW
Average static head	247 ft
Plant efficiency	67.00 %
Stored energy	23,600 MWh
Number of units	1

Construction time: 3 years
Construction cost: $159 per kW
Price level: 1988
First commercial power: January 1972
FERC project number:

River or water source: Salt River

Owner: Bureau of Reclamation
 Attn: D-3000
 P.O. Box 25007
 Denver, Colorado 80225

Designers:
 Salt River Project
 Bechtel Corporation

Plant Manager/Superintendent:
 Salt River Project
 P.O. Box 52025
 Pheonix, Arizona 85072-2025

 (602) 236-3854

DAM	UPPER RESERVOIR	LOWER RESERVOIR
Type	Existing concrete thin arch dam	Existing concrete thin arch dam.
Height (ft)	305	224
Crest length (ft)	660	380
Volume (yd³)	162,000	125,000
RESERVOIR		
Type	Existing Apache Lake constructed	Existing Canyon Lake, constructed
Surface area (acres)	2,669	939
Usable power storage (acre-ft)	116,985	25,699
Power pool fluctuation (ft)	18.0	5.5
Operating levels		
Maximum (ft)	1,913.5	1,660.9
Minimum (ft)	1,896.7	1,655.4
Drainage area (miles²)	0.0	
Seepage (ft³/s)		
SPILLWAY		
Design flood		
Return period (years)		
Flow (ft³/s)		
Capacity (ft³/s)	150,090	150,090
Type	One overflow and one tunnel spillway	Chute spillway
Gates		
Number	9 + 1	2
Type	9 radial & 1 fixed wheel	Fixed wheel
Width (ft)	25.92	49.87
Height (ft)	22.97	49.87
OUTLET WORKS		
Discharge capacity (ft³/s)		
Number of water passages	None	
Dimensions of water passages		
Height (ft)		
Width (ft)		
Diameter (ft)		
Type of gates		
Number of gates	None	

Horse Mesa #4 - Page 1 (English)

Plant Name: **HORSE MESA #4**

Plant location Tortilla Flat, AZ Maricopa County	Owner: Bureau of Reclamation Attn: D-3000 P.O. Box 25007 Denver, Colorado 80225

Rated capacity   97 MW
Average static head   75.3 m
Plant efficiency   67.00 %
Stored energy   23,600 MWh
Number of units   1
	Designers:
	   Salt River Project
	   Bechtel Corporation

Construction time: 3 years
Construction cost: $159 per kW
Price level: 1988
First commercial power: January 1972
FERC project number:
	Plant Manager/Superintendent:
	   Salt River Project
	   P.O. Box 52025
	   Pheonix, Arizona 85072-2025
	
	   (602) 236-3854

River or water source: Salt River	

DAM	UPPER RESERVOIR	LOWER RESERVOIR
Type	Existing concrete thin arch dam	Existing concrete thin arch dam.
Height (m)	93.0	68.3
Crest length (m)	201.2	115.8
Volume (m³)	123,858	95,797
RESERVOIR		
Type	Existing Apache Lake constructed	Existing Canyon Lake, constructed
Surface area (Mm²)	10.80	3.80
Usable power storage (Mm³)	144.300	31.700
Power pool fluctuation (m)	5.49	1.68
Operating levels		
Maximum (m)	583.25	506.25
Minimum (m)	578.10	504.57
Drainage area (Mm²)		
Seepage (m³/s)		
SPILLWAY		
Design flood		
Return period (years)		
Flow (m³/s)		
Capacity (m³/s)	4,250	4,250
Type	One overflow and one tunnel spillway	Chute spillway
Gates		
Number	9 + 1	2
Type	9 radial & 1 fixed wheel	Fixed wheel
•Width (m)	7.900	15.200
Height (m)	7.000	15.200
OUTLET WORKS		
Discharge capacity (m³/s)		
Number of water passages	None	
Dimensions of water passages		
Height (m)		
Width (m)		
Diameter (m)		
Type of gates		
Number of gates	None	

HORSE MESA #4

INTAKES	UPPER INTAKE	LOWER INTAKE
Number	1	1
Type	Penstock intake	Draft tube
Design discharge (ft³/s)		
Gross inlet area (ft²)	715	723
(at trash racks)		
Bar racks		
spacing (in)	7.25	7.25
shape		Rectangular
depth/thickness (in)	3.50 / 0.75	3.50 / 0.75
diameter (in)		
Emergency gates		
number	1	None
height/width (ft)	20.34 / 17.06	
type	Fixed wheel	N/A
Service gates		
number	None	None
height/width (ft)		
type	N/A	N/A
Bulkhead/stop logs(Y or N)	Y	Y
number of units serviced	1	3
Hoists		
number	1	3
capacity (tons)	248	10
type	Hydraulic	Electric

WATER PASSAGES	Upper Tunnel	Shaft	Lower Tunnel	Surge Tanks Upper	Lower	Penstocks	Tailrace Tunnel
Number						1	
Diameter (ft)						15.4	
Length (ft)						187	
Maximum velocity (ft/s)						30.2	
Concrete lining thickness (in)							
Total length of concrete sections (ft)							
Steel liner Thickness							
Minimum (in)						0.75	
Maximum (in)						1.00	
Material grade						A-36	
Total length of steel-lined sections (ft)						187	

Notes:

Horse Mesa #4 - Page 2 (English)

HORSE MESA #4

INTAKES	UPPER INTAKE	LOWER INTAKE
Number	1	1
Type	Penstock intake	Draft tube
Design discharge (m³/s)		
Gross inlet area (m²)	66.4	67.2
(at trash racks)		
Bar Racks:		
spacing (mm)	184	184
shape		Rectangular
depth/thickness (mm)	89 / 19	89 / 19
diameter (mm)		
Emergency gates		
number	1	None
height/width (m)	6.200 / 5.200	
type	Fixed wheel	N/A
Service gates		
number	None	None
height/width (m)		
type	N/A	N/A
Bulkhead/stop logs (Y or N)	Y	Y
number of units serviced	1	3
Hoists		
number	1	3
capacity (Mg)	225	9
type	Hydraulic	Electric

WATER PASSAGES	Upper Tunnel	Shaft	Lower Tunnel	Surge Tanks Upper	Lower	Penstocks	Tailrace Tunnel
Number						1	
Diameter (m)						4.70	
Length (m)						57.1	
Maximum velocity (m/s)						9.20	
Concrete lining thickness (m)							
Total length of concrete sections (m)							
Steel liner							
Thickness							
Minimum (mm)						19	
Maximum (mm)						25	
Material grade						A-36	
Total length of steel-lined sections (m)						57.1	

Notes:

HORSE MESA #4

POWERHOUSE and RELATED FEATURES

Powerhouse Structure
Type: Semi-outdoor concrete
Length: 83 ft   Width: 75 ft   Height: 93 ft

Guard Valves
Number: None   Diameter: ft
Type:

Transformers
Number: 1
Ratings: 112
Voltages: (kV) 115 / 13.2

Generator
Rating generating (MVA): 96.5   Rating pumping (MVA): 84.0
Insulation type:
Starting method: Pony motor
Starting equipment: 10,000 HP motor

Runners
Material: Cast steel
Minimum unit submergence: 20.4 ft
WR^2:
Manufacturer: Baldwin-Lima-Hamilton
Model test by: Baldwin-Lima-Hamilton

	Reversible Runners	Reversible Motor/Generator		
Number	1			1
Diameter (ft)	18.00	Rotor	Stator	25.92
rpm synchronous	150.0			150.0
rpm overspeed	187.5			187.5
Type	Francis	Synchronous, umbrella type		

Information on Runners

Condition:	Gross Head (ft) Generating	Gross Head (ft) Pumping	Capacity (MW) Generating	Capacity (MW) Pumping	Discharge (ft³/s) Generating	Discharge (ft³/s) Pumping	Turbine/Pump Eff.(%) Generating	Turbine/Pump Eff.(%) Pumping
Maximum head & maximum power	258	258	97					
Minimum head & maximum power	236	236			5,650	3,800		

Condition:	Net Head (ft) Generating	Net Head (ft) Pumping	Capacity (MW) Generating	Capacity (MW) Pumping	Discharge (ft³/s) Generating	Discharge (ft³/s) Pumping	Turbine/Pump Eff.(%) Generating	Turbine/Pump Eff.(%) Pumping
Rated head @ best gate	246	260	80			3,740		

HORSE MESA #4

POWERHOUSE and RELATED FEATURES

Powerhouse Structure
Type: Semi-outdoor concrete
Length: 25.3 m   Width: 22.9 m   Height: 28.3 m

Guard Valves
Number: None   Diameter: m
Type:

Transformers
Number: 1
Ratings: 112
Voltages: (kV) 115 / 13.2

Generator
Rating generating (MVA): 96.5   Rating pumping (MVA): 84.0
Insulation type:
Starting method: Pony motor
Starting equipment: 10,000 HP motor

Runners
Material: Cast steel
Minimum unit submergence: 6.22 m
WR^2:
Manufacturer: Baldwin-Lima-Hamilton
Model test by: Baldwin-Lima-Hamilton

	Reversible Runners	Reversible Motor/Generator		
Number	1			1
Diameter m	5.50	Rotor	Stator	7.900
rpm synchronous	150.0			150.0
rpm overspeed	187.5			187.5
Type	Francis	Synchronous, umbrella type		

Information on Runners

Condition:	Gross Head (m) Generating	Gross Head (m) Pumping	Capacity (MW) Generating	Capacity (MW) Pumping	Discharge (m³/s) Generating	Discharge (m³/s) Pumping	Turbine/Pump Eff.(%) Generating	Turbine/Pump Eff.(%) Pumping
Maximum head & maximum power	78.7	78.7	97					
Minimum head & maximum power	71.8	71.8			160.0	107.6		

Condition:	Net Head (m) Generating	Net Head (m) Pumping	Capacity (MW) Generating	Capacity (MW) Pumping	Discharge (m³/s) Generating	Discharge (m³/s) Pumping	Turbine/Pump Eff.(%) Generating	Turbine/Pump Eff.(%) Pumping
Rated head @ best gate	75.1	79.4	80			105.9		

Horse Mesa #4 - Page 3 (Metric)

HORSE MESA #4

Plant Data:
  Average GWh generating per year:
  Average GWh pumping per year:
  Starting time from standstill (s):
  Changeover time pumping to generating (min):
  Planned/scheduled time
    between major overhauls (years): 10
  Outage time required per unit
    during major overhauls (weeks): 9
  Representative plant availability (%): 87.3
  Representative planned outages (weeks per year): 2

Miscelleneous Notes:
  Horse Mesa Dam incorporates three conventional 11-MW units and one reversible 96.5-MW unit.

  The spillway capacity of the project is based on upstream regulation and discharge-capacity requirements, and it is not related to the pumped storage operation.

  Information provided on availability and unit outages are based on average values over the years from 1982 to 1989.

  Spillway gates are of two types, 9 radial gates and 1 fixed-wheel gate. The radial gates are 7.9 m (26 ft) wide by 7.0 m (23 ft) high. The fixed-wheel gate is 12.2 m (40 ft) wide by 13.6 m (45 ft) high.

Cavitation Experience:
  A total of 250 pounds of metal loss occurs per year.

Significant or Unique Problems:

List of Licenses Required:

ENVIRONMENTAL FEATURES

Recreation:
  Recreation opportunities include fishing, waterskiing, and camping at the campgrounds.

Fish and Wildlife:

Social:

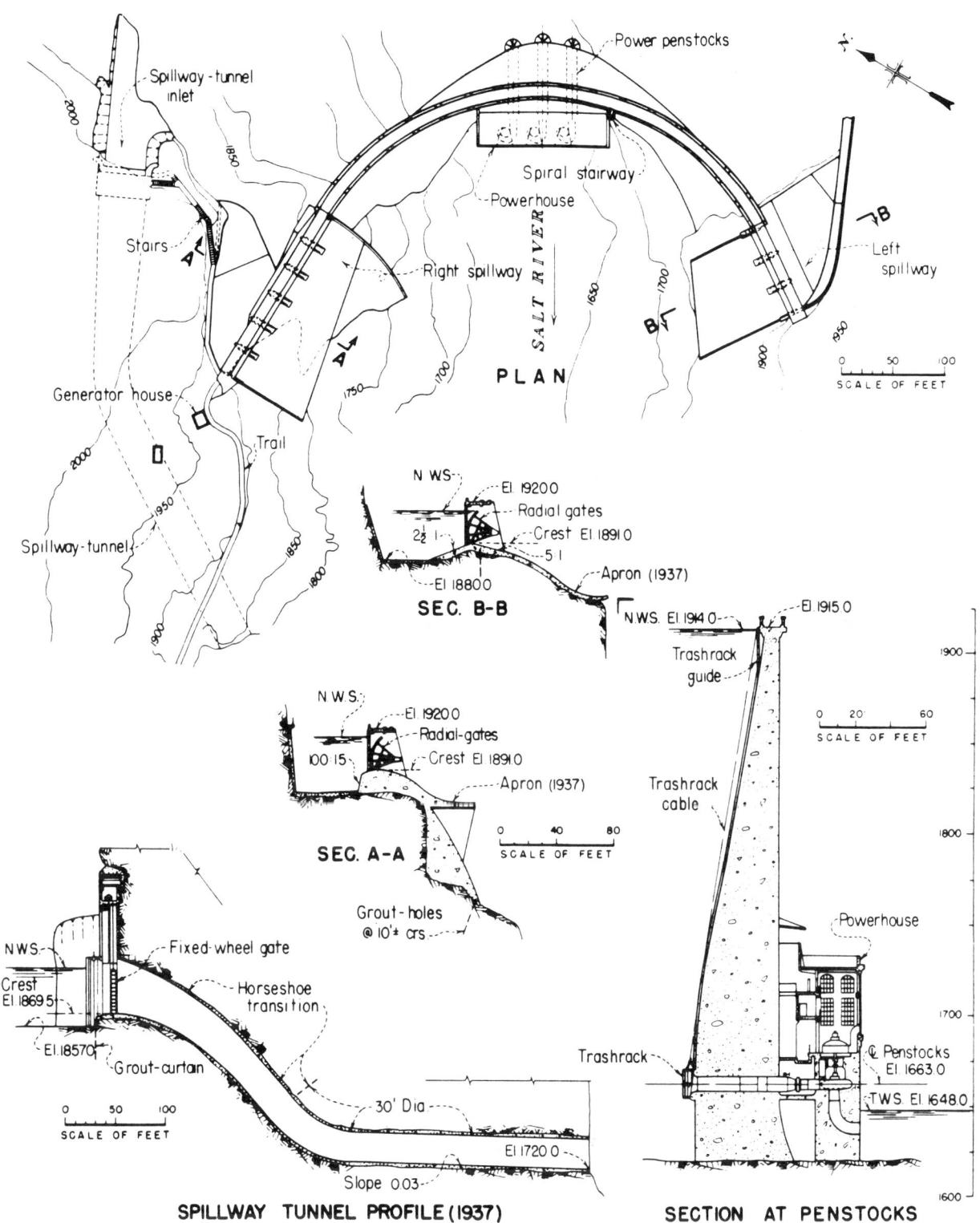

Horse Mesa Dam and Powerplant, Plan and Sections

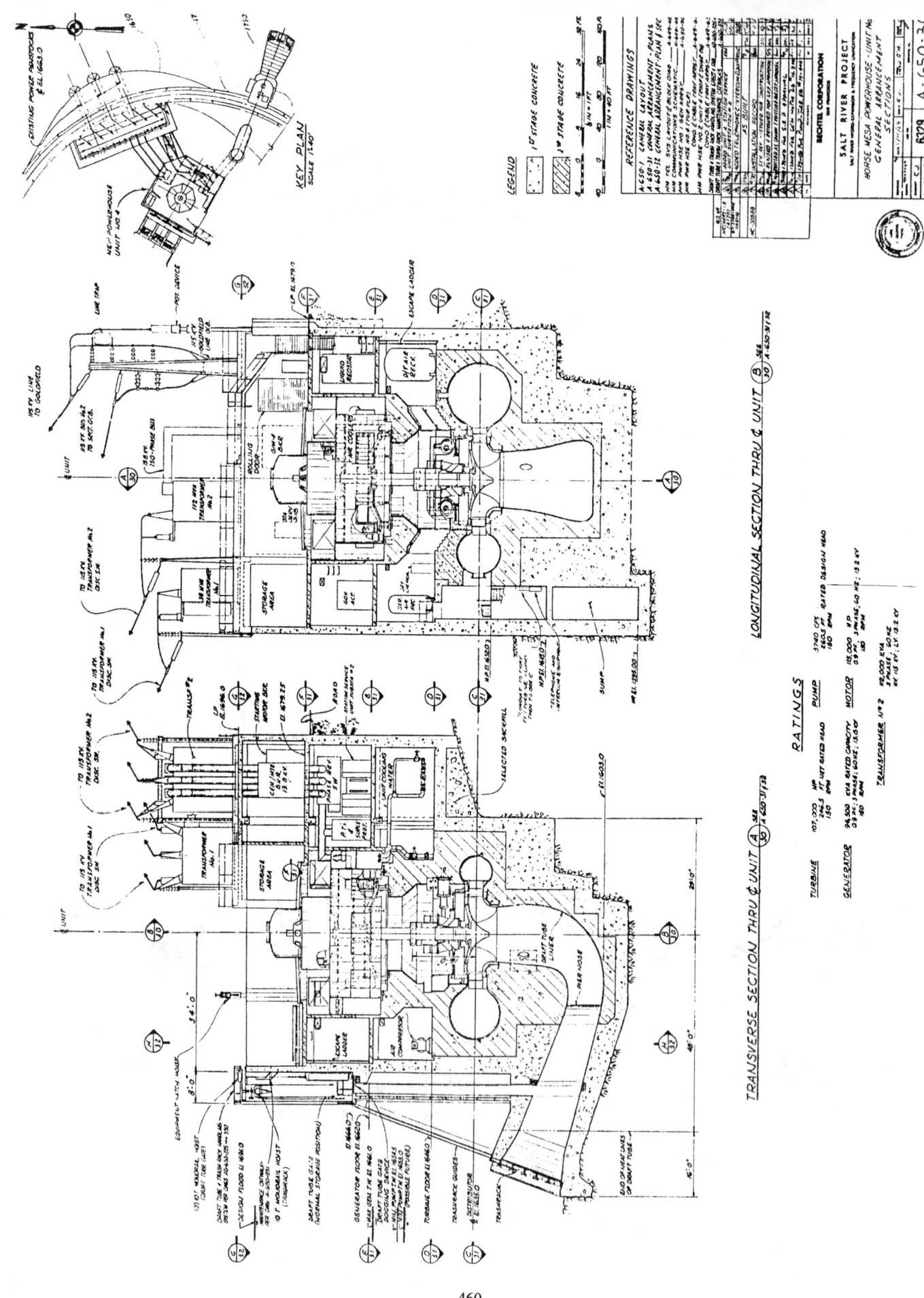

Plant Name: **JOCASSEE PUMPED STORAGE STATION**

Plant location:
  Salem, SC
  Pickens County

Rated capacity	610 MW
Average static head	325 ft
Plant efficiency	80.00 %
Stored energy	57,544 MWh
Number of units	4

Construction time: 6 years, 11 months
Construction cost: $164 per kW
Price level: 1973
First commercial power: December 1973
FERC project number: 2503

Owner: Duke Power Company
  P.O. Box 1006
  Charlotte, North Carolina 28201-1006

Designers:
  Duke Power Company, Design Engineering
  Chas. T. Main, Consultants

Plant Manager/Superintendent:
  Richard R. Miller
  HC-76 Box 170
  Salem, South Carolina 29676

  (803) 944-1455

River or water source: Whitewater, Thompson, and Keowee Rivers

	UPPER RESERVOIR	LOWER RESERVOIR
DAM		
Type	Zoned fill, Jocassee Dam	Homogenous earthfill, Keowee Dam Station
Height (ft)	385	170
Crest length (ft)	1,750	3,500
Volume (yd³)	13,000,000	2,043,000
RESERVOIR		
Type	Constructed, Lake Jocassee	Constructed, Lake Keowee
Surface area (acres)	7,565	18,372
Usable power storage (acre-ft)	215,700	391,700
Power pool fluctuation (ft)	30.0	25.0
Operating levels		
Maximum (ft)	1,110.0	800.0
Minimum (ft)	1,080.0	775.0
Drainage area (miles²)	148.0	439.0
Seepage (ft³/s)	3.390	1.330
SPILLWAY		
Design flood		
Return period (years)		
Flow (ft³/s)		
Capacity (ft³/s)	46,000	98,000
Type	Concrete chute	Concrete chute
Gates		
Number	2	4
Type	Taintor	Taintor
Width (ft)	38.00	38.00
Height (ft)	33.00	35.00
OUTLET WORKS		
Discharge capacity (ft³/s)		
Number of water passages	None	None
Dimensions of water passages		
Height (ft)		
Width (ft)		
Diameter (ft)		
Type of gates		
Number of gates	None	None

Plant Name: JOCASSEE PUMPED STORAGE STATION

Plant location Salem, SC Pickens County	Owner: Duke Power Company P.O. Box 1006 Charlotte, North Carolina 28201-1006
Rated capacity 610 MW Average static head 99.1 m Plant efficiency 80.00 % Stored energy 57,544 MWh Number of units 4	Designers: Duke Power Company, Design Engineering Chas. T. Main, Consultants
Construction time: 6 years, 11 months Construction cost: $164 per kW Price level: 1973 First commercial power: December 1973 FERC project number: 2503	Plant Manager/Superintendent: Richard R. Miller HC-76 Box 170 Salem, South Carolina 29676 (803) 944-1455
River or water source: Whitewater, Thompson, and Keowee Rivers	

	UPPER RESERVOIR	LOWER RESERVOIR
DAM		
Type	Zoned fill, Jocassee Dam	Homogenous earthfill, Keowee Dam Station
Height (m)	117.3	51.8
Crest length (m)	533.4	1,066.8
Volume (m³)	9,939,218	1,561,986
RESERVOIR		
Type	Constructed, Lake Jocassee	Constructed, Lake Keowee
Surface area (Mm²)	30.62	74.35
Usable power storage (Mm³)	266.064	483.158
Power pool fluctuation (m)	9.14	7.62
Operating levels		
Maximum (m)	338.33	243.84
Minimum (m)	329.18	236.22
Drainage area (Mm²)	383.320	1,137.011
Seepage (m³/s)	0.0960	0.0377
SPILLWAY		
Design flood		
Return period (years)		
Flow (m³/s)		
Capacity (m³/s)	1,303	2,775
Type	Concrete chute	Concrete chute
Gates		
Number	2	4
Type	Taintor	Taintor
Width (m)	11.582	11.582
Height (m)	10.058	10.668
OUTLET WORKS		
Discharge capacity (m³/s)		
Number of water passages	None	None
Dimensions of water passages		
Height (m)		
Width (m)		
Diameter (m)		
Type of gates		
Number of gates	None	None

Jocassee Pumped Storage Station - Page 1 (Metric)

JOCASSEE PUMPED STORAGE STATION

INTAKES	UPPER INTAKE	LOWER INTAKE
Number	2	4
Type	Circular, 8 sided	Rectangular
Design discharge (ft³/s)	7,115	
Gross inlet area (ft²) (at trash racks)	3,840	4,122
Bar racks		
spacing (in)	6.50	6.50
shape	Rectangular	Circular
depth/thickness (in)	4.00 / 0.75	
diameter (in)		1.50
Emergency gates		
number	24	None
height/width (ft)	9.50 / 19.00	
type	Stop logs (steel)	
Service gates		
number	2	6
height/width (ft)	27.50 / 39.63	22.00 / 14.00
type	Cylindrical	Rectangular
Bulkhead/stop logs (Y or N)	Y	Y
number of units serviced	2	2
Hoists		
number	2	1
capacity (tons)	120	15
type	120-T fixed hoist, 20-T rotating hoist	Gantry

WATER PASSAGES	Upper Tunnel	Shaft	Lower Tunnel	Surge Tanks Upper	Surge Tanks Lower	Penstocks	Tailrace Tunnel
Number		2	2			4	
Diameter (ft)		33.5	33.5			20.5	
Length (ft)		276	1,200			160	
Maximum velocity (ft/s)		18.0	18.0				
Concrete lining thickness (in)		12.00	12.00				
Total length of concrete sections (ft)		276	1,200				
Steel liner							
Thickness							
Minimum (in)						0.96	
Maximum (in)						1.20	
Material grade						A-516	
Total length of steel-lined sections (ft)						266	

Notes:
Penstocks: Diameter and length are average values.

Jocassee Pumped Storage Station - Page 2 (English)

JOCASSEE PUMPED STORAGE STATION

INTAKES	UPPER INTAKE	LOWER INTAKE
Number	2	4
Type	Circular, 8 sided	Rectangular
Design discharge (m³/s)	201.5	
Gross inlet area (m²) (at trash racks)	356.7	382.9
Bar Racks:		
spacing (mm)	165	165
shape	Rectangular	Circular
depth/thickness (mm)	102 / 19	
diameter (mm)		38
Emergency gates		
number	24	None
height/width (m)	2.896 / 5.791	
type	Stop logs (steel)	
Service gates		
number	2	6
height/width (m)	8.382 / 12.078	6.706 / 4.267
type	Cylindrical	Rectangular
Bulkhead/stop logs (Y or N)	Y	Y
number of units serviced	2	2
Hoists		
number	2	1
capacity (Mg)	109	14
type	120-T fixed hoist, 20-T rotating hoist	Gantry

WATER PASSAGES	Upper Tunnel	Shaft	Lower Tunnel	Surge Tanks Upper	Surge Tanks Lower	Penstocks	Tailrace Tunnel
Number		2	2			4	
Diameter (m)		10.21	10.21			6.25	
Length (m)		84.1	365.8			48.8	
Maximum velocity (m/s)		5.49	5.49				
Concrete lining thickness (m)		0.305	0.305				
Total length of concrete sections (m)		84.1	365.8				
Steel liner Thickness							
Minimum (mm)						24	
Maximum (mm)						31	
Material grade						A-516	
Total length of steel-lined sections (m)						81.1	

Notes:
Penstocks: Diameter and length are average values.

Jocassee Pumped Storage Station - Page 2 (Metric)

JOCASSEE PUMPED STORAGE STATION
POWERHOUSE and RELATED FEATURES

Powerhouse Structure
Type: Aboveground, concrete, outdoor
Length: 360 ft   Width: 131 ft   Height: 59 ft

Guard Valves
Number: None   Diameter: ft
Type:

Transformers
Number: 4 West. Class FOA
Ratings: 192
Voltages: (kV) 230 / 14.4

Generator
Rating generating (MVA): 170.0   Rating pumping (MVA): 168.0
Insulation type: Thermolastic Class B NEMA code BBBXV
Starting method: 4 units synchronous; 2 units pony motor
Starting equipment: Units 1,3&4 semi-sync, Unit 2 reduced voltage.

Runners
Material: Cast steel with 0.5mm (0.019 in) stainless steel overlay
Minimum unit submergence: 16.0 ft
WR²: 172,000,000 (lbf x ft²)
Manufacturer: Allis Chalmers
Model test by: Allis Chalmers

	Reversible Runners	Reversible Motor/Generator			
Number	4				4
Diameter (ft)	24.00	Rotor	34.01	Stator	34.12
rpm synchronous	120.0				120.0
rpm overspeed	138.0				138.0
Type	Francis	Umbrella, vertical shaft			

Information on Runners

Condition:	Gross Head (ft) Generating	Gross Head (ft) Pumping	Capacity (MW) Generating	Capacity (MW) Pumping	Discharge (ft³/s) Generating	Discharge (ft³/s) Pumping	Turbine/Pump Eff.(%) Generating	Turbine/Pump Eff.(%) Pumping
Maximum head & maximum power	335	335	193	156	6,710	5,000	92.0	78.0
Minimum head & maximum power	310	310	152	170		6,200	91.0	

Condition:	Net Head (ft) Generating	Net Head (ft) Pumping	Capacity (MW) Generating	Capacity (MW) Pumping	Discharge (ft³/s) Generating	Discharge (ft³/s) Pumping	Turbine/Pump Eff.(%) Generating	Turbine/Pump Eff.(%) Pumping
Rated head @ best gate	294	294	170	170		6,200	93.0	92.0

Jocassee Pumped Storage Station - Page 3 (English)

JOCASSEE PUMPED STORAGE STATION

POWERHOUSE and RELATED FEATURES

Powerhouse Structure
Type: Aboveground, concrete, outdoor
Length: 109.7 m   Width: 40.1 m   Height: 18.0 m

Guard Valves
Number: None   Diameter: m
Type:

Transformers
Number: 4 West. Class FOA
Ratings: 192
Voltages: (kV) 230 / 14.4

Generator
Rating generating (MVA): 170.0   Rating pumping (MVA): 168.0
Insulation type: Thermolastic Class B NEMA code BBBXV
Starting method: 4 units synchronous; 2 units pony motor
Starting equipment: Units 1,3&4 semi-sync, Unit 2 reduced voltage.

Runners
Material: Cast steel with 0.5mm (0.019 in) stainless steel overlay
Minimum unit submergence: 4.88 m
WR^2: 71,096,000 (Newtons x m^2)
Manufacturer: Allis Chalmers
Model test by: Allis Chalmers

	Reversible Runners	Reversible Motor/Generator		
Number	4			4
Diameter m	7.32	Rotor 10.367	Stator	10.401
rpm synchronous	120.0			120.0
rpm overspeed	138.0			138.0
Type	Francis	Umbrella, vertical shaft		

Information on Runners

Condition:	Gross Head (m)		Capacity (MW)		Discharge (m³/s)		Turbine/Pump Eff.(%)	
	Generating	Pumping	Generating	Pumping	Generating	Pumping	Generating	Pumping
Maximum head & maximum power	102.1	102.1	193	156	190.0	141.6	92.0	78.0
Minimum head & maximum power	94.5	94.5	152	170		175.6	91.0	

Condition:	Net Head (m)		Capacity (MW)		Discharge (m³/s)		Turbine/Pump Eff.(%)	
	Generating	Pumping	Generating	Pumping	Generating	Pumping	Generating	Pumping
Rated head @ best gate	89.6	89.6	170	170		175.6	93.0	92.0

Jocassee Pumped Storage Station - Page 3 (Metric)

JOCASSEE PUMPED STORAGE STATION

Plant Data:

Average GWh generating per year:	1,245
Average GWh pumping per year:	1,646
Starting time from standstill (s):	150
Changeover time pumping to generating (min):	7
Planned/scheduled time between major overhauls (years):	20
Outage time required per unit during major overhauls (weeks):	24
Representative plant availability (%):	92.0
Representative planned outages (weeks per year):	1

Miscelleneous Notes:

The Lower Dam is made up of two homogeneous fill embankments with the following characteristics:

Height (m)	51.8	45.7
Crest length (m)	1067	533
Volume (m³)	1,562,000	1,529,000
Height (ft)	170	150
Crest length (ft)	3500	1750
Volume (yd³)	2,043,000	2,000,000

The changeover time from pumping to generating is 7 minutes in auto mode and 4.5 minutes in the manual mode.

The construction time of 6 years and 11 months also includes Keowee Hydro, a conventional plant.

Cavitation Experience:
Very minor cavitation has occurred with the stainless steel.

Significant or Unique Problems:
Problems have includee O&M mods, baffle plates, wearing rings, thrust bearing upper bracket weld failures.

List of Licenses Required:
FERC, DEHC (sewage and water supply - potable)

ENVIRONMENTAL FEATURES

Recreation:
Four paved lake access areas with nine ramps and primitive campgrounds. Activities include swimming, waterskiing, and power sail-boating. Numerous fishing tournaments have been held.

Fish and Wildlife:
Fish present at the lake include small- and large-mouth bass, rainbow and brown trout, catfish, crappie, shalle cracker, and pike. Wildlife include turkey, deer, bear, wildcat, fox, eagles, falcons, boar, mountain lion, quail, rabbit, mink, and beaver.

JOCASSEE PUMPED STORAGE STATION

Social:
The project includes 500 acres with a bathhouse, county store, beach, and camping and rental cabins which are leased to State of South Carolina, which performs operation and maintainance. The area is a South Carolina State Park and is part of the mitigation for project which is known as Keowee-Toxaway.

Page 4 (Continued)

The Jocassee project—a hydro and nuclear power partnership

By R. B. Siebensohn*

A good combination of hydro and thermal power generation is found at the Keowee-Toxaway project near Clemson, South Carolina, USA, which is scheduled for initial operation in 1974. This power scheme, incorporating the largest pump-turbines ever built in the USA, consists of conventional and pumped-storage hydroelectric generation, as well as one of the largest nuclear generating stations in the country

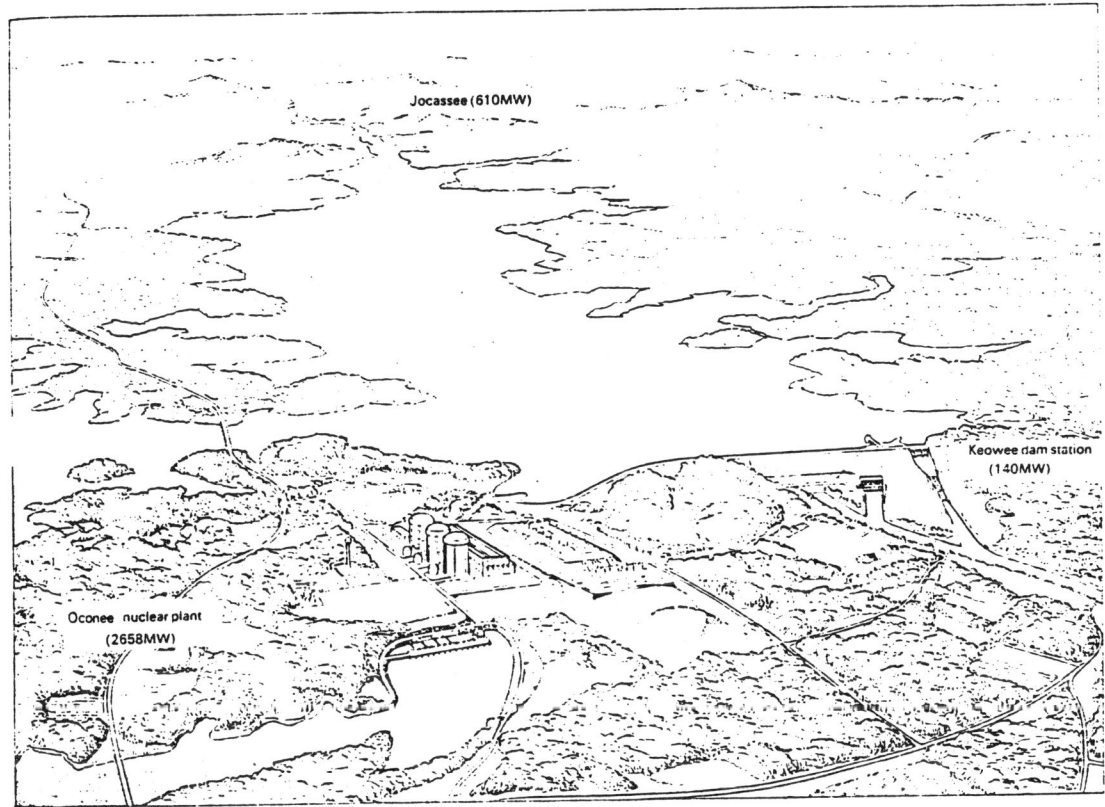

Fig. 1. An overall view of the Keowee-Toxaway project

THE Jocassee development, as part of the Keowee-Toxaway project, is formed by a 385ft high and 1750ft long rockfill dam, two saddle dikes, a gated spillway structure and an outdoor power-house containing four vertical reversible pump-turbines directly coupled to single-speed generator-motors.

The rated generating capacity of the plant will be 610MW under a net effective head of 294ft.

Two separate pressure tunnels, each 33·5ft in diameter, ll supply water to the pump-turbines through two tower-type intake structures in Lake Jocassee. The upper reservoir will have a surface area of 7565 acres at full pond elevation of 1110ft and a maximum drawdown of 30ft.

The plant will normally be used for peaking purposes and will have an average annual energy output from natural stream flow of about 81·7GWh. During periods of high reservoir level, the units may be operated on peak so as to utilize the full overload capability of the generator-motors. Due to the short duration of peak, it is expected that this capability will be considerably in excess of the normal thermal rating for continuous operation of the electric machinery.

* Project Engineer, Hydro-Turbine-Generator Division, Allis-Chalmers, USA

Reprinted, with permission, from International Water Power and Dam Construction, July/August 1970.

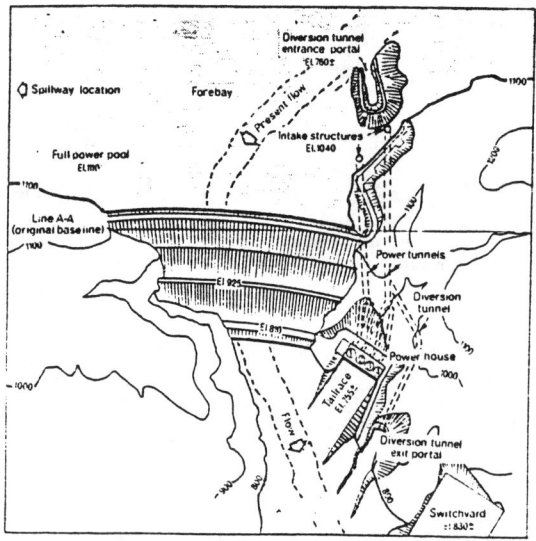

Fig. 2. The layout of the Jocassee dam and power station

Over the life of the plant, it is estimated that the units may be operated as generators 30 to 50 hours and as pumps up to 100 hours per week.

Model tests

An extensive model research programme was established to satisfy contractual requirements, to prov~ guaranteed pump-turbine performance values, and to d~ termine characteristic operating data necessary for the prototype design.

The test programme comprised measurement, recording and analysis of the following:

Capacity, efficiency and cavitation characteristics in the pumping and generating cycles;

Maximum required pump motor input;

Maximum turbine runaway speed at steady state;

Investigation of vibration behaviour, noise level and such other phenomena as axial hydraulic thrust;

Pressure distribution under shut-off head conditions;

Four-quadrant synoptic curve plots of shaft torque and speed versus discharge as shown in Fig. 3; and

Dynamic torque on wicket gates during operation in the pumping, generating and energy-dissipation ranges.

Transient phenomena during start-up and shut-down operation, speed and pressure rise, discharge and run-

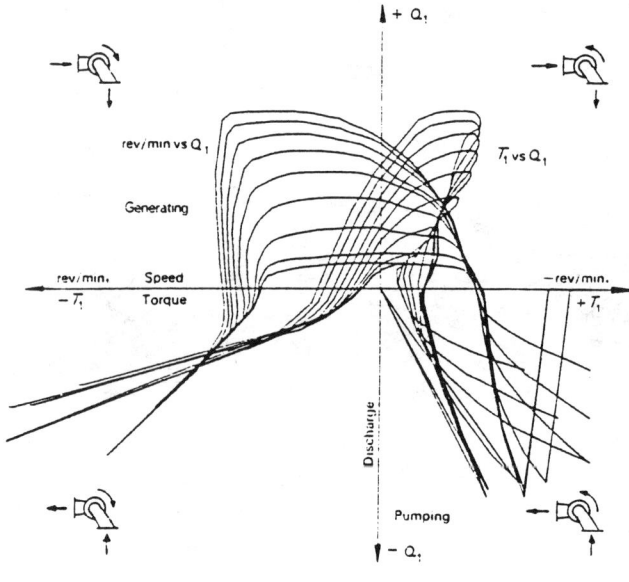

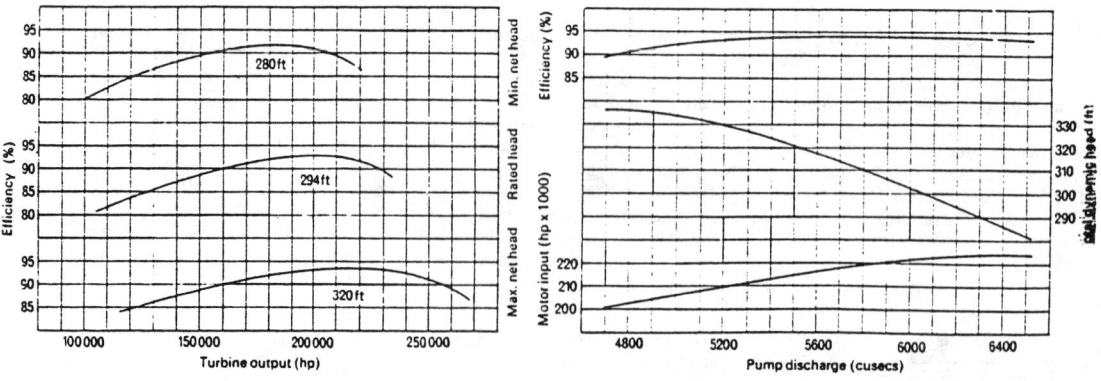

(Left) Fig. 3. Computer plotted synoptic curves of shaft torque and speed versus discharge
(Below left) Fig. 4 and (below) Fig. 5. Performance characteristics of Jocassee prototype pump-turbine units

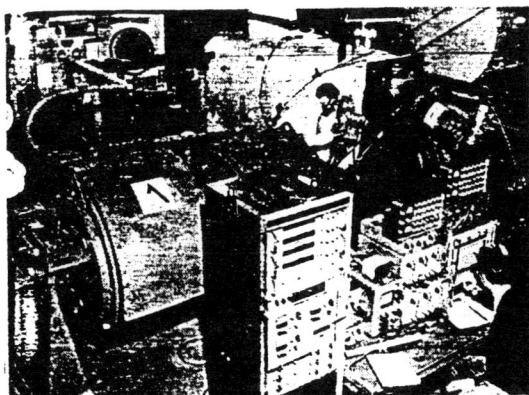

Arrangement of closed loop model test stand in Allis Chalmers' hydraulic laboratory

away speed as a result of load rejection or pump power failure were calculated with the help of a digital computer.

The tests were performed in a closed loop test stand in Allis-Chalmers' hydraulic laboratory. The model, exactly homologous with the pump-turbine and its water passages, had an impeller runner with six vanes and an outside diameter of about 19in. Arrangement of the model test equipment is shown in the photograph.

Performance characteristics of the prototype units, as given in Figs. 4 and 5, are calculated from the model test data applying the affinity laws for hydrodynamic similarity. Efficiencies for the pumping and generating cycles are stepped-up values from the model to prototype size using the modified Moody formula.

It is frequently claimed that cavitation for pump-bines is more critical during the pumping cycle than in the generating mode. This opinion cannot be generalised and only has validity when predominant interest in high pump performance leads to machines with oversized turbine runners.

In the case of the Jocassee units, however, the emphasis on generating performance prevailed over the setting required by pumping operation.

Pump-turbine design

With the rapid increase in unit size and output, it is no longer sufficient to rely only on empirical data, conventional calculation methods and mechanical design extrapolations to predict accurately the stresses and deflections in pump-turbine components. Typical of the advanced techniques employed is an analysis of main stress-carrying machine parts according to finite-element methods.

These methods, based upon the concept of replacing the actual continuous structure with a mathematical model comprised of structural elements of finite size, enable us to produce optimum designs with balanced loadings and deflection of the various components.

The analysis of the deformational behaviour of the large machine parts with a high degree of accuracy is particularly important since deflections unlike stresses do increase with unit size.

The plate-steel spiral cases are designed for 210lb/in^2 maximum operating pressure with 100% radiographing of all accessible weld joints. From experience gained on eral other jobs of comparable size, shop assembly of the spiral cases could be eliminated with an attendant cost saving.

Except for sections where the small end of the spiral joins the inlet section, the entire casings will be field fitted and welded. For structural reasons, the stay ring vanes will be fabricated from high-strength plate-steel ASTM 517, Grade 7 (brand name T-1).

The main guide-bearing is of the forced lubricated type with positive adjustability of oil flow and specially designed oil grooving for reversibility. It is equipped with a supplementary gravity oil supply tank providing the necessary lubrication during emergency shutdown sequence.

For high structural rigidity, the bearing housing is made integral with the head cover. The bearing shell is split vertically into two sections to facilitate dismantling. Thermostatic controlled heat exchangers maintain the oil temperature in the reservoir while the unit is idle.

The gate mechanism will be the only fast-closing emergency shut-off device in the power-plant. The gates will be integrally cast from ASTM Specification A27, Grade 70 steel. Castings are to be heat treated for improved impact properties.

As in conventional regulating systems, the gates will be actuated by the operating ring, links and levers positioned by a pair of hydraulic servomotors. The adjusting mechanism between the gate stem and operating ring will have shear pins so that in the event of blockage of one or more gates during closing, the pin will break and permit the remaining gates to close.

Each gate is to be equipped with a patented gate restraining mechanism. This device will prevent the gate from flailing between the gate stops during unstable hydraulic conditions should a shear pin fail. The gate mechanism will also have a pneumatic-operated shear pin failure detection system to actuate alarm signals and shut the unit down.

Since there are no shut-off valves between the penstocks and the units, the distributor is equipped with gate seals to reduce leakage into the wheel case during shutdown or depressed tail-water conditions. A floating seal ring in the main shaft packing box, specially designed for use on pump-turbines, accommodates the vertical and horizontal shaft movements during transient operating conditions.

Table I gives characteristic design and performance data, and Fig. 6 illustrates a cross section through the pump-turbine distributor and shows some of the machine design details.

The Francis-type impeller runners are larger overall physically than any previously built in the USA. They are some 15% larger than the world's most powerful pump-turbine runners currently in operation.

To facilitate shipping and handling, the runners will be split into halves for assembly at the job site. Finish machined weight of the runner assembly will be about 200t.

The runners are designed to minimise friction and windage losses and are composed of cast steel vanes, crowns and bands, assembled and welded together in the shop. Each runner will have welded stainless-steel inlays to protect against pitting due to cavitation. After finish machining, the runners will be statically shop balanced.

Generator-motors

The umbrella type generator-motors, supplied by the Westinghouse Electric Corp, will have one combined thrust-and-guide bearing supported on the lower generator bracket mounted immediately below the rotor.

The thrust bearing is designed for 2 170 000lb axial thrust occurring during steady-state operating conditions. The bearing has an automatic high pressure oil system

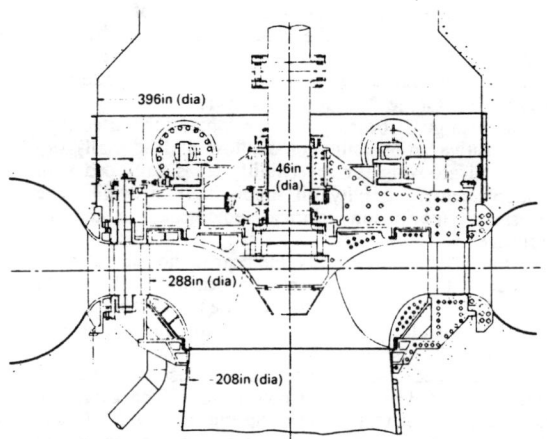

Fig. 6. The pump-turbine distributor section

to supply pressure oil to bearing surfaces during starting and stopping.

Provisions are made for a static excitation system and the generator-motors are suitable for synchronously or semi-synchronously starting another unit or being started with applied voltage through the 230kV system.

One of the units, in addition to being designed for synchronous starting, is designed for reduced-voltage starting by use of a three winding step-up transformer. All units may be operated as synchronous condensers to furnish reactive power to the system. As spinning reserve in the generating direction, the units are immediately available for power supply.

The speed of the pump-turbine units is controlled by mechanical-hydraulic governors of the cabinet actuator type as manufactured by the Woodward Governor Co and supplied by Allis-Chalmers. Nominal governor operating pressure is 350lb/in^2.

By the hydraulic transients analysis, the minimum gate closing time was determined to be 15s. Manual control of the units will be provided from the control room at the Jocassee power station.

Table I
Characteristic design and performance data
Pump-turbine characteristics

A. Pumping:
  - Rated total head — 294ft
  - Rated discharge — 6 200 cusecs
  - Head range of operation — 290—344ft
  - Maximum shut-off head — 470ft

B. Generating:
  - Rated net head — 294ft
  - Rated output — 228 000hp
  - Maximum net head — 320ft
  - Maximum output — 268 000hp
  - Minimum net head — 280ft
  - Minimum submergence — 12ft

C. Speeds:
  - Synchronous speeds, pumping and generating — 120 rev/min
  - Maximum runaway speed — 194 rev/min

D. Dimensions:
  - Impeller runner diameter — 288in
  - Spiral case entrance diameter — 216in
  - Turbine pit diameter — 396in
  - Main shaft diameter — 46in

Generator-motor characteristics

E. Motoring:
  - Rated capacity at 80°C — 233 000hp
  - Power factor — Unity
  - Voltage — 14 400

F. Generating:
  - Rated output at 60°/80°C — 170 000/196 000kva
  - Power factor — 90%
  - Flywheel effect — 172 000 000lbft^2

Conclusion

With their unprecedented dimensions the Jocassee units represent another advance in the development of large single stage Francis-type pump-turbines.

Generally, it might be observed that the trend toward higher unit capacities persists in a continuing effort to minimise installed cost per kW. The ultimate limits for unit size have not been reached and it seems that they will be determined by both economic and technical considerations.

Reference

1. CHACOUR S. "A High Precision Axi-Symmetric Triangular Element used in the Study of Hydraulic Turbine Components." ASME 1970 Spring Conference of Fluids Engineering Division, Detroit, Michigan, USA.

Plant Name: **KINZUA**

Plant location:
Warren, PA

Rated capacity	350 MW
Average static head	646 ft
Plant efficiency	66.70 %
Stored energy	3,920 MWh
Number of units	2

Construction time: 4 years, 6 months
Construction cost: $505 per kW
Price level: 1988
First commercial power: December 1970
FERC project number: 2280

River or water source: Allegheny River

Owner: Pennsylvania Electric Company
1001 Broad Street
Johnstown, Pennsylvania 15907

Designers:
Harza Engineering Company, Chicago, Illinois

Plant Manager/Superintendent:
Kinzua Plant Manager
P.O. Box 126
Warren, Pennsylvania 16365

(814) 533-8111

	UPPER RESERVOIR	LOWER RESERVOIR
DAM		
Type	Rolled fill - sandstone	Concrete and earthfill
Height (ft)	115	
Crest length (ft)	7,800	
Volume (yd³)	260,805,000	
RESERVOIR		
Type	Constructed	On stream, Allegheney Reservoir
Surface area (acres)	110	20,992
Usable power storage (acre-ft)	5,756	4,650
Power pool fluctuation (ft)	70.0	105.0
Operating levels		
Maximum (ft)	2,073.3	1,365.0
Minimum (ft)	2,003.3	1,260.0
Drainage area (miles²)	0.2	21,080.0
Seepage (ft³/s)		
SPILLWAY		
Design flood		
Return period (years)	Overpump	
Flow (ft³/s)	7,910	
Capacity (ft³/s)	7,910	
Type	Fuse plug for over pumping	
Gates		
Number	None	None
Type		
Width (ft)		
Height (ft)		
OUTLET WORKS		
Discharge capacity (ft³/s)		
Number of water passages	None	None
Dimensions of water passages		
Height (ft)		
Width (ft)		
Diameter (ft)		
Type of gates		
Number of gates	None	None

Kinzua - Page 1 (English)

Plant Name: **KINZUA**

Plant location
  Warren, PA

Owner: Pennsylvania Electric Company
        1001 Broad Street
        Johnstown, Pennsylvania 15907

Rated capacity   350 MW
Average static head   197.0 m
Plant efficiency   66.70 %
Stored energy   3,920 MWh
Number of units   2

Designers:
    Harza Engineering Company, Chicago, Illinois

Construction time: 4 years, 6 months
Construction cost: $505 per kW
Price level: 1988
First commercial power: December 1970
FERC project number: 2280

Plant Manager/Superintendent:
    Kinzua Plant Manager
    P.O. Box 126
    Warren, Pennsylvania   16365

    (814) 533-8111

River or water source: Allegheny River

	UPPER RESERVOIR	LOWER RESERVOIR
DAM		
Type	Rolled fill - sandstone	Concrete and earthfill
Height (m)	35.1	
Crest length (m)	2,377.4	
Volume (m³)	199,400,000	
RESERVOIR		
Type	Constructed	On stream, Allegheney Reservoir
Surface area (Mm²)	0.45	84.95
Usable power storage (Mm³)	7.100	5.736
Power pool fluctuation (m)	21.34	32.00
Operating levels		
Maximum (m)	631.94	416.05
Minimum (m)	610.61	384.05
Drainage area (Mm²)	0.445	54,597.255
Seepage (m³/s)		
SPILLWAY		
Design flood		
Return period (years)	Overpump	
Flow (m³/s)	224	
Capacity (m³/s)	224	
Type	Fuse plug for over pumping	
Gates		
Number	None	None
Type		
Width (m)		
Height (m)		
OUTLET WORKS		
Discharge capacity (m³/s)		
Number of water passages	None	None
Dimensions of water passages		
Height (m)		
Width (m)		
Diameter (m)		
Type of gates		
Number of gates	None	None

Kinzua - Page 1 (Metric)

KINZUA

INTAKES	UPPER INTAKE	LOWER INTAKE
Number	1	1
Type	Bell mouth with vortex	Tower designed for temperature control
Design discharge (ft³/s)	4,167	4,181
Gross inlet area (ft²) (at trash racks)	792	615
Bar racks		
spacing (in)		1.50
shape		
depth/thickness (in)		0.38 /
diameter (in)		
Emergency gates		
number	None	1
height/width (ft)		25.30 / 18.50
type		Closure bulkhead
Service gates		
number	None	2
height/width (ft)		23.49 / 16.50
type		Fixed wheel - vertical
Bulkhead/stop logs (Y or N)		
number of units serviced		
Hoists		
number	None	None
capacity (tons)		
type		Hydraulic

WATER PASSAGES	Upper Tunnel	Shaft	Lower Tunnel	Surge Tanks Upper	Lower	Penstocks	Tailrace Tunnel
Number		1	1			2	
Diameter (ft)		22.3	22.3			14.0	
Length (ft)		482	1,990			330	
Maximum velocity (ft/s)		24.0	24.0			24.0	
Concrete lining thickness (in)							
Total length of concrete sections (ft)		482	1,506				
Steel liner Thickness							
Minimum (in)			0.77			0.59	
Maximum (in)			0.92			1.65	
Material grade			ASTM A-537, Grade A			Several see drawings	
Total length of steel-lined sections (ft)			484			330	

Notes:

KINZUA

INTAKES	UPPER INTAKE	LOWER INTAKE
Number	1	1
Type	Bell mouth with vortex	Tower designed for temperature control
Design discharge (m³/s)	118.0	118.4
Gross inlet area (m²) (at trash racks)	73.6	57.1
Bar Racks:		
spacing (mm)		38
shape		
depth/thickness (mm)		10 /
diameter (mm)		
Emergency gates		
number	None	1
height/width (m)		7.711 / 5.639
type		Closure bulkhead
Service gates		
number	None	2
height/width (m)		7.159 / 5.029
type		Fixed wheel - vertical
Bulkhead/stop logs (Y or N)		
number of units serviced		
Hoists		
number	None	None
capacity (Mg)		
type		Hydraulic

WATER PASSAGES	Upper Tunnel	Shaft	Lower Tunnel	Surge Tanks Upper	Surge Tanks Lower	Penstocks	Tailrace Tunnel
Number		1	1			2	
Diameter (m)		6.81	6.81			4.27	
Length (m)		146.9	606.6			100.6	
Maximum velocity (m/s)		7.31	7.32			7.32	
Concrete lining thickness (m)							
Total length of concrete sections (m)		146.9	459.0				
Steel liner Thickness							
Minimum (mm)			20			15	
Maximum (mm)			23			42	
Material grade			ASTM A-537, Grade A			Several see drawings	
Total length of steel-lined sections (m)			147.5			100.6	

Notes:

KINZUA

POWERHOUSE and RELATED FEATURES

Powerhouse Structure
Type: Fully enclosed
Length: 230 ft Width: 70 ft Height: 138 ft

Guard Valves
Number: 2 Diameter: 9.5 ft
Type: Spherical

Transformers
Number: 3
Ratings: Group 1-215/241 group 2-29/32
Voltages: (kV) 230 / 138

Generator
Rating generating (MVA): 220.0 Rating pumping (MVA): 195.0
Insulation type:
Starting method: Direct start with pelton wheel or back-to-back start
Starting equipment:

Runners
Material: Steel
Minimum unit submergence: 72.0 ft
WR^2:
Manufacturer: Newport News
Model test by: Newport News

	Reversible Runners	Reversible Motor/Generator	
Number	2		2
Diameter (ft)	18.70	Rotor	Stator
rpm synchronous	225.0		
rpm overspeed			
Type	Francis		

Information on Runners

Condition:	Gross Head (ft) Generating	Gross Head (ft) Pumping	Capacity (MW) Generating	Capacity (MW) Pumping	Discharge (ft³/s) Generating	Discharge (ft³/s) Pumping	Turbine/Pump Eff.(%) Generating	Turbine/Pump Eff.(%) Pumping
Maximum head & maximum power	813	825			4,150	2,320		
Minimum head & maximum power	644	680				3,600		

Condition:	Net Head (ft) Generating	Net Head (ft) Pumping	Capacity (MW) Generating	Capacity (MW) Pumping	Discharge (ft³/s) Generating	Discharge (ft³/s) Pumping	Turbine/Pump Eff.(%) Generating	Turbine/Pump Eff.(%) Pumping
Rated head @ best gate	646	700	162	195		3,200		

KINZUA

POWERHOUSE and RELATED FEATURES

Powerhouse Structure
Type: Fully enclosed
Length: 70.1 m Width: 21.3 m Height: 42.1 m

Guard Valves
Number: 2 Diameter: 2.90 m
Type: Spherical

Transformers
Number: 3
Ratings: Group 1-215/241 group 2-29/32
Voltages: (kV) 230 / 138

Generator
Rating generating (MVA): 220.0 Rating pumping (MVA): 195.0
Insulation type:
Starting method: Direct start with pelton wheel or back-to-back start
Starting equipment:

Runners
Material: Steel
Minimum unit submergence: 21.95 m
WR^2:
Manufacturer: Newport News
Model test by: Newport News

	Reversible Runners	Reversible Motor/Generator	
Number	2		2
Diameter m	5.69	Rotor	Stator
rpm synchronous	225.0		
rpm overspeed			
Type	Francis		

Information on Runners

Condition:	Gross Head (m) Generating	Gross Head (m) Pumping	Capacity (MW) Generating	Capacity (MW) Pumping	Discharge (m³/s) Generating	Discharge (m³/s) Pumping	Turbine/Pump Eff.(%) Generating	Turbine/Pump Eff.(%) Pumping
Maximum head & maximum power	247.9	251.5			117.5	65.7		
Minimum head & maximum power	196.2	207.3				101.9		

Condition:	Net Head (m) Generating	Net Head (m) Pumping	Capacity (MW) Generating	Capacity (MW) Pumping	Discharge (m³/s) Generating	Discharge (m³/s) Pumping	Turbine/Pump Eff.(%) Generating	Turbine/Pump Eff.(%) Pumping
Rated head @ best gate	196.9	213.4	162	195		90.6		

Kinzua - Page 3 (Metric)

KINZUA

Plant Data:
   Average GWh generating per year: 1
   Average GWh pumping per year: 1
   Starting time from standstill (s):
   Changeover time pumping to generating (min):
   Planned/scheduled time
      between major overhauls (years):
   Outage time required per unit
      during major overhauls (weeks):   23
   Representative plant availability (%):   88.0
   Representative planned outages (weeks per year):   3

Miscelleneous Notes:
   The Kinzua plant combines the features of conventional and pumped storage for later release to the lower reservoir or Allegheny River. Pump turbine units 1 and 2 pump and generate between the upper and lower reservoirs, but special gating arrangements also permit unit 2 to generate to the Alleghany River.

   Pumped output is that portion of the output that is attributable wholly to the release of water through units 1 or 2 from the upper reservoir down to the level where pumping began. Conventional output is that portion of the output which is attributable to the further passage of water from the level at which pumping began to the level of the river through units 2 or 3.

   Unit 3, the conventional Francis unit, has a rating of 40,200 hp, compared to 217,000 hp for the reversible units. Data in the pumped storage database reflect the reversible units.

Cavitation Experience:
   Routine yearly runner cavitation occurs.

Significant or Unique Problems:

List of Licenses Required:

ENVIRONMENTAL FEATURES

Recreation:
   Recreation includes boating, fishing, and camping at Kinzua Dam.

Fish and Wildlife:

Social:

THE SENECA PUMPED-STORAGE PLANT

SEYMOUR B. ROCK
Senior Engineer
Civil and Mechanical Engineering Department
The Cleveland Electric Illuminating Company
Cleveland, Ohio

INTRODUCTION

The Seneca Pumped Storage Plant is now under construction in northwestern Pennsylvania. The three-unit plant will have a rated capacity of 380,000 kW and is scheduled for service in mid-1969. The power facilities are being installed adjacent to a concrete-earthfill dam recently completed by the Corps of Engineers across the Allegheny River near Warren, Pennsylvania, about 200 miles upstream from Pittsburgh (Fig. 1).

When the Corps of Engineers announced plans for the construction of the Kinzua Dam for purposes of flood control, low-flow augmentation and recreation, the Pennsylvania Electric Company (Penelec) initiated studies aimed at determining how best to utilize the power potential of the site.

Since there was an 800-ft plateau immediately adjacent to the left bank of the river at this location, it quickly became apparent that the site was well adapted to the construction of a pumped-storage plant.

In 1962, Penelec filed an application with the Federal Power Commission for a license to install a pumped-storage plant at this site.

Meanwhile, generation planning studies by The Cleveland Electric Illuminating Company (CEI) had been indicating the desirability of some form of peaking capacity for the CEI system. Since the feasibility of the Kinzua site was far superior to that of any potential pumped-storage site in the CEI service area, CEI joined Penelec in carrying out the final phases of the planning studies for what was later named the Seneca Pumped-Storage Plant (Fig. 2).

PLANNING CONSIDERATIONS

In carrying out these studies, the best alternative to the pumped-storage project appeared to be a mine-mouth steam plant located in western Pennsylvania. In evaluating the two choices, recognition was given to the principal advantages of pumped hydro:

1. Higher availability is expected from a pumped-storage plant compared with a steam plant. For a steam-electric station to provide capacity equal to that of a hydro station, it was estimated that its installed capacity must be greater than that of the hydro by at least 10 percent in order to provide for the greater outage time of the steam facility.
2. It was further estimated that the capital cost of a coal-fired steam plant would exceed that of a pumped-storage plant by approximately $10 per kW. Although the general level of construction costs has risen sharply since these studies were carried out, it is felt that this same capital cost differential in favor of pumped storage continues to prevail.

On the basis of the planning studies, it was concluded that the pumped-storage

Reprinted from Volume 29 Proceedings of the American Power Conference, 1967.

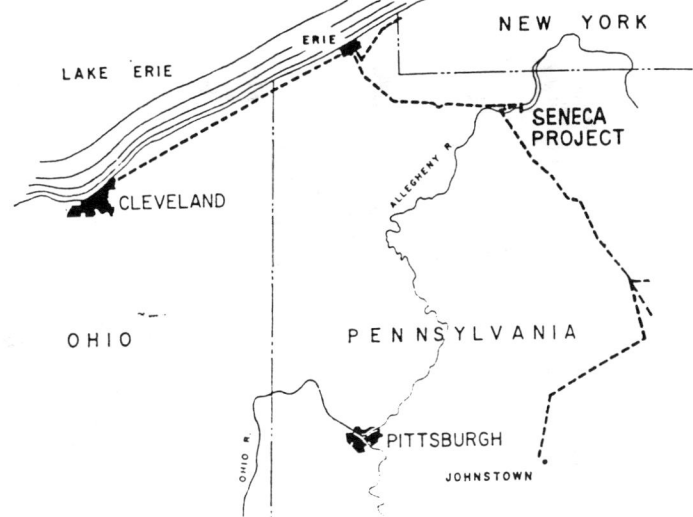

Fig. 1—Seneca project general area.

plant demonstrated a clear economic advantage over the alternate steam plant for both companies. For this reason, CEI joined Penelec in submitting an amended license application to the Federal Power Commission in May 1964.

The FPC issued its license for the Project in December 1965. The license conditions include the payment of an annual charge by CEI and Penelec to the United States for use of Government-owned facilities at the site. The license was promptly accepted by both companies and construction work initiated in April 1966.

While economically favorable on balance, there are, of course, important disadvantages associated with pumped storage. The pumping to generating cycle has an overall efficiency of about 75 percent or less, with transmission losses further reducing this figure. This means that the project consumes 4 kilowatthours of off-peak energy for every 3 kilowatthours produced for on-peak use.

A more significant penalty, however, is the large number of low incremental cost kilowatthours lost in the case where the alternate installation would have been a modern base-load steam unit.

This "lost opportunity" for unrestricted generation of low-cost kilowatthours is an important penalty associated with pumped storage. This is more true on systems with significant amounts of higher cost generating equipment still in service, although interconnections operate to reduce this penalty. This effect is expected to become less over the years as the efficiency spread in operative equipment narrows.

Another inherent limitation of pumped storage is that the economic benefits tend to decrease as the amount of pumped storage installed at any particular time increases. This is because of the increased energy associated with a given block of capacity as each succeeding increment of capacity is moved into the lower position in the load curve which it would usually assume.

In other words, each additional pumped-storage increment becomes less of a peaking unit and more of a base-load plant. This places increased emphasis on the adverse fuel cost characteristics of

Fig. 2—General view of Seneca plant project.

pumped hydro. If, however, the second increment is deferred until sufficient system growth has taken place, it is perfectly possible that this increment may be as advantageous as the first one was originally. The same would apply for succeeding increments, if installed at proper intervals of load growth.

Penelec is a member of the four-company General Public Utilities System which was previously committed to substantial blocks of pumped-storage capacity already in the construction and planning stages.

Since Cleveland Electric Illuminating had no peaking units on its system, it appeared to be in the mutual interests of both companies for CEI to assume a major portion of the initial ownership of the Seneca Plant. The initial ratio agreed to was 80 percent for Cleveland Electric Illuminating Company and 20 percent for Pennsylvania Electric Company.

The Harza Engineering Company of Chicago was retained to perform engineering and construction management services.

PLANT DESCRIPTION

Lower Intake Structure

A lower intake structure (Fig. 3) is being installed in the Allegheny Reservoir adjacent to the upstream face of the Kinzua Dam. Water flow to and from the reversible pump-turbines will be permitted by opening two fixed-wheel-type vertical lift intake gates, one for each of two conduits. These gates will be capable of withstanding unbalanced pressure against either face of the gate. Each gate will be

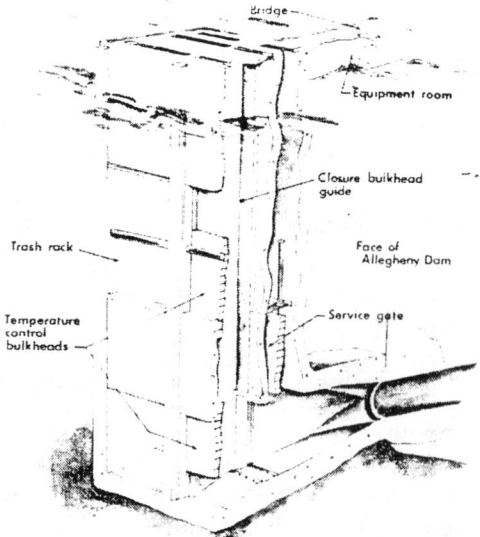

Fig. 3—Lower intake structure.

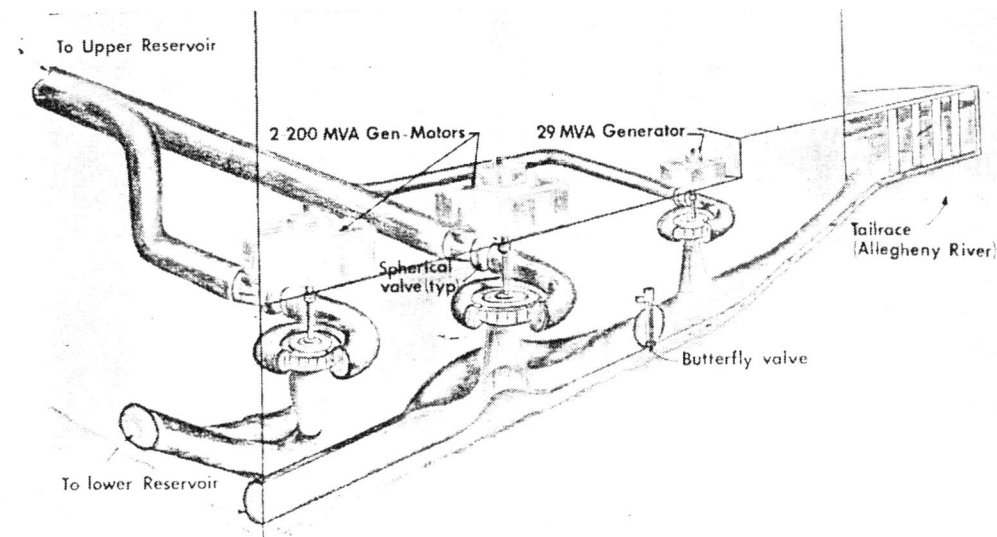

Fig. 4—Arrangement of units and water passages.

operated by, and connected to, a hydraulic hoist. When not in use, the gates will be kept in the "open" position just above the top of the conduit.

One closure bulkhead, designed to operate in slots upstream of the intake gate, will be provided. Placement of the closure bulkhead will permit access to, or removal of, the intake gate for inspection and maintenance. The bulkhead will be handled by means of a lifting chain and a roadway crane and will be placed and removed under balanced head conditions.

To permit the maximum degree of flexibility in controlling the temperature of water discharged by the Seneca units, two temperature control bulkheads, each with an upper and lower opening, will be provided. These allow selection of water entering the low-head conduits from either of two elevations in the Allegheny Reservoir. This will permit us to conform to the desires of regulatory authorities to withdraw water from particular levels for water temperature and quality control.

Low-Head Conduit

Two 15-ft low-head steel conduits carry water between the lower intake structure and the power station during both the pumping and generating cycles. These conduits pass through openings in the concrete section of the Kinzua Dam, which were left as blockouts when the dam was built.

Upper Reservoir

The upper reservoir is located on a plateau adjacent to the left abutment of the Kinzua Dam. The floor of the upper reservoir will be approximately 800 ft above the elevation of the Allegheny River downstream of the dam and approximately 700 ft above the water level behind the dam. In the horizontal direction, the centerline of the upper reservoir is about a half mile from the power station.

High-Head Conduit

A high-head conduit connects the upper reservoir to the power station. For the entire vertical run of this conduit, and a portion of the horizontal run, the conduit will be concrete lined with an inside diameter of 22 ft 4 in.

For several hundred feet of the horizontal run, where the depth of overburden is reduced, the conduits will be steel lined. The inside diameter of the main conduit steel liners will be 21 ft 6 in. The penstock branches to the individual units are, of course, smaller in diameter.

Power Station

Finally, there is the power station which houses two reversible pump-turbines and their associated generator-motors. In the generating direction, these units have a rated capacity of 175,000 kW each. Also included is a small third unit of approximately 30,000 kW capacity which can operate only in the generating direction.

Figure 4 illustrates the general arrangement of the three units in the power station, showing the conduits bringing water to and carrying water away from the three units. Unit No. 1, which is a reversible unit, recirculates water between the Allegheny Reservoir behind Kinzua Dam and the upper reservoir.

Under the terms of the license issued by the Federal Power Commission, CEI and Penelec have the option of utilizing for power generation any water which the Corps of Engineers might wish to discharge downstream past the Kinzua Dam. Two units have been designed to permit such utilization of the water.

Unit No. 3, the small unit which operates in the generating direction only, receives all of its water supply from the upper reservoir and discharges it to the Allegheny River downstream of the dam. All of the water supplied to this unit, therefore, will have been pumped to the upper reservoir by the large reversible units. Unit No. 3 has been sized so that it can efficiently utilize the 500-cfs minimum downstream release specified by the Corps, which is responsible for operation of the dam.

The plant also includes a second large reversible pump-turbine similar to Unit No. 1 except for the configuration of the draft tube. Unit No. 2 will also be capable of recirculating water between the Allegheny Reservoir and the upper reservoir exactly as does Unit No. 1. In addition to this, however, Unit No. 2 will also be capable of receiving water from the upper reservoir and discharging it to the Allegheny River downstream of the Kinzua Dam.

The purpose of this arrangement is to permit CEI and Penelec to use the hydraulic head made available by the Kinzua Dam. Discharging water downstream rather than back to the Allegheny Reservoir enhances the output of Unit No. 2 because of the additional head made available by discharging to an elevation averaging approximately 100 ft below that of the Allegheny Reservoir. Such downstream discharge can take place only during such periods when discharges in excess of the 500-cfs minimum specified previously are permissible.

Figure 5 shows diagrammatically the flows to and from each of the three units as well as their nominal design conditions.

DESIGN HIGHLIGHTS

Synchronous Starting

Unit No. 3 also has one other important job to do. It provides the starting torque to the reversible generator-motors when it is necessary to start them in the motor (or pumping) direction. The generator-motors will be started one at a time by connecting the armature of each to the Unit No. 3 generator while at

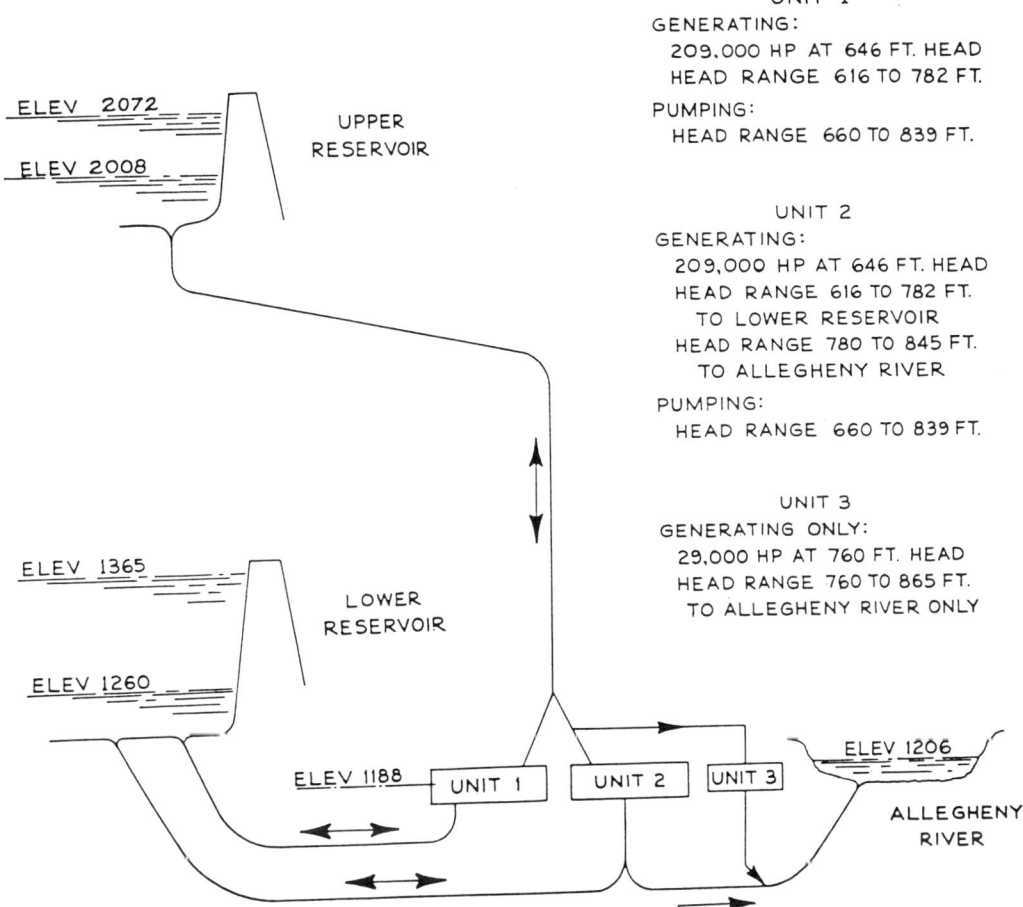

Fig. 5—Water flow diagram and unit design conditions.

standstill, exciting the fields of the generator and motor and using the Unit No. 3 turbine to accelerate its generator and the electrically-connected motor in synchronism up to rated speed.

Following is a brief outline of this synchronous starting procedure:

1. Water in the draft tube of the pump-turbine being started will be lowered by compressed air in order to clear the runner of water.
2. The Unit No. 3 generator and the generator-motor stator windings will be connected together via a 1200-ampere bus.
3. Excitation will be applied to the generator and generator-motor fields from separate static exciters fed from the plant auxiliary power system.
4. The wicket gates of Unit No. 3 turbine will be opened to a position sufficient to overcome the breakaway resistance of the two machines, but not so much that the Unit No. 3 generator will accelerate ahead of the motor being started. The turbine will then start and accelerate its generator and the generator-motor to rated speed.

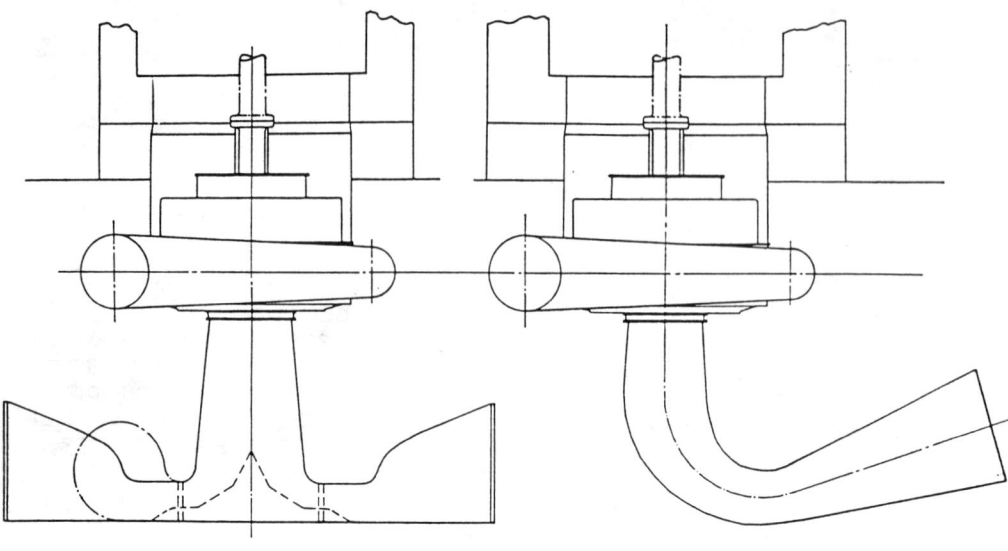

UNIT 2   UNIT 1

Fig. 6—Comparison of conventional and divided draft tube configurations.

5. The direct-connected exciter of the generator-motor will be brought up to no-load excitation voltage as the generator-motor accelerates. Excitation of the generator-motor will be transferred to its exciter at 90 percent speed.
6. The governor of Unit No. 3 will take control automatically when rated speed is achieved. Speed and voltage matching relays and an automatic synchronizer will act to first trip Unit No. 3 from the starting bus and then to close the main circuit breaker and synchronize the generator-motor to the CEI-Penelec system which will then supply pumping energy.
7. The tailwater depressing system will be shut down and the air release line opened. This will permit tailwater to rise into the impeller-runner chamber and cause the pump to develop full shutoff head. The motor will now begin to draw additional power from the system and, when the spherical valve and wicket gates open, the pump will begin to deliver water to the upper reservoir.

Divided Draft Tube

Another unusual feature of the Seneca Plant is the use of the divided draft tube on one of the reversible pump-turbines.

The divided draft tube makes it possible for the unit to discharge to either of two tailwater locations and elevations. Figure 6 shows the general appearance of the divided draft tube on Unit No. 2 as contrasted with the conventional elbow draft tube on Unit No. 1.

An extensive series of model tests was run on the proposed divided draft tube in order to optimize its design to verify that its performance characteristics would be satisfactory for commercial operation, and to compare its efficiency and capability with those of the conventional design.

Efficiency and capability penalties were found to be small so that these appear to suggest no economic loss of consequence. In the generating direction, operating limitations also appear to be negligible when discharging to Allegheny Reservoir. The principal adverse hydraulic characteristics observed were in the pumping direction.

More than 25 different design modifications were tested in the hydraulic laboratory in an attempt to achieve a uniformly-distributed flow through the draft tube and into the eye of the pump impeller. The principal modification incorporated into the design is a steel cone to be installed on the floor of the draft tube directly below the centerline of the impeller. Provision has been made for the introduction of compressed air into the draft tube through the tip of this cone. This may be of some benefit in suppressing cavitation and vortexing effects should these arise.

Vortices were observed extending from the vicinity of the tip of the cone and into the impeller under conditions of low NPSH and low water level in the upper reservoir. During most periods of pumping operation, these limitations should not prove severe. The low NPSH condition will occur for only a short period during the winter months when the level of water in the Allegheny Reservoir will be seasonally low. Furthermore, the low level condition in the upper reservoir may only occur for a matter of a few hours during the weekly pumping cycle, since we do not expect to empty the upper reservoir during the daily generating period as a rule.

Finally, we can specify that, during the relatively short period of adverse water level, pumping should be accomplished by Unit No. 1 alone or by Unit No. 1 and Unit No. 2 operating in parallel. Such parallel operation would act to increase the flow and friction loss in the high-head conduit, thus increasing the discharge head against which Unit No. 2 will be pumping. For these reasons, the operating limitations observed during the model testing program should impose no significant restrictions on the commercial feasibility of the divided draft tube Unit No. 2.

Asphalt-Lined Upper Reservoir

The upper reservoir, shown in Fig. 7, is estimated to cost a little over $6 million and will be capable of storing approximately 6000 acre-feet of water with a potential energy content of 4,200,000 kWh—equivalent to about 11 hours of generation at full capacity. Part of this water will be dead storage below the minimum operating level, which will serve to assure sufficient energy in case of emergency to provide cranking power to the CEI-Penelec system.

The configuration of the upper reservoir is based on a "balanced-cut-and-fill" principle whereby excavated materials are used for construction of the dike, with no net input or spoiling of materials.

The geology of the plateau includes a sandstone cap extending to a depth of about 190 ft below the floor of the upper reservoir. The remaining 600 ft down to the Allegheny River consists principally of shale and siltstone layers. Bedding planes of these formations dip slightly away from the river. This dip is favorable with respect to the stability of the hillside slope.

There are two sets of vertical joints in the plateau. The major set extends in a direction essentially parallel to the river,

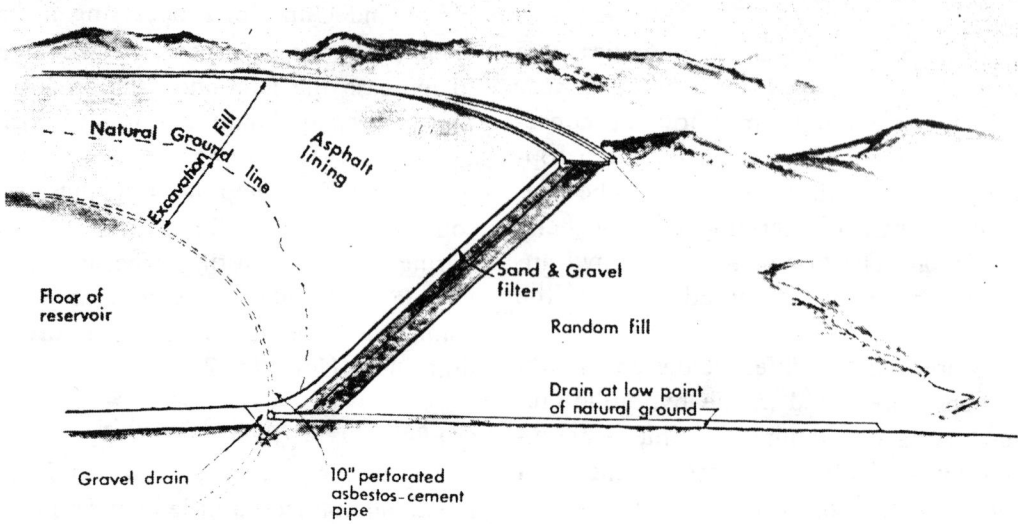

Fig. 7—General view of upper reservoir construction.

while the minor joint set is approximately perpendicular to the river.

The central core of the dike consists of random fill material, principally sandstone excavated from the floor of the upper reservoir. Procedures for placement of the dike material were carefully specified and inspected.

As it turned out, the material excavated from the floor contained a smaller percentage of fines than estimated on the basis of preliminary surveys. As a result, a lesser volume of in-place material from the floor was required to bring the dike up to its required elevation.

We were able to take advantage of this situation to add an additional 5 ft to the height of the dike while raising the original floor elevation. The raising of the dike height enhanced the average effective capacity of the plant by approximately 2800 kW. This additional capacity was obtained at an incremental construction cost below the average for the upper reservoir.

All floor and dike surfaces in contact with the water will be fully asphalt lined. Several alternatives for minimizing leakage of water from the upper reservoir were studied carefully. These included the utilization of an impervious core and the use of alternate lining materials, such as concrete and PVC. Figure 8 shows the design finally adopted.

On the inside surfaces of the dike, a 6-in. layer of sand and gravel filter materials will be placed on top of the random fill of the main portion of the dike. Above this will be a 3-in. drainage layer of porous asphaltic concrete. Above this 9-in. bed, two 1½-in. courses of asphaltic concrete will be laid. Finally, on the surface exposed to the water, a modified asphalt seal coat with asbestos fibers will be installed.

The sloping sides of the reservoir will be well drained in order to prevent the development of uplift pressures behind the lining during drawdown. The porous asphaltic concrete layer will lead to a gravel drain located in a trench around the entire toe of the inside slope. The

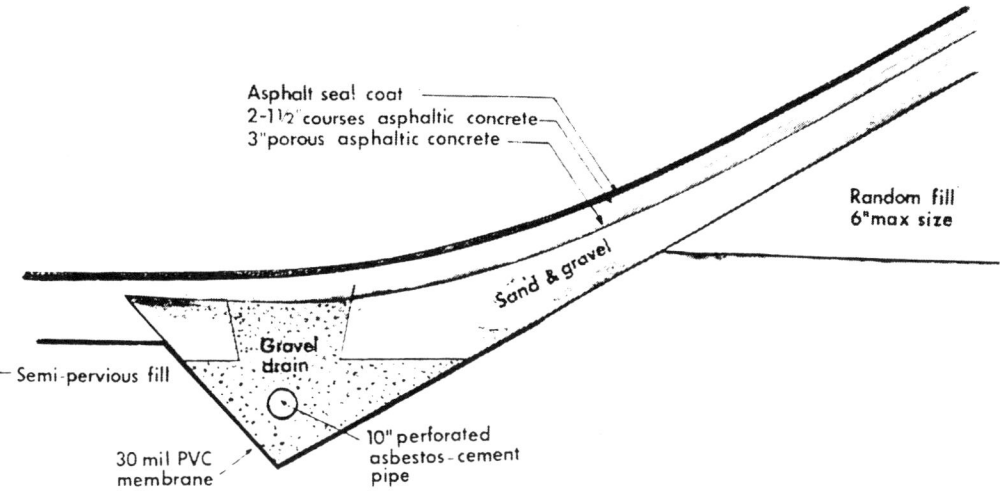

Fig. 8—Lining and drainage details—upper reservoir.

gravel drain will surround a 10-in. diameter perforated asbestos-cement pipe. Three 12-in. diameter asbestos-cement pipes spaced around the perimeter will connect the 10-in. perforated pipe to the outer toe of the dike.

The floor lining will consist of two 1½-in. courses of asphaltic concrete covered by an asphalt seal coat. As a base for this lining, a layer of compacted fill at least 1 ft thick will be installed. During excavation operations, which are now substantially completed, very careful attention was devoted to preventing blast damage to the rock which will remain in place beneath the floor of the upper reservoir.

One purpose of the semipervious fill layer is to ensure a relatively dense layer under the asphalt that will prevent the intrusion of water from the more permeable dike drainage system. No special drainage facilities will be installed beneath the floor. We are depending on the two vertical joint systems in the underlying rock and the permeability of the sandstone to percolate water down to the underlying water table.

OPTIMIZING PLANT DESIGN

Optimizing the design of a pumped-storage plant and its principal components poses some unusual problems. Essentially, these arise because the capacity of the plant depends not only on the size and configuration of the generating equipment itself, but on the energy storage capacity of the upper reservoir as well.

It is readily apparent that, when the upper reservoir is full, the plant's ability to meet the requirements imposed on it is at a maximum; when the upper reservoir is empty, the plant has no capability even though all units may be serviceable and ready to generate.

For this reason, then, it is essential that studies aimed at optimizing one of these plant design parameters properly reflect the influence of the others, as appropriate.

In optimizing the design of the Seneca Plant, we utilized a procedure developed by hydraulic engineers for the General Public Utilities System. Although this procedure was followed in optimizing the size and configuration of the upper reservoir, the most significant application was in developing the final design of the reversible pump-turbines.

In the present state-of-the-art, it is virtually impossible for the purchaser to specify an optimum design for the pump-turbine to be installed on his system. There are a great many design variables which affect the pumping and generating performance and the balance between them. In addition, there are a number of manufacturing, technical and cost considerations which also must be factored into the design.

The evaluation procedure we used made it possible for us to analyze several possible pump-turbine configurations and their associated performance characteristics. Results of these analyses were relayed to the manufacturer's design staff to guide the direction of their efforts.

The manufacturer, in turn, communicated to us his thinking with respect to feasible design objectives and associated cost data. An extensive program of model testing was carried out in order to verify calculations and confirm the commercial suitability of the configuration finally selected.

These cooperative efforts proved to be highly successful. We were able to develop a pump-turbine design far better suited to our system needs than the one originally proposed. Furthermore, this improvement was achieved at only a modest increase in contract price.

For this reason, the exchange of technical and economic data between the manufacturer and customer during the design and model testing phases of the contract should be encouraged. To the extent possible, adequate time should be allowed to permit realization of the full benefits of such exchanges.

BIBLIOGRAPHY

1. Ley, R. D. and Loane, E. S., "Symposium on Pumped Storage: General Planning of Pumped Storage," *Proc. ASCE, J. Power Div.*, 88, PO2, 211-31 (1962) July.

2. Loane, E. S. and Ley, R. D., "Economics of Pumped Storage," Paper presented at Edison Electric Institute Electrical System and Equipment Committee Meeting, St. Louis, Mo., September 29, 1964.

3. Ley, R. D., "The Economic Analysis of Pumped Storage Projects," *Proc. ASCE, J. Power Div.*, 92, PO2, 77-90 (1966) April.

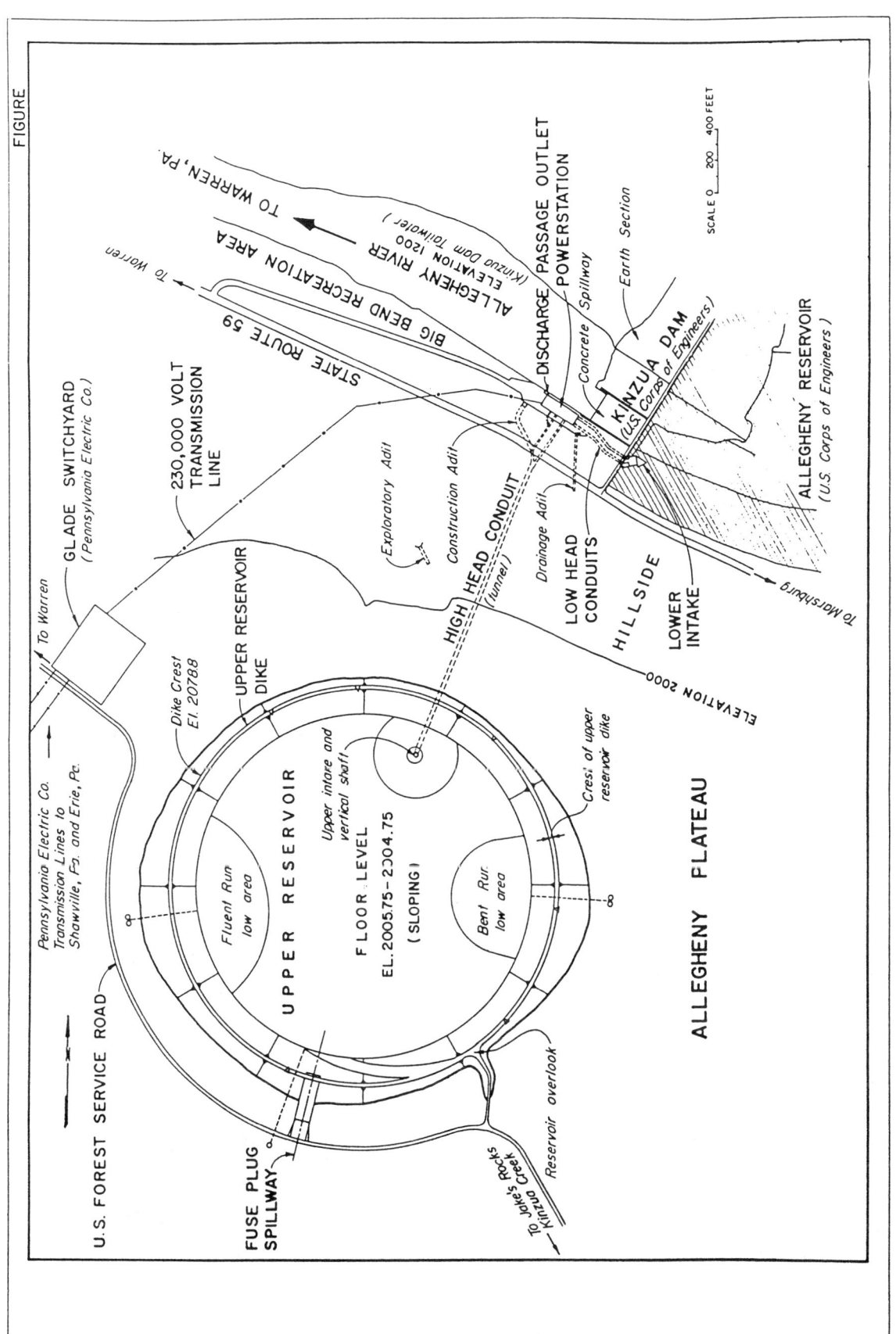
FIGURE

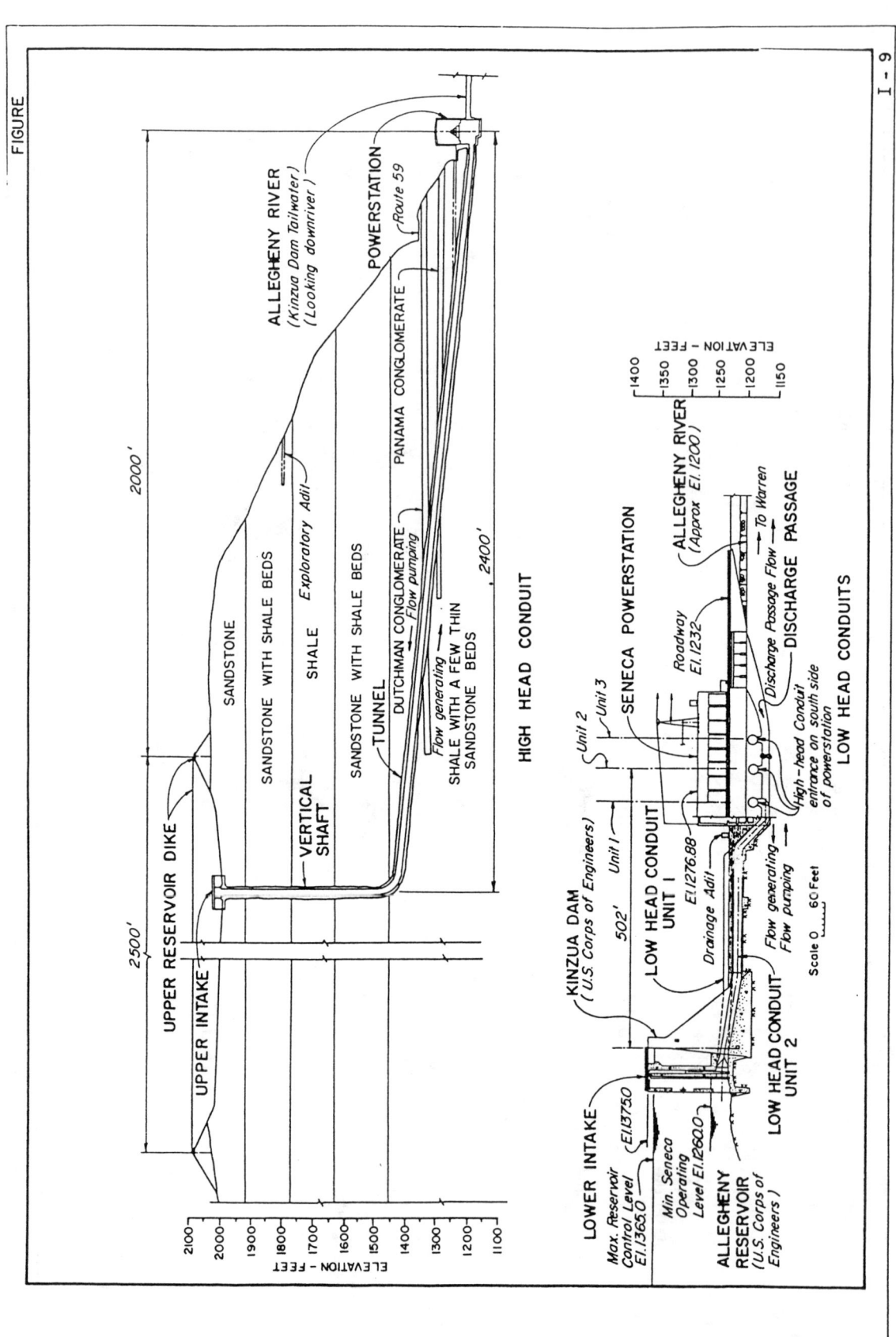

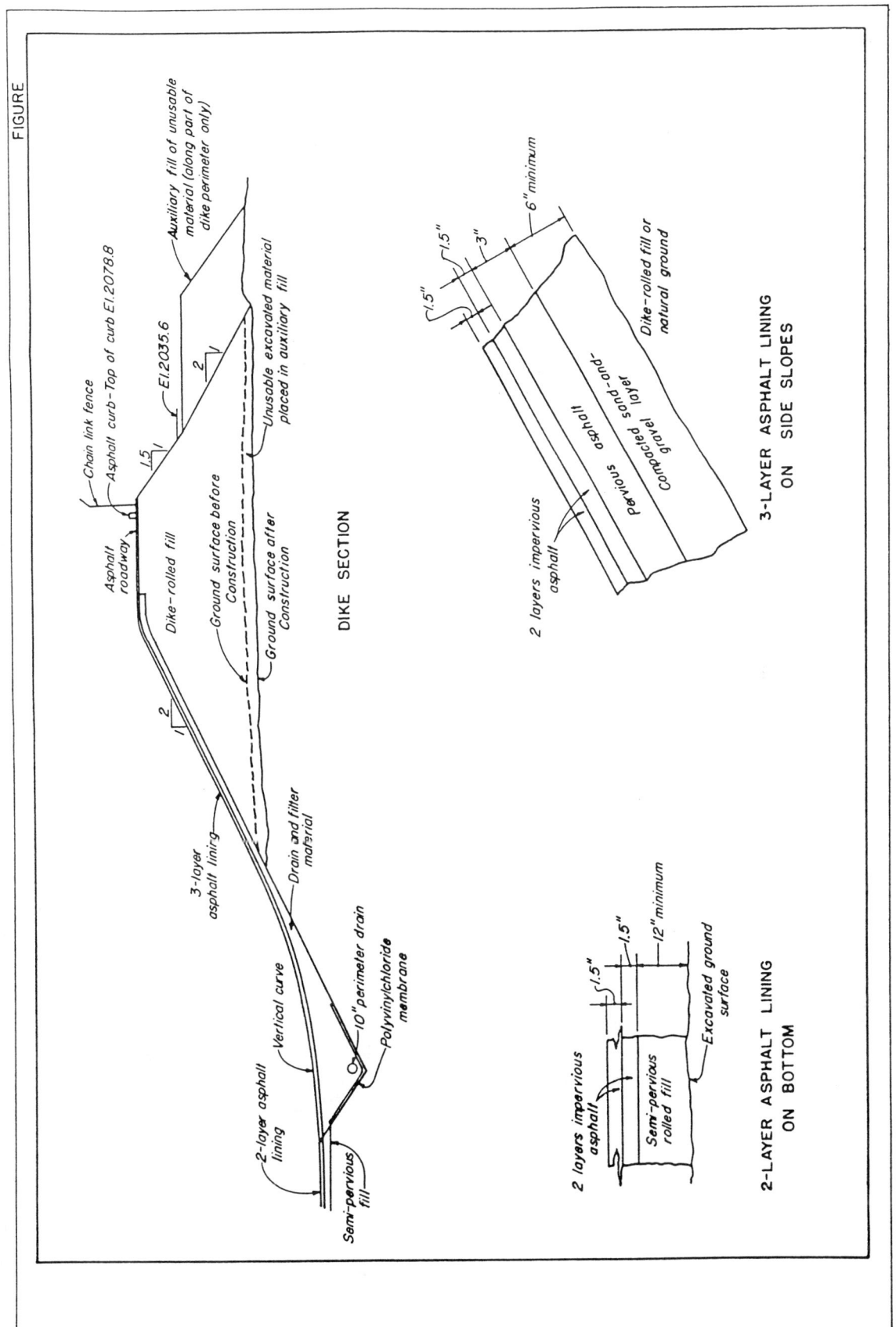

FIGURE

Center of upper intake
609,544.59N
1,656,886.51E

7°30'
45° (typ.)
3'-4"
Pier
₵ 4" steel drain pipe connect to header

₵ high head conduit
N22°30'37.85"E
to powerstation

₵ 6" steel drain header pipe, connect header pipe to high head conduit drain

22'-4"
41'-3"
12'-4"R
Crest Dia. 31'-9"
100'-0"
3'-4"
2'-0"
30°
El. 1981.80

PLAN

SCALE 0 — 12 FEET

2'-0"
100' Diameter
El. 2004.50
El. 2005.00
El. 2002.00
Short pier
15'-10½' Radius Weir
Long pier
19'-6"
4'-9"R
Ellipse
Crest El. 1988.5
Floor El. 1983.0
El. 1981.80
1'-9"
y x
Origin of ellipse
2'-0"
$\dfrac{x^2}{22.1681} + \dfrac{y^2}{71.4025} = 1$
6" header drain pipe
Top of vertical shaft El. 1975.05

SECTION

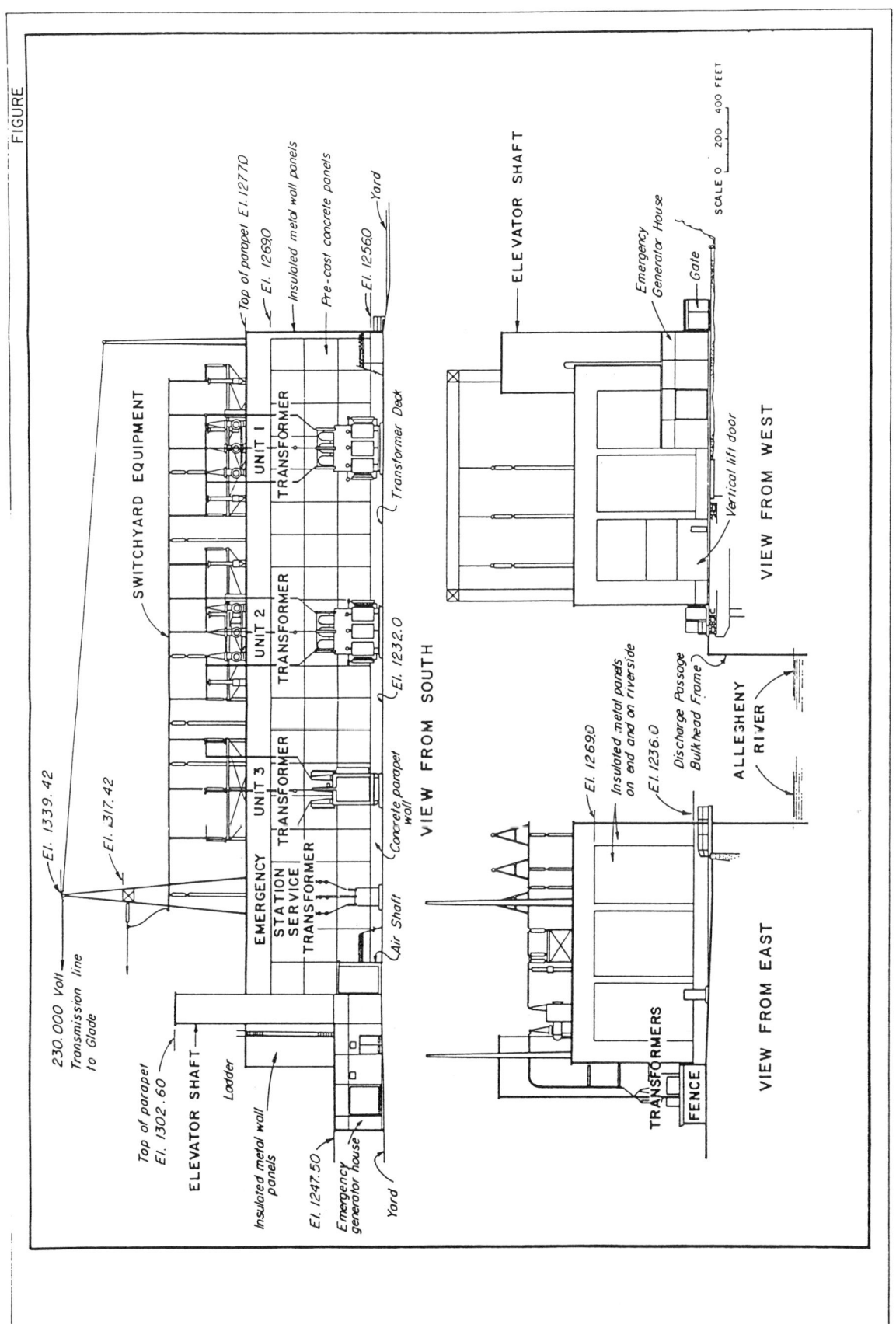

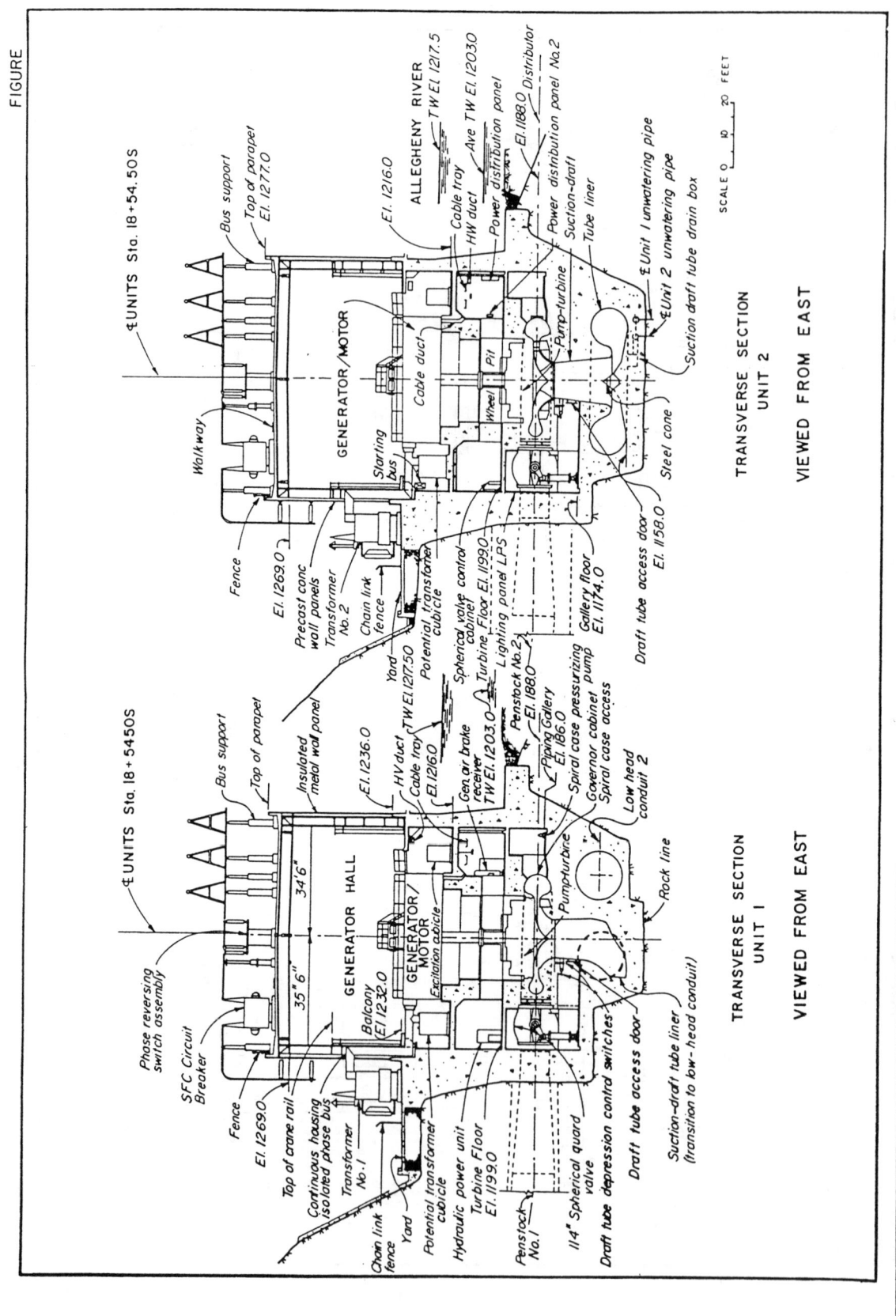

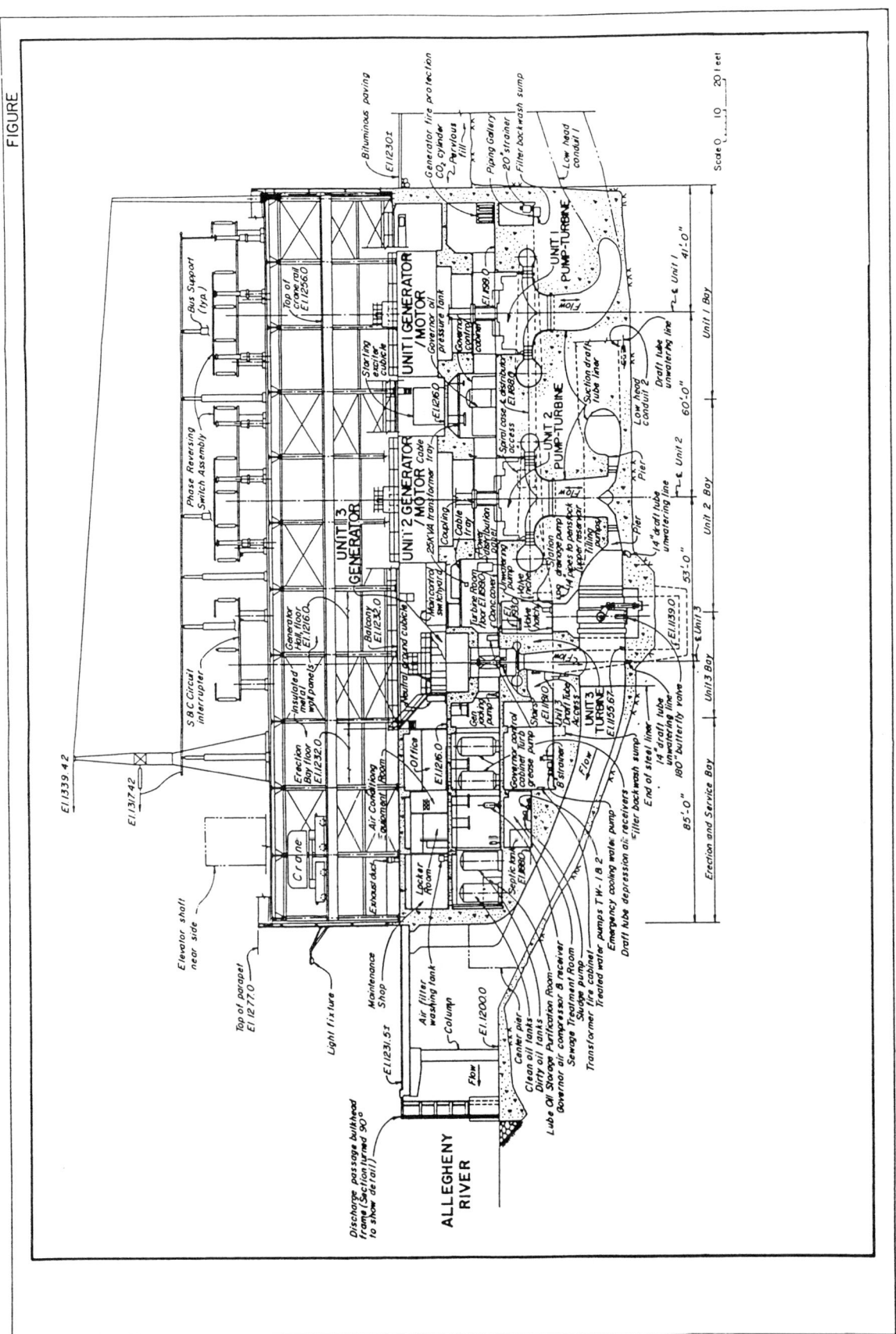

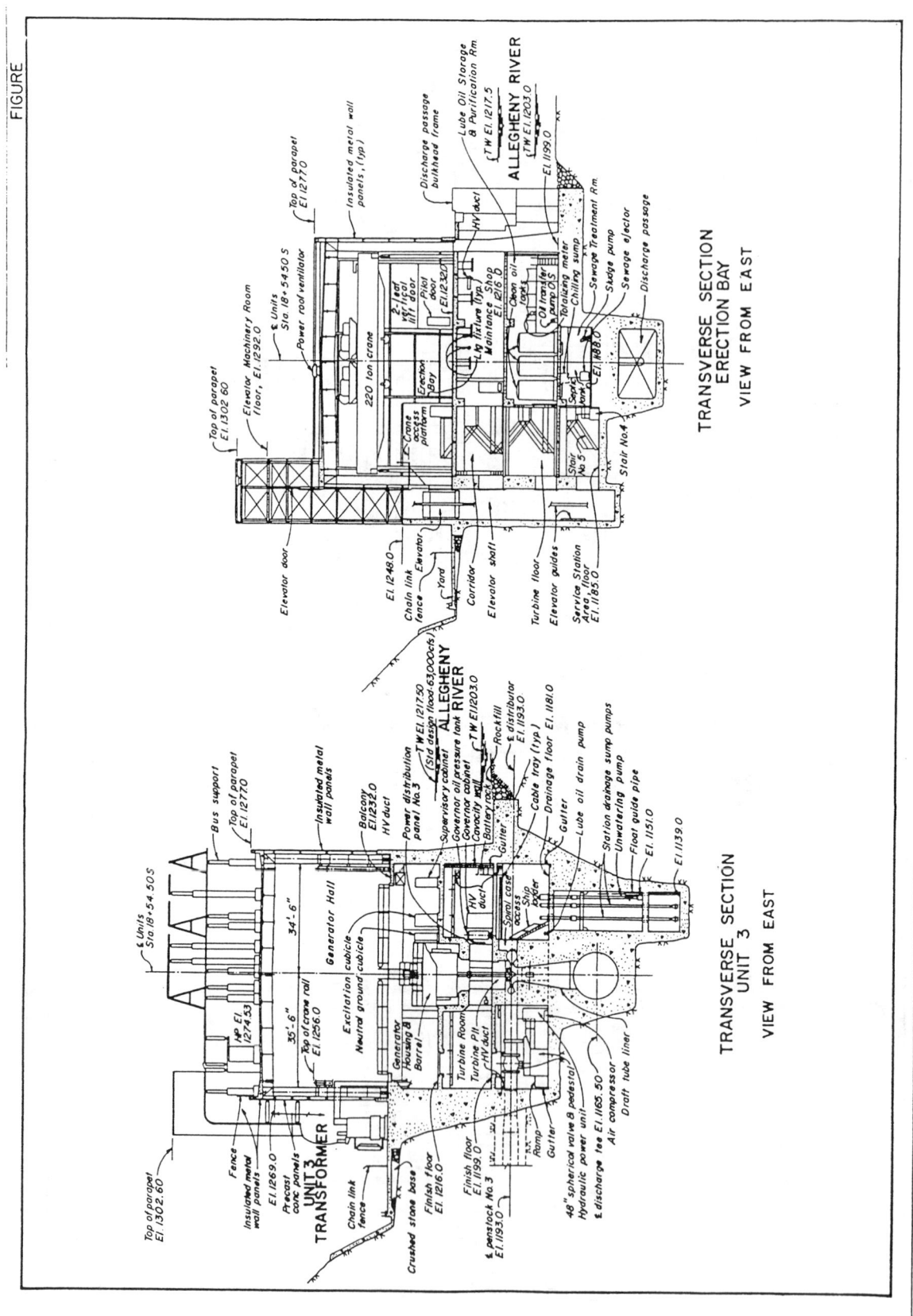

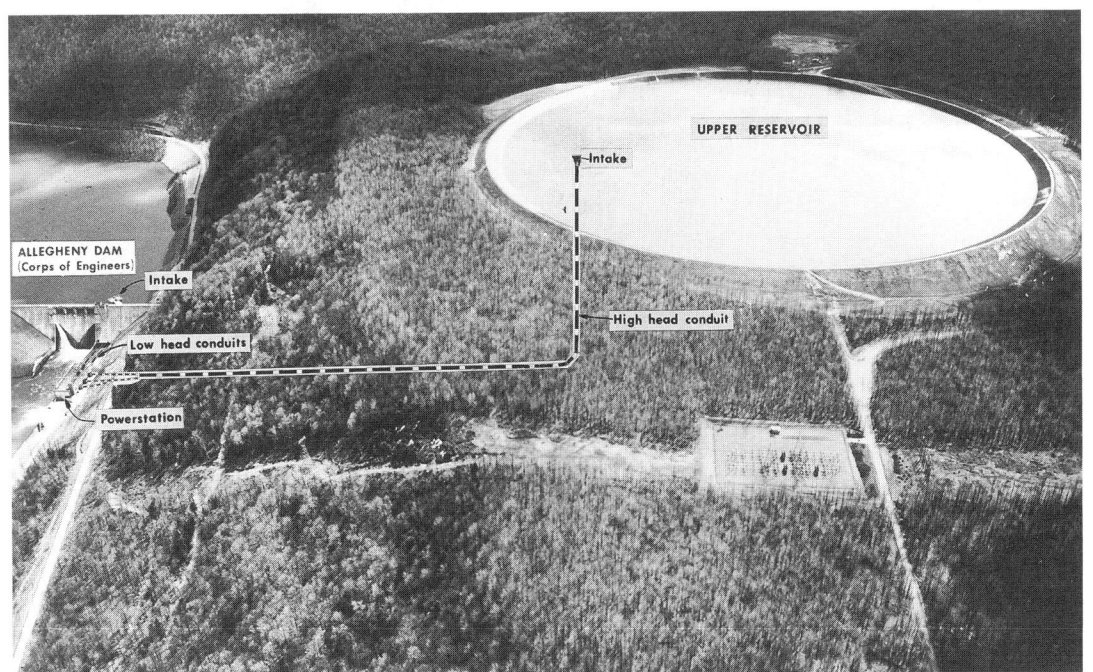

Aerial View of Upper Reservoir

Powerhouse - Units 1 and 2 Reversible, Unit 3 Conventional (foreground)

Kinzua Pumped Storage Plant

Lower Intake at Corps of Engineers' Allegheny Dam

Kinzua Pumped Storage Plant

Plant Name: **LEWISTON**

Plant location: 　Lewiston, NY 　Niagara County	Owner:　New York Power Authority 　　　　　123 Main Street 　　　　　White Plains, New York　10601
Rated capacity　　　　　240 MW Average static head　　　 75 ft Plant efficiency　　　　　% Stored energy　　　　4,800 MWh Number of units　　　　　12	Designers: 　　VHL, Hall and Rich, Division of Chas. T. Main, Inc. 　　Prudential Center 　　Boston, Mass. 02199
Construction time: 3 years, 9 months Construction cost: $333 per kW Price level: 1961 First commercial power: December 1961 FERC project number: 2216 River or water source: Niagara River	Plant Manager/Superintendent: 　　John Philips 　　Niagra Power Project 　　5777 Lewiston Road 　　Lewiston, New York　14092 　　(706) 285-3211

	UPPER RESERVOIR	LOWER RESERVOIR
DAM		
Type	Earth and rockfill	Concrete gravity
Height (ft)	55	389
Crest length (ft)	34,500	1,840
Volume (yd³)	9,250,000	1,104,000
RESERVOIR		
Type	Constructed, Lewiston Reservoir	Constructed
Surface area (acres)	1,900	
Usable power storage (acre-ft)	57,000	
Power pool fluctuation (ft)	35.0	28.0
Operating levels		
Maximum (ft)	655.0	565.0
Minimum (ft)	620.0	548.0
Drainage area (miles²)	3.0	
Seepage (ft³/s)		
SPILLWAY		
Design flood		
Return period (years)		
Flow (ft³/s)		
Capacity (ft³/s)		
Type	None	
Gates		
Number	None	None
Type		
Width (ft)		
Height (ft)		
OUTLET WORKS		
Discharge capacity (ft³/s)		
Number of water passages	None	None
Dimensions of water passages		
Height (ft)		
Width (ft)		
Diameter (ft)		
Type of gates		
Number of gates	None	None

Plant Name: **LEWISTON**

```
Plant location                              Owner:  New York Power Authority
  Lewiston, NY                                      123 Main Street
  Niagara County                                    White Plains, New York  10601

Rated capacity              240 MW
Average static head         22.9 m         Designers:
Plant efficiency               %                   VHL, Hall and Rich, Division of Chas. T. Main, Inc.
Stored energy           4,800 MWh                  Prudential Center
Number of units                12                  Boston, Mass. 02199

Construction time: 3 years, 9 months       Plant Manager/Superintendent:
Construction cost: $333 per kW                     John Philips
Price level: 1961                                  Niagra Power Project
First commercial power: December 1961              5777 Lewiston Road
FERC project number: 2216                          Lewiston, New York 14092
                                                   (706) 285-3211
River or water source: Niagara River
```

	UPPER RESERVOIR	LOWER RESERVOIR
DAM		
Type	Earth and rockfill	Concrete gravity
Height (m)	16.8	118.6
Crest length (m)	10,515.6	560.8
Volume (m³)	7,072,136	844,069
RESERVOIR		
Type	Constructed, Lewiston Reservoir	Constructed
Surface area (Mm²)	7.69	
Usable power storage (Mm³)	70.309	
Power pool fluctuation (m)	10.67	8.53
Operating levels		
Maximum (m)	199.64	172.21
Minimum (m)	188.98	167.03
Drainage area (Mm²)	7.689	
Seepage (m³/s)		
SPILLWAY		
Design flood		
Return period (years)		
Flow (m³/s)		
Capacity (m³/s)		
Type	None	
Gates		
Number	None	None
Type		
Width (m)		
Height (m)		
OUTLET WORKS		
Discharge capacity (m³/s)		
Number of water passages	None	None
Dimensions of water passages		
Height (m)		
Width (m)		
Diameter (m)		
Type of gates		
Number of gates	None	None

Lewiston - Page 1 (Metric)

LEWISTON

INTAKES	UPPER INTAKE	LOWER INTAKE
Number	12	12
Type	Concrete gravity with rectangular openings	Concrete gravity with recangular openings
Design discharge (ft^3/s)	5,000	5,000
Gross inlet area (ft^2) (at trash racks)	880	667
Bar racks		
spacing (in)	6.00	6.00
shape	Rectangular	Rectangular
depth/thickness (in)	3.00 / 1.00	3.00 / 1.00
diameter (in)		
Emergency gates		
number	None	
height/width (ft)	25.00 / 16.17	25.00 / 20.67
type	Stop log	Stop log
Service gates		
number	12	None
height/width (ft)	25.00 / 15.17	
type	Head gates	
Bulkhead/stop logs(Y or N)	Y	Y
number of units serviced		
Hoists		
number	One	One
capacity (tons)	65	150
type	Gantry	Gantry

WATER PASSAGES	Upper Tunnel	Shaft	Lower Tunnel	Surge Tanks Upper	Lower	Penstocks	Tailrace Tunnel
Number						12	
Diameter (ft)						24.0	
Length (ft)						152	
Maximum velocity (ft/s)						14.5	
Concrete lining thickness (in)							
Total length of concrete sections (ft)						60	
Steel liner Thickness							
Minimum (in)						0.47	
Maximum (in)						0.69	
Material grade						A-201-B	
Total length of steel-lined sections (ft)						92	

Notes:
Penstocks: Diameter varies from 7.31 m (24 ft) to 5.49 m (18 ft). Penstocks are embedded in mass concrete.

LEWISTON

INTAKES	UPPER INTAKE	LOWER INTAKE
Number	12	12
Type	Concrete gravity with rectangular openings	Concrete gravity with recangular openings
Design discharge (m³/s)	141.6	141.6
Gross inlet area (m²) (at trash racks)	81.8	62.0
Bar Racks:		
spacing (mm)	152	152
shape	Rectangular	Rectangular
depth/thickness (mm)	76 / 25	76 / 25
diameter (mm)		
Emergency gates		
number	None	
height/width (m)	7.620 / 4.928	7.620 / 6.300
type	Stop log	Stop log
Service gates		
number	12	None
height/width (m)	7.620 / 4.624	
type	Head gates	
Bulkhead/stop logs (Y or N)	Y	Y
number of units serviced		
Hoists		
number	One	One
capacity (Mg)	59	136
type	Gantry	Gantry

WATER PASSAGES	Upper Tunnel	Shaft	Lower Tunnel	Surge Tanks Upper	Surge Tanks Lower	Penstocks	Tailrace Tunnel
Number						12	
Diameter (m)						7.32	
Length (m)						46.3	
Maximum velocity (m/s)						4.42	
Concrete lining thickness (m)							
Total length of concrete sections (m)						18.3	
Steel liner Thickness							
Minimum (mm)						12	
Maximum (mm)						17	
Material grade						A-201-B	
Total length of steel-lined sections (m)						28.0	

Notes:
Penstocks: Diameter varies from 7.31 m (24 ft) to 5.49 m (18 ft). Penstocks are embedded in mass concrete.

Lewiston - Page 2 (Metric)

LEWISTON

POWERHOUSE and RELATED FEATURES

Powerhouse Structure
Type: Surface
Length: 917 ft   Width: 240 ft   Height: 160 ft

Guard Valves
Number: None   Diameter: ft
Type:

Transformers
Number: 4
Ratings: 100 each
Voltages: (kV) 240 / 13.2

Generator
Rating generating (MVA): 25.0   Rating pumping (MVA): 28.0
Insulation type:
Starting method: Depress water level, start units spinning, open valves and gates
Starting equipment: Air depression system

Runners
Material:
Minimum unit submergence: 4.0 ft
WR^2:
Manufacturer: Allis Chalmers
Model test by: Allis Chalmers

	Reversible Runners	Reversible Motor/Generator			
Number	12				12
Diameter (ft)	17.20	Rotor	24.92	Stator	30.67
rpm synchronous	112.5				112.5
rpm overspeed	202.0				202.0
Type	Modified Francis				

Information on Runners

	Gross Head (ft)		Capacity (MW)		Discharge (ft³/s)		Turbine/Pump Eff.(%)	
Condition:	Generating	Pumping	Generating	Pumping	Generating	Pumping	Generating	Pumping
Maximum head & maximum power	113	118	35	26	4,900	3,000	76.0	87.0
Minimum head & maximum power	66	76	15	21	4,400	4,100	65.0	80.0

	Net Head (ft)		Capacity (MW)		Discharge (ft³/s)		Turbine/Pump Eff.(%)	
Condition:	Generating	Pumping	Generating	Pumping	Generating	Pumping	Generating	Pumping
Rated head @ best gate	111	118	28	26	3,500	3,000	86.0	86.0

Lewiston - Page 3 (English)

LEWISTON

POWERHOUSE and RELATED FEATURES

Powerhouse Structure
Type: Surface
Length: 279.5 m Width: 73.2 m Height: 48.8 m

Guard Valves
Number: None Diameter: m
Type:

Transformers
Number: 4
Ratings: 100 each
Voltages: (kV) 240 / 13.2

Generator
Rating generating (MVA): 25.0 Rating pumping (MVA): 28.0
Insulation type:
Starting method: Depress water level, start units spinning, open valves and gates
Starting equipment: Air depression system

Runners
Material:
Minimum unit submergence: 1.22 m
WR^2:
Manufacturer: Allis Chalmers
Model test by: Allis Chalmers

	Reversible Runners	Reversible Motor/Generator		
Number	12			12
Diameter m	5.23	Rotor 7.595	Stator	9.347
rpm synchronous	112.5			112.5
rpm overspeed	202.0			202.0
Type	Modified Francis			

Information on Runners

Condition:	Gross Head (m) Generating	Gross Head (m) Pumping	Capacity (MW) Generating	Capacity (MW) Pumping	Discharge (m³/s) Generating	Discharge (m³/s) Pumping	Turbine/Pump Eff.(%) Generating	Turbine/Pump Eff.(%) Pumping
Maximum head & maximum power	34.4	36.0	35	26	138.8	85.0	76.0	87.0
Minimum head & maximum power	20.1	23.2	15	21	124.6	116.1	65.0	80.0

Condition:	Net Head (m) Generating	Net Head (m) Pumping	Capacity (MW) Generating	Capacity (MW) Pumping	Discharge (m³/s) Generating	Discharge (m³/s) Pumping	Turbine/Pump Eff.(%) Generating	Turbine/Pump Eff.(%) Pumping
Rated head @ best gate	33.8	36.0	28	26	99.1	85.0	86.0	86.0

Lewiston - Page 3 (Metric)

LEWISTON

Plant Data:
   Average GWh generating per year: 426
   Average GWh pumping per year: 712
   Starting time from standstill (s):
   Changeover time pumping to generating (min):   8
   Planned/scheduled time
      between major overhauls (years):
   Outage time required per unit
      during major overhauls (weeks):   12
   Representative plant availability (%):   93.0
   Representative planned outages (weeks per year):   2

Miscelleneous Notes:
The upper reservoir contains 2.4 million m³ (3.1 million yd³) of earthfill and 4.70 million m³ (6.15 million yd³) of rock.

The lower dam volume of 844,100 m³ (1,104,000 yd³) is concrete.

Turnaround times:
  Generating to pumping   8 min
  Pumping to generating   8 min
  Mode changes   730 per machine per year
The lower reservoir is the forebay to the Robert Moses Niagara Powerplant.

Rated capacity per unit is 28,000 hp at 22.9 m (75 ft) net head generating and 96.3 m³/s (3,400 ft³/s) at 25.9 m (85 ft) total dynamic head.

Cavitation Experience:
The turbine runners do cavitate and they are repaired once every two years when they are out for routine maintenance.

Significant or Unique Problems:

List of Licenses Required:
   FERC.

ENVIRONMENTAL FEATURES

Recreation:
Hyde Park was enlarged by 583 Km² (144 acres) deeded to the City of Niagara Falls from the Power Authority, and an existing 9-hole golf course was expanded to 18 holes. Another 18-hole course was rearranged. A modern clubhouse was built and an underpass conects the two courses.

An artificial hill for coasting near the clubhouse was constructed. Also, at Lewiston Historical Park, 174 Km² (43 acres) were provided to the Niagra Frontier State Park Commission. Recreational facilities were developed there while historic sites were preserved and

LEWISTON

reconstructed.

Fish and Wildlife:
The Power Authority has built a fishing facility at the upper reservoir with an approach and parking for 20 cars.

Social:
The Power Authority has provided financial assistance to the town of Lewiston to improve and upgrade its water supply.

Robert Moses Niagara Power Plant is in foreground, then open canal connecting it with reservoir pump-generating plant.

At entrance to Intake No. 1, Power Authority Chairman Robert Moses stands between Asa George, Assistant Chief Engineer of the Authority, left, and the author, Col. William S. Chapin, its Chief Engineer and General Manager.

Twin intakes from Niagara River are seen 21 hours after first water flowed. Each intake is 700 ft long and has 40 slotted openings taking water from 13 to 26 ft below the surface.

The Niagara Power Project

WILLIAM S. CHAPIN
General Manager and Chief Engineer
Power Authority of the State of New York
New York, N. Y.

Power started to flow on February 10 from two of the 13 generators at the Robert Moses Niagara Power Plant at Niagara Falls, N. Y. In this 2,190,000-kw plant, costing $720 million, an additional generator will go on the line every five weeks until all are operating. A reservoir pump-generating plant, a feature of the project, will be ready for operation late in the year, when sufficient generators in the main powerhouse are working to utilize all the available water.

This hydroelectric plant, largest in the Western world, was built by the Power Authority of the State of New York. It was put in service in three years from the receipt of a workable license from the Federal Power Commission, on January 30, 1958. Financing was done through the sale of self-liquidating bonds to private investors and does not involve credit or grants from the state or federal government.

The great plant was built by a number of the nation's outstanding constructors, working on separate sections to complete the work in record time. Contracts for equipment and for field work were let in large blocks, many of them in the $20 million to $100 million bracket. One firm handled $170 million in contracts. Table I lists the major contracts for the work.

In many ways the Niagara plant is

unique. It is by far the largest hydro plant ever built in a busy industrial and residential area. Power had to be developed from run-of-the-river flow, taking the water as it comes without regulation. Long transmission lines tie the plant into the St. Lawrence project and a state-wide grid, and provide for interchange with Canadian plants.

By treaty with Canada, flow over Niagara Falls is to be permanently maintained at a rate that will preserve, and even enhance, the beauty of the Falls. This requires that a flow of 100,000 cfs be maintained over the Falls during the daylight hours of the tourist season and 50,000 cfs at other times. This comes from an average flow of 202,500 cfs, the water available for power being divided between Canada and the United States.

Operation without a dam is practical because the Niagara and St. Lawrence Rivers draw their water from the 300,000-sq mile tributary area of the Great Lakes, which forms the finest reservoir in the world, feeding the rivers so uniformly that the maximum flow is only about twice the minimum flow. This ratio is in great contrast to most other large rivers: 25 to 1 for the Mississippi, 35 to 1 for the Columbia, and 31 to 1 for the Ottawa. Despite the uniformity of the flow, the non-uniformity of the electric demand, the added 50,000 cfs of water available during night hours, and the complete lack of upstream storage make pumped storage an economic necessity—one of the unusual features of the project.

Robert Moses, dynamic chairman

The Niagara Power Project is being constructed by the same forces that recently completed the $650 million St. Lawrence Power Project, which won the ASCE award as the Outstanding Civil Engineering Achievement of 1960. The Power Authority of the State of New York did not possess a secret weapon to enable it to overcome all the obstacles in its way in financing, designing and constructing this billion-dollar self-liquidating project in the short period of six years. It did have the great abilities of its dynamic chairman, Robert Moses, who reorganized it and injected new life into it in 1954. With characteristic vigor he gave overall direction to the financing and prosecution of the work on an exceptionally rapid schedule. Mr. Moses, however, gives credit to the staff and the consultants responsible for the design, construction and daily direction of the work. The role played by Uhl, Hall & Rich, the principal engineering consultants, in the design and supervision

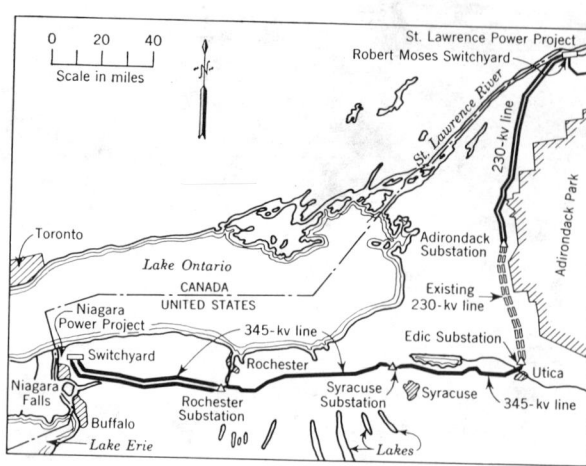

FIG. 1. A high-tension tie-line connects the Niagara and St. Lawrence power plants, increasing the firm capacity of both projects.

of this project cannot be overemphasized.

The Power Authority itself consists of five trustees and a small supervisory staff of competent, dedicated people. It would have been impossible for the Authority to get together the design and supervisory staff needed for work of this magnitude even had there been the desire to do so. Chairman Moses followed a practice of many years standing in directing large engineering projects, of employing the best engineering and architectural consultants available. The engineers were given latitude to carry on the work without interference from the owner's staff or others. But the responsibility for getting results on a precise schedule was imposed upon them.

Canada shares water

Canada, through The Hydro-Electric Power Commission of Ontario, shares half the flow at Niagara, after some minor adjustments to equitably apportion water diverted into the Great Lakes by Canada and out of them by the United States. For years unnecessary controversies, both international and local in character, delayed authorization for the full utilization of Niagara water. However, Canada took earlier advantage of the Niagra treaty than did the United States and, in 1955, completed its Sir Adam

TABLE I. Major contracts for Niagara Power Project

DESCRIPTION	CONTRACTOR	APPROX. COST MILLIONS OF $
Intake and 8,000 ft of covered conduit	Merritt-Chapman & Scott Corp.	72
Covered conduit, 9,000 ft	Balf, Savin & Winkelman	38
Covered conduit, 5,500 ft	Gull-De Felice	34
Open canal, 4,000 ft, construction of reservoir	Kiewit, Morrison-Knudsen, Perini, Walsh	46
Power plant, general contract	Merritt-Chapman & Scott	100
Hydraulic turbines	Baldwin-Lima-Hamilton Corp. and Newport News Shipbuilding	21
Generators	Westinghouse Electric Corp.	21
Penstocks	Chicago Bridge & Iron Co.	11
Transformers	Ferranti Electric, Inc.	4
Reservoir pump-generating plant	Tuscarora Contractors (Arundel, Dixon and Hunkin-Conkey)	40
Pump turbines	Allis-Chalmers	12
Motor-generators	Allis-Chalmers	8
Niagara switchyard	Emerson Gordon Electric Co. and Day & Zimmerman	11

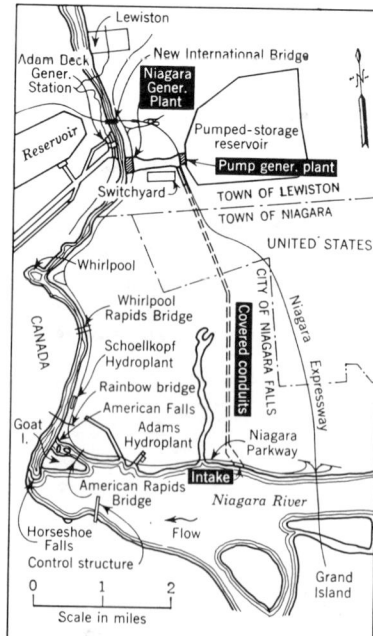

FIG. 2 Area map locates important features of project.

CIVIL ENGINEERING • April 1961

About 83 percent of total concrete had been placed in reservoir pump-generating plant in October 1960. Tailrace bridge piers and one lane of expressway deck are completed.

Suspended above unit No. 2 of main power plant, 595-ton rotor is about to be lowered into place inside a pit-mounted stator that provides only 5/8-in. clearance all around.

Beck Power Plant No. 2, which is only a little smaller than the United States Niagara Plant. (See "Canadian Power Development at Niagara Falls Nears Completion," by Otto Holden, M. ASCE, CIVIL ENGINEERING, Jan. 1956, vol. p. 6.)

Partial destruction of the Schoellkopf plant by a rock slide in June 1956 gave impetus to United States plans. However a clearcut go-ahead for the project could not be obtained until early 1958. The Schoellkopf Plant, and the G. S. Adams plant, that has been producing power from the Falls with the same equipment since 1895, will be retired since the new plant will more efficiently utilize the water available.

At the Falls the Niagara River drops spectacularly 160 ft. Before it reaches the Falls it drops some 60 ft through cataracts, and below the Falls it drops another 115 ft through several rapids. By taking the water from above the upper rapids, at about the level of Lake Erie, and discharging it near the level of Lake Ontario, 314 ft of the total 326-ft difference in water levels between the two lakes is utilized.

The bedrock structure in the area is comparatively simple, consisting of nearly horizontal layers of shale, limestone, dolomite and sandstone. These layers have remained stable since they were first deposited and are not folded or faulted. This situation simplified construction somewhat, but because the rock is known to be subject to some movement, and because of the failure that occurred at the Schoellkopf plant, excavation at the main Niagara generating plant was carried back at about a 45-deg angle.

Maximum power from the available Niagara water is obtained by diverting the water 2½ miles above the Falls. Each of the two intakes is a concrete structure 700 ft long with 48 vertically slotted openings, which take the water from 13 to 26 ft below the river's surface. The structure acts as a fender to keep floating ice moving steadily downstream without entering the power intakes. Gates 49 ft wide and 68 ft high control the flow of the water to two parallel cut-and-cover tunnels which, lined, are 46 ft wide and 66 ft high. (See "Twin Conduits for Niagara," H. P. Cerutti, CIVIL ENGINEERING, July 1960, p. 50.) These tunnels carry the water 4¼ miles through cuts up to 150 ft deep, to an open canal, which forms the Niagara generating plant forebay. The latter is 4,000 ft long, 500 ft wide, and 110 ft deep. This serves the reservoir pump and generating plant as well as the major plant.

Robert Moses Niagara Power Plant

Located in the Niagara Gorge about 5 miles below the Falls, the power plant will have 13 vertical-shaft Francis-type hydraulic turbines. These have welded plate-steel spiral casings with cast-steel runner components welded into a unit. Each unit

Niagara Generating Plant is seen at end of open canal, with Sir Adam Beck Generating Station of Hydro-Electric Power Commission of Ontario on far side of Niagara River. In foreground Bailey bridge carries road across the canal.

Transmission lines for Niagara Power Project are seen looking north from the vicinity of Gibson Substation. The switchyard is located just south of the Open Canal. See Fig. 2. High-tension line connnects Niagara and St. Lawrence power plants

uses 6,250 cfs of water at best gate and 7,500 cfs at full gate. The design capacity of each turbine is 200,000 hp at best efficiency, at a 300-ft head. The generators deliver 167,000 kva or 150,000 kw at 0.90 power factor, which puts them among the largest units ever built.

The powerhouse is a modified outdoor type; elimination of the conventional superstructure afforded substantial economy. The structure is 1,850 ft long, 389 ft high above the foundation, and 580 ft in width. A traveling crane of 630-ton capacity will service the units. It carries a heated housing sufficiently large to completely enclose a unit for minor service or for transport to the service building. The turbines discharge into the Niagara River just upstream from the existing Sir Adam Beck Power Plant on the Canadian side of the river.

Because of the large amount of power, which must be transported a very considerable distance for use, transformers in the plant take the 13,800-v output of the generators and step it up to 115 or 230 kv. At the nearby switchyard it can be stepped up further to 345 kv for distant transmission.

Reservoir pump-generating plant

A dual-purpose pump-generating plant will take the extra water available during nighttime flow and lift it as much as 100 ft for storage to meet the maximum power demands of daylight hours. Twelve pump-generator units are installed in a structure 974 ft long, 160 ft high, and having a width of 240 ft. The units are of the fixed-blade, non-feathering Francis type and have runners of 206-in. diameter, the largest that could be shipped in one piece. Each generator is capable of pumping 3,400 cfs at a net head of 85 ft and of producing 20,000 to 28,000 kw of power as the water flows back through the units. The same water then passes through the turbines of the main plant, augmenting the allocated natural flow during daylight hours.

The two generating plants will have an installed capacity of 2,190,000 kw, compared with 1,974,000 kw at Grand Coulee Dam and 1,824,000 kw at the St. Lawrence Project jointly operated by the New York State Power Authority and Ontario Hydro. The annual energy output is expected to be 13 billion kwhr.

An important element in the Niagara Project is the high-tension tie-line between the Niagara and St. Lawrence plants. The Authority is constructing two 345,000-v circuits between the new plant and Rochester and one

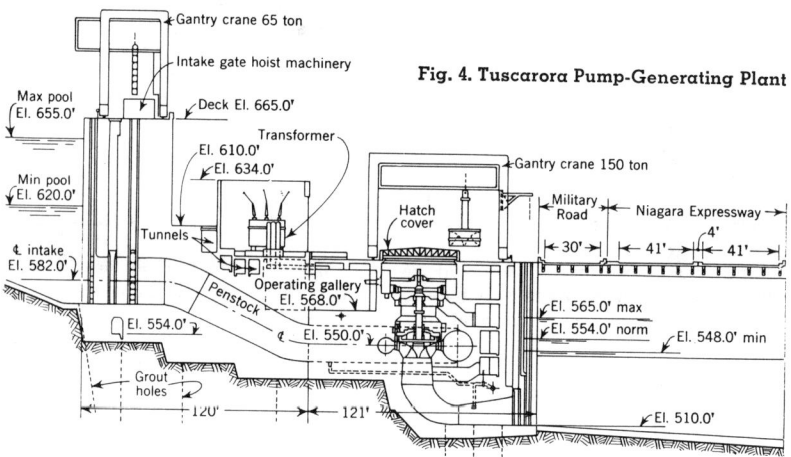

Fig. 3. Niagara Generating Plant is seen in cross section.

Fig. 4. Tuscarora Pump-Generating Plant

345,000-v circuit from Rochester to Utica, where a connection is made to the St. Lawrence plant by two existing 230,000-v lines. The tie-line increases the firm power capacity of each project, because normally water flow at the St. Lawrence Project is highest during the summer, the season during which the U.S.-Canadian treaty reduces the available water for power production at Niagara Falls.

The scars of construction are now rapidly disappearing at the Niagara plant as covering of the conduits is completed. The area traversed by the conduits was criss-crossed with streets, railroads and electrical transmission lines, all of which had to be maintained while an open cut as much as 150 ft deep was made. Subsurface utilities included not only the usual water, sewer and gas lines but also hydrogen, acetylene, chlorine and other chemical lines between industrial plants. Some 19 transmission lines and 100 homes had to be moved. The entire area is now being restored for industrial, residential and recreational use.

The incidental program in connection with the project, participated in by other agencies, includes the Niagara Parkway, the Niagara Expressway, a new bridge between Lewiston, N.Y., and Queenston, Ontario, and the long-awaited railroad grade-crossing eliminations. May 30, 1962, will mark the opening of the new international bridge and the expressway approach to it. One section of the Niagara Parkway was opened to traffic last summer. Other sections will receive traffic as they are completed. The year 1963 should see substantial completion of the entire integrated program at Niagara.

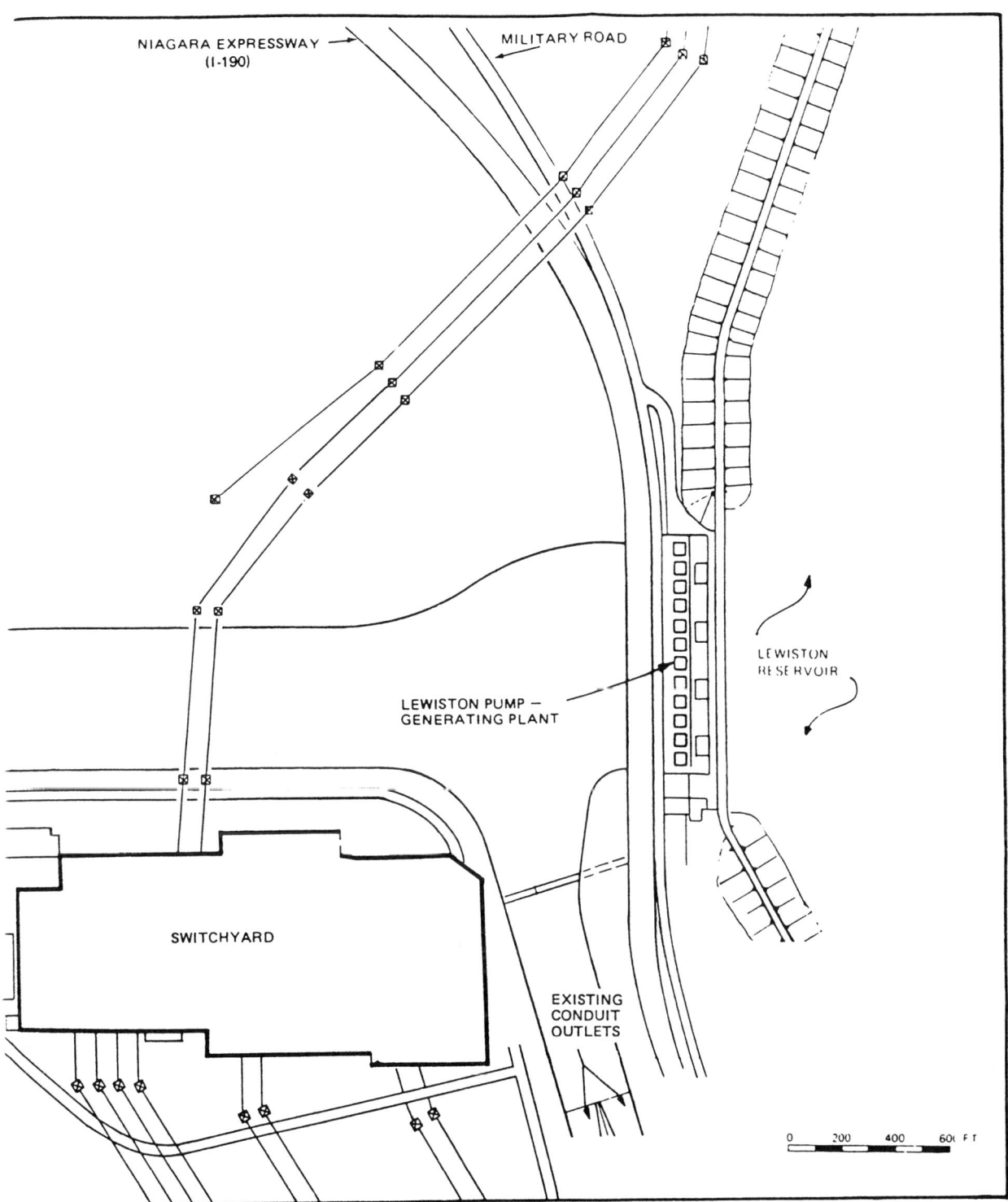

EXISTING PROJECT

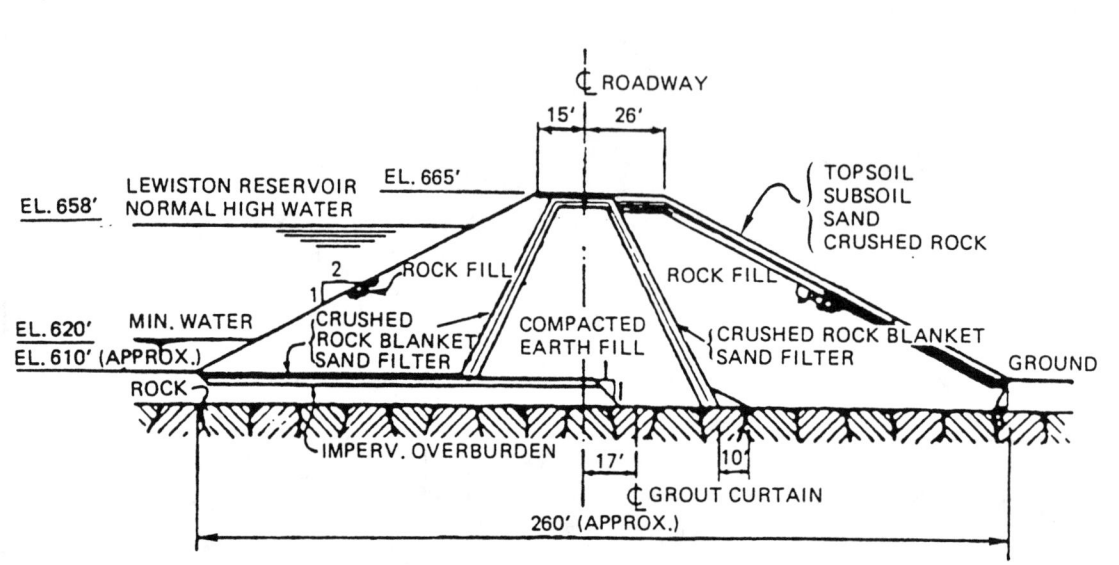

Note: Elevations are referenced to the U.S. Lake Survey Datum of 1935. An elevation of 100 feet USLS Datum would correspond to an elevation of 98.27 feet referenced to the International Great Lakes Datum, 1955, the datum presently in use.

DIKE CROSS SECTION

Aerial View

Powerhouse and Upper Reservoir

Lewiston Pumped Storage Plant

Lewiston Pump/Generating Plant

Bibliography

1. Davis, Calvin Victor and Kenneth E. Sorensen, <u>Handbook of Applied Hydraulics</u>, Third Edition, 1970, McGraw-Hill Book Company, pp. 25-17.

2. <u>Niagara Power Project</u> - <u>Data</u> - <u>Statistics</u>, Fourth Revision, April, 1965, Power Authority of the State of New York.

3. Hamlin, James B., "Experiences with Large Pump-Turbines at Niagara Falls," <u>Proceedings of the American Power Conference</u>, 1964, Vol. 26, page 704.

4. Chapin, William S., "The Niagara Power Project," <u>Civil Engineering</u>, April, 1961, ASCE, pp. 36-39.

Plant Name: **LUDINGTON**

Plant location:
  Ludington, MI
  Mason County
 	
Rated capacity   1,979 MW
Average static head   364 ft
Plant efficiency   71.50 %
Stored energy   17,200 MWh
Number of units   6
 	
Construction time: 3 years
Construction cost: $170 per kW
Price level: 1973
First commercial power: January 1973
FERC project number: 2680
 	
River or water source: Lake Michigan (Lower Reservoir)
	Owner: Consumers Power (51%) / Detroit Edison (49%)
	        212 W. Michigan Avenue
	        Ludington, Michigan 49201
	
	Designers:
	        Ebasco Services, Inc.
	        2 World Trade Center
	        New York, N.Y. 10048
	
	Plant Manager/Superintendent:
	        Richard Gerkowski / Consumers Power
	        3535 South Lake Shore Drive
	        Ludington, Michigan 49431
	
	        (616) 845-6264

DAM	UPPER RESERVOIR	LOWER RESERVOIR
Type	Asphalt faced, zoned earth embankment	None
Height (ft)	170	
Crest length (ft)	31,824	
Volume (yd³)	37,669,000	
RESERVOIR		
Type	Constructed	Lake Michigan
Surface area (acres)	840	14,336,248
Usable power storage (acre-ft)	52,210	
Power pool fluctuation (ft)	66.9	
Operating levels		
Maximum (ft)	942.3	579.7
Minimum (ft)	875.3	579.7
Drainage area (miles²)	1.3	
Seepage (ft³/s)	18.364	
SPILLWAY		
Design flood		
Return period (years)		
Flow (ft³/s)		
Capacity (ft³/s)		
Type	None	None
Gates		
Number	None	None
Type		
Width (ft)		
Height (ft)		
OUTLET WORKS		
Discharge capacity (ft³/s)		
Number of water passages	None	None
Dimensions of water passages		
Height (ft)		
Width (ft)		
Diameter (ft)		
Type of gates		
Number of gates	None	None

Plant Name: **LUDINGTON**

Plant location Ludington, MI Mason County	Owner: Consumers Power (51%) / Detroit Edison (49%) 212 W. Michigan Avenue Ludington, Michigan 49201

Rated capacity 1,979 MW
Average static head 111.0 m
Plant efficiency 71.50 %
Stored energy 17,200 MWh
Number of units 6

Designers:
Ebasco Services, Inc.
2 World Trade Center
New York, N.Y. 10048

Construction time: 3 years
Construction cost: $170 per kW
Price level: 1973
First commercial power: January 1973
FERC project number: 2680

Plant Manager/Superintendent:
Richard Gerkowski / Consumers Power
3535 South Lake Shore Drive
Ludington, Michigan 49431

(616) 845-6264

River or water source: Lake Michigan (Lower Reservoir)

	UPPER RESERVOIR	LOWER RESERVOIR
DAM		
Type	Asphalt faced, zoned earth embankment	None
Height (m)	51.8	
Crest length (m)	9,700.0	
Volume (m³)	28,800,000	
RESERVOIR		
Type	Constructed	Lake Michigan
Surface area (Mm²)	3.40	58,018.00
Usable power storage (Mm³)	64.400	
Power pool fluctuation (m)	20.40	
Operating levels		
Maximum (m)	287.20	176.70
Minimum (m)	266.80	176.70
Drainage area (Mm²)	3.370	
Seepage (m³/s)	0.5200	
SPILLWAY		
Design flood		
Return period (years)		
Flow (m³/s)		
Capacity (m³/s)		
Type	None	None
Gates		
Number	None	None
Type		
Width (m)		
Height (m)		
OUTLET WORKS		
Discharge capacity (m³/s)		
Number of water passages	None	None
Dimensions of water passages		
Height (m)		
Width (m)		
Diameter (m)		
Type of gates		
Number of gates	None	None

Ludington - Page 1 (Metric)

LUDINGTON

INTAKES	UPPER INTAKE	LOWER INTAKE
Number	1	6
Type	Horizontal, reinforced concrete	Draft tube
Design discharge (ft³/s)	12,660	11,001
Gross inlet area (ft²) (at trash racks)	817	1,324
Bar racks		
spacing (in)		14.00
shape		Circular
depth/thickness (in)		2.00 /
diameter (in)		
Emergency gates		
number	None	None
height/width (ft)		
type		
Service gates		
number	6	12
height/width (ft)	28.87 / 28.87	21.00 / 32.15
type	Vertical lift, fixed wheel	Slide
Bulkhead/stop logs (Y or N)	Y	
number of units serviced	1	
Hoists		
number	6	1
capacity (tons)	250	360
type	Mechanical, motor drive	Gantry crane

WATER PASSAGES	Upper Tunnel	Shaft	Lower Tunnel	Surge Tanks Upper	Lower	Penstocks	Tailrace Tunnel
Number						6	
Diameter (ft)						28.5	
Length (ft)						1,299	
Maximum velocity (ft/s)						27.9	
Concrete lining thickness (in)							
Total length of concrete sections (ft)							
Steel liner Thickness							
Minimum (in)						0.50	
Maximum (in)						1.44	
Material grade						A516 & A53 7	
Total length of steel-lined sections (ft)						1,299	

Notes:
Penstocks: Penstocks are encased in concrete at the embankment and powerhouse.

Ludington - Page 2 (English)

LUDINGTON

INTAKES	UPPER INTAKE	LOWER INTAKE
Number	1	6
Type	Horizontal, reinforced concrete	Draft tube
Design discharge (m³/s)	358.5	311.5
Gross inlet area (m²) (at trash racks)	75.9	123.0
Bar Racks:		
spacing (mm)		356
shape		Circular
depth/thickness (mm)		51 /
diameter (mm)		
Emergency gates		
number	None	None
height/width (m)		
type		
Service gates		
number	6	12
height/width (m)	8.800 / 8.800	6.400 / 9.800
type	Vertical lift, fixed wheel	Slide
Bulkhead/stop logs (Y or N)	Y	
number of units serviced	1	
Hoists		
number	6	1
capacity (Mg)	227	327
type	Mechanical, motor drive	Gantry crane

WATER PASSAGES	Upper Tunnel	Shaft	Lower Tunnel	Surge Tanks Upper	Surge Tanks Lower	Penstocks	Tailrace Tunnel
Number						6	
Diameter (m)						8.70	
Length (m)						396.0	
Maximum velocity (m/s)						8.50	
Concrete lining thickness (m)							
Total length of concrete sections (m)							
Steel liner Thickness							
Minimum (mm)						13	
Maximum (mm)						37	
Material grade						A516 & A537	
Total length of steel-lined sections (m)						396.0	

Notes:
Penstocks: Penstocks are encased in concrete at the embankment and powerhouse.

Ludington - Page 2 (Metric)

LUDINGTON

POWERHOUSE and RELATED FEATURES

Powerhouse Structure
Type: Semi-outdoor
Length: 576 ft   Width: 170 ft   Height: 106 ft

Guard Valves
Number: None   Diameter: ft
Type:

Transformers
Number: 9
Ratings: 257
Voltages: (kV) 345 / 20

Generator
Rating generating (MVA): 310.0   Rating pumping (MVA): 337.0
Insulation type:
Starting method: Back-to-back synchronous for 4 units and pony motors for 2 units
Starting equipment:

Runners
Material: Fabricated steel with stainless steel cladding
Minimum unit submergence: 26.5 ft
WR^2: 900,000,000 (lbf x ft^2)
Manufacturer: Hitachi
Model test by: Hitachi

	Reversible Runners	Reversible Motor/Generator
Number	6	6
Diameter (ft)	27.50	Rotor 47.01 Stator
rpm synchronous	112.5	112.5
rpm overspeed	163.0	163.0
Type	Francis	Outdoor, synchronous

Information on Runners

Condition:	Gross Head (ft) Generating	Gross Head (ft) Pumping	Capacity (MW) Generating	Capacity (MW) Pumping	Discharge (ft³/s) Generating	Discharge (ft³/s) Pumping	Turbine/Pump Eff.(%) Generating	Turbine/Pump Eff.(%) Pumping
Maximum head & maximum power	364	364	343	321	12,678	8,052	92.9	87.6
Minimum head & maximum power	295	295	269	315	12,360	11,509	89.7	92.7

Note: Data in the above table are based on design data.

Condition:	Net Head (ft) Generating	Net Head (ft) Pumping	Capacity (MW) Generating	Capacity (MW) Pumping	Discharge (ft³/s) Generating	Discharge (ft³/s) Pumping	Turbine/Pump Eff.(%) Generating	Turbine/Pump Eff.(%) Pumping
Rated head @ best gate	353	286	308	323	11,509	11,124	92.2	92.7

Note: Data in the above table are based on design data.

LUDINGTON

POWERHOUSE and RELATED FEATURES

Powerhouse Structure
Type: Semi-outdoor
Length: 175.5 m   Width: 51.8 m   Height: 32.3 m

Guard Valves
Number: None   Diameter: m
Type:

Transformers
Number: 9
Ratings: 257
Voltages: (kV) 345 / 20

Generator
Rating generating (MVA): 310.0   Rating pumping (MVA): 337.0
Insulation type:
Starting method: Back-to-back synchronous for 4 units and pony motors for 2 units
Starting equipment:

Runners
Material: Fabricated steel with stainless steel cladding
Minimum unit submergence: 8.08 m
WR^2: 372,014,000 (Newtons x m^2)
Manufacturer: Hitachi
Model test by: Hitachi

	Reversible Runners	Reversible Motor/Generator	
Number	6		6
Diameter m	8.38	Rotor 14.330	Stator
rpm synchronous	112.5		112.5
rpm overspeed	163.0		163.0
Type	Francis		Outdoor, synchronous

Information on Runners

Condition:	Gross Head (m)		Capacity (MW)		Discharge (m³/s)		Turbine/Pump Eff.(%)	
	Generating	Pumping	Generating	Pumping	Generating	Pumping	Generating	Pumping
Maximum head & maximum power	111.0	111.0	343	321	359.0	228.0	92.9	87.6
Minimum head & maximum power	90.0	90.0	269	315	350.0	325.9	89.7	92.7

Note: Data in the above table are based on design data.

Condition:	Net Head (m)		Capacity (MW)		Discharge (m³/s)		Turbine/Pump Eff.(%)	
	Generating	Pumping	Generating	Pumping	Generating	Pumping	Generating	Pumping
Rated head @ best gate	107.6	87.2	308	323	325.9	315.0	92.2	92.7

Note: Data in the above table are based on design data.

Ludington - Page 3 (Metric)

LUDINGTON

Plant Data:

Average GWh generating per year:	2,547
Average GWh pumping per year:	3,560
Starting time from standstill (s):	420
Changeover time pumping to generating (min):	12
Planned/scheduled time between major overhauls (years):	2
Outage time required per unit during major overhauls (weeks):	4
Representative plant availability (%):	94.0
Representative planned outages (weeks per year):	3

Miscelleneous Notes:
Seepage from the upper reservoir varies from 0.42 m³/s (15 ft³/s) to 0.62 m³/s (22 ft³/s).

The intake structure in the upper reservoir has six separate inlets. The data reported applies to each of the six inlets.

The hoist has a mechanical drive with two speeds, one for maintenance and one for emergency closure.

Cavitation Experience:
Nominal cavitation has occurred to the runners.

Significant or Unique Problems:
A significant design feature of the plant is the offshore jetty system which provides protection against wave action, littoral drifts, and the transport of sand through the units.

List of Licenses Required:
FERC, Corps Section 404 permit.

ENVIRONMENTAL FEATURES

Recreation:
Recreational facilities include the Lake Michigan Vista Point Scenic Overlook; a picnic area with pavillion, restrooms, and playground; and a trailer park with 48 campsites, showers, restrooms and playground. Periodic "open houses" and special tours are held.

Fish and Wildlife:
Yellow perch and forage species such as sculpins and darters are abundant near the plant offshore structures. Fishing near the plant provides for trout and salmon catches.

Social:
The plant is a landmark for boating navigation.

Ludington pumped-storage project wins 1973 outstanding CE achievement award

Situated on the eastern shore of Lake Michigan, the Ludington project is the biggest pumped-storage facility in the world. Employing the largest pump-turbine and generator-motors ever constructed, the facility takes water from Lake Michigan during off-peak hours and pumps it up to a huge man-made reservoir, 1 mile wide and 2 miles long. During peak-power periods, the water rushes down to Lake Michigan again, generating 1.9 million kw of electric power. The reservoir dike is innovative; it was built from local sand. To keep water from seeping into the dike's interior the dike's inside bank was paved with a unique asphalt sandwich containing its own drainage system.

CARL F. WHITEHEAD, M. ASCE
Manager of projects
Ebasco Services Inc.
New York, N.Y.

DONN RUOTOLO, M. ASCE
Project manager
Ebasco Services Inc.
New York, N.Y.

The Ludington project is the largest pumped storage installation in the world. And the first such project to receive ASCE's "Outstanding Civil Engineering Achievement Award." This massive plant, designed and constructed by Ebasco Engineering Corporation for Consumers Power Co. and Detroit Edison Co. will generate 1,872,000 kw of electric power. Estimated total cost is over $300 million.

Water from Lake Michigan (see Fig. 1) is pumped into a manmade upper reservoir, 344 ft (105 m) above, using off-peak electricity, and stored there until required for generation. The water is then released down penstocks and through turbines to generate electricity on its return to the lake. Three of six power units are now operational.

The project is vast. The 2.2-mile (3.5 km) long by 1-mile (1.6 km) wide upper reservoir, with a gross capacity of 82,000 acre-ft (101 million m³), is contained within a 6-mile (9.6 km) long embankment constructed of over 37 million yd³ (28.2 million m³) of fill, averaging 103 ft (31.4 m) in height. Inside slopes of the embankment are faced with 535,000 tons (485,000 mt) of hydraulic asphalt concrete, covering an area equivalent to 60 mi (97 km) of two-lane highway. The powerhouse penstock encasements and intake structure (see photo) required 440,000 yd³ (336,000 m³) of concrete, and 19,000 tons (17,200 mt) of reinforcing steel.

The Ludington Project contributes to the well-being of the Michigan area as a source of readily available power. The power plant is as pollution free as present technology permits. Local and regional areas have also benefited from the major construction force empoyed, the permanent operating staff needed at the plant, and the taxes payed by the project. A scenic overlook from the top of the upper reservoir will give visitors a magnificent view. There will also be picnic grounds and a trailer park with overnight accommodations.

Project planning

Initially, three problems confronted designers. The first: an encircling reservoir dike, up to 170 ft (52 m) high, would have to be constructed on a pervious foundation using local materials. Though it would be desirable to build the dike out of rock, it was not available in the vicinity. What was available was poorly graded fine sand with layers of clay and calcareous silty sand. Sand for the dike would come from excavating the reservoir. Additionally, a way would have to be found to keep

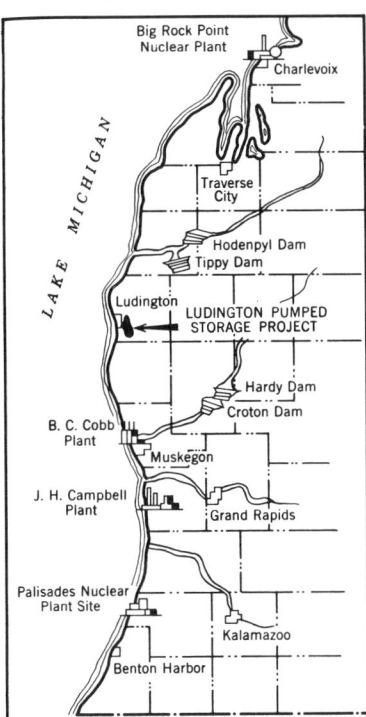

Fig. 1. The Ludington pumped-storage project, on the east shore of Lake Michigan, draws power from the Michigan power pool to pump water from the lake up to a man-made reservoir; and returns ⅔ of this power to the pool when the water falls back to the lake.

water from seeping through the bottom of the reservoir.

The second problem was to design the upper reservoir and lake-front structures to handle massive volumes of moving water (up to 80,000 cfs)—without serious scour. But first we had to do scale model studies of the upper reservoir and the lakefront structures.

The third problem was coordination of design and construction, to complete dike and powerhouse in a mere three years and four months. So, we designed and built a *two-stage* powerhouse, permitting early completion of exterior walls and slabs. Erection of turbine generators and placement of the second stage concrete could then be accomplished under a temporary roof without weather delays. Two-stage development also permitted use of a gantry crane to erect pump-turbines and generator-motors. Further, to minimize winter operations, we scheduled overtime during good weather. And we reduced and relocated craft manpower peaks (see March 73, pp. 72-75).

Model studies

An upper reservoir model (scale: 1/22) was constructed and tested at Alden Research Laboratories at Worcester Poly to analyze flow into the intake and jetting action of pump discharge into the upper reservoir. These studies developed the outline of the deflector apron at the intake (see Fig. 2 and photo A), which minimized the impact of the 20 fps (6.1 m/s) discharge velocities exiting from the six 28 ft (8.5 m) diam. penstocks. The studies also defined the areas where scour protection would be needed on the reservoir bottom; established filling criteria for an empty reservoir; and recommended a baffle wall (photo A) to eliminate possible vortexing at the intake of the upper reservoir.

Lakefront (lower reservoir) studies were done at the University of Michigan Hydraulics Laboratory to develop a protective harbor for the powerhouse tailrace channel, to limit wave heights at the powerhouse to within a safe range (5 ft max) while minimizing currents in the tailrace area. These studies developed the optimum arrangement of harbor and jetty-breakwater scheme (see photo B), suggested a rubble mound construction to save money, and determined the velocity distribution and scour potential in the tailrace channel (see "The Ludington Pumped Storage Project—Design Consideration and Water Flow Studies," ASCE National Water-Resources Conference 1971, Phoenix).

Developing a crack-resistant asphalt

Our goal was to develop an asphalt mix to cover the dike's inside slope that would be impervious to water, deformable, resistant to ice erosion, and free from cracking and fissuring because of heating or aging. J. R. Benson (Denver) and Lord Laboratories (Boulder, Colo.) did the initial research. Then, a half-scale test section was constructed near the site, an exact duplicate of the proposed embankment outer membrane, including pipes and pumps (see Fig. 3 a & b).

Constructing the embankment

Situated on a bluff, the upper reservoir has a foundation of loose-to-medium-dense glacial outwash deposits for about 30 ft (9.1 m) overlying exceptionally dense material. With rock located 800 ft (244 m) below surface, a vertical cutoff was impossible. Thus, the embankment face and reservoir bottom had to be sealed with an impervious membrane. Except for the impervious upstream face and drainage material, all materials for the 6-mile (9.7 km)-long embankment came from the reservoir. The embankment has four zones: random fill; a chimney drain of clean site sand; a site sand zone; and an impermeable upstream face of calcareous silty sand that provides a working surface for asphalt placing equipment, and insures embankment stability by preventing water from seeping into it (see Fig. 3a). A longitudinal drain and lateral drains are provided at the base of the dike's foundation (see Fig. 3b), to limit water seepage into the dike should water bypass the asphalt membrane.

Heterogeneity of the soils made it impractical to rely on a *performance spec* for constructing the dike. So, we conducted a test program to develop a *method spec* that would spell out: the most suitable compaction equipment for each zone; the thickness of each layer; number of passes; speed of machine; and optimum moisture content. Controlling layer thickness proved effective for ensuring quality compaction.

Because of the embankment's zoned construction and the need to obtain all materials from within the reservoir area, careful planning was needed for placement and compaction. The maximum fill placed in one day was 180,000 yd^3 (138,000 m^3). 37 million yd^3 (28 million m^3) was placed and compacted in three construction seasons (mid-April to mid-Dec).

To meet production schedules exceeding 1 million yd^3/wk (764,000 m^3/wk), the earthmoving contractor employed 95 24-yd twin-engine scrapers and an impressive array of loading, hauling, spreading and compaction equipment. Overlays were made of the entire reservoir and the daily operation of each spread planned in detail before actual work. Earthmoving was done in two 10-hour shifts, 6 days/wk.

During winter of 1975-1972, conditions were unfavorable for placement of embankment or lining. But freezing weather had stabilized offsite clay borrow pits and haul roads. So, the earthmoving contractor used the time to haul 2.5 million yd^3 (1.9 million m^3) of clay to stockpiles in the reservoir, within easy working distance after closure of the embankment.

Embankment has asphalt face

With only fine grained glacial material available, it would have been too costly to build a conventional dike with impervious core, filter zones and slope protection. The most economical section would be an impervious membrane on the reservoir-side of the dike face connected to the impervious bottom lining of the reservoir. We considered three impervious facings: reinforced concrete slab; hydraulic asphalt concrete; and steel plate. Besides being impervious, the embankment facing had to take deflection due to embankment settlement; adapt to large variations in temperature (−29 to +38°C); resist erosion from wave action and ice; and handle seepage through the outer layer.

Hydraulic asphalt concrete was selected for technical and economic advantages. From bottom to top, the asphalt pavement (see Fig. 3b) consists of a 3 in. (76 mm) layer of hydraulic asphalt concrete; 18 in. (460 mm) of well-graded 1½ in. (38 mm)-(max) aggregate, which serves as drain; a 2½ in. (64 mm) layer of binder course, serving as binder between overlying impervious asphalt and drain material; and two 2½ in. (64 mm) layers of impervious hydraulic asphalt concrete.

Unique drainage system

A unique safety feature of the asphalt sandwich is the drainage system (Fig. 3b). At the toe of the asphalt face and inside the 18-in. (460 mm) drain layer is a 10 in. (254 mm) *slotted* fiberglass pipe running around the entire inner perimeter of the embankment. To this pipe are connected 8 in. (203 mm) diam. fiberglass casings (no slots) at 150 ft (45.7 m) intervals; these run up the slope and terminate in a manhole at the top of the embankment. Any water finding its way into the drainage zone between asphalt layers is collected in the toe drain and pumped to the surface by submersible pumps inside the 8-in. casings. The pipe was specially designed to withstand the loading of a heavy roller or D8 tractor with minimum cover.

Should a crack occur in the surface pavement, possibly forcing engineers to lower the reservoir water level, water inside the 18-in. (460 mm) drain could exert outward pressure, lifting the upper asphalt layer. But the pumps will prevent this by pumping the water from inside the drainage layer. Their capacity and spacing can lower the water level in the drainage layer at a minimum rate of 8 ft/hr (2.4 m/hr), the maximum reservoir drawdown rate. The pumps also maintain a low water level in the drainage layer from minor seepage through the asphalt.

The size of the pumps, 8 in. (203 mm) casings and manhole on the top of the embankment allow pumps to be lowered into casings and pulled out for servicing. Operation of all pumps is monitored at points along the embankment and in the powerhouse. Water level recorders will chart water levels inside the face drain. To date no water has been detected in the drainage layer.

Asphalt placement

Four specially-designed bridge type pavers, the largest ever built, were made for the Ludington job by Rex-Arbau (Rex Chainbelt) to place the HAC subbase course, binder and the first HAC surface course (photo B). These huge rigs spanned 149 ft (45.4 m), paving one-half the slope distance each pass. Three smaller, specially designed paving machines, previously used only in Europe, were winched up the slope to place the upper HAC surface course. The contractor worked two 12-hour shifts, five days per week during the 10½ month construction period, placing 535,000 tons (485,000 mt) of asphalt concrete and 390,000 yd³ (298,000 m³) of crushed drain rock. Asphalt placing rates exceeding 6500 tons/day (5900 mt/day) (see Dec., CE 1972, pp. 54-57).

Reservoir lining

A clay lining was selected for the impervious upper reservoir bottom because of cost, availability and flexibility. Design thicknesses of the lining vary from 3 to 5 ft (0.9 to 1.5 m) in the middle of the reservoir to 8 ft (2.4 m) along the 5:1 slope of the embankment, and 10 ft (3.0 m) on the top of the asphalt berm. Actual thicknesses varied, depending on the ability to select clay from borrow pits and exposure of clay to frost prior to reservoir filling.

Most clay borrow pits were located within 5 miles (8.0 km) of the reservoir. About 1/6 of the clay was excavated from the reservoir. High clay moisture content made placement difficult, calling for special placement procedures. As clay lining was brought to finish grade, water was ponded over completed sections to prevent drying out and cracking prior to initial reservoir filling operations. To date, seepage through the clay has been greater than anticipated, but decreasing.

Intake

The upper reservoir intake (photo A) is a concrete structure having a separate inlet with vertical slide gates for each unit. These gates close at 2 ft/min. (10.2 mm/s) for maintenance operations and 8 ft/min. (40.8 mm/s) during emergencies—even with full-gate turbine flow.

Penstocks

There are six penstocks (Photo C), one for each unit, varying in diam. from 28.5 to 24 ft (8.7 to 7.3 m). Upper and lower sections, where backfill is greater than 4 ft (1.3 m), are encased in concrete. At the upper section, maximum fill over the encasement is 100 ft (30.5 m) (see Fig. 2). The remainder of the penstock, between the downstream toe of the embankment and the lower encasement, was completely backfilled, requiring stiffeners at 5 ft (1.5 m) intervals to withstand earth pressures.

To allow for creep or settlement, differential settlement joints were installed at the top and bottom of the slope. These joints can take 4 in. (102 mm) of longitudinal expansion or contraction and 2 in. (50.8 mm) of differential settlement. Because the penstocks are on a soil slope, a monitored drainage system was provided. Between penstocks, a gravel bed with collecting pipes drain rain penetration or ground seepage. Horizontal drains, extending up to 300 ft (91.4 m) into the slope, drain ground water deep in the hillside.

The penstocks are painted inside and out and have cathodic protection in the backfilled sections. Strain gages indicate penstock stress, most useful during backfilling and operations.

Penstock steel was delivered in 10 ft (3.0 m) long curved sections, then

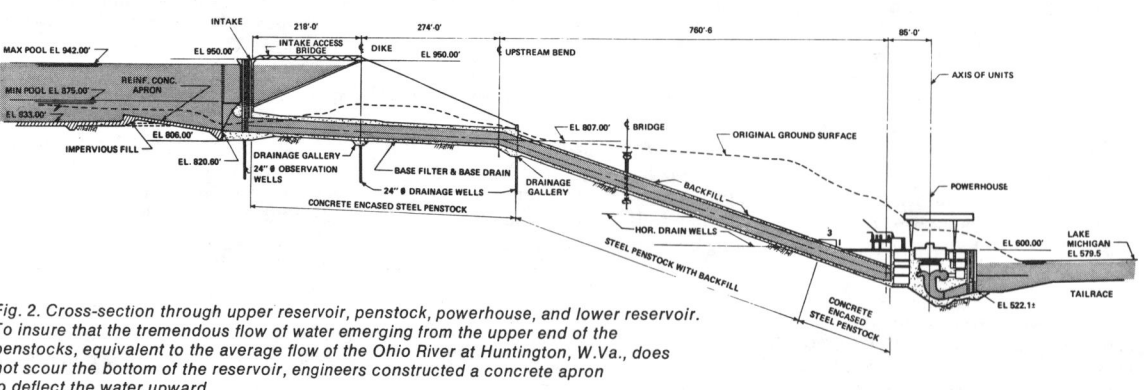

Fig. 2. Cross-section through upper reservoir, penstock, powerhouse, and lower reservoir. To insure that the tremendous flow of water emerging from the upper end of the penstocks, equivalent to the average flow of the Ohio River at Huntington, W.Va., does not scour the bottom of the reservoir, engineers constructed a concrete apron to deflect the water upward.

welded into 30 ft (9.1 m) can sections. These were slid down erection tracks along the 3:1 slope, and supported on temporary steel brackets and spread concrete footings spaced at 90 ft (27.4 m) until welding was complete. A continuous 120° concrete saddle was placed under each penstock. Just before the penstock sections were backfilled, steel supports were burned off to avoid local stress concentrations.

Powerhouse

The semi-outdoor powerhouse (photo C) is conventional—except that it bears on the dense clay under Lake Michigan. This foundation required a unit block design that isolates each unit in a separate 85 ft (25.9 m) wide block of concrete. A water stop scheme using double water stops with a drain in-between keeps total water leakage into the powerhouse, over 600 ft (183 m) long 170 ft (51.8 m) wide and 85 ft (25.9 m) below water level, to less than 5 gals/min (0.019 m³/min)!

A steel trestle was erected in the electrical bay at the rear of the powerhouse with its upper deck girders clear of the concrete roof slab (photo C). From this trestle two revolving gantry cranes were able to boom out 150 ft (45.7 m) to place the front wall of the powerhouse and also the massive thrust blocks and penstock encasement in the rear. Four-yard concrete buckets were transported by truck from the batch plant near the top of the access road to the powerhouse. The concrete tailrace retaining walls extending out from both ends of the powerhouse were placed by crawler cranes.

The first concrete in the powerhouse was placed in September 1970 and concrete was essentially completed in October 1972. A daily average of 1000 yd³ (764 m³) was sustained over one

Giant slope paver placing asphalt subbase. Asphalt is brought to the lower end of the "bridge" by side-dump trucks. These feed skip cars that run atop the bridge, in turn feeding an asphalt spreader within the bridge.

To protect the powerhouse from the waves of Lake Michigan (background), engineers constructed these two jetties and breakwater. Upper reservoir inlet-outlet is in foreground. Its upward-sloping concrete apron (Fig. 2) helps prevent water emerging from the ends of the six penstocks from scouring the bottom of the upper reservoir. Perforated baffle wall minimizes vortexing of water flowing into penstocks.

Fig. 3a. The inside slope of the dike is covered with a specially designed asphalt sandwich (Fig. 3b) designed to keep water from seeping into the dike's interior. Should water somehow get through or around this sandwich and into the dike's interior, the drainage system within the dike—chimney drain, longitudinal drain, and lateral drains —will carry it away.

Fig. 3b. Water penetrating the outer asphalt layers would move into the pavement drain, flow down to and collect in the 10 in. perforated pipe, and then be pumped up the slope to the surface by a submersible pump.

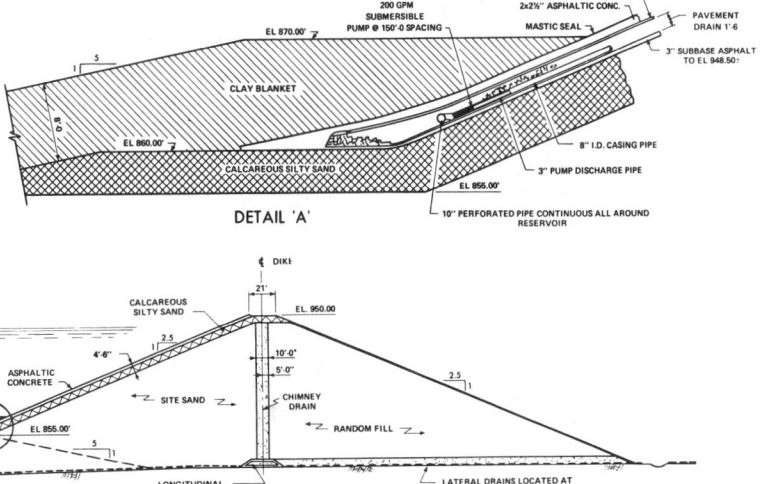

90-day stretch with a one-day peak at 2500 yd³ (1910 m³). Most concrete was placed on the swing and graveyard shifts to free the revolver cranes for daytime use for handling forms, reinforcing steel, embedded pump-turbine parts.

To maintain a winter concrete schedule, aggregate, delivered 150 miles (241 km) by bulk lake carriers, was stockpiled to carry over from November to the spring opening of the Great Lakes shipping season. The 180-yd³/hr (137 m³/hr) batch plant was winterized; provisions made to heat aggregate stockpiles and mix water. Exterior forms were insulated and steam-heated enclosures installed at the powerhouse to maintain optimum curing temperatures during cold weather. And using fly ash improved workability and reduced cement quantities. An innovative technique permitted placement of granular backfill around the powerhouse during winter weather. A brine solution was sprayed on backfill material to retard freezing.

Powerhouse equipment

The pump-turbines and generator-motor units (Hitachi) are the largest ever constructed, both in power and physical size. Rated at 450,000 hp (312,000 kw), each turbine has an inlet diameter of 24 ft (7.3 m), a runner diam. of 27.5 ft (8.3 m), and a generator-motor diameter of 68 ft (20.7 m). A 360 ton (327 mt) station gantry crane handles the impeller-runner but not the generator-motor rotor, due to the rotors large size and weight. The rotor is fabricated so that the runner can pass through the rim when the spider is removed. Stacking of the rotor rim and the stator was done in place.

Trash racks are not required at the pump intake. Draft tube trash racks were studied carefully for jetting velocities in the generating mode. At other installations fatigue failures have occurred at welded bar connections. So, round bars were used and, instead of being welded to the horizontal members, were fitted snugly with split supports which were in turn welded to the horizontal members.

Four water level devices prevent overpumping. These devices will shut down all units when reservoir water reaches a predetermined level.

Lakefront protection

To protect the powerhouse against waves from Lake Michigan, engineers built a north and a south jetty, extending 1,600 ft (488 m) into the lake, and a breakwater 1,850 ft (564 m) long, 2,700 ft (823 m) from and parallel to the shore (see cover & photo B). These structures will reduce 14 ft (4.3 m) waves at the breakwater to 4 ft (1.2 m) near the powerhouse. Location, shape and height of the jetties and breakwater were determined by model testing.

Near shore, the jetties are made of circular, steel-sheet pile cells; these were part of the cofferdam during powerhouse construction. The remainder of the jetties and the entire breakwater are rubble mound. A cofferdam of 27 40-ft (12.2 m) diam. sheetpile cells connected the two jetties, while the powerhouse was excavated to a depth of 85 ft (25.9 m).

Water in the 1,100 ft (335 m) wide tailrace channel, with minimum depth of 28.5 ft (8.7 m), has an average velocity of less than 3 fps (0.92 m/s) with all units operating either in pumping or generating mode. To protect small craft, channel outlets to the lake have been sized for a maximum 1.5 fps (0.46 m/s) inlet and outlet velocity. To minimize scour, the channel bottom immediately in front of the powerhouse is protected with a reinforced concrete slab; the remainder of the bottom is covered with riprap.

Esthetics

The Lakeshore Drive Highway Bridge presents a pleasing appearance. And the buried penstocks permit an uninterrupted view of the buff above the lake. The embankment's exterior slopes have been planted with local grasses and trees. Several acres just outside the reservoir have been reserved for recreation; facilities include an overnight trailer camp, rest rooms and showers, a children's playground, and picnic facilities. ▽

Water from the upper man-made reservoir (background) will rush down these six huge penstocks to the powerplant below (foreground). Note the six power units under construction, each to consist of a generator-motor coupled to a pump-turbine.

The key people on the Ebasco civil engineering team that have made the Ludington pumped-storage project a reality: sitting (l. to r.), G. Nerses, principal A-S engineer; R. L. Polvi, project supt., now construction manager; C. F. Whitehead, manager of projects; Donn Ruotolo, project manager; J. H. Womer, senior engineer, A-S dept. Standing (l. to r.), L. Gertler, principal C-H engineer; C. T. McCreedy, construction supt.; and J. F. Daly, resident engineer. Missing, E. Comninellis, principal engineer, C-H dept; and R. E. Mitchell, resident engineer.

Aerial View of Outlet Works

Aerial View of Reservoir, Lake Michigan in Background

Ludington Pumped Storage Plant

Ludington Pump/Generating Plant

Bibliography

1. Comninellis, E., "Ludington Pumped Storage Project," *Journal of the Power Division*, ASCE, Vol. 99, No. PO1, Proc. Paper 9696, May, 1973, pp. 069-088.

2. Seguin, R. L., "Ludington Pumped Storage Hydroelectric Generating Station, General Plant Construction and Electrical Details," Paper presented at the EEI Electrical System and Equipment Committee Meeting, Grand Rapids, Michigan, May 21-23, 1973.

3. Forgey, Harry L., "Symposium on the Ludington Pumped Storage Hydroelectric Generating Station," *Proceedings of the American Power Conference*, 1974, Vol. 36, pp. 797-841.

4. Whitehead, Carl F. and Donn Ruotolo, "Ludington Pumped-Storage Project Wins 1973 Outstanding CE Achievement Award," *Civil Engineering*, June, 1973, ASCE.

Plant Name: **MORMON FLAT #2**

Plant location:	Owner: Bureau of Reclamation
Tortilla Flat, AZ	Attn: D-3000
Maricopa County	P.O. Box 25007
	Denver, Colorado 80225
Rated capacity 47 MW	
Average static head 129 ft	Designers:
Plant efficiency 67.00 %	Salt River Project
Stored energy 2,761 MWh	Bechtel Corporation
Number of units 1	
Construction time: 3 years	Plant Manager/Superintendent:
Construction cost: $507 per kW	Salt River Project
Price level:	P.O. Box 52025
First commercial power: January 1971	Pheonix, Arizona 85072-2025
FERC project number:	
	(602) 236-3854
River or water source: Salt River	

DAM	UPPER RESERVOIR	LOWER RESERVOIR
Type	Existing concrete thin arch dam	Existing concrete thin arch dam
Height (ft)	224	207
Crest length (ft)	380	1,531
Volume (yd³)	60,000	120,000
RESERVOIR		
Type	Existing Canyon Lake constructed	Existing Saguaro Lake, constructed
Surface area (acres)	939	1,255
Usable power storage (acre-ft)	25,699	8,521
Power pool fluctuation (ft)	15.0	7.0
Operating levels		
Maximum (ft)	1,660.9	1,529.5
Minimum (ft)	1,645.0	1,522.3
Drainage area (miles²)	0.0	
Seepage (ft³/s)		
SPILLWAY		
Design flood		
Return period (years)		
Flow (ft³/s)		
Capacity (ft³/s)	150,090	
Type	Chute spillway	2 spillways
Gates		
Number	2	9
Type	Fixed wheel	Radial
Width (ft)	49.87	27.00
Height (ft)	49.87	22.97
OUTLET WORKS		
Discharge capacity (ft³/s)		1,808
Number of water passages	None	1
Dimensions of water passages		
Height (ft)		
Width (ft)		
Diameter (ft)		6.99
Type of gates		Butterfly valve
Number of gates	None	1

Plant Name: **MORMON FLAT #2**

Plant location
  Tortilla Flat, AZ
  Maricopa County
	Owner: Bureau of Reclamation
	       Attn: D-3000
	       P.O. Box 25007
	       Denver, Colorado 80225

Rated capacity   47 MW
Average static head   39.3 m
Plant efficiency   67.00 %
Stored energy   2,761 MWh
Number of units   1
	Designers:
	    Salt River Project
	    Bechtel Corporation

Construction time: 3 years
Construction cost: $507 per kW
Price level:
First commercial power: January 1971
FERC project number:
	Plant Manager/Superintendent:
	    Salt River Project
	    P.O. Box 52025
	    Phoenix, Arizona 85072-2025
	
	    (602) 236-3854

River or water source: Salt River	

	UPPER RESERVOIR	LOWER RESERVOIR
DAM		
Type	Existing concrete thin arch dam	Existing concrete thin arch dam
Height (m)	68.3	63.1
Crest length (m)	115.8	466.5
Volume (m³)	45,797	91,747
RESERVOIR		
Type	Existing Canyon Lake constructed	Existing Saguaro Lake, constructed
Surface area (Mm²)	3.80	5.08
Usable power storage (Mm³)	31.700	10.510
Power pool fluctuation (m)	4.57	2.13
Operating levels		
Maximum (m)	506.25	466.20
Minimum (m)	501.40	464.00
Drainage area (Mm²)		
Seepage (m³/s)		
SPILLWAY		
Design flood		
Return period (years)		
Flow (m³/s)		
Capacity (m³/s)	4,250	
Type	Chute spillway	2 spillways
Gates		
Number	2	9
Type	Fixed wheel	Radial
Width (m)	15.200	8.230
Height (m)	15.200	7.000
OUTLET WORKS		
Discharge capacity (m³/s)		51
Number of water passages	None	1
Dimensions of water passages		
Height (m)		
Width (m)		
Diameter (m)		2.130
Type of gates		Butterfly valve
Number of gates	None	1

Mormon Flat #2 - Page 1 (Metric)

MORMON FLAT #2

INTAKES	UPPER INTAKE	LOWER INTAKE
Number	1	1
Type	Penstock intake	Draft tube
Design discharge (ft³/s)	5,424	3,599
Gross inlet area (ft²)	805	807
(at trash racks)		
Bar racks		
spacing (in)	5.25	5.25
shape	Rectangular	Rectangular
depth/thickness (in)	3.50 / 0.75	3.50 / 0.75
diameter (in)		
Emergency gates		
number	1	None
height/width (ft)	16.40 / 20.01	
type	Fixed wheel	
Service gates		
number	None	None
height/width (ft)		
type		
Bulkhead/stop logs(Y or N)	Y	Y
number of units serviced	1	1
Hoists		
number	1	3
capacity (tons)	88	15
type	Hydraulic	Electric

WATER PASSAGES	Upper Tunnel	Shaft	Lower Tunnel	Surge Tanks Upper	Surge Tanks Lower	Penstocks	Tailrace Tunnel
Number						1	
Diameter (ft)						18.0	
Length (ft)						175	
Maximum velocity (ft/s)						21.3	
Concrete lining thickness (in)							
Total length of concrete sections (ft)							
Steel liner							
Thickness							
Minimum (in)						0.62	
Maximum (in)						0.75	
Material grade						A-36	
Total length of steel-lined sections (ft)						175	

Notes:

Mormon Flat #2 - Page 2 (English)

MORMON FLAT #2

INTAKES	UPPER INTAKE	LOWER INTAKE
Number	1	1
Type	Penstock intake	Draft tube
Design discharge (m³/s)	153.6	101.9
Gross inlet area (m²) (at trash racks)	74.8	75.0
Bar Racks:		
spacing (mm)	133	133
shape	Rectangular	Rectangular
depth/thickness (mm)	89 / 19	89 / 19
diameter (mm)		
Emergency gates		
number	1	None
height/width (m)	5.000 / 6.100	
type	Fixed wheel	
Service gates		
number	None	None
height/width (m)		
type		
Bulkhead/stop logs (Y or N)	Y	Y
number of units serviced	1	1
Hoists		
number	1	3
capacity (Mg)	80	14
type	Hydraulic	Electric

WATER PASSAGES	Upper Tunnel	Shaft	Lower Tunnel	Surge Tanks Upper	Surge Tanks Lower	Penstocks	Tailrace Tunnel
Number						1	
Diameter (m)						5.49	
Length (m)						53.3	
Maximum velocity (m/s)						6.50	
Concrete lining thickness (m)							
Total length of concrete sections (m)							
Steel liner							
Thickness							
Minimum (mm)						16	
Maximum (mm)						19	
Material grade						A-36	
Total length of steel-lined sections (m)						53.3	

Notes:

Mormon Flat #2 - Page 2 (Metric)

MORMON FLAT #2

POWERHOUSE and RELATED FEATURES

Powerhouse Structure
Type: Semi-outdoor concrete
Length: 94 ft   Width: 72 ft   Height: 93 ft

Guard Valves
Number: None   Diameter: ft
Type:

Transformers
Number: 1
Ratings: 78.4
Voltages: (kV) 115 / 13.8

Generator
Rating generating (MVA): 47.0   Rating pumping (MVA): 45.5
Insulation type:
Starting method: Synchronous, static AC frequency converter
Starting equipment: Amortisseur bar

Runners
Material: Cast steel
Minimum unit submergence: 13.3 ft
WR^2:
Manufacturer: Allis Chalmers
Model test by: Allis Chalmers

	Reversible Runners	Reversible Motor/Generator		
Number	1			1
Diameter (ft)	17.10	Rotor	Stator	
rpm synchronous	138.5			138.5
rpm overspeed	173.0			173.0
Type	Francis			

Information on Runners

Condition:	Gross Head (ft)		Capacity (MW)		Discharge (ft³/s)		Turbine/Pump Eff.(%)	
	Generating	Pumping	Generating	Pumping	Generating	Pumping	Generating	Pumping
Maximum head & maximum power	139	139	47					
Minimum head & maximum power	126	126			5,424	3,599		

Condition:	Net Head (ft)		Capacity (MW)		Discharge (ft³/s)		Turbine/Pump Eff.(%)	
	Generating	Pumping	Generating	Pumping	Generating	Pumping	Generating	Pumping
Rated head @ best gate	135	143	42	45		4,070		

MORMON FLAT #2

POWERHOUSE and RELATED FEATURES

Powerhouse Structure
Type: Semi-outdoor concrete
Length: 28.7 m   Width: 21.9 m   Height: 28.3 m

Guard Valves
Number: None   Diameter: m
Type:

Transformers
Number: 1
Ratings: 78.4
Voltages: (kV) 115 / 13.8

Generator
Rating generating (MVA): 47.0   Rating pumping (MVA): 45.5
Insulation type:
Starting method: Synchronous, static AC frequency converter
Starting equipment: Amortisseur bar

Runners
Material: Cast steel
Minimum unit submergence: 4.05 m
WR^2:
Manufacturer: Allis Chalmers
Model test by: Allis Chalmers

	Reversible Runners	Reversible Motor/Generator		
Number	1			1
Diameter m	5.20	Rotor	Stator	
rpm synchronous	138.5			138.5
rpm overspeed	173.0			173.0
Type	Francis			

Information on Runners

Condition:	Gross Head (m)		Capacity (MW)		Discharge (m³/s)		Turbine/Pump Eff.(%)	
	Generating	Pumping	Generating	Pumping	Generating	Pumping	Generating	Pumping
Maximum head & maximum power	42.3	42.3	47					
Minimum head & maximum power	38.4	38.4			153.6	101.9		

Condition:	Net Head (m)		Capacity (MW)		Discharge (m³/s)		Turbine/Pump Eff.(%)	
	Generating	Pumping	Generating	Pumping	Generating	Pumping	Generating	Pumping
Rated head @ best gate	41.1	43.6	42	45		115.3		

Mormon Flat #2 - Page 3 (Metric)

MORMON FLAT #2

Plant Data:
  Average GWh generating per year:
  Average GWh pumping per year:
  Starting time from standstill (s):
  Changeover time pumping to generating (min):
  Planned/scheduled time
    between major overhauls (years):   10
  Outage time required per unit
    during major overhauls (weeks):   9
  Representative plant availability (%):   85.7
  Representative planned outages (weeks per year):   2

Miscelleneous Notes:
  Mormon Flat Dam incorporates a 10-MW conventional unit and one 47-MW reversible unit.

  A second spillway was recently constructed at the project.

  Information on unit availability and unit outages are based on average values for the years 1982 through 1989.

Cavitation Experience:
  Cavitation causes 250 pounds of metal loss per year.

Significant or Unique Problems:

List of Licenses Required:

ENVIRONMENTAL FEATURES

Recreation:
  Recreation includes fishing, waterskiing, and the use of campgrounds.

Fish and Wildlife:

Social:

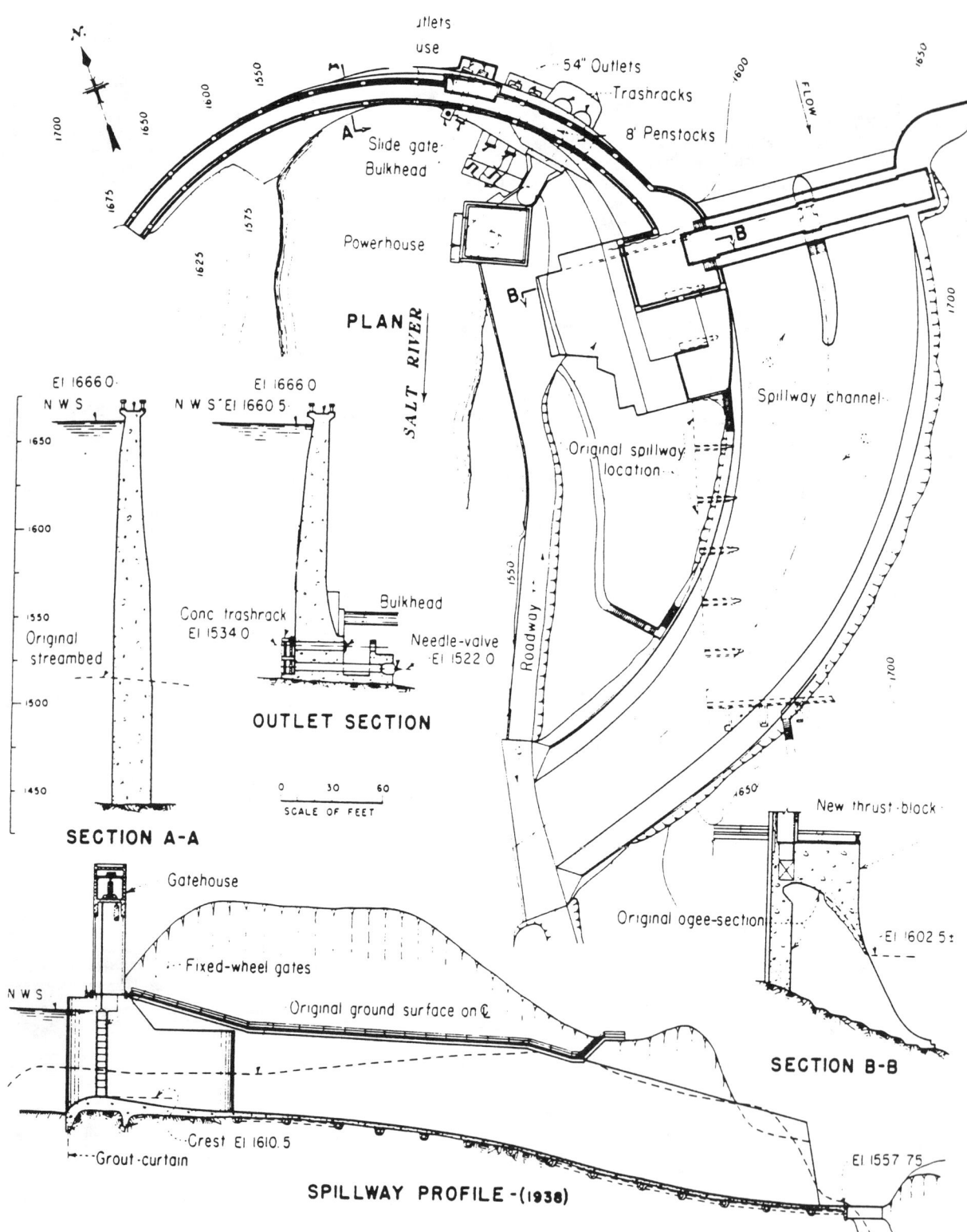

Mormon Flat Dam, Plan and Sections

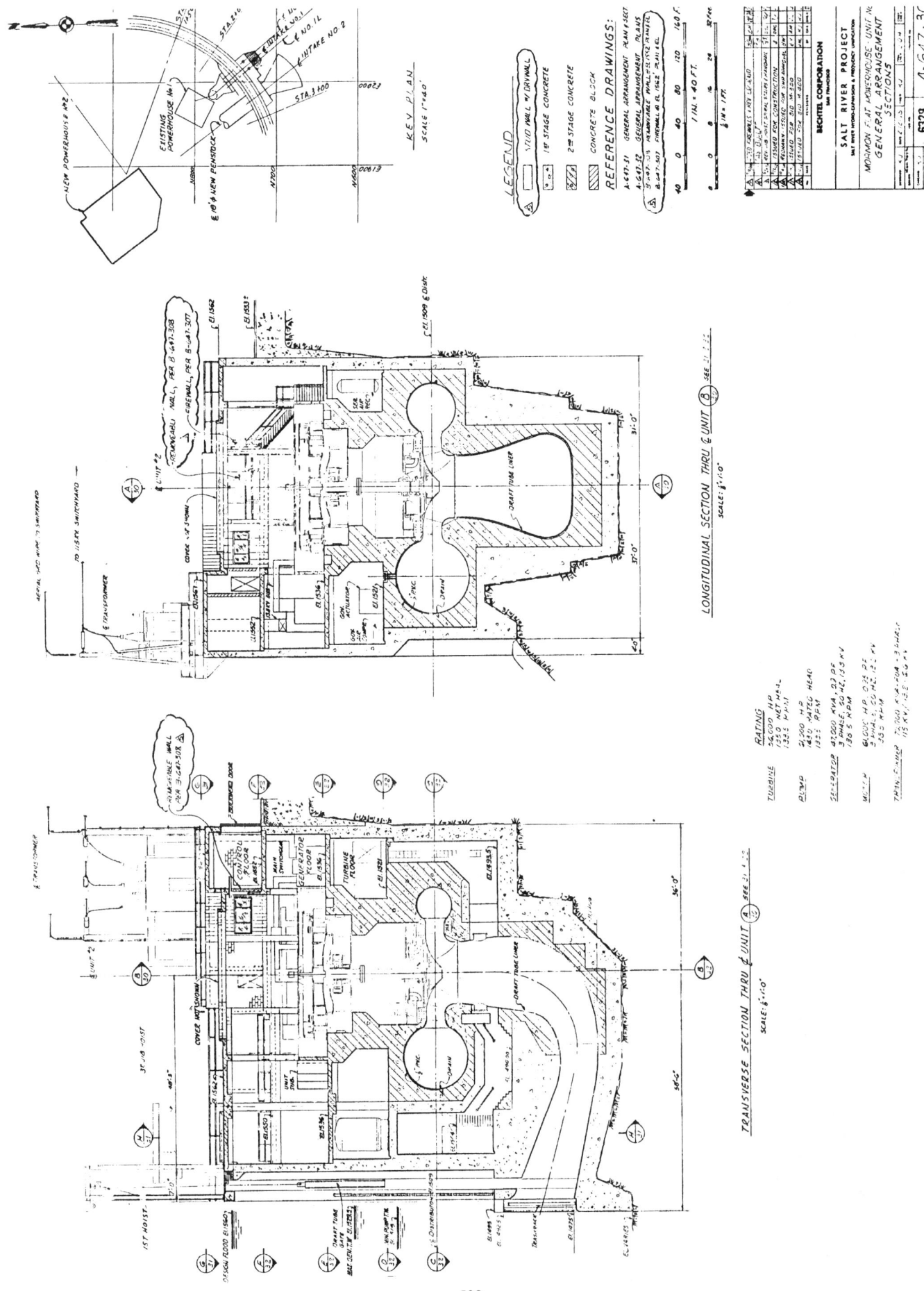

Plant Name: **MT. ELBERT**

Plant location: Leadville, CO Lake County	Owner: Bureau of Reclamation Attn: D-3000 P.O. Box 25007 Denver, Colorado 80225
Rated capacity 200 MW Average static head 448 ft Plant efficiency 81.30 % Stored energy 2,410 MWh Number of units 2	Designers: Bureau of Reclamation
Construction time: 10 years Construction cost: Price level: First commercial power: June 1981 FERC project number: River or water source: Trans Mountain Diversion	Plant Manager/Superintendent: Granite Star Route Leadville, Colorado 81228 (719) 486-2325

	UPPER RESERVOIR	LOWER RESERVOIR
DAM		
Type	Zoned earthfill embankment, Mt. Elbert forebay	Zoned earthfill embankment
Height (ft)	92	55
Crest length (ft)	2,650	3,140
Volume (yd³)	846,000	624,000
RESERVOIR		
Type	Constructed	On stream, Twin Lakes
Surface area (acres)	282	2,817
Usable power storage (acre-ft)	7,395	67,930
Power pool fluctuation (ft)	30.0	4.0
Operating levels		
Maximum (ft)	9,645.7	9,200.0
Minimum (ft)	9,590.0	9,168.7
Drainage area (miles²)	1.0	105.0
Seepage (ft³/s)	0.004	1.605
SPILLWAY		
Design flood		
Return period (years)		
Flow (ft³/s)	9,070	12,840
Capacity (ft³/s)		
Type	Morning glory	Morning glory 12.11 m (39.7 ft)
Gates		
Number	None	None
Type	None - stored inflow	None
Width (ft)		
Height (ft)		
OUTLET WORKS		
Discharge capacity (ft³/s)		3,465
Number of water passages	None	2
Dimensions of water passages		
Height (ft)		8.00
Width (ft)		6.50
Diameter (ft)		
Type of gates		Slide gates
Number of gates	None	2

Mt. Elbert - Page 1 (English)

Plant Name: **MT. ELBERT**

Plant location Leadville, CO Lake County	Owner: Bureau of Reclamation Attn: D-3000 P.O. Box 25007 Denver, Colorado 80225
Rated capacity 200 MW Average static head 136.6 m Plant efficiency 81.30 % Stored energy 2,410 MWh Number of units 2	Designers: Bureau of Reclamation
Construction time: 10 years Construction cost: Price level: First commercial power: June 1981 FERC project number:	Plant Manager/Superintendent: Granite Star Route Leadville, Colorado 81228 (719) 486-2325
River or water source: Trans Mountain Diversion	

	UPPER RESERVOIR	LOWER RESERVOIR
DAM		
Type	Zoned earthfill embankment, Mt. Elbert forebay	Zoned earthfill embankment
Height (m)	28.0	16.8
Crest length (m)	807.7	957.1
Volume (m³)	646,622	477,100
RESERVOIR		
Type	Constructed	On stream, Twin Lakes
Surface area (Mm²)	1.14	11.40
Usable power storage (Mm³)	9.122	83.790
Power pool fluctuation (m)	9.14	1.22
Operating levels		
Maximum (m)	2,940.01	2,804.16
Minimum (m)	2,923.03	2,794.62
Drainage area (Mm²)	2.590	271.950
Seepage (m³/s)	0.0001	0.0454
SPILLWAY		
Design flood		
Return period (years)		
Flow (m³/s)	257	364
Capacity (m³/s)		
Type	Morning glory	Morning glory 12.11 m (39.7 ft)
Gates		
Number	None	None
Type	None - stored inflow	None
Width (m)		
Height (m)		
OUTLET WORKS		
Discharge capacity (m³/s)		98
Number of water passages	None	2
Dimensions of water passages		
Height (m)		2.438
Width (m)		1.981
Diameter (m)		
Type of gates		Slide gates
Number of gates	None	2

Mt. Elbert - Page 1 (Metric)

MT. ELBERT

INTAKES	UPPER INTAKE	LOWER INTAKE
Number	2	4
Type	Concrete structure	Draft tube
Design discharge (ft³/s)	3,590	6,920
Gross inlet area (ft²) (at trash racks)	1,200	1,572
Bar racks		
spacing (in)	6.00	6.50
shape	Elliptical	Elliptical
depth/thickness (in)	2.00 / 0.62	2.00 / 0.62
diameter (in)		
Emergency gates		
number	2	None
height/width (ft)	12.50 / 15.00	
type	Wheel mounted	
Service gates		
number	None	None
height/width (ft)		
type		
Bulkhead/stop logs (Y or N)	Y	Y
number of units serviced	4	4
Hoists		
number	1	None
capacity (tons)		
type	Hydraulic hoist - temporary	

WATER PASSAGES	Upper Tunnel	Shaft	Lower Tunnel	Surge Tanks Upper	Surge Tanks Lower	Penstocks	Tailrace Tunnel
Number				2		2	
Diameter (ft)				40.0		15.0	
Length (ft)				50		3,000	
Maximum velocity (ft/s)						20.3	
Concrete lining thickness (in)							
Total length of concrete sections (ft)							
Steel liner Thickness							
Minimum (in)						0.63	
Maximum (in)						1.37	
Material grade						A516 GRADE 60	
Total length of steel-lined sections (ft)						3,000	

Notes:

MT. ELBERT

INTAKES	UPPER INTAKE	LOWER INTAKE
Number	2	4
Type	Concrete structure	Draft tube
Design discharge (m³/s)	101.7	196.0
Gross inlet area (m²) (at trash racks)	111.5	146.0
Bar Racks:		
spacing (mm)	152	165
shape	Elliptical	Elliptical
depth/thickness (mm)	51 / 16	51 / 16
diameter (mm)		
Emergency gates		
number	2	None
height/width (m)	3.810 / 4.572	
type	Wheel mounted	
Service gates		
number	None	None
height/width (m)		
type		
Bulkhead/stop logs (Y or N)	Y	Y
number of units serviced	4	4
Hoists		
number	1	None
capacity (Mg)		
type	Hydraulic hoist - temporary	

WATER PASSAGES	Upper Tunnel	Shaft	Lower Tunnel	Surge Tanks Upper	Surge Tanks Lower	Penstocks	Tailrace Tunnel
Number				2		2	
Diameter (m)				12.19		4.57	
Length (m)				15.2		914.4	
Maximum velocity (m/s)						6.19	
Concrete lining thickness (m)							
Total length of concrete sections (m)							
Steel liner Thickness							
Minimum (mm)						16	
Maximum (mm)						35	
Material grade						A516 GRADE 60	
Total length of steel-lined sections (m)						914.4	

Notes:

Mt. Elbert - Page 2 (Metric)

MT. ELBERT

POWERHOUSE and RELATED FEATURES

Powerhouse Structure
Type: Surface
Length: 146 ft   Width: 104 ft   Height: 182 ft

Guard Valves
Number: None   Diameter: ft
Type:

Transformers
Number: 2
Ratings: 142.4
Voltages: (kV) 230 / 11.5

Generator
Rating generating (MVA): 105.3   Rating pumping (MVA): 127.0
Insulation type: Mica
Starting method: Line starting
Starting equipment: Solid state (1 shared between units)

Runners
Material: Stainless Steel A296 CF8
Minimum unit submergence: 38.0 ft
WR^2: 6,525,000 (lbf x ft²)
Manufacturer: Allis Chalmers, Hitachi
Model test by: Allis Chalmers, Hitachi

	Reversible Runners	Reversible Motor/Generator	
Number	2		2
Diameter (ft)	18.80	Rotor	Stator
rpm synchronous	180.0		180.0
rpm overspeed	270.0		
Type	Francis		Vertical Shaft

Information on Runners

Condition:	Gross Head (ft) Generating	Gross Head (ft) Pumping	Capacity (MW) Generating	Capacity (MW) Pumping	Discharge (ft³/s) Generating	Discharge (ft³/s) Pumping	Turbine/Pump Eff.(%) Generating	Turbine/Pump Eff.(%) Pumping
Maximum head & maximum power	475	485	139	132	3,725	2,400	92.0	92.3
Minimum head & maximum power	400	430	103		3,325	3,400		

Condition:	Net Head (ft) Generating	Net Head (ft) Pumping	Capacity (MW) Generating	Capacity (MW) Pumping	Discharge (ft³/s) Generating	Discharge (ft³/s) Pumping	Turbine/Pump Eff.(%) Generating	Turbine/Pump Eff.(%) Pumping
Rated head @ best gate	400	430	103	132	3,325	3,400	92.0	

Mt. Elbert - Page 3 (English)

MT. ELBERT

POWERHOUSE and RELATED FEATURES

Powerhouse Structure
Type: Surface
Length: 44.5 m   Width: 31.7 m   Height: 55.5 m

Guard Valves
Number: None   Diameter: m
Type:

Transformers
Number: 2
Ratings: 142.4
Voltages: (kV) 230 / 11.5

Generator
Rating generating (MVA): 105.3   Rating pumping (MVA): 127.0
Insulation type: Mica
Starting method: Line starting
Starting equipment: Solid state (1 shared between units)

Runners
Material: Stainless Steel A296 CF8
Minimum unit submergence: 11.58 m
WR^2: 2,697,000 (Newtons x m^2)
Manufacturer: Allis Chalmers, Hitachi
Model test by: Allis Chalmers, Hitachi

	Reversible Runners	Reversible Motor/Generator		
Number	2	2		
Diameter m	5.74	Rotor	Stator	
rpm synchronous	180.0		180.0	
rpm overspeed	270.0			
Type	Francis		Vertical Shaft	

Information on Runners

Condition:	Gross Head (m)		Capacity (MW)		Discharge (m³/s)		Turbine/Pump Eff.(%)	
	Generating	Pumping	Generating	Pumping	Generating	Pumping	Generating	Pumping
Maximum head & maximum power	144.8	147.8	139	132	105.5	68.0	92.0	92.3
Minimum head & maximum power	121.9	131.1	103		94.2	96.3		

Condition:	Net Head (m)		Capacity (MW)		Discharge (m³/s)		Turbine/Pump Eff.(%)	
	Generating	Pumping	Generating	Pumping	Generating	Pumping	Generating	Pumping
Rated head @ best gate	121.9	131.1	103	132	94.2	96.3	92.0	

Mt. Elbert - Page 3 (Metric)

MT. ELBERT

Plant Data:

Average GWh generating per year:	
Average GWh pumping per year:	
Starting time from standstill (s):	
Changeover time pumping to generating (min):	5
Planned/scheduled time between major overhauls (years):	2
Outage time required per unit during major overhauls (weeks):	3
Representative plant availability (%):	90.0
Representative planned outages (weeks per year):	3

Miscelleneous Notes:

This is a multipurpose project, no cost estimate for the separable pumped storage portion is available.

The spillway for the lower reservoir is a 12.11-m (39.75-ft) diameter morning glory.

The drainage area on the upper reservoir is not applicable because it is part of a transmountain water supply. The reservoir is partially filled by flows through the tunnel and by minor inflow from the surrounding area.

The upper reservoir is completly lined with PVC.

Cavitation Experience:
None.

Significant or Unique Problems:
The problem of excessive seepage in the upper reservoir has been resolved with a polythene lining.

List of Licenses Required:
Congressional authorization, Corps Section 404 & 401 permits

ENVIRONMENTAL FEATURES

Recreation:
The Mt. Elbert Powerplant was built in an existing reservoir. This lower reservoir was originally two natural lakes, that were enlarged twice by constructing two separate dams. The latest reservoir enlargement increased tourism and recreational activities including boating, hiking, camping, fishing, and swimming. Over 20,000 people tour the powerplant's visitor center each year. Recreational development includes two boat ramps, two picnic areas, a campground and parking areas.

Fish and Wildlife:
Twin Lakes supports a healthy trout (especially Lake Trout) population. The forebay reservoir also supports trout.

MT. ELBERT

Social:
Prior to the latest reservoir enlargement, most of the buildings of the old towns of Interlaken and Twin Lakes were moved and restored. This project is part of an irrigation project.

Progress at the Mt Elbert pumped-storage powerplant in Colorado

By H. G. Arthur*

Now in its ninth year of construction, the Fryingpan-Arkansas development has been described as one of the most important water resources developments ever undertaken in the western United States. It comprises two hydro installations—the proposed Otero scheme and the Mt Elbert pumped-storage powerplant, currently under construction, which is the subject of this article

WHAT WILL BE the US Bureau of Reclamation's first installation of pump-turbines that are to supply much-needed peaking capacity is uniquely identified with current construction at the Mt Elbert pumped-storage powerplant which is part of BuRec's Fryingpan-Arkansas project in Colorado. The plant, near the state's highest mountain peak, 14 431ft-high Mt Elbert, is about 15 miles southwest of the mountain community of Leadville and is some 9200ft above sea level.

Scheduled to be placed in operation in December, 1976, the Mt Elbert plant will be the 54th hydro plant constructed and operated by BuRec in the western United States. Initially, it will house one 100MW reversible pump-turbine unit, with space being provided in the plant for installation of a second similar unit at a later date.

The site of the scheme is on the shore of Twin Lakes, an existing 111 000acre-ft reservoir that is to be enlarged to a capacity of 166 000acre-ft. A forebay reservoir with a capacity of about 10 000acre-ft is to be built on a hill above the powerplant.

The pump-turbines will be used to pump as much as 3400cusecs of water from Twin Lakes into the forebay reservoir during periods of low power demand, and the flow will be reversed through the turbines to meet daily peak power demands.

Speaking in January, 1972 when the contract for construction of the Mt Elbert plant was awarded, and citing the significance of the installation in terms of its environmental impact, the Commissioner of the Bureau of Reclamation stated that: "Pumped storage has developed through an era of increasing concern over environmental protection. It is an efficient and reliable method of furnishing badly-needed peaking power, which has a minimum of adverse effect on the quality of our living and environmental surroundings".

Fryingpan-Arkansas project

The Mt Elbert plant and the 11MW Otero power station, planned for construction later as a peaking plant, will be

* Director of Design and Construction, Bureau of Reclamation, US Department of the Interior, Denver, Colorado 80225, USA.

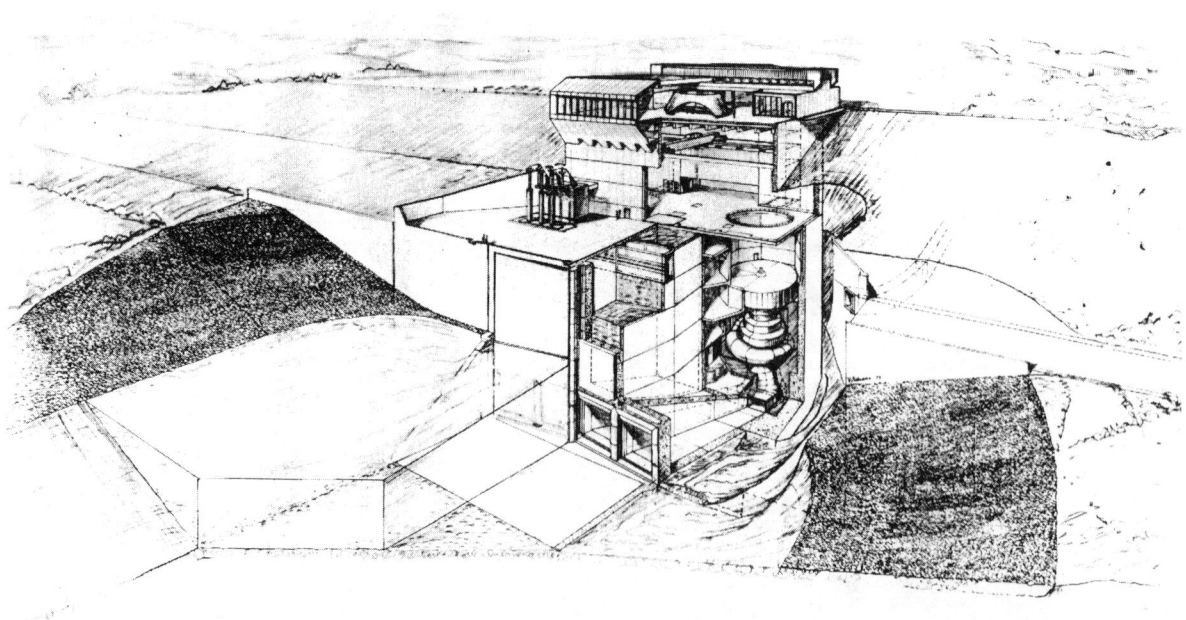

An artist's impression of the Mt Elbert pumped-storage plant on Twin Lakes reservoir

Reprinted, with permission, from International Water Power and Dam Construction, June 1974.

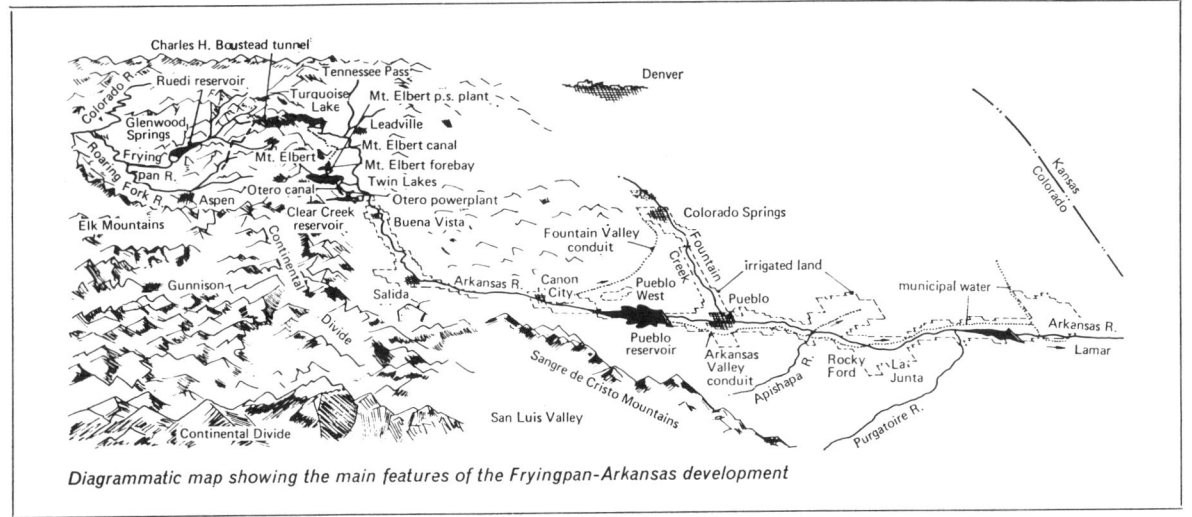
Diagrammatic map showing the main features of the Fryingpan-Arkansas development

the two power installations of the Fryingpan-Arkansas development. Now in its ninth year of construction, this development is a multi-purpose, trans-mountain diversion scheme in the central and southeastern part of Colorado.

Construction work on Ruedi dam, the first major feature of the scheme, was begun in July, 1964 and completed in July, 1968. At the ground-breaking ceremonies for the dam, the Fryingpan-Arkansas scheme was spoken of as one of the greatest water resources developments ever undertaken in the western United States.

Under the scheme water is diverted from the Colorado River basin on the western slope of the Rocky Mountains to the Arkansas River basin on the eastern slope of the Continental Divide.

The water diverted from the western slope, together with regulated flows from the Arkansas River, will provide supplementary irrigation, and municipal- and industrial-water supplies, and be used to generate hydroelectric power. Flood control is also an important aim of the scheme, together with fish and wildlife enhancement,

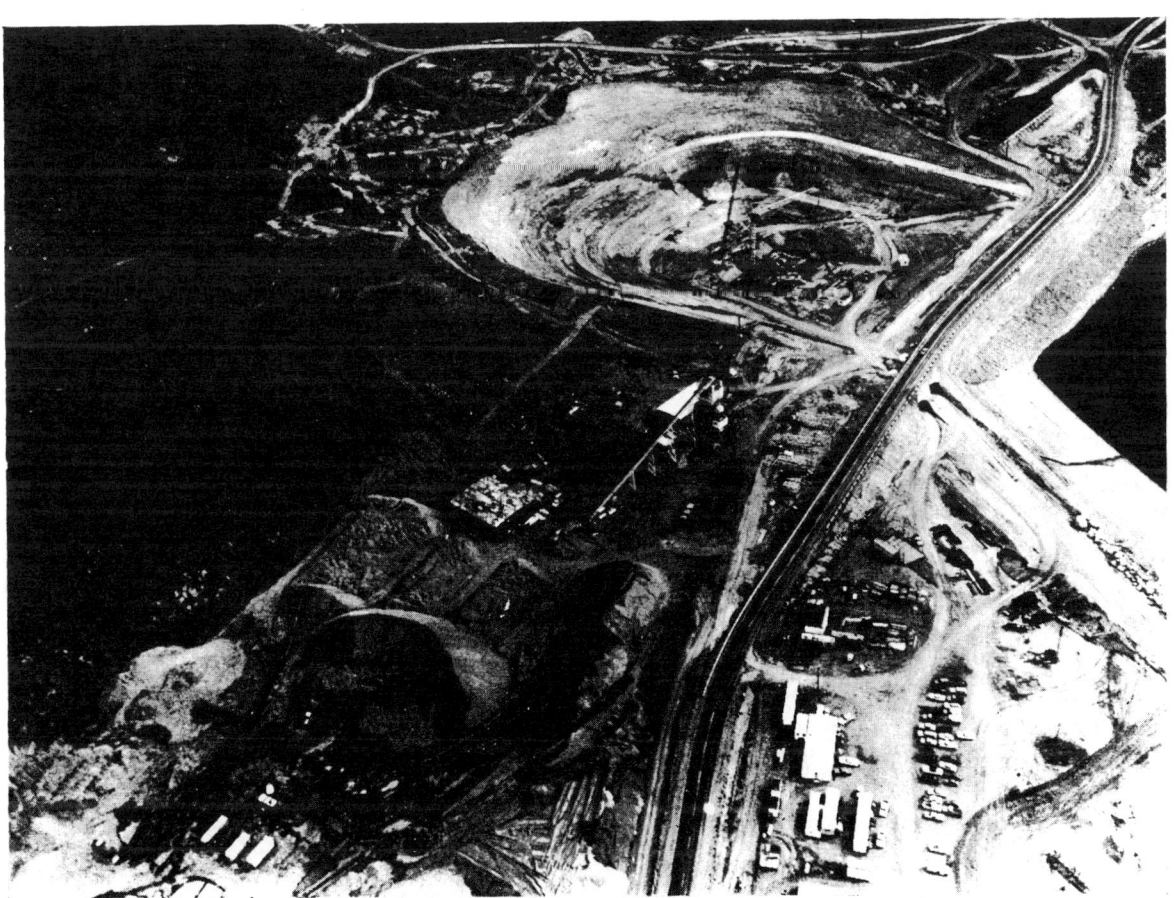
The site of the Mt Elbert powerplant (top right) as seen from the air during an early stage of construction (September 1972)

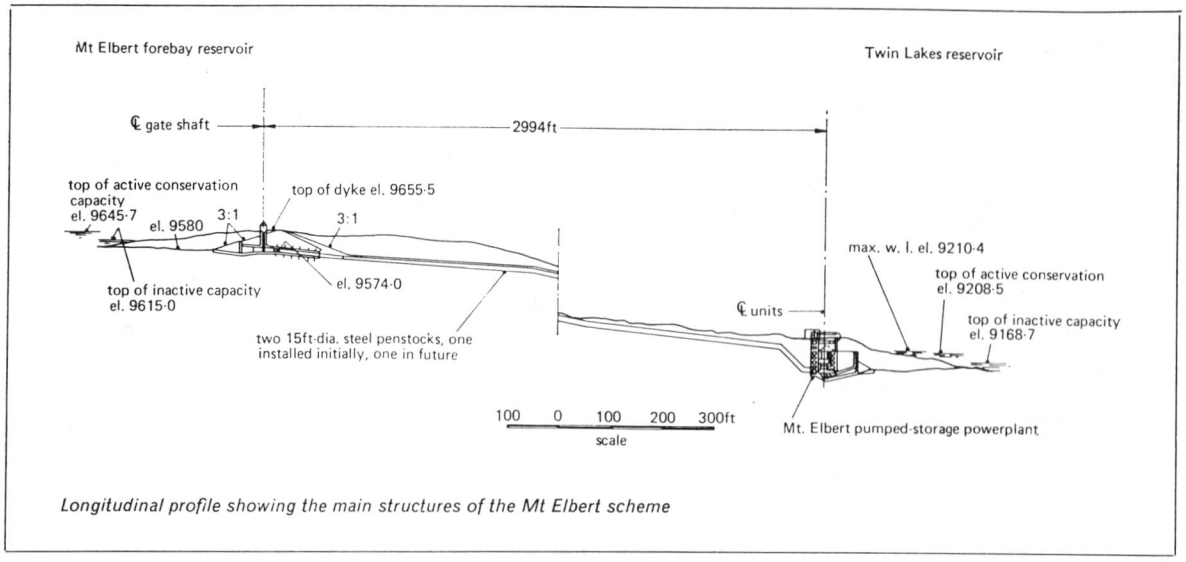
Longitudinal profile showing the main structures of the Mt Elbert scheme

recreational development, and sediment control.

The collection and diversion system is in a mountainous area above el. 10 000ft on the headwaters of the Fryingpan River, on the western slope of the Continental Divide. The service area of the project is in the Arkansas River basin in Colorado.

The Fryingpan River is a tributary of the Roaring Fork River which, in turn, is a tributary of the Colorado River. On the eastern slope, the headwaters of the Arkansas River originate above timber-line in the high mountains near Leadville, and the river then flows south and east through canyons and foothills to the high plains of eastern Colorado.

The scheme comprises five broad systems of water resources development: (i) Ruedi dam, which provides storage of water to replace that diverted to the eastern slope; (ii) the collection system on the western slope; (iii) the storage and diversion structures; (iv) the power system; and (v) the irrigation and municipal- and industrial-water supply systems. The last three systems are on the eastern slope.

The western slope collection system takes water from tributaries of the Fryingpan River at about el. 10 000ft, and comprises the North Side system, the South Side system, and the Charles H. Boustead tunnel. All three are integrated to divert an average of 69 200acre-ft of water annually from the western to the eastern slope.

The South Side system diverts water from the headwaters of Hunter Creek and tributaries of the Fryingpan River through 19·2 miles of tunnels and conduits to the central collection point at the west portal of the Boustead tunnel.

The North Side system also diverts water from other tributary headwaters of the Fryingpan River through a system of 4·1 miles of conduits and 7·2 miles of tunnels to the west portal of the Boustead tunnel.

The 5·4-mile-long Boustead tunnel then carries water from the western slope through the Continental Divide.

The major storage and diversion features on the eastern slope are Sugar Loaf dam, Twin Lakes dam, Otero canal, Elbert canal, Clear Creek dam, and Pueblo dam.

The Sugar Loaf dam increases the capacity of Turquoise Lake from 17 000 to 129 000acre-ft and the greatly enlarged reservoir stores and regulates all project water that is diverted from the western slope.

The Otero canal will deliver water from Twin Lakes through the Otero powerplant to the existing Clear Creek reservoir, which will then supply water to the Arkansas River.

Water released from Turquoise Lake will be carried by the Elbert canal to the forebay supplying the Mt Elbert powerplant.

The Fryingpan-Arkansas power system will include the two powerplants, switchyards and transmission lines and, as explained previously, the generation system will ultimately have a peaking capacity of 211MW. The two plants will be interconnected with other BuRec power facilities.

The principal storage feature of the Fryingpan-Arkansas project on the eastern slope will be the 357 000acre-ft Pueblo reservoir, impounded by the large concrete and earthfill Pueblo dam, which is being constructed on the Arkansas River 6 miles west of the city of Pueblo.

The project's service area of 280 000 acres of irrigable land will receive supplementary irrigation supplies from the water diverted from the western slope. These areas are within the Arkansas River basin and will be served by existing distribution systems.

Current plans call for the Fryingpan-Arkansas project to supply 20 000acre-ft (or about 7000 million gallons) of water annually to supplement the existing supplies of several eastern-slope municipalities. About 190 miles of pipeline will be laid to carry water to cities and towns in the Arkansas River valley. Another pipeline is also being planned to serve the city of Colorado Springs.

All reimbursable costs of the project are to be returned to the US Treasury within a 50-year period following completion of construction. Under Reclamation law, all costs allocated to hydro power and municipal- and industrial-water investments are to be returned with interest; revenues from the sale of power produced by the two powerplants will assist toward this repayment.

Costs allocated to flood control, fish and wildlife conservation, and recreational development are not reimbursable.

Construction of the Mt Elbert powerplant

Construction of the Mt Elbert powerplant began in January, 1972 under a $16 239 483 contract awarded to the Martin K. Eby Construction Co Inc, and Equipment Rental Sales Co Inc, operating as a joint venture under the name Eby and Co, Wichita, Kansas.

Work under the contract, which includes first-stage construction of the powerplant structure and installation of the elevators and cranes, was 90% completed by April

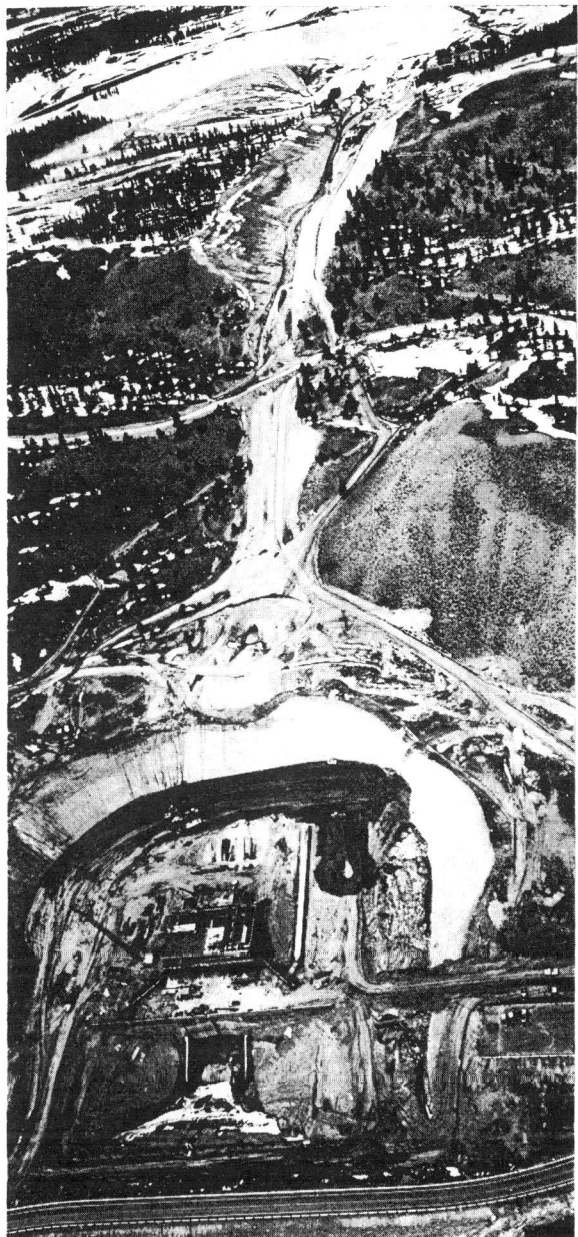
Overall view of the Mt Elbert powerplant and penstock areas looking up towards the forebay inlet-outlet structure

Table I—Pump-turbine specifications

Impeller-runner	
Maximum outside diameter (in)	220
Weight (lb)	150 000
Throat diameter (in)	136
Height (in)	87
Maximum radial thrust on impeller runner (lb)	46 200
Head cover	
Maximum diameter (in)	288
Weight of half section (lb)	117 000
Main shaft	
Diameter (in)	36
Weight (lb)	45 000
Spiral casing (largest section)	
Dimensions (in)	543 × 180 × 144
Weight (lb)	137 000
Pump-turbine fly-wheel effect	
WR^2 watered (lb-ft^2)	5 500 000
WR^2 dewatered (lb-ft^2)	5 000 000
Maximum pumping power	
Required over entire range of operation (hp)	178 500

1, 1974. Completion of the contract is called for by February 1, 1975.

The powerplant is a reinforced concrete structure, 146 × 213ft in plan and 192ft high. A total of 40 000yd³ of first-stage concrete will be placed for the main structure and a further 6000yd³ will be placed for the penstock anchor blocks, retaining walls, training structure, and other powerplant structures.

A large part of the main structure will be underground, requiring the excavation of 1·359 million cubic yards of earth material and a total backfill of 1·057 million cubic yards.

Work under the first-stage contract also includes construction of one complete 15ft-diameter steel penstock, approximately 3000ft long, and 15ft-diameter steel stubs for the future installation of a second penstock. For aesthetic reasons, the penstocks will be buried.

One stub will extend from the powerplant; the other will extend from the inlet-outlet structure being constructed in the forebay reservoir. The total length of the stubs will be about 840ft.

The second penstock will be completed later to convey water to the second 100MW reversible pump-turbine.

The suction draft tube for the first unit will be a concrete conduit extending 134ft from the centreline of the units to an open trapezoidal-channel or tailrace. This channel, which will discharge into Twin Lakes, will be approximately 350ft long.

Provisions for stoplogs are to be made at the end of the draft tube, and at the uphill end of the penstock a gated concrete structure containing two 12·5 × 15ft wheel-mounted emergency gates, one for each penstock, will be constructed.

Other work under the first-stage contract includes installation of two 225t overhead travelling cranes; draft-tube bulkhead gates; trashracks; piping, metalwork, and electrical conduit embedded in the first-stage concrete; and steel framing and crane girders.

The two cranes will be used for installation, maintenance and servicing of the plant's equipment.

The pump-turbine unit

In February, 1971, BuRec awarded a $1·7434 million contract for manufacture of the first Mt Elbert pump-turbine to the Allis-Chalmers Manufacturing Co, York, Pennsylvania.

Manufacturing work, representing 45% of the contract, is being carried out at the firm's factory at York. A further 38% of the contract work is being done for Allis-Chalmers by the Fuji Electric Co, at Kawasaki, Japan. The remainder of the work will be carried out at the powerplant site.

The pump-turbine will be a Francis machine with a vertical shaft, single-impeller runner, and fixed stayvanes. The specifications are given in Table I.

In terms of rated capacity, BuRec's requirement is that when the pump-turbine is pumping at a speed of 180rev/min over a total head range of 468 to 485ft there must be an average capacity of not less than 80% of the average

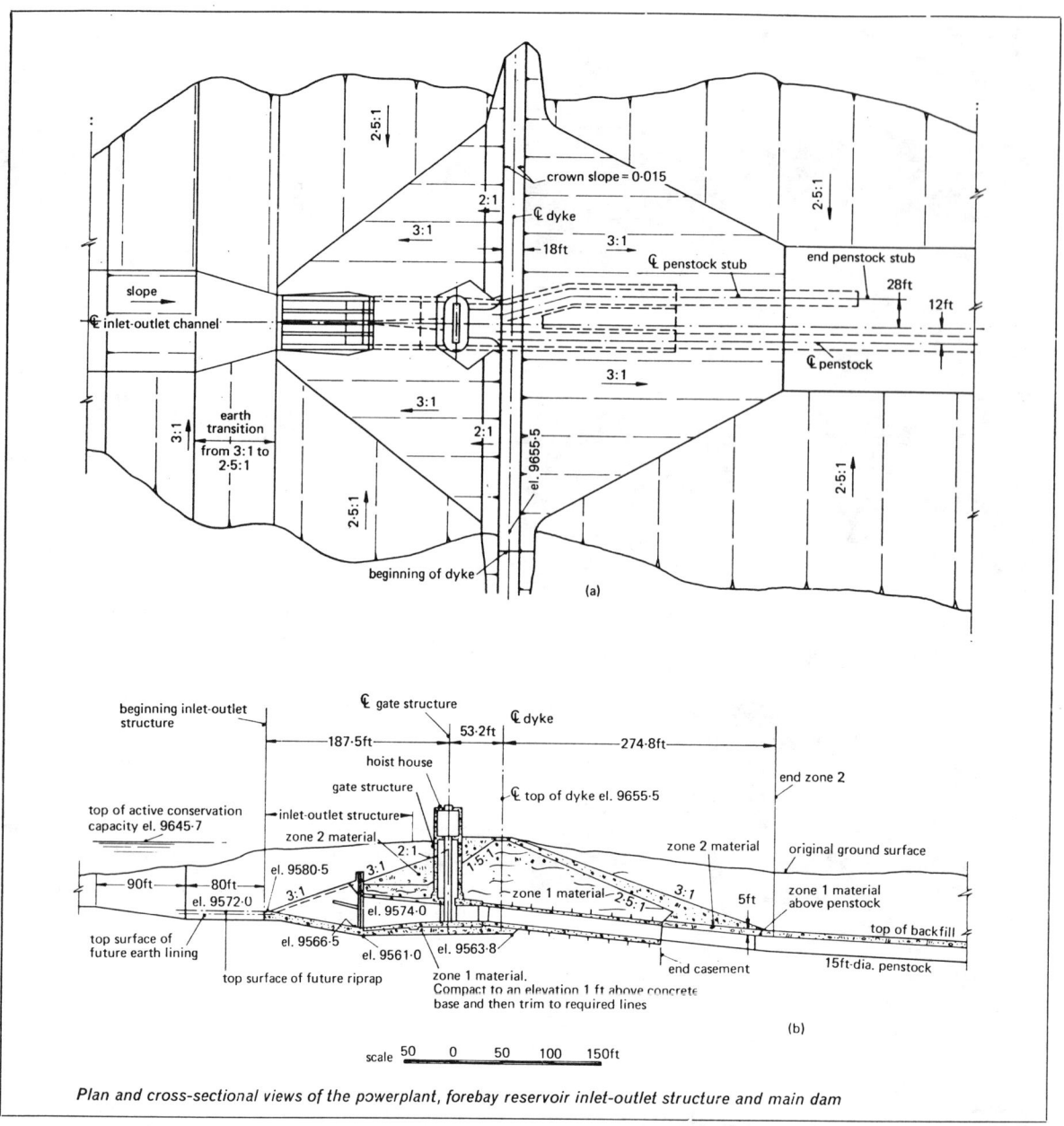

Plan and cross-sectional views of the powerplant, forebay reservoir inlet-outlet structure and main dam

turbine discharge when developing 138 000hp over an effective head range of 448 to 466ft.

When it is operating as a turbine at a speed of 180rev/min, the pump-turbine is to have a capacity of not less than 138 000hp at full gate opening when it is operating under an effective head of 405ft.

As part of the contract Allis-Chalmers fabricated a test model of the pump-turbine that is homologous with the prototype and upon which acceptance of the hydraulic design for manufacture of the prototype was based.

The model was complete with casing extension, spiral casing, staying, wicket gates, impeller-runner, suction draft tube, tailwater side trashrack, and about six diameters of straight penstock or pump-discharge pipe.

The throat diameter of the model impeller-runner was 10in. Features affecting hydraulic thrust were accurately modelled to permit prediction of prototype thrusts and transparent sections were provided in the upper portion of the draft tube liner to allow observation of cavitation effects.

In addition, piezometer taps were installed in the spiral casing and draft tube to record pressure and other data.

In accepting the hydraulic design for manufacture of the prototype, BuRec engineers placed particular emphasis on the hydraulic performance of the model and its ability to operate smoothly and quietly under all normal operating conditions.

The model tests indicated that the unit should meet or exceed all specified performance requirements.

The motor-generator

Manufacture of the motor-generator to operate with the first Mt Elbert pump-turbine unit, is proceeding under a $2 220 421 contract awarded in November, 1971 to the Westinghouse Electric Corp. Installation of the machine is scheduled for completion by November 1, 1975.

Progress on the powerplant structure as on February 27, this year

Specifications for the unit when it is operating at el. 9164ft are given in Table II.

The motor-generator is to be capable of continuous operation as a generator at 115% rated kVA and as a motor at 105% rated horsepower, both at the rated voltage, power factor and frequency values.

It is estimated that the motor-generator will be subjected to 300 motor and 300 generator starts each year. Accordingly, the design requirements are that the machine must be capable of safely withstanding not less than 1000 starts, without major maintenance (downtime) of more than 1 day, on either the static-starting system or its controls.

Another design requirement is that the machine must synchronize within 10min after initiation of the starting cycle.

Table II—Motor-generator specifications		
	Motor	Generator
Rating	170 000hp	105 263kVA
Speed (rev/min)	180	180
Power factor (%)	95	95
Frequency (Hz)	60	60
Number of phases	3	3
Voltage between phases (kV)	11·5	11·5
Excitation voltage (V)	250	250

Armature winding is wye-connected, suitable for either grounded or ungrounded neutral service

The powerplant will normally generate power between 8am and 10pm unless there is a need for reserve capacity on the interconnected power system. Pumping operation will usually be scheduled for the remainder of the 24hr day and at weekends.

Operating Procedures

Pump-turbine. The water surface at the pump intake on the Twin Lakes reservoir will vary between el. 9168·7 and 9208·5ft asl. At the forebay reservoir inlet-outlet structure, the water surface will vary between a minimum of el. 9615·0ft and a maximum of el. 9646·8ft.

Head losses exterior to the pump-turbine, but including discharge-line friction, trashrack, entrance and discharge losses, are estimated to be 15ft for the 3400cusec discharge.

For pumping operation, the pump-turbine unit will be started with the wicket gates closed and the impeller dewatered. After the unit is synchronized, the wicket gates will be opened.

For generation, the unit will be started and brought up to synchronous speed as a conventional turbine-driven generator, with speed control by the wicket gates and governor.

BuRec requirements are that the pump-turbine operates satisfactorily as a pump over the entire range of heads possible between the limiting water-surface elevations mentioned earlier, taking into account head losses, but without throttling and with the submergence provided. The total pumping head will vary between 420 and 485ft.

Additionally, the pump-turbine must operate satisfactorily as a turbine over the entire range of heads possible

Progress on the forebay reservoir inlet-outlet structure as in August 1973

between the same limiting water-surface elevations, taking into account head losses. The effective head on the turbine will vary between 390 and 475ft.

Motor-generator. The motor-generator will have a system using static components for startup in the pumping direction. Alternating current is taken from the incoming transmission system and converted to a controlled DC potential. This power is then re-converted to a variable frequency which is directly applied to the stator of the machine.

The unit will startup in the motor direction at low frequency, this frequency being increased to 60Hz, and at synchronous speed the unit is connected to the transmission system and pumping begins.

For operation as a motor, the method of starting will be by means of the static starting system, with the wicket gates closed and the suction draft tube dewatered. After the unit is synchronized, the wicket gates will be opened.

For operation as a generator, the unit will be started and brought up to synchronous speed as a conventional hydropower generator, speed control being provided by the wicket gates and governor. The machine will then be synchronized with the power system.

The powerplant is designed for remote operation and also for local attended operation from the unit control panel on the main board in the control room.

Starting, stopping, and running control of the unit will be carried out by local-manual or local-automatic means at the unit control and by supervisory means at the remote centre.

Control of the excitation and voltage-regulating equipment, and all unit auxiliaries, will be incorporated into the local-manual, local-automatic, and supervisory control schemes.

The operating procedure will be essentially as follows, and the sequence will be carried out automatically after the initial selection:

Motor operation (impeller dewatered, wicket gates closed)
(i) Motor operation will be selected.
(ii) Auxiliaries will be started.
(iii) The field will be applied from the static excitation system.
(iv) The voltage regulator will be applied.
(v) The circuit breakers of the static starting system will close to connect the motor to the power system and bring the unit up to speed.
(vi) The unit will be synchronized with the power system by an automatic synchronizer.
(vii) The motor-generator circuit breaker will close to connect the machine to the power system at the same time that the circuit breakers of the static starting system are opened.
(viii) The wicket gates will be opened and pumping will begin.

Generator operation (hydraulic start with penstock full)
(i) Generator operation will be selected.
(ii) Auxiliaries will be started.
(iii) Wicket gates will be partially opened to bring unit up to speed.
(iv) The field will be applied automatically.
(v) The voltage regulator will be applied.
(vi) The unit will be synchronized with the power system by an automatic synchronizer.
(vii) The motor-generator circuit breaker will close to connect the machine to the power system.
(viii) The unit will be operated under electric governor control of the wicket gates.

Under emergency loading conditions, the motor-generator is to be capable of being loaded as a generator from zero to full load in 30s.

An important feature of the unit is its capacity to achieve rapid changeover from pumping to generating operation within approximately 5min.

The unit controls will be such that when the motor-generator is operating in the pumping mode, operation of the proper control switch will automatically cause the machine to transfer rapidly from pumping to generating.

For shutdown of the unit, the normal sequence will be closure of the wicket gates, followed by opening of the motor-generator circuit breaker. Under emergency shutdown, the wicket gates will be automatically closed by the governor in 15s.

After the main circuit breaker has opened, the static starting equipment can again be connected to the machine terminals and to the power-line system. This will allow the stored energy in the rotating machine to be returned to the external line.

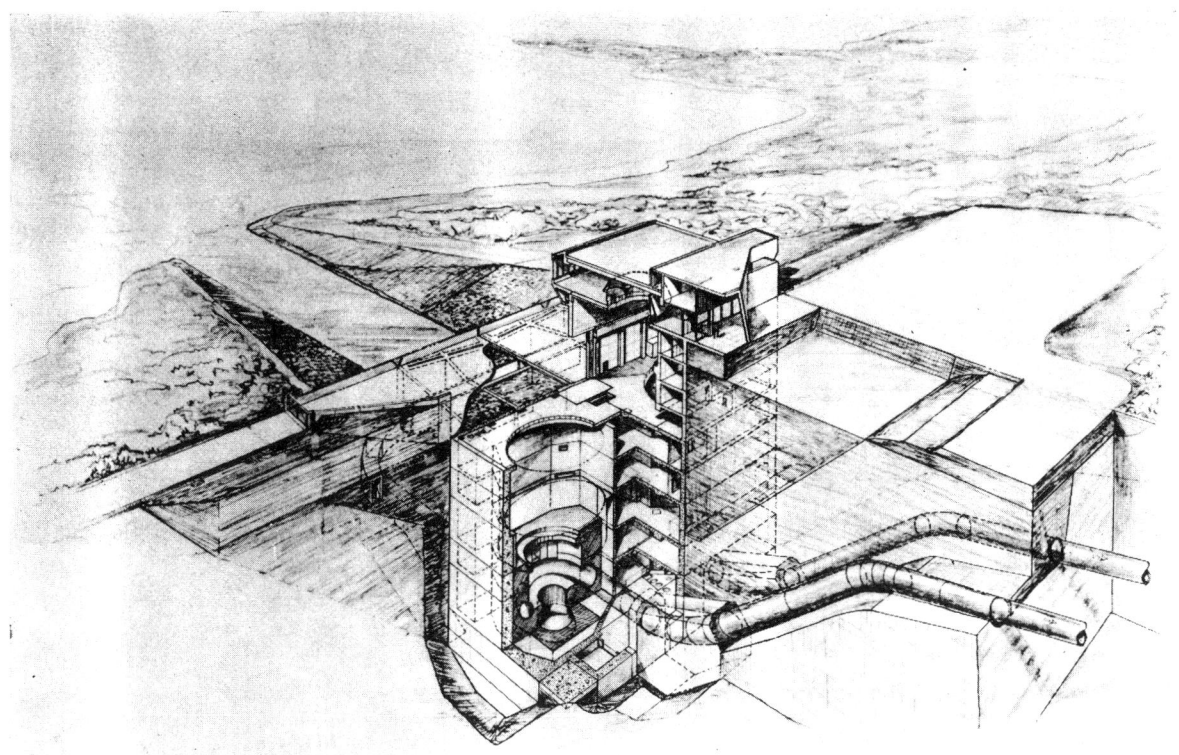

Another artist's impression of the Mt Elbert powerplant, this time showing the penstocks (the top one being to the future unit)

Completion Plans

Current proposals envisage awarding a contract for the completion of the powerplant in mid-1974. This work will include placement of the second-stage concrete for installation of the first pump-turbine unit, installation of the unit and associated equipment, and of certain electrical and mechanical equipment not installed under the current construction contract.

The completion contract will also include finishing of the building interior, and construction of the plant switchyard, which is to be built near the forebay on a relatively flat area of glacial moraine.

A 230kV cable will connect the switchyard with the power transformer and will extend through 800ft of cable tunnel from the powerplant. The remainder of the cable to the switchyard is to be buried underground.

The 230/11·5kV, 142·4MVA, three-phase power transformer will be installed outside the powerplant to step up the generated voltage to 230kV. A second transformer having the same rating will be installed later when the plant is expanded to its full capacity of 200MW.

Other work planned under the completion contract will include installation of batteries, oil-purification equipment and storage tanks, and shop equipment for machine repairs. The shop will also serve the needs of other project facilities.

Environmental considerations

As the Mt Elbert powerplant site is in a major recreational area of the Colorado Mountains (at present there are about 100 000 visitors annually), the Bureau of Reclamation has expended considerable effort to ensure that the plant will blend as unobtrusively as possible with its surroundings and that the impact of construction activities on the environment is closely controlled.

Specifications for construction of the powerplant include provisions for the contractor to carry out his activities in a manner that will prevent any unnecessary destruction, scarring, or defacing of the natural surroundings in the vicinity of his work.

These specifications also include provisions for the contractor to control air pollution resulting from his construction work. Burning of woody vegetation is not permitted and waste water from aggregate processing, concrete batching, or other construction activities must not enter streams or water courses without appropriate control methods.

Sanitary wastes are to be disposed of by burial at approved sites or by other approved methods.

BuRec engineers have called for the powerplant penstocks to be buried as an aesthetic consideration; this is also compatible with engineering and economic analyses.

The design decision to have the transformer circuit carried underground by means of an HV cable from the powerplant to the switchyard was also a contribution to preserving the environment of the powerplant site.

The switchyard itself incorporates BuRec's latest designs to render the facility functionally and aesthetically pleasing; switchyard structures will have rigid structural steel configurations, instead of the multi-member latticed-frame design.

All completed structures connected with the powerplant, as well as the plant itself, will be fully landscaped.

In dedicating the Charles H. Boustead tunnel in June, 1972, the US Secretary of the Interior, Rogers Morton, stressed the concern of the Department of the Interior over the impact of the Fryingpan-Arkansas project structures, such as the Mt Elbert installation, on the environment. He stated that the project "is unique for its environmental considerations. The project demonstrates that man can meet his needs for water and energy while at the same time he preserves the beauty of nature and the balance of the ecosystem."

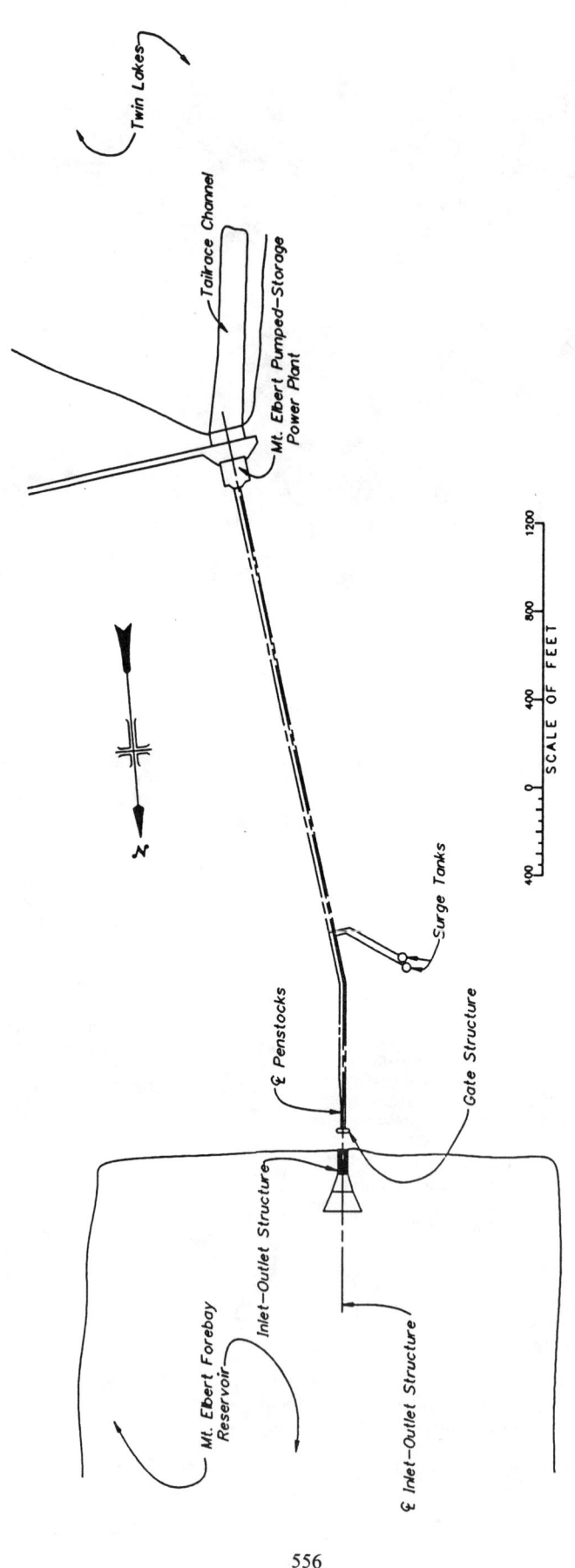

MT. ELBERT PUMPED-STORAGE POWER PLANT PLAN

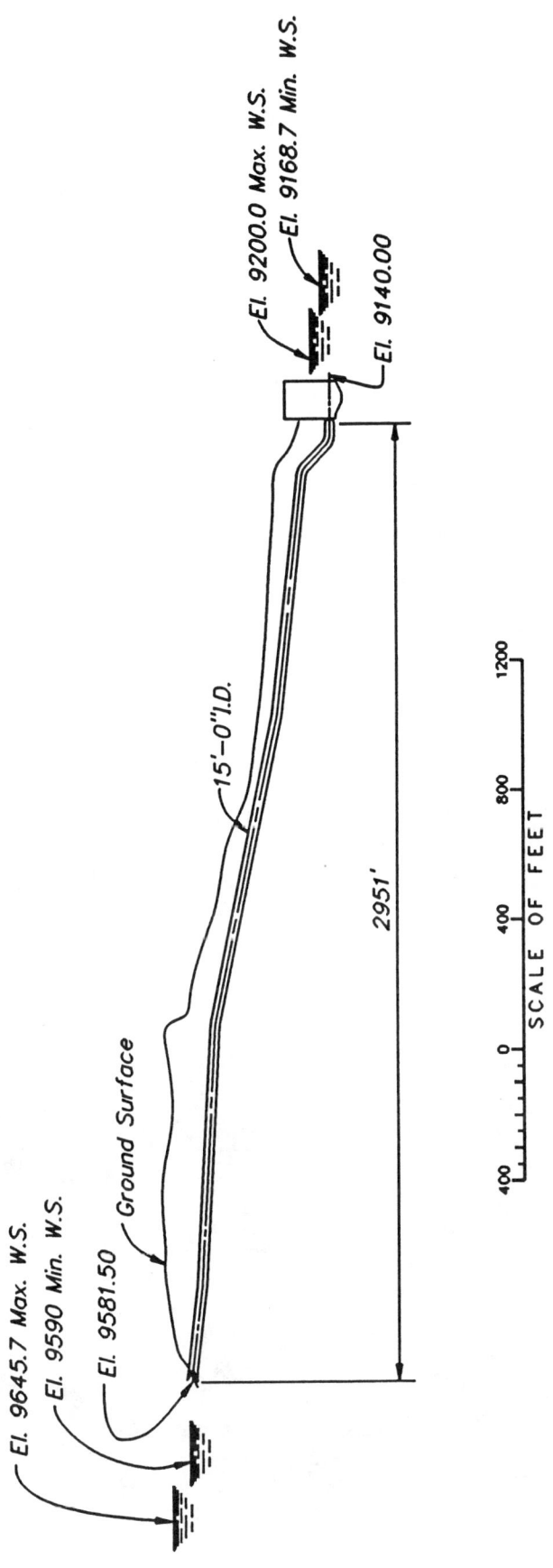

MT. ELBERT PUMPED-STORAGE POWER PLANT PROFILE

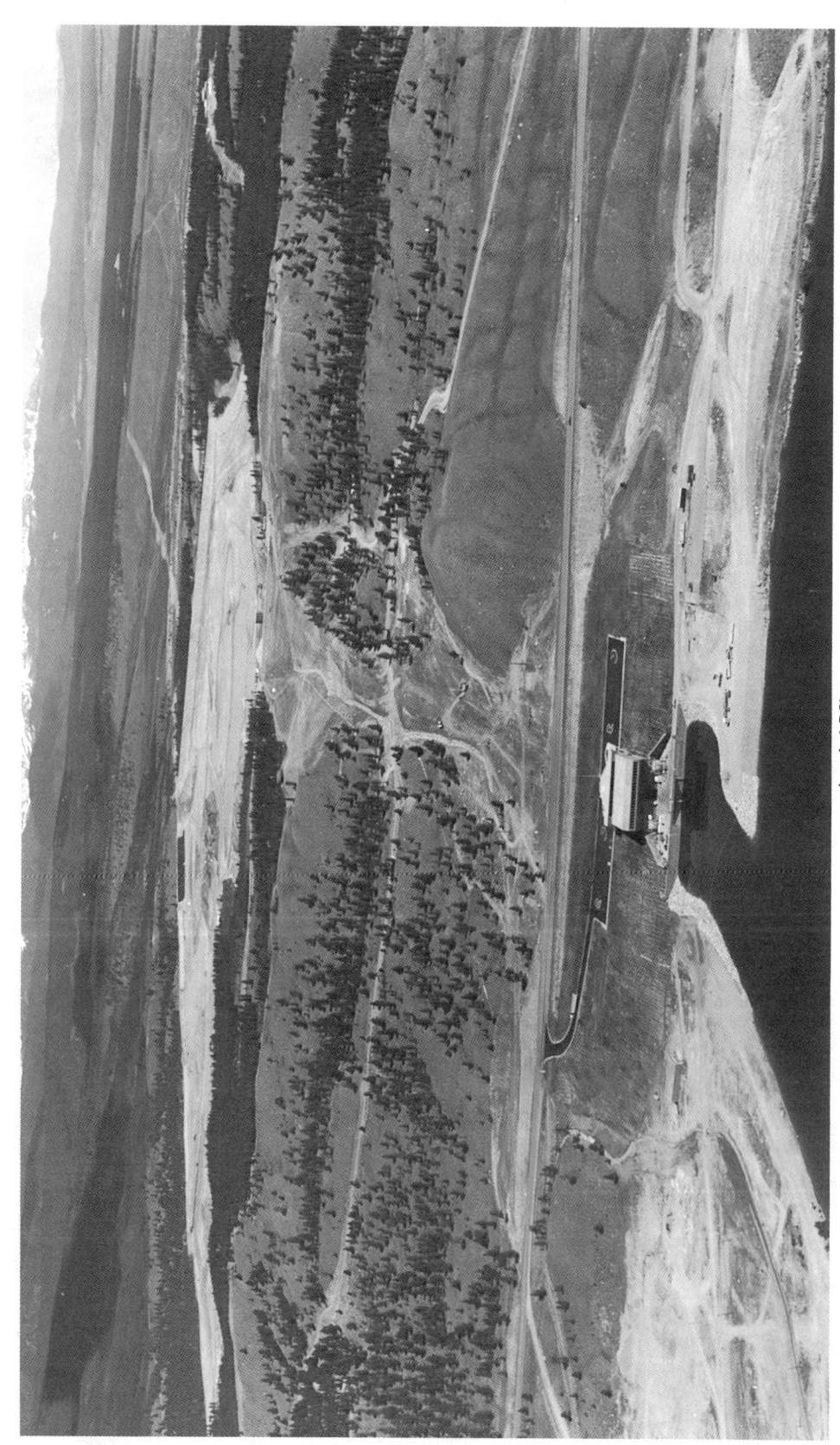

Aerial View

Mt. Elbert Pumped Storage Plant

Plant Name: **MUDDY RUN**

Plant location:
  Drumore, PA
  Lancaster County
	Owner: Philadelphia Electric Company
	2301 Market Street
	Philadelphia, Pennsylvania 19101

Rated capacity 800 MW Average static head 353 ft Plant efficiency 71.40 % Stored energy 11,100 MWh Number of units 8	Designers: Philadelphia Electric Company

Construction time: 2 years, 6 months
Construction cost: $90 per kW
Price level: 1967
First commercial power: April 1967
FERC project number: 2355
 	
River or water source: Susquehanna River
	Plant Manager/Superintendent:
	  Plant Manager, Muddy Run Pumped Storage
	  Rd #1
	  Drumore, Pennsylvania 17518
	
	  (301) 457-4161

	UPPER RESERVOIR	LOWER RESERVOIR
DAM		
Type	Earth and rockfill	Concrete gravity
Height (ft)	250	102
Crest length (ft)	4,400	4,649
Volume (yd^3)	5,179,000	667,000
RESERVOIR		
Type	Constructed	On stream, Conwingo Pond
Surface area (acres)	960	2,656
Usable power storage (acre-ft)	33,158	80,990
Power pool fluctuation (ft)	50.0	4.5
Operating levels		
Maximum (ft)	520.0	109.0
Minimum (ft)	470.0	105.5
Drainage area (miles^2)	9.2	27,000.0
Seepage (ft^3/s)		
SPILLWAY		
Design flood		
Return period (years)		
Flow (ft^3/s)	13,000	1,170,000
Capacity (ft^3/s)		1,170,000
Type	Concrete, ungated	
Gates		
Number	None	50
Type		Stoney type crest
Width (ft)		38.00
Height (ft)		22.50
OUTLET WORKS		
Discharge capacity (ft^3/s)		
Number of water passages	None	None
Dimensions of water passages		
Height (ft)		
Width (ft)		
Diameter (ft)		
Type of gates		
Number of gates	None	None

Plant Name: **MUDDY RUN**

Plant location Drumore, PA Lancaster County	Owner: Philadelphia Electric Company 2301 Market Street Philadelphia, Pennsylvania 19101
Rated capacity 800 MW Average static head 107.6 m Plant efficiency 71.40 % Stored energy 11,100 MWh Number of units 8	Designers: Philadelphia Electric Company
Construction time: 2 years, 6 months Construction cost: $90 per kW Price level: 1967 First commercial power: April 1967 FERC project number: 2355	Plant Manager/Superintendent: Plant Manager, Muddy Run Pumped Storage Rd #1 Drumore, Pennsylvania 17518 (301) 457-4161
River or water source: Susquehanna River	

	UPPER RESERVOIR	LOWER RESERVOIR
DAM		
Type	Earth and rockfill	Concrete gravity
Height (m)	76.2	31.1
Crest length (m)	1,341.1	1,417.0
Volume (m³)	3,960,000	510,000
RESERVOIR		
Type	Constructed	On stream, Conwingo Pond
Surface area (Mm²)	3.89	10.75
Usable power storage (Mm³)	40.900	99.900
Power pool fluctuation (m)	15.24	1.37
Operating levels		
Maximum (m)	158.50	33.22
Minimum (m)	143.26	32.16
Drainage area (Mm²)	23.828	69,930.070
Seepage (m³/s)		
SPILLWAY		
Design flood		
Return period (years)		
Flow (m³/s)	368	33,131
Capacity (m³/s)		33,131
Type	Concrete, ungated	
Gates		
Number	None	50
Type		Stoney type crest
Width (m)		11.582
Height (m)		6.858
OUTLET WORKS		
Discharge capacity (m³/s)		
Number of water passages	None	None
Dimensions of water passages		
Height (m)		
Width (m)		
Diameter (m)		
Type of gates		
Number of gates	None	None

Muddy Run - Page 1 (Metric)

MUDDY RUN

INTAKES	UPPER INTAKE	LOWER INTAKE
Number	4	8
Type	Concrete structure, tower	Draft tube
Design discharge (ft³/s)	6,800	
Gross inlet area (ft²) (at trash racks)	2,224	
Bar racks		
spacing (in)	6.00	
shape		
depth/thickness (in)	3.50 / 0.63	
diameter (in)		
Emergency gates		
number	None	None
height/width (ft)		
type		
Service gates		
number	1	None
height/width (ft)		
type	Circular gate, 7.16 m (23.5 ft) diameter	
Bulkhead/stop logs (Y or N)		
number of units serviced		
Hoists		
number	4	None
capacity (tons)		
type	Roller chain type	

WATER PASSAGES	Upper Tunnel	Shaft	Lower Tunnel	Surge Tanks Upper	Surge Tanks Lower	Penstocks	Tailrace Tunnel
Number		4	4			8	
Diameter (ft)		24.5	24.5			14.0	
Length (ft)		350	440			500	
Maximum velocity (ft/s)		18.5	18.5			28.5	
Concrete lining thickness (in)		18.00	18.00				
Total length of concrete sections (ft)		350	440				
Steel liner Thickness							
Minimum (in)						0.88	
Maximum (in)						1.30	
Material grade						ASTM A-36	
Total length of steel-lined sections (ft)						400	

Notes:

Muddy Run - Page 2 (English)

MUDDY RUN

INTAKES	UPPER INTAKE	LOWER INTAKE
Number	4	8
Type	Concrete structure, tower	Draft tube
Design discharge (m³/s)	192.6	
Gross inlet area (m²) (at trash racks)	206.6	
Bar Racks:		
spacing (mm)	152	
shape		
depth/thickness (mm)	89 / 16	
diameter (mm)		
Emergency gates		
number	None	None
height/width (m)		
type		
Service gates		
number	1	None
height/width (m)		
type	Circular gate, 7.16 m (23.5 ft) diameter	
Bulkhead/stop logs (Y or N)		
number of units serviced		
Hoists		
number	4	None
capacity (Mg)		
type	Roller chain type	

WATER PASSAGES	Upper Tunnel	Shaft	Lower Tunnel	Surge Tanks Upper	Surge Tanks Lower	Penstocks	Tailrace Tunnel
Number		4	4			8	
Diameter (m)		7.47	7.47			4.27	
Length (m)		106.7	134.1			152.4	
Maximum velocity (m/s)		5.64	5.64			8.69	
Concrete lining thickness (m)		0.457	0.457				
Total length of concrete sections (m)		106.7	134.1				
Steel liner Thickness							
Minimum (mm)						22	
Maximum (mm)						33	
Material grade						ASTM A-36	
Total length of steel-lined sections (m)						121.9	

Notes:

Muddy Run - Page 2 (Metric)

MUDDY RUN

POWERHOUSE and RELATED FEATURES

Powerhouse Structure
Type: Outdoor type with open generator deck
Length: 600 ft   Width: 140 ft   Height: 80 ft

Guard Valves
Number: None   Diameter: ft
Type:

Transformers
Number: 4
Ratings: 250
Voltages: (kV) 220 / 13.8

Generator
Rating generating (MVA): 111.0   Rating pumping (MVA): 111.0
Insulation type:
Starting method: Pony motor
Starting equipment: Reduced voltage taps on transformers

Runners
Material: ASTM A-27 gr 65-35 cast buckets welded
Minimum unit submergence: 20.0 ft
WR^2:
Manufacturer: Baldwin-Lima-Hamilton
Model test by: Baldwin-Lima-Hamilton

	Reversible Runners	Reversible Motor/Generator		
Number	8			8
Diameter (ft)	17.90	Rotor	Stator	
rpm synchronous	180.0			180.0
rpm overspeed				
Type	Modified Francis		Umbrella, outdoor	

Information on Runners

Condition:	Gross Head (ft)		Capacity (MW)		Discharge (ft³/s)		Turbine/Pump Eff.(%)	
	Generating	Pumping	Generating	Pumping	Generating	Pumping	Generating	Pumping
Maximum head & maximum power	415	415			4,370	2,600		
Minimum head & maximum power	361	361				3,400		

Condition:	Net Head (ft)		Capacity (MW)		Discharge (ft³/s)		Turbine/Pump Eff.(%)	
	Generating	Pumping	Generating	Pumping	Generating	Pumping	Generating	Pumping
Rated head @ best gate	353	427	103	115		2,610		

Muddy Run - Page 3 (English)

MUDDY RUN

POWERHOUSE and RELATED FEATURES

Powerhouse Structure
Type: Outdoor type with open generator deck
Length: 182.9 m   Width: 42.7 m   Height: 24.4 m

Guard Valves
Number: None   Diameter: m
Type:

Transformers
Number: 4
Ratings: 250
Voltages: (kV) 220 / 13.8

Generator
Rating generating (MVA): 111.0   Rating pumping (MVA): 111.0
Insulation type:
Starting method: Pony motor
Starting equipment: Reduced voltage taps on transformers

Runners
Material: ASTM A-27 gr 65-35 cast buckets welded
Minimum unit submergence: 6.10 m
WR^2:
Manufacturer: Baldwin-Lima-Hamilton
Model test by: Baldwin-Lima-Hamilton

	Reversible Runners	Reversible Motor/Generator	
Number	8		8
Diameter m	5.46	Rotor	Stator
rpm synchronous	180.0		180.0
rpm overspeed			
Type	Modified Francis		Umbrella, outdoor

Information on Runners

Condition:	Gross Head (m) Generating	Gross Head (m) Pumping	Capacity (MW) Generating	Capacity (MW) Pumping	Discharge (m³/s) Generating	Discharge (m³/s) Pumping	Turbine/Pump Eff.(%) Generating	Turbine/Pump Eff.(%) Pumping
Maximum head & maximum power	126.5	126.5			123.7	73.6		
Minimum head & maximum power	110.0	110.0				96.3		

Condition:	Net Head (m) Generating	Net Head (m) Pumping	Capacity (MW) Generating	Capacity (MW) Pumping	Discharge (m³/s) Generating	Discharge (m³/s) Pumping	Turbine/Pump Eff.(%) Generating	Turbine/Pump Eff.(%) Pumping
Rated head @ best gate	107.6	130.2	103	115		73.9		

Muddy Run - Page 3 (Metric)

MUDDY RUN

Plant Data:
 Average GWh generating per year: 1,354
 Average GWh pumping per year: 1,937
 Starting time from standstill (s):
 Changeover time pumping to generating (min):
 Planned/scheduled time
  between major overhauls (years):
 Outage time required per unit
  during major overhauls (weeks):   35
 Representative plant availability (%):   91.0
 Representative planned outages (weeks per year):

Miscelleneous Notes:
 None.

Cavitation Experience:

Significant or Unique Problems:

List of Licenses Required:

ENVIRONMENTAL FEATURES

Recreation:
 Recreation incudes a park with fishing and camping facilities.

Fish and Wildlife:

Social:

Muddy Run - Page 4

MUDDY RUN PUMPED-STORAGE PROJECT

STANLEY MOYER
Assistant Chief Mechanical Engineer

and

W. HAINES DICKINSON
Senior Engineer

Mechanical Engineering Division
Philadelphia Electric Company
Philadelphia, Pennsylvania

The Muddy Run Pumped-Storage Generating Plant was planned to supply peaking capacity to meet Philadelphia Electric Company requirements. Scheduled for completion during 1967, this $75,000,000 project will add a unique, new type of generation, rated at 800,000 kW, to the Company's 1966 year-end capability of approximately 3,660,000 kW.

HISTORY AND JUSTIFICATION

Generating capacity studies were commenced in 1960 to select the most economical plant addition that would enable the Company to meet its load requirements for the next decade.

Since summer peak loads of higher magnitude and longer duration than former winter peaks had begun to establish maximum capacity requirements, the need was apparent to investigate, from an economic standpoint, the use of generating units designed and priced specifically for the varying demands of peaking service. The base-load requirements had been met over the past decade by the installation of approximately 1,400,000 kW in modern reheat units.

Five types of peaking units were available for comparison at the time. They were steam peaking generators, diesel generators, gas turbine generators, conventional hydro generators and pumped-storage hydro generators. Capital and operating cost data were readily available for the first three types of plant from our own experience. The Company's experience with power boilers and steam heat boilers of various ratings provided reliable costs for steam peaking. Three 2000-kW diesel units and two gas turbines rated at 22,000 and 20,000 kW had been installed on the system previously, so costs were available.

The capital costs for conventional hydro generating capacity can vary greatly. Costs for a completely new project could be quite high, but additional units in an existing plant can be the most economical type of installation. Background gained through the 1964 installation of four new units in the Conowingo Hydro Plant evidenced the capital savings that can be realized. However, new sites for conventional hydro plants were almost nonexistent in our territory and so conventional hydro was not considered for further study.

Capital costs for a pumped-storage installation are very much dependent on the availability of a site. An investigation of possible plant sites was therefore started. The following criteria were established to aid in the selection of a pumped-storage site.

1. The site was to be adjacent to an existing reservoir, lake or river of sufficient size to serve as a lower reservoir. This would reduce total construction costs by eliminating the costs associated with building a lower reservoir.

Reprinted from Volume 29 Proceedings of the American Power Conference, 1967.

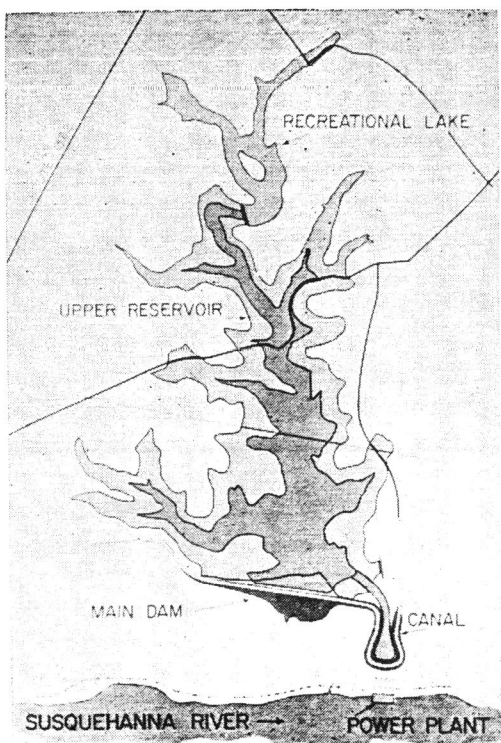

Fig. 1—Muddy Run layout.

2. The acreage required for a hydro project is great and consequently the cost of the land can be significant. A price of about $1000 per acre was to be a maximum. Site preparation costs were to be minimized by the avoidance of areas requiring extensive relocation of existing public facilities.
3. A natural ravine or depression was sought to permit the construction of an upper storage reservoir with the minimum investment in dams and dikes. The minimum water storage in the upper reservoir was to be capable of generating at least 1000 megawatthours.
4. The difference in elevation or head between the upper and lower reservoirs was to be at least 300 feet to make the development economically feasible.
5. The total length of the conduit system necessary to carry the water from the upper reservoir to the plant was not to exceed about 12 times the available head on the plant. The cost of longer conduits would become proportionately too expensive in the overall project.

The Muddy Run site met all of the criteria and proved more economical than several other sites considered. Preliminary plant layouts were developed and a total cost estimate prepared. The capital cost developed was entered into the overall comparative economic study. A current summary of capital costs appears later.

The economic comparison of the several alternative peaking-type generating sources was made by determining for each source (a) the total annual cost for one kilowatt of capacity and 500 kilowatthours of energy, and (b) the total annual cost for one kilowatt of capacity and 1500 kilowatthours of energy.

Capacity cost determinations consisted of the fixed charges for the installed plant capital expenditures (including land and transmission lines required for connection with the existing power system), fixed operating and maintenance expenses, administrative expenses, the cost of fuel inventory, and an unavailability allowance in the case of the steam plant.

The energy costs in the case of the pumped-storage plant were based on the average cost of off-peak energy, and the energy costs for the diesel, gas turbine and steam plants were based on their respective fuels. The evaluation also allowed credit in the case of steam for additional hours of generation, transmission credit for the diesels and gas turbines, and a spinning reserve credit for the pumped-storage plant. The use of Muddy Run units for spinning reserve will reduce the operation of steam units for this purpose and result in an estimated saving of about $250,000 per year.

The following tabulation shows the capacity and energy costs used in the evaluation, with Muddy Run based on off-peak energy:

	Capacity Cost $/kW	Energy Cost Mills/kWh
Muddy Run	89	2.6
Diesels (oil)	100	8.9
Gas Turbines	110	13.0
Steam Peaking (coal)	120	2.6

The study indicated that based on the anticipated operation at a 17 percent load factor, 1500 hours per year, the total annual costs for one kilowatt of capacity and 1500 kilowatthours of energy were:

Muddy Run	$14.25
Steam Peaking	16.62
Diesels	16.71
Gas Turbines	19.61

The economic effect of the operation of Muddy Run on the Conowingo plant, 12 miles downstream, was also considered. The optimum coordination of Muddy Run at certain river flows will result in a higher average weekly head in the Conowingo reservoir and in the availability of more weekday water flow at the Conowingo plant due to the increased water storage capacity. It is estimated that the resulting increase in effective peaking capacity at Conowingo when averaged over a year will be about ten megawatts.

Some of the advantages and disadvantages of pumped-hydro that could only be partially weighed during the course of the economic study are as follows:

Advantages
1. Fast startup.
2. Load leveling for steam units on the system.
3. Self-modernizing through improved basic efficiency of the system, including additions of nuclear units.
4. Low operating and maintenance costs.
5. Automatic unattended operation.

Disadvantages
1. Site location likely to be remote from load center, resulting in high transmission costs and losses.
2. A large plant is required to best achieve low investment.
3. Fuel cost penalty. This results from deferring efficient thermal generation. When peaking generation is installed, instead of efficient base-load units, it incurs greater use of less efficient units.

The study clearly indicated that the most economical choice to provide additional capacity was the Muddy Run Pumped-Storage Plant. As a consequence, the Board of Directors of the Philadelphia Electric Company authorized the construction of the project in December 1962. Federal Power Commission License No. 2355 was issued on November 9, 1964, for a term of 50 years.

GENERAL PROJECT DESCRIPTION

The Muddy Run Pumped-Storage Generating Plant is a project of the recirculating type, located about 18 miles south of the city of Lancaster, Pennsylvania, on the eastern shore of the Susquehanna River. The Company's existing Conowingo pond serves as the lower reservoir for the Muddy Run Project. At the normal water surface elevation of 109 feet, the useful volume of the Cono-

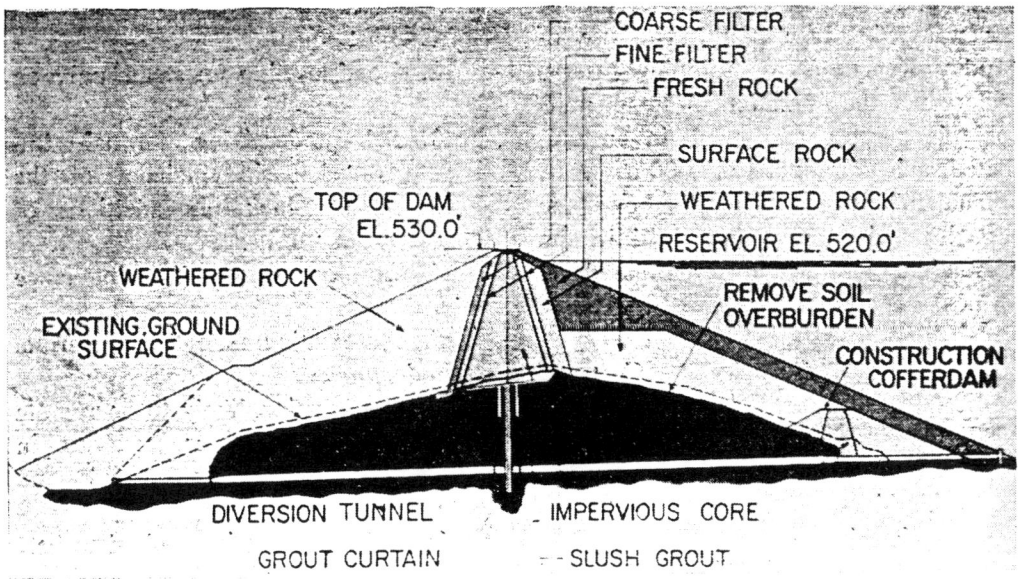

Fig. 2—Muddy Run dam.

wingo pond is about 2½ times the useful volume of the upper reservoir.

The upper reservoir (Fig. 1) is created by an earth and rock fill impoundment constructed across a deep rural valley in which was flowing a small stream called Muddy Run. In area the full pond water surface is nearly 1000 acres including a 100 acre finger which has been separately dammed for recreational uses. The 33,200 acre-feet of water in the top 50 feet of the reservoir available for power generation represent about 55 percent of the total storage and provide for 11,100,000 kilowatthours of energy. The full reservoir water level elevation is 520 feet above sea level and establishes a gross head of 411 feet.

The upper reservoir is located in an area of deeply weathered Wissahickon schist. The main dam is a zoned earth and rock fill structure about 4400 feet long at the crest with a maximum height of 250 feet and contains about 5,250,000 cubic yards of material. The upstream and downstream shells (Fig. 2) are primarily of variably weathered schist with fresh material used to face the upstream shell in the drawdown area. The impervious core material consists of a fine micaceous silt taken from borrow areas on the site. Field compaction of core material was very successful, attaining at least 95 percent of modified Proctor maximum density. The core is blanketed both upstream and downstream with graded two-stage filters, each 8 feet thick. To complete the impervious barrier, a grout curtain was injected from the bottom of the core trench at least 20 feet into the impermeable fresh rock.

The water passages begin in the upper reservoir with a canal, 180 feet wide at the base by 80 feet deep, which extends toward the powerhouse along a high ridge of ground. The natural rock through which the canal was excavated provides a competent cross section for the first 1300 feet of length. However, the rest of the canal (approximately 1000 feet), because of the open jointing in the exposed rock and because the ridge becomes

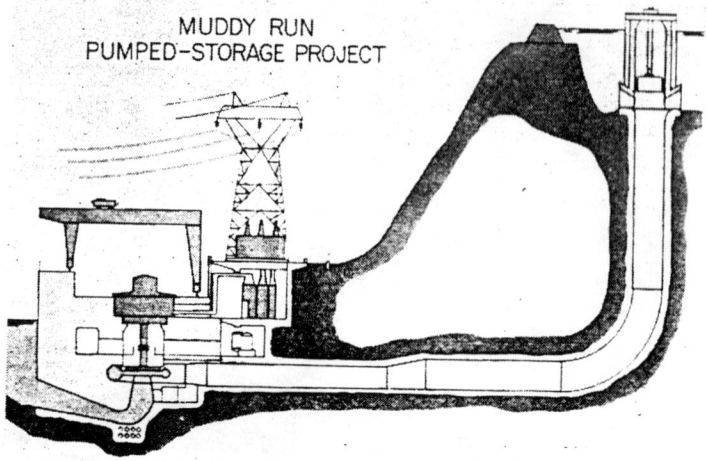

Fig. 3—Cross section.

quite narrow, is lined on the sides and bottom with reinforced concrete placed over porous pipe drains and anchored into the rock with grouted dowels.

At the powerhouse end of the canal, four concrete intake structures (Fig. 3) are installed and house the cylinder gates. The cylinder gates are 23.5 feet in diameter, and each gate provides complete shutoff to the shaft-tunnel system leading to its corresponding pair of pump-turbines. Except for the wicket gates associated with the units, the cylinder gates are the only means of shutoff, and were dsigned to close under full water flow in the event of an emergency. The normal closing time is approximately five minutes.

From the cylinder gates, the water conduit system goes underground, extending approximately 1300 feet to the plant. Four 24.5-foot diameter concrete lined shafts extend vertically down into the hillside. A 90-degree bend at the bottom of each shaft changes the direction and starts a horizontal run to the plant. About 500 feet ahead of the plant each tunnel bifurcates into two smaller 14-foot diameter steel-lined tunnels, each leading to a pump-turbine. Calculations indicated that there would be a 25 percent pressure rise on full-load rejection of two units simultaneously, resulting in 235 psi of internal pressure, which was the basis of design for the penstock and spiral case.

A concrete nongated overflow spillway located on the right side of the canal is sized to pass the peak runoff from a storm producing 26.5 inches of rainfall in six hours with the power plant shut down.

The powerhouse (Fig. 4), located on the Conowingo reservoir, is of the outdoor type and contains about 120,000 cubic yards of concrete. The plant structure is 600 feet long and 140 feet wide. The plant includes a service bay area at one end, and is serviced by a 250-ton gantry crane.

The eight reversible pump-turbines were designed and manufactured to develop 138,000 hp under the minimum net head of 353 feet. The output will be considerably greater at high heads, and will be limited by the service factor of the generator-motors. The pumping discharge will vary from approximately 3400 cubic feet per second at minimum head to 2600 cubic feet per second at the maximum head. The operating speed is 180 rpm in both directions.

The eight generator-motors were designed, manufactured and erected by Westinghouse. As a generator, each unit

TABLE I
COST BREAKDOWN

	$/kW	%
Land and public road relocations	3.40	3.6
Power plant structures	12.50	13.3
Dam and waterways	34.00	36.3
Pump-turbines, generator-motors and associated equipment	18.90	20.2
Accessory electrical equipment	4.75	5.1
Transmission facilities	3.50	3.8
Miscellaneous	1.60	1.7
Direct Cost	78.65	84.0
Engineering, contingency, interest during construction and general overhead	15.10	16.0
Gross Cost	93.75	100.0

is rated at 100,000 kW, 0.9 power factor, 13.8 kV, 3 phase, 60 cycles. As a motor, each unit will develop a maximum of 154,000 hp.

The units are paired electrically as well as hydraulically. Each of the four 250-MVA main transformers is connected through low-side breakers to two generator-motors. These transformers step up the generator voltage to 220 kilovolts for transmission out of the plant to the Peach Bottom area four miles downstream, where a new substation provides a tie to the system.

Starting in the pumping mode is accomplished at a reduced voltage, allowing the unit to accelerate to rated speed at 6.6 kV supplied by half voltage taps on two of the four main transformers.

As the project is nearing completion, a review of actual costs is appropriate. The present total cost is estimated at $75,000,000 for a rated capacity of 800,000 kW. The project was awarded to The Arundel Corporation of Baltimore slightly more than three years ago on a lump sum basis after competitive bidding. Since then, several changes have been made to the design and scope of work which necessitated modifications to the contract price. The price per kilowatt is currently $5 higher than the cost originally estimated for the economic study.

A cost breakdown by major accounts based on the nameplate rating of 800,000 kW is shown in *Table I*.

STARTUP EXPERIENCE

On Saturday, March 11, 1967, 28 months after the start of construction at Muddy Run, the initial unit was rotated for the first time. The second unit was rotated for the first time two days later. Our intention here is only to highlight the startup experience, reserving the details for presentation at a later date.

The temporary startup instrumentation was supplied, installed and operated by the Company's Testing Division. Variable speed, strip chart recording equipment was used to record on 14 channels various items connected with the pump-turbine generator-motors. Pressure transducers were used to measure penstock pressures during load reject tests and linear transducers measured servomotor travel. Shaft runout, vertical movement of the unit, speed and power were also recorded. Recording oscillographs were utilized to secure operating characteristics of the generator-motor and control relaying.

The initial operation of the first two units was accomplished with local manual controls. A minimum operating force is being provided from the Conowingo Hydro Plant to serve until all units are in service. The ultimate control system for Muddy Run provides for remote automatic, local automatic or local manual operation of all units in both the pumping and generating modes. The design of Muddy Run was based on the premise that the plant will be unattended following the completion of construction and the satisfactory operation of all units. A

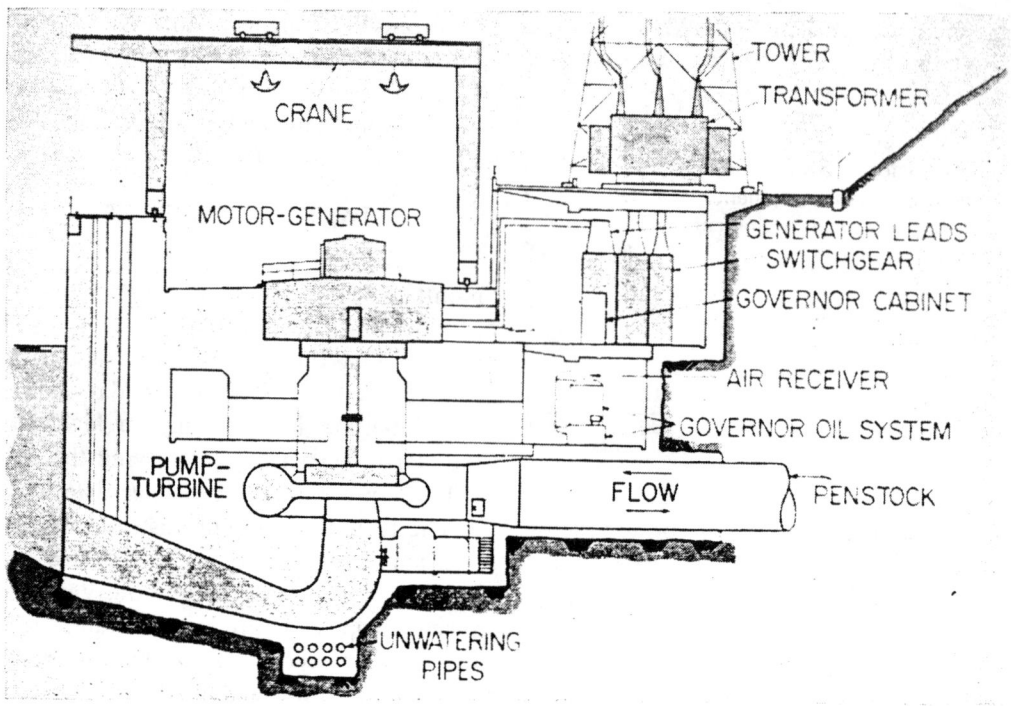

Fig. 4—Muddy Run powerhouse.

microwave system is being installed to enable future operations to be monitored and controlled from the existing manned Conowingo Hydro Plant located approximately 12 miles downstream.

The initial wet runs of the first two units occurred six days after the first dry run in the pumping mode. Considerable pumping was necessary initially since the upper reservoir has only a minor inflow from Muddy Run and a vast quantity of water for dead storage had to be pumped up to the upper reservoir before there could be any operation of the units as turbine-generators.

There was concern expressed regarding the initial pumping operations since it had to be done against zero head. Suggestions were made for filling the vertical shaft from external sources to create a discharge head before the first pumping operation with the main unit, but all suggestions seemed most complicated. It was therefore decided to attempt the initial pumping against zero head by opening the wicket gates only slightly and pumping the water up the shaft. The normal pumping head will vary over a range of 368 to 427 feet.

The first wet pumping run was accomplished without incident. The wicket gates were opened about 2 percent as soon as the unit was primed. As the discharge head increased, the wicket gates were opened wider. In a matter of minutes, the shaft was full and the period of concern was over. From the experience gained, it appears acceptable to fill the shaft initially with the main unit on future plants of similar head and penstock arrangement.

After each 20-foot increase in water surface elevation in the upper pond an inspection of the dam and readings of all the instrumentation incorporated in the dam were secured. Slope indicators, which are electronic devices utilized to detect structural movement of the various construction zones in either the longitudinal or transverse directions, and settlement indicators, which are electronic devices designed to measure the amount of vertical settlement occurring at various elevations in the dam, were incorporated in the embankment. Piezometers to measure water pressures existing at various depths inside the dam, and survey monuments across the crest of the dam providing a base for vertical and horizontal alignment complete the instrumentation in the dam. In addition, there are relief wells and measuring weirs installed on the downstream side to provide further data regarding water seepage through the dam. The reading, plotting and checking of all these devices was done during daylight and generally required two or three days for each lift.

The initial generation occurred without major incident two weeks after the first wet pumping run. By this time, the dead storage in the upper reservoir had been filled with water and a depth of about five feet of water was available in the power portion.

The first two units at Muddy Run were declared in commercial operation at 7:00 a.m., April 10, 1967, for loads up to 110,000 kilowatts each. Projecting this rating to the entire plant would result in a total capacity of 880,000 kilowatts and a unit price of $85.25 per kilowatt.

CONCLUSIONS

The Muddy Run Project will be completed this year. Six years of system planning studies, project design and construction are nearing a satisfactory conclusion with the initial operation of the first two units successfully accomplished. From our experience on this project we have two suggestions to pass along to designers of similar projects in the future.

We recommend to the civil engineer that he keep the design of earth and/or rock fill embankments flexible enough to accept field changes which will permit the use of all the material available at the site after actual excavation has completely disclosed its nature.

To the mechanical and hydraulic engineers we recommend studies of the possibility of including a cylinder gate at the speed ring of the pump-turbines supplementing or replacing the orthodox wicket gates.

Muddy Run Pump/Generating Plant

Bibliography

1. Moyer, Stanley and W. Haines Dickinson, "Muddy Run Pumped-Storage Project," *Proceedings of the American Power Conference*, 1967, Vol. 29, pp. 699-706.

2. Dickinson, W. Haines, "Muddy Run Pumped Storage: Big Capacity at Low Cost," *Electrical World*, July 4, 1966, pp. 31-33.

Plant Name: **NORTHFIELD MOUNTAIN**

Plant location:
  Northfield & Erving, MA
  Franklin County
	Owner: Connecticut Light and Power Co./Western Mass. Electric Co.
	       c/o Northeast Utilities
	       P.O. Box 270
	       Hartford, Connecticut 06101-0270

Rated capacity 1,080 MW Average static head 745 ft Plant efficiency 74.60 % Stored energy 10,500 MWh Number of units 4	Designers: Stone and Webster Engineering Corp.

Construction time: 4 years, 9 months
Construction cost: $145 per kW
Price level: 1972
First commercial power: November 1972
FERC project number: 2485
	Plant Manager/Superintendent:
	    Plant Manager
	    Route 63
	    Erving, Massachusetts 06101
	
	    ( ) -

River or water source: Connecticut River	

	UPPER RESERVOIR	LOWER RESERVOIR
DAM		
Type	Central core rockfill	Existing concrete Turner Falls Dam
Height (ft)	145	55
Crest length (ft)	5,600	
Volume (yd³)	2,700,000	
RESERVOIR		
Type	Constructed	On stream, on the Connecticut River, Turner Falls Pond
Surface area (acres)	275	2,000
Usable power storage (acre-ft)	14,350	12,745
Power pool fluctuation (ft)	84.5	9.0
Operating levels		
Maximum (ft)	1,004.5	185.0
Minimum (ft)	920.0	176.0
Drainage area (miles²)	0.5	7,163.0
Seepage (ft³/s)	2.000	
SPILLWAY		
Design flood		
Return period (years)		
Flow (ft³/s)	11,400	210,000
Capacity (ft³/s)		
Type	Overflow	Gated
Gates		
Number	None	7
Type	None	3 Radial, 4 Bascule
Width (ft)		120.00
Height (ft)		13.25
OUTLET WORKS		
Discharge capacity (ft³/s)	60	
Number of water passages	None	None
Dimensions of water passages		
Height (ft)		
Width (ft)		
Diameter (ft)		
Type of gates		
Number of gates	None	None

Plant Name: **NORTHFIELD MOUNTAIN**

Plant location
 Northfield & Erving, MA
 Franklin County

Rated capacity	1,080 MW
Average static head	227.1 m
Plant efficiency	74.60 %
Stored energy	10,500 MWh
Number of units	4

Construction time: 4 years, 9 months
Construction cost: $145 per kW
Price level: 1972
First commercial power: November 1972
FERC project number: 2485

River or water source: Connecticut River

Owner: Connecticut Light and Power Co./Western Mass. Electric Co.
 c/o Northeast Utilities
 P.O. Box 270
 Hartford, Connecticut 06101-0270

Designers:
 Stone and Webster Engineering Corp.

Plant Manager/Superintendent:
 Plant Manager
 Route 63
 Erving, Massachusetts   06101

 ( ) -

	UPPER RESERVOIR	LOWER RESERVOIR
DAM		
Type	Central core rockfill	Existing concrete Turner Falls Dam
Height (m)	44.2	16.8
Crest length (m)	1,706.9	
Volume (m³)	2,064,299	
RESERVOIR		
Type	Constructed	On stream, on the Connecticut River, Turner Falls Pond
Surface area (Mm²)	1.11	8.09
Usable power storage (Mm³)	17.701	15.721
Power pool fluctuation (m)	25.76	2.74
Operating levels		
Maximum (m)	306.17	56.39
Minimum (m)	280.42	53.65
Drainage area (Mm²)	1.295	18,552.189
Seepage (m³/s)	0.0566	
SPILLWAY		
Design flood		
Return period (years)		
Flow (m³/s)	323	5,947
Capacity (m³/s)		
Type	Overflow	Gated
Gates		
Number	None	7
Type	None	3 Radial, 4 Bascule
Width (m)		36.576
Height (m)		4.039
OUTLET WORKS		
Discharge capacity (m³/s)	2	
Number of water passages	None	None
Dimensions of water passages		
Height (m)		
Width (m)		
Diameter (m)		
Type of gates		
Number of gates	None	None

Northfield Mountain - Page 1 (Metric)

NORTHFIELD MOUNTAIN

INTAKES	UPPER INTAKE	LOWER INTAKE
Number	1	1
Type	Horizontal, side intake	Horizontal
Design discharge (ft³/s)	20,000	20,000
Gross inlet area (ft²) (at trash racks)	4,055	3,300
Bar racks		
spacing (in)	6.00	6.00
shape		
depth/thickness (in)	0.75 / 0.75	0.75 / 0.75
diameter (in)		
Emergency gates		
number	None	None
height/width (ft)		
type		
Service gates		
number	None	None
height/width (ft)		
type		
Bulkhead/stop logs (Y or N)		Y
number of units serviced		10
Hoists		
number	None	1
capacity (tons)		
type		Mobile crane w/ jacking device

WATER PASSAGES	Upper Tunnel	Shaft	Lower Tunnel	Surge Tanks Upper	Surge Tanks Lower	Penstocks	Tailrace Tunnel
Number		1		4	1	4	1
Diameter (ft)		31.0		20.0		14.8	33.0
Length (ft)		850		185		340	5,130
Maximum velocity (ft/s)		26.5				33.5	22.6
Concrete lining thickness (in)							
Total length of concrete sections (ft)							
Steel liner Thickness							
Minimum (in)						1.00	
Maximum (in)						2.00	
Material grade						A 36	
Total length of steel-lined sections (ft)						340	

Notes:
Tailrace Tunnel: Horseshoe shape.

Northfield Mountain - Page 2 (English)

NORTHFIELD MOUNTAIN

INTAKES	UPPER INTAKE	LOWER INTAKE
Number	1	1
Type	Horizontal, side intake	Horizontal
Design discharge (m³/s)	566.3	566.3
Gross inlet area (m²)	376.7	306.6
(at trash racks)		
Bar Racks:		
spacing (mm)	152	152
shape		
depth/thickness (mm)	19 / 19	19 / 19
diameter (mm)		
Emergency gates		
number	None	None
height/width (m)		
type		
Service gates		
number	None	None
height/width (m)		
type		
Bulkhead/stop logs (Y or N)		Y
number of units serviced		10
Hoists		
number	None	1
capacity (Mg)		
type		Mobile crane w/ jacking device

WATER PASSAGES	Upper Tunnel	Shaft	Lower Tunnel	Surge Tanks Upper	Surge Tanks Lower	Penstocks	Tailrace Tunnel
Number		1		4	1	4	1
Diameter (m)		9.45		6.10		4.51	10.06
Length (m)		259.1		56.4		103.6	1,563.6
Maximum velocity (m/s)		8.08				10.21	6.89
Concrete lining thickness (m)							
Total length of concrete sections (m)							
Steel liner Thickness							
Minimum (mm)						25	
Maximum (mm)						51	
Material grade						A 36	
Total length of steel-lined sections (m)						103.6	

Notes:
Tailrace Tunnel: Horseshoe shape.

Northfield Mountain - Page 2 (Metric)

NORTHFIELD MOUNTAIN

POWERHOUSE and RELATED FEATURES

Powerhouse Structure
Type: Underground
Length: 328 ft   Width: 70 ft   Height: 155 ft

Guard Valves
Number: 4   Diameter: 9.5 ft
Type: Spherical

Transformers
Number: 2, forced oil cooled
Ratings: 500
Voltages: (kV) 345 / 13.8

Generator
Rating generating (MVA): 235.0   Rating pumping (MVA): 217.0
Insulation type: Class B
Starting method: Synchronous start system using unit No. 3
Starting equipment: Pony motor with liquid rheostat

Runners
Material: Cast steel with stainless overlay
Minimum unit submergence: 106.0 ft
WR²:
Manufacturer: Baldwin-Lima-Hamilton
Model test by: Baldwin-Lima-Hamilton

	Reversible Runners	Reversible Motor/Generator	
Number	4		4
Diameter (ft)	17.20	Rotor	Stator
rpm synchronous	257.0		257.0
rpm overspeed	390.0		390.0
Type	Francis	Ac synch. vertical, 28 poles wound rotor	

Information on Runners

Condition:	Gross Head (ft) Generating	Gross Head (ft) Pumping	Capacity (MW) Generating	Capacity (MW) Pumping	Discharge (ft³/s) Generating	Discharge (ft³/s) Pumping	Turbine/Pump Eff.(%) Generating	Turbine/Pump Eff.(%) Pumping
Maximum head & maximum power	828	828	307		5,152			
Minimum head & maximum power	735	735				3,850		

Condition:	Net Head (ft) Generating	Net Head (ft) Pumping	Capacity (MW) Generating	Capacity (MW) Pumping	Discharge (ft³/s) Generating	Discharge (ft³/s) Pumping	Turbine/Pump Eff.(%) Generating	Turbine/Pump Eff.(%) Pumping
Rated head @ best gate	745	740	260	268		3,600		

Northfield Mountain - Page 3 (English)

NORTHFIELD MOUNTAIN

POWERHOUSE and RELATED FEATURES

Powerhouse Structure
Type: Underground
Length: 100.0 m   Width: 21.3 m   Height: 47.2 m

Guard Valves
Number: 4   Diameter: 2.90 m
Type: Spherical

Transformers
Number: 2, forced oil cooled
Ratings: 500
Voltages: (kV) 345 / 13.8

Generator
Rating generating (MVA): 235.0   Rating pumping (MVA): 217.0
Insulation type: Class B
Starting method: Synchronous start system using unit No. 3
Starting equipment: Pony motor with liquid rheostat

Runners
Material: Cast steel with stainless overlay
Minimum unit submergence: 32.31 m
WR^2:
Manufacturer: Baldwin-Lima-Hamilton
Model test by: Baldwin-Lima-Hamilton

	Reversible Runners	Reversible Motor/Generator	
Number	4		4
Diameter m	5.25	Rotor	Stator
rpm synchronous	257.0		257.0
rpm overspeed	390.0		390.0
Type	Francis	Ac synch. vertical, 28 poles wound rotor	

Information on Runners

Condition:	Gross Head (m)		Capacity (MW)		Discharge (m³/s)		Turbine/Pump Eff.(%)	
	Generating	Pumping	Generating	Pumping	Generating	Pumping	Generating	Pumping
Maximum head & maximum power	252.4	252.4	307		145.9			
Minimum head & maximum power	224.0	224.0				109.0		

Condition:	Net Head (m)		Capacity (MW)		Discharge (m³/s)		Turbine/Pump Eff.(%)	
	Generating	Pumping	Generating	Pumping	Generating	Pumping	Generating	Pumping
Rated head @ best gate	227.1	225.6	260	268		101.9		

Northfield Mountain - Page 3 (Metric)

NORTHFIELD MOUNTAIN

Plant Data:
Average GWh generating per year:	1,178
Average GWh pumping per year:	1,619
Starting time from standstill (s):	
Changeover time pumping to generating (min):	
Planned/scheduled time between major overhauls (years):	14
Outage time required per unit during major overhauls (weeks):	18
Representative plant availability (%):	
Representative planned outages (weeks per year):	

Miscelleneous Notes:

The pumping/generating ratio is stated as 1.36; this is an average of the range of 1.33 to 1.41.

The lower reservoir has seven gates, of which three are radial type and four are bascule type. The information on sizes is for the four Bascule gates. The three radial gates are 12.2 m (40 ft) high and 12.2 m (40 ft) wide.

The pumped storage plant used the existing Turners Falls Dam as the lower reservoir.

The design discharge for the intakes was 566.4 m³/s (20,000 ft³/s) in the generating mode and 396.4 m³/s (14,000 ft³/s) in the pumping mode.

The tailrace tunnel is a horseshoe shape.

The length of the shafts (259 m (850 ft)) does not include the length of the manifolds, penstocks, and transitions. If these are included, the total length is 570 m (1,870 ft).

The length of the tailrace tunnel (1,564 m (5,130 ft)) does not include the length of the manifolds and transitions. If these are included, the total length is 1,700 m (5,575 ft).

The minimum unit submergence of 32.3 m (106 ft) is from the tailwater to distributor centerline.

The pump is rated at 359,000 hp at the minimum TDH of 225 m (740 ft). The turbine is rated at 227 m (745 ft) net head.

390 RPM is the maximum allowable overspeed of the turbine and generator; under steady state conditions, the speed drops to 337 RPM.

The diameter of the turbine is 524.5 cm (206.5 in.) to the tip. The throat diameter is 312.4 cm (123.0 in.).

Cavitation Experience:
Cavitation removes 50 to 100 pounds of steel per year per unit.

Significant or Unique Problems:

Northfield Mountain - Page 4

NORTHFIELD MOUNTAIN

There have been no major construction or operating problems.

The underground powerhouse flooded during precommercial testing because of inadvertent operation of the valve control system. Upthrust/downthrust and oil pressure system refinements are as described in a paper by F. Harty.

List of Licenses Required:
FERC, water quality certification.

ENVIRONMENTAL FEATURES

Recreation:
Hiking/skiing trails, skating, riverboat tours, and a visitor center are available.

Fish and Wildlife:

Social:

Journal of the
POWER DIVISION
Proceedings of the American Society of Civil Engineers

NORTHFIELD MOUNTAIN PUMPED STORAGE PROJECT[a]

By S. Hale Lull,[1] and Antonio Ferreira,[2] M. ASCE

INTRODUCTION

The 1,000 Mw pumped storage project will be constructed by the Northeast Utilities operating companies. The Connecticut Light and Power Company, The Hartford Electric Light Company, and Western Massachusetts Electric Company together with the Northeast Utilities Service Company comprise the newly formed Northeast Utilities. Western Massachusetts Electric Company serves the electrical needs of a major portion of the western part Massachusetts. It operates generation, transmission, and distribution facilities in the service territory covering the cities of Pittsfield, Greenfield, and Springfield, Mass., and surrounding areas. The Connecticut Light and Power Company and The Hartford Electric Light Company are two of the three major utilities operating in Connecticut. The combined peak load of the three utilities is approximately 2,290 Mw, and the peak occurs normally in December.

The operating companies of Northeast Utilities, together with The United Illuminating Company, the third major utility in Connecticut, are the member companies of the Connecticut Valley Electric Exchange (CONVEX) which provides central-dispatching services for the four utilities, and through which they effect economy—interchange transactions with each other and with neighboring utilities.

Prior to the corporate affiliation of three of its members in July, 1966, the CONVEX companies were jointly planning their generation and transmission expansion programs, and coordinating these activities closely with the remaining utilities of New England through participation in the activities of the

---
Note.—Discussion open until April 1, 1969. To extend the closing date one month, a written request must be filed with the Executive Secretary, ASCE. This paper is part of the copyrighted Journal of the Power Division, Proceedings of the American Society of Civil Engineers, Vol. 94, No. PO2, November, 1968. Manuscript was submitted for review for possible publication on December 8, 1967.

[a] Presented at the October 16-20, 1967, ASCE National Meeting on Water Resources Engineering, held at New York, N.Y.

[1] Vice-Pres., System Planning, Northeast Utilities Service Co., Hartford, Conn.

[2] Chf. Hydr. Engr., Northeast Utilities Service Co., Hartford, Conn.

Planning Committee of the Electric Coordinating Council of New England (ECCNE).

PLANNING STUDIES

Planning studies performed in 1964 disclosed the need on the part of the CONVEX companies of additional generation over and above their then existing plans for 1971. The ECCNE Planning Committee studies indicated a similar need for the New England utilities as a whole. Generation simulation studies performed by CONVEX engineers indicated clearly the desirability of a source of short-hour, or peaking, power on their system by 1971.

It is well known in the industry that New England utilities, situated as they are in an area of high transportation costs, have of economic necessity been among the forerunners in the utilization of nuclear-power production. This action has not only tended to circumvent excessive fuel-transportation costs but has made, and to a greater extent in the future will make, available low incremental cost off-peak generation.

Faced with these parameters, the CONVEX planning engineers performed economic analyses which indicated that development of the Northfield Mountain Pumped Storage site at 1000 Mw was the most economical method of meeting their needs for added capacity in 1971, and that such an installation was even more readily adaptable to the needs of the larger New England load as a whole.

LICENSE APPLICATION

For this reason, on January 14, 1966, the three operating companies of Northeast Utilities jointly made application to the Federal Power Commission for a 50-yr license to construct, operate, and maintain the Northfield Project. The filing was made pursuant to the issuance of a preliminary permit to Western Massachusetts Electric Company on March 4, 1965, for the purpose of maintaining priority of application for license during the period when pertinent feasibility studies were in progress. The companies are currently (1967) awaiting receipt of the necessary license so they may proceed with the actual construction of the project.

SCOPE OF PROJECT

The Northfield Mountain Pumped Storage Project is a multipurpose hydroelectric project located on the Connecticut River at a point just after it passes from New Hampshire into Massachusetts. Included in the project is the pumped hydrogenerating plant, a recreation plan, and a water-supply proposal. The hydroelectric portion of the project is currently estimated to cost approximately $75,000,000 for the 1000 Mw installation exclusive of transmission facilities. This results in a cost per kilowatt of approximately $75.

Herein are outlined the major design features of the multipurpose Northfield Mountain Project and to delineate the philosophies which resulted in this specific design.

OPERATING FEATURES

Design Philosophy.—Recognizing the needs of both the Northeast Utilities system and the New England system as a whole for added short-hour generation to meet their rapidly fluctuating peak loads and in an effort to profit to a small degree by the lesson learned from the November 9, 1965 blackout, the Northfield Mountain hydroelectric project is designed with considerable attention placed on the speed of start of its units from either a spinning-in-air mode or from a standstill. A properly designed pumped hydroplant can become one of the greatest assets which an electric utility has in restoring power to a system after a major shutdown e.g., the 1965 blackout.

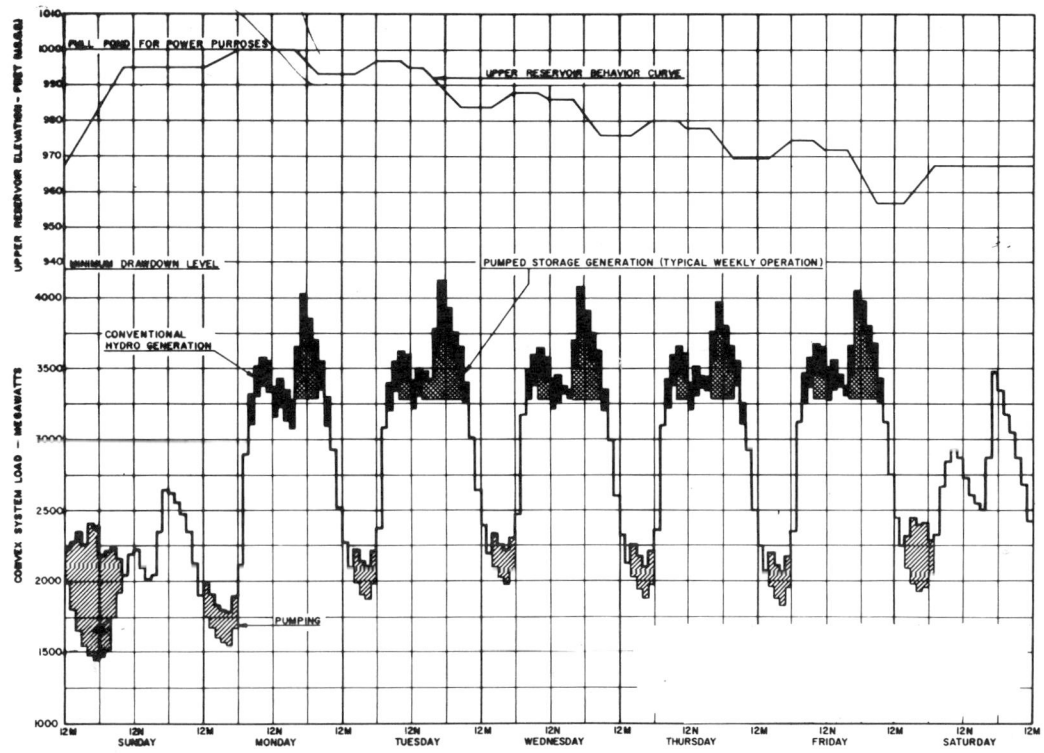

FIG. 1.—ECONOMY OPERATION, NORTHFIELD MOUNTAIN PUMPED STORAGE PROJECT, TYPICAL DECEMBER WEEK 1973, DISPATCHED ON CONVEX SYSTEM

Northeast Utilities seeks to profit from the Northfield Mountain pumped hydro installation by the utilization of its facilities to assure: (1) Realization of the full capability of the plant as firm capacity to meet its load; (2) emergency capacity available to the system on short notice; and (3) to the extent consistent with these goals, the operation of the facility for maximum economy.

The project has been designed with its upper reservoir as the limiting factor in the production of energy. This reservoir will be sufficiently large to permit 8500 Mw of generation in addition to the storage which will be reserved for water-supply purposes.

To assure optimum utilization of the project for emergency capacity, it is anticipated that enough water to permit approximately 2500 Mwh of generation will be retained in the upper reservoir at all times to meet possible emergencies.

Operating Cycle.—The remaining storage, equivalent to approximately 6000 Mwh, is sufficient to permit operation of the project on a modified weekly cycle. Thus, a portion of the water releases made during the day will be pumped back to the upper reservoir the following night, with the remaining portion being restored over the weekend. Fig. 1 illustrates the method by which the project would operate if it were dispatched to meet the CONVEX system load during a typical winter week. The peak hours of the five weekdays will be met by existing conventional hydrogeneration, as shown by the solid black area in Fig. 1. The pump-storage generation is shown during the five weekdays as a double cross-hatched area immediately below the conventional hydrogeneration. The remaining load indicated in white on Fig. 1, must be carried by existing fossil-fueled and nuclear units. The pumping performed during the week is shown as cross-hatched areas partially filling the off-peak valleys throughout the week. Only a portion of the pumping is performed dur-

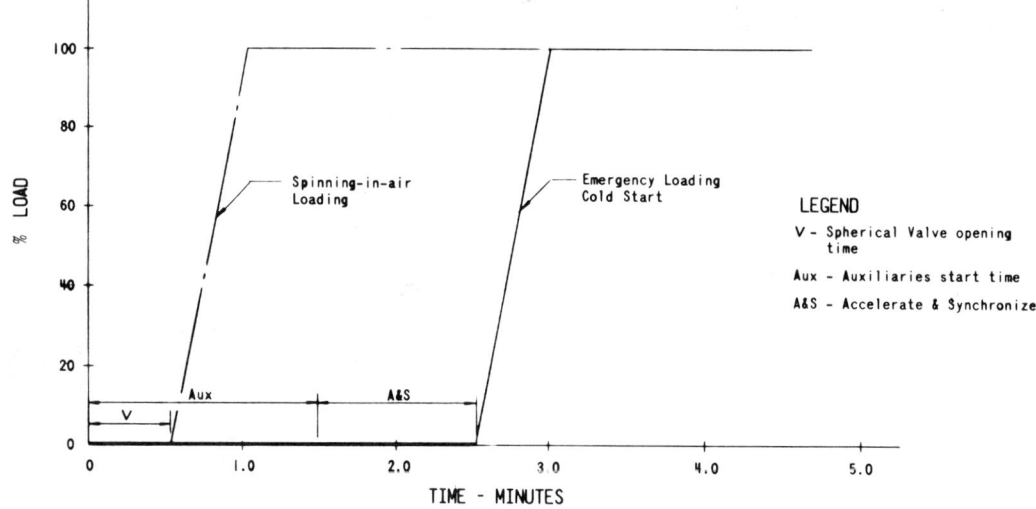

FIG. 2.—GENERATOR EMERGENCY LOADING SEQUENCE, NORTHFIELD MOUNTAIN PUMPED STORAGE PROJECT

ing the early morning hours of the day, with the bulk of the pumping occurring over the weekend.

With the project dispatched against the entire load of New England, as it undoubtedly will be by the time the project is placed in service, peaks of much greater magnitude and shorter duration will be experienced than exist on the smaller CONVEX load. Under these conditions the Northfield Mountain Project will operate at full capacity more frequently than is indicated in Fig. 1.

At the top of Fig. 1 can be seen the upper-reservoir behavior curve. This curve is indicative of the flexibility of the relatively large upper reservoir being provided at Northfield. The upper reservoir is full on Monday morning, with its level gradually receding through the week and being finally restored over the weekend. With flexibility available in the timing of the pumping cycle, permitting acceptance or rejection of available pumping costs, the project is not limited to a day-to-day operating cycle. The pumping operation shown in Fig. 1 consists of limited pumping throughout the week, because it takes advantage of the lower weekend-pumping costs to restore the reservoir to its maximum elevation by Monday. As the reservoir-behavior curve indicates, only approximately two thirds of the available storage was utilized for power production purposes during the week.

Generator Loading Cycles.—Despite the large physical and electrical size of the generator motors, great flexibility is possible in operating as generators. Operation of the units will call for both normal and emergency cycles.

In an emergency start (Fig. 2), approximately 1 and 1/2 min are required for opening of the spherical valves and starting of auxiliaries. The units then will be accelerated to speed and synchronized in approximately 1 min. Loading of the units will be accomplished in less than 30 sec, resulting in a fully loaded condition within 3 min of cold start.

In a normal cold start, a procedure similar to the emergency cold start will be followed except that loading above minimum (approximately 100 Mw) will be at a variable and controllable rate between 5 Mw and 100 Mw per min.

An alternate and considerably faster on-the-line emergency loading is contemplated by operating the units in a spinning-in-air mode, generate direction. This eliminates auxiliary and synchronizing times. Spinning-in-air operation takes place with the unit unwatered and the spherical valve closed. Loading can be initiated as soon as the valve is opened. A spherical valve time of less than 30 sec is being provided. Loadings in the spinning-in-air mode from initiation of valve opening to full load will be accomplished in approximately 1 min, as shown in Fig. 2. While it is recognized that an increase in input loading may be experienced immediately prior to speed-no-load gate, the sequence shown depicts generation loading only for purposes of simplification.

Once the units are on the line, they will be operated by an automatic-joint controller locally within the plant. Studies have shown that the most economical loading of whatever units are in service results from equally distributed loads. The joint-control unit will operate to adjust all governor settings alike. Load level, or governor setting, will be continually adjusted by means of a signal to the local joint controller initiated by an existing remotely located computer. This computer, a General Electric GE-PAC, is located in the CONVEX dispatch center 80 miles away. It will set station-load levels in accordance with economic-generation requirements. Computer impulses, and consequently station-load settings, will be adjusted every 4 sec to follow system-load changes. Remote-start push buttons for emergency or normal conditions will be provided both remotely and locally.

The ability, under emergency conditions, to have these units fully loaded from a complete shutdown within 3 min makes them capable of functioning as a portion of the assured reserve capacity of the New England utilities.

Pump Loading Cycle.—The normal mode of pump start calls for unwatering of the turbine runner by admission of compressed air, accelerating the unit to synchronous speed, synchronizing and subsequently loading to optimum gate. Various methods of acclerating the unit to synchronous speed were investigated, including full-voltage start, reduced-voltage start and wound-rotor motor start. The method selected was the wound-rotor motor start.

This scheme calls for a wound-rotor motor mounted on the main shaft directly above the generator-motor. A constant torque is provided by rotor-resistance adjustments accomplished by means of liquid rheostats which are automatically adjusted during the startup cycle. Inrushes are small and voltage drops generally negligible with this method of starting. The wound-rotor motors are rated 13,000 hp and will be designed for operation at 4.16 kv. The resultant starting time by this method is approximately 8 min to 10 min.

Once the units are brought up to speed and synchronized, pump loading takes place. Because the units are inherently inefficient at low-gate openings, loading to optimum gate (approximately 235 Mw) will be accomplished as fast as system conditions permit. The pumps will operate at constant load, and will not be under computer control. Local and remote push buttons for start or stop of units will be provided.

Note that manual trip out of one or more units at Northfield is an effective means of load shedding during emergency conditions. Thus, regardless of their mode of operation (generating, pumping, or shut down) the Northfield Mountain pumped hydro units will be a tremendous asset to the northeast in the event of a major emergency.

DESCRIPTION OF PROJECT FEATURES

Location.—The principal project structures will be located in the towns of Northfield and Erving, in Franklin County, Mass. (Fig. 3). The upper reservoir

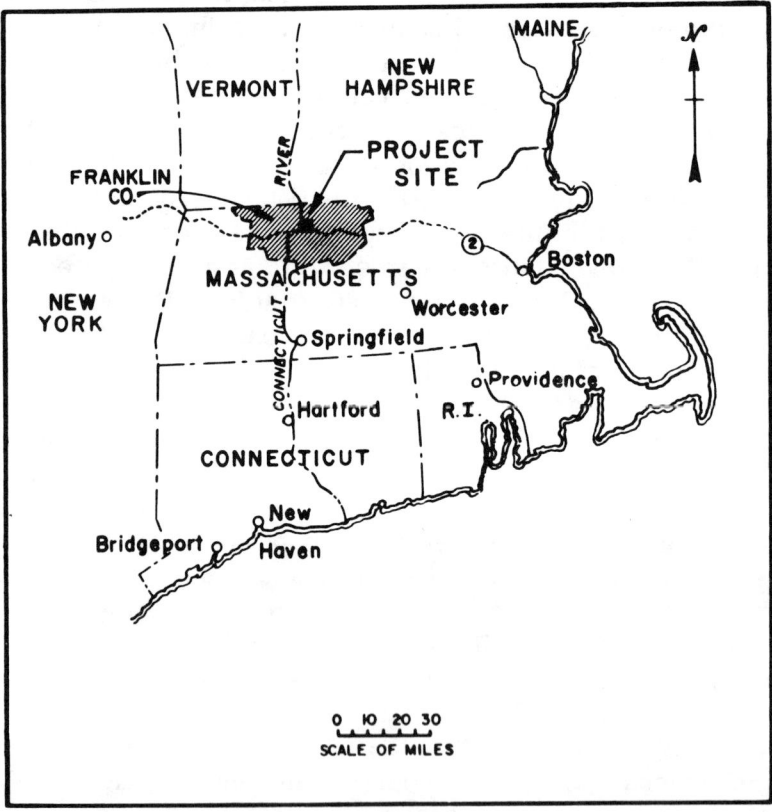

FIG. 3.—LOCATION MAP

will be constructed atop Northfield Mountain approximately 3/4 miles from the east bank of the Connecticut River. The tailrace canal will be constructed on the east bank of the Connecticut River approximately 5-1/2 river miles above the Turners Falls dam of Western Massachusetts Electric Company's existing run-of-river hydroelectric installation.

Underground Configuration.—To obtain for the units the high degree of speed regulation (15 sec governor time) and governor stability that are necessary to meet the system conditions of assured reserve, and so that the units can be capable of responding to rapid changes of system demand, the pressure-conduit length has been kept short. To accomplish this, the powerhouse has been located underground adjacent to the upper reservoir. The resulting long tailrace tunnel will be subject to surges which will be controlled within necessary limits by the surge-shaft system located immediately downstream of the pump-turbines.

Notwithstanding these operating-design parameters, the topography at the project site is unfavorable to the more conventional arrangement in which the powerhouse is located at the end of the flow line. The intervening ground surface between the upper and lower reservoir descends relatively rapidly from about El. 1,000 ft at the upper reservoir intake location to El. 400 within a span of 3,600 ft. In the remaining 2,400 ft, the surface descends by a further 150 ft to El. 250 at the tailrace canal. Except adjacent to the upper reservoir,

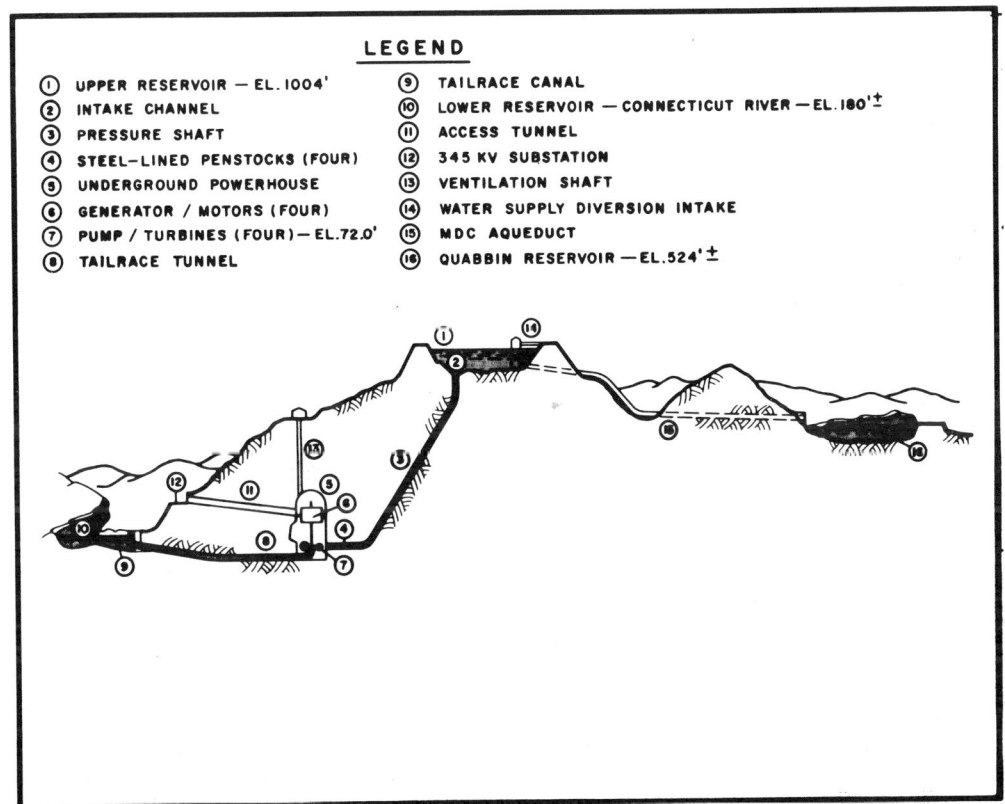

FIG. 4.—CROSS SECTION, NORTHFIELD MOUNTAIN PUMPED STORAGE PROJECT

rock cover over an unreinforced pressure tunnel located at the practicable depth below the ground surface would be inadequate to withstand the hydraulic loads. Because of these lower ground elevations near the lower reservoir, a surge tank located within effective distance of the powerhouse would have an unsupported height in excess of 700 ft. Such a structure was not considered economically feasible. Preliminary studies of two alternate configurations having the powerhouse located at the end of the flow line, with and without surge tanks, have been carried out in sufficient detail to confirm that their costs would be higher than that of the underground configuration adopted. The studies also showed that, in the absence of a surge tank, the desired operating criteria could not be met.

Civil and Hydraulic Features.

Upper Reservoir (Item 1, Fig. 4).—The upper reservoir and apurtenant structures will be constructed on Northfield Mountain in the Town of Erving. The main dam will be constructed across Briggs Brook at the south end of the reservoir. This brook originates within the small drainage area on the mountain top plateau. The dam is currently designed as a compacted rock-fill embankment with an impervious earth core. The top of the dam will be at El. 1,010, the top 4 ft of which has been allocated for water-supply storage. Maximum height of the dam above Briggs Brook, will be approximately 150 ft with a crest length of approximately 5,600 ft. Supplementary dams of similar rock-fill construction, with impervious core, will total an additional 5,600 ft in length, and will close of saddles along the crest of the Briggs Brook watershed on the rim of Northfield Mountain. The top of the rock-fill dam will be sufficiently wide for a roadway. There will be an 8-ft high chain-link fence surrounding the perimeter of the entire upper reservoir for the protection of the public and wildlife. Runoff from rainfall will normally be discharged through the turbines, but a high level ungated spillway designed to discharge 200 cfs down Packard Brook will be provided at the east side of the main dam.

At the maximum operating elevation for power purposes, El. 1,000, the reservoir will have a total stored volume of 17,050 acre-ft and a surface area of 286 acres. At the minimum operating El. of 938, the stored volume will be 4,300 acre-ft with a surface area of 134 acres. The maximum drawdown level of 62 ft provides a usable storage volume for power purposes of 12,750 acre-ft with an estimated energy content of 8,500 Mwhr.

Lower Reservoir (Item 10, Fig. 4).—The pond behind the dam at Turners Falls which is presently part of Western Massachusetts Electric Company's licensed project No. 1889 will serve as the lower pool into which the water discharged during the generating cycle will be stored. To completely store the entire usable contents of the upper reservoir without interfering with the operation of the existing run-of-river hydroelectric installation, which currently uses the Turners Falls Pond as its headwater pool, it will be necessary to increase the volume of the pond. Bascule-type spillway gates and tainter gates will replace the existing wooden flashboards on the Turners Falls dam, and will provide the necessary volume increase in that the new gates will be approximately 5-1/2 ft higher than the existing 7-1/3 ft-high flashboards. The pond will fluctuate between a minimum El. of 176, and a normal maximum of approximately El. 183.

The lower reservoir will remain a part of the licensed project No. 1889 of Western Massachusetts Electric Company, but it is expected that the Federal Power Commission (FPC) order amending this existing license will provide for the use of the Turners Falls pond by the pumped storage project. The canal-headgate house on the Turners Falls side of the dam will also undergo some minor modifications, principally the raising of the operating-floor levels consistent with the slight increase in operating elevations of the lower reservoir.

Intake Channel (Item 2, Fig. 4).—A channel excavated in rock will connect the main body of the upper reservoir to the pressure shaft intake on the west flank of Northfield Mountain. Material excavated from the channel will be used to construct the rock-fill dams. A low-level concrete weir with a stop log equipped sluiceway will be provided across the low portion of the channel bottom to retain sufficient water to refill the pressure shaft after unwatering for maintenance purposes, and also for initial filling upon completion of construction. The ungated pressure shaft intake will consist of a reinforced concrete bell-mouthed transition structure designed to reduce hydraulic entrance losses. A trash rack structure will be provided to prevent trash from being drawn into the intake. Racks and intake will be totally submerged at all reservoir operating elevations. Model studies of the intake channel, currently in progress, indicate that a roof will be necessary over the mouth of the intake to reduce the formation of vortices. The exact shape and size of the intake will not be determined until these model tests have been completed.

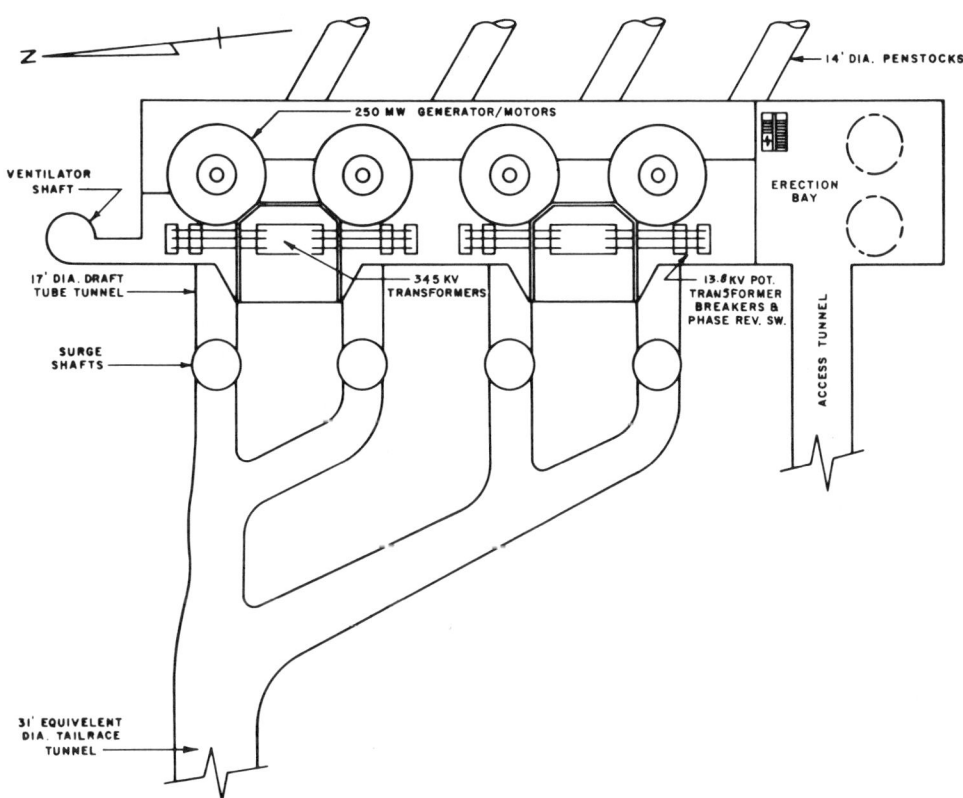

FIG. 5.—PLAN VIEW, UNDERGROUND POWERHOUSE, NORTHFIELD MOUNTAIN PUMPED STORAGE PROJECT

Pressure Shaft (Item 3, Fig. 4).—The pressure shaft will consist of a 31 ft diam reinforced concrete lined shaft. It will be excavated in bedrock from the intake channel invert at an inclination of approximately 50° from the horizontal. Approximately 600 ft upstream of the powerhouse, the shaft will enter a nearly horizontal reinforced concrete-lined tunnel section which will run on approximately a 10% slope toward the powerhouse. This section will split into four steel lined tunnel branches, (Item 4, Fig. 4) each 14 ft in diam, leading to the four turbine-spiral cases. Each branch will reduce to a 10 ft diam as it enters the powerhouse, and a hydraulically-operated spherical valve will be provided just upstream of each spiral-case entrance. In the steel-lined tunnel branches, concrete will be placed between the liner and the rock.

Powerhouse (Item 5, Figs. 4 and 5).—The powerhouse, approximately 300 ft long by 70 ft wide, by 120 ft high (from center line of turbine distributor to roof), will be in a chamber excavated in the bedrock of the mountain. It will contain the four 250-Mw generator-motors, the pump-turbines, spherical valves, governors, traveling cranes, and other necessary auxiliary electrical and mechanical equipment. A service bay will be provided in the south end of the powerhouse for erection and maintenance purposes. The control room will be located in the powerhouse under the service bay. Connecting the powerhouse and the upper slope of the mountain will be a vertical ventilation shaft, approximately 15 ft in diam, containing a vertical-stair system. The powerhouse will have floors at or near El. 80, El. 99, El. 112, and El. 123, with the main floor being at the latter elevation. The turbine-room floor will be at El. 80. The center line of the pump-turbine distributor has been set at El. 72. The arched powerhouse roof will be reinforced concrete lined, if such support is found to be structurally necessary. The powerhouse chamber walls above the level of the main floor will not be lined.

Two 175-ton capacity traveling electric bridge cranes, each with a 15-ton capacity auxiliary hoist, will be provided. During construction, the cranes will be operated at high speed. A lower speed for equipment erection and maintenance purposes will be specified for these units.

Surge Chambers.—A concrete lined surge-chamber system excavated in the rock will be provided at the ends of the pump-turbine-draft tubes to control surges induced in the long tailrace tunnel by normal fluctuation in the generating and pumping rates and following sudden shutdown because of line faults during either mode of operation. Each of the four vertical chambers will be 20 ft in diam by 180 ft high. Additional volume will be provided by means of two tunnels connecting the vertical chambers, one at the bottom, and one approximately in the middle. Slide gates will be provided at the junction of the draft tubes with the risers to the surge tanks to enable the pump-turbines to be unwatered. These will be operated by individual electric hoists located at the top of the surge tanks. Access to the hoists will be through a gallery excavated from the vertical-access shaft. The gallery will also serve as a vent for the surge tanks.

Tailrace Tunnel (Item 8, Fig. 4).—A single concrete-lined tailrace tunnel will be joined to each of the four draft tubes at the surge chamber riser location by means of transition galleries of varying diameters. The tunnel will be horseshoe shaped and approximately 32 ft wide by 31 ft high, and will be concrete lined for hydraulic purposes. It will be pitched for drainage from the surge-chamber location to a low point near the exit, and then rise at a steeper grade to the tailrace portal. A means of pumping out the tailrace tunnel for inspection and maintenance will be provided at this low point. A concrete control structure will be provided at the tunnel entrance. The structure will support trash racks and will have provisions for inserting stop logs, which will be handled by truck crane as no hoist will be provided at this location. Near the entrance portal, the tunnel will be flared to approximately twice the tunnel cross-sectional area to reduce exit and entrance velocities.

Access Tunnel (Item 11, Fig. 4).—Access to and ventilation within the powerhouse will be provided by means of a 24 ft wide by 24 ft high inclined tunnel ascending to the surface from the south end of the powerhouse chamber at El. 123 to the lower west slope of Northfield Mountain at approximately El. 387. This access tunnel will be unlined, and will have no roof support. The tunnel invert will have an asphalt pavement. It will be provided with gutters and drains, the drainage running into the powerhouse sump and then into the tailrace.

Tailrace Canal (Item 9, Fig. 4).—Connecting the Connecticut River and the tailrace-tunnel exit will be a canal approximately 500 ft long excavated on the east bank of the Connecticut River. The canal will be bell-mouthed at its entrance into the river to provide low velocities of flow into and out of the lower reservoir. A floating barrier will be provided across the mouth of the canal to prevent entry of boats, and to keep floating trash and debris from being drawn into the canal during the pumping cycle. Along the entire canal perimeter, a chain-link fence will be erected for the safety of the public and the protection of wildlife.

Transformer Location.—As shown in Fig. 5, on the downstream side of the powerhouse two side-wall vaults will be excavated, between units No. 1 and 2 and between units No. 3 and 4, to house the step-up transformers. The floor of the transformer vaults will be at El. 112 ft. From the two transformers, power will be transmitted through two 345 kv pipe-type cables installed in the access tunnel leading to the Northfield Mountain Substation which will be located on the lower west slope of the mountain in the vicinity of the access tunnel adit.

Pump/Turbines and Spherical Valves (Item 7, Fig. 4).—There will be four 257 rpm pump-turbines of sufficient size, capacity, and physical character to operate as turbines to produce 250 Mw output from the generator-motors, and to operate as pumps to lift the water from the lower reservoir to the upper reservoir. The pump-turbines will operate at net generating heads ranging from 720 ft to 823 ft. Expected maximum output at 823 ft net head will be approximately 270 Mw. When operated as pumps at minimum head, each unit will require an input of approximately 310,000 hp. They will deliver to the upper reservoir, flow rates ranging from 3,500 cfs to 2,000 cfs depending upon upper reservoir levels. The turbine equipment will include steel-draft-tube liners, steel-spiral cases, steel-pit liners, and wicket gates operated by actuator-type governors.

Compressed-air equipment will be provided to depress the water in the draft tube below the bottom of the runner during pump start up and for spinning-in-air operation.

Based on model study results which indicated no significant variation in performance between a 120-in. and 114-in. diam spherical valve, the smaller valve will be provided. During the pumping mode, a slight reduction in efficiency was indicated with the reduced inlet section at higher heads. However, a marked increase in pumping efficiency was noted during the model tests for the reduced-inlet section at heads of 810 ft and below. The valves will be hydraulically operated by a high-pressure oil system charged by an insert gas-control system rather than by compressed air. The insert gas-charge system was selected to eliminate the possibility of flash-point explosions caused by the heat of sudden compression. The accumulator tanks will have sufficient oil capacity to permit several opening and closing cycles. The bypass system for the spherical valve is being designed internally within the valve as part of the design effort to minimize exposed high-pressure piping or valving within the powerhouse cavern.

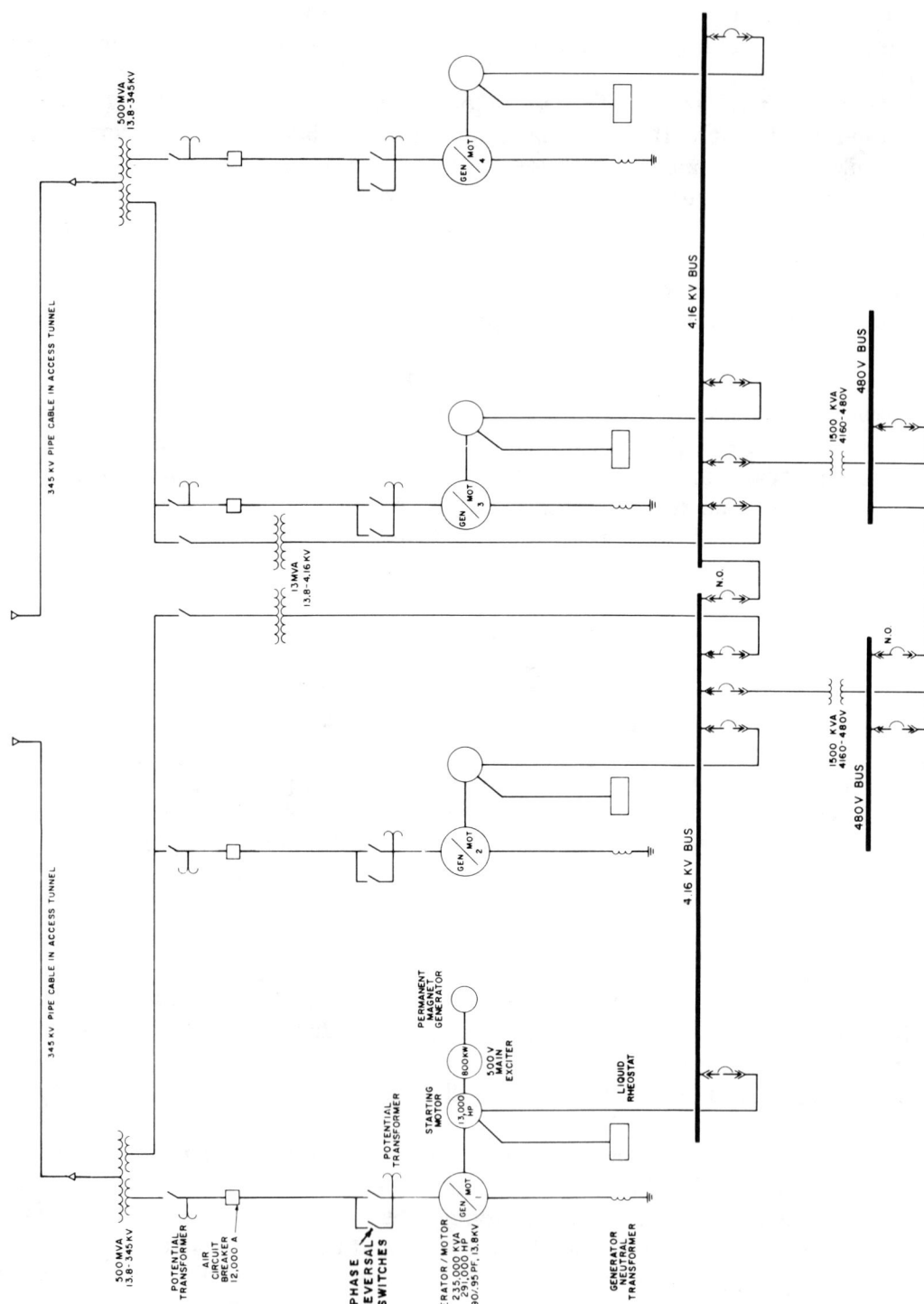

FIG. 6.—ONE-LINE DIAGRAM, NORTHFIELD MOUNTAIN PUMPED STORAGE PROJECT

Electrical Features.

General.—The electrical-plant layout was conceived with an eye toward space economy as well as cost economy because the project is to be an underground installation. For this reason, major equipment, such as generator-motors and power transformers, were optimally rated to fit anticipated conditions of operation. Full plant capacity of 1000 Mw will be achieved by utilizing the 80°C rise continuous rating on the generators. Two double-secondary power transformers were selected, and voltage insulation levels will be reduced consistent with system study data. Fig. 6 shows the one-line diagram of the station electrical arrangements.

Because the Northfield project will be one of the largest pumped hydro-installations in the world when completed in 1971, ratings of equipment will fall into a category of industry firsts in several instances. The generator-motors will be the largest installed as of the anticipated date of completion.

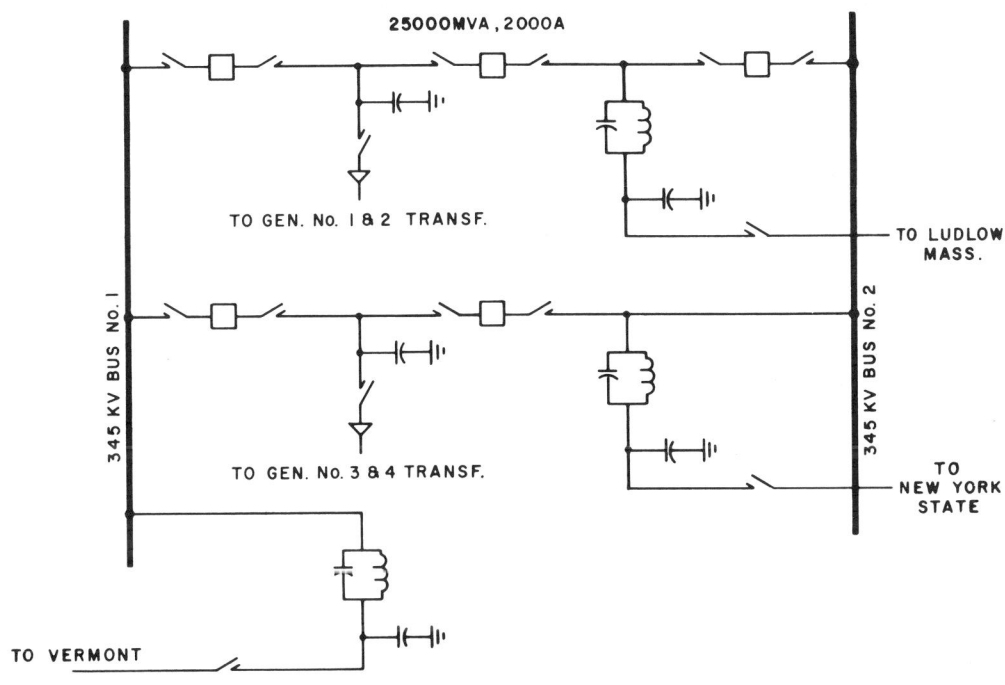

FIG. 7.—ONE-LINE DIAGRAM, NORTHFIELD MOUNTAIN SUBSTATION

The generator circuit breakers with a rating of 12,000 amp will be the highest-rated devices of this type installed in the United States. The main power transformers, in rating as well as size, will be the largest units installed in the United States in an underground location.

Generator/Motors (Item 6, Fig. 4).—The generator-motors will be vertical shaft, direct-coupled type. Each generator will be rated 270,000 kva/348,000 hp, 80°C rise, 0.90-0.95 pf, 257 rpm, 13.8 kv, 3 phase. The units will be capable of 282,000 kva at 0.9 pf as a generator, and will have a maximum pump rating of 368,000 hp at unity power factor, 80°C rise. Each unit will be provided with excitation equipment, pressure-lubricated thrust and guide bearings, brakes, cooling, and a fire-protection system.

The generator-motors will be provided with a piggy-back starting motor located on the main shaft above each unit. This motor, rated 13,000 hp, will be utilized in bringing units up to speed in the pump direction to minimize system-voltage disturbance on starting. The rotors of each generator-motor unit will weigh approximately 1,000,000 lb, will have peripheral speeds of approximately 18,000 fpm, and a WR^2 of 57,000,000 lb-ft^2. Stator segments will be provided in four sections. Upper and lower guide bearings will be furnished, and thrusts in excess of 2,400,000 lb are anticipated on the thrust bearing which will be located below the rotor.

Main Power Transformers.—The main power transformers will be dual-secondary transformers rated 500 Mva, 345-13.8-13.8 kv, 3 phase, 60 cycle, water-cooled, 65°C rise. These transformers will be located in sound-reducing vaults located between each pair of units. Each generator-motor will feed into one 13.8 kv winding of the double-secondary transformers through an individual circuit breaker.

Miscellaneous Equipment.—Isolated-phase bus duct and associated disconnecting and the phase-reversing switches utilized in changing from generate to pump modes will be rated 12,000 amp. The 345 kv connection between the underground transformers and the switching station above ground will be accomplished by means of pipe-type cables. Each circuit will consist of three cables in a single steel pipe, approximately 3,500 ft long, which will be routed through the access tunnel.

345 kv Switching Station.—The 345 kv switching station will be located on the hillside adjacent to the access tunnel adit (Item 12, Fig. 4) and its general configuration is represented in the one-line diagram, Fig. 7. Output of the station will be transmitted to points north in Vermont, west in New York State, and south in Massachusetts. The switching station will utilize spoil available from the access tunnel and cavern excavations for fill material. Circuit breakers will be of the air-blast type, and insulators throughout will be gray in color to provide a pleasing appearance.

MULTIPURPOSE FEATURES

Recreation.—Recognizing both the desirability and their obligation to provide for comprehensive development of the natural resources of the stretch of the Connecticut River to be licensed to them, the operating companies of Northeast Utilities have proposed a multipurpose development at Northfield Mountain. The potential of the 20 mile long lower reservoir, formed by the Turners Falls dam across the Connecticut River, for development for recreational purposes was obvious. Early in its planning, Northeast Utilities retained a recreation consultant to develop a comprehensive plan for recreation facilities to be developed on or adjacent to the lower reservoir of the pumped hydro project. The plan which evolved contemplates the development of nine different geographical areas in the vicinity of the project. Seven of these areas are located on the existing reservoir, licensed to Western Massachusetts Electric Company in connection with the Turners Falls development, which will become a joint use facility with the development of the pumped hydro plant. The remaining two areas are associated specifically with the proposed pumped hydro development. Included within these nine areas are facilities to permit swimming, picnicking, camping, boat camping, winter sports, hunting, fishing, bridle and hiking trails, horse and snow machine rentals, and vacationing in rental cabins.

It is proposed that the recreation facilities associated with the Northfield Mountain project be jointly developed by the operating companies and by the

Commonwealth of Massachusetts. The companies' recreation specialist is currently working with representatives of the Department of Natural Resources, Commonwealth of Massachusetts, to produce specific plans for the development of the nine recreation areas in connection with the project. It is anticipated that the utilities will be ready, upon receipt of their license to construct the Northfield project, to go forward with the development of the recreation plans in accordance with a program jointly worked out with the Commonwealth.

Fish Studies.—The operating companies of Northeast Utilities have also agreed to participate in a study of the Connecticut River designed to establish the potential of the river to support a program of enhancement of fishery resources. Participating in the study are state fish and wildlife agencies representing the four states abutting the Connecticut River as well as the U.S. Bureau of Sport Fisheries and Wildlife and the U.S. Bureau of Commercial Fisheries.

The studies seek to: (1) Establish a proper statement of intent with respect to enhancement of fishery resources of the Connecticut River basin; (2) develop a research program to establish the feasibility of proposed policy; and (3) develop the design of facilities to effectuate the program. The primary objective of the program is to restore to the river the shad runs which existed prior to the advent of hydro power, and the development of the system of dams on the river. Attention is also being given to other types of anadromous fish, and there is some hope that a partial restoration of the original salmon run on the Connecticut River can be made.

The results of this comprehensive study of the Connecticut River will, among other things, form the basis for determination of need, capacity and design of possible fish passage and protective devices at the Northfield Mountain and the Turners Falls developments.

Municipal Water Supply.—Again, recognizing the importance of the multipurpose-project concept, the operating companies of Northeast Utilities have also proposed to utilize the pumping and storage facilities of the Northfield project to supply Connecticut River water at times of high freshet river flows to the Quabbin Reservoir of the Massachusetts Metropolitan District Commission, which supplies water to the Boston metropolitan area.

Specifically the proposal presented to the Commonwealth was that the Northeast Utilities would be willing, whenever the Connecticut River flow exceeded 15,000 cfs, to use its facilities to pump 50 mcf (375,000,000 gal) of water daily to the upper reservoir for diversion to the Quabbin Reservoir of the Massachusetts Metropolitan District Commission. The diversion would be accomplished by means of an aqueduct to be provided by the Commonwealth. This water supply function of the project will require some specific additions to the project structures. Various project factors determine the effectiveness and the limitations of the quantities and rates of water which can be transferred.

The units, as pumps, will each develop over 300,000 hp. They will be capable of pumping rates from 2,000 cfs to 3,500 cfs each. If the average pumping rate is taken as 2,500 cfs per unit, the four unit total would be 10,000 cfs, or 75,000 gal per sec.

Quabbin Reservoir (Item 16, Fig. 4), 10 miles southeasterly of the Northfield Mountain upper reservoir, has a maximum El. of 524 ft. Thus, there is almost 500 ft of gravity head between the project's upper-storage reservoir and the Quabbin Reservoir. By raising the elevation of the upper reservoir by 4 ft, an additional 50,000,000 cu ft of water can be impounded. An additional water supply intake structure (Item 14, Fig. 4), separate from the intake structure for the power plant, must also be constructed in the upper reservoir for this purpose.

Recognizing the variance in flows of the Connecticut River, an analysis was made to determine the proper timing for such transferrals of water from the Connecticut River to Quabbin. The Western Massachusetts Electric Company's hydroelectric facility at Turners Falls is the largest user of water along the river downstream from the project. The maximum usable flow of this facility is in the order of 15,000 cfs. When the flow of the river exceeds this amount, the excess is spilled at the Turners Falls dam, and is also spilled by all other downstream users, ending as wasted water at Long Island Sound. For this reason, 15,000 cfs was selected as the minimum flow at which water could be transferred to Quabbin with no detrimental effect to any downstream

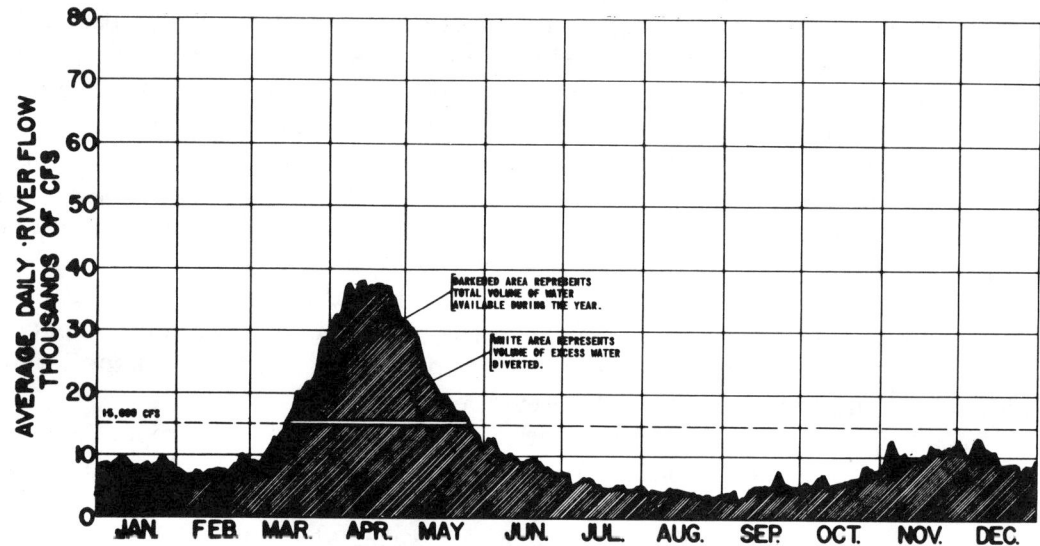

FIG. 8.—AVERAGE ANNUAL HYDROGRAPH, CONNECTICUT RIVER AT TURNERS FALLS, MASS. (BASED ON AVERAGE DAILY FLOWS 1915 THROUGH 1965)

user. As a matter of interest, some decrease in river content at high flows could be of definite advantage from a flood-control point of view. However, the net decrease is so minor that no flood-control benefits are actually attributed to it.

Fig. 8 indicates the average daily flows of the Connecticut River for the period of record 1915 through 1965, and shows how the flows of the Connecticut River vary considerably from one season to another. As expected, the record indicates that the flows of the river are high during the spring as a result of the combination of the spring rains coinciding with the snow-melt runoff of the watershed, and indicates that the river flow is low during the late summer months and early fall. This method of depicting average daily river flows indicates approximately 70 days with average daily flows greater than 15,000 cfs.

With all four units operating as pumps and using the average pumping rate of 2,500 cfs per unit (equivalent to 75,000 gal per sec), the 50 mcf (375,000,000 gal) of water which can be impounded in the additional 4 ft of reservoir height can be pumped to the upper reservoir in approximately 1.4 hr.

Normally, the flows of the Connecticut River at Turners Falls begin to exceed 15,000 cfs sometime about the middle of March; they reach an average of almost 40,000 cfs for approximately 20 days, and the flows begin to recede below 15,000 cfs towards the end of May. These are average conditions. The dry year of 1965 had only 26 days of over 15,000 cfs flows, and in the wet year of 1954 the 15,000 cfs flows were equaled or exceed for 142 days. Based on average conditions, the number of days used in the proposal was 70 days per year.

The water is, in effect, being pumped out of the pond behind the dam at Turners Falls. While the pumping is in progress, the waters of the Connecticut River are continuously being discharged through the hydroelectric facility at Turners Falls at the rate of 15,000 cfs. The total pumping rate of 10,000 cfs sustained for 1.4 hr, when it is redistributed on a 24-hr basis by the ponding and regulating abilities of the Turners Falls dam, results in a reduction in average-daily flow of less than 600 cfs.

During this 70-day period of a so-called average year, a total of over 26 billion gal of water can be transferred to the Quabbin Reservoir. The total volume of water flowing to the ocean during such an average year, as shown by the total area under the curve, would be in the order of 2.73 trillion gal. Thus, the transferred water represents less than 1% of this volume. Dividing the 26 billion gallons of transferred water by 365 days results in an annual average increase in the Quabbin usage of 72,000,000 gal per day—approximately one quarter of the present usage.

The proposal as presented contemplates that the utilities will increase the height of the reservoir and construct the separate intake structure. They expect to be reimbursed for the additional construction to the project, and for the pumping cost. The design and construction of the aqueduct from the project boundary to the Quabbin Reservoir (Item 15, Fig. 4) will be the responsibility of the Metropolitan District Commission.

CONCLUSIONS

Upon receipt of the FPC license which is expected sometime towards the latter part of this year, construction of the tunnels will begin. The plant will be ready for commercial operation in the fall of 1971.

The topography, geology, and physical location with respect to the load centers of New England and of the project site all combine to make the Northfield project an ideal comprehensive development, well suited for the planned multipurpose utilization.

The development will add 1,000 Mw of needed peaking capacity to the New England system. It will contribute substantially to the esthetic environment of the stretch of Connecticut valley in which it will be located by virtue of the added recreational areas to be opened up as part of the project. The solid rock formation of the mountain enables the entire waterway and generating system to be placed underground, thus facilitating design for high speed starting of units. Finally; the physical relationship of the project's upper reservoir to the Quabbin Reservoir makes possible a water-supply arrangement which will benefit a large portion of the population of the Commonwealth of Massachusetts.

LIST OF PROJECT ARTICLES

NORTHFIELD MOUNTAIN PUMPED STORAGE PROJECT

1. Unique Electrical Design Features of the Northfield Mountain Pumped Storage Project, F.L. Brennan and G.A. St. Onge, IEEE Paper presented at the 1969 Winter Power Meeting, New York, January 26-31, 1969.

2. Electrical Features of the Northfield Mountain Pumped Storage Project, R.P. Samolis, IEEE Paper 69-TP 147 PWR, Presented at the 1969 Winter Power Meeting, New York, January 26-31, 1969.

3. Northfield Mountain Pumped Storage Project, R.W. Gunwaldsen and A. Ferreira, Civil Engineering, May 1971.

4. Multipurpose Aspects of the Northfield Mountain Pumped Storage Project, A. Ferreira, Proceedings, International Conference on Pumped Storage Development and its Environmental Effects, AWRA, Milwaukee, WI, September 1971.

5. Hydraulic Structures at Northfield Mountain Pumped Storage Project, F.R. Harty, Jr., Paper presented at the 19th Annual Specialty Conference, Hydraulic Division ASCE, August 1971.

6. Initial Operating and Testing Experience, Northfield Mountain Pumped Storage Project, A. Ferreira, Proceedings of the American Power Conference, Vol. 35, 1973.

7. Hydraulic Effects Associated with Preliminary Operation of the Northfield Mountain Pumped Storage Project, F.R. Harty, Jr., Paper presented at the ASCE Hydraulics Division Specialty Conference, Knoxville, TN, August 1974.

8. Measurement of Tilt from Hydrodynamic Loads Powerhouse Cavern of the Northfield Mountain Pumped Storage Project, I. Simon and W.F. Swiger, ASCE Power Division, March 1975.

9. Startup and Preliminary Operation of Northfield Mountain Pumped Storage Project, R.G. Chevalier and R.P. Samolis, 37th Annual Meeting of the American Power Conference, Chicago, IL, April 1975.

10. Operating Experience of Northfield Mountain Pumped Storage Project, L.W. Noyes and R.G. Chevalier, 37th Annual Meeting of the American Power Conference, Chicago, IL, April 1975.

11. Planning and Design of Northfield Mountain Pumped Storage Project, A. Ferreira and W.E. Fisher, 37th Annual Meeting of the American Power Conference, Chicago, IL, April 1975.

12. Flooding and Restoration of Northfield Mountain Pumped Storage Project, R.P. Samolis and A. Ferreira, 37th Annual Meeting of the American Power Conference, Chicago, IL, April 1975.

13. Northfield Mountain PUmped Storage Project Performance, W.F. Swiger, P.A. Wild, and T.J. Lamb, ASCE, Journal of Geotechnical Division, Vol. 106, No. 6, June 1980.

14. 16 Years Operating and Maintenance of the 1,080 MW Northfield Mountain Pumped Storage Plant, A. Ferreira, Institution of Civil Engineers, International Conference on Pumped Storage, London, United Kingdom, 1990.

Plant Name: RACCOON MOUNTAIN

Plant location:
Chattanooga, TN
Marion County

Rated capacity	1,530 MW
Average static head	896 ft
Plant efficiency	76.90 %
Stored energy	31,400 MWh
Number of units	4

Construction time: 8 years, 6 months
Construction cost: $222 per kW
Price level: 1981
First commercial power: December 1978
FERC project number: None

River or water source: Tennessee River at river mile 444.6

Owner: Tennessee Valley Authority
Mr 6N 36A
Chattanooga, Tennessee 37401

Designers:
TVA Office of Engineering

Plant Manager/Superintendent:
J.M. Epps
TVA F+H PR
Chattanooga, Tennessee 37402

(615) 751-3391

	UPPER RESERVOIR	LOWER RESERVOIR
DAM		
Type	Earth and rockfill	Concrete with earthfill embankments, Nickajack Dam
Height (ft)	230	81
Crest length (ft)	8,500	3,766
Volume (yd³)	9,351,000	
RESERVOIR		
Type	Constructed	On stream, constructed, Nickajack Reservoir
Surface area (acres)	528	10,371
Usable power storage (acre-ft)	36,340	20,673
Power pool fluctuation (ft)	142.0	2.0
Operating levels		
Maximum (ft)	1,672.0	634.0
Minimum (ft)	1,530.0	632.0
Drainage area (miles²)	1.4	21,872.6
Seepage (ft³/s)		
SPILLWAY		
Design flood		
Return period (years)		
Flow (ft³/s)		500,000
Capacity (ft³/s)		
Type	None	
Gates		
Number	None	10
Type		Radial
Width (ft)		40.00
Height (ft)		40.00
OUTLET WORKS		
Discharge capacity (ft³/s)		
Number of water passages	None	None
Dimensions of water passages		
Height (ft)		
Width (ft)		
Diameter (ft)		
Type of gates		
Number of gates	None	None

Plant Name: **RACCOON MOUNTAIN**

Plant location
  Chattanooga, TN
  Marion County
	Owner: Tennessee Valley Authority
	  Mr 6N 36A
	  Chattanooga, Tennessee 37401

Rated capacity 1,530 MW Average static head 273.1 m Plant efficiency 76.90 % Stored energy 31,400 MWh Number of units 4	Designers: TVA Office of Engineering

Construction time: 8 years, 6 months
Construction cost: $222 per kW
Price level: 1981
First commercial power: December 1978
FERC project number: None
	Plant Manager/Superintendent:
	  J.M. Epps
	  TVA F+H PR
	  Chattanooga, Tennessee 37402
	
	  (615) 751-3391

River or water source: Tennessee River at river mile 444.6	

	UPPER RESERVOIR	LOWER RESERVOIR
DAM		
Type	Earth and rockfill	Concrete with earthfill embankments, Nickajack Dam
Height (m)	70.1	24.7
Crest length (m)	2,590.8	1,148.0
Volume (m³)	7,149,000	
RESERVOIR		
Type	Constructed	On stream, constructed, Nickajack Reservoir
Surface area (Mm²)	2.14	41.97
Usable power storage (Mm³)	44.825	25.500
Power pool fluctuation (m)	43.28	0.61
Operating levels		
Maximum (m)	509.63	193.24
Minimum (m)	466.34	192.63
Drainage area (Mm²)	3.630	56,650.000
Seepage (m³/s)		
SPILLWAY		
Design flood		
Return period (years)		
Flow (m³/s)		14,158
Capacity (m³/s)		
Type	None	
Gates		
Number	None	10
Type		Radial
Width (m)		12.192
Height (m)		12.192
OUTLET WORKS		
Discharge capacity (m³/s)		
Number of water passages	None	None
Dimensions of water passages		
Height (m)		
Width (m)		
Diameter (m)		
Type of gates		
Number of gates	None	None

RACCOON MOUNTAIN

INTAKES	UPPER INTAKE	LOWER INTAKE
Number	1	1
Type	Horseshoe-shaped vertical concrete tower	Reinforced concrete
Design discharge (ft³/s)	18,000	21,000
Gross inlet area (ft²) (at trash racks)	27,648	3,080
Bar racks		
spacing (in)		
shape		
depth/thickness (in)		
diameter (in)		
Emergency gates		
number	1	None
height/width (ft)	10.00 /	
type	Spherical valves 3.048 m (10 ft) in diameter	
Service gates		
number	None	4
height/width (ft)		5.00 / 5.56
type		Slide
Bulkhead/stop logs (Y or N)	N	N
number of units serviced		
Hoists		
number	None	4
capacity (tons)		20
type		Stationary hoist crane

WATER PASSAGES	Upper Tunnel	Shaft	Lower Tunnel	Surge Tanks Upper	Surge Tanks Lower	Penstocks	Tailrace Tunnel
Number		1	1		1	4	1
Diameter (ft)		35.0	35.0		44.0	17.8	35.0
Length (ft)		900	975		275	150	1,575
Maximum velocity (ft/s)		21.0	21.0			20.3	21.0
Concrete lining thickness (in)		12.00	12.00		18.00		
Total length of concrete sections (ft)		900	975				1,575
Steel liner Thickness							
Minimum (in)						0.95	
Maximum (in)						1.12	
Material grade						SA 517 Grade B	
Total length of steel-lined sections (ft)						115	

Notes:
Shaft: Connects directly to intake structure.
Lower Tunnel: Length does not include short transitions from 2 @ 7.47 m (24.5 ft) to 4 @ 5.33 m (17.5 ft).
Tailrace Tunnel: Length is from center line of units to discharge structure.

Raccoon Mountain - Page 2 (English)

RACCOON MOUNTAIN

INTAKES	UPPER INTAKE	LOWER INTAKE
Number	1	1
Type	Horseshoe-shaped vertical concrete tower	Reinforced concrete
Design discharge (m³/s)	509.7	594.7
Gross inlet area (m²) (at trash racks)	2,568.6	286.1
Bar Racks:		
spacing (mm)		
shape		
depth/thickness (mm)		
diameter (mm)		
Emergency gates		
number	1	None
height/width (m)	3.048 /	
type	Spherical valves 3.048 m (10 ft) in diameter	
Service gates		
number	None	4
height/width (m)		1.524 / 1.695
type		Slide
Bulkhead/stop logs (Y or N)	N	N
number of units serviced		
Hoists		
number	None	4
capacity (Mg)		18
type		Stationary hoist crane

WATER PASSAGES	Upper Tunnel	Shaft	Lower Tunnel	Surge Tanks Upper	Surge Tanks Lower	Penstocks	Tailrace Tunnel
Number		1	1		1	4	1
Diameter (m)		10.67	10.67		13.41	5.43	10.67
Length (m)		274.3	297.2		83.8	45.7	480.1
Maximum velocity (m/s)		6.40	6.40			6.19	6.40
Concrete lining thickness (m)		0.305	0.305		0.457		
Total length of concrete sections (m)		274.3	297.2				480.1
Steel liner Thickness							
Minimum (mm)						24	
Maximum (mm)						28	
Material grade						SA 517 Grade B	
Total length of steel-lined sections (m)						35.1	

Notes:
Shaft: Connects directly to intake structure.
Lower Tunnel: Length does not include short transitions from 2 @ 7.47 m (24.5 ft) to 4 @ 5.33 m (17.5 ft).
Tailrace Tunnel: Length is from center line of units to discharge structure.

RACCOON MOUNTAIN

POWERHOUSE and RELATED FEATURES

Powerhouse Structure
Type: Underground, rock, reinforced concrete and steel
Length: 490 ft Width: 72 ft Height: 157 ft

Guard Valves
Number: 4 Diameter: 10.0 ft
Type: Spherical

Transformers
Number: 2, 3-Phase
Ratings: 4 - 3PH 415
Voltages: (kV) 161 / 23

Generator
Rating generating (MVA): 391.5 Rating pumping (MVA): 399.0
Insulation type: Thermosetting
Starting method: Inverter, converter, back-to-back, direct control at powerhouse
Starting equipment:

Runners
Material: Stainless steel
Minimum unit submergence: 107.0 ft
WR²:
Manufacturer: Allis Chalmers
Model test by: Allis Chalmers

	Reversible Runners	Reversible Motor/Generator	
Number	4		4
Diameter (ft)	16.60	Rotor	Stator
rpm synchronous	300.0		300.0
rpm overspeed	455.0		455.0
Type	Francis		Siemons water cooled

Information on Runners

Condition:	Gross Head (ft)		Capacity (MW)		Discharge (ft³/s)		Turbine/Pump Eff.(%)	
	Generating	Pumping	Generating	Pumping	Generating	Pumping	Generating	Pumping
Maximum head & maximum power	1,042	1,042	400	360				
Minimum head & maximum power	900		300	360				

Condition:	Net Head (ft)		Capacity (MW)		Discharge (ft³/s)		Turbine/Pump Eff.(%)	
	Generating	Pumping	Generating	Pumping	Generating	Pumping	Generating	Pumping
Rated head @ best gate	1,020	1,000	400	360	5,020	3,850		

Raccoon Mountain - Page 3 (English)

RACCOON MOUNTAIN

POWERHOUSE and RELATED FEATURES

Powerhouse Structure
Type: Underground, rock, reinforced concrete and steel
Length: 149.4 m   Width: 21.9 m   Height: 47.9 m

Guard Valves
Number: 4   Diameter: 3.05 m
Type: Spherical

Transformers
Number: 2, 3-Phase
Ratings: 4 - 3PH 415
Voltages: (kV) 161 / 23

Generator
Rating generating (MVA): 391.5   Rating pumping (MVA): 399.0
Insulation type: Thermosetting
Starting method: Inverter, converter, back-to-back, direct control at powerhouse
Starting equipment:

Runners
Material: Stainless steel
Minimum unit submergence: 32.61 m
WR²:
Manufacturer: Allis Chalmers
Model test by: Allis Chalmers

	Reversible Runners	Reversible Motor/Generator		
Number	4			4
Diameter m	5.06	Rotor	Stator	
rpm synchronous	300.0			300.0
rpm overspeed	455.0			455.0
Type	Francis		Siemons water cooled	

Information on Runners

Condition:	Gross Head (m)		Capacity (MW)		Discharge (m³/s)		Turbine/Pump Eff.(%)	
	Generating	Pumping	Generating	Pumping	Generating	Pumping	Generating	Pumping
Maximum head & maximum power	317.6	317.6	400	360				
Minimum head & maximum power	274.3		300	360				

Condition:	Net Head (m)		Capacity (MW)		Discharge (m³/s)		Turbine/Pump Eff.(%)	
	Generating	Pumping	Generating	Pumping	Generating	Pumping	Generating	Pumping
Rated head @ best gate	310.8	304.8	400	360	142.2	109.0		

RACCOON MOUNTAIN

Plant Data:
Average GWh generating per year:	1,800
Average GWh pumping per year:	2,225
Starting time from standstill (s):	300
Changeover time pumping to generating (min):	5
Planned/scheduled time between major overhauls (years):	
Outage time required per unit during major overhauls (weeks):	
Representative plant availability (%):	84.0
Representative planned outages (weeks per year):	4

Miscelleneous Notes:
None.

Cavitation Experience:
Runner cavitation occurs and heavy repair is required every 2 years.

Significant or Unique Problems:
Problems include vibration, the loss of seven sets of turbine ring seals, and numerous turbine bearing failures.

List of Licenses Required:
None.

ENVIRONMENTAL FEATURES

Recreation:
A boat ramp, handicapped ramps at the lower pool, and a picnic area are available. Visitor tours are also conducted.

Fish and Wildlife:
The top lake is closed to the public. No hunting, fishing, and access to lower pool is allowed.

Social:
None.

RACCOON MOUNTAIN PUMPED-STORAGE PLANT

FINAL DESIGN REPORT

General Design Features

The principal features of the project (figures 1 through 3) are a lower reservoir formed by TVA's Nickajack Dam, nearly 20 miles downstream from the site, discharge structure, waterway tunnels, surge chamber, access tunnels to an underground transformer vault and powerplant chamber, intake structure, upper reservoir formed by an earth and rock-filled dam and switchyards located on top of the mountain near the east end of the upper reservoir. An access road from the northwest side of the lower portal and discharge area stems from a state highway.

Controlling Design Features

The lower reservoir is controlled by Nickajack Dam and established a normal operating water surface from elevation 632.0 to elevation 634.0. The upper reservoir maximum water surface is elevation 1672.0, and the normal minimum water surface is elevation 1530.0. The useful storage between these two elevations of 36,340 acre-feet provides approximately 20 hours of generation at plant capacity of 1530 MW before refilling the upper reservoir by pumping. With the 4-units operating as pumps, the normal reservoir operating range of useful storage can be refilled in approximately 27 hours. The plant was designed to be remotely controlled from the Chickamauga Power Service Center, but since commissioning, the plant has been operated as a manned plant.

Following are the principal components and dimensions of the structures, conduits, and dams as finally designed (figure 4 and exhibits 1 through 4):

1. An enclosure structure was provided to aid navigation. This was accomplished by driving nineteen 32-foot-diameter by 45-foot-high gravel-filled deflector cells and six 19-foot-diameter by 37-foot-high gravel-filled closure cells. A concrete guide wall was placed atop the cells and ramps were provided for fisherman access to the cells.

2. A 4-bay trashrack structure, 71 feet wide by 55 feet high, located at the river end of the discharge structure.

3. A 3-bay flow distribution structure, 49 feet wide by 37 feet high, located within the flaring walls of the discharge structure.

4. A stoplog structure, 43 feet wide by 82 feet high, located at the mouth of the 35-foot-diameter discharge tunnel.

5. Access Tunnels

    The main access tunnel is a horseshoe-shaped tunnel, 30 feet wide by 24 feet 6 inches high. From the transformer vault to the powerplant chamber, the tunnel is 41 feet wide by 31 feet high with an arched roof. The total length of the tunnel is approximately 1,366 feet long.

    The ventilation and emergency exit tunnel is a horseshoe-shaped tunnel, 15 feet wide by 21 feet high. The total length of the tunnel is approximately 1,225 feet.

The cable and visitors access tunnel is a horseshoe-shaped tunnel, 18 feet wide by 21 feet high. The total length of the tunnel is approximately 482 feet. The 23-foot-diameter vertical elevator and cable shaft is approximately 1,070 feet high.

The surge chamber vent and access tunnel is a horseshoe-shaped tunnel 18 feet wide by 21 feet high. The total length of the tunnel is approximately 1,127 feet long.

6. An underground surge chamber located on the low pressure side of the pump/turbines; top section, 52 feet wide by 25 feet high by 305 feet long; lower chamber 45 feet wide by 33 feet high by 275 feet long; 4 separate shafts, one for each unit, 41 feet in diameter, 90 feet high; 4 risers, 9 feet wide by 16 feet long by 51 feet high which connect the unit draft tubes and the surge chamber.

7. An underground transformer vault, 55 feet wide by 48 feet high by 385 feet long with a flat roof.

8. An underground powerplant chamber, including control and erection bays, 72 feet wide by 490 feet long with four reversible hydraulic pump/turbine units. Each unit has a rated generating capacity of 382,500 kW at 1,020-foot net head, and a rated pumping capacity of 3,850 CFS at 1000-foot total head. Maximum excavated height in the erection bay, 69 feet; in the area between units, 102 feet; in the area of the spherical values, 138 feet; and in the unit area beneath the draft tubes, 165 feet. The powerplant chamber has a

crowned roof. There are four individual bus tunnels, 28 feet wide by 70 feet long, that join the transformer vault and the powerplant chamber with a maximum excavated height of 52 feet.

9. Waterways

On the discharge side of the reversible pump/turbines, the four circular elbow-type, steel-lined draft tubes connect to 17-foot 6-inch-diameter concrete-lined tunnels. The tunnels enter a collector tunnel that transitions from 17 feet 6 inches to 24 feet 6 inches to 30 feet 6 inches to a 35-foot-diameter tunnel. The total length of the discharge tunnels is approximately 1,575 feet.

On the intake side of the reversible pump/turbine, four 115-foot-long steel penstock liners transition from a 10 foot diameter to a 17 foot 6 inch diameter. The penstock liners connect to 17-foot 6-inch-diameter concrete-lined tunnels that transition to 24 feet 6 inches to a 35-foot-diameter tunnel. The 35-foot-diameter tunnel continues for approximatley 975 feet and connects to a 35-foot-diameter elbow. A 900-foot vertical concrete-lined shaft, 35 feet in diameter, transitions to a 50-foot-diameter opening at the intake structure. The total length of the intake waterways, including the vertical shaft, is approximately 2,450 feet.

10. A horseshoe-shaped, multiport, silo-type intake structure, 220 feet high, with a 92-foot-inside diameter is located directly above the intake shaft.

11. An intake channel was excavated approximatey 1550 feet into the north side of the rim. At its deepest section, the channel is 180 feet wide and 245 feet high.

12. The west rockfill dam with sloping earth core is approximately 3900 feet long, maximum height 230 feet.

13. The south rockfill dam with sloping earth core is approximately 4500 feet long, maximum height 100 feet.

14. A rolled fill saddle dam to close a low spot at the juncture of the east and north rims is approximately 750-feet-long, maximum height 40 feet.

15. A retention dam across an inlet at the east side of the reservoir is approximately 250 feet long, maximum height 25 feet.

16. A 38 feet high rectangular control structure located at the upstream toe of the west dam housing a 4-foot by 4-foot slide gate.

17. A 4-foot-diameter pipe encased in a cut and cover trench at the low point of the west dam.

18. A multilevel 161-kV switchyard, intertie transformer yard, and 500-kV switchyard is located at the northeast side of the upper reservoir.

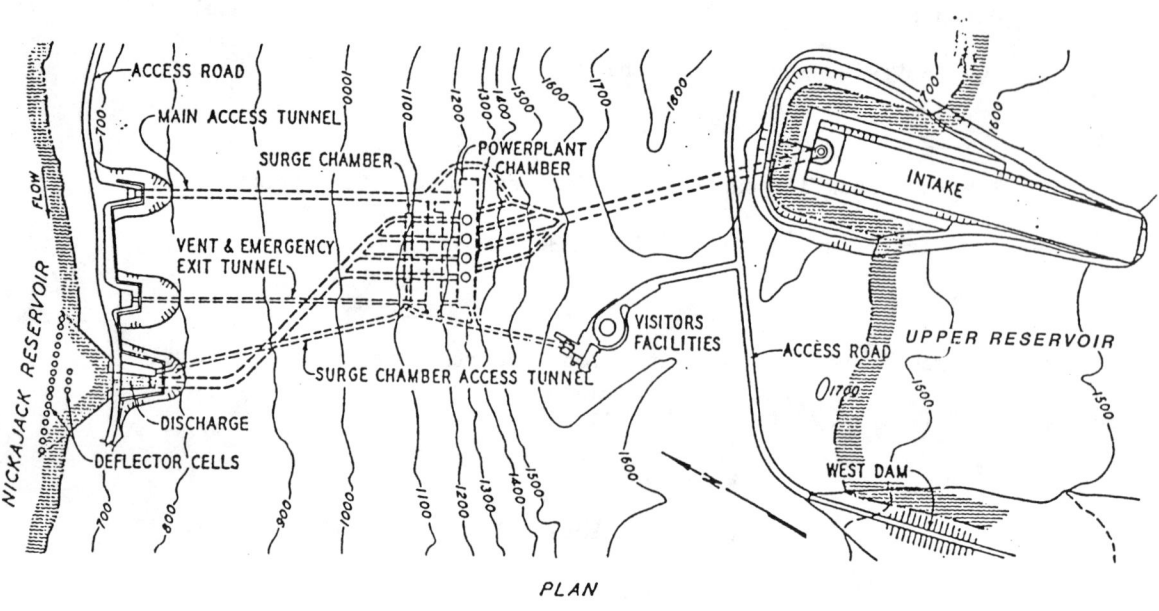

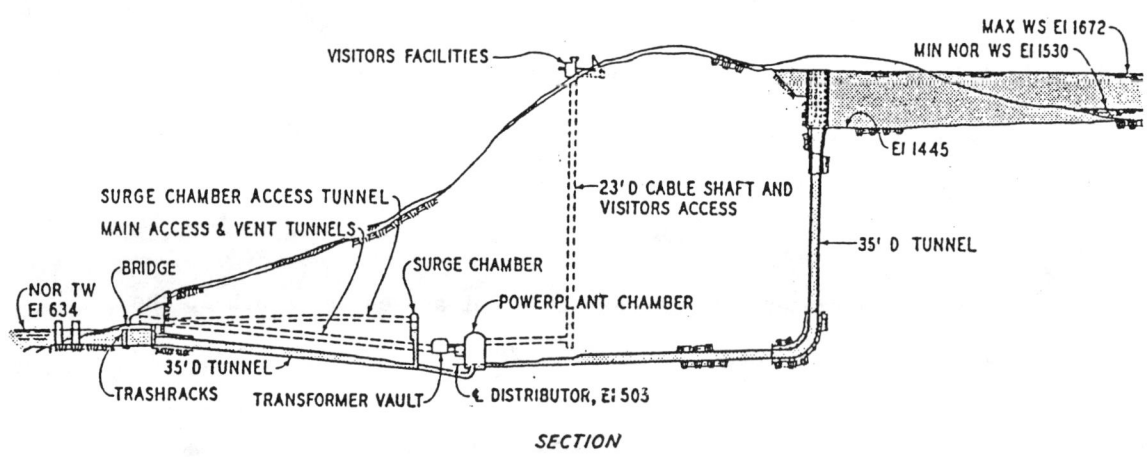

PLAN AND SECTION - POWER FACILITIES

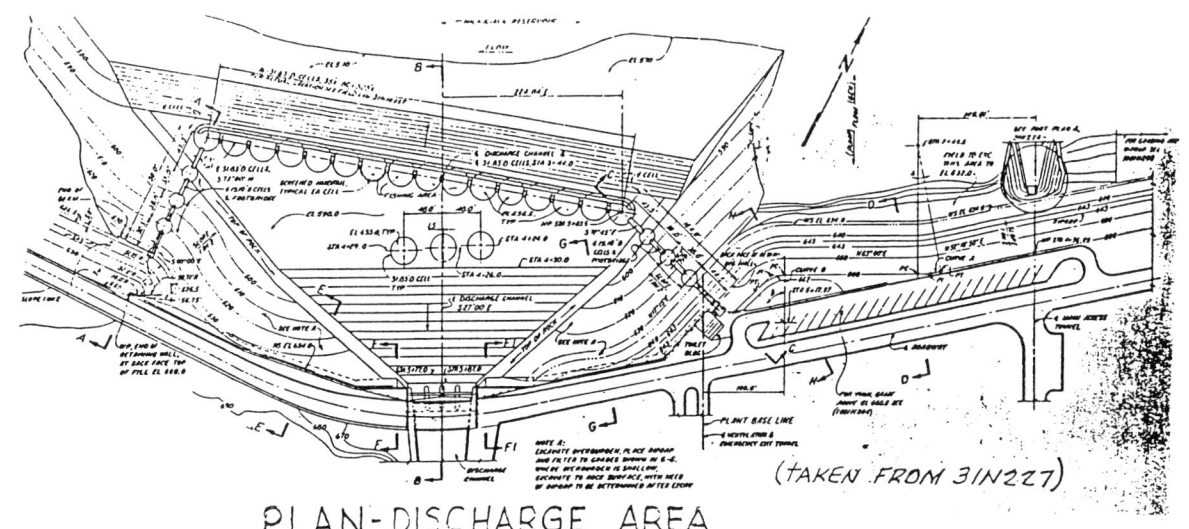

PLAN-DISCHARGE AREA

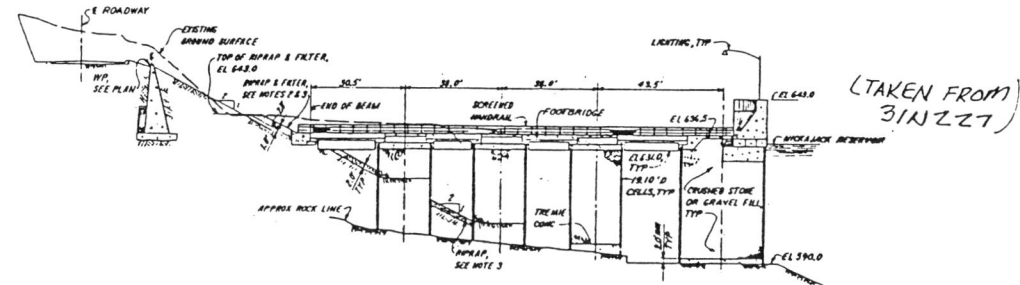

SECTION THRU DISCHARGE CELLS AND BRIDGE

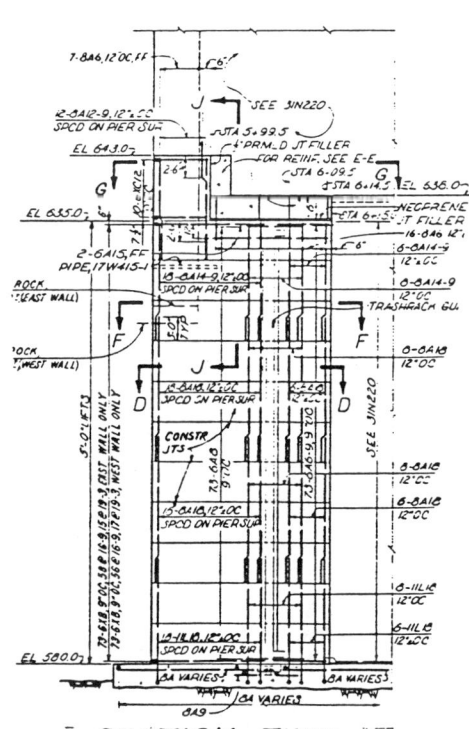

SECTION THRU TRASHRACK STRUCTURE

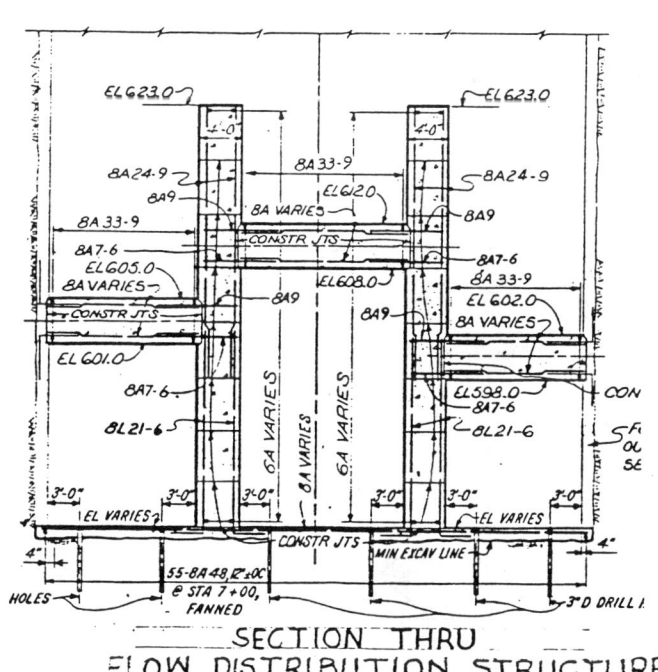

SECTION THRU FLOW DISTRIBUTION STRUCTURE

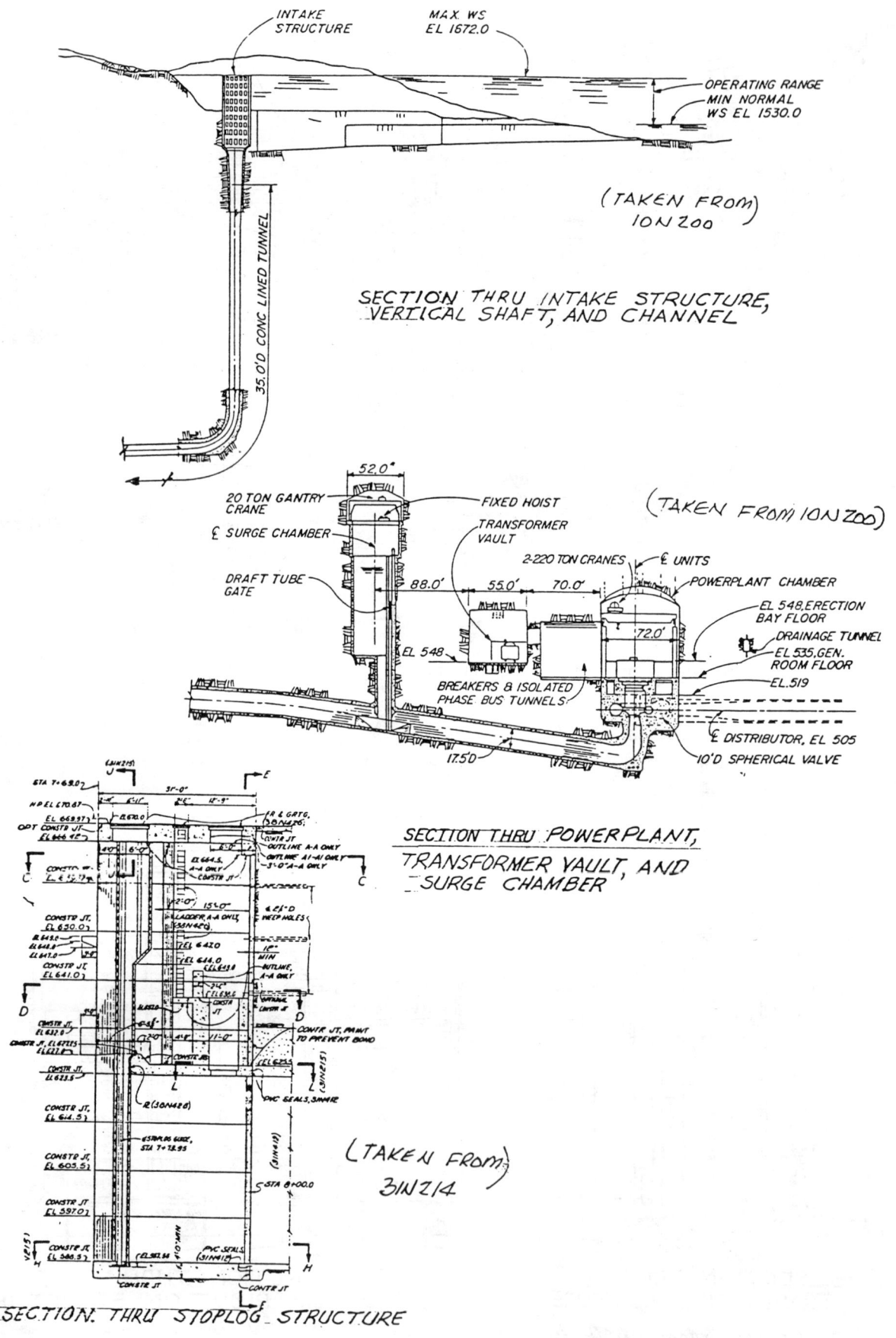

PRINCIPLE COMPONENTS

Upper Reservoir

Raccoon Mountain Pumped Storage Plant

Aerial View of Upper Reservoir and Outlet Works

Raccoon Mountain Pumped Storage Plant

Plant Name: **ROCKY MOUNTAIN**

Plant location:
  Armuchee, GA
  Floyd County
 	
Rated capacity   760 MW
Average static head 650 ft
Plant efficiency   83.50 %
Stored energy   5,700 MWh
Number of units   3
 	
Construction time:
Construction cost:
Price level:
First commercial power: December 1995
FERC project number: 2725
 	
River or water source: Heath Creek
	Owner: Oglethorpe Power Corporation
	       2100 East Exchange Place
	       P.O. Box 1349
	       Tucker, Georgia 30085
	
	Designers:
	  Harza Engineering Company
	  Tunnels And Upper Intake: Southern Company Services, Inc.
	
	Plant Manager/Superintendent:
	  Plant Manager
	  Rocky Mountain Pumped Storage
	  306 Antioch Road
	  Armuchee, Georgia 30105
	  (404) 236-3780

	UPPER RESERVOIR	LOWER RESERVOIR
DAM		
Type	Zoned earthfill	Concrete gravity, zoned earthfill
Height (ft)	80	72
Crest length (ft)	12,800	650
Volume (yd³)	10,738,000	571,000
RESERVOIR		
Type	Constructed	Constructed
Surface area (acres)	210	600
Usable power storage (acre-ft)	10,200	15,690
Power pool fluctuation (ft)	51.0	20.0
Operating levels		
Maximum (ft)	1,392.0	710.5
Minimum (ft)	1,341.0	681.0
Drainage area (miles²)	0.0	15.0
Seepage (ft³/s)		
SPILLWAY		
Design flood		
Return period (years)		PMF
Flow (ft³/s)		24,900
Capacity (ft³/s)	13,500	24,900
Type	Fuse plug	Overflow
Gates		
Number		2
Type		Radial
Width (ft)		22.50
Height (ft)		23.55
OUTLET WORKS		
Discharge capacity (ft³/s)		359
Number of water passages		1
Dimensions of water passages		
Height (ft)		
Width (ft)		
Diameter (ft)		4.00
Type of gates		Jet flow
Number of gates	None	1

Rocky Mountain - Page 1 (English)

Plant Name: **ROCKY MOUNTAIN**

Plant location Armuchee, GA Floyd County	Owner: Oglethorpe Power Corporation 2100 East Exchange Place P.O. Box 1349 Tucker, Georgia 30085	
Rated capacity 760 MW Average static head 198.1 m Plant efficiency 83.50 % Stored energy 5,700 MWh Number of units 3	Designers: Harza Engineering Company Tunnels And Upper Intake: Southern Company Services, Inc.	
Construction time: Construction cost: Price level: First commercial power: December 1995 FERC project number: 2725	Plant Manager/Superintendent: Plant Manager Rocky Mountain Pumped Storage 306 Antioch Road Armuchee, Georgia 30105 (404) 236-3780	
River or water source: Heath Creek		

DAM	UPPER RESERVOIR	LOWER RESERVOIR
Type	Zoned earthfill	Concrete gravity, zoned earthfill
Height (m)	24.4	22.0
Crest length (m)	3,901.4	198.0
Volume (m³)	8,210,000	436,400
RESERVOIR		
Type	Constructed	Constructed
Surface area (Mm²)	0.85	2.43
Usable power storage (Mm³)	12.582	19.353
Power pool fluctuation (m)	15.54	6.10
Operating levels		
Maximum (m)	424.28	216.56
Minimum (m)	408.74	207.57
Drainage area (Mm²)		38.850
Seepage (m³/s)		
SPILLWAY		
Design flood		
Return period (years)		PMF
Flow (m³/s)		705
Capacity (m³/s)	382	705
Type	Fuse plug	Overflow
Gates		
Number		2
Type		Radial
Width (m)		6.858
Height (m)		7.178
OUTLET WORKS		
Discharge capacity (m³/s)		10
Number of water passages		1
Dimensions of water passages		
Height (m)		
Width (m)		
Diameter (m)		1.219
Type of gates		Jet flow
Number of gates	None	1

ROCKY MOUNTAIN

INTAKES	UPPER INTAKE	LOWER INTAKE
Number	1	3
Type	Covered morning glory	Draft tube
Design discharge (ft³/s)	18,477	6,126
Gross inlet area (ft²)	3,848	724
(at trash racks)		
Bar racks		
spacing (in)		9.00
shape		Rectangular
depth/thickness (in)		4.00 / 0.75
diameter (in)		
Emergency gates		
number	None	None
height/width (ft)		
type		
Service gates		
number	None	6
height/width (ft)		16.77 / 17.78
type		Vertical lift
Bulkhead/stop logs (Y or N)	N	
number of units serviced		
Hoists		
number	None	6
capacity (tons)		30
type		Wire rope

WATER PASSAGES	Upper Tunnel	Shaft	Lower Tunnel	Surge Tanks Upper	Surge Tanks Lower	Penstocks	Tailrace Tunnel
Number		1	1			3	
Diameter (ft)		35.0	35.0			19.0	
Length (ft)		567	1,935			482	
Maximum velocity (ft/s)		19.2	19.2			21.8	
Concrete lining thickness (in)		20.04	20.04			24.00	
Total length of concrete sections (ft)		564	1,935			118	
Steel liner Thickness							
Minimum (in)						1.50	
Maximum (in)						2.75	
Material grade		None	None			A516/38	
Total length of steel-lined sections (ft)						349	

Notes:
Penstocks: Penstock length varies from 140.2 m (460 ft) to 153.3m (503 ft). See also Page 4 notes.

ROCKY MOUNTAIN

INTAKES	UPPER INTAKE	LOWER INTAKE
Number	1	3
Type	Covered morning glory	Draft tube
Design discharge (m³/s)	523.2	173.5
Gross inlet area (m²)	357.5	67.3
(at trash racks)		
Bar Racks:		
spacing (mm)		229
shape		Rectangular
depth/thickness (mm)		102 / 19
diameter (mm)		
Emergency gates		
number	None	None
height/width (m)		
type		
Service gates		
number	None	6
height/width (m)		5.110 / 5.420
type		Vertical lift
Bulkhead/stop logs (Y or N)	N	
number of units serviced		
Hoists		
number	None	6
capacity (Mg)		27
type		Wire rope

WATER PASSAGES	Upper Tunnel	Shaft	Lower Tunnel	Surge Tanks Upper	Surge Tanks Lower	Penstocks	Tailrace Tunnel
Number		1	1			3	
Diameter (m)		10.67	10.67			5.79	
Length (m)		172.8	589.8			147.0	
Maximum velocity (m/s)		5.85	5.85			6.63	
Concrete lining thickness (m)		0.508	0.508			0.610	
Total length of concrete sections (m)		172.0	589.8			36.0	
Steel liner Thickness							
Minimum (mm)						38	
Maximum (mm)						70	
Material grade		None	None			A516/38	
Total length of steel-lined sections (m)						106.3	

Notes:
Penstocks: Penstock length varies from 140.2 m (460 ft) to 153.3m (503 ft). See also Page 4 notes.

ROCKY MOUNTAIN

POWERHOUSE and RELATED FEATURES

Powerhouse Structure
Type: Surface - indoor
Length: 323 ft Width: 158 ft Height: 173 ft

Guard Valves
Number: 3 Diameter: 10.7 ft
Type: Spherical

Transformers
Number: 3 phase, 2 winding
Ratings: 237/316/395
Voltages: (kV) 230 / 19.2

Generator
Rating generating (MVA): 314.0 Rating pumping (MVA): 319.3
Insulation type: Class B
Starting method: Starting motor
Starting equipment: Static variable frequency converter

Runners
Material: 13-5 stainless steel casting
Minimum unit submergence: 71.0 ft
WR^2: 136,000,000 (lbf x ft²)
Manufacturer: Hitachi
Model test by: Hitachi

	Reversible Runners	Reversible Motor/Generator			
Number	3				3
Diameter (ft)	18.70	Rotor	26.41	Stator	26.57
rpm synchronous	225.0				225.0
rpm overspeed	337.5				337.5
Type	Francis				Semi-umbrella

Information on Runners

Condition:	Gross Head (ft)		Capacity (MW)		Discharge (ft³/s)		Turbine/Pump Eff.(%)	
	Generating	Pumping	Generating	Pumping	Generating	Pumping	Generating	Pumping
Maximum head & maximum power	690	717	321	301	6,159	4,361		
Minimum head & maximum power	613	634	263	315	5,739	5,170		

Note: Data in the above table are based on model tests.

Condition:	Net Head (ft)		Capacity (MW)		Discharge (ft³/s)		Turbine/Pump Eff.(%)	
	Generating	Pumping	Generating	Pumping	Generating	Pumping	Generating	Pumping
Rated head @ best gate	613	613	253	263	5,700	5,170		

Note: Data in the above table are based on model tests.

Rocky Mountain - Page 3 (English)

ROCKY MOUNTAIN

POWERHOUSE and RELATED FEATURES

Powerhouse Structure
Type: Surface - indoor
Length: 98.3 m   Width: 48.3 m   Height: 52.7 m

Guard Valves
Number: 3   Diameter: 3.25 m
Type: Spherical

Transformers
Number: 3 phase, 2 winding
Ratings: 237/316/395
Voltages: (kV) 230 / 19.2

Generator
Rating generating (MVA): 314.0   Rating pumping (MVA): 319.3
Insulation type: Class B
Starting method: Starting motor
Starting equipment: Static variable frequency converter

Runners
Material: 13-5 stainless steel casting
Minimum unit submergence: 21.64 m
WR^2: 56,215,000 (Newtons x m²)
Manufacturer: Hitachi
Model test by: Hitachi

	Reversible Runners	Reversible Motor/Generator	
Number	3		3
Diameter m	5.70	Rotor 8.050	Stator 8.100
rpm synchronous	225.0		225.0
rpm overspeed	337.5		337.5
Type	Francis		Semi-umbrella

Information on Runners

Condition:	Gross Head (m)		Capacity (MW)		Discharge (m³/s)		Turbine/Pump Eff.(%)	
	Generating	Pumping	Generating	Pumping	Generating	Pumping	Generating	Pumping
Maximum head & maximum power	210.3	218.5	321	301	174.4	123.5		
Minimum head & maximum power	186.8	193.1	263	315	162.5	146.4		

Note: Data in the above table are based on model tests.

Condition:	Net Head (m)		Capacity (MW)		Discharge (m³/s)		Turbine/Pump Eff.(%)	
	Generating	Pumping	Generating	Pumping	Generating	Pumping	Generating	Pumping
Rated head @ best gate	186.8	186.8	253	263	161.4	146.4		

Note: Data in the above table are based on model tests.

Rocky Mountain - Page 3 (Metric)

ROCKY MOUNTAIN

Plant Data:
Average GWh generating per year:
Average GWh pumping per year:
Starting time from standstill (s):
Changeover time pumping to generating (min):
Planned/scheduled time
   between major overhauls (years):
Outage time required per unit
   during major overhauls (weeks):
Representative plant availability (%):
Representative planned outages (weeks per year):

Miscelleneous Notes:
Construction of this project began in May of 1991.

The gross inlet area of 67.32 m² (724.6 ft²) is a total for the two inlet areas of 33.66 m² (362.3 ft²) each. The lower intake includes three trashrack panels per opening that are non-interchangeable.

The design discharges in the tables for the intakes are for the generating mode. The design discharges for the pumping mode for the single upper and triple lower intake are 439.14 m³/s (15,500 f³/s) and 146.38 m³/s (5,170 f³/s), respectivly.

The length of the three penstocks varies from 140.2 m (42.7 ft) to 153.42 m (503.3 ft). The length of the reinforced sections varies from 26.42 m (86.67 ft) to 46.94 m (154 ft). The lengths of the steel lined sections varies from 26.52 (87 ft) to 46.94 m (154 ft).

The total fill in the embankment for the lower dam is 436,400 m³ (570,800 yd³). This is made up of 37,300 m³ (48,800 yd³) of concrete and 399,100 m³ (522,000 yd³) of earthfill.

The manufacturer of the guard valves was Mitsubishi.

The diameter of the pump-turbine is listed as 5.70 m (224") to the tip. It is 4.1 m (161") to the eye. The stator diameter of 8.1 m (26.6 ft) is an inside diameter.

Cavitation Experience:
The project is not in operation yet.

Significant or Unique Problems:
Exploration of the main dam, left abutment, indicated limestone solution channels that require an approximate 200-foot-deep cutoff trench.

List of Licenses Required:
FERC, Georgia Dept Natural Resources, COE Section 401 & 404.

ENVIRONMENTAL FEATURES

ROCKY MOUNTAIN

Recreation:
　　Recreationists enjoy the picnic areas, beach, boat ramps, fishing, swimming, playground, and trails and a camp in the upper reservoir area.

Fish and Wildlife:
　　Preservation area is at the project site.

Social:
　　Highway improvement was included as a part of the project.

Rocky Mountain project will provide peak power for Georgia

By D. R. Murphy and W.R. Ivarson, Senior Project Manager* and Engineering Co-ordinator**

An 800 MW pumped-storage project is under construction in Georgia, USA, which will improve the reliability of the system operated by the Oglethorpe Power Corporation. The system which the project will serve (when complete in 1995) had a 1989 peak load of 3315 MW and an annual energy consumption of almost 14 000 GWh. Over the past five years, the system has exhibited an average demand growth of 6 per cent annually. The main features of the project design are described here.

The Rocky Mountain pumped-storage project is under construction at Heath Creek in Floyd County, Georgia, USA. The site is approximately 10 miles (16 km) northwest of Rome, Georgia, and 8 miles (12.8 km) east of the Georgia-Alabama state line. The scheme will include: an upper reservoir with an area of 221 acres (90 ha) on Rock Mountain; a lower operating reservoir with an area of 600 acres (243 ha) in Heath Creek and two auxiliary pools of 400 and 205 acres (162 and 83 ha); approximately 2700 ft (823 m) of tunnel and penstock; a nominal 800 MW powerplant; and, a 2.7 miles (4.3 km) transmission line and associated switching substation. A project plan is shown in Fig. 1.

Project history

In the late 1960s, the Georgia Power Corporation began studying the feasibility of a pumped-storage project at Rock Mountain. An initial licence application was filed with the Federal Power Commission (FPC) in October 1972. Studies continued, and a revised application was filed in January 1974. The various state and federal agencies were consulted, and an environmental impact statement, written by the FPC staff, was issued in September 1975 and finalized in May 1976. An order issuing a major licence was issued in January 1977.

Design studies began immediately on receipt of the licence, and construction began in late 1978. At the end of 1979, the main dam site of the lower reservoir was relocated approximately one mile (1.6 km) downstream, to avoid foundation problems. Also in late 1979, exploratory and drainage adit construction began. Between 1980 and 1987, work proceeded with the powerhouse excavation, county road relocation, penstock, tunnel, shaft and upper reservoir intake/outlet and powerhouse access bridge construction, and miscellaneous water main improvements.

In the early 1980s, Oglethorpe Power recognized the need for pumped storage in the generation mix and began a programme to identify potential sites. In the mid 1980s, Georgia Power delayed its plans for the construction of the Rocky Mountain project. After lengthy negotiations, the utilities reached an agreement, under which Oglethorpe would complete the design and construction, and operate the project. In December 1987, the utilities made application to the Federal Energy Regulatory Commission (FERC formerly FPC) for the transfer of the licence; this transfer was granted in January 1988. An agreement between Oglethorpe Power Corporation and Georgia Power Corporation was reached in December 1988, with Oglethorpe Power Corporation taking responsibility for the project and pursuing the development of the project since that time.

*Oglethorpe Power Corporation, 2100 East Exchange Place, Tucker, Georgia 30085, USA; and,
**Engineering Co-ordinator for the Rocky Mountain Project, Harza Engineering Co, 150 South Wacker Drive, Chicago, Illinois 60606, USA.

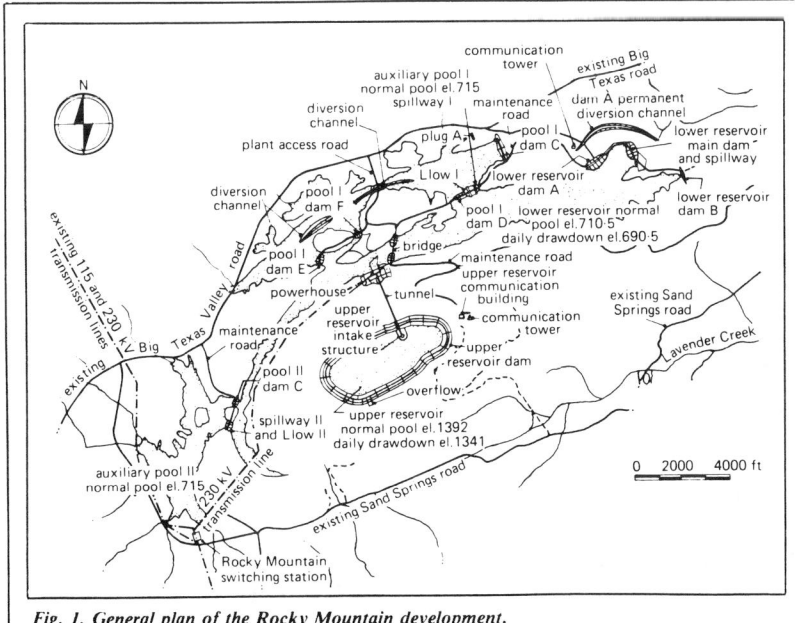

Fig. 1. General plan of the Rocky Mountain development.

Project design

There have been some significant changes with respect to electrical equipment because of the length of time the project has been under development. The initial design of the pump starting equipment provided for the use of separate pony motors to limit electric current requirements, while accelerating the units to a significant rotational speed. This scheme has been replaced by static frequency converters to enable the main motors to be used without separate ones for rotational starting. Also, the redesign of the computer control and instrumentation equipment has taken advantage of

Reprinted, with permission, from International Water Power and Dam Construction, February 1991.

the advances in that equipment over the years. The major project features, however, have remained similar.

Dams and reservoirs

The upper reservoir will be impounded by a continuous earth- and rockfill dam approximately 12 800 ft (3900 m) long, 75 ft (23 m) high on average, the crest elevation being 1400 ft (427 m) above sea level. The dam will be founded on an interbedded shale and sandstone formation. The reservoir will have a surface area of 221 acres (90 ha) at the maximum operating pool elevation of 1390 ft (424 m) and will have 9300 acre-ft (11.5 × 10⁶ m³) of storage between this elevation and the normal minimum pool elevation of 1341 ft (409 m).

The lower operating reservoir will be impounded by: a 550 ft (168 m)-long main earthfill dam with its crest at elevation 721 ft (220 m); a 480 ft- (146 m)-long concrete gravity structure, with a gated spillway to provide controlled discharge for pool elevations of 710.5 ft (216.6 m) and above; and a cut-off trench excavated approximately 700 ft (213 m) into the right abutment and backfilled with zoned earthfill up to el. 721 ft (220 m). Flow through the main spillway will be controlled by two tainter gates, each 24 ft (7.3 m) high by 23 ft (7 m) wide. An earthfill dam (dam A) about 1260 ft (384 m) long at el. 721 ft (220 m) will close a gap in the ridge, forming the reservoir barrier to the northwest of the main dam. An earth dike (dam B) to be constructed across a depression southeast of the main dam will be approximately 700 ft (213 m) long, with its crest at el. 720 ft (220 m). The reservoir will have a surface area of approximately 600 acres (243 ha) at normal full pool elevation, 710.5 ft (216.6 m), and contain 13 000 acre-ft (16 × 10⁶ m³) of active storage between els. 710.5 and 681 ft (216.6 and 208 m).

Two auxiliary pools will be provided adjacent to the lower reservoir and will have surface areas of 400 and 205 acres (262 and 83 ha) at their normal full pool elevation of 715 ft (218 m). The auxiliary pools will provide dry weather storage for the project, as the catchment area is small and inflow will be exceeded by evaporation losses, seepage losses and release requirements during extreme low flows. They will also be used for recreation in association with other facilities at the site. The auxiliary pools will be impounded by five earthfill dams and two ungated spillways flowing into the lower reservoir. Their combined usable storage will be 5800 acre-ft (7.2 × 10⁶ m³).

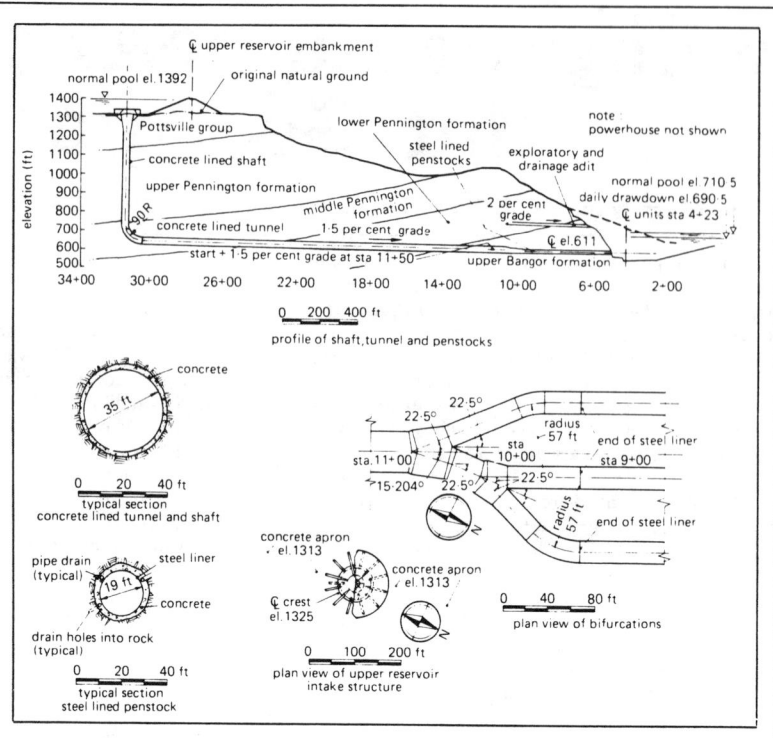

Fig. 2. Tunnel and shaft profile and typical sections.

The main dam south abutment and cutoff trench; drilling and grouting is in progress.

Upper reservoir intake/outlet

The upper reservoir intake/outlet is a circular 140 ft- (43 m)-diameter concrete structure, with a crest elevation of 1317 ft (402 m) and a vortex-suppression roof at el. 1341 ft (409 m).

Tunnel and penstocks

The waterway connecting the powerhouse to the upper and lower reservoirs comprises: a 567 ft- (173 m)-long, 35 ft- (10.7 m)-diameter, concrete-lined shaft; a 1935 ft- (590 m)-long, 35 ft- (10.7 m)-diameter horizontal tunnel; two concrete lined bifurcations; three 19 ft- (5.8 m)-diameter reinforced concrete-lined penstock connections of varying lengths; and, three 19 ft- (5.8 m)-diameter steel penstocks, each approximately 470 ft- (143 m)-long. The tunnel and shaft are shown in Fig. 2.

Powerhouse

The powerhouse will be a conventional surface (concrete) structure containing three units. Each unit will consist of a reversible pump/turbine rated at 263 MW at 613 ft (187 m) minimum net head, directly connected to a motor/generator rated at 314 MVA, with a 0.9 power factor (at 60°C rise). A hydraulically operated 128 in- (325 cm) diameter spherical valve will be installed at the junction of each penstock and spiral case extension. The full gate discharge of the three generating units will be approximately 18 500 ft^3/s (524 m^3/s) at the maximum operating head. The three unit transformers (each three-phase) and associated pull-off structure will be installed on the powerhouse roof. Fig. 3 shows the powerhouse configuration.

Transmission facilities

The unit transformers will be connected to an existing single-circuit 230 kV transmission system at a new switching substation to be constructed southwest of the auxiliary pool II. The transmission line will begin immediately behind the powerhouse near the base of Rock Mountain, and will extend approximately 2.7 miles (4.3 km) to the new substation.

Recreational facilities

In accordance with conditions of the FERC licence, recreational facilities will be provided for the general public at the project. It is estimated that there will be about 95 000 visits to the project and recreation facilities annually by the year 2000. The recreational opportunities provided at the site, primarily associated with and in the vicinity of the auxiliary pools, will include: overnight camping, picnicking, a beach suitable for swimming, non-motorized boating, hiking trails with viewing points, and a visitor's centre with displays regarding the area's resources and the Rocky Mountain project.

Construction schedule and risk management

General policies governing the establishment of the construction contract, the selection of contractors, and other details of the contracting process are based on Rural Electrification Administration (REA) guidelines and requirements as well as on accepted industry practice. All contracts for the project are of the fixed price type with lump sum or unit price bid items as far as possible. Oglethorpe Power Corporation will purchase the major equipment for the project under separate contracts, and supply this to the general construction contractor for installation.

Oglethorpe Power Corporation and its Engineer, Harza Engineering Company, chose to isolate areas of difficult or uncertain conditions of construction from the main general construction contract, so as to reduce the risk of problems which could affect completion of the project or cause complications relating to claims within the main contract. One of the first steps in risk reduction was to test the facilities already constructed and on which operation of the project would depend. The most important feature was the tunnel and shaft. Oglethorpe Power and Harza wanted to know what leakage rate could be expected from the tunnel when pressurized, the route of leakage water, what the piezometric head would be along leakage routes, and the behaviour of the tunnel when dewatered after a period of operation. It was important to know this before entering into the main general construction contract. To test the tunnel, steel bulkheads were installed at the portal end of the steel penstocks. The tunnel and shaft were filled with water to el. 1300 ft (396 m), which was almost to the top of the shaft. The resulting head on the tunnel was 681 ft (210 m), as opposed to the completed project maximum

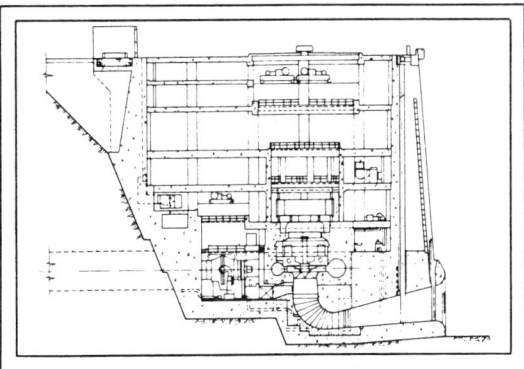

Fig. 3. Longitudinal section of the powerhouse.

normal operating net head of 702 ft (214 m). This test was carried out in late 1989 and early 1990, and demonstrated the competence of the tunnel. The maximum observed leakage rate was 302 gal/min (1143 l/m), and the increased water pressure in the rock mass surrounding the tunnel was within acceptable limits. More than 60 per cent of this leakage was accounted for by observation and measurement on the ground surface.

In addition to tunnel testing, contracts for all or part of the construction of the two dams were let separately ahead of the main contract. The south abutment of the lower reservoir main dam contained cavernous limestone features which required an extensive cutoff trench. A saddle dike in the vicinity (dam B) also required an extensive cutoff trench because of the same geological formation. This work was carried out in 1989 and 1990 under two separate construction contracts. Construction of dam B will continue into 1991.

The main general construction contract was signed in October 1990. Equipment procurement contracts are still pending for a number of the major equipment components, including: the spillway gates and hoists, governors for the pump turbines, switchgear and motor control centres, transformers, static frequency converters, cranes, draft tube gates, and trashracks. Two additional general construction contracts for the transmission line, substation and recreational facilities will be bid in 1991 and 1994 respectively.

As a further method of minimizing the risk of an unsuccessful or untimely start-up, the construction schedule was carefully planned to provide an adequate water supply at the time of project completion. The main construction contractor is required to complete the auxiliary reservoirs dams and spillways by 1 December 1991, so as to begin impounding. Completion of all the work for lower reservoir filling is to be carried out by 1 December 1993. The three units are scheduled to be fully operational by 15 December 1994, 15 February 1995, and 15 April 1995, respectively, with all construction activities except for recreational facilities scheduled for completion by 31 May 1995. The total project cost (to completion) at the time of writing is estimated to be approximately US$700 million.

Conclusion

Oglethorpe Power is a largely residential system with a capacity factor well below 50 per cent. This project proved to be the most economical solution to the various supply options, which included off-system power purchases, independent power producer purchases and the installation of combustion turbines.

GENERAL PLAN

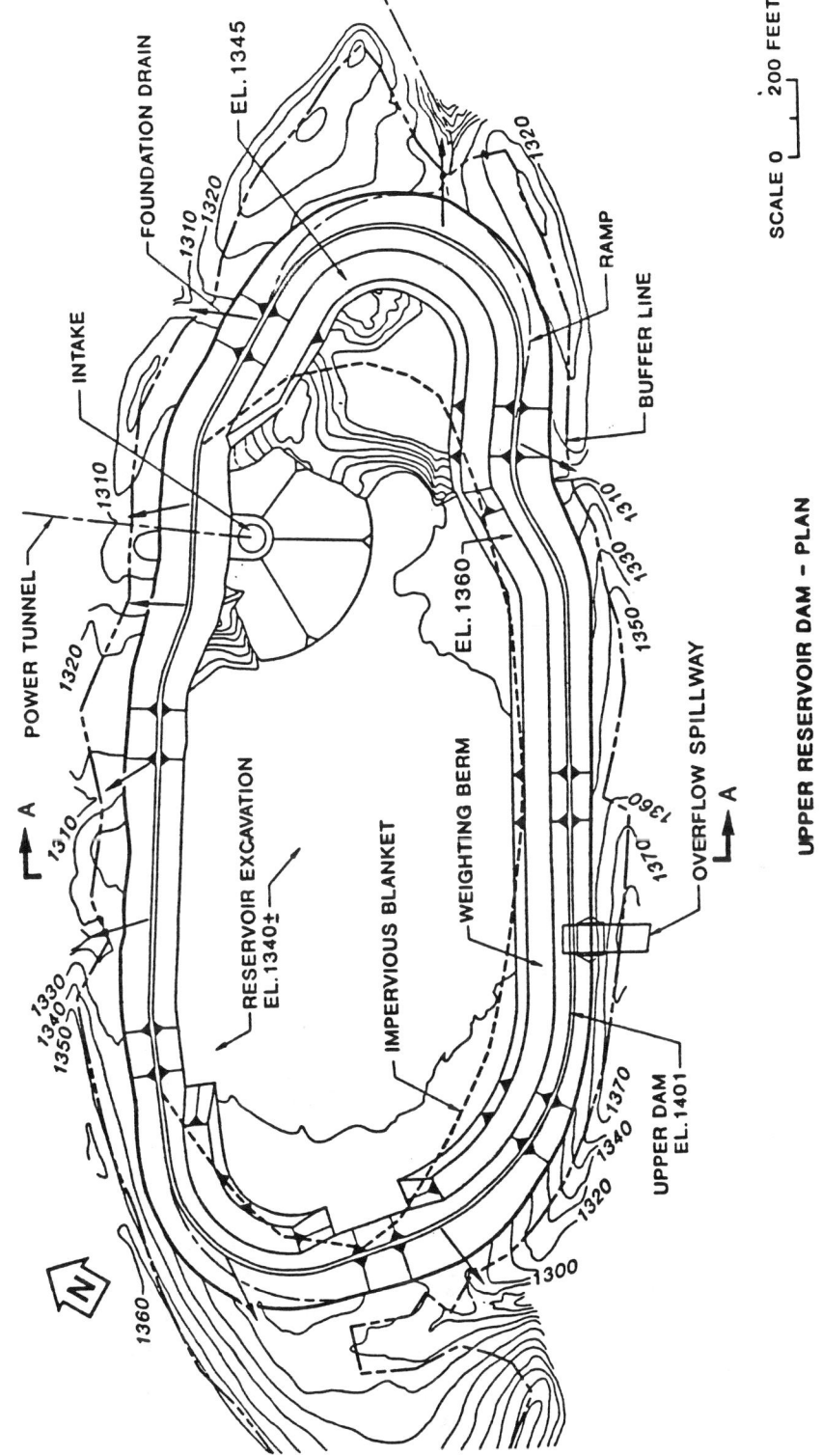

UPPER RESERVOIR DAM – PLAN

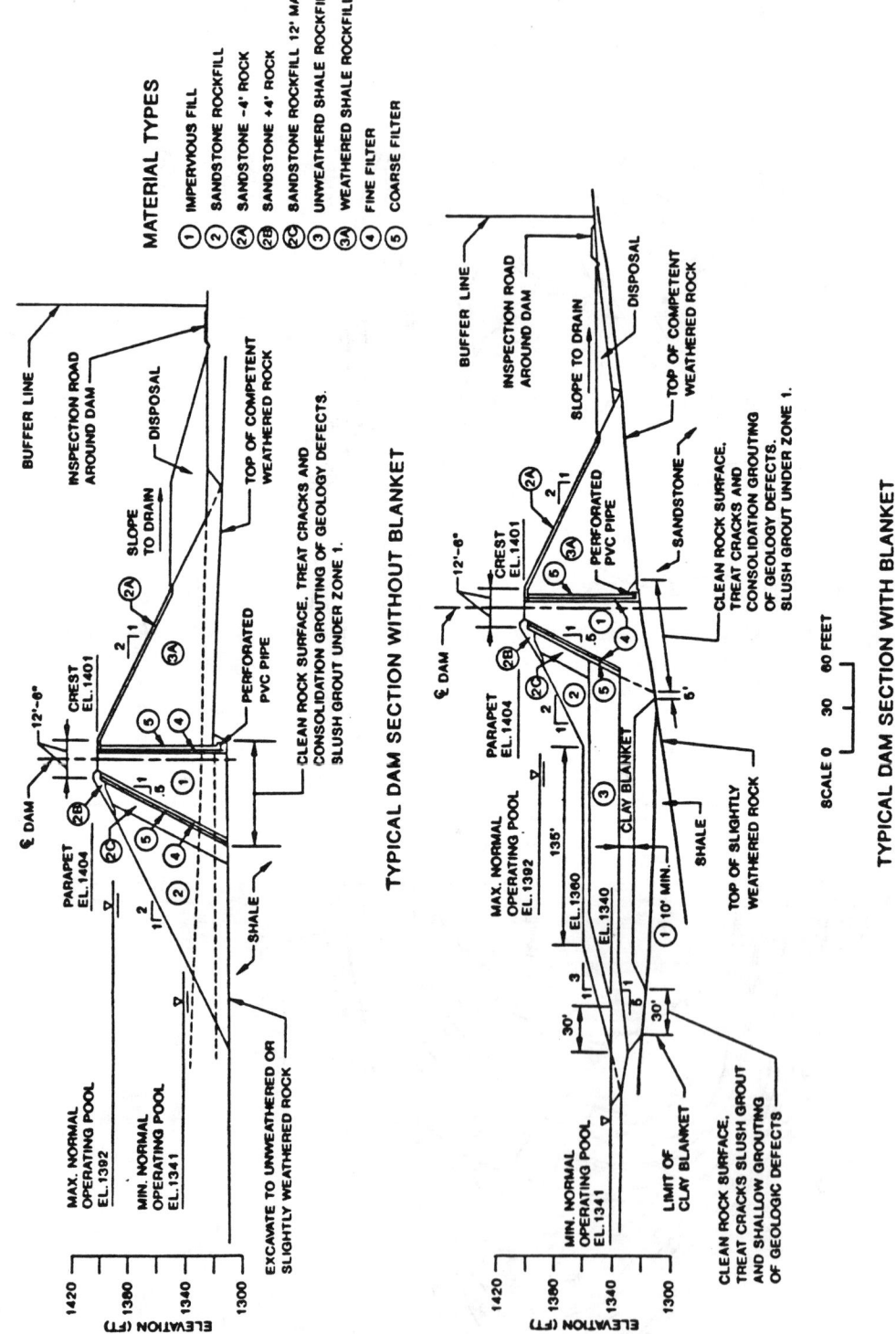

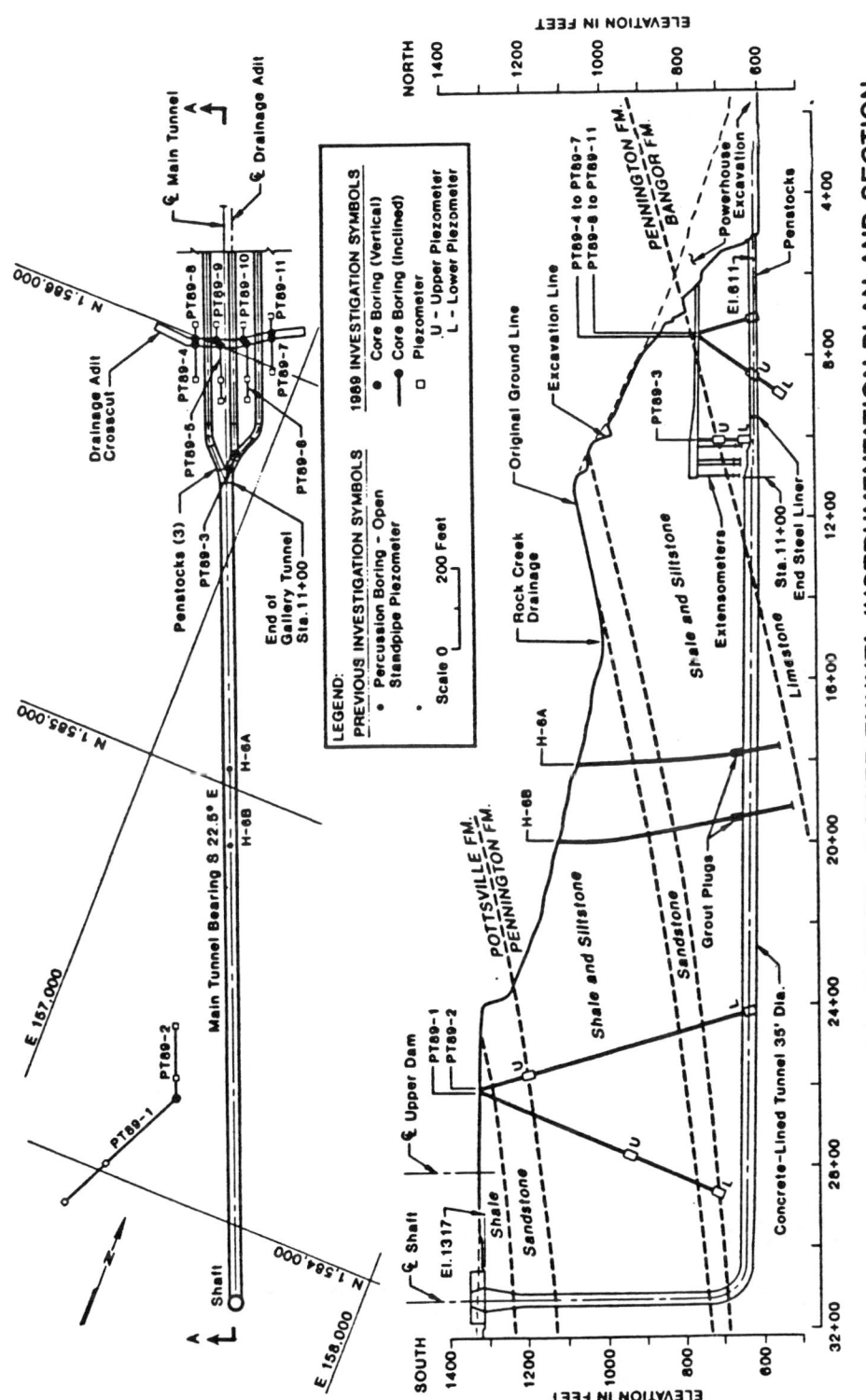

TUNNEL FILLING TEST - POWER TUNNEL INSTRUMENTATION PLAN AND SECTION

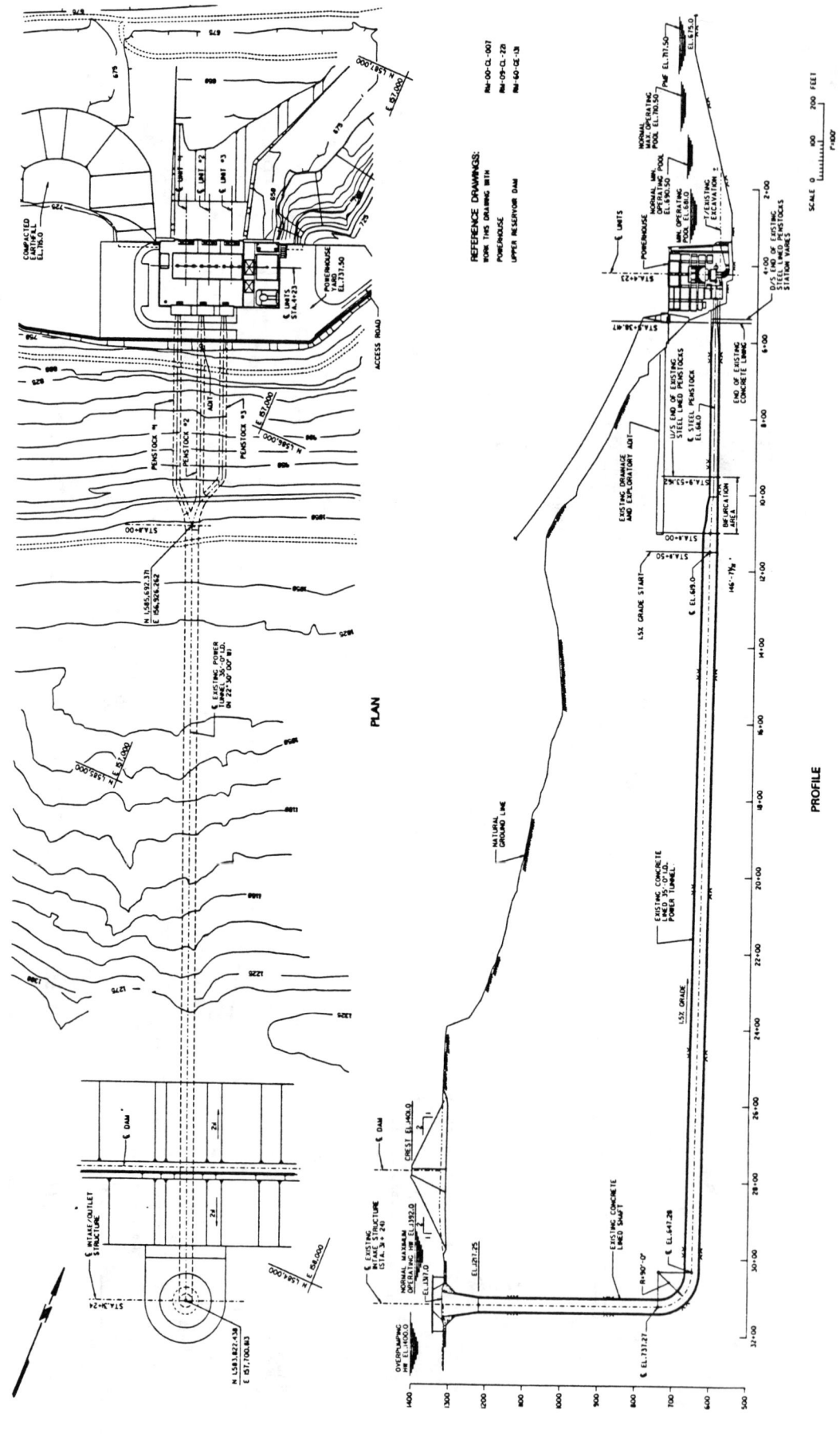

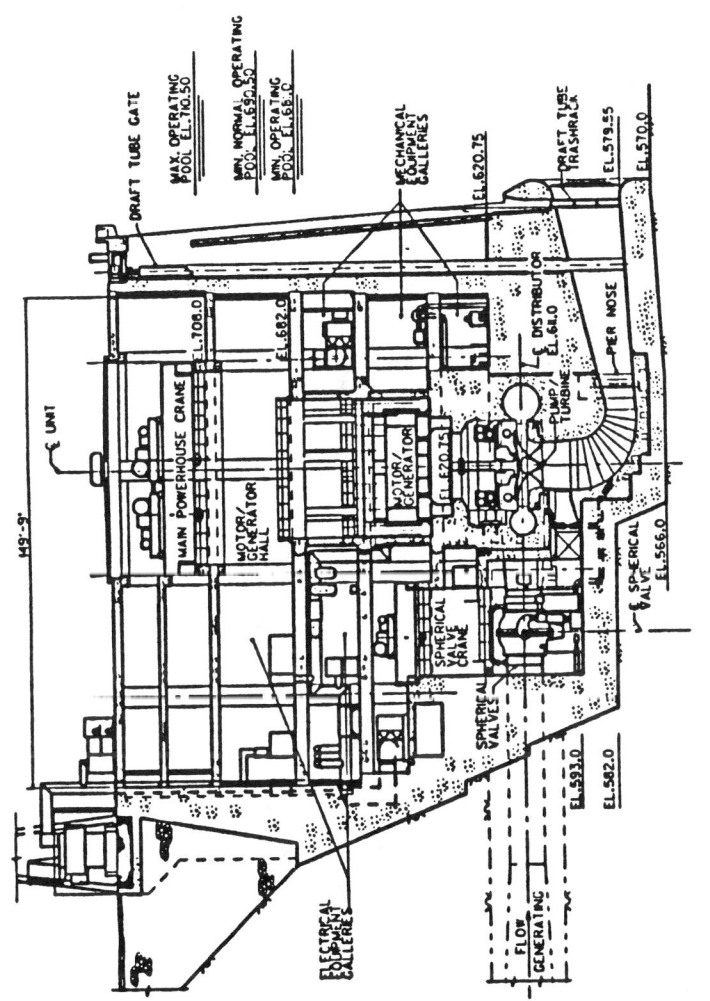

Powerstation Transverse Section

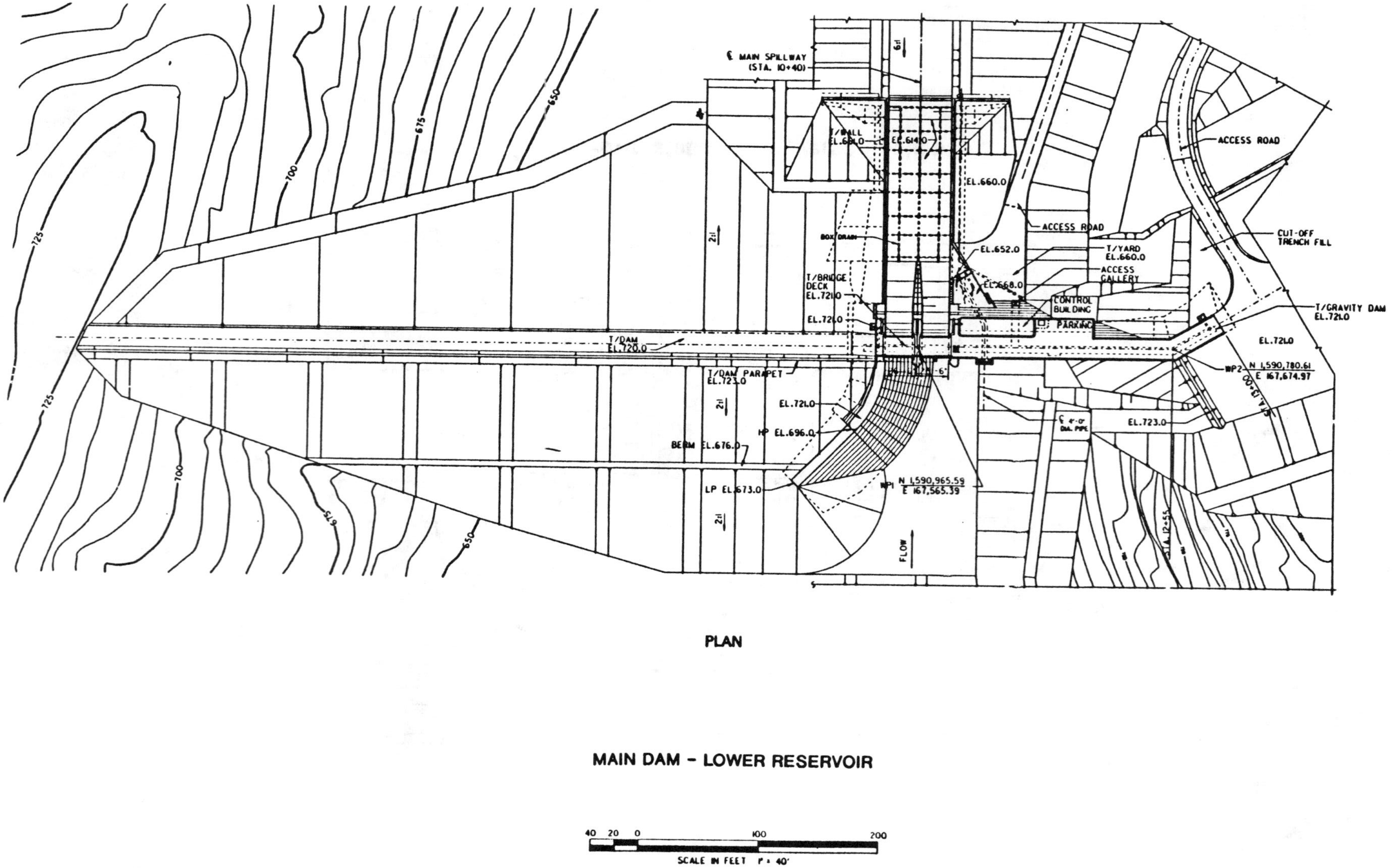

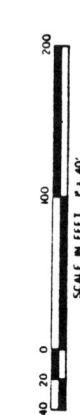

MAIN DAM – LOWER RESERVOIR

PLAN

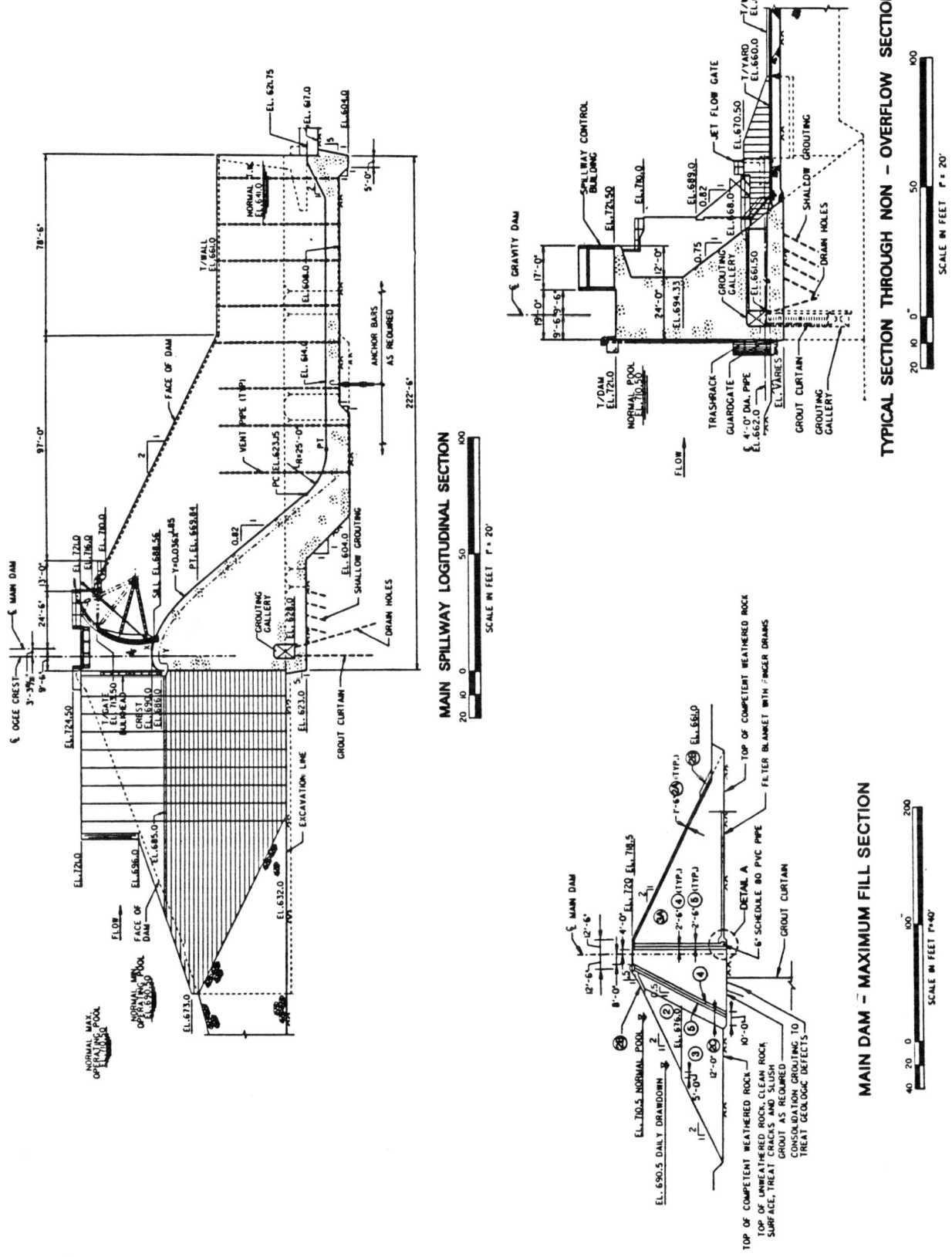

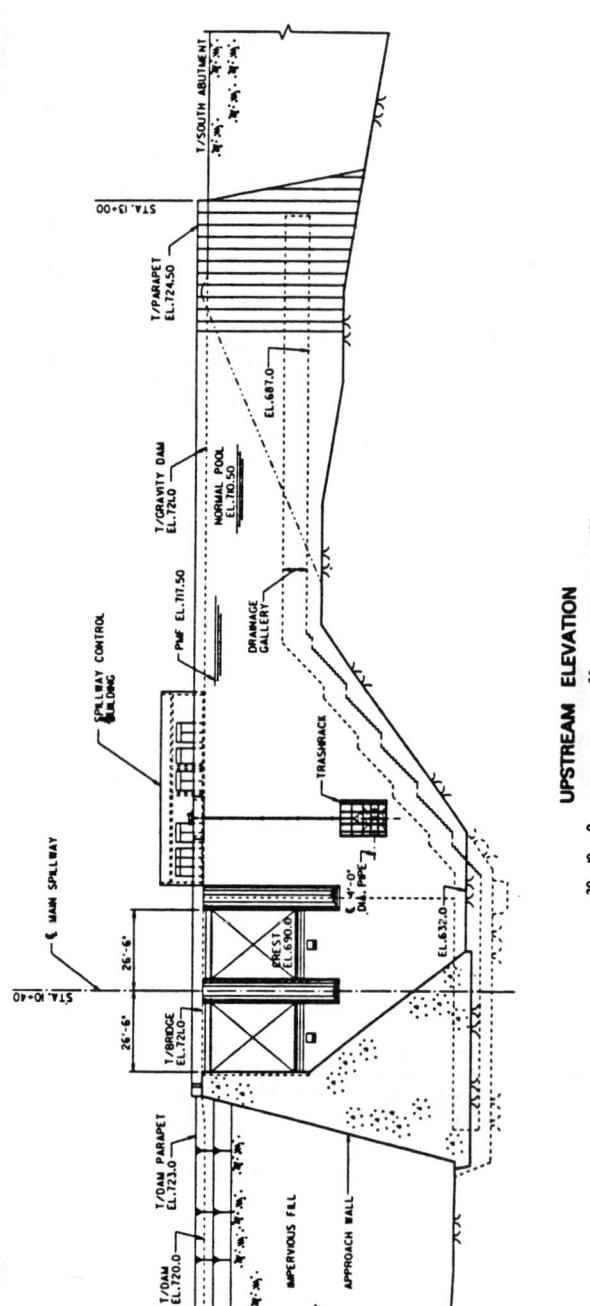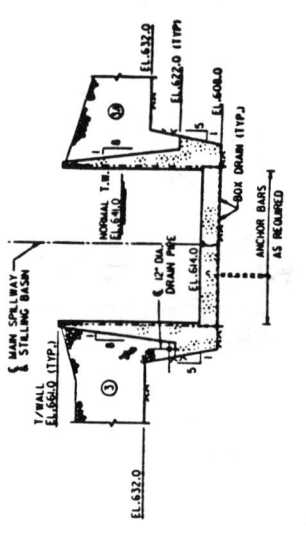

ROCKY MOUNTAIN PROJECT

Bibliography

1. "Rocky Mountain Project - Overview," D. Murphy, W.J. Bogdovitz, N.M. Hernandez.

2. "Rocky Mountain Project - Lower Tunnel Filling Test," D.A. Frey, D.E. Kleiner, B.C. Trammell.

3. "Rocky Mountain Project - Conceptual Design of Upper Reservoir," D.E. Kleiner, K.L. Wong, J. Boknecht.

4. "Rocky Mountain Project - PMF," N.L. Schickedanz, J.E. Lindell, B.C. Trammell, J. Boknecht.

5. "Rocky Mountain Project - Seepage Cut-Off in Solutioned Limestone of the Main Dam," A.H. Stukey, K.L. Wong, G. Taylor.

6. "Rocky Mountain Project - Intake/Discharge Channel Flow Simulation Study," C.Y. Wei, J. Boknecht.

7. "Rocky Mountain-Pump/Turbines and Related Mechanical Equipment," J.H.T. Sun, B.C. Goss.

All above papers: Proceedings, WATERPOWER '91, July 24-26, 1991, Denver, Colorado.

8. "Shaft and Tunnel Lining Techniques Used at the Rocky Mountain Pumped Storage Project," T.R. Harrell, W.D. Pressley. Proceedings, Vol. 2, Rapid Excavation and Tunnelling Conference, New York, N.Y., June 1985.

9. "675 MW at Rocky Mountain," R.A. Corso, USCOLD News, July 1977.

Plant Name: **ROCKY RIVER**

Plant location: New Milford, CT Litchfield County	Owner: Connecticut Light and Power Company P.O. Box 270 Hartford, Connecticut 06101
Rated capacity 32 MW Average static head 225 ft Plant efficiency % Stored energy 26,787 MWh Number of units 1	Designers: U.G.I. Contracting Company Philadephia, Pennsylvania
Construction time: 3 years, 4 months Construction cost: $232 per kW Price level: 1928 First commercial power: January 1929 FERC project number: 2576 River or water source: Housatonic River	Plant Manager/Superintendent: John M. MacDonald P.O. Box 270 New Milford, Connecticut 06776 (203) 355-6527

	UPPER RESERVOIR	LOWER RESERVOIR
DAM		
Type	Earth with wooden core wall	No lower reservoir, on Housatonic River
Height (ft)	100	
Crest length (ft)	952	
Volume (yd³)	368,000	
RESERVOIR		
Type	Constructed	On stream
Surface area (acres)	5,420	
Usable power storage (acre-ft)	53,480	
Power pool fluctuation (ft)	30.0	
Operating levels		
Maximum (ft)	430.0	200.0
Minimum (ft)	400.0	200.0
Drainage area (miles²)	35.0	
Seepage (ft³/s)		
SPILLWAY		
Design flood		
Return period (years)		
Flow (ft³/s)		
Capacity (ft³/s)		
Type	None	
Gates		
Number	None	None
Type		
Width (ft)		
Height (ft)		
OUTLET WORKS		
Discharge capacity (ft³/s)		
Number of water passages		None
Dimensions of water passages		
Height (ft)		
Width (ft)		
Diameter (ft)		
Type of gates		
Number of gates		None

Plant Name: **ROCKY RIVER**

Plant location
  New Milford, CT
  Litchfield County
	Owner: Connecticut Light and Power Company
	        P.O. Box 270
	        Hartford, Connecticut 06101

Rated capacity   32 MW
Average static head   68.6 m
Plant efficiency   %
Stored energy   26,787 MWh
Number of units   1
	Designers:
	   U.G.I. Contracting Company
	   Philadephia, Pennsylvania

Construction time: 3 years, 4 months
Construction cost: $232 per kW
Price level: 1928
First commercial power: January 1929
FERC project number: 2576
 	
River or water source: Housatonic River
	Plant Manager/Superintendent:
	   John M. MacDonald
	   P.O. Box 270
	   New Milford, Connecticut 06776
	
	   (203) 355-6527

	UPPER RESERVOIR	LOWER RESERVOIR
DAM		
Type	Earth with wooden core wall	No lower reservoir, on Housatonic River
Height (m)	30.5	
Crest length (m)	290.2	
Volume (m³)	281,203	
RESERVOIR		
Type	Constructed	On stream
Surface area (Mm²)	21.93	
Usable power storage (Mm³)	65.967	
Power pool fluctuation (m)	9.14	
Operating levels		
Maximum (m)	131.06	60.96
Minimum (m)	121.92	60.96
Drainage area (Mm²)	90.650	
Seepage (m³/s)		
SPILLWAY		
Design flood		
Return period (years)		
Flow (m³/s)		
Capacity (m³/s)		
Type	None	
Gates		
Number	None	None
Type		
Width (m)		
Height (m)		
OUTLET WORKS		
Discharge capacity (m³/s)		
Number of water passages		None
Dimensions of water passages		
Height (m)		
Width (m)		
Diameter (m)		
Type of gates		
Number of gates		None

Rocky River - Page 1 (Metric)

ROCKY RIVER

INTAKES	UPPER INTAKE	LOWER INTAKE
Number	1	2
Type	Circular concrete tower intake	Treadwell
Design discharge (ft³/s)	1,620	1,620
Gross inlet area (ft²)	1,170	119
(at trash racks)		
Bar racks		
spacing (in)	3.00	4.00
shape		
depth/thickness (in)		
diameter (in)		
Emergency gates		
number	None	None
height/width (ft)		
type		
Service gates		
number	1	2
height/width (ft)	18.00 / 18.50	8.00 / 5.00
type	Broom gate	Steel, slide
Bulkhead/stop logs (Y or N)	Y	Y
number of units serviced	3	
Hoists		
number	1	1
capacity (tons)	75	125
type	Electric, 16 HP motor	Electric powerhouse Niles crane

WATER PASSAGES	Upper Tunnel	Shaft	Lower Tunnel	Surge Tanks Upper	Surge Tanks Lower	Penstocks	Tailrace Tunnel
Number				1		1	1
Diameter (ft)				28.0		15.0	8.0
Length (ft)				76		1,677	
Maximum velocity (ft/s)							
Concrete lining thickness (in)							
Total length of concrete sections (ft)							
Steel liner Thickness							
Minimum (in)						0.50	
Maximum (in)						0.63	
Material grade						A36	
Total length of steel-lined sections (ft)						672	

Notes:
Penstocks: Penstocks are wood stave, steel, and concrete.

Rocky River - Page 2 (English)

ROCKY RIVER

INTAKES	UPPER INTAKE	LOWER INTAKE
Number	1	2
Type	Circular concrete tower intake	Treadwell
Design discharge (m³/s)	45.9	45.9
Gross inlet area (m²) (at trash racks)	108.7	11.1
Bar Racks:		
spacing (mm)	76	102
shape		
depth/thickness (mm)		
diameter (mm)		
Emergency gates		
number	None	None
height/width (m)		
type		
Service gates		
number	1	2
height/width (m)	5.486 / 5.639	2.438 / 1.524
type	Broom gate	Steel, slide
Bulkhead/stop logs (Y or N)	Y	Y
number of units serviced	3	
Hoists		
number	1	1
capacity (Mg)	68	113
type	Electric, 16 HP motor	Electric powerhouse Niles crane

WATER PASSAGES	Upper Tunnel	Shaft	Lower Tunnel	Surge Tanks Upper	Surge Tanks Lower	Penstocks	Tailrace Tunnel
Number				1		1	1
Diameter (m)				8.53		4.57	2.44
Length (m)				23.2		511.2	
Maximum velocity (m/s)							
Concrete lining thickness (m)							
Total length of concrete sections (m)							
Steel liner Thickness							
Minimum (mm)						13	
Maximum (mm)						16	
Material grade						A36	
Total length of steel-lined sections (m)						204.8	

Notes:
Penstocks: Penstocks are wood stave, steel, and concrete.

Rocky River - Page 2 (Metric)

ROCKY RIVER

POWERHOUSE and RELATED FEATURES

Powerhouse Structure
Type: Above ground
Length: 164 ft   Width: 36 ft   Height: 107 ft

Guard Valves
Number: 2   Diameter: 4.5 ft
Type: Butterfly

Transformers
Number: 2
Ratings: 500
Voltages: (kV) 1200 / 13.8

Generator
Rating generating (MVA): 314.0   Rating pumping (MVA):
Insulation type: Class B
Starting method: 1/2 voltage swap over after up-to-speed
Starting equipment:

Runners
Material: Bronze
Minimum unit submergence: 5.8 ft
WR^2: 9,375,000 (lbf x ft²)
Manufacturer: Worthington
Model test by: Worthington

	Reversible Runners	Reversible Motor/Generator	
Number	2	3	
Diameter (ft)	4.50	Rotor	Stator
rpm synchronous	327.0		225.0
rpm overspeed	634.0		337.5
Type	Francis		Semi-umbrella

Information on Runners

Condition:	Gross Head (ft) Generating	Gross Head (ft) Pumping	Capacity (MW) Generating	Capacity (MW) Pumping	Discharge (ft³/s) Generating	Discharge (ft³/s) Pumping	Turbine/Pump Eff.(%) Generating	Turbine/Pump Eff.(%) Pumping
Maximum head & maximum power	230	230	32	7	2,080	500	80.0	80.0
Minimum head & maximum power	210	210	31	6	2,080	500	80.0	91.0

Note: Data in the above table are based on operating data.

Condition:	Net Head (ft) Generating	Net Head (ft) Pumping	Capacity (MW) Generating	Capacity (MW) Pumping	Discharge (ft³/s) Generating	Discharge (ft³/s) Pumping	Turbine/Pump Eff.(%) Generating	Turbine/Pump Eff.(%) Pumping
Rated head @ best gate	240	240	32	7	1,870	250	70.0	78.0

Note: Data in the above table are based on operating data.

ROCKY RIVER

POWERHOUSE and RELATED FEATURES

Powerhouse Structure
Type: Above ground
Length: 50.0 m Width: 11.0 m Height: 32.6 m

Guard Valves
Number: 2 Diameter: 1.37 m
Type: Butterfly

Transformers
Number: 2
Ratings: 500
Voltages: (kV) 1200 / 13.8

Generator
Rating generating (MVA): 314.0 Rating pumping (MVA):
Insulation type: Class B
Starting method: 1/2 voltage swap over after up-to-speed
Starting equipment:

Runners
Material: Bronze
Minimum unit submergence: 1.77 m
WR^2: 3,875,000 (Newtons x m^2)
Manufacturer: Worthington
Model test by: Worthington

	Reversible Runners	Reversible Motor/Generator		
Number	2			3
Diameter m	1.37	Rotor	Stator	
rpm synchronous	327.0			225.0
rpm overspeed	634.0			337.5
Type	Francis			Semi-umbrella

Information on Runners

Condition:	Gross Head (m)		Capacity (MW)		Discharge (m^3/s)		Turbine/Pump Eff.(%)	
	Generating	Pumping	Generating	Pumping	Generating	Pumping	Generating	Pumping
Maximum head & maximum power	70.1	70.1	32	7	58.9	14.2	80.0	80.0
Minimum head & maximum power	64.0	64.0	31	6	58.9	14.2	80.0	91.0

Note: Data in the above table are based on operating data.

Condition:	Net Head (m)		Capacity (MW)		Discharge (m^3/s)		Turbine/Pump Eff.(%)	
	Generating	Pumping	Generating	Pumping	Generating	Pumping	Generating	Pumping
Rated head @ best gate	73.2	73.2	32	7	53.0	7.1	70.0	78.0

Note: Data in the above table are based on operating data.

ROCKY RIVER

Plant Data:

Average GWh generating per year:	12
Average GWh pumping per year:	9
Starting time from standstill (s):	300
Changeover time pumping to generating (min):	300
Planned/scheduled time between major overhauls (years):	20
Outage time required per unit during major overhauls (weeks):	24
Representative plant availability (%):	98.0
Representative planned outages (weeks per year):	1

Miscelleneous Notes:

This is a seasonal pumped storage project with one generator and two pumps.

The penstock for this plant was reported to be 511 m (1,677 ft) long. Upstream from the surge tank, the wood stove penstock is 306.9 m (1,007 ft) long with an inside diameter of 4.6 m (15 ft). Downstream from the surge tank, the steel penstock has variable diameters of 3.96 m (13 ft), 3.66 m (12 ft), and 3.35 m (11 ft).

Cavitation Experience:

Very little cavitation has occurred.

Significant or Unique Problems:

Because of ice suspended in the water, pumping cannot occur during certain cold weather temperatures. The ise plugs up the cooling water strainers and also plugs up the racks.

List of Licenses Required:

FERC.

ENVIRONMENTAL FEATURES

Recreation:

The lake is ringed with homes and is heavily used for boating and fishing. Picknicking, fishing and nature trails are provided at Dikes Point.

Fish and Wildlife:

The lake has many species of fish and the open areas have deer, wild turkeys, and many other forms of wildlife.

Social:

Aerial View of Outlet Works and Powerhouse

Rocky River Pumped Storage Plant

Plant Name: **RUSSEL DAM**

Plant location: Elberton, GA		Owner: Savannah District, Corps of Engineers
		100 W. Oglethorpe Avenue
		Savannah, Georgia 31401
Rated capacity	360 MW	Designers:
Average static head	144 ft	
Plant efficiency	%	
Stored energy	MWh	
Number of units	4	
Construction time:		Plant Manager/Superintendent:
Construction cost:		
Price level:		
First commercial power: June 1987		
FERC project number:		
River or water source: Savannah River		

DAM	UPPER RESERVOIR	LOWER RESERVOIR
Type	Concrete	Zoned fill, concrete and earth
Height (ft)	195	170
Crest length (ft)	1,834	5,680
Volume (yd³)	4,499,000	9,550,000
RESERVOIR		
Type	Constructed	Constructed
Surface area (acres)	26,653	71,100
Usable power storage (acre-ft)	126,864	1,045,000
Power pool fluctuation (ft)	5.0	18.0
Operating levels		
Maximum (ft)		
Minimum (ft)		
Drainage area (miles²)	283.0	6,144.0
Seepage (ft³/s)	1.000	
SPILLWAY		
Design flood		
Return period (years)		
Flow (ft³/s)	801,500	1,015,000
Capacity (ft³/s)	801,500	1,015,000
Type	Ogee concrete gravity	
Gates		
Number	10	23
Type	Tainter	Tainter
Width (ft)	50.00	40.00
Height (ft)	45.00	35.00
OUTLET WORKS		
Discharge capacity (ft³/s)		
Number of water passages	None	None
Dimensions of water passages		
Height (ft)		
Width (ft)		
Diameter (ft)		
Type of gates		
Number of gates	None	None

Plant Name: RUSSEL DAM

Plant location: Elberton, GA

Owner: Savannah District, Corps of Engineers
100 W. Oglethorpe Avenue
Savannah, Georgia 31401

Rated capacity   360 MW
Average static head   43.9 m
Plant efficiency   %
Stored energy   MWh
Number of units   4

Designers:

Construction time:
Construction cost:
Price level:
First commercial power: June 1987
FERC project number:

Plant Manager/Superintendent:

River or water source: Savannah River

	UPPER RESERVOIR	LOWER RESERVOIR
DAM		
Type	Concrete	Zoned fill, concrete and earth
Height (m)	59.4	51.8
Crest length (m)	559.0	1,731.3
Volume (m³)	3,439,734	7,301,502
RESERVOIR		
Type	Constructed	Constructed
Surface area (Mm²)	107.86	287.74
Usable power storage (Mm³)	156.485	1,288.996
Power pool fluctuation (m)	1.52	5.49
Operating levels		
Maximum (m)		
Minimum (m)		
Drainage area (Mm²)	732.971	15,912.976
Seepage (m³/s)	0.0283	
SPILLWAY		
Design flood		
Return period (years)		
Flow (m³/s)	22,696	28,742
Capacity (m³/s)	22,696	28,742
Type	Ogee concrete gravity	
Gates		
Number	10	23
Type	Tainter	Tainter
Width (m)	15.240	12.192
Height (m)	13.716	10.668
OUTLET WORKS		
Discharge capacity (m³/s)		
Number of water passages	None	None
Dimensions of water passages		
Height (m)		
Width (m)		
Diameter (m)		
Type of gates		
Number of gates	None	None

Russel Dam - Page 1 (Metric)

RUSSEL DAM

INTAKES	UPPER INTAKE	LOWER INTAKE
Number	4	
Type		
Design discharge (ft³/s)	30,000	
Gross inlet area (ft²)		
(at trash racks)		
Bar racks		
spacing (in)		
shape		
depth/thickness (in)		
diameter (in)		
Emergency gates		
number		
height/width (ft)		
type		
Service gates		
number		
height/width (ft)		
type		
Bulkhead/stop logs(Y or N)		
number of units serviced		
Hoists		
number		
capacity (tons)		
type		

WATER PASSAGES	Upper Tunnel	Shaft	Lower Tunnel	Surge Tanks Upper	Lower	Penstocks	Tailrace Tunnel
Number						4	
Diameter (ft)						26.0	
Length (ft)							
Maximum velocity (ft/s)							
Concrete lining thickness (in)							
Total length of concrete sections (ft)							
Steel liner							
Thickness							
Minimum (in)							
Maximum (in)							
Material grade							
Total length of steel-lined sections (ft)							

Notes:

RUSSEL DAM

INTAKES	UPPER INTAKE	LOWER INTAKE
Number	4	
Type		
Design discharge (m³/s)	849.5	
Gross inlet area (m²) (at trash racks)		
Bar Racks:		
spacing (mm)		
shape		
depth/thickness (mm)		
diameter (mm)		
Emergency gates		
number		
height/width (m)		
type		
Service gates		
number		
height/width (m)		
type		
Bulkhead/stop logs (Y or N)		
number of units serviced		
Hoists		
number		
capacity (Mg)		
type		

WATER PASSAGES	Upper Tunnel	Shaft	Lower Tunnel	Surge Tanks Upper	Lower	Penstocks	Tailrace Tunnel
Number						4	
Diameter (m)						7.93	
Length (m)							
Maximum velocity (m/s)							
Concrete lining thickness (m)							
Total length of concrete sections (m)							
Steel liner							
Thickness							
Minimum (mm)							
Maximum (mm)							
Material grade							
Total length of steel-lined sections (m)							

Notes:

RUSSEL DAM

POWERHOUSE and RELATED FEATURES

Powerhouse Structure
Type:
Length: 0 ft   Width: 0 ft   Height: 0 ft

Guard Valves
Number: None   Diameter: ft
Type:

Transformers
Number:
Ratings:
Voltages: (kV)

Generator
Rating generating (MVA):   Rating pumping (MVA):
Insulation type:
Starting method:
Starting equipment:

Runners
Material:
Minimum unit submergence: 0.0 ft
WR²:
Manufacturer: Vevey Dominion
Model test by:

	Reversible Runners	Reversible Motor/Generator	
		Rotor	Stator
Number	4		
Diameter (ft)			
rpm synchronous	120.0		
rpm overspeed			
Type	Francis		

Information on Runners

Condition:	Gross Head (ft)		Capacity (MW)		Discharge (ft³/s)		Turbine/Pump Eff.(%)	
	Generating	Pumping	Generating	Pumping	Generating	Pumping	Generating	Pumping
Maximum head & maximum power								
Minimum head & maximum power								

Condition:	Net Head (ft)		Capacity (MW)		Discharge (ft³/s)		Turbine/Pump Eff.(%)	
	Generating	Pumping	Generating	Pumping	Generating	Pumping	Generating	Pumping
Rated head @ best gate	157	161	93	88				

Russel Dam - Page 3 (English)

RUSSEL DAM

POWERHOUSE and RELATED FEATURES

Powerhouse Structure
Type:
Length: m Width: m Height: m

Guard Valves
Number: None Diameter: m
Type:

Transformers
Number:
Ratings:
Voltages: (kV)

Generator
Rating generating (MVA): Rating pumping (MVA):
Insulation type:
Starting method:
Starting equipment:

Runners
Material:
Minimum unit submergence: m
WR^2:
Manufacturer: Vevey Dominion
Model test by:

	Reversible Runners	Reversible Motor/Generator	
		Rotor	Stator
Number	4		
Diameter m			
rpm synchronous	120.0		
rpm overspeed			
Type	Francis		

Information on Runners

Condition:	Gross Head (m)		Capacity (MW)		Discharge (m³/s)		Turbine/Pump Eff.(%)	
	Generating	Pumping	Generating	Pumping	Generating	Pumping	Generating	Pumping
Maximum head & maximum power								
Minimum head & maximum power								

Condition:	Net Head (m)		Capacity (MW)		Discharge (m³/s)		Turbine/Pump Eff.(%)	
	Generating	Pumping	Generating	Pumping	Generating	Pumping	Generating	Pumping
Rated head @ best gate	47.9	49.1	93	88				

Russel Dam - Page 3 (Metric)

RUSSEL DAM

Plant Data:
   Average GWh generating per year:
   Average GWh pumping per year:
   Starting time from standstill (s):
   Changeover time pumping to generating (min):
   Planned/scheduled time
      between major overhauls (years):
   Outage time required per unit
      during major overhauls (weeks):
   Representative plant availability (%):
   Representative planned outages (weeks per year):

Miscelleneous Notes:
   The upper dam is constructed of 764,000 m³ (999,000 yd³) of concrete and 2,676,000 m³ (3,500,000 yd³) of earthen materials.

   The lower dam is constructed of 803,000 m³ (1,050,000 yd³) of concrete and 2,676,000 m³ (3,500,000 yd³) of earthen materials.

Cavitation Experience:

Significant or Unique Problems:

List of Licenses Required:

ENVIRONMENTAL FEATURES

Recreation:
   Recreation facilities include a fishing pier, boat ramps, and a visitor center.

Fish and Wildlife:
   Bar screens being designed for fish protection.

Social:

Plant Name: **SALINA**

Plant location:
Salina, OK
Mayes County

Rated capacity 260 MW
Average static head 225 ft
Plant efficiency 70.00 %
Stored energy 2,650 MWh
Number of units 6

Construction time: 2 years, 5 months
Construction cost: $1,000 per kW
Price level: 1971
First commercial power: May 1968
FERC project number: 2524

Owner: Grand River Dam Authority
P.O. Box 409
Vanita, Oklahoma 74301-0409

Designers:
Benham Holoway, Tulsa, Oklahoma

Plant Manager/Superintendent:
Darrel B. Burroughs
Box 234
Langley, Oklahoma 74350

(918) 434-5920

River or water source: Saline Creek, Hudson Lake, and Neosho River

	UPPER RESERVOIR	LOWER RESERVOIR
DAM		
Type	Rockfill and earth	Rockfill earth concrete constructed
Height (ft)	200	
Crest length (ft)	2,300	
Volume (yd³)	3,700,000	
RESERVOIR		
Type	Constructed	Constructed, Lake Hudson
Surface area (acres)	785	
Usable power storage (acre-ft)		
Power pool fluctuation (ft)	20.0	17.0
Operating levels		
Maximum (ft)	865.0	636.0
Minimum (ft)	845.0	619.0
Drainage area (miles²)	0.0	
Seepage (ft³/s)	28.620	
SPILLWAY		
Design flood		
Return period (years)		
Flow (ft³/s)		
Capacity (ft³/s)		
Type	None	
Gates		
Number		
Type	None	
Width (ft)		
Height (ft)		
OUTLET WORKS		
Discharge capacity (ft³/s)		
Number of water passages	None	
Dimensions of water passages		
Height (ft)		
Width (ft)		
Diameter (ft)		
Type of gates		
Number of gates	None	

Salina - Page 1 (English)

Plant Name: **SALINA**

Plant location
Salina, OK
Mayes County

Rated capacity	260 MW
Average static head	68.6 m
Plant efficiency	70.00 %
Stored energy	2,650 MWh
Number of units	6

Construction time: 2 years, 5 months
Construction cost: $1,000 per kW
Price level: 1971
First commercial power: May 1968
FERC project number: 2524

Owner: Grand River Dam Authority
P.O. Box 409
Vanita, Oklahoma 74301-0409

Designers:
Benham Holoway, Tulsa, Oklahoma

Plant Manager/Superintendent:
Darrel B. Burroughs
Box 234
Langley, Oklahoma 74350

(918) 434-5920

River or water source: Saline Creek, Hudson Lake, and Neosho River

	UPPER RESERVOIR	LOWER RESERVOIR
DAM		
Type	Rockfill and earth	Rockfill earth concrete constructed
Height (m)	61.0	
Crest length (m)	701.0	
Volume (m³)	2,828,854	
RESERVOIR		
Type	Constructed	Constructed, Lake Hudson
Surface area (Mm²)	3.18	
Usable power storage (Mm³)		
Power pool fluctuation (m)	6.10	5.18
Operating levels		
Maximum (m)	263.65	193.85
Minimum (m)	257.56	188.67
Drainage area (Mm²)		
Seepage (m³/s)	0.8104	
SPILLWAY		
Design flood		
Return period (years)		
Flow (m³/s)		
Capacity (m³/s)		
Type	None	
Gates		
Number		
Type	None	
Width (m)		
Height (m)		
OUTLET WORKS		
Discharge capacity (m³/s)		
Number of water passages	None	
Dimensions of water passages		
Height (m)		
Width (m)		
Diameter (m)		
Type of gates		
Number of gates	None	

SALINA

INTAKES	UPPER INTAKE	LOWER INTAKE
Number	6	
Type		
Design discharge (ft³/s)		
Gross inlet area (ft²)	154	
(at trash racks)		
Bar racks		
spacing (in)	3.75	
shape	Rectangular	
depth/thickness (in)	5.00 / 0.50	
diameter (in)		
Emergency gates		
number	None	
height/width (ft)		
type		
Service gates		
number	None	
height/width (ft)		
type		
Bulkhead/stop logs (Y or N)	Y	
number of units serviced	6	
Hoists		
number	1	
capacity (tons)		
type	Overhead	

WATER PASSAGES	Upper Tunnel	Shaft	Lower Tunnel	Surge Tanks Upper	Lower	Penstocks	Tailrace Tunnel
Number						6	
Diameter (ft)						14.0	
Length (ft)						640	
Maximum velocity (ft/s)							
Concrete lining thickness (in)							
Total length of concrete sections (ft)							
Steel liner Thickness							
Minimum (in)						0.50	
Maximum (in)						0.75	
Material grade						Steel	
Total length of steel-lined sections (ft)						640	

Notes:

SALINA

INTAKES	UPPER INTAKE	LOWER INTAKE
Number	6	
Type		
Design discharge (m³/s)		
Gross inlet area (m²) (at trash racks)	14.3	
Bar Racks:		
spacing (mm)	95	
shape	Rectangular	
depth/thickness (mm)	127 / 13	
diameter (mm)		
Emergency gates		
number	None	
height/width (m)		
type		
Service gates		
number	None	
height/width (m)		
type		
Bulkhead/stop logs (Y or N)	Y	
number of units serviced	6	
Hoists		
number	1	
capacity (Mg)		
type	Overhead	

WATER PASSAGES	Upper Tunnel	Shaft	Lower Tunnel	Surge Tanks Upper	Surge Tanks Lower	Penstocks	Tailrace Tunnel
Number						6	
Diameter (m)						4.27	
Length (m)						195.1	
Maximum velocity (m/s)							
Concrete lining thickness (m)							
Total length of concrete sections (m)							
Steel liner							
Thickness							
Minimum (mm)						13	
Maximum (mm)						19	
Material grade						Steel	
Total length of steel-lined sections (m)						195.1	

Notes:

SALINA

POWERHOUSE and RELATED FEATURES

Powerhouse Structure
Type:
Length: 390 ft Width: 92 ft Height: 61 ft

Guard Valves
Number: 6 Diameter: 13.4 ft
Type: Butterfly

Transformers
Number: 6
Ratings:
Voltages: (kV) 161

Generator
Rating generating (MVA): 48.0 Rating pumping (MVA): 48.0
Insulation type:
Starting method: Cross line
Starting equipment: Cross line

Runners
Material: Steel
Minimum unit submergence: 5.0 ft
WR^2:
Manufacturer: Allis Chalmers
Model test by: Allis Chalmers

	Reversible Runners	Reversible Motor/Generator			
Number	6				6
Diameter (ft)	15.00	Rotor	18.40	Stator	18.48
rpm synchronous	1/1.4				1/1.4
rpm overspeed	197.0				197.0
Type	Francis				Westinghouse

Information on Runners

	Gross Head (ft)		Capacity (MW)		Discharge (ft³/s)		Turbine/Pump Eff.(%)	
Condition:	Generating	Pumping	Generating	Pumping	Generating	Pumping	Generating	Pumping
Maximum head & maximum power	245	245	43	48	2,800	2,000	83.0	86.0
Minimum head & maximum power	225	215	30	48	2,000	2,000	80.0	86.0

	Net Head (ft)		Capacity (MW)		Discharge (ft³/s)		Turbine/Pump Eff.(%)	
Condition:	Generating	Pumping	Generating	Pumping	Generating	Pumping	Generating	Pumping
Rated head @ best gate	225	245	42	48	2,800	1,900	83.0	86.0

Salina - Page 3 (English)

SALINA

POWERHOUSE and RELATED FEATURES

Powerhouse Structure
Type:
Length: 118.9 m   Width: 28.0 m   Height: 18.7 m

Guard Valves
Number: 6   Diameter: 4.09 m
Type: Butterfly

Transformers
Number: 6
Ratings:
Voltages: (kV) 161

Generator
Rating generating (MVA): 48.0   Rating pumping (MVA): 48.0
Insulation type:
Starting method: Cross line
Starting equipment: Cross line

Runners
Material: Steel
Minimum unit submergence: 1.52 m
WR^2:
Manufacturer: Allis Chalmers
Model test by: Allis Chalmers

	Reversible Runners	Reversible Motor/Generator		
Number	6			6
Diameter m	4.57	Rotor 5.607	Stator	5.632
rpm synchronous	171.4			171.4
rpm overspeed	197.0			197.0
Type	Francis			Westinghouse

Information on Runners

Condition:	Gross Head (m) Generating	Gross Head (m) Pumping	Capacity (MW) Generating	Capacity (MW) Pumping	Discharge (m³/s) Generating	Discharge (m³/s) Pumping	Turbine/Pump Eff.(%) Generating	Turbine/Pump Eff.(%) Pumping
Maximum head & maximum power	74.7	74.7	43	48	79.3	56.6	83.0	86.0
Minimum head & maximum power	68.6	65.5	30	48	56.6	56.6	80.0	86.0

Condition:	Net Head (m) Generating	Net Head (m) Pumping	Capacity (MW) Generating	Capacity (MW) Pumping	Discharge (m³/s) Generating	Discharge (m³/s) Pumping	Turbine/Pump Eff.(%) Generating	Turbine/Pump Eff.(%) Pumping
Rated head @ best gate	68.6	74.7	42	48	79.3	53.8	83.0	86.0

Salina - Page 3 (Metric)

SALINA

Plant Data:

Average GWh generating per year:	59
Average GWh pumping per year:	102
Starting time from standstill (s):	120
Changeover time pumping to generating (min):	5
Planned/scheduled time between major overhauls (years):	20
Outage time required per unit during major overhauls (weeks):	52
Representative plant availability (%):	
Representative planned outages (weeks per year):	

Miscelleneous Notes:
None.

Cavitation Experience:
Cavitation repair was required after 20 years.

Significant or Unique Problems:
Problems have included excessive unit run-out, along with significant guide vane vibration and movement.

List of Licenses Required:
FERC.

ENVIRONMENTAL FEATURES

Recreation:
Boat ramps are readily available and the lake provide good fishing for bass, crappie, and other fish. No powered motor boats are allowed.

Fish and Wildlife:

Social:

CASE STUDY IN THE REHABILITATION
OF A PUMPED/STORAGE INSTALLATION
SALINA POWERHOUSE
GRAND RIVER DAM AUTHORITY

JAMES L. KEPLER *

KEELING T. McGAUGHEY **

Abstract

This paper presents a specific case study in the rehabilitation of a pump/turbine for the Grand River Dam Authority at its Salina plant. The study addresses problems such as cavitation repair, bushing replacement, runner seal replacement and unit alignment. State-of-the-art design improvements, coupled with experience factors, were key elements needed to assure a successful project rehabilitation.

*   Senior Project Manager/Sales Engineer, Rehabilitation Services, Voith Hydro, Inc., P. O. Box 712, York, PA, 17405.

** Assistant General Manager of Engineering and Operations, Grand River Authority, 707 South Wilson, P.O. Box 409, Vinita, OK 74301-0409.

Background

The pump/turbines at the Salina Pumped Storage Project (See Figures 1 & 2) were manufactured by Allis-Chalmers Corporation, now Voith Hydro, Inc., in 1967. The powerhouse contains six (6) right hand, 130" eye diameter, Francis Type reversible pump/turbines.

Figure 1 - Salina Project

Operating Conditions

	Generating	Pumping
Rated Output	60,000 HP	2320 cfs
Rated Net Head	225 FT	245 FT
Rated Speed	171.4 RPM	171.4 RPM
Operation Duties	19,157 Hours	22,121 Hours

The Salina Project is located in Oklahoma near the town of Salina, approximately sixty (60) miles northeast of Tulsa.

Unit Condition

From the initial 1967 installation, the operating characteristics of these machines had deteriorated considerably. Excessive unit run-out was noted, along with significant wicket gate vibration and movement.

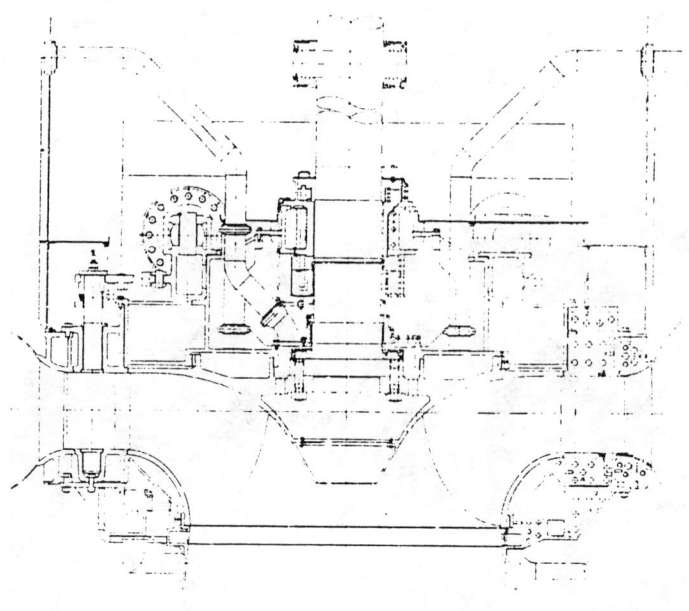

Figure 2 - Section of Pump/Turbine

Major problem areas noted before disassembly were: broken lower rotating wearing ring, excessive gate end seal wear, excessive gate vertical seal clearance, wicket gate bushing wear and runner hydraulic surface irregularities. Major reasons for deterioration over the past twenty-two (22) years can be attributed to the deposit of approximately 2,000 pounds, per unit, of repair weld on the runner buckets and band. (See Figure 3) These weld overlay repairs were made without surface contour grinding to the original hydraulic shape.

The additional weld material and rough surfaces contributed significantly in propagating additional cavitation damage, leading to mechanical and hydraulic imbalance. In addition, the wicket gate vertical seal joints had been repair welded in-place, causing gate warpage which contributed to increased gate leakage and gate end seal wear.

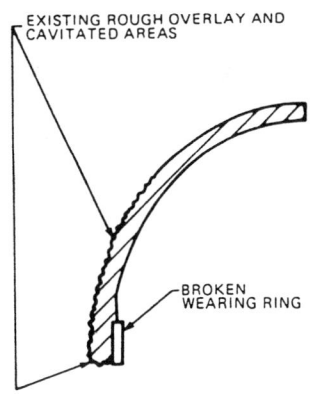
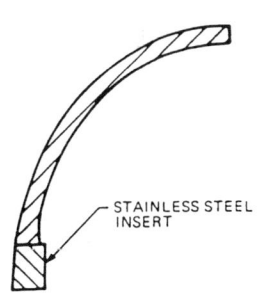

Figure 3 - Runner Band before Repair

Figure 4 - Runner Band after Repair

Rehabilitation

Runner

Due to the existing condition of the runner, it was necessary to ship it to the Voith shop, where hydraulic engineers checked and monitored hydraulic surface contour repairs, correct any hydraulic vent deviations and monitor restoration of vane profiles. Additionally, edge shape, vane entrance and discharge angles were checked and restored.

Pump/turbines should in most cases be balanced dynamically. This was another reason for the return of the runner to the shop where balancing corrections could easily be accomplished.

In the case of Salina, it was decided to cut-off the entire lower section of the cavitated runner band (appoximately 7 inches), and install a solid stainless steel section, forming an integral rotating wearing ring.(See Figure 4) Major high stress areas, i.e. bucket to band and crown attachment fillets, were all given thorough non-destructive examinations.

The major portion of the runners' hydraulic surface areas had cavitation repairs completed by application of stainless steel overlay and all surfaces were ground to an acceptable hydraulic shape. Finally, the runner was dynamically balanced on Voith's in-house balancing equipment, to an ISO G-16 standard.

Addition of the solid stainless steel insert, hydraulic surface contour grinding, as well as the increase in stainless steel weld overlay on the water passages will all greatly assist in the reduction of future cavitation/damage.

Figure 5 - Runner As Received For Rehabilitation

Head Covers and Bottom Ring

The outer and inner head covers and bottom ring were also returned to the shop where it was determined that the head cover and bottom ring gate bushing bores had deteriorated, requiring line-boring. This line boring process would assure reconditioned gate bores that were plumb with the center of the unit and parallel with each other. Just as importantly, the line boring would assist in proper gate to gate and gate end sealing. If the above mentioned seal points were not making precise contact, excessive water leakage would occur during shutdown, as well as motoring operations.

Again, due to the deteriorated condition, a complete new set of gate end seal assemblies was provided.(See Figure 9)

Previous excessive gate end seal leakage caused cavitation and high velocity water erosion on the water passage surfaces of the head cover and bottom ring. These surfaces were repaired with stainless steel, and the surface ground to a smooth hydraulic contour.

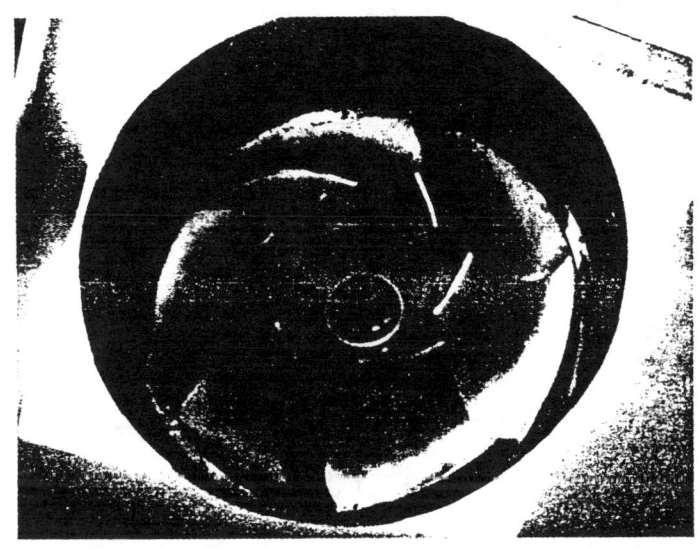

Figure 6 - Runner after Repair

Following the wicket gate hole line boring process, a new set of bronze (SAE 660) gate bushings was installed in the outer head cover and bottom ring.(See Figure 9) These bushings were manufactured to the original tolerances, with exception of the middle and lower bushings. These bushings were designed to contain a seal, located very close to the water passage flow and contained in the bushing itself.(See Figure 7) The purpose of this design is to protect the wicket gate journals, sealing out river water contaminants, and at the same time containing and preventing grease lubrication from being washed out of the bearing areas. These bushings were manufactured with very tight tolerances, especially in the wall thickness, ±.001". These tolerances are required to assure bearing alignment and proper surface journal contact of all three gate bushings.

Wicket Gates

Wicket gates (See Figures 8 & 9) were another problem at Salina, where the bearing journals and the gate seal surfaces were all out of tolerance due to wear, i.e. they were not diametrically square and perpendicular.

It is very important when rehabilitating wicket gates, that all final machined surface tolerances be strictly maintained with respect to the centerline of the wicket gate journal diameters. Bearing journals and vertical seal joints should be perpendicular and gate ends square with the centerline of the gate. The Salina gate bearing journals and vertical seal areas were overlayed with stainless steel and machined to acceptable tolerances and surface finish.

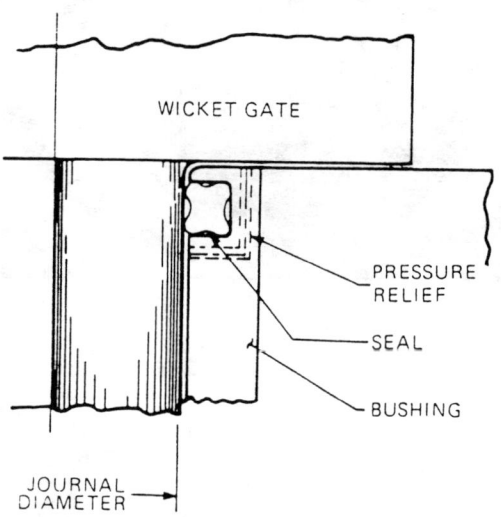

Figure 7 - Gate Bushing Seal

While in the shop all hydraulic surfaces that were cavitated or erroded were repaired with stainless steel and ground to the correct hydraulic contour.

Quality control always plays an important part of rehabilitation. The wicket gates were given a magnetic particle (M.T.) and dye penetrant (P.T.) examination in the high stressed areas of the gate, i.e. journal to leaf juncture. This area was also inspected after undercut machining and prior to bearing journal stainless overlay.

Turbine Guide Bearing

The turbine guide bearing's babbitted surface was found to be cracked and areas of babbitt broken out of the bearing surface. The bearing was totally cleaned, blasted, re-babbitted and machined, restoring it to the original design.

It was concluded that shaft run-out and unit mechanical and hydraulic imbalance were major causes of bearing babbitt deterioration.

Wearing Ring/Seal Rings

Increased wearing ring clearances had permitted increased water leakage into the cavity between the runner and head cover, and between the runner and discharge ring.(See Figure 2) The increase in water leakage and containment between the head cover and runner had also increased the thrust load of the rotating parts. With an increase in load, coupled with unit vibration and runout, thrust bearing temperatures increased. Considering all of the above mentioned reasons, coupled with disassembly costs, it was decided to replace all wearing rings. Ring replacement materials are stainless steel, with mating rings of different stainless qualities, i.e. 400 Series and 300 Series.

The Salina wearing rings were all replaced, machined and checked to run in center with each corresponding ring and to the center of the turbine bearing.

The pump/turbines lower stationary wearing ring mounting surface, which is located in the discharge ring, was checked in the field and found to be out of center with the remaining embedded parts. It was also out of parallel with the center of the unit plumb line. As a result of this condition, it was necessary to field machine the wearing ring mounting surface back into tolerance. The new ring was then manufactured and installed to accept the newly machined dimensions.

Figure 8 - Wicket Gate Repair

Gate Ring

The gate ring was also found to have excessive wear on the bearing pads. Excessive wear in this area permitted the gate ring to "skate", limiting full and repeated wicket gate vertical seal contact or squeeze. These bearing pads were all replaced on the Salina project, permitting the gate ring to properly operate on center.

Gate Linkage

Before disassembly, significant wicket gate vibration and movement was noted. The gate linkage bushings were all found to be worn and eccentric, exceeding allowable design tolerances. Excessive wear and gate linkage movement limit full gate closure and in addition create a dynamic loading condition, thus promoting increased gate shear pin breakage. Grand River Dam Authority decided to replace these bushings during their outage, providing for more stable gate control.

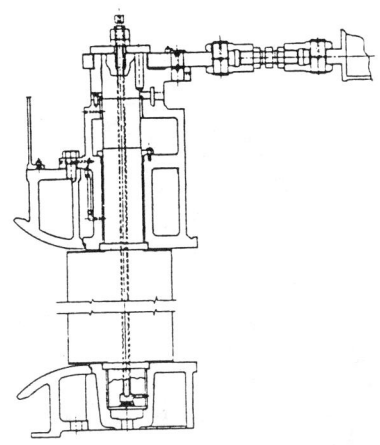

Figure 9 - Gate Assembly

Servomotors

Grand River Dam Authority also chose to use this rehabilitation period to disassemble their gate servomotors, checking for cylinder condition and also replacing servomotor piston rings and packings.

Field Responsibilities

Before disassembly, clearance measurements and checks were recommended by Voith (See Figure 2). Post disassembly checks with only the head-cover and bottom ring doweled in place were also performed. Installing a center of unit plumb line at this stage of partial assembly, coupled with the use of an accurate transit, will indicate whether or not the unit or powerhouse has moved or shifted. In the case of Grand River Dam Authority's Salina project, it was found that the specific unit being rehabilitated had shifted, which had moved the discharge ring, headcover, and generator stator out of level. This condition made it necessary to field machine the bottom ring and head cover mounting surface. The stator also had to be re-leveled, re-centered and doweled.

Figure 10 - Runner Received at Powerhouse

Re-assembly is the most critical of field operations. The head cover, bottom ring and four wicket gates were first installed, checking for center and level. At the same time wicket gate end clearance, uniformity and accuracy were ascertained.

After all of the turbine components were installed, excluding bearings and packing boxes, four plumb lines were placed in the turbine pit. These plumb lines were used to check and adjust for rotating parts, center and plumb. Following complete installation and prior to start-up, all clearances i.e. bearing, wearing ring, air gap etc. were checked and recorded. This data will be used as a guide and reference standard for Grand River Dam Authority during future years of operation.

Results

As reported by Grand River Dam Authority and Voith's field service technical representative, the rehabilitated unit went back on line with positive results. Due to all of the rehabilitation work, the unit is presently producing four additional reclaimed megawatts of power. Grand River Dam Authority also reports a twelve second improvement time in start-up. Additionally the unit goes on and off line and operates over-all much more efficiently and smoother than before.

GENERAL COMPARISONS

	Before	After
Gate Mechanical Movement	.125"-.375"	.007"-.012"
Shaft Runout	.020"-.035"	.0015"-.002"
Output	41 MW	45 MW
Vertical Gate Seal	.010"-.030"	.002" MAX
Gate End Seal	.010"-.020"	.000"
Pump Start-Up Time	39 seconds	27 seconds

Conclusion

Major rehabilitation efforts will always produce positive results. Many power companies consider over-all down time and costs to be excessive and not worth the time. However, considering the normal down time of six to seven months for this magnitude of work, the payback and reduced maintenance over the following ten to fifteen years is favorable.

Salina Pump/Generating Plant

Bibliography

1. Kepler, J. L. and K. T. McGaughey, "Pump-Turbine Refurbishment at the Salina Powerhouse," _Water Power & Dam Construction_, April, 1990, pp. 26-28.

2. Kepler, J. L. and K. T. McGaughey, "Case Study in the Rehabilitation of a Pumped/Storage Installation, Salina Powerhouse, Grand River Dam Authority," _Waterpower '89_, ASCE, pp. 1597-1608.

Plant Name: **SMITH MOUNTAIN**

Plant location:		Owner:	Appalacian Power Company
Sandy Level, VA			P.O. Box 2021
Pittsylvania/Bedford County			Roanoke, Virginia 24022-2121
Rated capacity	240 MW		
Average static head	195 ft	Designers:	
Plant efficiency	73.00 %		Ebasco Services
Stored energy	3,332 MWh		2 World Trade Center
Number of units	3		New York, New York 10048
Construction time: 6 years		Plant Manager/Superintendent:	
Construction cost: $48 per kW			Dale Fisher
Price level: 1966			Route 1
First commercial power: December 1965			Sandy Level, Virginia 24161
FERC project number: 2210			
			(703) 985-2586
River or water source: Roanoke River			

	UPPER RESERVOIR	LOWER RESERVOIR
DAM		
Type	Concrete arch	Concrete gravity
Height (ft)	235	94
Crest length (ft)	820	981
Volume (yd³)	191,000	109,000
RESERVOIR		
Type	Constructed, Smith Mountain Reservoir	Constructed, Leesville Reservoir
Surface area (acres)	20,608	3,262
Usable power storage (acre-ft)	157,764	37,779
Power pool fluctuation (ft)	8.0	13.0
Operating levels		
Maximum (ft)	795.0	613.0
Minimum (ft)	787.0	600.0
Drainage area (miles²)	1,023.9	1,505.0
Seepage (ft³/s)		
SPILLWAY		
Design flood		
Return period (years)		
Flow (ft³/s)	200,000	242,000
Capacity (ft³/s)	50,010	161,990
Type	Free overflow	Gated
Gates		
Number	None	4
Type		Radial
Width (ft)		49.87
Height (ft)		35.10
OUTLET WORKS		
Discharge capacity (ft³/s)		74
Number of water passages	None	1
Dimensions of water passages		
Height (ft)		
Width (ft)		
Diameter (ft)		0.98
Type of gates		Valve
Number of gates	None	2

Smith Mountain - Page 1 (English)

Plant Name: **SMITH MOUNTAIN**

Plant location Sandy Level, VA Pittsylvania/Bedford County	Owner: Appalacian Power Company P.O. Box 2021 Roanoke, Virginia 24022-2121
Rated capacity 240 MW Average static head 59.4 m Plant efficiency 73.00 % Stored energy 3,332 MWh Number of units 3	Designers: Ebasco Services 2 World Trade Center New York, New York 10048
Construction time: 6 years Construction cost: $48 per kW Price level: 1966 First commercial power: December 1965 FERC project number: 2210 River or water source: Roanoke River	Plant Manager/Superintendent: Dale Fisher Route 1 Sandy Level, Virginia 24161 (703) 985-2586

	UPPER RESERVOIR	LOWER RESERVOIR
DAM		
Type	Concrete arch	Concrete gravity
Height (m)	71.6	28.6
Crest length (m)	249.9	299.0
Volume (m³)	146,317	83,610
RESERVOIR		
Type	Constructed, Smith Mountain Reservoir	Constructed, Leesville Reservoir
Surface area (Mm²)	83.40	13.20
Usable power storage (Mm³)	194.600	46.600
Power pool fluctuation (m)	2.44	3.96
Operating levels		
Maximum (m)	242.32	186.84
Minimum (m)	239.88	182.88
Drainage area (Mm²)	2,652.000	3,898.000
Seepage (m³/s)		
SPILLWAY		
Design flood		
Return period (years)		
Flow (m³/s)	5,663	6,853
Capacity (m³/s)	1,416	4,587
Type	Free overflow	Gated
Gates		
Number	None	4
Type		Radial
Width (m)		15.200
Height (m)		10.700
OUTLET WORKS		
Discharge capacity (m³/s)		2
Number of water passages	None	1
Dimensions of water passages		
Height (m)		
Width (m)		
Diameter (m)		0.300
Type of gates		Valve
Number of gates	None	2

Smith Mountain - Page 1 (Metric)

SMITH MOUNTAIN

INTAKES	UPPER INTAKE	LOWER INTAKE
Number	3	4
Type	Bell mouth	Bell mouth
Design discharge (ft³/s)	5,000	
Gross inlet area (ft²) (at trash racks)		
Bar racks		
spacing (in)	4.00	
shape	Rectangular	
depth/thickness (in)	3.50 / 0.37	
diameter (in)		
Emergency gates		
number	3	4
height/width (ft)	24.38 / 22.64	
type	Wheel	Wheel
Service gates		
number	None	None
height/width (ft)		
type		
Bulkhead/stop logs (Y or N)	N	N
number of units serviced		
Hoists		
number	3	4
capacity (tons)	100	100
type	Motor operated, wire rope	Motor operated, wire rope

WATER PASSAGES	Upper Tunnel	Shaft	Lower Tunnel	Surge Tanks Upper	Lower	Penstocks	Tailrace Tunnel
Number						3	
Diameter (ft)						26.0	
Length (ft)						170	
Maximum velocity (ft/s)							
Concrete lining thickness (in)							
Total length of concrete sections (ft)							
Steel liner Thickness							
Minimum (in)							
Maximum (in)							
Material grade							
Total length of steel-lined sections (ft)							

Notes:
Penstocks: 1 @ 7.92 m (26 ft) and 2 @ 6.10 m (20 ft) dia. Length varies, longest is about 51.8 m (170 ft).

Smith Mountain - Page 2 (English)

SMITH MOUNTAIN

INTAKES	UPPER INTAKE	LOWER INTAKE
Number	3	4
Type	Bell mouth	Bell mouth
Design discharge (m³/s)	141.6	
Gross inlet area (m²) (at trash racks)		
Bar Racks:		
spacing (mm)	102	
shape	Rectangular	
depth/thickness (mm)	89 / 9	
diameter (mm)		
Emergency gates		
number	3	4
height/width (m)	7.430 / 6.900	
type	Wheel	Wheel
Service gates		
number	None	None
height/width (m)		
type		
Bulkhead/stop logs (Y or N)	N	N
number of units serviced		
Hoists		
number	3	4
capacity (Mg)	91	91
type	Motor operated, wire rope	Motor operated, wire rope

WATER PASSAGES	Upper Tunnel	Shaft	Lower Tunnel	Surge Tanks Upper	Lower	Penstocks	Tailrace Tunnel
Number						3	
Diameter (m)						7.93	
Length (m)						51.8	
Maximum velocity (m/s)							
Concrete lining thickness (m)							
Total length of concrete sections (m)							
Steel liner Thickness							
Minimum (mm)							
Maximum (mm)							
Material grade							
Total length of steel-lined sections (m)							

Notes:
Penstocks: 1 @ 7.92 m (26 ft) and 2 @ 6.10 m (20 ft) dia. Length varies, longest is about 51.8 m (170 ft).

SMITH MOUNTAIN

POWERHOUSE and RELATED FEATURES

Powerhouse Structure
Type: Outdoor
Length: 351 ft Width: 98 ft Height: 40 ft

Guard Valves
Number: None Diameter: ft
Type:

Transformers
Number: 3
Ratings: 80
Voltages: (kV) 138 / 13.8

Generator
Rating generating (MVA): 69.5 Rating pumping (MVA): 81.5
Insulation type:
Starting method: Reduced voltage, transformer tap
Starting equipment: Reduced voltage, transformer tap

Runners
Material: Cast steel
Minimum unit submergence: 4.0 ft
WR^2:
Manufacturer: Allis Chalmers
Model test by: Allis Chalmers

	Reversible Runners	Reversible Motor/Generator		
Number	3			3
Diameter (ft)	22.30	Rotor	Stator	
rpm synchronous	105.9			105.9
rpm overspeed				
Type	Francis			Umbrella

Information on Runners

	Gross Head (ft)		Capacity (MW)		Discharge (ft³/s)		Turbine/Pump Eff.(%)	
Condition:	Generating	Pumping	Generating	Pumping	Generating	Pumping	Generating	Pumping
Maximum head & maximum power	195	195	75	77	5,350	4,188	85.0	90.0
Minimum head & maximum power	174	174	65	78	4,976	4,640	89.0	88.0

Note: Data in the above table are based on design data.

	Net Head (ft)		Capacity (MW)		Discharge (ft³/s)		Turbine/Pump Eff.(%)	
Condition:	Generating	Pumping	Generating	Pumping	Generating	Pumping	Generating	Pumping
Rated head @ best gate	180	197	70	75		4,220	90.0	93.0

Note: Data in the above table are based on field test.

SMITH MOUNTAIN

POWERHOUSE and RELATED FEATURES

Powerhouse Structure
Type: Outdoor
Length: 107.0 m   Width: 29.7 m   Height: 12.0 m

Guard Valves
Number: None   Diameter: m
Type:

Transformers
Number: 3
Ratings: 80
Voltages: (kV) 138 / 13.8

Generator
Rating generating (MVA): 69.5   Rating pumping (MVA): 81.5
Insulation type:
Starting method: Reduced voltage, transformer tap
Starting equipment: Reduced voltage, transformer tap

Runners
Material: Cast steel
Minimum unit submergence: 1.22 m
WR^2:
Manufacturer: Allis Chalmers
Model test by: Allis Chalmers

	Reversible Runners	Reversible Motor/Generator		
Number	3			3
Diameter m	6.80	Rotor	Stator	
rpm synchronous	105.9			105.9
rpm overspeed				
Type	Francis			Umbrella

Information on Runners

Condition:	Gross Head (m)		Capacity (MW)		Discharge (m³/s)		Turbine/Pump Eff.(%)	
	Generating	Pumping	Generating	Pumping	Generating	Pumping	Generating	Pumping
Maximum head & maximum power	59.5	59.5	75	77	151.5	118.6	85.0	90.0
Minimum head & maximum power	53.0	53.0	65	78	140.9	131.4	89.0	88.0

Note: Data in the above table are based on design data.

Condition:	Net Head (m)		Capacity (MW)		Discharge (m³/s)		Turbine/Pump Eff.(%)	
	Generating	Pumping	Generating	Pumping	Generating	Pumping	Generating	Pumping
Rated head @ best gate	54.9	60.0	70	75		119.5	90.0	93.0

Note: Data in the above table are based on field test.

Smith Mountain - Page 3 (Metric)

SMITH MOUNTAIN

Plant Data:
```
Average GWh generating per year:                382
Average GWh pumping per year:                   388
Starting time from standstill (s):              180
Changeover time pumping to generating (min):     14
Planned/scheduled time
    between major overhauls (years):              1
Outage time required per unit
    during major overhauls (weeks):              12
Representative plant availability (%):         32.0
Representative planned outages (weeks per year): 14
```

Miscelleneous Notes:

Smith Mountain includes three reversible Francis units and two conventional Francis units. Data in the tables reflect the reversible units.

The three reversible units are not the same. Two are 6.8 m (22 ft) in diameter and operate at 105.9 RPM. The third unit is 8.0 m (26 ft) in diameter and operates at 90.0 RPM.

The capacity of the units at Smith Mountain are:
- 2 70-MW reversible
- 1 100-MW reversible
- 2 160-MW conventional

Cavitation Experience:

Significant or Unique Problems:

The intake trashracks failed in the pumping mode. Both reversible units have had wicket gate failures resulting in extensive runner damage.

List of Licenses Required:

ENVIRONMENTAL FEATURES

Recreation:

Recreational facilities include numerous picnic areas, over 25 marinas, a visitor center, wildlife trails, and a 20.23 Mm² (5,000-acre) wildlife management area.

Fish and Wildlife:

The lake is stocked with gamefish, such as striped largemouth and smallmouth bass and muskel. The 20.23 Mm² (5,000 acre) wildlife refuge is managed by the Department of Game and Fisheries.

Social:

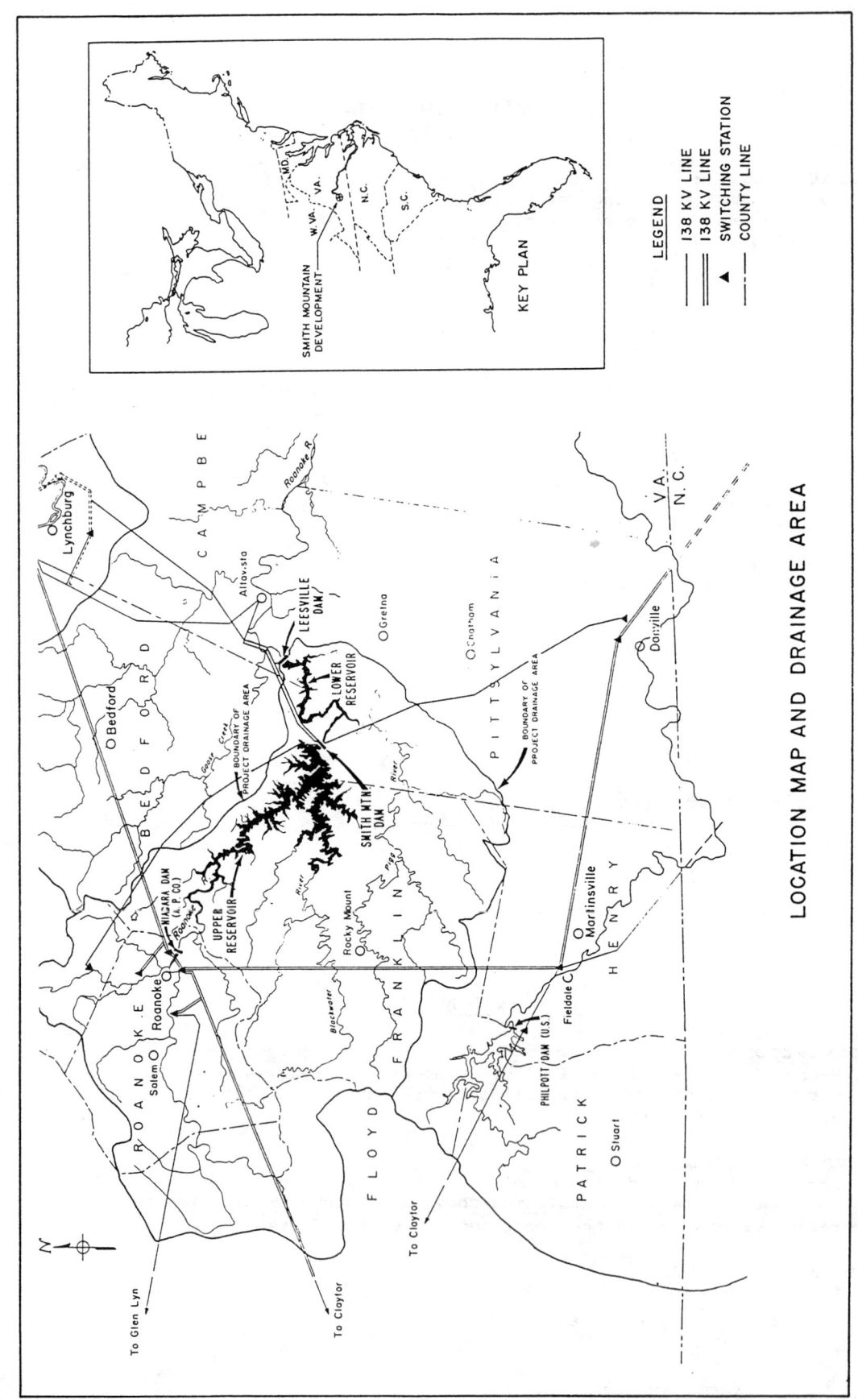

FIGURE 1

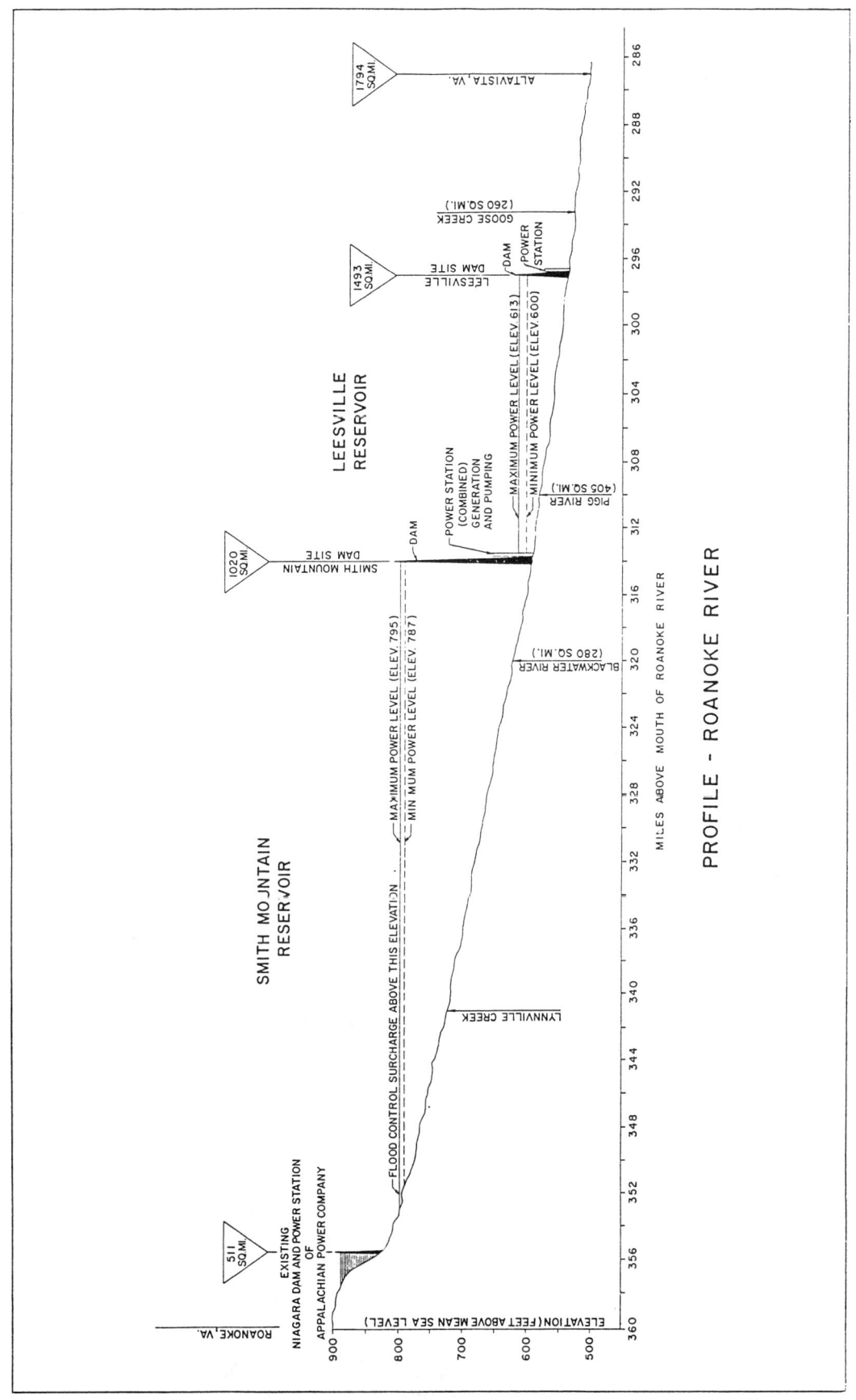

FIGURE 2

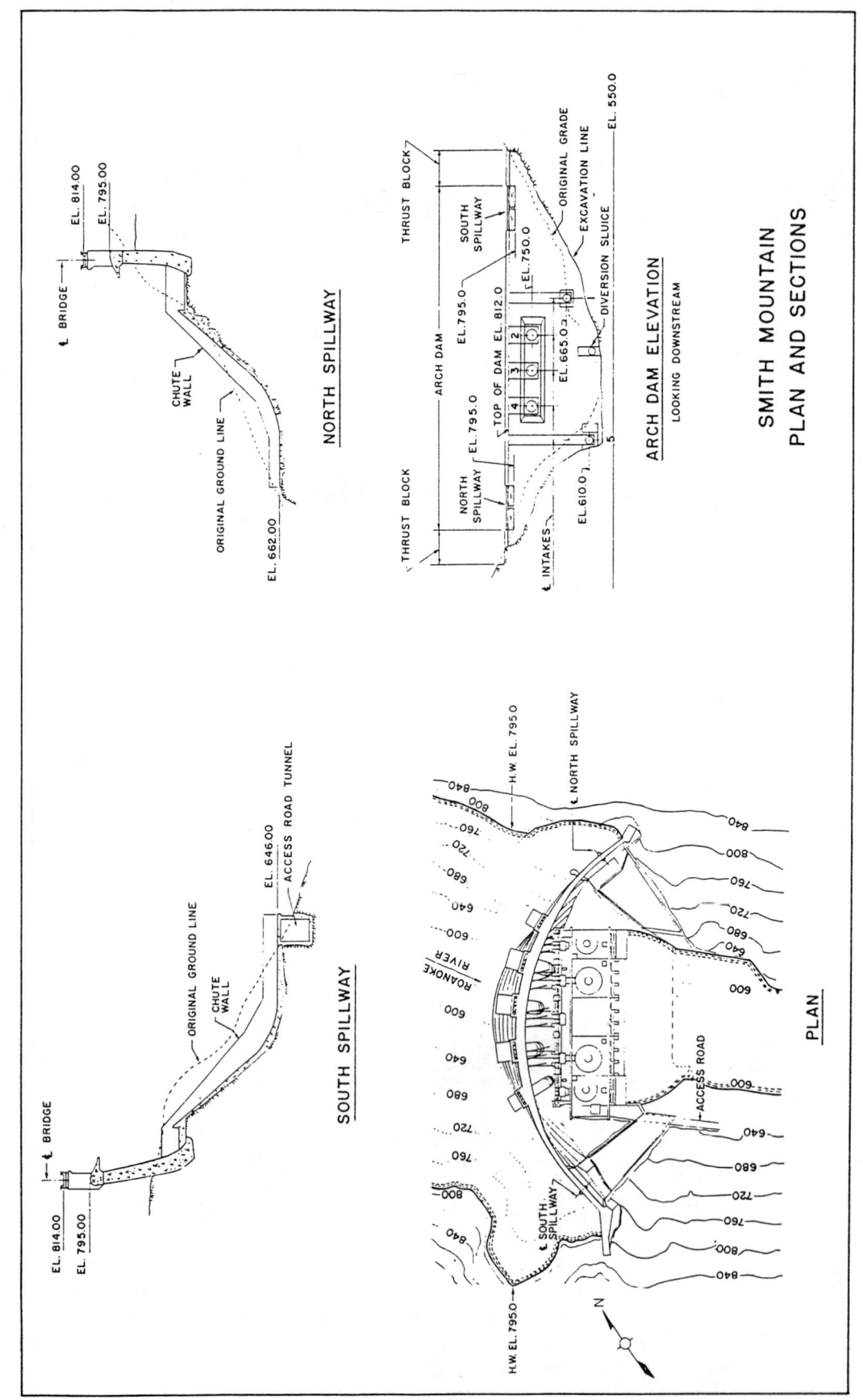

FIGURE 3

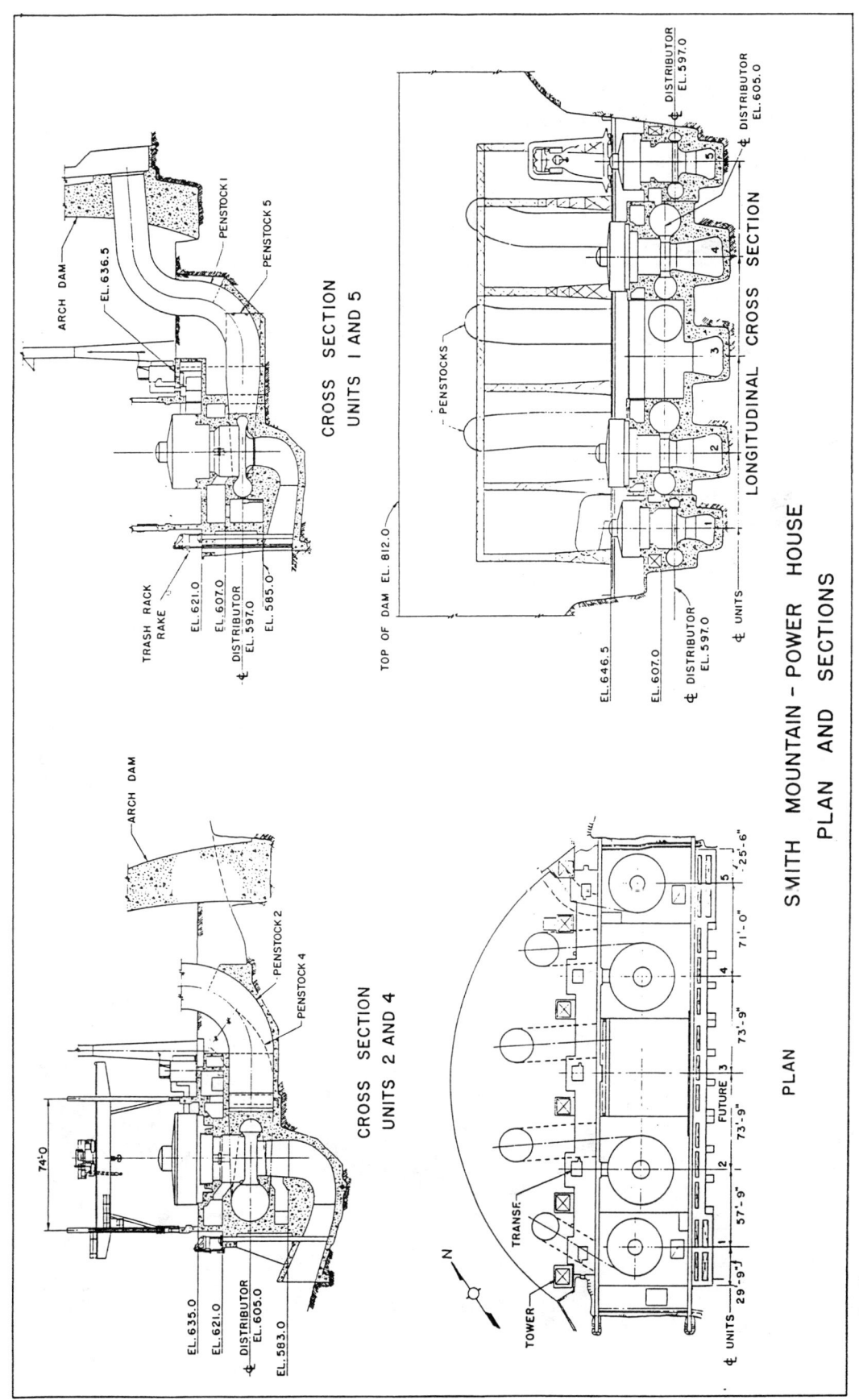

FIGURE 4

FIGURE 5

Smith Mountain Dam and Powerhouse

Smith Mountain Pumped Storage Plant

Smith Mountain Pump/Generating Plant

Bibliography

1. Hroncich, M. J. and J. M. Mullarkey, "Symposium on Pumped Storage: Run-of-River Pumped Storage - Smith Mountain Project," *Journal of the Power Division*, ASCE, Vol. 88, No. PO2, July, 1962, pp. 183-209.

Plant Name: **TAUM SAUK**

Plant location:
St. Louis, MO
Reynolds County

Rated capacity	350 MW
Average static head	790 ft
Plant efficiency	50.00 %
Stored energy	2,750 MWh
Number of units	2

Construction time: 3 years, 7 months
Construction cost: $105 per kW
Price level: 1963
First commercial power: January 1963
FERC project number:

River or water source: East Fork of the Black River

Owner: Union Electric Company
1901 Chateau
P.O. Box 149
St. Louis, Missouri 63166

Designers:
Sverdup and Parcel

Plant Manager/Superintendent:

	UPPER RESERVOIR	LOWER RESERVOIR
DAM		
Type	Rockfill, concrete face, asphalt floor	Ogee weir, concrete, gravity
Height (ft)	92	60
Crest length (ft)	6,660	390
Volume (yd^3)	3,750,000	22,000
RESERVOIR		
Type	Constructed	Constructed
Surface area (acres)	55	200
Usable power storage (acre-ft)	4,350	6,350
Power pool fluctuation (ft)	88.0	15.0
Operating levels		
Maximum (ft)	1,597.0	750.0
Minimum (ft)	1,505.0	735.0
Drainage area (miles^2)	0.1	91.0
Seepage (ft^3/s)	11.000	
SPILLWAY		
Design flood		
Return period (years)		
Flow (ft^3/s)		120,000
Capacity (ft^3/s)		120,000
Type		Ogee weir
Gates		
Number		
Type		
Width (ft)		
Height (ft)		
OUTLET WORKS		
Discharge capacity (ft^3/s)	8,800	2,000
Number of water passages	1	2
Dimensions of water passages		
Height (ft)		8.00
Width (ft)		10.00
Diameter (ft)	27.00	
Type of gates		Sluice
Number of gates	None	2

Taum Sauk - Page 1 (English)

Plant Name: **TAUM SAUK**

Plant location St. Louis, MO Reynolds County	Owner: Union Electric Company 1901 Chateau P.O. Box 149 St. Louis, Missouri 63166
Rated capacity 350 MW Average static head 240.8 m Plant efficiency 50.00 % Stored energy 2,750 MWh Number of units 2	Designers: Sverdup and Parcel
Construction time: 3 years, 7 months Construction cost: $105 per kW Price level: 1963 First commercial power: January 1963 FERC project number:	Plant Manager/Superintendent:
River or water source: East Fork of the Black River	

DAM	UPPER RESERVOIR	LOWER RESERVOIR
Type	Rockfill, concrete face, asphalt floor	Ogee weir, concrete, gravity
Height (m)	28.0	18.3
Crest length (m)	2,030.0	118.9
Volume (m³)	2,867,082	16,820
RESERVOIR		
Type	Constructed	Constructed
Surface area (Mm²)	0.22	0.81
Usable power storage (Mm³)	5.366	7.833
Power pool fluctuation (m)	26.82	4.57
Operating levels		
Maximum (m)	486.77	228.60
Minimum (m)	458.72	224.03
Drainage area (Mm²)	0.223	235.690
Seepage (m³/s)	0.3115	
SPILLWAY		
Design flood		
Return period (years)		
Flow (m³/s)		3,398
Capacity (m³/s)		3,398
Type		Ogee weir
Gates		
Number		
Type		
Width (m)		
Height (m)		
OUTLET WORKS		
Discharge capacity (m³/s)	249	57
Number of water passages	1	2
Dimensions of water passages		
Height (m)		2.438
Width (m)		3.048
Diameter (m)	8.230	
Type of gates		Sluice
Number of gates	None	2

Taum Sauk - Page 1 (Metric)

TAUM SAUK

INTAKES	UPPER INTAKE	LOWER INTAKE
Number	1	
Type	Cylindrical vertical reinforced concrete	
Design discharge (ft³/s)	8,793	
Gross inlet area (ft²) (at trash racks)	581	
Bar racks		
spacing (in)		
shape		
depth/thickness (in)		
diameter (in)		
Emergency gates		
number	None	
height/width (ft)		
type		
Service gates		
number	None	
height/width (ft)		
type		
Bulkhead/stop logs (Y or N)	N	
number of units serviced		
Hoists		
number	1	
capacity (tons)	300	
type	Semi-Gantry	

WATER PASSAGES	Upper Tunnel	Shaft	Lower Tunnel	Surge Tanks Upper	Surge Tanks Lower	Penstocks	Tailrace Tunnel
Number		1	1			2	
Diameter (ft)		27.2	25.5			13.5	
Length (ft)		431	6,571			248	
Maximum velocity (ft/s)		15.0	17.2			31.0	
Concrete lining thickness (in)		6.00					
Total length of concrete sections (ft)		35	4,764				
Steel liner Thickness							
Minimum (in)			0.63				
Maximum (in)			1.88			2.00	
Material grade			T-1,T-1A,A-201			T-1	
Total length of steel-lined sections (ft)			1,800			248	

Notes:
Lower Tunnel: Concrete-lined section is horseshoe shaped. Steel section diameter is 5.49 m (18 ft).

TAUM SAUK

INTAKES	UPPER INTAKE	LOWER INTAKE
Number	1	
Type	Cylindrical vertical reinforced concrete	
Design discharge (m³/s)	249.0	
Gross inlet area (m²) (at trash racks)	54.0	
Bar Racks:		
spacing (mm)		
shape		
depth/thickness (mm)		
diameter (mm)		
Emergency gates		
number	None	
height/width (m)		
type		
Service gates		
number	None	
height/width (m)		
type		
Bulkhead/stop logs (Y or N)	N	
number of units serviced		
Hoists		
number	1	
capacity (Mg)	272	
type	Semi-Gantry	

WATER PASSAGES	Upper Tunnel	Shaft	Lower Tunnel	Surge Tanks Upper	Surge Tanks Lower	Penstocks	Tailrace Tunnel
Number		1	1			2	
Diameter (m)		8.28	7.77			4.12	
Length (m)		131.4	2,002.8			75.6	
Maximum velocity (m/s)		4.57	5.24			9.45	
Concrete lining thickness (m)		0.152					
Total length of concrete sections (m)		10.7	1,452.1				
Steel liner Thickness							
Minimum (mm)			16				
Maximum (mm)			48			51	
Material grade			T-1,T-1A,A-201			T-1	
Total length of steel-lined sections (m)			548.6			75.6	

Notes:
Lower Tunnel: Concrete-lined section is horseshoe shaped. Steel section diameter is 5.49 m (18 ft).

TAUM SAUK

POWERHOUSE and RELATED FEATURES

Powerhouse Structure
Type: Semi-outdoor
Length: 150 ft   Width: 88 ft   Height: 80 ft

Guard Valves
Number: 2   Diameter: 9.0 ft
Type: Spherical

Transformers
Number: 2
Ratings: 230
Voltages: (kV) 138 / 13.5 3 phase FOA

Generator
Rating generating (MVA): 204.0   Rating pumping (MVA): 179.0
Insulation type: Asphaltic Mica
Starting method: Pony motor
Starting equipment: Pony motor

Runners
Material: Welded A27, 70-36
Minimum unit submergence: 32.0 ft
WR^2:
Manufacturer: Allis Chalmers
Model test by: Allis Chalmers

	Reversible Runners	Reversible Motor/Generator	
Number	2		2
Diameter (ft)	21.30	Rotor	Stator
rpm synchronous	200.0		200.0
rpm overspeed	220.0		220.0
Type	Francis		Salient POLF

Information on Runners

Condition:	Gross Head (ft) Generating	Gross Head (ft) Pumping	Capacity (MW) Generating	Capacity (MW) Pumping	Discharge (ft³/s) Generating	Discharge (ft³/s) Pumping	Turbine/Pump Eff.(%) Generating	Turbine/Pump Eff.(%) Pumping
Maximum head & maximum power	875	875	225	175	4,400	1,775		
Minimum head & maximum power	764	764	175	210	2,530	2,675		

Condition:	Net Head (ft) Generating	Net Head (ft) Pumping	Capacity (MW) Generating	Capacity (MW) Pumping	Discharge (ft³/s) Generating	Discharge (ft³/s) Pumping	Turbine/Pump Eff.(%) Generating	Turbine/Pump Eff.(%) Pumping
Rated head @ best gate	790	764	220	179		2,650		

Taum Sauk - Page 3 (English)

TAUM SAUK

POWERHOUSE and RELATED FEATURES

Powerhouse Structure
Type: Semi-outdoor
Length: 45.7 m   Width: 26.8 m   Height: 24.4 m

Guard Valves
Number: 2   Diameter: 2.74 m
Type: Spherical

Transformers
Number: 2
Ratings: 230
Voltages: (kV) 138 / 13.5 3 phase FOA

Generator
Rating generating (MVA): 204.0   Rating pumping (MVA): 179.0
Insulation type: Asphaltic Mica
Starting method: Pony motor
Starting equipment: Pony motor

Runners
Material: Welded A27, 70-36
Minimum unit submergence: 9.75 m
WR²:
Manufacturer: Allis Chalmers
Model test by: Allis Chalmers

	Reversible Runners	Reversible Motor/Generator		
Number	2			2
Diameter m	6.48	Rotor	Stator	
rpm synchronous	200.0			200.0
rpm overspeed	220.0			220.0
Type	Francis			Salient POLF

Information on Runners

Condition:	Gross Head (m)		Capacity (MW)		Discharge (m³/s)		Turbine/Pump Eff.(%)	
	Generating	Pumping	Generating	Pumping	Generating	Pumping	Generating	Pumping
Maximum head & maximum power	266.7	266.7	225	175	124.6	50.3		
Minimum head & maximum power	232.9	232.9	175	210	71.6	75.8		

Condition:	Net Head (m)		Capacity (MW)		Discharge (m³/s)		Turbine/Pump Eff.(%)	
	Generating	Pumping	Generating	Pumping	Generating	Pumping	Generating	Pumping
Rated head @ best gate	240.8	232.9	220	179		75.0		

Taum Sauk - Page 3 (Metric)

TAUM SAUK

Plant Data:
- Average GWh generating per year:
- Average GWh pumping per year:
- Starting time from standstill (s):
- Changeover time pumping to generating (min):
- Planned/scheduled time
  between major overhauls (years):
- Outage time required per unit
  during major overhauls (weeks):
- Representative plant availability (%):
- Representative planned outages (weeks per year):

Miscelleneous Notes:
None.

Cavitation Experience:
Moderate.

Significant or Unique Problems:

List of Licenses Required:

ENVIRONMENTAL FEATURES

Recreation:

Fish and Wildlife:

Social:

Taum Sauk Pumped-Storage Power Project

By **Edward A. Rudulph, F. ASCE**, Manager, Taum Sauk Project,
Union Electric Company, St. Louis, Mo.

The pumped-storage hydroelectric Taum Sauk Project of the Union Electric Company of St. Louis, Mo. has two reversible pump turbines, each with a rated generating capacity of 175 megawatts (mw). Pumping height and power head vary from 764 to 875 ft. To resist the high water pressure, USS "T-1" steel is used for partial lining of the tunnels. A concrete-gravity dam creates a lower pool, which impounds 6,000 acre-ft of water. At the upstream end of the lower pool a canal was excavated inland to a desirable site for the power station. A tunnel leads from the power station to the upper reservoir, which is formed by a rock-filled dike on a hilltop. The upper reservoir will contain 4,360 acre-ft of water, equivalent to 2,750,000 kwhr at the output terminals of the power plant.

As pumps, each of the pump-turbines will range in discharge from 1,775 cfs (against a total dynamic head of 875 ft) to 2,675 cfs (against a head of 764 ft). These are the maximum and minimum operating heads. Power consumption for the two conditions is expected to be 175 mw and 210 mw respectively, the higher power being required at the lower head. As generators, each unit will deliver 225 mw at the maximum net head of 820 ft and 175 mw at minimum net head of 692 ft. The project uses 3 kw of power for each 2 kw it returns to the system.

The Taum Sauk Project utilizes a group of small streams that come together to form the East Fork of the Black River in Missouri, with a drainage area of about 80 sq miles. The site is about 90 miles in a direct line from St. Louis. When generating, Taum Sauk will serve the load near it. The heavy loss in transmission is from steam plants in St. Louis to Taum Sauk for pumping.

Adjacent to the stream, and the pool formed by the lower dam, rises one of the high porphyry knobs characteristic of the Missouri Ozarks. A reservoir of sufficient size could be built at the top of this knob by a rock dike around an area of 38 acres. The dike is 6,000 ft in circumference and 82 ft high above the reservoir floor. It has a top width of 12 ft with natural slopes, as dumped, of about 1 vertical of 1.3 horizontal on both the inside and outside faces. It is faced on the inside with a concrete slab 10 in. thick reinforced at the middle of the slab with No. 7 bars spaced 12 in. on centers both ways. The slab was built in panels 60 ft in maximum width, extending from bottom to top of the rock fill. Copper strips seal the joints. At the top, the slab extends above the dike into a vertical parapet wall 10 ft high and 1 ft thick. The floor of the reservoir is sealed with a 4-in. layer of asphaltic concrete built up in two 2-in. layers.

The two pools are connected by waterways consisting of a channel excavated in rock for a distance of 1,600 ft from the upstream end of the lower pool, at the end of which the reversible pumping and generating station is located. Upstream from the power station, a penstock 248 ft long connects with a 25.5-ft by 25.5-ft tunnel of horseshoe section 6,572 ft long. For 1,800 ft at its lower end, this tunnel has a lining 18,5 ft in diameter. The rest of the tunnel is unlined and terminates at the bottom of a vertical shaft 431 ft high and 27.17 ft in diameter, which extends to the floor of the upper pool.

To keep the cost of the tunnel lining and penstock to a minimum, USS "T-1" and T-1 Type A steel is used for the first time in the United States in penstocks and tunnel lining. Protection of each water-wheel is provided by a 108-in. spherical turbine inlet valve at the spiral-casing inlet of each of the two units.

Deep submergence of the pump impeller (turbine runner) is required to limit cavitation, especially when the unit is operating as a pump. For this reason, draft-tube gates are provided to permit full unwatering of the units for inspection and repair.

The reversible pump-turbines and spherical valves were built by Allis-Chalmers Manufacturing Company and the generator motors by General Electric Company. The general contract for the dam and reservoir and the installation of equipment is held by Fruin-Colnon Contracting Company and Utah Construction & Mining Company, joint venturers. The subcontractor for the penstock and tunnel lining was the Nooter Corp.

Structural and mechanical engineering was done by the Sverdrup and Parcel Engineering Company. Electrical engineering and over-all coordination of the project were handled by the owner, the Union Electric Company.

This article was originally published in CIVIL ENGINEERING, a publication of the American Society of Civil Engineers, Vol. 33, No. 1, pages 40-43, January, 1963.

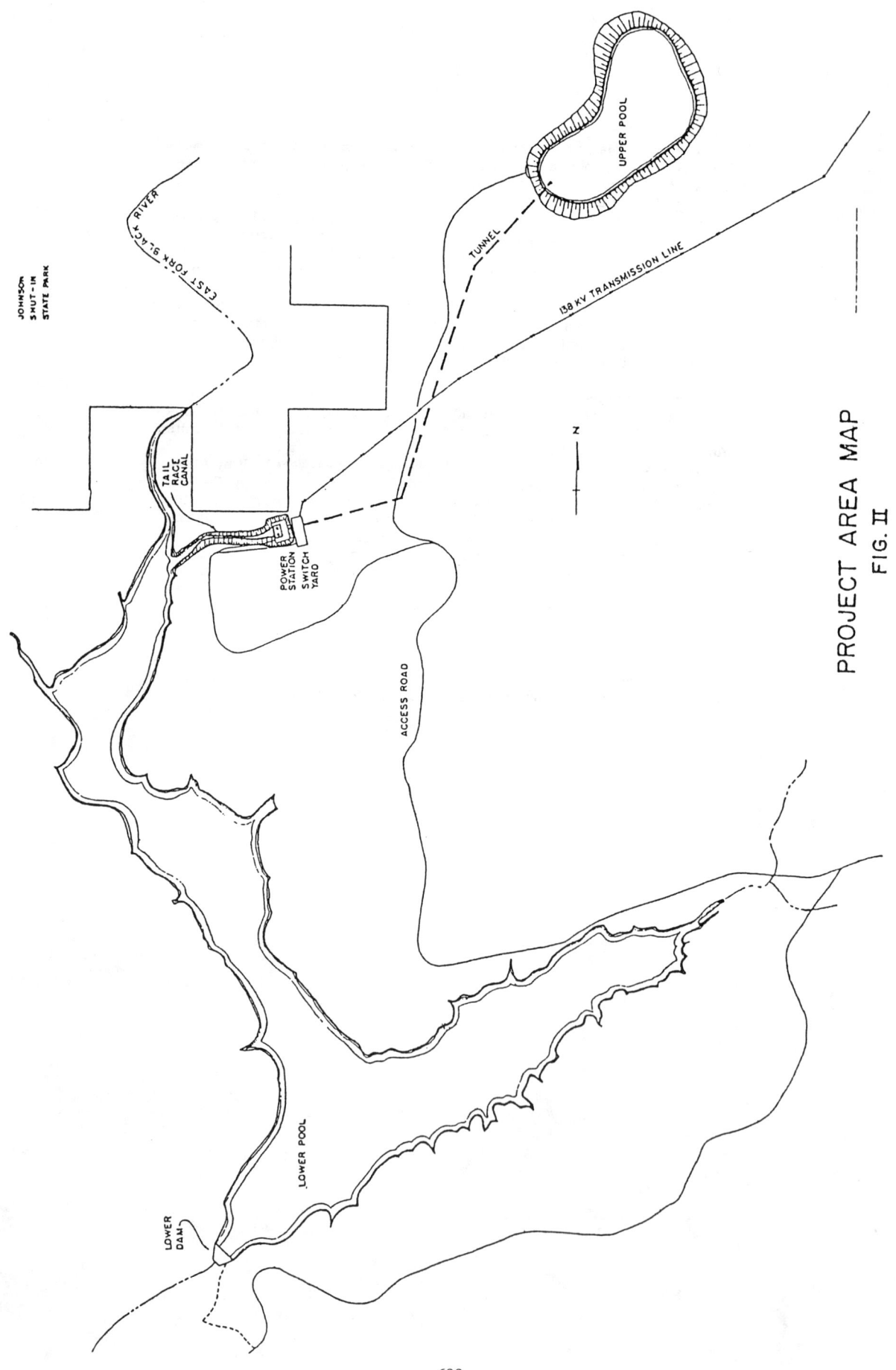

PROJECT AREA MAP
FIG. II

BIBLIOGRAPHY

1. Gamble, George P., "Union Electric's Taum Sauk Pumped Storage Hydro Plant," Power Engineering. In two parts. 55-58, November 1960 and 54-56, December 1960.

2. Gamble, George P., "Pumped Storage For Peaking Service. Development and Description of the Taum Sauk Project," American Power Conference. Proceedings. Vol. XXIII, 687-699, 1961.

3. Gamble, George P., "Taum Sauk Will Provide 350 Mw of Peaking Hydro," Electrical World, 50-53 May 8, 1961.

4. Gamble, George P. and Meyerand, R.G., "The Taum Sauk Pumped Storage Project," Machinery Lloyd (Overseas Edition) 36-41, March 16, 1963.

5. Jonas, R.G. and Meagher, M.R., "Automatic Control For the Taum Sauk Project," AIEE District Conference Paper, No. DPA 62-758, 1962.

6. Meyerand, R.G., "Summary of Electric System and Equipment Committee Meeting, February 12-14. EEI Bulletin, 123-124 April, 1961.

7. Meyerand, R.G., "Engineering Behind the Automation Scheme at Taum Sauk Hydro Plant," Power Engineering. In 2 parts. 41-44, November 1962 and 44-45, December 1962.

8. Rudulph, E.A., "Taum Sauk Pumped Storage Power Project," Civil Engineering, 40-43, January 1963.

9. Vencill, G.J., "The Taum Sauk Development" - Part 2, Electric Light and Power 44-47, May 1, 1961.

10. Vencill, G.J., "Factors Governing Selection of Large Pumped Storage Project to Serve Utility Peak Loads." Presented by George P. Gamble at the World Power Conference, 6th. Australia, 1962.

11. Whitlow, George S., "The Taum Sauk Project" - Part 1, Electric Light and Power 46-49, April 15, 1961.

12. Gamble, George P., and Rudulph, Edward A., Pumped Storage for Peaking Service, the Taum Sauk Project - A Step Forward," American Power Conference, Chicago, Illinois, April 1964.

Plant Name: **THERMALITO POWER PLANT**

Plant location:
Oroville, CA
Butte County

Rated capacity 115 MW
Average static head 93 ft
Plant efficiency %
Stored energy MWh
Number of units 4

Construction time: 4 years, 1 month
Construction cost: $210 per kW
Price level: 1968
First commercial power: February 1968
FERC project number: 2100

River or water source: Thermalito Power Canal

Owner: California Department of Water Resources
1416 Ninth Street
Sacramento, California 95965

Designers:
California Deptartment of Water Resources
Division of Design and Construction

Plant Manager/Superintendent:
Rolland Williams - Chief, Oroville Division
460 Glen Drive
Oroville, California 95965

(916) 534-2413

	UPPER RESERVOIR	LOWER RESERVOIR
DAM		
Type	Zoned earthfill	Zoned earthfill
Height (ft)	71	39
Crest length (ft)	16,000	42,000
Volume (yd³)	1,580,000	5,020,000
RESERVOIR		
Type	Constructed	Constructed
Surface area (acres)	630	4,500
Usable power storage (acre-ft)	9,936	2,888
Power pool fluctuation (ft)	4.0	13.5
Operating levels		
Maximum (ft)	226.0	136.5
Minimum (ft)	222.0	123.0
Drainage area (miles²)	0.1	
Seepage (ft³/s)		5.0
SPILLWAY		
Design flood		
Return period (years)		
Flow (ft³/s)		
Capacity (ft³/s)		
Type		
Gates		
Number		
Type		None
Width (ft)		
Height (ft)		
OUTLET WORKS		
Discharge capacity (ft³/s)	10,000	
Number of water passages	1	14
Dimensions of water passages		
Height (ft)	16.25	
Width (ft)	17.92	
Diameter (ft)		
Type of gates	Radial	See notes
Number of gates	1	None

Plant Name: **THERMALITO POWER PLANT**

Plant location Oroville, CA Butte County	Owner: California Department of Water Resources 1416 Ninth Street Sacramento, California 95965
Rated capacity 115 MW Average static head 28.4 m Plant efficiency % Stored energy MWh Number of units 4	Designers: California Deptartment of Water Resources Division of Design and Construction
Construction time: 4 years, 1 month Construction cost: $210 per kW Price level: 1968 First commercial power: February 1968 FERC project number: 2100	Plant Manager/Superintendent: Rolland Williams - Chief, Oroville Division 460 Glen Drive Oroville, California 95965 (916) 534-2413
River or water source: Thermalito Power Canal	

	UPPER RESERVOIR	LOWER RESERVOIR
DAM		
Type	Zoned earthfill	Zoned earthfill
Height (m)	21.6	11.9
Crest length (m)	4,876.8	12,801.6
Volume (m³)	1,207,997	3,838,067
RESERVOIR		
Type	Constructed	Constructed
Surface area (Mm²)	2.55	18.21
Usable power storage (Mm³)	12.256	3.562
Power pool fluctuation (m)	1.22	4.12
Operating levels		
Maximum (m)	68.89	41.61
Minimum (m)	67.67	37.49
Drainage area (Mm²)	0.259	12.950
Seepage (m³/s)		
SPILLWAY		
Design flood		
Return period (years)		
Flow (m³/s)		
Capacity (m³/s)		
Type		
Gates		
Number		None
Type		
Width (m)		
Height (m)		
OUTLET WORKS		
Discharge capacity (m³/s)	283	
Number of water passages	1	14
Dimensions of water passages		
Height (m)	4.953	
Width (m)	5.461	
Diameter (m)		
Type of gates	Radial	See notes
Number of gates	1	None

Thermalito Power Plant - Page 1 (Metric)

THERMALITO POWER PLANT

INTAKES	UPPER INTAKE	LOWER INTAKE
Number	4	
Type		Afterbay pool
Design discharge (ft³/s)	4,200	
Gross inlet area (ft²) (at trash racks)		
Bar racks		
spacing (in)	6.00	
shape	Rectangular	
depth/thickness (in)	2.25 / 0.50	
diameter (in)		
Emergency gates		
number	None	None
height/width (ft)		
type		
Service gates		
number	None	None
height/width (ft)		
type		
Bulkhead/stop logs (Y or N)	Y	
number of units serviced	4	4
Hoists		
number	1	1
capacity (tons)	75	25
type	Gantry crane	Electric Gantry crane

WATER PASSAGES	Upper Tunnel	Shaft	Lower Tunnel	Surge Tanks Upper	Surge Tanks Lower	Penstocks	Tailrace Tunnel
Number						4	
Diameter (ft)						21.0	
Length (ft)						118	
Maximum velocity (ft/s)						11.3	
Concrete lining thickness (in)						24.00	
Total length of concrete sections (ft)							
Steel liner Thickness							
Minimum (in)							
Maximum (in)							
Material grade							
Total length of steel-lined sections (ft)						118	

Notes:
Penstocks: 3 @ 6.40 m (21 ft) diameter, 1 @ 7.32 m (24 ft) diameter.

THERMALITO POWER PLANT

INTAKES	UPPER INTAKE	LOWER INTAKE
Number	4	
Type		Afterbay pool
Design discharge (m³/s)	118.9	
Gross inlet area (m²) (at trash racks)		
Bar Racks:		
spacing (mm)	152	
shape	Rectangular	
depth/thickness (mm)	57 / 13	
diameter (mm)		
Emergency gates		
number	None	None
height/width (m)		
type		
Service gates		
number	None	None
height/width (m)		
type		
Bulkhead/stop logs (Y or N)	Y	
number of units serviced	4	4
Hoists		
number	1	1
capacity (Mg)	68	23
type	Gantry crane	Electric Gantry crane

WATER PASSAGES	Upper Tunnel	Shaft	Lower Tunnel	Surge Tanks Upper	Surge Tanks Lower	Penstocks	Tailrace Tunnel
Number						4	
Diameter (m)						6.40	
Length (m)						36.0	
Maximum velocity (m/s)						3.44	
Concrete lining thickness (m)						0.610	
Total length of concrete sections (m)							
Steel liner							
Thickness							
Minimum (mm)							
Maximum (mm)							
Material grade							
Total length of steel-lined sections (m)						36.0	

Notes:
Penstocks: 3 @ 6.40 m (21 ft) diameter, 1 @ 7.32 m (24 ft) diameter.

THERMALITO POWER PLANT

POWERHOUSE and RELATED FEATURES

Powerhouse Structure
Type: Aboveground
Length: 350 ft   Width: 151 ft   Height: 137 ft

Guard Valves
Number: None   Diameter: ft
Type:

Transformers
Number: 4
Ratings: 4 total: 1 @ 26.75; 3 @ 24.40
Voltages: (kV) 230 / 13.2

Generator
Rating generating (MVA): 115.0   Rating pumping (MVA): 87.0
Insulation type:
Starting method: Pump unwatered across the line
Starting equipment: Pump unwatered across the line

Runners
Material: Cast steel ASTM A-27 grade 7036
Minimum unit submergence: 0.0 ft
WR^2:
Manufacturer: Allis Chalmers
Model test by: Allis Chalmers

	Reversible Runners	Reversible Motor/Generator	
Number	3		3
Diameter (ft)	17.20	Rotor	Stator
rpm synchronous	112.5		112.5
rpm overspeed	169.0		169.0
Type	Francis		

Information on Runners

Condition:	Gross Head (ft) Generating	Gross Head (ft) Pumping	Capacity (MW) Generating	Capacity (MW) Pumping	Discharge (ft³/s) Generating	Discharge (ft³/s) Pumping	Turbine/Pump Eff.(%) Generating	Turbine/Pump Eff.(%) Pumping
Maximum head & maximum power	101	101	30	29	3,780	2,770	90.9	92.6
Minimum head & maximum power	85	85	26	28	3,900	3,470	90.0	92.5

Note: Data in the above table are based on design & model data.

Condition:	Net Head (ft) Generating	Net Head (ft) Pumping	Capacity (MW) Generating	Capacity (MW) Pumping	Discharge (ft³/s) Generating	Discharge (ft³/s) Pumping	Turbine/Pump Eff.(%) Generating	Turbine/Pump Eff.(%) Pumping
Rated head @ best gate	85		26					

Thermalito Power Plant - Page 3 (English)

THERMALITO POWER PLANT

POWERHOUSE and RELATED FEATURES

Powerhouse Structure
Type: Aboveground
Length: 106.7 m   Width: 46.0 m   Height: 41.8 m

Guard Valves
Number: None   Diameter: m
Type:

Transformers
Number: 4
Ratings: 4 total: 1 @ 26.75; 3 @ 24.40
Voltages: (kV) 230 / 13.2

Generator
Rating generating (MVA): 115.0   Rating pumping (MVA): 87.0
Insulation type:
Starting method: Pump unwatered across the line
Starting equipment: Pump unwatered across the line

Runners
Material: Cast steel ASTM A-27 grade 7036
Minimum unit submergence: m
WR^2:
Manufacturer: Allis Chalmers
Model test by: Allis Chalmers

	Reversible Runners	Reversible Motor/Generator		
Number	3			3
Diameter m	5.23	Rotor	Stator	
rpm synchronous	112.5			112.5
rpm overspeed	169.0			169.0
Type	Francis			

Information on Runners

Condition:	Gross Head (m) Generating	Gross Head (m) Pumping	Capacity (MW) Generating	Capacity (MW) Pumping	Discharge (m³/s) Generating	Discharge (m³/s) Pumping	Turbine/Pump Eff.(%) Generating	Turbine/Pump Eff.(%) Pumping
Maximum head & maximum power	30.8	30.8	30	29	107.0	78.4	90.9	92.6
Minimum head & maximum power	25.9	25.9	26	28	110.4	98.3	90.0	92.5

Note: Data in the above table are based on design & model data.

Condition:	Net Head (m) Generating	Net Head (m) Pumping	Capacity (MW) Generating	Capacity (MW) Pumping	Discharge (m³/s) Generating	Discharge (m³/s) Pumping	Turbine/Pump Eff.(%) Generating	Turbine/Pump Eff.(%) Pumping
Rated head @ best gate	25.9		26					

Thermalito Power Plant - Page 3 (Metric)

THERMALITO POWER PLANT

Plant Data:

Average GWh generating per year:	
Average GWh pumping per year:	
Starting time from standstill (s):	300
Changeover time pumping to generating (min):	60
Planned/scheduled time between major overhauls (years):	1
Outage time required per unit during major overhauls (weeks):	2
Representative plant availability (%):	73.0
Representative planned outages (weeks per year):	2

Miscelleneous Notes:

The Thermalito powerplant has three modified Francis turbines and one Kaplan unit. The tables describe the Francis units. The Kaplan unit has the following characteristics:

Diameter	4.318 m	(14.167 ft)
Spacing	21.3 m	(70 ft)
Rating as Turbine	45,000 hp	
RPM synchronous	138.5 rpm	
Generation		
Maximum discharge	127.4 m³/s	(4500 ft³/s)
@ head	25.9 m	(85 ft)
Material		
Cast steel ASTM A148 Grade 80-50 with 18-8 stainless steel overlay.		

Turbine design and model tests by Allis Chalmers.

The plant has four penstocks. The three leading to the pumped storage units have a diameter of 6.4 m (21 ft). The other has a diameter of 7.31 m (24 ft).

The plant has 14 water passages that make up the outlet works, described as follows:

Discharge capacity m³/s	481	65	1.4	14.1	34
Number of water passages	1	4	1	3	5
Dimensions of water passages					
Height m	4.3	1.8			
Width m	4.3	2.1			
Diameter mm			762	1,830	2,440
Type of gates	Radial	Slide	Slide	Slide	Slide
Number of gates	5	4	1	3	5
Discharge capacity ft³/s	17,000	2,300	50	500	1,200
Number of water passages	1	4	1	3	5
Dimensions of water passages					
Height ft	14	6			
Width ft	14	7			
Diameter in.			30	72	96
Type of gates	Radial	Slide	Slide	Slide	Slide
Number of gates	5	4	1	3	5

THERMALITO POWER PLANT

Cavitation Experience:
The Thermalito powerplant has minimal cavitation while running at a constant head.

Significant or Unique Problems:
Debris plugs the intakes.

List of Licenses Required:
FERC.

ENVIRONMENTAL FEATURES

Recreation:
Facilities include a public boat ramp, picnic area, and a swimming beach. Recreationists enjoy boating, fishing, waterskiing, swimming, and hunting.

Fish and Wildlife:
Trout, bass, catfish, panfish, and ducks are present in the area.

Social:

Page 4 (Continued)

Aerial View

Powerhouse and Outlet Works

Thermalito Pumped Storage Plant

Plant Name: **WALLACE DAM**

Plant location:
  Eatonton, GE
  Putman-Handcock County
	Owner: Georgia Power Company
	P.O. Box 4545
	Atlanta, Georgia 30302

Rated capacity 209 MW Average static head 95 ft Plant efficiency % Stored energy 213,000 MWh Number of units 4	Designers: Southern Company Services, Birmingham, Alabama
Construction time: 5 years, 11 months Construction cost: $1,074 per kW Price level: 1988 First commercial power: December 1979 FERC project number: 2413 River or water source: Oconee	Plant Manager/Superintendent: J. R. Pope P.O. Box 630 Eatonton, Georgia 31024-0630 (404) 526-3608

	UPPER RESERVOIR	LOWER RESERVOIR
DAM Type	Concrete, earth, non-overflow	Concrete and earth, non-overflow.
Height (ft)	117	100
Crest length (ft)	2,395	2,766
Volume (yd^3)	990,000	947,000
RESERVOIR Type	Constructed, Oconee Lake	Existing, Lake Sinclair
Surface area (acres)	19,050	15,400
Usable power storage (acre-ft)	756,694	724,120
Power pool fluctuation (ft)	5.0	2.0
Operating levels		
Maximum (ft)	435.0	340.0
Minimum (ft)	430.0	338.0
Drainage area (miles^2)	1,830.0	2,910.0
Seepage (ft^3/s)		
SPILLWAY		
Design flood		
Return period (years)	PMF	PMF
Flow (ft^3/s)	268,000	479,000
Capacity (ft^3/s)	268,000	479,000
Type	Gated ogee	Gated ogee
Gates		
Number	5	24
Type	Radial lift	Radial lift
Width (ft)	42.00	30.00
Height (ft)	44.00	18.00
OUTLET WORKS		
Discharge capacity (ft^3/s)		
Number of water passages	None	None
Dimensions of water passages		
Height (ft)		
Width (ft)		
Diameter (ft)		
Type of gates		
Number of gates	None	None

Plant Name: **WALLACE DAM**

Plant location Eatonton, GE Putman-Handcock County	Owner: Georgia Power Company P.O. Box 4545 Atlanta, Georgia 30302
Rated capacity 209 MW Average static head 29.0 m Plant efficiency % Stored energy 213,000 MWh Number of units 4	Designers: Southern Company Services, Birmingham, Alabama
Construction time: 5 years, 11 months Construction cost: $1,074 per kW Price level: 1988 First commercial power: December 1979 FERC project number: 2413 River or water source: Oconee	Plant Manager/Superintendent: J. R. Pope P.O. Box 630 Eatonton, Georgia 31024-0630 (404) 526-3608

	UPPER RESERVOIR	LOWER RESERVOIR
DAM Type	Concrete, earth, non-overflow	Concrete and earth, non-overflow.
Height (m)	35.7	30.5
Crest length (m)	730.0	843.1
Volume (m³)	756,694	724,120
RESERVOIR Type	Constructed, Oconee Lake	Existing, Lake Sinclair
Surface area (Mm²)	77.09	62.32
Usable power storage (Mm³)	933.374	893.194
Power pool fluctuation (m)	1.52	0.61
Operating levels		
Maximum (m)	132.59	103.63
Minimum (m)	131.06	103.02
Drainage area (Mm²)	4,739.705	7,536.908
Seepage (m³/s)		
SPILLWAY Design flood		
Return period (years)	PMF	PMF
Flow (m³/s)	7,589	13,564
Capacity (m³/s)	7,589	13,564
Type	Gated ogee	Gated ogee
Gates		
Number	5	24
Type	Radial lift	Radial lift
Width (m)	12.802	9.144
Height (m)	13.411	5.486
OUTLET WORKS		
Discharge capacity (m³/s)		
Number of water passages	None	None
Dimensions of water passages		
Height (m)		
Width (m)		
Diameter (m)		
Type of gates		
Number of gates	None	None

Wallace Dam - Page 1 (Metric)

WALLACE DAM

INTAKES	UPPER INTAKE	LOWER INTAKE
Number	6	4
Type	Concrete	Concrete
Design discharge (ft³/s)	6,400	8,000
Gross inlet area (ft²)	1,863	1,662
(at trash racks)		
Bar racks		
spacing (in)	14.50	14.50
shape	V-round H-elliptical	V-round, H-elliptical
depth/thickness (in)	22.00 / 1.00	22.00 / 1.00
diameter (in)	4.00	4.00
Emergency gates		
number	None	None
height/width (ft)		
type		
Service gates		
number	2	3
height/width (ft)	32.35 / 17.11	16.15 / 23.50
type	Slipe	Slipe
Bulkhead/stop logs (Y or N)	Y	
number of units serviced	2	
Hoists		
number	1	1
capacity (tons)	50	35
type	Gantry	Gantry

WATER PASSAGES	Upper Tunnel	Shaft	Lower Tunnel	Surge Tanks Upper	Lower	Penstocks	Tailrace Tunnel
Number						6	6
Diameter (ft)						25.5	
Length (ft)						95	77
Maximum velocity (ft/s)						15.7	21.2
Concrete lining thickness (in)							
Total length of concrete sections (ft)						83	77
Steel liner							
Thickness							
Minimum (in)							
Maximum (in)							
Material grade							
Total length of steel-lined sections (ft)						12	

Notes:
Tailrace Tunnel: Diameter is reported as 23.266 m (76.33 ft) by 4.648 m (15.25 ft).

WALLACE DAM

INTAKES	UPPER INTAKE	LOWER INTAKE
Number	6	4
Type	Concrete	Concrete
Design discharge (m³/s)	181.2	226.5
Gross inlet area (m²)	173.1	154.4
(at trash racks)		
Bar Racks:		
spacing (mm)	368	368
shape	V-round H-elliptical	V-round, H-elliptical
depth/thickness (mm)	559 / 25	559 / 25
diameter (mm)	102	102
Emergency gates		
number	None	None
height/width (m)		
type		
Service gates		
number	2	3
height/width (m)	9.861 / 5.216	4.921 / 7.163
type	Slipe	Slipe
Bulkhead/stop logs (Y or N)	Y	
number of units serviced	2	
Hoists		
number	1	1
capacity (Mg)	45	32
type	Gantry	Gantry

WATER PASSAGES	Upper Tunnel	Shaft	Lower Tunnel	Surge Tanks Upper	Surge Tanks Lower	Penstocks	Tailrace Tunnel
Number						6	6
Diameter (m)						7.77	
Length (m)						29.0	23.5
Maximum velocity (m/s)						4.79	6.46
Concrete lining thickness (m)							
Total length of concrete sections (m)						25.3	23.5
Steel liner Thickness							
Minimum (mm)							
Maximum (mm)							
Material grade							
Total length of steel-lined sections (m)						3.7	

Notes:
Tailrace Tunnel: Diameter is reported as 23.266 m (76.33 ft) by 4.648 m (15.25 ft).

WALLACE DAM

POWERHOUSE and RELATED FEATURES

Powerhouse Structure
Type: Concrete
Length: 531 ft   Width: 100 ft   Height: 90 ft

Guard Valves
Number: None   Diameter: ft
Type:

Transformers
Number: 2
Ratings:
Voltages: (kV) 230 / 14.4

Generator
Rating generating (MVA): 52.2   Rating pumping (MVA): 63.0
Insulation type: Thermoelastic epoxy
Starting method: Reduced voltage, transformer tap
Starting equipment:

Runners
Material: Steel
Minimum unit submergence: 7.0 ft
WR^2:
Manufacturer: Allis Chalmers
Model test by: Allis Chalmers

	Reversible Runners	Reversible Motor/Generator	
Number	4		4
Diameter (ft)	21.90	Rotor	Stator
rpm synchronous	85.7		85.7
rpm overspeed	105.0		105.0
Type	Modified Francis		Synchronous motor

Information on Runners

Condition:	Gross Head (ft) Generating	Gross Head (ft) Pumping	Capacity (MW) Generating	Capacity (MW) Pumping	Discharge (ft³/s) Generating	Discharge (ft³/s) Pumping	Turbine/Pump Eff.(%) Generating	Turbine/Pump Eff.(%) Pumping
Maximum head & maximum power	96	96			8,000	6,400	92.0	
Minimum head & maximum power	89	89			8,000	6,400	91.0	

Note: Data in the above table are based on model tests.

Condition:	Net Head (ft) Generating	Net Head (ft) Pumping	Capacity (MW) Generating	Capacity (MW) Pumping	Discharge (ft³/s) Generating	Discharge (ft³/s) Pumping	Turbine/Pump Eff.(%) Generating	Turbine/Pump Eff.(%) Pumping
Rated head @ best gate	96	96	53	63	7,200	7,800	91.8	93.0

Note: Data in the above table are based on field test.

Wallace Dam - Page 3 (English)

WALLACE DAM

POWERHOUSE and RELATED FEATURES

Powerhouse Structure
Type: Concrete
Length: 161.8 m Width: 30.5 m Height: 27.4 m

Guard Valves
Number: None Diameter: m
Type:

Transformers
Number: 2
Ratings:
Voltages: (kV) 230 / 14.4

Generator
Rating generating (MVA): 52.2 Rating pumping (MVA): 63.0
Insulation type: Thermoelastic epoxy
Starting method: Reduced voltage, transformer tap
Starting equipment:

Runners
Material: Steel
Minimum unit submergence: 2.13 m
WR^2:
Manufacturer: Allis Chalmers
Model test by: Allis Chalmers

	Reversible Runners	Reversible Motor/Generator	
Number	4		4
Diameter m	6.68	Rotor	Stator
rpm synchronous	85.7		85.7
rpm overspeed	105.0		105.0
Type	Modified Francis		Synchronous motor

Information on Runners

Condition:	Gross Head (m)		Capacity (MW)		Discharge (m³/s)		Turbine/Pump Eff.(%)	
	Generating	Pumping	Generating	Pumping	Generating	Pumping	Generating	Pumping
Maximum head & maximum power	29.3	29.3			226.5	181.2	92.0	
Minimum head & maximum power	27.1	27.1			226.5	181.2	91.0	

Note: Data in the above table are based on model tests.

Condition:	Net Head (m)		Capacity (MW)		Discharge (m³/s)		Turbine/Pump Eff.(%)	
	Generating	Pumping	Generating	Pumping	Generating	Pumping	Generating	Pumping
Rated head @ best gate	29.3	29.3	53	63	203.9	220.9	91.8	93.0

Note: Data in the above table are based on field test.

Wallace Dam - Page 3 (Metric)

WALLACE DAM

Plant Data:
   Average GWh generating per year:
   Average GWh pumping per year:
   Starting time from standstill (s):   240
   Changeover time pumping to generating (min):   6
   Planned/scheduled time
     between major overhauls (years):   5
   Outage time required per unit
     during major overhauls (weeks):   9
   Representative plant availability (%):   96.0
   Representative planned outages (weeks per year):   1

Miscelleneous Notes:
   Wallace Dam includes 6 units, 4 reversible units and 2 conventional units.

   The upper dam volume of 578,534 m^3 (756,694 yd^3) includes 302,514 m^3 (395,673 yd^3) of concrete and 276,020 m^3 (361,021 yd^3) of earth.

   The lower dam volume of 553,630 m^3 (724,120 yd^3) includes 161,858 m^3 (211,703 yd^3) of concrete and 391,771 m^3 (512,317 yd^3) of earth.

Cavitation Experience:
   Cavitation has progressed beyond the stainless steel overlay. Required amount of rebuilding with stainless steel has gradually reduced to now include only minor areas at each overhaul.

Significant or Unique Problems:
   Runner band failure, excessive gate stem bushing wear on pump units, stator coil failure on pump during initial start-up.

List of Licenses Required:
   FERC and Corps Section 404 permit during construction.

ENVIRONMENTAL FEATURES

Recreation:
   Recreational usage exceeds 600,000 visitors per year. Includes boating, hiking, camping, fishing, and swimming. project has caused extensive recreational development.

Fish and Wildlife:
   Active Osprey are nesting in the reservoir and Bald Eagles nest at the Dam. A wildlife refuge was created from the project. There are no fishery issues.

Social:

WALLACE DAM

Project has increased tourism and rec development. Taxes collected have increased. Farmland converted to residential and recreatonal land near lake.

WALLACE DAM DATA

PURPOSE

To provide low cost electricity at peak loads by means of the pump storage method of generation.

To provide public recreational facilities throughout the area.

Wallace Dam is also strategically located to provide system stability and flexibility in operation. It will provide spinning reserve centrally located with a maximum of 324 MW, valuable area protection, and emergency start-up power for Plant Branch, located on Lake Sinclair. In addition it will provide sites for future thermal plants.

DESCRIPTION

The Wallace Dam site is located on the Oconee River in the Altamaha Drainage Basin, 1.5 miles north of Georgia Highway 16 between Eatonton and Sparta, Georgia, at the headwaters of Lake Sinclair, and 1/4 mile south of the intersection of the Greene, Putnam, and Hancock County lines.

PLANT DATA

DAM

Type:	Gravity Concrete/Earth Dam
Top Elevation (Plant Datum):	445
Stream Bed Elevation (Plant Datum):	328
Maximum Height Above Stream Bed (ft.):	117
Total Length (ft.):	2,395
Full Pond Operating Head (ft.):	95

CONCRETE STRUCTURES

Structure	Concrete Quantity (cu.yds.)	Length (ft.)
West Nonoverflow	70,141	300
East Nonoverflow	35,943	226
Spillway	55,559	266
Headworks	115,460	531
Powerhouse	118,570	–
Total Concrete Stucture	395,673	1,323

EARTH DAMS

Dams	Earth Fill Quantity (cu.yds.)	Rock Fill Quantity (cu.yds.)	Length (ft.)
West Earth Dam	202,808	41,465	347
East Earth Dam	138,368	21,779	725
Total Main Earth Dams	341,176	63,344	1,072
Saddle Dike	19,845	5,606	900
Total Earth Dams	361,021	68,950	1,972

EXCAVATION

Structure	Common (cu.yds.)	Rock (cu.yds.)
West Nonoverflow	156,670	8,736
Spillway	36,414	8,280
East Nonoverflow	9,405	2,226
Powerhouse & Headworks	62,200	218,146
Total Concrete Dam	264,689	237,388
Tailrace	8,770,000	-
Total Excavation	9,034,689	237,388

MISCELLANEOUS INFORMATION

Application for License Filed	April 29, 1966
Original License Order Issued	August 6, 1969
Application for Amendment Filed	September 4, 1969
Order Amending License Order Issued	June 19, 1970
License Accepted	July 15, 1970
Application for Modification of Effective Date	September 24, 1970
Order Modifying License Order Issued	February 18, 1971
Effective Date of License	June 1, 1970
Began Construction of Access Road and Staging Area	September 30, 1971
Began Construction of Plant Area	June 8, 1972
First Concrete Placed	October 30, 1972
Construction Suspended	December 20, 1972
Construction Resumed	May 7, 1974
Construction Suspended	January 6, 1975
Construction Resumed	May 13, 1976
First Unit in Commercial Operation	December, 1979
Estimated Completion	May, 1980

UNIT DATA

Pump Turbines and Turbines manufactured by Allis-Chalmers.
Generator Motors and Generators manufactured by General Electric.
Governors manufactured by Woodward Governor Company.

Number of Pump Turbine/Generator Motor Units	4
Nameplate Rating -	
- Each Reversible Pump (hp)	83,000
- Each Generator (MW)	52.2
Number of Turbine/Generator Units	2
Nameplate Rating Each Turbine/Generator (MW)	56.25
Total Nameplate Capacity (MW)	321.3
Total Dependable Capacity (MW)	321.3
Average Annual Energy -	
- Natural (kWH)	128,000,000
- Total (kWH)	341,000,000
Generator Voltage (kV)	14.4
Transmission Voltage (kV)	230.0
Exciter Rated Capacity -	
- Pump Turbine (kV)	446
- Conventional (kv)	348
Exciter Voltage -	
- Pump Turbine (kV)	371
- Conventional (kV)	364
Shaft Velocity -	
- Pump Turbine (rpm)	85.8
- Conventional (rpm)	120.0

LAKE DATA

OCONEE LAKE DATA

RESERVOIR

Official Name:	Oconee Lake
Drainage Area Above Dam Site (sq. mi.):	1,830
Normal Full Pond Elevation (ft. above Plant Datum):	435.0
Full Reservoir Storage (acre-ft.):	370,000
Full Reservoir Storage (billions of gallons):	120.5
Full Reservoir Shoreline (miles):	374
Surface Area (acres):	19,050
Normal Daily Drawdown (ft.):	1.5

Reservoir extends into Greene, Hancock, Morgan, and Putnam Counties.

MITIGATION AND RECREATIONAL ENHANCEMENTS

Land provided to State Game and Fish as -
- Mitigation for inundated wildlife habitat (ac.): 10,300
- Fish plots (50 plots-5 acres each) (ac. total): 250
- Wildlife habitat (ac.): 1,250
- Recreational Areas (16 areas) (ac. total): 3,215

SINCLAIR LAKE

Sinclair Lake is formed by the Oconee, Little River, and other small tributaries which are retained by Sinclair Dam, which is located on the Oconee River approximately 20 miles downstream of Wallace Dam. Near Milledgeville, Georgia, the lake encompasses approximately 15,000 acres within Baldwin, Putnam, Hancock and Jones Counties, which provides hydropower for two hydro turbine generators at the Sinclair Dam Plant.

Sinclair Lake is also the tailpond for Wallace Dam and acts as a holding pond for pumped storage at Wallace Dam.

Normal full pond elevation is 340.0 feet above sea level, with a normal daily drawdown of approximately 1.8 feet.

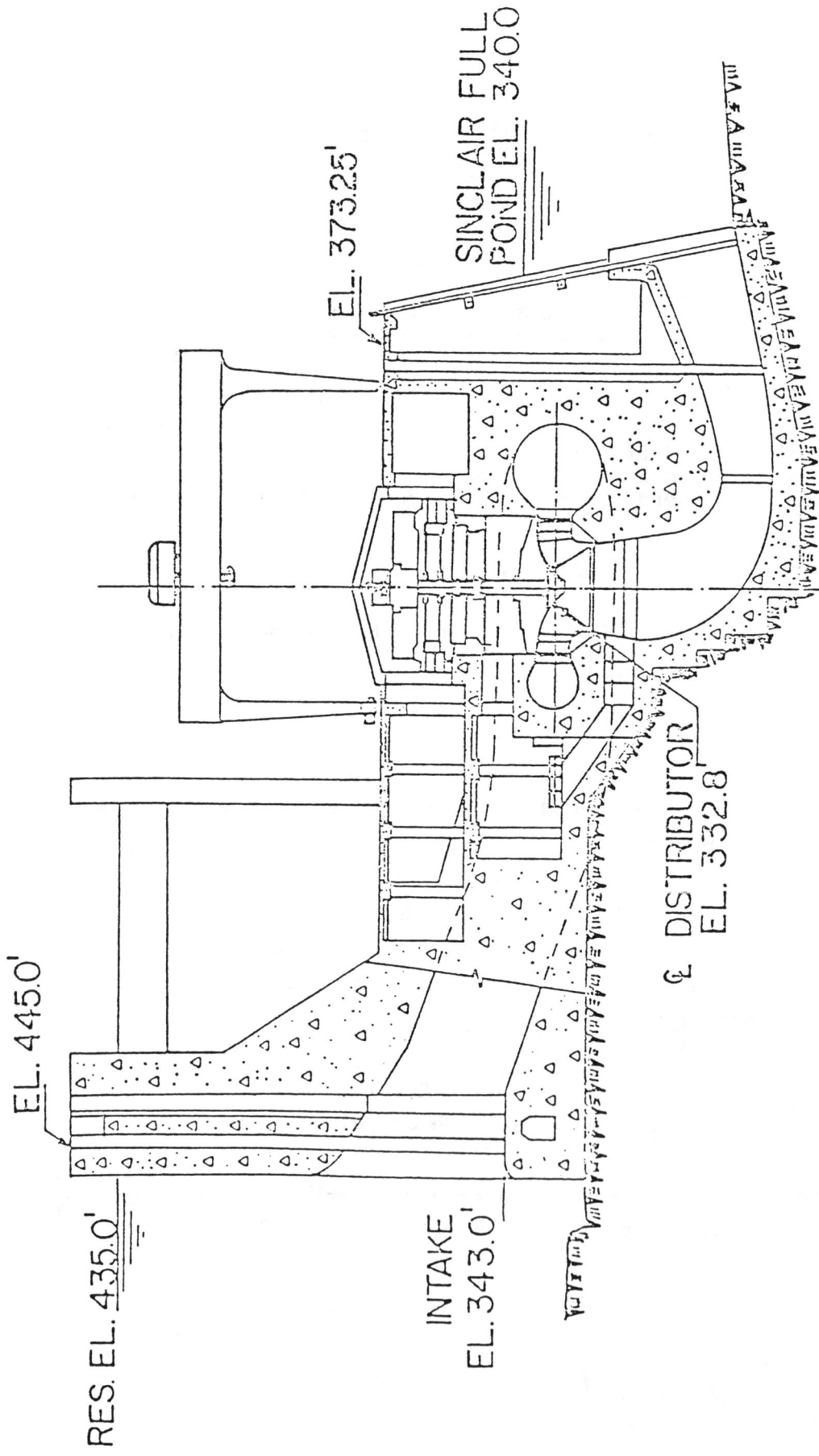

FIGURE 2. SECTION OF PUMP-TURBINE UNIT

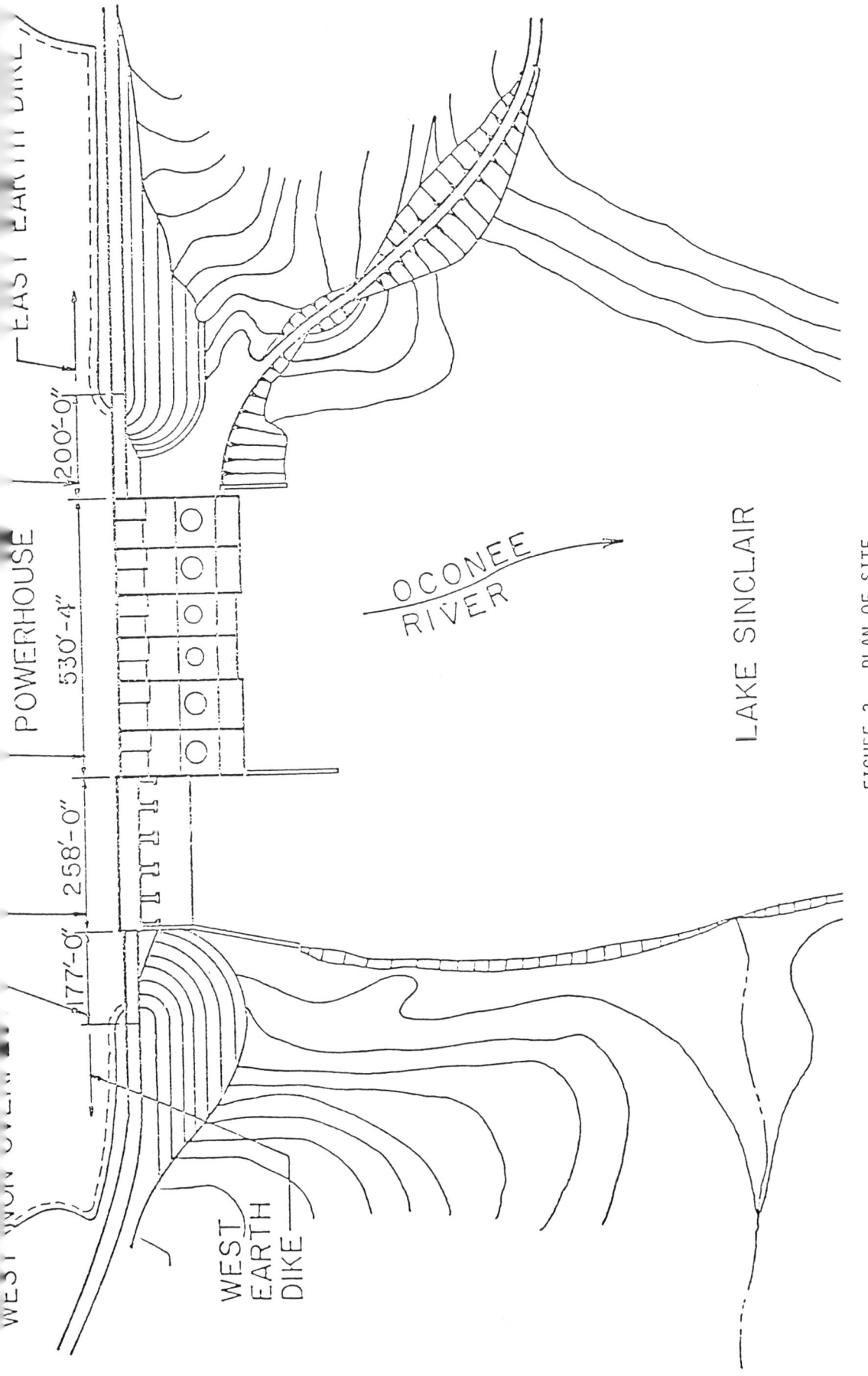

FIGURE 3. PLAN OF SITE

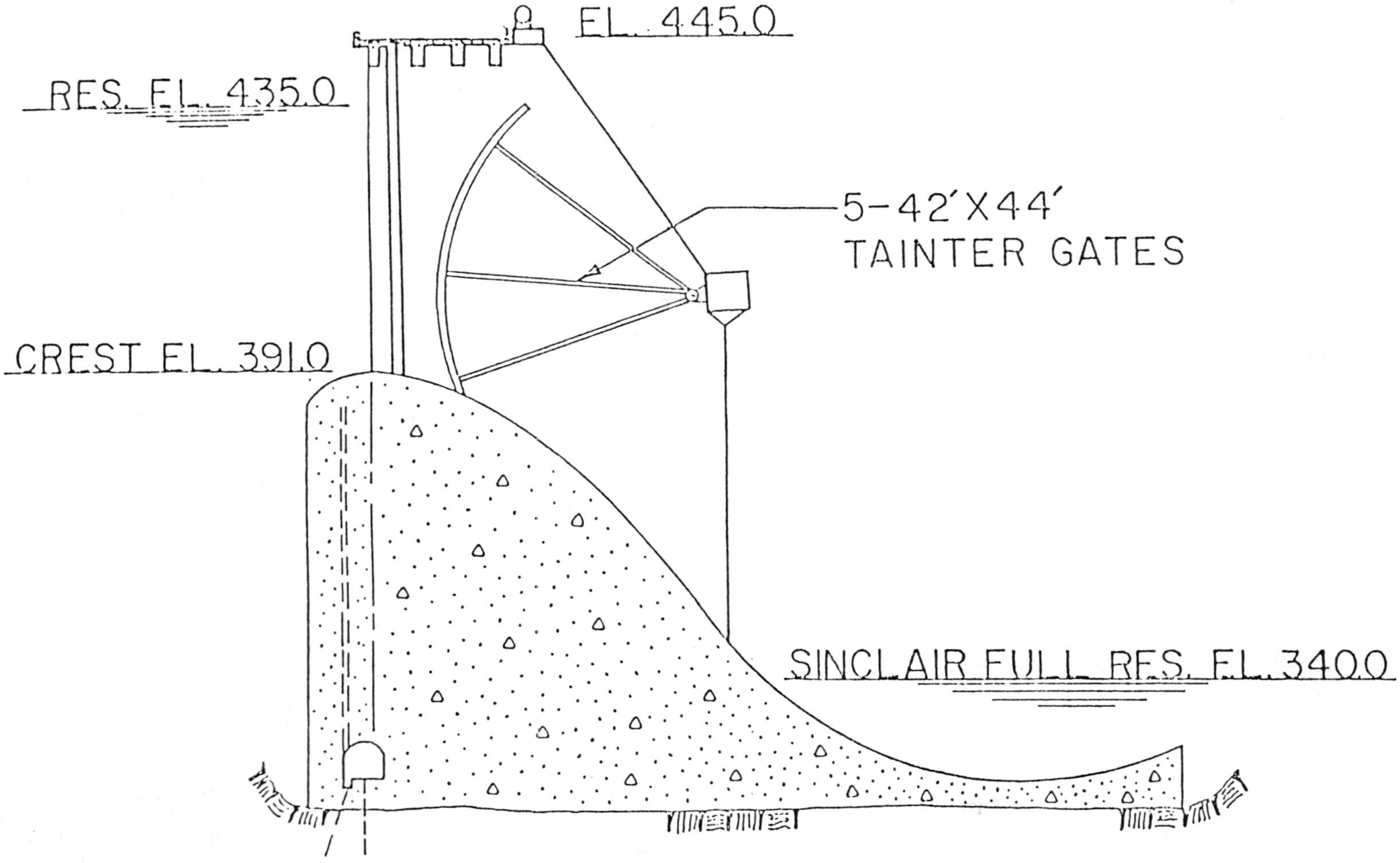

FIGURE 4. SPILLWAY SECTION

Wallace Dam and Powerhouse

Wallace Pumped Storage Plant

BIBLIOGRAPHY

1. "Optimization of the Operation of Wallace Dam and Sinclair Dam as a Pumped Storage Development," Master's Thesis by James Owen Patrick, Georgia Institute of Technology, Atlanta, Georgia.

2. "Wallace Dam," by Major H. Thompson, Jr., Journal of the Power Division, Proceedings of ASCE, Vol. 98, No. PO2, October 1972, pages 333-348. Paper #9259.

3. "Unsteady Seepage Analysis of Wallace Dam," by Mustafa M. Aral, M. ASCE and Morris L. Maslia, A.M. ASCE, Journal of Hydraulic Engineering, Vol. 109, No. 6, June 1983. Paper #18045.

Plant Name: **YARDS CREEK STATION**

Plant location:
Blairstown, NJ
Warren County

Owner: Jersey Central P&L Co. & Public Service Electric & Gas Co.
Madison Avenue At Punch Bowl Road
Morristown, New Jersey 07960

Rated capacity	360 MW
Average static head	732 ft
Plant efficiency	66.67 %
Stored energy	2,800 MWh
Number of units	3

Designers:
Ebasco Services, Inc.

Construction time: 2 years, 6 months
Construction cost: $92 per kW
Price level: 1965
First commercial power: September 1965
FERC project number: 2309

Plant Manager/Superintendent:
Fred Kunich, Manager
Yards Creek Station, Jersey Central P&L
P.O Box L
Blairstown, New Jersey 07825
(908) 362-6163

River or water source: Yards Creek

DAM	UPPER RESERVOIR	LOWER RESERVOIR
Type	Rockfill with earth core	Earthfill, till core
Height (ft)	70	55
Crest length (ft)	9,600	1,300
Volume (yd³)	26,700,000	8,700,000
RESERVOIR		
Type	Constructed	On stream, constructed
Surface area (acres)	154	282
Usable power storage (acre-ft)	4,650	4,650
Power pool fluctuation (ft)	49.0	23.5
Operating levels		
Maximum (ft)	1,555.0	818.5
Minimum (ft)	1,506.0	795.0
Drainage area (miles²)	0.2	4.3
Seepage (ft³/s)	0.989	
SPILLWAY		
Design flood		
Return period (years)		
Flow (ft³/s)		11,000
Capacity (ft³/s)		11,000
Type	None	Concrete ogee
Gates		
Number	None	None
Type		
Width (ft)		
Height (ft)		
OUTLET WORKS		
Discharge capacity (ft³/s)		318
Number of water passages	None	2
Dimensions of water passages		
Height (ft)		
Width (ft)		
Diameter (ft)		2.50
Type of gates		Cone valve
Number of gates		2

Plant Name: **YARDS CREEK STATION**

Plant location Blairstown, NJ Warren County	Owner: Jersey Central P&L Co. & Public Service Electric & Gas Co. Madison Avenue At Punch Bowl Road Morristown, New Jersey 07960
Rated capacity 360 MW Average static head 223.0 m Plant efficiency 66.67 % Stored energy 2,800 MWh Number of units 3	Designers: Ebasco Services, Inc.
Construction time: 2 years, 6 months Construction cost: $92 per kW Price level: 1965 First commercial power: September 1965 FERC project number: 2309	Plant Manager/Superintendent: Fred Kunich, Manager Yards Creek Station, Jersey Central P&L P.O Box L Blairstown, New Jersey 07825 (908) 362-6163
River or water source: Yards Creek	

	UPPER RESERVOIR	LOWER RESERVOIR
DAM		
Type	Rockfill with earth core	Earthfill, till core
Height (m)	21.3	16.8
Crest length (m)	2,926.1	396.2
Volume (m³)	20,413,624	6,651,630
RESERVOIR		
Type	Constructed	On stream, constructed
Surface area (Mm²)	0.62	1.14
Usable power storage (Mm³)	5.736	5.736
Power pool fluctuation (m)	14.94	7.16
Operating levels		
Maximum (m)	473.96	249.48
Minimum (m)	459.03	242.32
Drainage area (Mm²)	0.622	11.137
Seepage (m³/s)	0.0280	
SPILLWAY		
Design flood		
Return period (years)		
Flow (m³/s)		311
Capacity (m³/s)		311
Type	None	Concrete ogee
Gates		
Number	None	None
Type		
Width (m)		
Height (m)		
OUTLET WORKS		
Discharge capacity (m³/s)		9
Number of water passages	None	2
Dimensions of water passages		
Height (m)		
Width (m)		
Diameter (m)		0.762
Type of gates		Cone valve
Number of gates		2

YARDS CREEK STATION

INTAKES	UPPER INTAKE	LOWER INTAKE
Number	1	3
Type	Conventional bell mouth concrete	Divided draft tube
Design discharge (ft³/s)	6,000	2,000
Gross inlet area (ft²) (at trash racks)	2,900	1,000
Bar racks		
spacing (in)	3.00	6.00
shape	Rectangular	Rectangular
depth/thickness (in)	3.00 / 0.38	3.00 / 0.50
diameter (in)		
Emergency gates		
number	None	None
height/width (ft)		
type		
Service gates		
number	6	6
height/width (ft)	10.00 / 12.00	12.00 / 12.00
type	Steel-faced stop log panels	Steel panels
Bulkhead/stop logs(Y or N)		
number of units serviced		
Hoists		
number	1	2
capacity (tons)	100	327
type	Mobile crane	Gantry crane

WATER PASSAGES	Upper Tunnel	Shaft	Lower Tunnel	Surge Tanks Upper	Surge Tanks Lower	Penstocks	Tailrace Tunnel
Number	1					3	
Diameter (ft)	20.0					10.0	
Length (ft)	1,500					2,000	
Maximum velocity (ft/s)	26.0					29.0	
Concrete lining thickness (in)	12.00						
Total length of concrete sections (ft)	1,500						
Steel liner Thickness							
Minimum (in)	0.82					0.98	
Maximum (in)	0.98					2.00	
Material grade						Luken penstock steel	
Total length of steel-lined sections (ft)	200					2,000	

Notes:

YARDS CREEK STATION

INTAKES	UPPER INTAKE	LOWER INTAKE
Number	1	3
Type	Conventional bell mouth concrete	Divided draft tube
Design discharge (m³/s)	169.9	56.6
Gross inlet area (m²) (at trash racks)	269.4	92.9
Bar Racks:		
spacing (mm)	76	152
shape	Rectangular	Rectangular
depth/thickness (mm)	76 / 10	76 / 13
diameter (mm)		
Emergency gates		
number		
height/width (m)	None	None
type		
Service gates		
number	6	6
height/width (m)	3.048 / 3.658	3.658 / 3.658
type	Steel-faced stop log panels	Steel panels
Bulkhead/stop logs (Y or N)		
number of units serviced		
Hoists		
number	1	2
capacity (Mg)	91	297
type	Mobile crane	Gantry crane

WATER PASSAGES	Upper Tunnel	Shaft	Lower Tunnel	Surge Tanks Upper	Surge Tanks Lower	Penstocks	Tailrace Tunnel
Number	1					3	
Diameter (m)	6.10					3.05	
Length (m)	457.2					609.6	
Maximum velocity (m/s)	7.93					8.84	
Concrete lining thickness (m)	0.305						
Total length of concrete sections (m)	457.2						
Steel liner Thickness							
Minimum (mm)	21					25	
Maximum (mm)	25					51	
Material grade						Luken penstock steel	
Total length of steel-lined sections (m)	61.0					609.6	

Notes:

Yards Creek Station - Page 2 (Metric)

YARDS CREEK STATION

POWERHOUSE and RELATED FEATURES

Powerhouse Structure
Type: Concrete below grade - surface
Length: 155 ft   Width: 64 ft   Height: 70 ft

Guard Valves
Number: 3   Diameter: 7.0 ft
Type: Spherical

Transformers
Number: 3
Ratings: 129/144 @ 55/65 degree rise
Voltages: (kV) 230 / 14.4

Generator
Rating generating (MVA): 125.0   Rating pumping (MVA): 140.0
Insulation type: NECCA Bond (National Elect. Coil Co.)
Starting method: Current limiting reactor across the line
Starting equipment: Current limiting reactor across the line

Runners
Material: 13-6 cast stainles steel
Minimum unit submergence: 25.0 ft
WR^2: 31,000,000 (lbf x ft²)
Manufacturer: Baldwin-Lima-Hamilton
Model test by: Baldwin-Lima-Hamilton

	Reversible Runners	Reversible Motor/Generator		
Number	3			3
Diameter (ft)	18.40	Rotor	18.50	Stator
rpm synchronous	240.0			240.0
rpm overspeed	364.0			364.0
Type	Francis			

Information on Runners

Condition:	Gross Head (ft) Generating	Gross Head (ft) Pumping	Capacity (MW) Generating	Capacity (MW) Pumping	Discharge (ft³/s) Generating	Discharge (ft³/s) Pumping	Turbine/Pump Eff.(%) Generating	Turbine/Pump Eff.(%) Pumping
Maximum head & maximum power	760	760	122			1,800	84.0	84.0
Minimum head & maximum power	688	688	112	138		1,900	82.0	

Note: Data in the above table are based on field tests.

Condition:	Net Head (ft) Generating	Net Head (ft) Pumping	Capacity (MW) Generating	Capacity (MW) Pumping	Discharge (ft³/s) Generating	Discharge (ft³/s) Pumping	Turbine/Pump Eff.(%) Generating	Turbine/Pump Eff.(%) Pumping
Rated head @ best gate	656	732	112			2,145		

Yards Creek Station - Page 3 (English)

YARDS CREEK STATION

POWERHOUSE and RELATED FEATURES

Powerhouse Structure
Type: Concrete below grade - surface
Length: 47.2 m   Width: 19.5 m   Height: 21.3 m

Guard Valves
Number: 3   Diameter: 2.13 m
Type: Spherical

Transformers
Number: 3
Ratings: 129/144 @ 55/65 degree rise
Voltages: (kV) 230 / 14.4

Generator
Rating generating (MVA): 125.0   Rating pumping (MVA): 140.0
Insulation type: NECCA Bond (National Elect. Coil Co.)
Starting method: Current limiting reactor across the line
Starting equipment: Current limiting reactor across the line

Runners
Material: 13-6 cast stainles steel
Minimum unit submergence: 7.62 m
WR^2: 12,814,000 (Newtons x m^2)
Manufacturer: Baldwin-Lima-Hamilton
Model test by: Baldwin-Lima-Hamilton

	Reversible Runners	Reversible Motor/Generator	
Number	3		3
Diameter m	5.61	Rotor 5.639	Stator
rpm synchronous	240.0		240.0
rpm overspeed	364.0		364.0
Type	Francis		

Information on Runners

Condition:	Gross Head (m)		Capacity (MW)		Discharge (m³/s)		Turbine/Pump Eff.(%)	
	Generating	Pumping	Generating	Pumping	Generating	Pumping	Generating	Pumping
Maximum head & maximum power	231.6	231.6	122			51.0	84.0	84.0
Minimum head & maximum power	209.6	209.6	112	138		53.8	82.0	

Note: Data in the above table are based on field tests.

Condition:	Net Head (m)		Capacity (MW)		Discharge (m³/s)		Turbine/Pump Eff.(%)	
	Generating	Pumping	Generating	Pumping	Generating	Pumping	Generating	Pumping
Rated head @ best gate	199.9	223.1	112			60.7		

Yards Creek Station - Page 3 (Metric)

YARDS CREEK STATION

Plant Data:

Average GWh generating per year:	537
Average GWh pumping per year:	807
Starting time from standstill (s):	360
Changeover time pumping to generating (min):	9
Planned/scheduled time between major overhauls (years):	10
Outage time required per unit during major overhauls (weeks):	9
Representative plant availability (%):	94.0
Representative planned outages (weeks per year):	2

Miscelleneous Notes:

The outlet works for the lower dam includes the following outlet works facilities:

    two - 0.76 m (2.5 ft) cone valves,
    one - .3 m (1 ft) butterfly valve, and
    one - .2 m (8 in.) butterfly valve.

Cavitation Experience:

There is minimal cavitation occurence.

Significant or Unique Problems:

Early problems now resolved include: pump-turbine runner cracking, M.G. winding insulation failures, main transformer failures, pump-turbine head cover cracking, and 230 kV air circut breaker leaks. The remaining problem is wicket gate vibration.

List of Licenses Required:

FERC, NJDEP.

ENVIRONMENTAL FEATURES

Recreation:

Recreation is passive, occurs during the daytime, and is year around. Facilities include a visitor center, nature overlooks, interpretive panels, hiking trails, a picnic area, and restrooms. No reservoir boating, swimming, or fishing is allowed. Hunting is allowed in specified areas per State game laws.

Fish and Wildlife:

There is no fishing in the reservoir and hunting is limited. The lower reservoir area is a wildlife refuge.

Social:

Group guided tours of the station are conducted.

YARDS CREEK PUMPED STORAGE PROJECT
GENERAL PROJECT DESCRIPTION

By
M. G. Salzman, Consultant Engineer
Ebasco Services Incorporated

LOCATION

The Yards Creek Pumped Storage Project is located in the Townships of Blairstown and Pahaquarry in Warren County, New Jersey. The high-level reservoir is on Kittatinny Mountain just southeast of the Delaware River and about 5 miles east of Delaware Water Gap. The low-level reservoir is on Yards Creek, on a small tributary of Paulins Kill which flows into Delaware River. The pumping and generating station is located about 4 miles northwest of Blairstown, New Jersey.

COMPONENT PARTS

The following component parts make up the project (see Figure 1).

1) Upper Reservoir - Dam and Dikes - Kittatinny Mountain.

2) Lower Reservoir - Dam, Spillway and Auxiliary Dike on Yards Creek.

3) Auxiliary Reservoir - Dam, Spillway and Auxiliary Supply Canal from Yards Creek.

4) Connecting Waterways - consisting of intake channel in Upper Reservoir, pressure tunnel and penstock with trifurcation to three distributor penstocks with spherical valves leading to pump-turbine units; draft-tubes and tailrace channel to Lower Reservoir.

5) Pumping and Generating Station - adjacent to Lower Reservoir, containing three vertical reversible pump-turbine units rated at 150,500 hp each as turbines under 656 feet minimum effective head, direct connected to generator/motors rated at 125,000 kva at 0.9 power factor as generators, with normal speed of 240 rpm.

6) Road System - including approximately 2 miles of new access road from the pumping and generating station around the escarpment of Kittatinny Mountain to the Upper Reservoir and the relocation of a portion of the township road which is below the high water level of the Lower Reservoir.

UPPER RESERVOIR - DAMS AND DIKES

This reservoir is located in a depressed swampy area on the top of the Kittatinny Range. There is practically no contributary drainage into the reservoir but the north swamp area drains into a ravine and tributary (designated tributary "Y") which flows east and south into Yards Creek.

The reservoir is formed by a noncontinuous series of rock-fill dikes around the borders of the swamps; these connect the high rock outcrops. They will impound an area of 164 acres at normal full pond elevation 1,555, reduced to 35 acres at minimum power elevation 1,506. The total storage capacity of the reservoir will be 4,900 acre-feet and the drawdown of 49 feet provides a usable storage of 4,650 acre-feet.

The maximum height of the north dam section across the tributary "Y" is 70 feet and the over-all length of dam and dikes is 9,000 feet. Approximately 1,900 feet of length along the west dike is designed to be the common dike between the present Yards Creek Upper Reservoir and the future upper level reservoir for the Delaware River pumped-storage development. A stub connection on this common dam will facilitate construction of the future dikes. The design of the dike is suitable for the variable drawdowns on either side of the dike.

LOWER RESERVOIR - DAM, DIKE AND SPILLWAY

This reservoir is formed by an earthfill dam constructed across Yards Creek approximately 1,500 feet above Mount Vernon Road, and by an auxiliary dike across a saddle at the southeastern side of the reservoir. Surface area is 300 acres at spillway crest elevation 818.5, with a storage capacity of 5,420 acre-feet. The power storage of 4,650 acre-feet, and an additional 270 acre-feet seasonal storage, require a drawdown to elevation 795, leaving a dead storage pool of 500 acre-feet.

The main dam is 1,300 feet long and connects at the right abutment with a concrete spillway, 100 feet wide, which is founded on shale. The spillway has a 10-foot wide crest in a notch at elevation 818.5 and a 90-foot wide ogee section at elevation 823.5. The spillway can discharge 6,000 cfs with the reservoir at elevation 830.8. The spillway chute is concrete-lined channel with a sloping floor terminating 165 feet downstream from the crest and discharging into the creek bed. The material from the approach channel excavation for the spillway furnished material for the berms at the toes of the main dam.

An outlet structure with trash racks and two 3-foot-diameter pipes was provided with two Howell-Bunger outlet valves located in a valve house which will permit regulated discharges of up to 200 cfs under conditions of maximum drawdown.

The maximum height of the main dam above stream bed is 56 feet and the structure was founded on sand and gravel with a central core of till and a slurry cut-off trench. The top of the dam is elevation 832.5 with an 11-foot width of roadway.

The auxiliary dike is 2,200 feet long and has a maximum height of 40 feet above the valley floor. The design of the dike is similar to that of the main dam including the rock berms at the toes.

AUXILIARY RESERVOIR - DAM, SPILLWAY AND SUPPLY CANAL

The Auxiliary Reservoir is located northeasterly of the Lower Reservoir and has a normal water elevation 868 with possible drawdown to minimum elevation 852.5. This will provide supplemental seasonal storage of about 480 acre-feet, compensating for evaporation, and releases from the project. The dam is 500 feet long and 20 feet high with a 10-foot-wide crest at elevation 870. The dam has a central core of clayey till with outer shell of sandy till.

The spillway consists of 100-foot-wide cut through the west abutment with a concrete sill and riprap. The excavation for the spillway provided a large part of the material used in the dam. The design capacity of the spillway is 250 cfs. The dam is also provided with two 18-inch outlet pipes with a screened inlet and control gates.

The Auxiliary Reservoir can obtain part of its water supply from the drainage area above it and part by an auxiliary supply canal from Yards Creek. This canal is approximately 1,500 feet long and can carry a maximum discharge of 10 cfs. The normal low water flow of Yards Creek will flow into the Auxiliary Reservoir and over the spillway into the Lower Reservoir. A discharge limiting structure at the entrance to the supply canal on Yards Creek will prevent floods from entering and damaging the canal.

CONNECTING WATERWAYS

Figure 1 shows the connecting waterways between the Upper Reservoir and the Lower Reservoir, and the profile of the reservoirs and connecting waterways, indicating variable static head differentials during the pumping and generating cycles with maximum 760 feet and minimum 687.5 feet, respectively.

The intake channel in the Upper Reservoir is some 1,500 feet long with a 35-foot base width. The intake is a conventional bell-mouth concrete structure with top elevation 1,560 and invert elevation 1,480. Trash racks and stop log slots are provided but no head gates or hoists are installed.

Comparative economic studies were made of combined tunnel and penstock plans with alternative all-tunnel schemes; the over-all economics dictated that the combined inclined tunnel through the escarpment should emerge into a single penstock with trifurcation into three separate penstocks above and near the station. The concrete-lined 20-foot diameter tunnel is approximately 1,500 feet long and partly steel-lined for 200 feet near the outlet connecting the 19 to 18-foot penstock, 1,960 feet long, extending to the trifurcation which branches into three 10-foot diameter penstocks, each about 270 feet long through tunnels leading to the spherical valves and pump-turbine units.

PUMPING AND GENERATING STATION

The station structure (Figure 2) was located on the basis of an economic study which balanced the cost of tailrace channel excavation with the cost of penstock and lost head, considering accessibility from the relocated road. It is a semioutdoor type with an outside overhead gantry crane and removable roof over the units. Each of the three penstocks will be connected to a spherical valve ahead of the three pump turbine units. The draft tubes are provided with trash racks.

The centerline of the spiral case and distributor is at elevation 770, the required 25-foot pump submergence below minimum reservoir elevation 795. The operating floor is at elevation 777.5 and the upper deck at elevation 792. Removable hatch covers are provided for access to the spherical valves, sump pumps, etc. An erection bay is located at the south end of the station for maintenance work on the generator/motor or pump/turbine parts. An electrical service bay is located in back of the generator area below the transformer deck at elevation 832.5.

The main plant structure is 140 feet long and 63.5 feet wide with the draft tube invert at elevation 751.5.

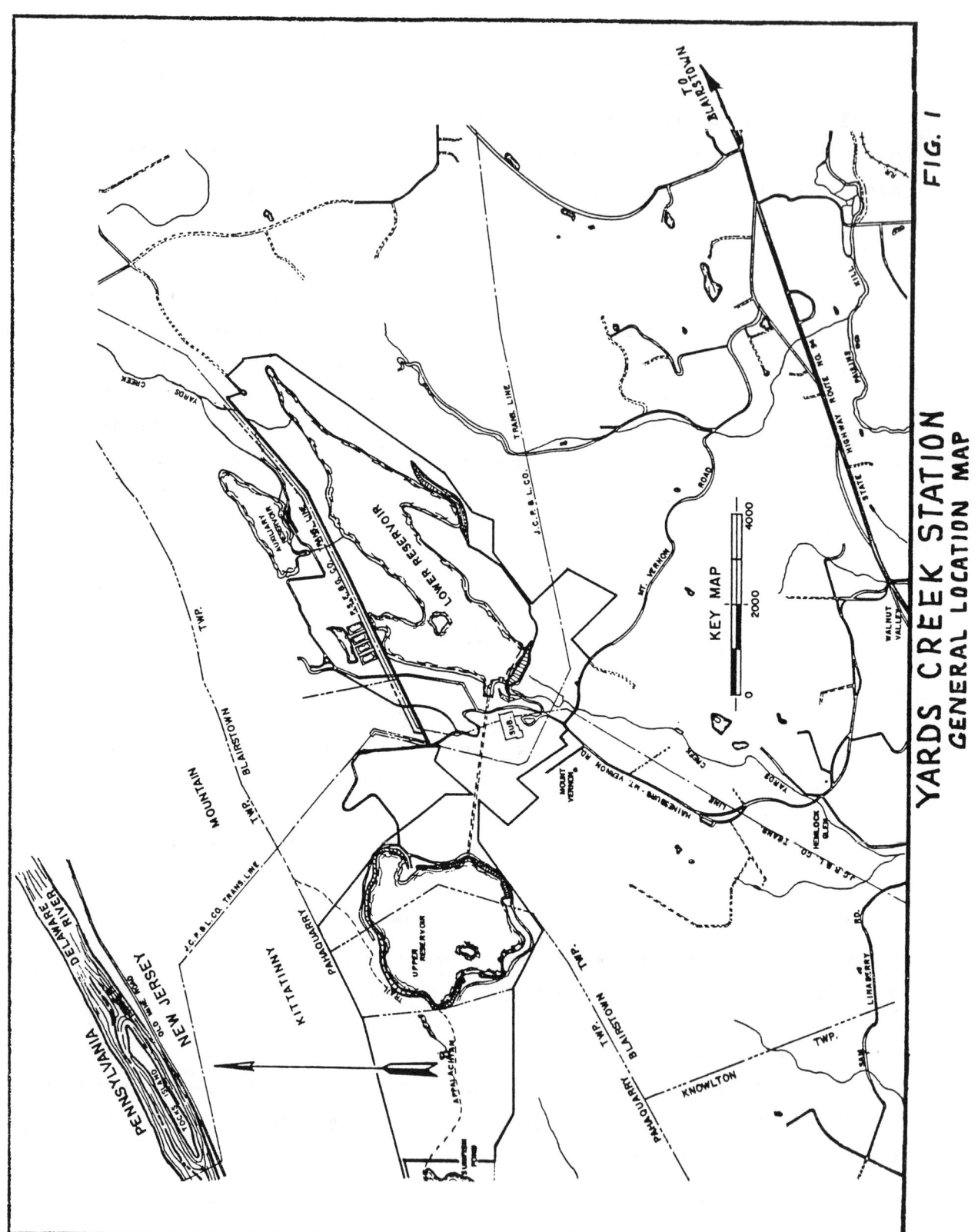

YARDS CREEK STATION
GENERAL LOCATION MAP

FIG. 1

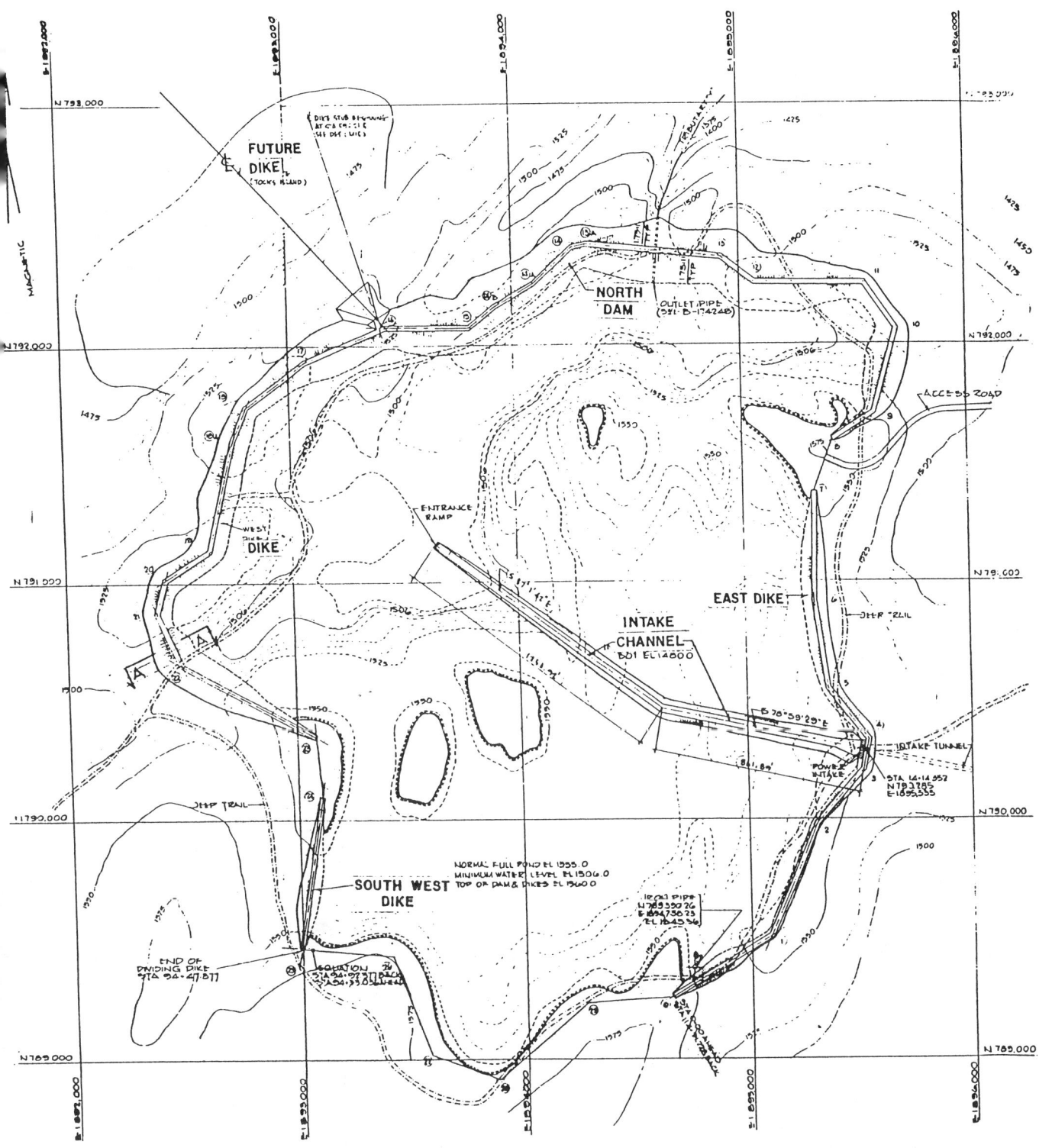

PLAN

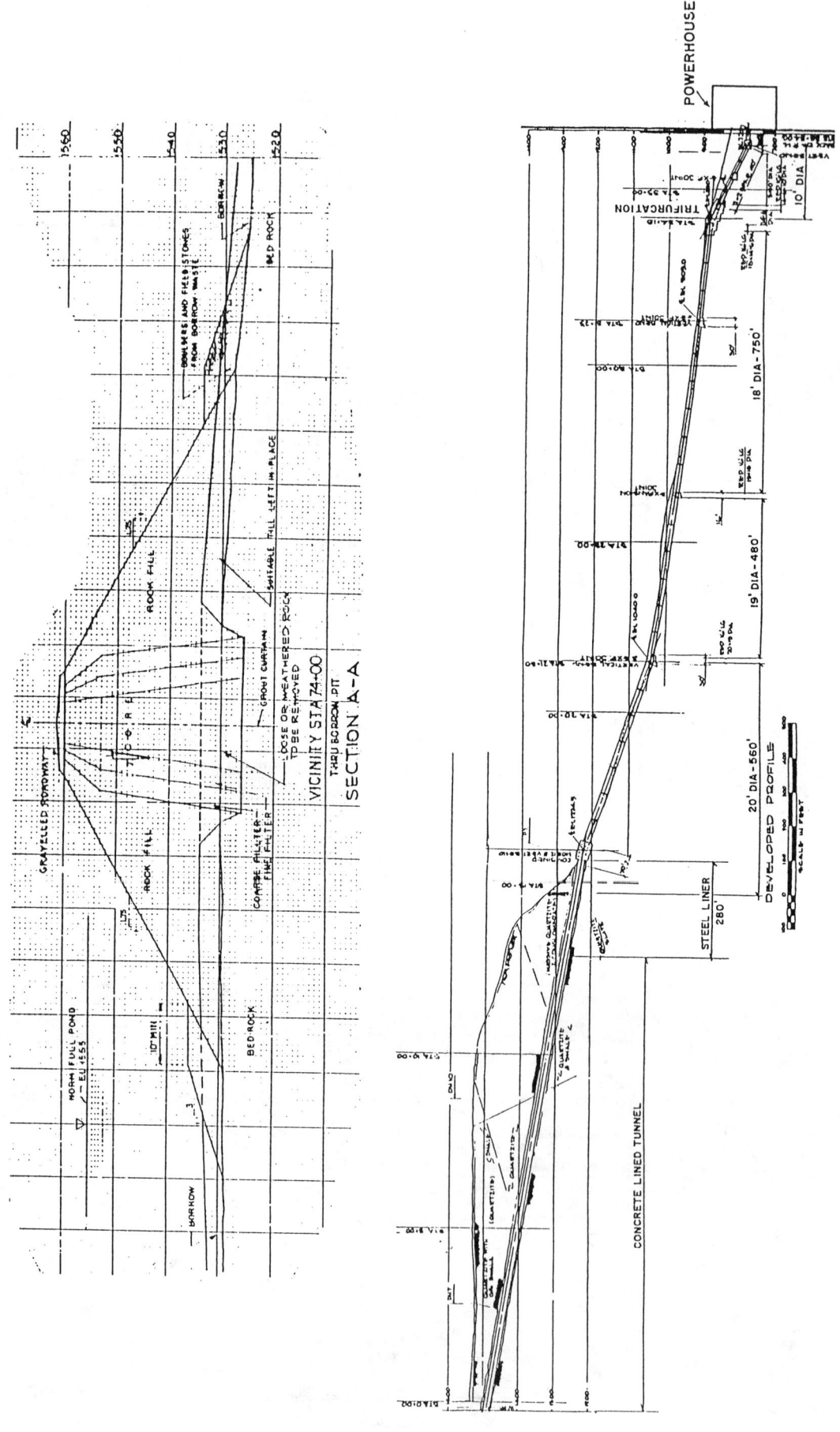

Aerial View

Powerhouse and Switchyard

Yards Creek Pumped Storage Plant

Yards Creek Pump/Generating Plant

Bibliography

1. Ley, R. D. and W. E. Liepe, "FPC Approves: Constuction Begins on Yards Creek Pumped-Storage Plant," *Electrical World*, June 17, 1963, pp. 92-94.

INDEX

Conduits summary data 28-30, 71-73

Environmental features
  Bad Creek 93-94, 100
  Balsam Meadow 112, 121-123
  Bath County 133
  Bear Swamp 161
  Blenheim-Gilboa 182
  Cabin Creek 197
  Carters 216
  Castaic 243
  Clarence Cannon 264
  De Gray 288-289
  Edward Hyatt 312
  Fairfield 342
  Flatiron 377
  Gianelli San Luis 392-393
  Grand Coulee 406-407
  Helms 421
  Hiwassee-Unit 2 445
  Horse Mesa #4 458
  Jocassee 467-468
  Kinzua 479
  Lewiston 507-508
  Ludington 523
  Mormon Flat #2 537
  Mt. Elbert 546-547
  Muddy Run 565
  Northfield Mountain 582
  Raccoon Mountain 608
  Rocky Mountain 625-626
  Rocky River 646
  Russel Dam 654
  Salina 661
  Smith Mountain 681
  Thermalito 707
  Wallace Dam 715-716
  Yards Creek Station 733
Environmental features summary data 43-53

Installed capacity, growth in 5

Lower dams, reservoirs, and spillways summary data 34-39, 77-81
Lower intakes
  Bad Creek 89, 90
  Balsam Meadow 108, 109
  Bath County 128, 129, 141-142
  Bear Swamp 157, 158
  Blenheim-Gilboa 177, 178
  Cabin Creek 193, 194
  Carters 212, 213
  Castaic 239, 240
  Clarence Cannon 260, 261
  De Gray 284, 285
  Fairfield 337, 338
  Gianelli San Luis 388, 389
  Grand Coulee 402, 403
  Helms 417, 418
  Hiwassee-Unit 2 441, 442
  Horse Mesa #4 454, 455
  Jocassee 463, 464
  Kinzua 475, 476
  Lewiston 503, 504
  Ludington 519, 520
  Mormon Flat #2 533, 534
  Mt. Elbert 542, 543
  Muddy Run 651, 562
  Northfield Mountain 577, 578
  Raccoon Mountain 604, 605
  Rocky Mountain 621, 622
  Rocky River 642, 643
  Russel Dam 650, 651
  Salina 657, 658
  Smith Mountain 677, 678
  Taum Sauk 691, 692
  Thermalito 702, 703
  Wallace Dam 711, 712
  Yards Creek Station 729, 730
Lower intakes summary data 40-42, 82-84
Lower reservoirs
  Bad Creek 87, 88, 97
  Balsam Meadow 106, 107
  Bath County 126, 127, 136-137
  Bear Swamp 15, 156
  Blenheim-Gilboa 175, 176, 188
  Cabin Creek 191, 192
  Carters 210, 211
  Castaic 237, 238
  Clarence Cannon 258, 259
  De Gray 282, 283
  Edward Hyatt 306, 307
  Fairfield 335, 336
  Flatiron 370, 371
  Gianelli San Luis 386, 387
  Grand Coulee 400, 401
  Helms 415, 416
  Hiwassee-Unit 2 439, 440
  Horse Mesa #4 452, 453
  Jocassee 461, 462
  Kinzua 473, 474
  Lewiston 501, 502
  Ludington 517, 518
  Mormon Flat #2 531, 532
  Mt. Elbert 540, 541
  Muddy Run 559, 560
  Northfield Mountain 575, 576
  Raccoon Mountain 602, 603
  Rocky Mountain 619, 620
  Rocky River 640, 641
  Russel Dam 648, 649
  Salina 655, 656
  Smith Mountain 675, 676
  Taum Sauk 689, 690
  Thermalito 700, 701
  Wallace Dam 709, 710
  Yards Creek Station 727, 728, 735-736

Plant data
  Bad Creek 93
  Balsam Meadow 112
  Bath County 132
  Bear Swamp 161
  Blenheim-Gilboa 181
  Cabin Creek 197
  Carters 216
  Castaic 243
  Clarence Cannon 264
  De Gray 288
  Edward Hyatt 312
  Fairfield 341
  Flatiron 376
  Gianelli San Luis 392
  Grand Coulee 406
  Helms 421
  Hiwassee-Unit 2 445
  Horse Mesa #4 458
  Jocassee 467, 468
  Kinzua 479
  Lewiston 507
  Ludington 523
  Mormon Flat #2 537
  Mt. Elbert 546
  Muddy Run 565
  Northfield Mountain 581
  Raccoon Mountain 608
  Rocky Mountain 625
  Rocky River 646
  Russel Dam 648, 649
  Russel Dam 654
  Salina 661
  Smith Mountain 681
  Taum Sauk 695, 696
  Thermalito 706
  Walace Dam 715
  Yards Creek Station 733
Plant efficiency 6
Plant summary data 54-61
Powerhouses
  Bad Creek 91, 92, 97
  Balsam Meadow 110, 111, 116
  Bath County 130, 131, 139-141
  Bear Swamp 159, 160, 165-167
  Blenheim-Gilboa 179, 180, 189
  Cabin Creek 195, 196
  Carters 214, 215, 228-233
  Castalc 241, 242, 248
  Clarence Cannn 262, 263
  De Gray 286, 287, 295-297
  Edward Hyatt 310, 311
  Fairfield 339, 340
  Flatiron 374, 375
  Gianelli San Luis 390, 391
  Grand Coulee 404, 405
  Helms 419, 420, 423-424, 426-427
  Hiwassee-Unit 2 443, 444
  Horse Mesa #4 456, 457
  Jocassee 465, 466
  Kinzua 477, 478
  Lewiston 505, 506
  Ludington 521, 522
  Mormon Flat #2 535, 536
  Mt. Elbert 544, 545
  Muddy Run 563, 564
  Northfield Mountain 579, 580
  Raccoon Mountain 606, 607
  Rocky River 644, 645
  Russel Dam 652, 653
  Salina 659, 660
  Smith Mountain 679, 680
  Taum Sauk 693, 694

Thermalito 704, 705
Wallace Dam 713, 714
Yards Creek Station 731, 732
Project data
   Bad Creek 87, 88
   Balsam Meadow 106, 107
   Bath County 126, 127
   Bear Swamp 155, 156
   Blenheim-Gilboa 175, 176, 183
   Cabin Creek 191, 192
   Carters 210, 211
   Castaic 237, 238
   Clarence Cannon 258, 259
   De Gray 282, 283
   Edward Hyatt 306, 307
   Fairfield 335, 336
   Flatiron 370, 371
   Gianelli San Luis 386, 387
   Grand Coulee 400, 401
   Helms 415, 416
   Hiwassee-Unit 2 439, 440
   Horse Mesa #4 452, 453
   Jocassee 461, 462
   Kinzua 473, 474
   Lewiston 501, 502
   Ludington 517, 518
   Mormon Flat #2 531, 532
   Mt. Elbert 540, 541
   Muddy Run 559, 560
   Northfield Mountain 575, 576
   Raccoon Mountain 602, 603
   Rocky Mountain 619, 620
   Rocky River 640, 641
   alina 655, 656
   Smith Mountain 675, 676
   Taum Sauk 689, 690
   Thermalito 700, 701
   Wallace Dam 709, 710
   Yards Creek Station 727, 728
Project description
   Bad Creek 95-104
   Balsam Meadow 113-124
   Bath County 134-152
   Bear Swamp 162-173
   Cabin Creek 200-208
   Carters 217-235
   Castaic 244-256
   Clarence Cannon 265-281
   De Gray 290-305
   Edward Hyatt 313-327
   Fairfield 343-368
   Flatiron 378-384
   Gianelli San Luis 394-398
   Grand Coulee 408-411
   Helms 422-436
   Hiwassee-Unit 2 446-451
   Jocassee 469-472
   Kinzua 480-500
   Lewiston 509-515
   Ludington 524-529
   Mt. Elbert 548-558
   Muddy Run 566-573
   Northfield Mountain 583-599
   Raccoon Mountain 609-618

Rocky Mountain 627-638
Salina 662-673
Smith Mountain 682-687
Taum Sauk 696-697
Wallace Dam 717-726
Yards Creek Station, 734-741
Pump-turbines summary data 31-33, 74-76
Pumped storage plant summary data 11-18
Pumped storage plants map 4
Pump/turbine rehabilitation, Salina case study 662-673

Upper dams, reservoirs, and spillways summary data 19-23, 62-66
Upper intakes
   Bad Creek 89, 90
   Balsam Meadow 108, 109
   Bath County 128, 129, 141-142
   Bear Swamp 157, 158, 163
   Blenheim-Gilboa 177, 178, 185
   Cabin Creek 193, 194
   Carters 212, 213
   Castaic 239, 240
   Clarence Cannon 260, 261
   De Gray 284, 285
   Edward Hyatt 308,309
   Fairfield 337, 338
   Flatiron 372, 373
   Gianelli San Luis 388, 389
   Grand Coulee 402, 403
   Helms 417, 418
   Hiwassee-Unit 2 441, 442
   Horse Mesa #4 454, 455
   Jocassee 463, 464
   Kinzua 475, 476
   Lewiston 503, 504
   Ludington 519, 520
   Mormon Flat #2 533, 534
   Mt. Elbert 542, 543
   Muddy Run 561, 562
   Raccoon Mountain 604, 605
   Rocky Mountain 621, 622
   Rocky River 642, 643
   Russel Dam 650, 651
   Salina 657, 658
   Smith Mountain 677, 678
   Taum Sauk 691, 692
   Thermalito 702, 703
   Wallace Dam 711, 712
   Yards Creek Station 729, 730
Upper intakes summary data 24-27, 67-70
Upper reservoirs
   Bad Creek 87, 88, 96
   Balsam Meadow 106, 107
   Bath County 126, 127, 137
   Bear Swamp 155, 156
   Blenheim-Gilboa 175, 176, 184
   Cabin Creek 191, 192
   Carters 210, 211
   Castaic 237, 238
   Clarence Cannon 258, 259
   De Gray 282, 283
   Edward Hyatt 306, 307
   Fairfield 335, 336

Flatiron 370, 371
Gianelli San Luis 386, 387
Grand Coulee 400, 401
Helms 415, 416
Hiwassee-Unit 2 439, 440
Horse Mesa #4 452, 453
Jocassee 461, 462
Kinzua 473, 474
Lewiston 501, 502
Ludington 517, 518
Mormon Flat #2 531, 532
Mt. Elbert 540, 541
Muddy Run 559, 560, 569
Northfield Mountain 575, 576
Northfield Mountain 577, 578
Racoon Mountain 602, 603
Rocky Mountain 619, 629
Rocky River 640, 641
Russel Dam 648, 649
Salina 655, 656
Smith Mountain 675, 676
Taum Sauk 689, 690
Thermalito 700, 701
Wallace Dam 709, 710
Yards Creek Station 727, 728, 734-735

Water passages
   Bad Creek 89, 90, 98-99
   Balsam Meadow 108, 109, 115
   Bath County 128, 129, 138-139
   Bear Swamp 157, 158
   Blenheim-Gilboa 177,178
   Cabin Creek 193, 194, 222-226
   Carters 212, 213
   Castaic 239, 240
   Clarence Cannon 260, 261
   De Gray 284, 285
   Edward Hyatt, 308, 309
   Fairfield 337, 338
   Flatiron 372, 373
   Gianelli San Luis 388, 389
   Grand Coulee 402, 403
   Helms 417, 418, 423
   Hiwassee-Unit 2 441, 442
   Horse Mesa #4 454, 455
   Jocassee 463, 464
   Kinzua 475, 476
   Lewiston 503, 504
   Ludington 519, 520
   Mormon Flat #2 533, 534
   Mt. Elbert 542, 543
   Muddy Run 561, 562, 570-571
   Northfield Mountain 577, 578
   Raccoon Mountain 604, 605
   Rocky Mountain 621, 622
   Rocky River 642, 643
   Russel Dam 650, 651
   Salina 657, 658
   Smith Mountain 677, 678
   Taum Sauk 691, 692
   Thermalito 702, 703
   Wallace Dam 711, 712
   Yards Creek Station 729, 730